Lysosomal Storage Disorders

Lysosomal Storage Disorders

John A. Barranger, M.D., Ph.D.
Mario A. Cabrera-Salazar, M.D.

Springer

John A. Barranger, MD, PhD
Department of Human Genetics
University of Pittsburgh
Pittsburgh, PA 15261
USA
jbarrang@helix.hgen.pitt.edu

Mario A. Cabrera-Salazar, MD
Genzyme Corporation
Framingham, MA 01701–9322
USA
Mario.Cabrera@genzyme.com

ISBN 978-0-387-70908-6 ISBN 978-0-387-70909-3 (eBook)

Library of Congress Control Number: 2007924717

Cover Illustration: Airplane (2001). The cover of this book is a painting from Giovane Araujo Curatolo; A patient from Brazil affected by MPS1. This painting was made as part of the Expression of Hope program; created as a mean to generate awareness and understanding of the incredible strength and courage of the thousands of patients living with lysosomal storage disorders.

For more information please visit http://www.expressionofhope.com

Printed on acid-free paper.

9 8 7 6 5 4 3 2 1

springer.com

To all those colleagues and friends
who made this text as good
as it is, I remain in debt.
Thank you for your thoughtful contributions.

John A. Barranger

To my parents Jorge and Sonia
and Luis Antonio, my grandfather;
a teacher who mastered the science
of educating generations.

Mario A. Cabrera-Salazar

To all patients with lysosomal storage
disorders, thank you for all the lessons
given to us

Preface

Scientific progress has been rapid in lysosomal biology during the last six decades. Its application to human disease is nothing less than spectacular. In no other group of disorders has knowledge and clinical utility progressed so speedily. Recall that the organelle was described in just 1955. Since then, the biochemical alterations and storage materials were described, the enzyme deficiencies discovered, the gene coding of these glycoproteins cloned and thousands of mutations defined. These advances have resulted in highly improved diagnosis for more than 50 diseases. For five diseases, including the most common lysosomal storage disorder, molecular therapy is a reality, extremely effective and very safe. This higher plateau of medical approaches to human disease is something to which all translational scientists aspire and only a few actually witness. The relief of pain and suffering is a tribute to the ideas and work of many dedicated investigators. Much of that work is presented in this text.

Despite our ability to treat some of these diseases through enzyme replacement therapy (ERT) and, accurately define different diseases that look alike, there is much to be learned about lysosomal disease. With each step up the barrier to knowledge, a new point of view is attained, a fresh perspective. Much is seen more clearly and many "allegories of the cave" are dispelled forever. Yet, our new view demands a new vision drawing us to find better definitions of what we see. This is how it has been with lysosomal diseases. No sooner had the little vesicle been described when a defect in it was shown to cause a human disease. With each advance came the hope for better diagnosis and visions of therapy. Nowhere was its value more evident than in the clear separation of the mucopolysaccharidoses into reliable diagnostic categories according to the enzyme that was deficient in them. Long-standing debates and controversies were settled forever.

Still though, as in all lysosomal storage disease, what makes the difference in the spectrum of iduronidase deficient disease from severe Hurler syndrome to mild Scheie disease remains to be defined. There is no doubt that multiple genes are involved in these complex diseases. Just when we thought that they were monogeneic, whole new visions of the diseases have emerged. One even suggests that lysosomal storage may not be the primary pathogenetic step, but rather, protesomal activation in an attempt to rid the cell of improperly folded lysosomal enzyme. The universality of this theory remains to be demonstrated. However, this new approach may provide another tool in the physician's bag to define and treat disease. It is with this enthusiasm that we should look forward to new definitions of the lysosome and lysosomal disease. This text provides a thoughtful introduction to a wide variety of data, concepts, and approaches to lysosomal storage disorders.

John A. Barranger
University of Pittsburgh, Pittsburgh, PA
May 2007

Contents

Contributors

John A. Barranger
Department of Human Genetics, University of Pittsburgh, Pittsburgh, PA 15261, USA, jbarrang@helix.hgen.pitt.edu

Alessandra Biffi
San Raffaele Telethon Institute for Gene Therapy and Vita Salute University, H. San Raffaele Scientific Institute, 20132 Milan, Italy, biffi.alessandra@hsr.it

Roscoe O. Brady
Developmental and Metabolic Neurology Branch, National Institute of Neurological Disorders and stroke, National Institutes of Health, Bethesda, MD 20892-1260, USA, bradyr@ninds.nih.gov

Roscoe O. Brady Jr.
Department of Psychiatry, Massachusetts General Hospital, Boston, MA 02114, USA, rbrady@partners.org

Douglas Brooks
Sansom Institute, School of Pharmacy and Medical Sciences, University of South Australia, North Adelaide, SA 5001, Australia; Lysosomal Diseases Research Unit, Department of Genetic Medicine, Children Youth and Women's Health Service, North Adelaide, SA 5006, Australia, Doug.Brooks@unisa.edu.au

Edward Burton
Department of Neurology, University of Pittsburgh School of Medicine, Pittsburgh, PA 15213, USA, burtonea@upmc.edu

Mario A. Cabrera-Salazar
Genetic Disease Science, Genzyme Corporation, Framingham, MA 01701-9322, USA, Mario.Cabrera@genzyme.com

Gustavo Charria-Ortiz
Department of Neurology, University of Miami School of Medicine, Miami, FL 33136, USA, Gcharria@med.miami.edu

Seng H. Cheng
Genetic Diseases Science, Genzyme Corporation, Framingham, MA 01701-9322, USA, seng.cheng@genzyme.com

Dan Clark
The Burnham Institute for Medical Research, La Jolla, CA 92037, USA, daclark@ucsd.edu

Lorne A. Clarke
Department of Medical Genetics, University of British Columbia, Child and Family Research Institute, Vancouver, BC, Canada V5Z4H4, lclarke@cw.bc.ca

Ronald G. Crystal
Department of Genetic Medicine, Weill Medical College of Cornell University, New York, NY 10021, USA, rgcryst@med.cornell.edu

Leonard D'Aiuto
Department of Surgery and Medicine, Molecular Medicine Institute University of Pittsburgh School of Medicine, Pittsburgh, PA 15219, USA, daiutol@msx.upmc.edu

Beverly L. Davidson
Departments of Internal Medicine, Neurology, Physiology & Biophysics, University of Iowa, Iowa City, IA, 52242 USA, beverly-davidson@uiowa.edu

Christian de Duve
The Rockefeller University, 1230, York Avenue, New York, NY 10021, USA and de Duve Institute, 75, Avenue Hippocrate, 1200 Brussels, Belgium, deduve_private@icp.ucl.ac.be

Robert J. Desnick
Department of Human Genetics, Mount Sinai School of Medicine, New York, NY10029, USA, rjdesnick@mssm.edu

William A. Gahl
Section on Human Biochemical Genetics, Medical Genetics Branch, National Human Genome Research Institute, NIH, Bethesda, MD 20892-1851, USA, bgahl@helix.nih.gov

Andrea Gambotto
Department of Surgery and Medicine, Molecular Medicine Institute University of Pittsburgh School of Medicine, Pittsburgh, PA 15219, USA, gambottoa@msx.upmc.edu

Michael H. Gelb
Department of Biochemistry, University of Washington, Seattle, WA 98195, USA, gelb@chem.washington.edu

Volkmar Gieselmann
Institut fur Physiologische ChemieUniversitat Bonn, Germany; Rheinische Friedrich-Wilhelms-Universität Bonn, Nussallee 11, D-53115 Bonn, Germany, gieselmann@institut.physiochem.uni-bonn.de

Joseph C. Glorioso III
Department of Molecular Genetics and Biochemistry, University of Pittsburgh School of Medicine, Pittsburgh, PA 15261, USA, glorioso@pitt.edu

Rodolfo Gonzalez
Center for Neuroscience and Aging Research, Burnham Institute for Medical Research, La Jolla, CA 92037, USA, rudyg@burnham.org

Neil R. Hackett
Department of Genetic Medicine, Weill Medical College of Cornell University, New York, NY 10021, USA, geneticmedicine@mail.med.cornell.edu

Amanda Helip-Wooley
Section on Human Biochemical Genetics, Medical Genetics Branch, National Human Genome Research Institute, NIH, Bethesda, MD 20892, USA, ahwooley@mail.nih.gov

John J. Hopwood
Lysosomal Diseases Research Unit, Department of Genetic Medicine, Department of Pediatrics, Children Youth and Women's Health Service, North Adelaide, Sa 5006, Australia, john.hopwood@adelaide.edu.au

Rahul Jandial
Department of Neurosurgery, UCSD School of Medicine, La Jolla, CA 92093, USA, rahuljandial@hotmail.com

Mylvaganam Jeyakumar
Department of Biochemistry, Glycobiology Institute, University of Oxford, Oxford OX1 3QU, UK, mylvaganam.jeyakumar@bioch.ox.ac.uk

Stephen M. Kaminksy
Department of Genetic Medicine, Weill Medical College of Cornell University, New York, NY 10021, USA, skaminsky@iavi.org

Seon Hee Kim
Department of Molecular Genetics and Biochemistry and Molecular Medicine Institute, University of Pittsburgh School of Medicine, Pittsburgh, PA 15261, USA

Robert Kleta
Section on Human Biochemical Genetics, Medical Genetics Branch, National Human Genome Research Institute, NIH, Bethesda, MD 20892, USA, kletar@mail.nih.gov

Dawn Laney
Emory Lysosomal Storage Disease Center, Emory University, Atlanta, GA 30322, USA, dlaney@genetics.emory.edu

Jean-Pyo Lee
The Burnham Institute for Medical Research, and Department of Pediatrics, University of California San Diego, La Jolla, CA 92037, USA
jeanpyol@ucsd.edu

Junko Matsuda
Institute of Glycotechnology, Future Science and Technology Joint Research Center, Tokai University, Hiratsuka, Japan 259-1292, junko@clin.med.tokushima-u.ac.jp

Margaret McGovern
Department of Human Genetics, Mount Sinai School of Medicine, New York, NY 10029, USA, margaret.mcgovern@mssm.edu

Scott McKercher
The Burnham Institute for Medical Research, La Jolla, CA 92037-1062, USA, SMckercher@burnham.org

Peter Meikle
Lysosomal Diseases Research Unit, Department of Genetic Medicine, Department of Pediatrics, Children Youth and Women's Health Service, North Adelaide, SA 2006, Australia, peter.meikle@adelaide.edu.au

Adriana M. Montaño
Department of Pediatrics, Cardinal Glennon Children's Hospital, Saint Louis University, St. Louis, MO 63110-2586, USA, amontano@slu.edu

Angela Montecalvo
Department of Surgery and Medicine, Molecular Medicine Institute
University of Pittsburgh School of Medicine, Pittsburgh, PA 15219, USA, anm45@pitt.edu

Kim Mooney
Genetic Counseling, Biomarin Pharmaceutical, Novato, CA 94949, USA, Kim.Mooney@biomarin.com

Cindy Morgan
University of California San Francisco, Stanford Lysosomal Disease Center, San Francisco, CA 94143, USA, morganc@peds.ucsf.edu

Josef Mueller
The Burnham Institute for Medical Research, La Jolla, CA 92037, USA; Zentrum für integrative Psychiatrie, UKSH School of Medicine, Kiel 24118, Germany, fmueller@burnham.org

Viv Muller
Lysosomal Diseases Research Unit, Department of Genetic Medicine, Children Youth and Women's Health Service, North Adelaide, SA 5006, Australia, viv.muller@cywhs.sa.gov.au

Luigi Naldini
San Raffaele Telethon Institute for Gene Therapy and Vita Salute University, H. San Raffaele Scientific Institute, Milan 20132, Italy, luigi.naldini@hsr.it

Tatsuo Nishioka
Department of Pediatrics, Cardinal Glennon Children's Hospital, Saint Louis University, St. Louis, Missouri 63110-2586, USA, nishiokat@slu.edu

Denise Y.J. Norato
Medical School, Life Sciences Center, Catholic University of Campinas, 13081-970 São Paulo, Brazil, norato@uol.com.br

Erin O'Rourke
Department of Human Genetics, University of Pittsburgh School of Medicine, Genzyme Corporation, Pittsburgh, PA 02142, USA, erin.orourke@genzyme.com

Tadao Orii
Department of Pediatrics, Gifu University School of Medicine, Seki, Gifu 501-3936, Japan

Kook In Park
Department of Pediatrics, Yonsei University College of Medicine, Seoul 120-749, Korea, kipark@yumc.yonsei.ac.kr

Emma Parkinson-Lawrence
Lysosomal Diseases Research Unit, Department of Genetic Medicine, Children Youth and Women's Health Service, North Adelaide, SA 5006, Australia, Emma.Parkinson-Lawrence@unisa.edu.au

David A. Pearce
Center for Aging and Developmental Biology, Department of Biochemistry and Biophysics, University of Rochester School of Medicine, Rochester, NY 14642, USA, david_kornack@urmc.rochester.edu

Frances M. Platt
Department of Biochemistry, Glycobiology Institute, University of Oxford, Oxford OX1 3QU, UK, frances.platt@pharm.ox.ac.uk

Paul D. Robbins
Department of Molecular Genetics and Biochemistry, Molecular Medicine Institute, University of Pittsburgh School of Medicine, Pittsburgh, PA 15261, USA, probb@pitt.edu

Edward H. Schuchmann
Department of Human Genetics, Mount Sinai School of Medicine, New York, NY 10029, USA, edward.schuchman@mssm.edu

C. Ronald Scott
Department of Pediatrics, University of Washington, Seattle, WA 98195, USA, crscott@u.washington.edu

Thomas N. Seyfried
Department of Biology, Boston College, Chestnut Hill, MA 02467, USA, thomas.seyfried@bc.edu

Calogera M. Simonaro
Department of Human Genetics, Mount Sinai School of Medicine, New York, NY 10029, USA, calogera.simonaro@mssm.edu

Evan Y. Snyder
The Burnham Institute for Medical Research, and Department of Pediatrics, University of California San Diego, La Jolla, CA 92037,USA, esnyder@burnham.org

Dolan Sondhi
Department of Genetic Medicine, Weill Medical College of Cornell University, New York, NY 10021, USA, dos2011@med.cornell.edu

Jennifer Sullivan
Metabolic Clinic, Duke University Medical Center, Durham, NC 27009, USA, Sulli019@mc.duke.edu

Kunihiko Suzuki
Institute of Glycotechnology, Future Science and Technology Joint Research Center, Tokai University, Hiratsuka 259-1292, Japan, kuni@suisui.ne.jp

Rossane M. Taylor
University of Sydney, Department of Animal Science, Faculty of Veterinary Science, Sydney, NSW 2006, Australia, rosannet@vetsci.usyd.edu.au

Shunji Tomatsu
Department of Pediatrics, Cardinal Glennon Children's Hospital, St. Louis, MO 63110-2586 USA, tomatsus@slu.edu

Frantisek Turecek
Department of Chemistry, University of Washington, Seattle, WA 98195, USA, turecek@chem.washington.edu

Chris Turner
Lysosomal Diseases Research Unit, Department of Genetic Medicine, Children Youth and Women's Health Service, North Adelaide, SA 5006, Australia, christopher.turner@adelaide.edu.au

Neal J. Weinreb
University Research Foundation for Lysosomal Storage Diseases and Northwest Oncology Hematology Associates PA, Coral Springs, FL 33065, USA, bonneal@hotmail.com

James. E. Wraith
Royal Manchester Children's Hospital, Manchester M27 4HA, UK, ed.wraith@cmmc.nhs.uk

FROM LYSOSOMES TO STORAGE DISEASES AND BACK: A PERSONAL REMINISCENCE

Christian de Duve

When the editors of this book invited me to write an introduction, my first reaction was that many others were more qualified to introduce a topic that has enjoyed so many recent developments of major importance, including singularly successful clinical applications, unusual for genetic diseases. The task, I felt, would be much better fulfilled by one of the pioneers of the field, several of whom are still in the forefront of research today. Then, it occurred to me that, as a witness to the circuitous pathway whereby lysosomes and storage diseases were first brought together, I had a story to tell that could perhaps be of interest, and hold some instructive aspects, especially to the younger generations. In order to stay within the boundaries of my competence, my story is restricted to events with which I have been personally associated. To put those events within their proper framework, I must, with due apologies, go back to the beginning of my own career as a scientist.

Seventy years ago, when, as a young medical student, I first entered a physiology research laboratory, chance put me in a group that was investigating the action of insulin on liver, a much disputed question at the time. Twelve years later, after gaining medical and chemical degrees at the Catholic University of Louvain, in Belgium, and completing my training as a biochemist in Sweden and the United States, I started my first, modest laboratory at my Belgian alma mater, with one overwhelming ambition: to elucidate the hepatic action of insulin, still a tantalizing unsolved problem. I was joined in this endeavor by a young MD, Henri-Géry Hers, who had already worked with me as a student, helping to rediscover glucagon. Two medical students, Jacques Berthet and Lucie Dupret (who was to become Mrs. Berthet) completed our small group.

Our first achievement consisted in the characterization of glucose 6-phosphatase, a hepatic enzyme involved in the formation of glucose from glycogen and suspected of interfering with the effect of insulin. I have recounted elsewhere (de Duve, 1969) how we were led to study the intracellular distribution of this enzyme by centrifugal fractionation techniques and, in the course of that work, ran into the intriguing phenomenon of latency displayed by another phosphatase, the unspecific acid phosphatase. Lured by this challenging observation, I put aside insulin, for what I thought would be only a short time, and embarked on a quest that led, in 1955, to the identification of a new group of cytoplasmic particles characterized by their content in hydrolytic enzymes with an acid pH optimum and believed to be involved in phenomena of intracellular digestion (de

Duve et al., 1955). These particles, to which I gave the name of lysosomes, together with the peroxisomes that came to be recognized a little later, raised so many fascinating vistas that I never returned to insulin (de Duve, 2004).

Thanks to our work and to that of many others, the main functions of lysosomes were quickly identified. By 1963, at a meeting at the Ciba Foundation in London, which brought together the main investigators in this rapidly developing field, I was able to summarize these functions as concerned with the digestion of extracellular materials taken up by endocytosis and in that of intracellular materials segregated by the newly described phenomenon of cellular autophagy, two terms that I coined for the occasion (de Duve, 1963). I further pointed out that, contrary to unicellular protists, the cells of higher animals lack the ability to discharge the contents of their used lysosomes to the outside by the phenomenon of cellular defecation, so that the lysosomes progressively turn into residual bodies, loaded with the debris of internalized materials that, for one reason or another, have failed to be digested and cleared. Jokingly referred to as "chronic cellular constipation" (de Duve, 1964a), this inability of our cells to unload the contents of their lysosomes attracted my early attention by its possible harmful consequences, in cellular aging, for example. Several of our experiments dealt with reproducing the syndrome of lysosomal overloading by the administration of endocytizable undigestible substances (Wattiaux, Wibo, and Baudhuin, 1965; de Duve and Wattaux, 1966) and, later, by exposing cells to inhibitory antibodies against lysosomal enzymes (Tulkens, Trouet, and Van Hoof, 1970). In my Ciba Foundation talk, I pointed to "congenital or acquired enzymic deficiency affecting the lysosomes" as one of the causes that could be responsible for enhanced residue accumulation (de Duve, 1963) and I was able to cite brand new findings of the Hers group on glycogen storage (Hers, 1963) as a possible example of this mechanism.

It is to be noted that, contrary to my other coworkers, who had accompanied me in my new venture, Hers, eager to start his own independent group, remained faithful to carbohydrate metabolism. In this field, he rapidly made several important contributions, dealing first with the metabolism of fructose and, later, with the regulation of glycogen breakdown and synthesis (Hers, 1983). In 1957, Hers became interested in a group of inborn diseases called glycogenoses and characterized by abnormal deposits of glycogen in a number of tissues. He was prompted to do this by the untimely death of Gerty Cori, a celebrated biochemist who had shared the 1947 Nobel Prize in medicine with her husband Carl for their work on glycogen metabolism. In her last years, Gerty Cori had undertaken to identify the enzymatic defects responsible for the various kinds of glycogenoses which, interestingly, are most often due to the deficiency of a nonlysosomal, metabolic enzyme, such as phosphorylase, amylo-1,6-glucosidase, or glucose 6-phosphatase.

After Gerty Cori's death, Hers decided to take over this line of research and, for this purpose, organized the analysis of biopsy samples of those rare diseases collected from various clinical centers. In the course of this work, he ran into a few cases of the mysterious condition first described in 1932 by the Dutch pediatrician J. C. Pompe and classified by Cori as glycogen storage disease type II. Confirming her findings, Hers found that all the known enzymes of glycogen metabolism were present in normal amounts in the pathological material, an observation consistent with the fact that the patients show a normal hyperglycemic response to epinephrine and to glucagon. Yet, defying the biochemical knowledge of the time, their tissues are severely overloaded with glycogen, leading to grave disabilities and, often, early death.

The solution, as often happens in science, came from serendipity. Interested in the possible deficiency, in another glycogenosis (type III), of a hypothetical glucosyl transferase postulated by other workers, Hers developed an assay for such an enzyme, based on the use of 14C-maltose as glucosyl donor and of glycogen as acceptor. He did indeed observe the expected activity but found it to be normal in type III samples and, furthermore, not to be due to the postulated transglucosylase, but to an unspecific α-glucosidase with an acid pH optimum (Hers, 1963). By chance, the assays were extended to three samples from patients with type II glycogenosis, all of which were found to entirely lack this enzyme. Thus the deficiency responsible for type II glycogenosis was identified in the course of a study aimed at type III.

It so happened that this discovery was made in a laboratory where lysosomes were "in the air." In this environment, the possibility that the newly discovered acid glucosidase could belong to lysosomes immediately came to Hers's mind and it did not take his group long to confirm that the enzyme was indeed situated in lysosomes (Lejeune, Thinès-Sempoux, and Hers, 1963). The first congenital lysosomal defect was thus discovered. But how could the observed enzyme deficiency be related to abnormal glycogen deposition?

Here is where Hers displayed remarkable insight. He reasoned that the glucosidase could be normally involved in the breakdown of glycogen molecules that become segregated within lysosomes by cellular autophagy. Should this hypothesis be correct, he further reasoned, the abnormal glycogen deposits should be present within membrane-bounded structures related to lysosomes, where they would escape digestion because of the absence of the required glucosidase and would, at the same time, be inaccessible to the metabolic breakdown system situated in the cytosol. This prediction was confirmed morphologically, in collaboration with Pierre Baudhuin, when a new case of type II glycogenosis became available for study (Bauduin, Hers, and Loeb, 1964).

Hers did not stop there. In a landmark editorial published in *Gastroenterology* (Hers, 1965), he generalized the concept of inborn lysosomal disease, defined its principal criteria, and proposed that it could explain a number of congenital mucopolysaccharidoses, lipidoses, and other storage conditions. In collaboration with François Van Hoof, he set out to test this hypothesis and found it indeed to be applicable to some mucopolysaccharidoses (Van Hoof and Hers, 1968). Hers also appreciated from the start that lysosomal defects were likely to be uniquely accessible to enzyme replacement therapy, because extracellular proteins, when taken up by cells, are normally directed to lysosomes for digestion (de Duve, 1964b). Attempts, first with an enzyme preparation from aspergillus (Bauduin et al., 1964) and later, with human enzyme purified from placenta (de Barsy et al., 1973) gave negative results, however, and, considering the small number of patients who might benefit from such therapy, this line of research was not pursued.

The story, as told so far, has dealt only with the work carried out in my Louvain laboratory, although, as I wish to emphasize, entirely without my participation. In fact, I was in the United States at the time and heard of the new findings only on my return. I found them fascinating but remember reacting with caution to Hers's generalization. His concept fitted so logically within our whole picture of the functions of lysosomes that I could not help fearing that there might be "a catch somewhere." Also conducive to caution was the fact that lysosomes are rarely involved in glycogenoses, which, as I mentioned above, are mostly due to the deficiency of a metabolic enzyme. The possibility that the same could be true in other storage diseases obviously had to be kept in mind. This, oddly enough, has turned out not to be the case. One of the strangest aspects of this strange story is that the generalized concept of a lysosomal disease was inspired by the most atypical among the genetic storage conditions.

The impact of this contribution on the evolution of the field is difficult to assess. Storage diseases were, at that time, being investigated in a number of laboratories. Electron-microscope examinations had, in a number of cases, revealed the abnormal deposits to be situated within membrane-bounded entities, which had even sometimes been identified as lysosome-related by acid-phosphatase cytochemistry. On the other hand, there was considerable interest among biochemists in the chemical structures of complex lipids and polysaccharides, including those that make up the stored materials, and the biosynthesis and breakdown of these substances were being actively deciphered. Several investigators were turning to pathological specimens in their studies. Although excessive synthesis was the favored explanation of the abnormalities, the history of the field shows that interest was beginning to veer in the direction of a degradation defect. In fact, the identification of several such defects almost coincided with Hers's discovery or followed it very closely. No doubt, the lysosomal concept would have emerged eventually, although perhaps in a slow and piecemeal fashion.

In this respect, the clear way in which Hers delineated a common pathogenic storage mechanism within our existing picture of lysosome function indubitably gave the field a considerable impetus. His concept provided a powerful unifying framework for a wide diversity of findings, it helped investigators interpret their results and plan further experiments, and it directly inspired many research projects. Several lysosomal enzymes were actually discovered thanks to this effort. A fitting conclusion to this phase in the history of storage diseases was provided in 1973 by the publication, under the joint editorship of Hers and Van Hoof, of a comprehensive work in which all the main investigators concerned with the field were brought together under the umbrella of *Lysosomes and Storage Diseases* (Hers and Van Hoof, 1973).

Since then, many important advances have been made, as this book eloquently shows. I must leave it to other, more authoritative voices to tell the rest of the story, ending my part with a reflection: in science, one does not always find what one is looking for; but what one finds sometimes turns out to be more interesting than what one was looking for. My colleague Hers and I both found this out independently. So have many other investigators. The history of science is littered with serendipitous discoveries. But for serendipity to bear fruit, investigators must enjoy enough freedom and flexibility to be able to follow the opportunity offered by chance. This message often escapes administrators, who insist on rigid programs and disallow any departure from an agreed project.

REFERENCES

Baudhuin, P., Hers, H.G., and Loeb, H. (1964). An electron-microscopic and biochemical study of Type II glycogenosis. *Lab. Invest.* **13**: 1140–1152 (1964).

de Barsy, T., Jacquemin, P., Van Hoof, F., and Hers, H.G. (1973). Enzyme replacement in Pompe disease: an attempt with purified human α-glucosidase. *Enzyme Therapy and Genetic Diseases, Birth Defects, Original Article Series*, **9**: 184–190.

de Duve, C. (1963). The lysosome concept. In: de Reuck, V.S. and Cameron, M.P. (Eds.), *Ciba Foundation Symposium on Lysosomes*. London: J. & A. Churchill, pp. 1–35.

de Duve, C. (1964a). From cytases to lysosomes. *Fed. Proc.* **23**: 1045–1049.
de Duve, C. (1964b). Born-again glucagon. *FASEB J.* **5**: 979–981.
de Duve, C. (1969). The lysosome in retrospect. In: Dingle, J.T. and Fell, H.B. (Eds.), *Lysosomes in Biology and Pathology*, Vol. 1. Amsterdam: North-Holland, pp. 3–40.
de Duve, C. (2004). My love affair with insulin. *J. Biol. Chem.* **27**:9, 21679–21688.
de Duve, C. and Wattiaux, R. (1966). Functions of lysosomes. *Annu. Rev. Physiol.* **28**: 435–492.
de Duve, C., Pressman, B.C., Gianetto, R., Wattiaux, R., and Appelmans, F. (1955). Tissue fractionation studies. 6. Intracellular distribution patterns of enzymes in rat-liver tissue. *Biochem. J.* **60**: 604–617.
Hers, H.G. (1963). α-Glucosidase deficiency in generalized glycogen storage disease (Pompe's disease). *Biochem. J.* **86**: 11–16.
Hers, H.G. (1965). Inborn lysosomal diseases. *Gastroenterology*, **48**: 625–633.
Hers, H.G. (1983). From fructose to fructose 2,6-bisphosphate, with a detour through lysosomes and glycogen. In: Semenza, G. (Ed.), *Selected Topics in the History of Biochemistry*, Amsterdam: Elsevier Science, pp. 71–101.
Hers, H.G. and Van Hoof, F. (1973). *Lysosomes and Storage Diseases*. New York: Academic Press.
Lejeune, N., Thinès-Sempoux, D., and Hers, H.G. (1963). Tissue fractionation studies. 16. Intracellular distribution and properties of α-glucosidases in rat liver. *Biochem. J.* **86**: 16–21.
Tulkens, P., Trouet, A., and Van Hoof, F. (1970). Immunological inhibition of lysosome function. *Nature*, **228**: 1282–1285.
Van Hoof, F. and Hers, H.G. (1968). The abnormalities of lysosomal enzymes in mucopolysacccharidoses. *Eur. J. Biochem.* **7**: 34–44.
Wattiaux, R., Wibo, M., and Baudhuin, P. (1963). Influence of the injection of Triton WR-1339 on the properties of rat-liver lysosomes. In: Reuck, A.V.S. and Cameron, M.P. (Eds.), *Ciba Foundation Symposium on Lysosomes*, London: J. & A. Churchill, pp. 176–200.

LYSOSOMAL BIOGENESIS AND DISEASE

Doug Brooks[1,2] and Emma Parkinson-Lawrence[1,2]

PERSPECTIVE

This chapter introduces key concepts in the area of lysosomal biogenesis, which were initially derived from the study of lysosomal storage disorders, but more recently developed from molecular studies on vesicular traffic and the cell biology of specific endosomal–lysosomal proteins. The dynamics of the endomembrane system is discussed and includes the concepts of biosynthesis, vesicular traffic, protein processing, secretion, enzyme uptake, and the degradation of macromolecular substrates to their constituents in the endosome–lysosome network. Each section highlights potential areas of dysfunction in endosome–lysosome proteins or their processing/transport machinery, and relates this to known disease states, to both enhance discussion of key areas of lysosomal biogenesis and introduce the other chapters of this book. We hypothesise that a defect at any point in the processes of endosome–lysosome biogenesis and function is susceptible to the effects of mutation and can therefore result in a genetic disease. Even for defects in the same gene, different mutations may have dramatically different effects based on how the message and gene product interact with the processing machinery and organelle milieu. In most cases mutations will have direct and obvious functional and clinical significance, but for other defects the effects may be more subtle with either only long-term significance or obvious effects at particular stages of development. Thus, an error in protein processing or vesicular traffic may be just as important to lysosomal function as a mutation in the coding sequence of a degradive lysosomal hydrolase, as iterated in this book.

[1]Lysosomal Diseases Research Unit, Department of Genetic Medicine, Children Youth and Women's Health Service, 72 King William Rd, North Adelaide, South Australia 5006, Australia,
[2]Department of Paediatrics, University of Adelaide, Adelaide, South Australia 5005, Australia.
Assoc. Prof. Doug Brooks, Lysosomal Diseases Research Unit, Department of Genetic Medicine, Children Youth and Women's Health Service, 72 King William Rd, North Adelaide, South Australia 5006, Australia. Phone: (61-8) 8161-7341; Fax: (61-8) 8161-7100; E-mail: doug.brooks@adelaide.edu.au

1 TRANSCRIPTION AND TRANSLATION OF ENDOSOME-LYSOSOME PROTEINS

1.1 Transcription of Message Encoding Endosome–Lysosome Protein

Transcription is the process that involves the synthesis of messenger RNA (mRNA), transfer RNA, ribosomal RNA, and other structural/regulatory RNA, by RNA polymerases from DNA templates in the nucleus (Figure 1). Each mRNA molecule carries the information required to exit the nucleus and encode for the subsequent synthesis of a specific protein on ribosomes, located either directly in the cytosol (for cytosolic proteins) or at the cytosolic face of the rough endoplasmic reticulum (RER: for membrane and soluble proteins of the endomembrane system). Proteins involved in endosome–lysosome biogenesis and function are mainly synthesised cotranslationally on the RER, but some components of the vesicular machinery are synthesised either in the cytosol or possibly on the cytosolic face of the RER.

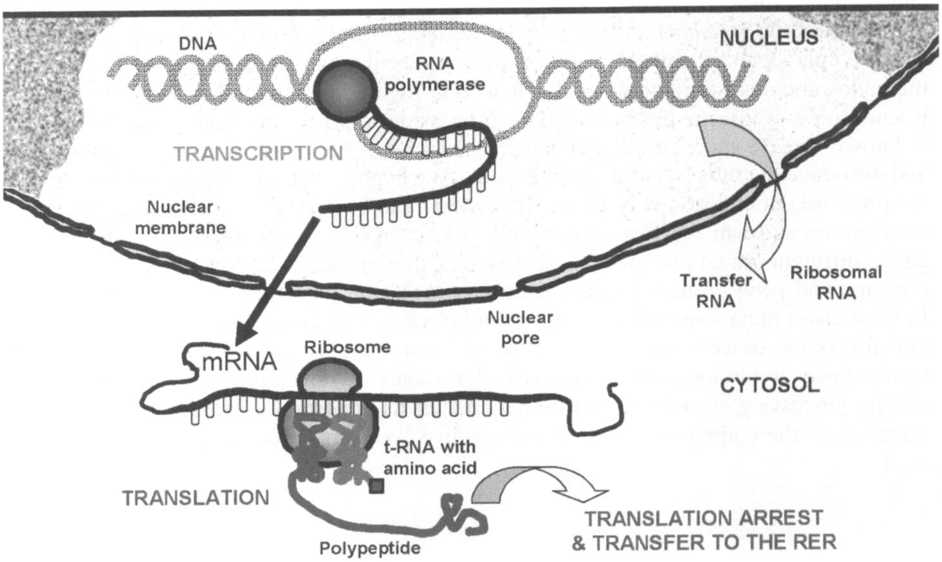

Figure 1. A schematic depicting the synthesis of mRNA from DNA in the nucleus (transcription), its export from the nucleus into the cytosol and the initiation of protein synthesis (translation) on ribosomes in the cytosol.

1.2 Aberrations Affecting mRNA

The production of mRNA and its transport to the cytosol does not necessarily ensure that a protein product will be normally synthesised. In mammalian cells, the polyadenylation signal found in the 3' untranslated region (UTR) is believed to enhance both the translation and stability of mRNA (Sachs, Sarnow & Hentze, 1997; Preiss & Hentz,

1998), as well as its transport from the nucleus to the cytoplasm (Hilleren et al., 2001). In two Fabry patients (a type of lysosomal storage disorder, LSD), frame-shift mutations in the 3' region of the α-galactosidase A gene, 1277delAA(del2) and 1284delACTT(del4), were shown to destroy the termination codon and del2 also altered the polyadenylation signal (Yasuda et al., 2003). Transcripts lacking termination codons are normally degraded by the cytoplasmic exosome (Frischmeyer et al., 2002; van Hoof et al., 2002). These mutations did not cause mRNA decay, but rather the formation of either a unique 3' structure or aberrant 3' end that resulted in inefficient translation.

Conversely, a patient with Hurler–Scheie syndrome (an attenuated form of mucopolysaccharidosis I, MPS I), with a T346M mutation (homozygous), had reduced levels of α-L-iduronidase mRNA consistent with mRNA decay (Lee-Chen et al., 1999). Frame-shift or nonsense mutations usually result in very low levels of mRNA due to a specific message degradation process (e.g. Menon & Neufeld, 1994; Myerowitz & Costigan, 1988). In the LSD, Tay Sachs disease, β-hexosaminidase frame-shift mutations have been described which result in premature termination of mRNA transcripts (Myerowitz & Costigan, 1988). One of the most common mutations found in Tay Sachs disease is a 4 bp insertion within exon 11 (1278ins4) of the hexosaminidase A gene, which results in undetectable levels of nonsense mRNA despite normal transcription (Boles & Proia, 1995; Paw & Neufeld, 1988). This suggested either an inherent instability of the mRNA or possibly the existence of a cellular process which has the capacity to recognise nonfunctional mRNA prior to the translocation process (Gieselmann, 1995). Premature stop codon mutations present as a major reason for the onset of pathophysiology in other LSD patients. For most LSD, the percentage of patients with at least one premature stop codon mutation approximates 20%, but in some disorders the percentage of patients is much higher (e.g. up to 95% of MPS I patients, Hein et al., 2004). Both the number of premature stop codon mutations (15 different α-L-iduronidase premature stop codon mutations) and the frequency of specific alleles identified in MPS I patients is high (Hein et al., 2004). For example, the two α-L-iduronidase premature stop codon mutations Q70X and W402X account for up to 70% of disease alleles in some populations (Scott et al., 1992a,b; Bunge et al., 1995, 1998). The impact of premature stop codons on protein biosynthesis tends to be dramatic, generally contributing to a very severe clinical presentation in patients, but to some extent this depends on the type and thus fidelity of the specific stop codon sequence that is introduced into the coding sequence (Hein et al., 2004).

1.3 Translation of Endosome–Lysosome Proteins

Proteins synthesised and translocated at the RER (Figure 2), include both integral membrane and lumenal proteins of either the RER, the Golgi apparatus, the endosomal network, lysosomes, or proteins destined for the secretory pathway (cell surface and extracellular space). Proteins destined for synthesis into either the limiting membrane or the lumen of the RER, have a specific hydrophobic signal sequence. Following the initial synthesis of this signal sequence on ribosomes in the cytosol (Figure 1), elongation of the polypeptide chain is arrested until docking with the RER has been completed (Figure 2). The first step in this docking process is the interaction of a signal recognition particle (SRP) with the ribosome signal sequence and binding of this complex to both SRP receptors and ribosome receptors located in the RER membrane (Neuhof et al., 1998; Raden & Gilmore, 1998; Menetret et al., 2000). This also involves the interaction with

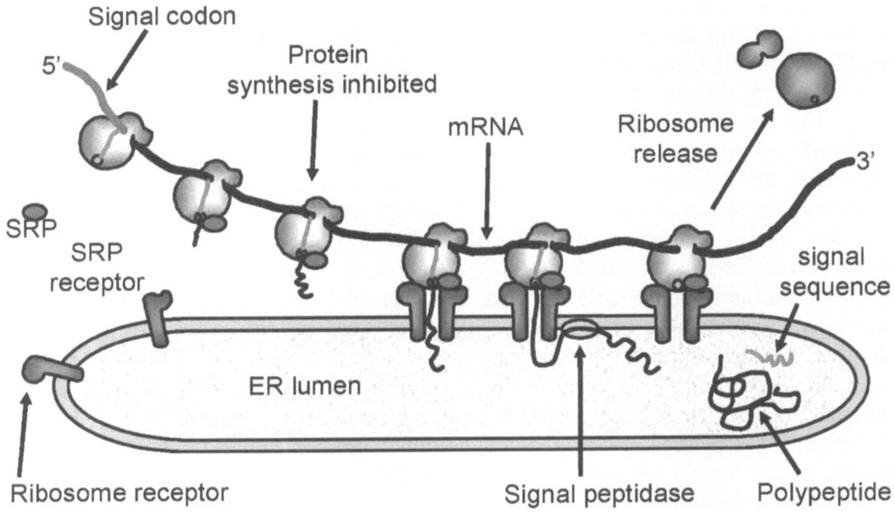

Figure 2. Ribosomes and mRNA, with a short polypeptide signal sequence, dock with both the ribosome and signal recognition particle receptors on the RER. This docking process then allows polypeptide synthesis to proceed through interaction with the Sec61 complex (translocation channel).

the translocation channel (Sec61 complex), which together facilitate the resumption of polypeptide biosynthesis (Gorlich & Rapoport, 1993). Regulatory elements operate at both ends of the translocation channel to control opening and closing of channel gates, and other elements control entry of polypeptide sequence into the phospholipid bi-layer of the RER (Meacock, Greenfield & High, 2000).

2 PROTEIN FOLDING AND PROCESSING OF ENDOSOME–LYSOSOME PROTEINS

2.1 Synthesis and Folding of Protein in the RER

The RER is a specialised subcellular compartment that provides a selective environment for complex folding, modification reactions, and oligomeric assembly of proteins (Helenius et al., 1992), which are either destined for secretion or delivery to the plasma membrane or through specific targeting for delivery to organelles in the endosome–lysosome pathway (Brooks, 1997; Barral et al., 2004). In the lumen of the RER, protein translocation, early protein processing, and folding of the nascent polypeptide proceed in a concerted process.

Molecular chaperones are involved in driving the folding of a nascent polypeptide chain in a series of controlled binding and release cycles catalysing the assembly of the polypeptide towards the lowest free energy state, representing the fully folded molecule.

Molecular chaperones use ATP to drive this folding process, consuming this energy source to enable the protein to traverse a series of energy barriers that are required to catalyse the transition of the polypeptide to specific folding intermediates. Molecular chaperones are also involved in presenting the nascent polypeptide chain in an appropriate conformation to allow specific processing events. For example, suitable sites for N-linked glycosylation (asparagine containing motif; NXS, NXT sequences) must be exposed to allow the transfer of high mannose oligosaccharides to the protein backbone (Goldberg, Gabel & Kornfeld, 1984; von Figura & Hasilick, 1986; Storrie, 1988).

N-linked glycosylation is an important initial processing event for lysosomal proteins as these structures impart stability to the folded protein and are used in chaperone mediated folding of the protein. These structures also act as the critical recipients of targeting signals (for soluble lysosomal proteins), that will either direct traffic to the endosome–lysosome pathway, or if absent, result in traffic out of the cell by the default secretory pathway. N-linked glycosylation is a complex process in its own right. It requires the initial sequential assembly of two N-acetylglucosamine (GlcNAc) sugars and five mannose (Man) sugars attached to dolichol via a pyro-phosphate linkage on the outside of the RER membrane (Hirschberg & Snider, 1987). This $GlcNAc_2$-Man_5 structure is then enzymically transferred to the lumen of the RER where further synthesis is required to complete a high mannose oligosaccharide, which is then capped with three sequential glucose sugar residues ($GlcNAc_2$-Man_9-glucose). It is only at this stage that the completed oligosaccharide structure can be attached to the consensus asparagine motifs within the linear sequence of the nascent polypeptide chain.

Folding and processing events have a significant role in driving the import of the nascent polypeptide chain into the RER, but this process is thought to be reversible for misfolded protein. The correct outcome for folding is the attainment of the appropriate tertiary and quaternary structure and thus functionality of the protein. Errors occur in the normal process of biosynthesis, translocation, processing, and folding. Proteins that are unable to attain their correct three-dimensional structure are retained in the RER where they associate with molecular chaperones for extended periods, in an attempt to fold. If correct folding is not achieved, a quality control process is evoked (Hurtley & Helenius, 1989), which selects the incorrectly folded protein for degradation by "endoplasmic reticulum-associated degradation" (ERAD; McCracken & Brodsky, 2003; Cohen & Kelly, 2003). This involves the complete unfolding of the protein and retrograde transport of the polypeptide to the cytosolic face of the RER for degradation by the proteosome machinery. The quality control and ERAD processes have very important implications for the processing and handling of mutant gene products (Lodish, 1988; Lippincott-Schwartz et al., 1988).

2.2 Disorders Involving Altered Folding of Endosome–Lysosome Proteins

Endosome–lysosome proteins are generally synthesised as large/high molecular weight precursors, which are folded, processed, and then transported through the endomembrane system for further glyco- and proteolytic-processing. Mutations in genes encoding endosome–lysosome proteins, such as missense/point mutations and small deletions and/or insertions generally result in incorrect folding of the mutant polypeptide. The mutant protein may fold to some degree and is often catalytically active, but is either unstable or does not pass the quality control process, which normally allows exit from the RER and entry into the Golgi for traffic to the endosome–lysosome system. The recent resolution of the crystal structure for some endosome–lysosome proteins, together with homology modelling and structure–function analysis, has afforded predictions about the influence of mutations based on their location and interaction within a protein of interest. Mutant proteins that are unable to attain a native structure are retained in the RER and may undergo ERAD, resulting in intracellular loss of the protein. This inability of mutant endosome–lysosome proteins to leave the RER and traffic to the lysosome is a major contributing factor to the onset of pathology in many LSD patients.

The immunochemical analysis of mucopolysaccharidosis (MPS) patient samples has revealed that most patients have reduced levels of protein and activity associated with the specific LSD (e.g. Brooks et al., 1991; Brooks, 1993). Salla disease (SD) and infantile sialic acid storage disorder (ISSD) result from mutations in the SLC17A5 gene, which encodes for sialin, the lysosomal sialic acid transporter. In SD, the attenuated form of the disorder, residual amounts of the mutant protein have been shown to reach the lysosome implying loss of enzyme in an early biosynthetic compartment (Naganawa et al., 2000). Moreover, in ISSD patients (the severe form of the disorder), most of the mutant protein has been shown to be retained either in the RER or the ER-Golgi intermediate compartment (ERGIC; Naganawa et al., 2000; Itoh et al., 2002; Aula et al., 2002; Wreden, Wlizla & Reimer, 2005). Mutations in arylsulfatase A (deficient in metachromatic leukodystrophy, MLD; Henseler et al., 1996; Poeppel et al., 2005 and references therein) and sulfamidase (deficient in MPS IIIA; Muschol et al., 2004) have also been shown to result in protein misfolding and retention in the RER, leading to degradation of the mutant protein. Several α-galactosidase A (deficient in Fabry disease) mutations have been reported to produce mutant protein that was retained in the RER and subsequently degraded by the proteosome machinery (Yasuda et al., 2003). A study of missense mutations in α-mannosidase (deficient in α-mannosidosis) also showed that misfolded protein was retained in the RER (Hansen et al., 2004). The mutations were mainly located within the core domain of the protein and affected either hydrogen bond formation or disulfide bridge formation (Heikinheimo et al., 2003), showing some similarity to the structural effects seen for the mutations in α-galactosidase A (Yasuda et al., 2003).

Most of the naturally occurring mutations in aspartylglucosaminuria (AGU) (deficiency in aspartylglucosaminidase) cause aberrant folding of the precursor polypeptide and inhibit subsequent activation by proteolytic cleavage (Peltola et al., 1996 and references therein). For example, a serine-to-proline substitution at position 72 did not ablate enzyme activity but specifically prevented proteolytic cleavage and activation in the RER. Structural changes caused by missense mutations are the most common cause of AGU disease (Saarela et al., 2001). These mutations have been shown to cause altered folding and destabilisation of the AGA polypeptide by either introducing bulky residues to limited spaces, affecting the location of active site residues, and/or preventing disulfide bond formation.

The lysosomal hydrolase acid β-glucosidase undergoes a series of maturation steps; a 64 kDa polypeptide with high mannose-type oligosaccharide chains is processed to a sialylated 69 kDa form in the Golgi with subsequent processing to a 59 kDa polypeptide (Bergmann & Grabowski, 1989). In one nonneuronopathic Ashkenazi Jewish patient, three non-Jewish type 1 as well as two type 2 neuronopathic Gaucher patients, only the 64 kDa form of acid β-glucosidase was detected. This suggested that the mutant protein had not been transported past the RER to the Golgi where sialylation and further processing occurs. In a severe neuronopathic type 2 Gaucher disease patient, a G202R mutation resulted in a mutant protein with only a slight reduction in enzyme activity when compared to the wild-type protein (Zimmer et al., 1999). However, localisation studies showed that the acid β-glucosidase protein remained within the RER as a mannose-rich polypeptide. Similarly, for mutations in acid α-glucosidase (functional deficiency results in Glycogenosis type II or Pompe Disease) there are a number of examples where mutant protein is retained in a pre-Golgi compartment, consistent with mutant protein retention in the RER (Reuser et al., 1985, 1987; Montalvo et al., 2004). Mutations in β-hexosaminidase (α-subunit) causing classical Tay Sachs disease have also been reported to generate misfolded polypeptide which was retained in the RER and degraded (Paw et al., 1990; Hechtman et al., 1989). The retention and degradation of mutant β-hexosaminidase (β-subunit) in an early biosynthetic compartment has also been reported in cases of Sandhoff disease (Dlott et al., 1990). In cases where small amounts of mutant protein escape the RER and traffic to the endosome–lysosome system, the residual activity is associated with attenuated forms of these disorders.

2.3 A Disorder Involving Aberrant Processing of Endosome-Lysosome Proteins

In the RER, lysosomal and other sulphatases undergo a specific post-translational modification, involving the conversion of a cysteine residue to a Cα-formylglycine. The Cα-formylglycine (Fgly) is part of the conserved hexapeptide L/V-Fgly-X-P-S-R motif (Selmer et al., 1996). From the crystal structures of arysulphatase A and B and mutation analysis, a critical cysteine was identified, which is modified to form the active site Cα-formylglycine residue in these sulphatase enzymes (Brooks et al., 1995; Bond et al., 1997; Lukatela et al., 1998). In the LSD, multiple sulfatase deficiency (MSD) the activity of all sulphatases is substantially decreased (Kolodny & Fluharty, 1995). This deficiency is caused by mutations in the resident RER protein, SUMF1, which normally catalyses the cysteine to Cα-formylglycine processing modification (Dierks et al., 2003; Cosma et al., 2003).

2.4 Vesicular Traffic from the RER to the Golgi

Having passed the quality control process in the RER, an endosome–lysosome glycoprotein must be transferred to the Golgi for further processing. Transport of proteins, between different compartments of the endomembrane system, follows a similar basic principle. This involves the formation of a small vesicular compartment from the donor organelle, vesicle traffic, then docking and fusion of the vesicle with the target organelle, which allows cargo delivery (Figure 3). The mechanism for vesicle formation is similar at different compartments, but specificity is derived from the precise molecular machinery involved at each compartment interface (briefly described at later points in this chapter). For traffic between the RER and the Golgi compartments, COPII and COPI coated vesicles are respectively involved in either anterograde or retrograde vesicular transport. These COP vesicles derive their names from the vesicular coat structures that are used in vesicle formation (Figure 4; reviewed in Lee et al., 2004). As this is the first point for the discussion of vesicular traffic in this chapter, the essential elements of the process are described as a generalised overview (see also Figure 5 and Rothman, 1994).

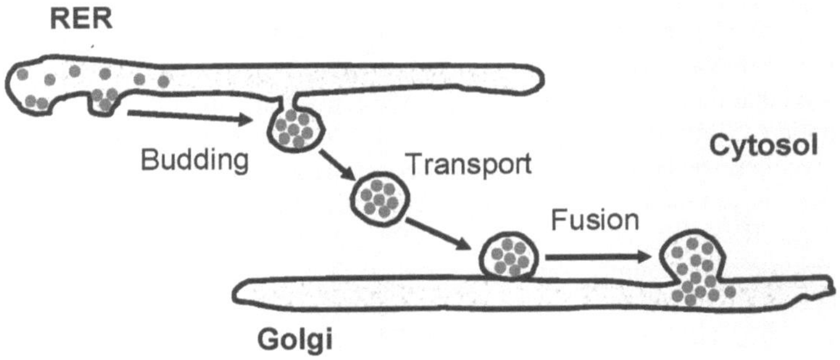

Figure 3. Recruitment of cargo (red) and the formation of vesicles at the RER, for transport and delivery of the cargo by vesicular traffic to the Golgi compartment. At the Golgi membrane, fusion of the vesicle with the target membrane allows the transfer of the transport vesicle's contents into the Golgi compartment.

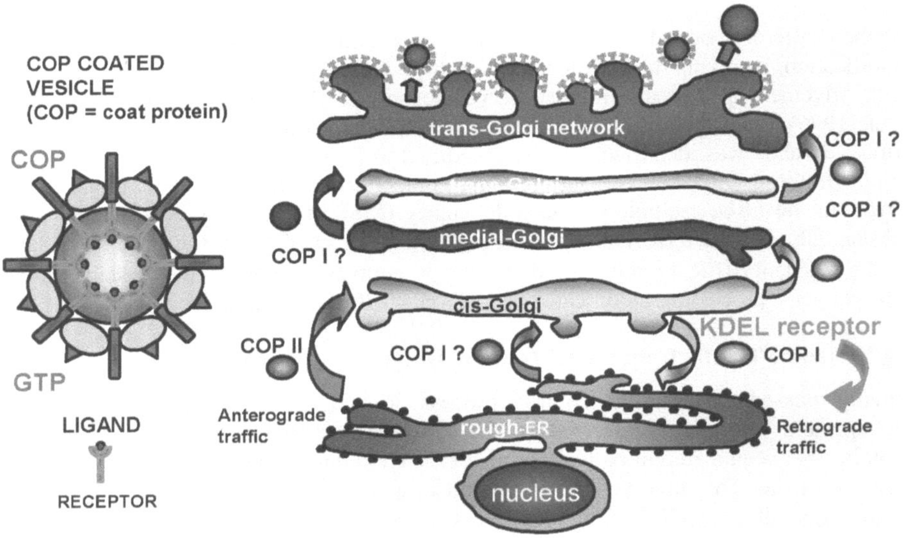

Figure 4. The left-hand figure shows a COP-coated vesicle, with internal ligand bound to its receptor. COP-coated vesicles are used for transport between the RER and the Golgi.

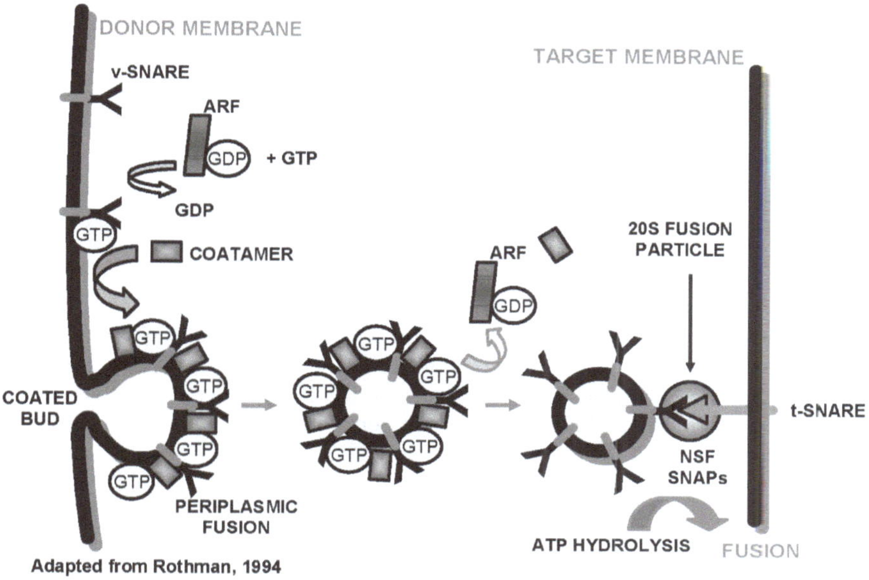

Figure 5. Essential elements of the vesicular machinery for: vesicle formation; vesicle transport, and the fusion process, which are required for the transfer of cargo between organelle compartments. Recruitment of SNARE molecules and coatamer constituents on the donor organelle results in membrane distortion and vesicle formation. Hydrolysis of GTP drives the dissociation of the coat proteins, revealing the vesicle for transport to the target organelle. Interaction of SNARE molecules is involved in the fusion of the vesicle with the target membrane, which is facilitated by ATP hydrolysis.

The first step in vesicle transport involves the recruitment of a specific cargo(s). This requires the segregation of cargo proteins destined for transport, into an area of membrane destined for vesicle formation (e.g. RER exit site). This may involve the interaction of cargo with a specific membrane protein receptor and localisation of this complex with other membrane proteins (or receptors and ligands) destined for transport. The recruitment and localisation process is achieved by the interaction of cytoplasmic sorting signals on the *trans*-membrane proteins with cytosolic coat proteins. The accessibility of an appropriately modified cytoplasmic sorting signal (e.g. dileucine or tyrosine motif) and the recruitment and activation of small GTP-binding proteins (e.g. Sar1 for COPII or Arf1 for COPI), drives the initiation of coat assembly. Coat formation is important in facilitating the distortion of the donor membrane and the subsequent curvature of the membrane that will eventually form a transport vesicle. At different points along the endomembrane system, interactions of coat proteins with elements of the cytoskeleton have also been shown to be important for this membrane distortion and vesicle formation. The recruitment of additional proteins (often referred to as adaptors) and scaffolding (e.g. clathrin at the TGN) form a complex that is the vesicular coat. The recruitment process must also include SNARE proteins that will be involved in the fusion of the vesicle with the target membrane (see below).

The next major step in the process of vesicle formation is budding from the donor membrane, which is achieved by constricting the membrane at the point where it will be pinched off. For example, dynamin has been well characterised for its involvement in the GTP-dependent excision of clathrin-coated vesicles from the cell surface. At this point two important processes must occur, GTP-dependent hydrolysis to remove the vesicular coat and interaction of the vesicle via motor proteins with the microtubule system, which then facilitates the movement of vesicles to the target organelle compartment.

The final step in delivery of the transport vesicle is its interaction with the target organelle. This is mediated first by tethering factors, which initiate a long-range interaction between the membranes destined for fusion and second by SNARE molecules (Figure 5, vesicle or v-SNARES and target or t-SNARES) that facilitate short-range interaction and the fusion process (for details on the SNARE hypothesis see Rothman, 1994). The fusion of the vesicle with the target compartment involves an ATP-dependent hydrolysis mechanism.

For protein traffic from the RER to the Golgi, delivery is thought to occur via an intermediate compartment called the ERGIC, which may have a role in both recovery of specific RER proteins and selective transfer of proteins destined for Golgi processing. Notably, defects in RER to Golgi traffic are yet to be defined as genetic diseases and are probably embryonic lethal due to the essential need for transfer of protein from the RER to the Golgi for either secretion or transfer to the cell surface or traffic to the endocytic system.

2.5 Endosome–Lysosome Protein Modification in the Golgi

The Golgi complex has two major specialist functions for endosome–lysosome proteins, glyco-processing and protein sorting. Glycosylation is the most common form of post-translational processing for proteins of the endomembrane system (Leblond & Bennett, 1977; Hirschberg & Snider, 1987; Lis & Sharon, 1993). Golgi processing of oligo-saccharides involves the removal or addition of saccharide residues and the addition of variable amounts of mannose, N-acetylgalactosamine, galactose, fucose, sialic acid, and repeat units of N-acetyl-lactosamine, in a concerted series of oligosaccharide transfer reactions (e.g. see Figure 6). The oligosaccharide transferases involved in attaching specific glycosylation residues have a restricted compartment distribution within the Golgi complex, giving rise to ordered sequential processing (Figure 7). Protein sorting, which occurs in the *trans*-Golgi, directs transport to either different vesicular compartments of the endomembrane system or to the cell surface. For soluble endosome–lysosome proteins, this sorting process is heavily reliant on the glycosylation structures attached within the RER and modified within distal compartments of the Golgi apparatus.

The most critical glycoprocessing event for soluble endosome–lysosome proteins is the attachment of the mannose 6-phosphate targeting signal to N-linked high mannose oligosaccharides (Figure 8). This occurs in the *cis*-Golgi compartment and is catalysed by the enzyme N-acetylglucosamine 1-phosphotransferase. This transfer reaction attaches the phosphate to mannose residues on the N-linked oligosaccharide, leaving a terminal glucosamine sugar covering the phosphate. The terminal glucosamine residue must be removed by an "uncapping enzyme" in the *trans*-Golgi (Figure 8), to allow recognition of the mannose 6-phosphate structure by mannose 6-phosphate receptors, facilitating protein targeting to the endosome–lysosome system.

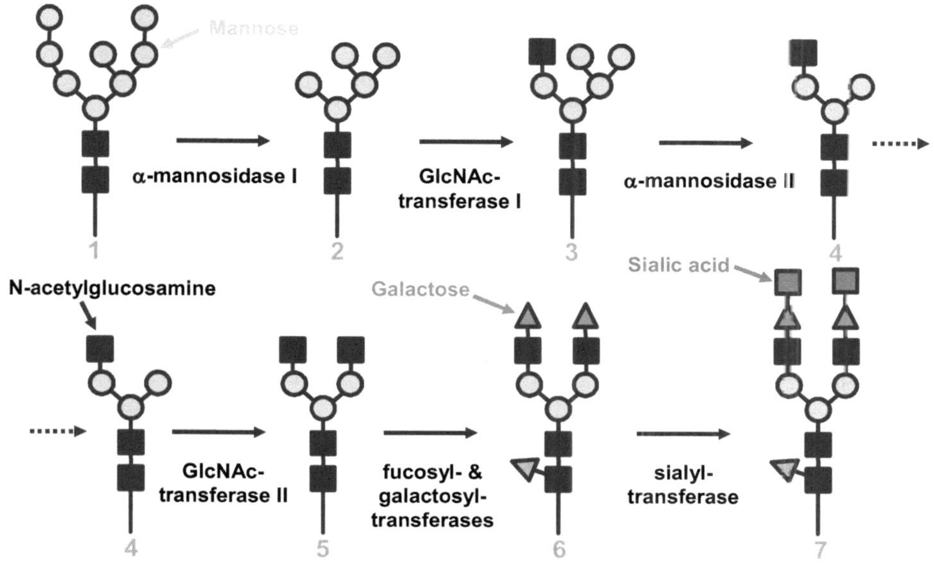

Figure 6. Some of the oligosaccharide modification events for N-linked high mannose-oligosaccharides being processed during their transition through the Golgi.

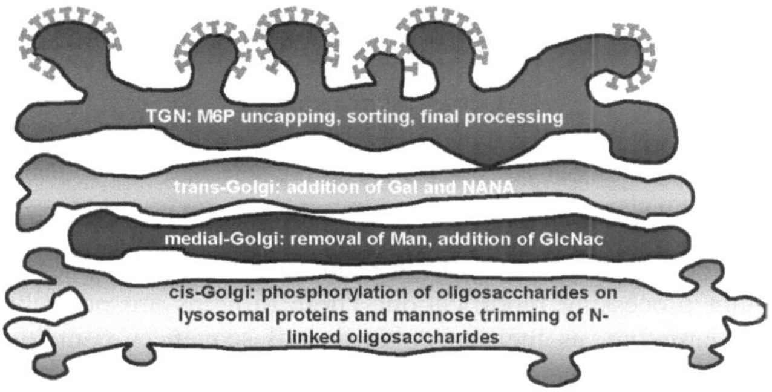

Figure 7. Different compartments in the Golgi apparatus and some of their main functional role(s) in relation to endosome–lysosome protein processing.

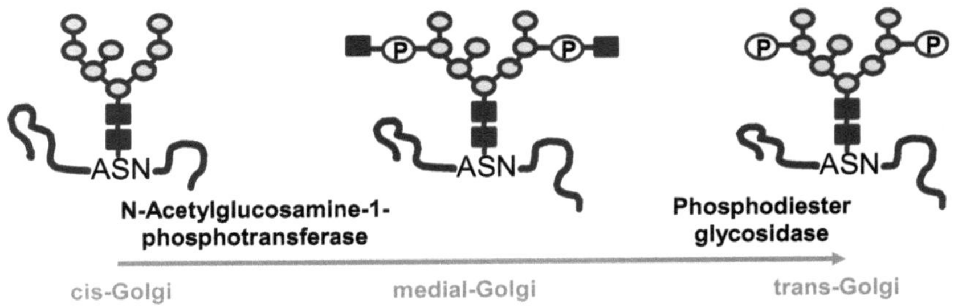

Figure 8. Diagrammatic representation of the key processing events involved in the attachment and exposure of mannose 6-phosphate signals, which are critical for the targeting of soluble endosome–lysosome proteins.

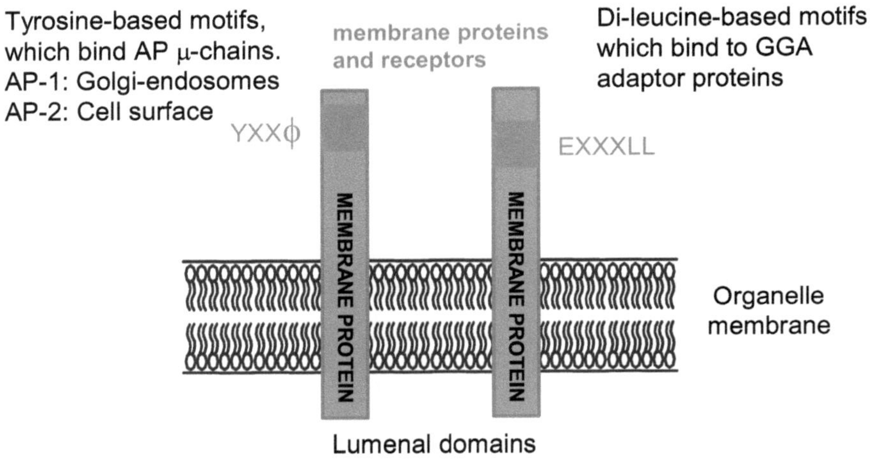

Figure 9. Targeting signals that are located in the cytoplasmic tail sequences of endosome–lysosome membrane proteins.

For integral membrane proteins of the endosome–lysosome system, glycosylation is also an important processing modification. These proteins are not reliant on mannose 6-phosphate for targeting, as their cytoplasmic tail sequences contain specific signals (Figure 9), which direct their traffic to endosomes and lysosomes. Some integral membrane proteins are heavily glycosylated, which may reflect either a structural, functional, or protective role for these glyco-moieties. Lysosomal membrane proteins often have N-linked glycosylation, which may account for up to 50% or more of the molecular size of these glycoproteins (e.g. LAMP-1 and LAMP-2). Other types of glycosylation, such as O-linked glycosylation are also present on integral endosome–lysosome membrane proteins. For both soluble and integral membrane endosome–lysosome proteins, the precise nature of the glycosylation at any given consensus site can exhibit considerable variability (Baenziger, 1994), providing an element of structural and biological uniqueness.

2.6 Lysosomal Storage Disorders Involving Altered Golgi Processing

The mannose 6-phosphorylation of N-linked oligosaccharide(s) on endosome–lysosome proteins occurs within the *cis*-Golgi, providing the unique targeting signal to direct the enzyme to the endosome–lysosome system (Figure 8). Mucolipidosis II/III (inclusion-cell disease, I-cell disease) results from a deficiency in the activity of UDP-N-acetylglucosamine: glycoprotein N-acetylglucosamine 1-phosphotransferase (Haslik, Waheed & von Figura, 1981; Creek & Sly, 1984) and results in a failure to add the mannose 6-phosphate targeting signal to soluble endosome–lysosome proteins. Soluble enzymes that lack mannose 6-phosphate residues cannot bind to mannose 6-phosphate receptors, in the *trans*-Golgi network (Reitman, Varki & Kornfeld, 1981), and are subsequently lost from the affected cell by the default secretory pathway (Wiesman, Vassella & Herschkowitz, 1971). Although cells produce normal amounts of active enzyme, this protein cannot be targeted to the endosome–lysosome system (see below) causing a high level of these enzymes in circulation. In addition, the mistargeted protein cannot be recaptured by mannose 6-phosphate receptors at the cell surface due to the absence of the mannose 6-phosphate recognition signal on the circulating enzyme. This results in a cellular deficiency in the majority of soluble lysosomal enzymes and causes the onset of the very severe clinical manifestations associated with I-cell disease (Kornfeld, 1986). In theory, a deficiency of the "uncapping enzyme" in the *trans*-Golgi could also generate an I-cell phenotype, based on the inability to expose the mannose 6-phosphate targeting signal due to a failure to remove the terminal N-acetylglucosamine sugar residue (see below for details of this enzyme activity), but a disorder relating to this defect is yet to be described in the literature.

3 ENDOSOME–LYSOSOME PROTEIN SORTING IN THE TRANS-GOLGI

As described above, the mannose 6-phosphate targeting signals that were added to the soluble glycoprotein in the *cis*-Golgi are uncapped by the action of N-acetylglucosamine 1-phosphodiester-α-N-acetylglucosaminidase (or uncovering enzyme), in the *trans*-Golgi. This final glycoprocessing step allows the interaction of the glycoprotein with mannose 6-phosphate receptors in the *trans*-Golgi, which then facilitates traffic to endosome–lysosome organelles (Figure 10; reviewed in: Kornfeld & Mellman, 1989; von Figura, 1991; Kornfeld, 1992; Hille-Rehfeld, 1995). The traffic of mannose 6-phosphate receptors and other integral membrane proteins to endosomes is based on the presence of a cytoplasmic tail sequence(s) (e.g. a dileucine or a tyrosine motif, Figure 9). The correct modification of this tail sequence (e.g. phosphorylation) is a key step in initiating vesiculation at the *trans*-Golgi. Clathrin-coated vesicles are formed at the *trans*-Golgi network in a similar manner to that described above for COP-coated vesicles in the RER-Golgi region (see Figure 10).

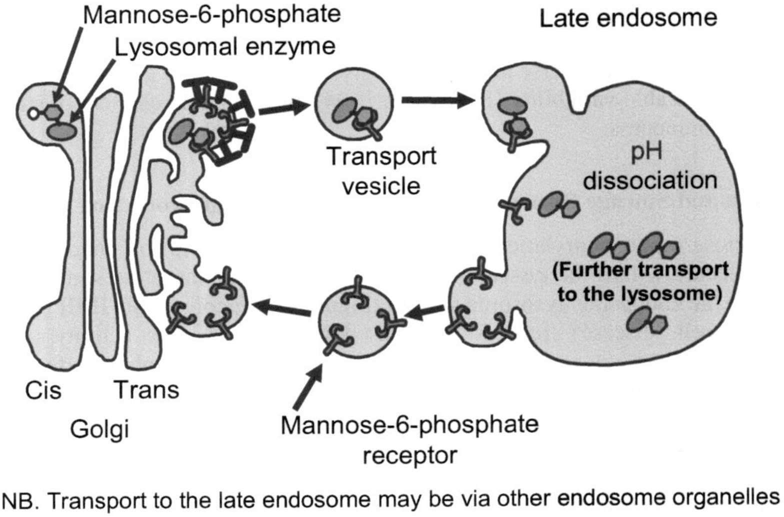

Figure 10. Mannose 6-phosphate targeting of soluble endosome–lysosome proteins. Uncapping of the mannose 6-phosphate targeting signal in the *trans*-Golgi results in interaction of the lysosomal enzyme with mannose 6-phosphate receptors. Vesicular transport of the receptor–ligand complex to the endosome compartment and dissociation of the ligand from the mannose 6-phosphate receptor in the acidic environment of the endosome, allows the recycling of the mannose 6-phosphate receptors to the Golgi compartment.

As for COP-coated vesicles, GTP-binding proteins (e.g. Arf, ADP ribosylation factor) and phosphoinositides initiate coat assembly in the *trans*-Golgi, acting as docking sites for adaptor proteins. For soluble lysosomal enzymes exiting the *trans*-Golgi compartment, the GGA (Golgi localised, Gamma-ear-containing Arf-binding) proteins have a critical role in coat assembly, mediating interactions between cytoplasmic targeting signals (e.g. on the mannose 6-phosphate receptor cytoplasmic tail sequence) and clathrin scaffolding. There is some evidence to suggest that lysosomal membrane proteins (e.g. LAMP-1 and LAMP-2) are packaged into different vesicular carriers (Karlsson & Carlsson, 1998), but LAMP-1 has also been reported in clathrin-coated vesicles (Honig et al., 1996). Vesicular traffic of lysosomal proteins from the *trans*-Golgi compartment is directed towards endosomes (Figures 10 and 11).

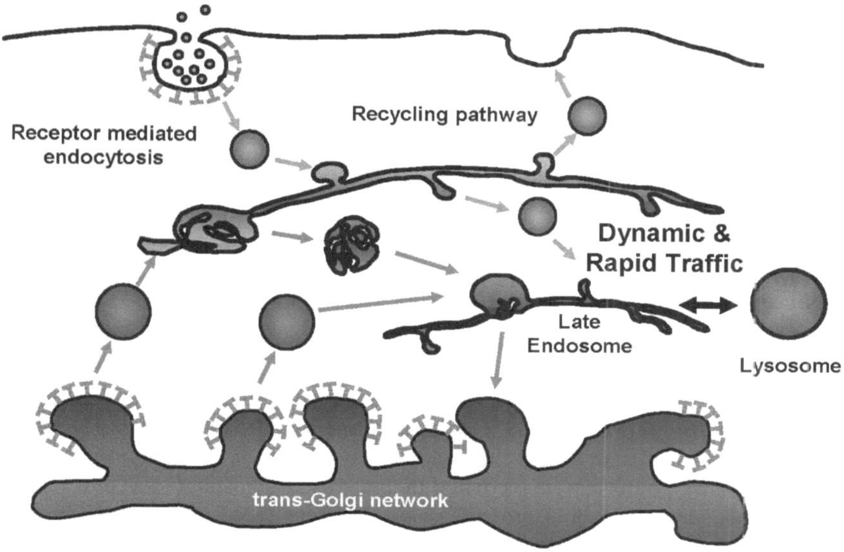

Figure 11. Pathways of vesicular traffic between the Golgi and endosome compartments (purple) and the endocytic pathway (green). Both pathways converge in endosomes and can interact with the terminal compartment in the pathway, the lysosome.

4 ENDOSOME COMPARTMENTS

4.1 Endocytic Pathway

Two major intracellular pathways of molecular traffic, coming from either the biosynthetic compartment(s) or the cell surface, converge in endosomes (Figure 11). Endosomes are by definition a series of tubular and vesicular compartments involved in the traffic of material from the cell surface (Figure 11). In the endocytic pathway a series of different compartments have been identified, which represent critical points (static windows) along this intracellular route. Early endosomes are the first major compartment reached following internalisation and delivery of material from the cell surface. This compartment has a role in the sorting of receptors and ligands for delivery either back to the cell surface or to other elements of the endocytic system. Delivery from the cell surface into early endosomes can involve a variety of different uptake mechanisms including clathrin-coated vesicles (e.g. AP-2 adaptor system), as well as nonclathrin-mediated uptake, such as phagocytosis, caveolae-mediated uptake, and pinocytosis (reviewed in Nichols & Lippincott-Schwartz, 2001).

Late endosomes (prelysosomal compartment) comprise a complex tubuloreticular structure often with internal membranes that are rich in cation-independent mannose 6-phosphate receptors. The passage of endocytic tracers from the cell surface and their entry into early endosomes can be followed and these markers are sequentially delivered to late endosome structures. The acidification of late endosomes (pH less than 5.5) is significantly lower than either the *trans*-Golgi or early endosomes, consistent with the role of this compartment in dissociating ligands from their receptors (e.g. mannose 6-phosphate receptor). The late endosome is also thought to have a major role in sorting receptors for recycling back to either the *trans*-Golgi network or the cell surface. Late endosomes have a high concentration of mannose 6-phosphate receptors, consistent with this role and these receptors are not passed to lysosomes. The late endosome is most probably the intracellular site for the degradation of most macromolecular substrates by lysosomal hydrolases (see below for further details).

4.2 Delivery of Soluble Endosome–Lysosome Proteins to Endosomes

Newly synthesised endosome–lysosome proteins can enter the endocytic pathway via either early or late endosomes (Figure 11). It is likely that the late endosome or prelysosomal compartment is the major entry point for glycoproteins exiting the *trans*-Golgi, due to the close proximity of these two compartments (Griffiths et al., 1988). Soluble lysosomal proteins can also access the endocytic system by uptake from the cell surface. Both of these entry points can be mediated by mannose 6-phosphate receptor recruitment of soluble endosome–lysosome proteins. Two different types of mannose 6-phosphate receptors operate in the endomembrane system (Pfeffer, 1988). The 46 kDa cation-dependent mannose 6-phosphate receptor (Dahms et al., 1987, 1989) has a major role in the delivery of soluble endosome–lysosome proteins out of the *trans*-Golgi network. The 300 kDa cation-independent mannose 6-phosphate receptor (Griffiths et al., 1988) has a major role in the delivery of glycoproteins from the cell surface into endosomes. Both of these mannose 6-phosphate receptors are type I membrane proteins (have one membrane spanning domain), which traffic to and are enriched in the prelysosomal compartment, but are not detected in lysosomes.

4.3 Delivery of Lysosomal Membrane Proteins to Endosomes

There is some uniformity for the traffic of soluble lysosomal proteins out of the *trans*-Golgi because of the interaction with mannose 6-phophate receptors. Although in reality there are two possible mannose 6-phosphate receptors and different potential affinities for these receptors based on the degree of enzyme phosphorylation. In contrast, endosome–lysosome membrane proteins have different possible sorting signals within the cytoplasmic tail sequences (Figure 9 and Table 1). Traffic of membrane proteins to the cell surface and endosome compartments has been reported, reflecting two possible paths to endosomes–lysosomes (e.g. Fukuda, 1991). In some cases (e.g. acid phosphatase), recycling between the cell surface and endosomes appears to be a major trafficking pathway, before processing and delivery to the lysosomal compartment. Clearly some lysosomal membrane proteins traffic beyond the late endosome, showing a high concentration in lysosomal organelles, clearly distinguishing their delivery from mannose 6-phosphate receptors, which do not pass beyond the late endosome compartment. Several lysosomal targeting sequences have been identified for lysosomal membrane proteins and have been shown to reside in the C-terminal cytosolic tail of lysosomal integral membrane proteins (Table 1). Thus the roles of tyrosine motifs (Lobel et al., 1989; Williams & Fukuda, 1990; Chen et al., 1990; Jing et al., 1990; Peters et al., 1990; Collawn et al.,

1990) and dileucine motifs (Sandoval et al., 1994; Ogata & Fukuda, 1994) have been established for the targeting of lysosomal membrane proteins.

LAMP, lysosomal associated membrane protein; LAP: lysosomal acid phosphatase; LIMP: lysosomal integral membrane protein; CD-M6PR: cation-dependent mannose 6-phosphate receptor; CI-M6PR: cation-independent mannose 6-phosphate receptor. The dots (...) refer to parts of the amino acid sequence which have not been listed. Targeting signals are in bold.

The phosphorylation of the mannose 6-phosphate receptor, at two specific serine residues appears to be closely correlated with the formation of clathrin-coated vesicles and exit from the Golgi (Meresse et al., 1990). This phosphorylation reaction appears to be specific to the *trans*-Golgi network, as internalization at the cell surface of mannose 6-phosphate receptor in coated vesicles does not result in the phosphorylation of the same serine residues. Thus, control of membrane protein exit from the *trans*-Golgi and other sites in the endosome–lysosome network may be controlled by specific modification events, in proximity to the known targeting signals. The events involved in the formation of clathrin-coated vesicles, their budding from the *trans*-Golgi network and fusion with the endosomal compartment are mediated by similar molecular events to that for transport between Golgi compartments, but there is evidence of specific adaptor protein involvement (Bonifacino & Traub, 2003).

Alternate targeting mechanisms for soluble lysosomal proteins (i.e. nonmannose 6-phosphate receptor mechanisms), seemed unlikely as the lysosomal storage disorder, I-cell disease was reported to result in the traffic of soluble lysosomal proteins into the secretory pathway. However, the targeting of cathepsin D in a cell line from an I-cell disease patient (Glickman & Kornfeld, 1993) suggested that there may be membrane-associated protein targeting mechanisms for some hydrolases.

Table 1. Cytoplasmic targeting motifs in endosome–lysosome membrane proteins

Protein	Cytoplasmic Tail Sequence	Cellular Location
CD-MPR	**MSS : ...**VGDDQLGEESEERDDH**LL**PM	Golgi & Endosomes
CI-MPR	**MSS : ...**DDRVG**L**VRGE...DDSDED**LL**HI	Cell surface, Endosomes
LAMP-1	**MSS :** RKRSHA**GY**QTI	Endosomes, Lysosomes
LAMP-2	**MSS :** LKHHHA**GY**EQF	Endosomes, Lysosomes
LAMP-3 (LIMP-I)	**MSS :** VKSIRS**GY**EVM	Endosomes, Lysosomes
LAP	**MSS :** RMQAQPP**GY**RHVADGEDHA	Cell surface, Endosomes Lysosomes
LIMP II	**MSS :** RGQGSTDEGTADERAP**LI**RT	Endosomes

4.4 Defects in the Endosome Pathway

Hermansky–Pudlak syndrome (HPS) is a group of genetic disorders characterized by defects in specific organelles of the endosome–lysosome system, particularly affecting melanosome and platelet dense granule formation (Starcevic, Nazarian & Dell'Angelica, 2002). In humans there are four forms, of which HPS type 2 results from a defect in the AP-3 sorting adaptor complex (Dell'Angelica et al., 1999). AP-3 is a coat protein involved in vesicle formation and traffic between the TGN and endosomes with an apparent role in sorting functions in endosomes (Boman, 2001; Robinson & Bonifacino,

2001; Ihrke et al., 2004; Peden et al., 2004). In AP-3 deficient humans there appears to be a specific defect in melanosome biogenesis (site of synthesis and storage of melanin pigment in melanocytes; Huizing et al., 2001 and references for M. Marks). AP-3 also appears to be involved in traffic from the TGN to lysosomes via an endosome-independent pathway (Rous et al., 2002) and facilitates the lysosomal targeting of CLN3. CLN3 mutations cause the LSD Batten disease (Kyttälä et al., 2005).

4.5 Role of the Endocytic Pathway in Enzyme Replacement Therapy

In the early 1960s, studies on the uptake of horseradish peroxidase by rat kidney epithelial cells identified an endocytic compartment, between the cell surface and lysosomes (Straus, 1964; Helenius et al., 1983). At various time points after endocytosis the internalised horseradish peroxidase was observed in different organelle populations en route to lysosomes. These studies suggested a functional role for these newly identified compartments, for the delivery of material endocytosed at the cell surface towards the lysosomal compartment for degradation. The endocytic pathway can be exploited to deliver exogenously delivered lysosomal glycoproteins to the lysosome (Figure 12). For enzyme replacement therapy, enzyme administered to a patient's circulatory system will result in contact with cell surface receptors on different cell types.

The cation-independent mannose 6-phosphate receptor, which is usually found in high concentration at the cell surface of most cell types, can be used to import lysosomal proteins into the endocytic network, provided that the glycoprotein has the correct targeting signal (i.e. mannose 6-phosphate on the N-linked oligosaccharide). Following transfer to an early endosome, the internalized glycoprotein would be trafficked to a late endosome compartment where it would be dissociated from the mannose 6-phosphate receptor by the acidic conditions. This would then normally allow for transfer of the glycoprotein to the lysosomal compartment. For lysosomal storage disorders it is likely that the target for the

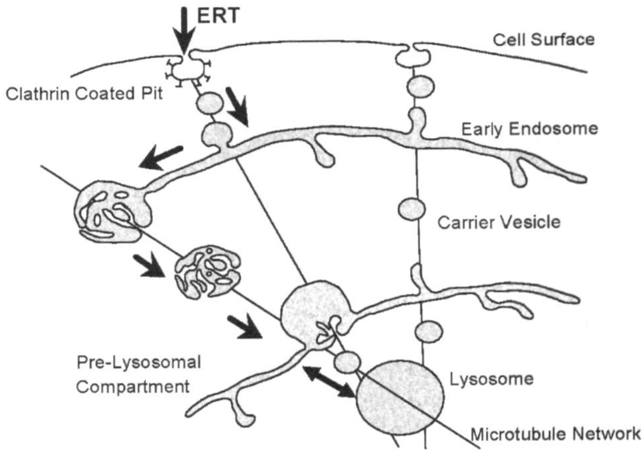

Figure 12. Endocytic pathway and uptake of enzyme for enzyme replacement therapy to access late endosome and lysosome compartments.

correction of storage is a hybrid endosome–lysosome compartment as discussed below. Endocytosis of a replacement protein should therefore access the endosome–lysosome system to correct the enzyme deficiency.

5 LYSOSOMAL COMPARTMENT

Lysosomes are single-membrane-bound organelles that contain a range of acid hydrolases (de Duve et al., 1955), including proteases, glycosidases, sulfatases, phosphatases, and lipases (Figure 13). The lysosome is the terminal compartment in both the endomembrane and biosynthetic and endocytic pathways, and interacts with endosomes carrying cargo from these destinations, as well as endosome compartments from other intracellular pathways (e.g. autophagosomes). Lysosomal organelles are heterogeneous in the morphological criteria of size, shape, and composition. As defined by Storrie (1988): the lysosome is an acidic compartment (which contains a proton pump ATPase); it must be the principal domicile of a 'mature, fully processed lysosomal protein' (the presence alone of a lysosomal protein within an organelle structure does not necessarily establish that organelle as a lysosome); it is the terminal compartment in the endocytic pathway and is distinguished from endosomes by the absence of mannose 6-phosphate receptors; lysosomes must behave as a high density organelle in fractionation experiments.

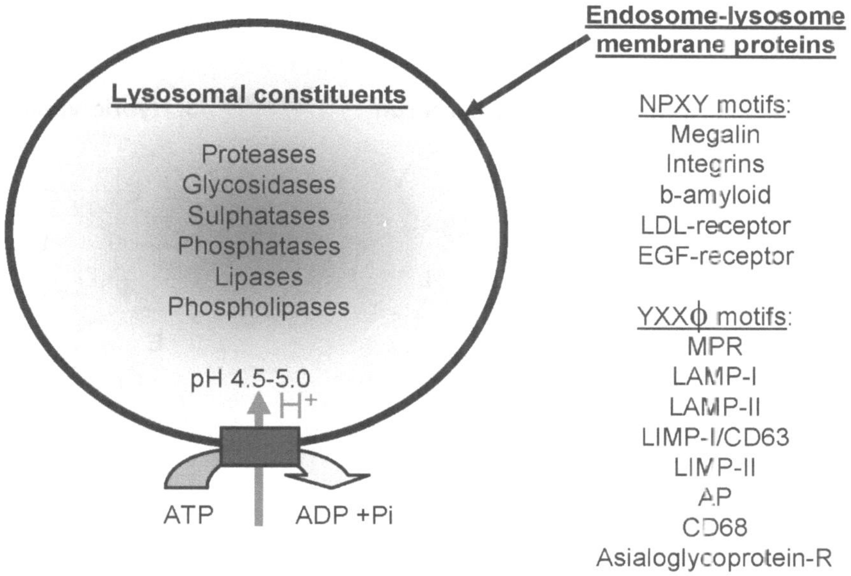

Figure 13. The lysosome and some of its major constituents.

6 DEGRADIVE COMPARTMENTS

6.1 Endosome–Lysosome Interactions

Recent evidence indicates that several different types of interaction can occur between endosomes and lysosomes (Figure 14; reviewed in Luzio et al., 2000, 2003) and it is suggested that this may relate to different functional requirements. First, vesicles may be formed from endosomes and then traffic to the lysosome. This type of vesicular transfer may have a functional role in segregating lysosomal enzymes, for traffic to and storage in the lysosomal compartment (e.g. newly synthesised lysosomal proteins coming from the biosynthetic compartments). The second type of interaction has been referred to as "kiss and run" where the two compartments come into close proximity for a short duration, forming a transient pore for the transfer of contents. A plausible reason for this transfer mechanism could involve the passage of either low molecular weight materials between the two organelles or possibly even ion transfer. Finally, endosomes and lysosomes may fuse to form a hybrid organelle. The recovery of lysosomal membrane and contents from this organelle has been reported by Bright et al., 1997.

The fusion of lysosomes and endosomes could represent the way in which these two organelles interact to degrade substrates and effectively form the main degradive compartment of the cell (Griffiths, 1996). Thus the low pH of the endosome could provide limited degradive potential and upon fusion with lysosomes, the increased acidity and influx of additional lysosomal hydrolases could provide for increased enzyme

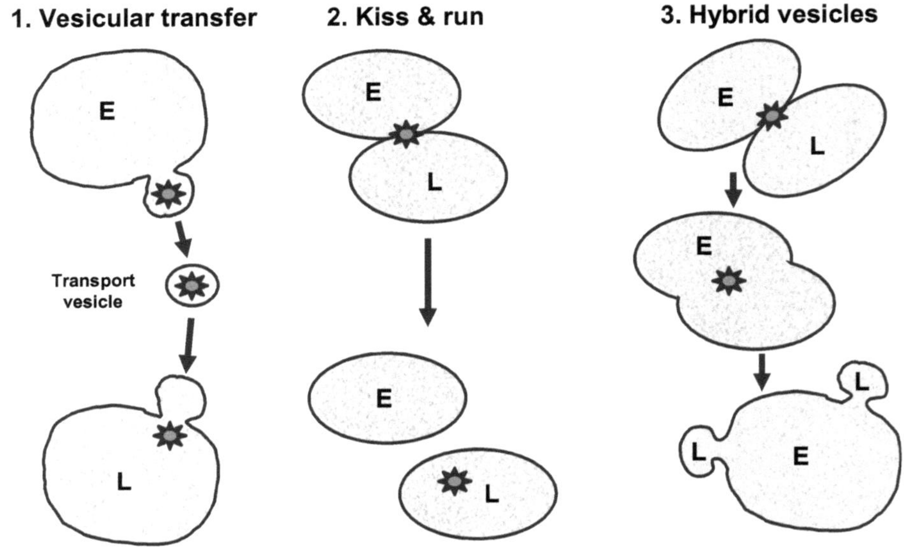

Adapted from the articles of Luzio and colleagues

Figure 14. Diagrammatic representation of three possible interactions between the late endosome and lysosome compartments.

catalytic activity and efficient turnover of substrates. Effectively, the lysosome could act as a reservoir for acid hydrolases, which then fuses with endosomes when required, providing an optimum catalytic environment for substrate breakdown. These events may not all occur at the same stage. For example, an endosome with a high content of mannose 6-phosphate receptor and attached ligand may acidify to release the cargo and allow the mannose 6-phosphate receptor to recycle back to the *trans*-Golgi, before it fuses with a lysosome to engage in a degradive phase to metabolize substrate(s).

6.2 Endosome–Lysosome Degradation and Processing

As described above, the most likely organelle for the degradation and turnover of macromolecular constituents is a hybrid endosome–lysosome compartment. However, the cleavage of macromolecular substrates such as, for example, proteoglycans begins in an early endosome compartment. Following internalisation from the cell surface proteoglycans are subject to protease action, which releases the long chain polysaccharide from the protein core, and endo-hydrolase digestion to generate polysaccharide fragments of approximately 5000 Da. The partially degraded polysaccharide chains are then delivered to late endosomes–lysosomes for exo-hydrolase digestion, which occurs rapidly and is specifically directed at the nonreducing end of the polysaccharide chains.

The process of macro-autophagy is responsible for delivery of organelles and other cytoplasmic material into the endomembrane system for breakdown and reutilisation. Membrane engulfment or pinocytosis (similar to that from the cell surface in macrophages) is used to capture material from the cytoplasm, which is destined for endosome–lysosome degradation. This process allows the cell to turn over intracellular organelles, or other molecular components which do not normally have access to the endomembrane system and is a potential source of extra small molecules and energy in times of cellular stress. Clearly, endosomes generated by the engulfment of intracellular material may be expected to be on a distinct intracellular pathway, compared to that described for endocytosis from the cell surface.

6.3 A Defect in Endosome–Lysosome Interaction

The lysosomal storage disorder mucolipidosis IV is characterised by the generalised accumulation of phospholipids, sphingolipids, and acid mucopolysaccharides in endosomes-lysosomes (Berman et al., 1974; Merin et al., 1975; Newell, Matalon & Meyers, 1975; Tellez-Nagel et al., 1976; Goebel, Kohischutter & Lenard, 1982; Folkerth et al., 1995). Although the activity of lysosomal hydrolases is not affected in this LSD, storage appears to result from aberrant sorting and or traffic along the late endosome–lysosome pathway (Bargal & Bach 1997; Chen, Bach, and Pagano, 1998). The defective gene is MCOLN1, which encodes the integral membrane protein Mucolipin-1 (Bargal et al., 2000; Bassi et al., 2000; Sun et al., 2000), which has been characterised as a Ca^{2+} permeable channel (LaPlante et al., 2002; Raychowdhury et al., 2004). Endocytotic and exocytotic processes are regulated by calcium. The efflux of calcium from the late endosome and lysosome compartments is required for the fusion of these organelles (Pryor et al., 2000). Studies on mutants of the ML-1 *C.elegans* orthologue, *cup-5*, suggest either a defect in the re-formation of lysosomes from hybrid endosome–lysosome organelles or an increased propensity to form hybrid organelles (Piper & Luzio, 2004).

6.4 Storage Disorders

Well over 50 different LSD have been recognized, making them a very significant group of genetic disorders, with a combined incidence of around 1 in 7700 live births (Meikle et al., 1999). The aetiology of the disorders involving soluble lysosomal hydrolases has been well described with the enzymology and sequence of most of these proteins defined. LSD have a number of common features including the storage of undegraded substrates in endosome–lysosome organelles and the onset of severe clinical symptoms, which tend to be multisystemic and progressive in dysfunction. The nature of the storage compartment in each LSD is likely to vary, but in most cases probably involves similar elements of the endosome–lysosome pathway. How this storage results in pathogenesis and how different storage products impart different clinical phenotypes in this set of disorders is yet to be elucidated but is the focus of current intense research. This set of disorders is a focus of this book.

7 SUMMARY

None of the static window pictorials or descriptions of the specific points along the pathway of endomembrane biogenesis, in this chapter, do justice to the dynamics of the endosome–lysosome system. The rapidity of these processes and the constant flux between different compartments is truly astounding. It is not surprising that a defect in any point along this complex and dynamic pathway can result in the group of disorders collectively labelled as LSD. Moreover, it is also obvious that each LSD has an impact on a number of functional aspects within the endomembrane system. The chapters that follow describe some of the classic defects in lysosomal hydrolases and their clinical consequences, which are the devastating LSD. The disorders of biogenesis and vesicular traffic, which are closely related to the more classic LSD involving soluble lysosomal hydrolases, are now giving functional recognition to other important components of the endosome–lysosome system.

REFERENCES

Aula, N., Jalanko, A., Aula, P., and Peltonen, L., 2002, Unraveling the molecular pathogenesis of free sialic acid storage disorders: Altered targeting of mutant sialin, *Mol Genet Metab.* **77**: 99.

Baenziger, J.U., 1994, Protein-specific glycosyltransferases: How and why they do it!, *FASEB J.* **8**: 1019.

Bargal, R. and Bach, G., 1997, Mucolipidosis type IV: Abnormal transport of lipids to lysosomes, *J Inher Metab Dis.* **20**: 625.

Bargal, R., Avidan, N., Asher, B., Olender, Z., Zeigler, M., Frumkin, A., Raas-Rothschild, A., Glusman, G., Lancet, D., and Bach, G., 2000, Identification of the gene causing mucolipidosis type IV, *Nature Genet.* **26**: 118.

Barral, J.M., Broadley, S.A., Schaffar, G., and Hartl, F.U., 2004, Roles of molecular chaperones in protein misfolding diseases, *Semin Cell Dev Biol.* **15**: 17.

Bassi, M.T., Manzoni, M., Monti, E., Pizzo, M.T., Ballabio, A., and Borsani, G., 2000, Cloning of the gene encoding a novel integral membrane protein, mucolipin, and identification of the two major founder mutations causing mucolipidosis type IV, *Am J Hum Genet.* **67**: 1110.

Bergmann, J.E. and Grabowski, G.A., 1989, Posttranslational processing of human lysosomal acid beta-glucosidase: A continuum of defects in Gaucher disease type 1 and type 2 fibroblasts, *Am J Hum Genet.* **44**: 741.

Berman, E.R., Livni, N., Shapira, E., Merin, S., and Levij, I.S., 1974, Congenital corneal clouding with abnormal systemic storage bodies: A new variant of mucolipidosis, *J Pediatr.* **84**: 519.

Boles, D.J. and Proia, R.L., 1995, The molecular basis of HEXA mRNA deficiency caused by the most common Tay-Sachs disease mutation, *Am J Hum Genet.* **56**: 716-724.

Boman, A.L., 2001, GGA proteins: new players in the sorting game, *J Cell Sci.* **114**: 3413.

Bond, C.S., Clements, P.R., Ashby, S.J., Collyer, C.A., Harrop, S.J., Hopwood, J.J., and Guss, J.M., 1997, Structure of a human lysosomal sulfatase, *Struct.* **5**: 277.

Bonifacino, J.S. and Traub, L.M., 2003, Signals for sorting of transmembrane proteins to endosomes and lysosomes, *Annu Rev Biochem.* **72**: 395.

Bright, N.A., Reaves, B.J., Mullock, B.M., and Luzio, J.P., 1997, Dense core lysosomes can fuse with late endosomes and are re-formed from the resultant hybrid organelles, *J Cell Sci.* **110**: 2027.

Brooks, D.A., 1993, Immunochemical analysis of lysosomal enzymes in mucopolysaccharidosis type I and type VI patients, *J Inher Metab Dis.* **16**: 3.

Brooks, D.A., 1997, Protein processing: a role in the pathophysiology of genetic disease, *FEBS Lett.* **409**: 115.

Brooks, D.A., McCourt, P.A.G., Gibson, G.J., Ashton, L.J., Shutter, M., and Hopwood, J.J., 1991, Analysis of N-acetylgalactosamine-4-sulfatase protein and kinetics in mucopolysaccharidosis type VI patients, *Am J Hum Genet.* **48**: 710.

Brooks, D.A., Robertson, D.A., Bindloss, C., Litjens, T., Anson, D., Peters, C., Morris, C.P., and Hopwood, J.J., 1995, Two site directed mutations abrogate enzyme activity but have different effects on conformation and cellular content of N-acetylgalactosamine 4-sulfatase protein, *Biochem J.* **307**: 457.

Bunge, S., Clements, P.R., Byers, S., Kleijer, W.J., Brooks, D.A., and Hopwood J.J., 1998, Genotype–phenotype correlations in mucopolysaccharidosis type I using enzyme kinetics, immunoquantification and in vitro turnover studies, *Biochim Biophys Acta.* **1407**: 249.

Bunge, S., Kleijer, W.J., Steglich, C., Beck, M., Schwinger, E., and Gal, A., 1995, Mucopolysaccharidosis type I: identification of 13 novel mutations of the α-L-iduronidase gene, *Hum Mutat.* **6**: 91.

Chen, C-S., Bach, G., and Pagano, R.E., 1998, Abnormal transport along the lysosomal pathway in mucolipidosis, type IV disease, *Proc Natl Acad Sci.* **95**: 6373.

Cohen, F.E. and Kelly, J.W., 2003, Therapeutic approaches to protein-misfolding diseases, *Nature.* **426**: 905.

Collawn, J.F., Stangel, M., Kuhn, L.A., Esekogwu, V., Jing, S.Q., Trowbridge, I.S., and Tainer, J.A., 1990, Transferrin receptor internalization sequence YXRF implicates a tight turn as the structural recognition motif for endocytosis, *Cell.***63**: 1061.

Cosma, M.P., Pepe, S., Annunziata, I., Newbold, R.F., Grompe, M., Parenti, G., and Ballabio, A., 2003, The multiple sulfatase deficiency gene encodes an essential and limiting factor for the activity of sulfatases, *Cell.* **113**: 445.

Creek, K.E. and Sly, W.S., 1984, *Lysosomes in Biology and Pathology*, Amsterdam: Elsevier/North-Holland, p. 63.

Dahms, N.M., Lobel, P., and Kornfeld, S., 1989, Mannose-6-phosphate receptors and lysosomal targeting, *J Biol Chem.* **264**: 12115.

Dahms, N.M., Lobel, P., Breitmeyer, J., Chirgwin, J.M. and Kornfeld, S., 1987, 46 kd mannose-6-phosphate receptor: cloning, expression, and homology to the 215 kd mannose-6-phosphate receptor, *Cell.* **50**: 181.

de Duve, C., Pressman, B.C., Gianetto, R., Wattiaux, R., and Appelmans, F., 1955, Tissue fractionation studies. 6. Intracellular distribution patterns of enzymes in rat-liver tissue, *Biochem J.* **60**: 604.

Dell'Angelica, E.C., Shotelersuk, V., Aguilar, R.C., Gahl, W.A., and Bonifacino, J.S., 1999, Altered trafficking of lysosomal proteins in Hermansky-Pudlak syndrome due to mutations in the β3A subunit of the AP-3 adaptor, *Mol Cell.* **3**: 11.

Dierks, T., Schmidt, B., Borissenko, L.V., Peng, J., Preusser, A., Mariappan, M., and von Figura, K., 2003, Multiple sulfatase deficiency is caused by mutations in the gene encoding the human C(alpha)-formylglycine generating enzyme, *Cell.* **113**: 435.

Dlott, B., d'Azzo, A., Quon, D.V., and Neufeld, E.F., 1990, Two mutations produce intron insertion in mRNA and elongated beta-subunit of human beta-hexosaminidase, *J Biol Chem.* **265**: 17921.

Folkerth, R.D., Alroy, J., Lomakina, I., Skutelsky, E., Raghavan, S.S., and Kolodny, E.H., 1995, Mucolipidosis IV: Morphology and histochemistry of an autopsy case, *J Neuropathol Exp Neurol.* **54**: 154.

Frischmeyer, P.A., van Hoof, A., O'Donnell, K., Guerrerio, A.L., Parker, R., and Dietz, H.C., 2002, An mRNA surveillance mechanism that eliminates transcripts lacking termination codons, *Science.* **295**: 2258.

Fukuda, M., 1991, Lysosomal membrane glycoproteins. structure, biosynthesis, and intracellular trafficking, *J Biol Chem.* **266**: 21327.

Gieselmann, V., 1995, Lysosomal storage diseases, *Biochim et Biophys Acta.* **1270**: 103.

Glickman, J.N. and Kornfeld, S., 1993, Mannose-6-phosphate-independent targeting of lysosomal enzymes in I-cell disease B lymphoblasts, *J Cell Biol.* **123**: 99.

Goebel, H.H., Kohischutter, A., and Lenard, H.G., 1982, Morphological and chemical biopsy findings in mucolipidosis IV, *Clin Neuropathol.* **1**: 73.

Goldberg, D., Gabel, C., and Kornfeld, S., 1984, *Lysosomes in Biology and Pathology 7* Amsterdam: North Holland, p. 45.

Gorlich, D. and Rapoport, T.A., 1993, Protein translocation into proteoliposomes reconstituted from purified components of the endoplasmic reticulum membrane, *Cell.* **75**: 615.

Griffiths, G., Hoflack, B., Simons, K., Mellman, I., and Kornfeld, S., 1988, The mannose-6-phosphate receptor and the biogenesis of lysosomes, *Cell.* **52**: 329.

Griffiths, G.M., 1996, Secretory lysosomes - A special mechanism of regulated secretion in haemopoietic cells, *Trends Cell Biol.* **6**: 329.

Hansen, G., Berg, T., Riise Stensland, H.M., Heikinheimo, P., Klenow, H., Evjen, G., Nilssen, O., and Tollersrud, O.K., 2004, Intracellular transport of human lysosomal alpha-mannosidase and alpha-mannosidosis-related mutants, *Biochem J.* **381**: 537.

Haslik, A., Waheed, A., and von Figura, K., 1981, Enzymatic phosphorylation of lysosomal enzymes in the presence of UDP-N-acetylglucosamine. Absence of the activity in I-cell fibroblasts, *Biochem Biophys Res Commun.* **98**: 761.

Hechtman, P., Boulay, B., Bayerlan, J., and Andermann, E., 1989, The mutation mechanism causing juvenile-onset Tay Sachs disease among Lebanese, *Clin Genet.* **35**: 364.

Heikinheimo, P., Helland, R., Leiros, H.-K., Karlsen, S., Evjen, G., Ravelli, R., Schoen, G., Ruigrok, R., Tollersrud, O.K., and McSweeney, S., 2003, The structure of bovine lysosomal α-mannosidase suggests a novel mechanism for low pH activation, *J Mol Biol.* **327**: 631.

Hein, L.K., Bawden, M., Muller, V.J., Sillence, D., Hopwood, J.J., and Brooks, D.A., 2004, α-L-iduronidase premature stop codons and potential read-through in mucopolysaccharidosis I patients, *J Mol Biol.* **338**: 453.

Helenius, A., Marquardt, T., and Braakman, I., 1992, The endoplasmic reticulum as a protein folding compartment, *Trends Cell Biol.* **2**: 227.

Helenius, A., Mellman, I., Wall, D., and Hubbard, A., 1983, Endosomes, *Trends Biochem Sci.* **July**: 245.

Henseler, M., Klein, A., Reber, M., Vanier, M.T., Landrieu, P., and Sandhoff, K., 1996, Analysis of splice-site mutation in the sap-precursor gene of a patient with metachromatic leukodystrophy, *Am J Hum Genet* **58**: 65.

Hille-Rehfeld, A., 1995, Mannose-6-phosphate receptors in sorting and transport of lysosomal enzymes, *Biochim Biophys Acta.* **1241**: 177.

Hilleren, P., McCarthy, T., Rosbash, M., Parker, R., and Jensen, T.H., 2001, Quality control of mRNA 3'-end processing is linked to the nuclear exosome, *Nature.* **413**: 538.

Hirschberg, C.B. and Snider, M.D., 1987, Topography of glycosylation in the rough endoplasmic reticulum and golgi apparatus, *Ann Rev Biochem.* **56**: 63.

Huizing, M., Sarangarajan, R., Strovel, E., Zhao, Y., Gahl, W.A., and Boissy, R.E., 2001, AP-3 mediates tyrosinase but not TRP-1 trafficking in human melanocytes, *Mol Biol Cell.* **12**: 2075.

Hurtley, S.M. and Helenius, A., 1989, Protein oligomerization in the endoplasmic reticulum, *Ann Rev Cell Biol.* **5**: 277.

Ihrke, G., Kyttälä, A., Russell, M.R., Rous, B.A., and Luzio, J.P., 2004, Differential use of two AP-3-mediated pathways by lysosomal membrane proteins, *Traffic* **5**: 946.

Itoh, K., Naganawa, Y., Matsuzawa, F., Aikawaw, S., Doi, H., Sasagasako, N., Yamada, T., Kira, J., Kobayashi, T., Pshezhetsky, A.V., and Sakuraba, H., 2002, Novel missense mutations in human lysosomal silaidase gene in sialidosis patients and prediction of structural alterations of mutant enzymes, *J Hum Genet.* **47**: 29.

Jing, S.Q., Spencer, T., Miller, K., Hopkins, C., and Trowbridge, I.S., 1990, Role of the human transferrin receptor cytoplasmic domain in endocytosis: Localization of a specific signal sequence for internalization, *J Cell Biol.* **110**: 283.

Karlsson, K. and Carlsson, S.R., 1998, Sorting of lysosomal membrane glycoproteins lamp-1 and lamp-2 into vesicles distinct from mannose-6-phosphate receptor/ gamma-adaptin vesicles at the trans-golgi network, *J Biol Chem.* **273**: 18966.

Kolodny, E.H. and Fluharty, A.L., 1995, In Scriver, C.R., Beaudet, A.L., Sly, W.S., and Valle, D. (Eds.), *The Metabolic and Molecular Bases of Inherited Disease*, New York: McGraw-Hill, p. 2693.

Kornfeld, R. and Kornfeld, S., 1985, Assembly of asparagine-linked oligosaccharides, *Ann Rev Biochem.* **54**: 631.

Kornfeld, S., 1986, Trafficking of lysosomal enzymes in normal and disease states, *J Clin Invest.* **77**: 1.

Kornfeld, S., 1992, Structure and function of the mannose-6-phosphate/insulinlike growth factor II receptors, *Ann Rev Biochem.* **61**: 307.

Kornfeld, S. and Mellman, I., 1989, The biogenesis of lysosomes, *Ann Rev Cell Biol.* **5**: 483.

Kyttälä, A., Yliannala, K., Schu, P., Jalanko, A., and Luzio, J.P., 2005, AP-1 and AP-3 facilitate lysosomal targeting of batten disease protein CLN3 via its dileucine motif, *J Biol Chem.* **280**: 10277.

LaPlante, J.M., Falardeau, J., Sun, M., Kanazirska, M., Brown, E.M., Slaugenhaupt, S.A., and Vassilev, P.M., 2002, Identification and characterisation of the single channel function of human mucolipin-1 implicated in mucolipidosis type IV, a disorder affecting the lysosomal pathway, *FEBS Lett.* **532**: 183.

Leblond, C.P. and Bennett, G., 1977, Role of the Golgi apparatus in terminal glycosylation. In Brinkley, B.R. and Porter, K.R. (Eds.), *International Cell Biology. International congress on Cell Biology*, 1976–1977, New York: Rockefeller University Press, pp. 326.

Lee, M.C., Miller, E.A., Goldberg, J., Orci, L., and Schekman, R., 2004, Bi-directional protein transport between the ER and Golgi, *Annu Rev Cell Dev Biol.* **20**: 87.

Lee-Chen, G.J., Lin, S.P., Tang, Y.F., and Chin, Y.W., 1999, Mucopolysaccharisosis type I: characterization of novel mutations affecting α-L-iduronidase activity, *Clin Genet.* **56**: 66.

Lippincott-Schwartz, J., Bonifacio, J.S., Yuan, L.C., and Klausner, R.D., 1988, Degradation from the endoplasmic reticulum: Disposing of newly synthesized proteins, *Cell.* **54**: 209.

Lis, H. and Sharon, N., 1993, Protein glycosylation: structural and functional aspects, *Eur J Biochem.* **218**: 1.

Lobel, P., Fujimoto, K., Ye, R.D., Griffiths, G., and Kornfeld, S., 1989, Mutations in the cytoplasmic domain of the 275 kD mannose-6-phosphate receptor differentially alter lysosomal enzyme sorting and endocytosis, *Cell.* **57**: 787.

Lodish, H.F., 1988, Transport of secretory and membrane glycoproteins from the rough endoplasmic reticulum to the Golgi. A rate limiting step in protein maturation and secretion, *J Biol Chem.* **263**: 2107.

Lukatela, G., Krauss, N., Theis, K., Selmer, T., Gieselmann, V., von Figura, K., and Saenger, W., 1998, Crystal structure of human arylsulfatase A: The aldehyde function and the metal ion at the active site suggest a novel mechanism for sulfate ester hydrolysis, *Biochem.* **37**: 3654.

Luzio, J.P., Poupon, V., Lindsay, M.R., Mullock, B.M., Piper, R.C., and Pryor, P.R., 2003, Membrane dynamics and the biogenesis of lysosomes, *Mol Membr Biol.* **20**: 141.

Luzio, J.P., Rous, B.A., Bright, N.A., Pryor, P.R., Mullock, B.M., and Piper, R.C., 2000, Lysosome-endosome fusion and lysosome biogenesis, *J Cell Sci.* **113**: 1515.

McCracken, A.A. and Brodsky, J.L., 2003, Evolving questions and paradigm shifts in endoplasmic-reticulum-associated degradation (ERAD), *Bioessays.* **25**: 868.

Meacock, S.L., Greenfield, J.J., and High, S., 2000, Protein targeting and translocation at the endoplasmic reticulum membrane-through the eye of a needle?, *Essays Biochem.* **36**: 1.

Meikle, P.J., Hopwood, J.J., Clague, A.E., and Carey W.F., 1999, Prevalence of lysosomal storage disorders, *JAMA*. **281**: 249.

Menetret, J.F., Neuhof, A., Morgan, D.G., Plath, K., Radermacher, M., Rapoport, T.A., and Akey, C.W., 2000, The structure of ribosome-channel complexes engaged in protein translocation, *Mol Cell*. **6**: 1219.

Menon, K.P. and Neufeld, E.F., 1994, Evidence for degradation of mRNA encoding alpha-L-iduronidase in Hurler fibroblasts with premature termination alleles, *Cell Mol Biol*. **40**: 999.

Meresse, S. and Hoflack, B., 1993, Phosphorylation of the cation-independent mannose-6-phosphate receptor is closely associated with its exit from the *trans*-Golgi network, *J Cell Biol*. **120**: 67.

Meresse, S., Ludwig, T., Frank, R., and Hoflack, B., 1990, Phosphorylation of the cytoplasmic domain of the bovine cation-independent mannose-6-phosphate receptor. Serines 2421 and 2492 are the targets of a casein kinase II associated to the Golgi-derived HAI adaptor complex, *J Biol Chem*. **265**: 18833.

Merin, S., Livni, N., Berman, E.R., and Yatziv, S., 1975, Mucolipidosis IV: ocular systemic and ultrastructural findings, *Invest Ophthal*. **14**: 437.

Montalvo, A.L.E., Cariati, R., Deganuto, M., Guerci, V., Garcia, R., Ciana, G., Bembi, B., and Pittis, M.G., 2004, Glycogenosis type II: Identification and expression of three novel mutations in the acid α-glucosidase gene causing the infantile form of the disease, *Mol Genet Metab*. **81**: 203.

Muschol, N., Storch, S., Ballhausen, D., Beesley, C., Westermann, J.-C., Gal, A., Ullrich, K., Hopwood, J.J., Winchester, B., and Braulke, T., 2004, Transport, enzymatic activity, and stability of mutant sulfamidase (SGSH) identified in patients with mucopolysaccharidosis type IIIA, *Hum Mutat*. **23**: 559.

Myerowitz, R., and Costigan, F.C., 1988, The major defect in Ashkenazi Jews with Tay-Sachs disease is an insertion in the gene for the alpha-chain of beta-hexosaminidase, *J Biol Chem*. **263**: 18587.

Naganawa, Y., Itoh, K., Shimmoto, M., Takiguchi, K., Doi, H., Nishizawa, Y., Kobayashi, T., Kamei, S., Lukong, K.E., Pshezhetsky, A.V., and Sakuraba, H., 2000, Molecular and structural studies of japanese patients with sialidosis type I, *J Hum Genet*. **45**: 241.

Neuhof, A., Rolls, M.M., Jungnickel, B., Kalies, K.U., and Rapoport, T.A., 1998, Binding of signal recognition particle gives ribosome/nascent chain complexes a competitive advantage in endoplasmic reticulum membrane interaction, *Mol Biol Cell*. **9**: 103.

Newell, F.W., Matalon, R., and Meyers, S., 1975, A new mucolipidosis with psychomotor retardation, corneal clouding and retina degeneration, *Am J Opyhal*. **80**: 440.

Nichols, B.J., and Lippincott-Schwartz, J., 2001, Endocytosis without clathrin coats, *Trends Cell Biol*. **11**: 406.

Ogata, S. and Fukuda, M., 1994, Lysosomal targeting of limp II membrane glycoprotein requires a novel leu-ile motif at a particular position in its cytoplasmic tail, *J Biol Chem*. **269**: 5210.

Paw, B.H. and Neufeld, E.F., 1988, Normal transcription of the beta-hexosaminidase alpha-chain gene in the Ashkenazi Tay-Sachs mutation, *J Biol Chem*. **263**: 3012.

Paw, B.H., Moskowitz, S.M., Uhrhammer, N., Wright, N., Kaback, M.M., and Neufeld, E.F., 1990, Juvenile GM2 gangliosidosis caused by substitution of histidine for arginine at position 499 or 504 of the alpha-subunit of beta-hexosaminidase, *J Biol Chem.* **265**: 9452.

Peden, A.A., Oorschot, V., Hesser, B.A., Austin, C.D., Scheller, R.H., and Klumperman, J., 2004, Localization of the AP-3 adaptor complex defines a novel endosomal exit site for lysosomal membrane proteins, *J Cell Biol.* **164**: 1065.

Peltola, M., Tikkanen, R., Peltonen, L., and Jalanko, A., 1996, Ser72Pro active-site disease mutation in human lysosomal aspartylglucosaminidase: Abnormal intracellular processing and evidence for extracellular activation, *Hum Mol Genet.* **5**: 737.

Peters, C., Braun, M., Weber, B., Wendland, M., Schmidt, B., Pohlmann, R., Waheed, A., and von Figura, K., 1990, Targeting of a lysosomal membrane protein: a tyrosine-containing endocytosis signal in the cytoplasmic tail of lysosomal acid phosphatase is necessary and sufficient for targeting to lysosomes, *EMBO J.* **9**: 3497.

Piper, R.C. and Luzio, J.P., 2004, CUPpling calcium to lysosomal biogenesis, *Trends Cell Biol.* **14**: 471.

Poeppel, P., Habetha, M., Marcão, A., Büssow, H., Berna, L., and Gieselmann, V., 2005, Missense mutations as a cause of metachromatic leukodystrophy. Degradation of arylsulfatse A in the endoplasmic reticulum, *FEBS J.* **272**: 1179.

Preiss, T. and Hentze, M.W., 1998, Dual function of the messenger RNA cap structure in poly(A)-tail-promoted translation in yeast, *Nature.* **392**: 516.

Pryor, P.R., Mullock, B.M., Bright, N., Gray, S.R., and Luzio, J.P., 2000, The role of intraorgenellar Ca^{2+} in late endosome-lysosome heterotypic fusion and in the reformation of lysosomes from hybrid organelles, *J Cell Biol.* **149**: 1053.

Raden, D. and Gilmore, R., 1998, Signal recognition particle-dependent targeting of ribosomes to the rough endoplasmic reticulum in the absence and presence of the nascent polypeptide-associated complex, *Mol Biol Cell.* **9**: 117.

Raychowdhury, M.K., Gonzalez-Perrett, S., Montalbetti, N., Timpanaro, G.A., Chasan, B., Goldmann, W.H., Stahl, S., Cooney, A., Goldin, E., and Cantiello, H.F., 2004, Molecular pathophysiology of mucolipidosis type IV: pH Dysregulation of the mucolipin-1 cation channel, *Hum Mol Genet.* **13**: 617.

Reitman, M.L., Varki, A., and Kornfeld, S., 1981, Fibroblasts from patients with I-cell disease and pseudo-Hurler polydystrophy are deficient in uridine 5'-diphosphate-N-acetylglucosamine: Glycoprotein N-acetylglucosaminylphosphotransferase activity, *J Clin Invest.* **67**: 1574.

Reuser, A.J.J., Kroos, M., Oude Elferink, R.P.J., and Tager, J.M., 1985, Defects in synthesis, phosphorylation, and maturation of acid α-glucosidase in glycogenosis type II, *J Biol Chem.* **260**: 8336.

Reuser, A.J.J., Kroos, M., Willemsen, R., Swallow, D., Tager, J.M., and Galjaard, H., 1987, Clinical diversity in glycogenesis type II Biosynthesis and in situ localization of acid α-glucosidase in mutant fibroblasts, *J Clin Invest.* **79**: 1689.

Robinson, M.S. and Bonifacino, J.S., 2001, Adaptor-related proteins, *Curr Opin Cell Biol.* **13**: 444.

Rothman, J.E., 1994, Mechanisms of intracellular protein transport, *Nature.* **372**: 55.

Rous, B.A., Reaves, B.J., Ihrke, G., Briggs, J.A., Gray, S.R., Stephens, D.J., Banting, G., and Luzio, J.P., 2002, Role of adaptor complex AP-3 in targeting wild-type and mutated CD63 to lysosomes, *Mol Biol Cell.* **13**: 1071.

Saarela, J., Minna, L., Oinonen, C., von Schantz, C., Jalanko, A., Rouvinen, J., and Peltonen, L., 2001, Molecular pathogenesis of a disease: structural consequences of aspartylglucosaminuria mutations, *Hum Mol Genet.* **10**: 983.

Sachs, A.B., Sarnow, P., and Hentze, M.W., 1997, Starting at the beginning, middle, and end: translation initiation in eukaryotes, *Cell* **89**: 831.

Sandoval, I.V., Arredondo, J.J., Alcalde, J., Gonzalez, N.A., Vandekerckhove, J., Jimenez, M.A., and Rico, M., 1994, The residues Leu(Ile)475-Ile(Leu, Val, Ala)476, contained in the extended carboxyl cytoplasmic tail, are critical for targeting of the resident lysosomal membrane protein LIMP II to lysosomes, *J Biol Chem.* **269**: 6622.

Scott, H.S., Litjens, T., Hopwood, J.J., and Morris, C.P., 1992a, A common mutation for mucopolysaccharidosis type I associated with a severe Hurler syndrome phenotype, *Hum Mutat.* **1**: 103.

Scott, H.S., Litjens, T., Nelson, P.V., Brooks, D.A., Hopwood, J.J., and Morris, C.P., 1992b, α-L-iduronidase mutations (Q70X and P533R) associate with severe Hurler syndrome phenotype, *Hum Mutat.* **1**: 333.

Selmer, T., Hallmann, A., Schmidt, B., Sumper, M., and von Figura, K., 1996, The evolutionary conservation of a novel protein modification, the conversion of cysteine to serine semialdehyde in arylsulfatase from Volvox carteri, *Eur J Biochem* **238**: 341.

Starcevic, M., Nazarian, R., and Dell'Angelica, E.C., 2002, The molecular machinery for the biogenesis of lysosome-related organelles: lessons from hermansky-pudlak syndrome, *Sem Cell and Dev Biol.* **13**: 271.

Storrie, B., 1988, Assembly of lysosomes: Perspectives from comparative molecular cell biology, *Int Rev Cytol.* **111**: 53.

Straus, W., 1964, Cytochemical observations on the relationship between lysosomes and phagosomes in kidney and liver by combined staining for acid phosphatase and intravenously injected horseradish peroxidase, *J Cell Biol.* **20**: 497.

Sun, M., Goldin, E., Stahl, S., Falardeau, J.L., Kennedy, J.C., Acierno, J.S., Bove, C., Kaneski, C.R., Nagle, J., Bromely, M.C., Colman, M., Schiffmann, R., and Slaugenhaupt, S.A., 2000, Mucolipidosis type IV is caused by mutations in a gene encoding a novel transient receptor potential channel, *Hum Mol Genet.* **9**: 2471.

Tellez-Nagel, I., Rapin, I., Iwamoto, T., Johnson, A.A., Norton, W.T., and Nitowsky, H., 1976, Mucolipidosis IV: Clinical ultrastructural, histochemical and chemical studies of a case, including a brain biopsy, *Arch Neurol.* **33**: 828.

Van Hoof, A., Frischmeyer, P.A., Dietz, H.C., and Parker, R., 2002, Exosome-mediated recognition and degradation of mRNAs lacking a termination codon, *Science.* **295**: 2262.

von Figura, K., 1991, Molecular recognition and targeting of lysosomal proteins, *Curr Opinion Cell Biol.* **3**: 642.

von Figura, K., and Hasilik, A., 1986, Lysosomal enzymes and their receptors, *Ann Rev Biochem.* **55**: 167.

Wiesman, U., Vassella, F., and Herschkowitz, N., 1971, "I-cell" disease: Leakage of lysosomal enzymes into extracellular fluids, *N Engl J Med.* **285**: 1090.

Williams, M.A., and Fukuda, M., 1990, Accumulation of membrane glycoproteins in lysosomes requires a tyrosine residue at a particular position in the cytoplasmic tail, *J Cell Biol.* **111**: 955.

Wreden, C.C., Wlizla, M., and Reimer, R.J., 2005, Varied mechanisms underlie the free sialic acid storage disorders, *J Biol Chem.* **280**: 1408.

Yasuda, M., Shabbeer, J., Benson, S.D., Maire, I., Burnett, R.M., and Desnick, R.J., 2003, Fabry disease: characterization of alpha-galactosidase A double mutations and the D313Y plasma enzyme pseudodeficiency allele, *Hum Mutat.* **22**: 486.

Yasuda, M., Shabbeer, J., Osawa, M., and Desnick, R.J., 2003, Fabry disease: novel α-Galactosidase A 3'-terminal mutations result in multiple transcripts due to aberrant 3'-end formation, *Am J Hum Genet.* **73**: 162.

Zimmer, K.P., le Coutre, P., Aerts, H.M., Harzer, K., Fukuda, M., O'Brien, J.S., and Naim, H.Y., 1999, Intracellular transport of acid beta-glucosidase and lysosome-associated membrane proteins is affected in Gaucher's disease (G202R mutation), *J Pathol.* **188**: 407.

THE CONCEPT OF TREATMENT IN LYSOSOMAL STORAGE DISEASES

Roscoe O. Brady

Even though sporadic reports of individual patients who were later classified having a lysosomal storage disease began to appear toward the end of the 19th century (Tay, 1881; Gaucher, 1882; Sachs, 1887), little was known about the true nature of storage materials in the LSDs until the early third of the 20th century. In 1907, F. Marchand reported that a hyaline-like material was stored in the "so-called idiopathic splenomegaly" of the Gaucher type. However, he erroneously believed that it was not a lipid because the material did not react with osmic acid. The accumulating substance was identified as a cerebroside by Lieb in 1924. Cerebrosides consist of three components: two are lipids (sphingosine and fatty acid), and the third is a carbohydrate. Galactocerebroside had been known since the beginning of this century to be the preponderant lipid of the brain on a weight basis. Lieb believed that this cerebroside accumulated in the organs and tissues of patients with Gaucher disease. However, the optical rotation of an aqueous solution of the sugar derived from the accumulating cerebroside was incompatible with its being galactose. In 1934, Aghion reported that the sugar moiety of the accumulating cerebroside was glucose rather than galactose. This finding was confirmed by many investigators, thereby conclusively establishing that the principal accumulating lipid in patients with Gaucher disease is glucocerebroside.

Progress in identifying the underlying cause of Gaucher or any other lysosomal storage disorder was very slow. Because of the difference in the nature of the hexose in the accumulating glucocerebroside in Gaucher disease and that in galactocerebroside in the brain, Thannhauser (1950) mentioned that a "galactose test" appeared to be normal in Gaucher disease. In order to clarify whether there might be an abnormality of hexose metabolism, my colleagues and I carried out a galactose tolerance test in a patient with type 1 Gaucher disease in 1956. The test revealed conclusively that galactose was normally handled in the tissues of patients with Gaucher disease.

At this point, we were faced with a considerable dilemma. Did patients with Gaucher disease make glucocerebroside instead of galactocerebroside in their organs and systemic tissues? Did they overproduce glucocerebroside, a normally minor sphingolipid component? Or was there another explanation? My colleague Eberhard Trams and I undertook investigations on the biosynthesis of cerebrosides in surviving spleen tissue slices obtained from patients with Gaucher disease who were undergoing splenectomy, a common procedure to palliate their distended abdomens and improve anemia and thrombocytopenia that are hallmarks of the disorder. We found that there was no abnormality

Developmental and Metabolic Neurology Branch, National Institute of Neurological Disorders and Stroke National Institutes of Health, Bethesda, MD 20892-1260. Tel (301) 496-3285; Fax: (301) 496-9480; e-mail rb57v@nih.gov

in the formation of either galactocerebroside or glucocerebroside in tissues derived from the patients. We therefore postulated that it was likely that there was a defect in glucocerebroside catabolism in Gaucher disease (Trams and Brady, 1960). It took several years to demonstrate the enzymatic defect in Gaucher disease. My colleagues and I tried to find an enzyme that catalyzed the hydrolytic cleavage of either fatty acid or glucose (or both) from unlabeled glucocerebroside isolated from spleen tissue from patients with Gaucher disease. The bioassays were not sufficiently sensitive to detect either of these biodegradative enzymes.

Because the chemical synthesis of glucocerebroside had not been accomplished, we attempted to label it throughout the molecule by incubating it with tritium gas under pressure in a sealed vessel for a week, the so-called Wilzbach technique. We were unable to obtain ^{3}H-glucocerebroside with sufficiently low background radioactivity to obtain meaningful enzymatic data.

About this time, Dr. David Shapiro at the Weizmann Institute of Science in Rehovot, Israel published his paper on the chemical synthesis of sphingomyelin. It had been known for a number of years that sphingomyelin was the offending accumulating lipid in the metabolic disorder known as Niemann–Pick disease (Klenk, 1934). I wrote to Dr. Shapiro and suggested that I come to the Weizmann Institute and with his guidance, prepare sphingomyelin radioactively labeled with ^{14}C in various portions of the molecule. Dr. Shapiro replied that he did not have radioactive counting facilities in his laboratory and suggested that he come to the United States and that we prepare labeled glucocerebroside to determine the metabolic defect in Gaucher disease. I obtained a small travel grant for Dr. Shapiro, and he and Julian Kanfer working in my laboratory synthesized two preparations of 14C-labeled glucocerebroside. The first was labeled in the fatty acid portion of the molecule, and the second in the glucose moiety.

We could not find any evidence of an enzyme in mammalian tissues that catalyzed the hydrolysis of the fatty acid moiety of glucocerebroside, but every tissue we examined had significant activity of the enzyme glucocerebrosidase that catalyzed the hydrolytic cleavage of glucose from this sphingolipid (Brady, Kanfer, and Shapiro, 1965a). When we examined glucocerebrosidase activity in human spleen tissues, we found a very significant reduction in the tissues from patients with Gaucher disease compared with controls (Brady et al., 1965b, 1966a). It should be noted that all surviving patients with Gaucher disease exhibited some, albeit significantly reduced, glucocerebrosidase activity.

In 1966, Drs. Kanfer and Shapiro and I labeled sphingomyelin with ^{14}C and discovered the enzyme sphingomyelinase in many mammalian tissues that catalyzes the hydrolytic cleavage of phosphocholine from sphingomyelin (Kanfer et al., 1966). Sphingomyelinase activity was found to be dramatically reduced in tissues obtained from patients with Niemann–Pick disease from that in human control tissue samples (Brady et al., 1966b). This deficiency is the metabolic defect in Niemann–Pick disease.

At the time these investigations were undertaken, it was unclear whether glucocerebrosidase should be designated as a lysosomal enzyme. The highest specific activity was found in spleen tissue. When rat and human spleen specimens were disrupted in two volumes of 0.1 M potassium phosphate buffer, most of the glucocerebrosidase activity was present in the 100,000 × g high-speed supernatant solution (Brady, Kanfer, and Shapiro, 1965a). The pH optimum for glucocerebrosidase was 6.0. A subsequent investigation comparing the subcellular distribution of glucocerebrosidase and galactocerebrosidase revealed the presence of an enzyme in rat intestinal tissue that appeared to catalyze the hydrolysis of both glucocerebroside and galactocerebroside

(Brady et al., 1965c). Homogenization of this tissue was carried out in nine volumes of a 0.25 M sucrose solution. The bulk of this enzymatic activity was recovered in subcellular particles that sedimented between 600 × g and 8400 × g. In an examination of the tissue distribution of sphingomyelinase, the liver was found to contain the preponderance of its activity. Subcellular distribution of sphingomyelinase was examined following homogenization in nine volumes of 0.25 M sucrose solution. Most of the sphingomyelinase activity was found in the particles sedimenting between 600 × g and 9000 × g although there was significant sphingomyelinase activity in higher fractions (Kanfer et al., 1966). The pH optimum for sphingomyelinase was 5.0.

An investigation of the activities of sphingolipid hydrolases in leukemic leukocytes revealed that glucocerebrosidase activity was highest in the subcellular fraction that contained most of the acid phosphatase activity that had been used as a marker for cell fractions containing enzymes associated with lysosomes. It was deduced that glucocerebrosidase and other sphingolipid hydrolases were lysosomal enzymes (Kampine et al., 1967). Direct confirmation of the lysosomal localization was obtained for glucocerebrosidase, galactocerebrosidase, and sphingomyelinase (Weinreb, Brady, and Tappel, 1968).

The discovery of the enzymatic defects in Gaucher disease and in Niemann–Pick disease provided a basis for understanding the metabolic abnormalities in other sphingolipid storage disorders including prediction of the specific enzymatic defects in Fabry disease, Tay–Sachs disease, and generalized (G_{M1}) gangliosidosis (Brady, 1966c). In that overview, I speculated on potential therapeutic strategies to treat patients with disorders of this type. It seemed quite elementary that if an enzyme were insufficiently active, one might attempt to purify it and inject it into patients to see if it would provide therapeutic benefit. A few years previously, Christian de Duve had speculated on the possibility of enzyme replacement for patients with lysosomal storage disorders to which the sphingolipidoses had subsequently been shown to belong as indicated in the previous paragraph. The first investigation along this line that my colleagues and undertook was the intravenous injection of hexosaminidase A that had been isolated from human urine into an infant with the Sandhoff form (O-variant) of Tay–Sachs disease. These patients accumulate ganglioside G_{M2} in the brain and other organs along with globoside (N-acetylgalactosaminylgalactosylgalactosylglucosylceramide) in peripheral tissues and in the blood. We found that there was a significant reduction of globoside in the circulation shortly after infusing hexosaminidase A (Johnson and Brady, 1972). None of the injected enzyme reached the brain. The patient had a mild pyrexia following infusion. There was no change in the patient's clinical condition.

In order to further reduce the possibility of sensitizing patients to a foreign protein, I decided to investigate the possibility that human placenta might contain the requisite sphingolipid hydrolyzing enzymes. The first of these that we purified from this source was ceramidetrihexosidase, now called α-galactosidase A (Johnson and Brady, 1972) that is deficient in patients with Fabry disease (Brady et al., 1967). We found that intravenous administration of small quantities of this enzyme to two patients with Fabry disease led to a rapid reduction of ceramidetrihexoside (globotriaosylceramide, Gb3) in the circulation (Brady et al., 1973). We were not permitted to perform organ biopsies at that time. Many years passed until additional enzyme replacement (ERT) trials were performed in patients with Fabry disease (Schiffmann et al., 2001; Schiffmann et al., 2002; Eng et al., 2001).

I learned that glucocerebrosidase was also present in placental tissue. My colleagues and I partially purified this enzyme from this source (Pentchev et al., 1973). We felt it, too, was sufficiently pure to administer it to patients with Gaucher disease. We carried out single intravenous injections of placental glucocerebrosidase in a young male with type 3 (chronic neuronopathic) Gaucher disease and in a splenectomized female with type 1 (non-neuronopatic) Gaucher disease. We obtained permission to perform percutaneous liver needle biopsies before and 24 hours following the infusion of the glucocerebrosidase. We observed a 26% decrease in the quantity of glucocerebroside in the post-infusion biopsies in both recipients (Brady et al., 1974). As in the patients with Fabry disease, there was a decrease in the quantity of glucocerebroside in the blood of both Gaucher patients following administration of glucocerebrosidase. In contrast with the return of blood Gb3 to pre-infusion levels within 72 hours in the two Fabry patients, the reduction of glucocerebroside in the circulation of the Gaucher patients lasted many weeks.

This finding provided considerable incentive to continue to investigate ERT in patients with lysosomal disorders. A long period of time was required to develop a satisfactory procedure to obtain sufficient quantities of glucocerebrosidase for efficacy trials (Furbish et al., 1977) An especially important development occurred when John Barranger and Scott Furbish discovered that glucocerebrosidase was a glycoprotein. A few years earlier, Ashwell and Morell discovered the first mammalian lectins (Van den Hamer et al., 1970). Ashwell and Kawasaki first defined a mannose-specific lectin in chicken liver. Other investigators confirmed this finding shortly thereafter (Achord et al., 1978).

Collaboration between Barranger and Ashwell led to the discovery that the mannose-specific lectin on Kupffer cells could be used as a drug delivery system for glucocerebrosidase (Barranger et al., 1978; Steer et al., 1978; Furbish et al., 1978). As a result, a single patient was investigated in a pilot study in 1984 using the mannose terminated glucocerebrosidase. The clinical results in this child were immediate and significant. Hemoglobin rose from 4.5 g/dL to 8.0 g/dL in a period of only ten weeks. Normalization of hematologic indices, resolution of organomegaly and bone marrow clearance of stored glucocerebrosidase were later documented (Barranger et al., 1989; Barton et al., 1990). Subsequently, a formal trial in 12 patients was conducted and the results were similarly convincing (Barton et al., 1991).

Macrophage-targeted placental glucocerebrosidase was approved by the U.S. Food and Drug Administration for the treatment of patients with Gaucher disease on April 5, 1991. The enzyme was subsequently produced recombinantly in Chinese hamster ovary cells using a cDNA based on the glucocerebrosidase gene first cloned by Barranger and Ginns (Ginns et al., 1984). Recombinant glucocererbrosidase was also modified so that its oligosaccharide side chains terminate with mannose. This preparation was biologically equivalent to mannose-terminal placental glucocerebrosidase (Grabowski et al., 1995). It was approved for the treatment of patients with Gaucher disease by the U.S. Food and Drug Administration in 1994. More than 4000 patients are now receiving this therapy. Encouraged by the success of ERT in Gaucher disease, patients with Fabry disease, and a number of additional metabolic storage disorders are now being treated by ERT (see Chapters 20, 21, 24, and 30, this book).

The proposal that ERT might benefit patients with metabolic storage disorders was initially met with great skepticism. The long period of time required to make it effective is in itself a significant lesson in medical research. In addition, it seems doubtful if successful ERT for lipid storage disorders could have been accomplished without the concept of long-term difficult research that underlay the establishment of the United States National Institutes of Health (NIH). Many times outside evaluators of my program were told that little if any progress had been made. Nevertheless, they were very supportive and urged that the research be continued to its eventual benefit to many patients. I feel I was fortunate to have been employed at the NIH during the time these fundamental investigations took place.

REFERENCES

Achord DT, Brot FE, Bell CE, Sly WS. Human beta-glucuronidase: In vivo clearance and in vitro uptake by a glycoprotein recognition system on reticuloendothelial cells. *Cell.* 1978; Sep;15(1): 269–278.

Aghion, H. *La maladie de Gaucher dans l'enfance.* 1934; Thèse, Paris.

Barton NW, Brady RO, Dambrosia JM, DiBisceglie AM, Doppelt SH, Hill SC, et al. Replacement therapy for inherited enzyme deficiency macrophage-targeted glucocerebrosidase for Gaucher's disease. *N Engl J Med.* 1991; 324: 1464–1470.

Barton NW, Furbish FS, Murray GJ, Garfield M, Brady RO. Therapeutic response to intravenous infusions of glucocerebrosidase in a patient with Gaucher disease. *Proc Natl Acad Sci USA.* 1990; 87: 1913–1916.

Barranger JA, Ohashi T, Hong CM, et al. Molecular pathology and therapy of Gaucher disease. *Jpn J Inherit Metabol Dis.* 1989; 51: 45–71.

Barranger JA, Pentchev PG, Furbish FS, Steer CJ, Jones EA, Brady RO. Studies of lysosomal function: I. Metabolism of some complex lipids by isolated hepatocytes and Kupffer cells. *Biochem Biophys Res Commun.* 1978 Aug 14; 83(3): 1055–1060.

Barranger JA, Rapoport SI, Fredericks WR, Pentchev PG, MacDermot KD, Steusing JK, Brady RO. Modification of the blood–brain barrier: increased concentration and fate of enzymes entering the brain. *Proc Natl Acad Sci USA.* 1979 Jan; 76(1): 481–485.

Brady, RO. Sphingolipidoses. *N Engl J Med.* 1966c; 275: 312–318.

Brady RO, Gal AE, Bradley RM, Martensson, E, Warshaw AL, Laster L. Enzymatic defect in Fabry's disease. Ceramidetrihexosidase deficency. *N Engl J Med.* 1967; 276: 1163–1167.

Brady RO, Gal AE, Kanfer JN, Bradley RM. The metabolism of cerebrosides. III. Purification and properties of a glucosyl- and galactosylceramide-cleaving enzyme from rat intestinal tissue. *J. Biol. Chem.* 1965; 240: 3766–3770.

Brady RO, Kanfer JN, Bradley RM, Shapiro D. Demonstration of a deficiency of glucocerebroside-cleaving enzyme in Gaucher's disease. *J. Clin. Inves.* 1966a; 45: 1112–1115.

Brady RO, Kanfer JN, Mock MB, Fredrickson DS. The metabolism of sphingomyelin. II. Evidence of an enzymatic deficiency in Niemann–Pick disease. *Proc. Natl. Acad. Sci. USA.* 1966b; 55: 366–369.

Brady RO, Kanfer J, Shapiro D. The metabolism of glucocerebrosides. I. Purification and properties of a glucocerebroside-cleaving enzyme from spleen tissue. *J. Biol. Chem.* 1965a; 240: 39–42.

Brady RO, Kanfer JN, Shapiro D. Metabolism of glucocerebrosides. II. Evidence of an enzymatic deficiency in Gaucher's disease. *Biochem. Biophys. Res. Commun.* 1965b; 18: 221–225.

Brady RO, Pentchev PG, Gal AE, Hibbert SR, Dekaban AS, Replacement therapy for inherited enzyme deficiency: Use of purified glucocerebrosidase in Gaucher's disease. *N Engl J. Med.* 1974; 291: 989–993.

Brady RO, Tallman JF, Johnson WG, Gal AE, Leahy WE, Quirk JM, Dekaban AS, Replacement therapy for inherited enzyme deficiency: Use of purified ceramidetrihexosidase in Fabry's disease. *N Engl J Med.* 1973; 289: 9–14.

Eng CM, Guffon N, Wilcox WR, Germain DP, Lee P, Waldek S, Caplan L, Linthorst GE, Desnick RJ. International Collaborative Fabry Disease Study Group, Safety and efficacy of recombinant human alpha-galactosidase A–replacement therapy in Fabry's disease. *N Engl J Med.* 2001; 345: 9–16.

Furbish FS, Blair HE, Shiloach J, Pentchev PG, Brady RO. Enzyme replacement therapy in Gaucher's disease: Large-scale purification of glucocerebrosidase suitable for human administration. *Proc Natl Acad Sci USA.* 1977; 74: 3560–3563.

Furbish FS, Steer CJ, Barranger JA, Jones EA, Brady RO. The uptake of native and desialylated glucocerebrosidase by rat hepatocytes and Kupffer cells. Biochem Biophys Res Commun. 1978 Apr;14; 81(3): 1047–1053.

Furbish FS, Steer CJ, Krett NL, Barranger, JA. Uptake and distribution of placental glucocerebrosidase in rat hepatic cells and effects of sequential deglycosylation. *Biochim Biophys Acta.* 1981; 673: 425–434.

Gaucher PCE. De l'épithélioma primitif de la rate. 1882: Thèse de Paris.

Ginns EI, Choudary PV, Martin BM, Winfield S, Stubblefield B, Mayor J, Merkle-Lehman D, Murray GJ, Bowers LA, Barranger JA. Isolation of cDNA clones for human beta-glucocerebrosidase using the lambda gt11 expression system. *Biochem Biophys Res Commun.* 1984 Sep 17; 123(2): 574–580.

Grabowski GA, Barton NW, Pastores G, Dambrosia JM, Banerjee TK, McKee MA et al. Enzyme therapy in Gaucher disease Type 1: Comparative efficacy of mannose-terminated glucocerebrosidase from natural and recombinant sources. *Ann Int Med.* 1995; 122: 33–39.

Johnson WG, Brady RO. Ceramidetrihexosidase from human placenta. *Methods Enzymol.* 1972; XXVIII: 849–856.

Johnson WG, Desnick RJ, Long DM, Sharp HL, Krivit W, Brady B, Brady RO. Intravenous injection of purified hexosaminidase A into a patient with Tay–Sachs disease. *Enzyme Therapy in Genetic Diseases.* In: Desnick RJ, Bernlohr RW, Krivit W. eds. Baltimore: Williams and Wilkins: 1973: 120–124. Birth Defects Original Article Series, IX.

Kampine JP, Brady RO, Yankee RA, Kanfer JN, Shapiro D, Gal AE. Sphingolipid hydrolases in leukemic leukocytes. *Cancer Res.* 1967; 27: 1312–1315.

Kanfer JN, Young OM, Shapiro D, Brady RO. The metabolism of sphingomyelin. I. Purification and properties of a sphingomyelin-cleaving enzyme from rat liver tissue. *J Biol Chem.* 1966; 241: 1081–1084.

Klenk E. Uber die nature der phosphatide der milz bei Niemann–Pickschen Krankheit. *Z Physiol Chem.* 1934; 229: 151–156.

Lieb H. Cerebrosidespeicherung bei Splenomegalie Typus Gaucher. *Ztschr Physiol Chem.* 1924; 140: 305–313.

Marchand F. Über Sogennante idiopathische Splenomegalie (Typus Gaucher). *Munchen med Wchnschr.* 1907; 54: 1102–1103.

Pentchev PG, Brady RO, Gal AE, Hibbert SR. Replacement therapy for inherited enzyme deficiency: sustained clearance of accumulated glucocerebroside in Gaucher's disease following infusion of purified glucocerebrosidase. *J Molec Med.* 1975; 1: 73–78.

Pentchev PG, Brady RO, Hibbert SR, Gal AE, Shapiro D. Isolation and characterization of glucocerebrosidase from human placental tissue. *J Biol Chem.* 1973; 248: 5256–5261.

Sachs B. On arrested cerebral development with special reference to cortical pathology. *J Nerv Ment Dis.* 1887; 14: 541–553.

Schiffmann R, Kopp JB, Austin HA, Sabnis S, Moore DF, Weibel T, Balow JE, Brady RO. Enzyme replacement therapy in Fabry disease. A randomized controlled trial. *JAMA* 2001; 285: 2743–2749.

Schiffmann R, Murray GJ, Treco D, Daniel P, Sellos-Moura M, Myers M, et al. Infusion of α-galactosidase A reduces tissue globotriaosylceramide storage in patients with Fabry disease. *Proc Natl Acad Sci USA.* 2000; 97: 365–370.

Steer CJ, Furbish FS, Barranger JA, Brady RO, Jones EA.The uptake of agalacto-glucocerebrosidase by rat hepatocytes and Kupffer cells. *FEBS Lett.* 1978 Jul 15; 91(2): 202–205.

Tay W. Symmetrical changes in the region of the yellow spot in each eye of an infant. *Trans Ophthal Soc UK.* 1881: 1: 55–57.

Thannhauser SJ. Diseases of cellular lipid metabolism. In: Christian HA (Ed.) Lipidoses, New York: Oxford University Press; 1950: 49.

Trams EG, Brady RO. Cerebroside synthesis in Gaucher's disease. *J Clin Invest.* 1960; 39: 1546–1560.

Van Den Hamer CJ, Morell AG, Scheinberg IH, Hickman J, Ashwell G. Physical and chemical studies on ceruloplasmin. IX. The role of galactosyl residues in the clearance of ceruloplasmin from the circulation. *J Biol Chem.* 1970 Sep 10; 245(17): 4397–402.

Weinreb NJ, Brady RO, Tappel AL. The lysosomal localization of sphingolipid hydrolases. *Biochim Biophys Acta.* 1968; 159: 141–146.

COMPLEX LIPID CATABOLISM

Roscoe O. Brady, M.D.[1] and Roscoe O. Brady, Jr., M.D., Ph.D.[2]

There are three principal classes of materials whose orderly biodegradation is required for salutary homeostasis of humans. These are glucose polymers such as glycogen, complex carbohydrates such as mucopolysaccharides, and complex lipids. Pathological changes that involve the last group usually involve lipids whose characteristic component is the long chain amino alcohol sphingosine (Figure 1).

D-erythro-trans-2-amino-4-octacecene-1-3-diol

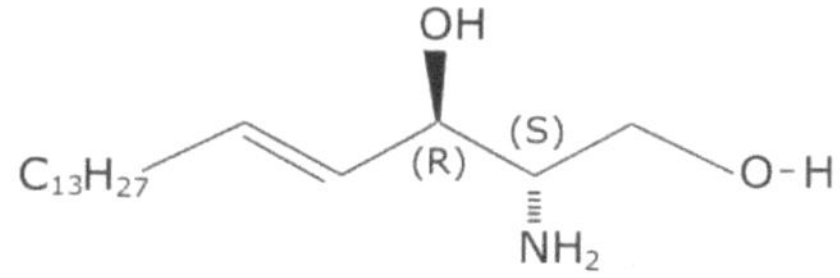

Figure 1. Sphingosine.

Sphingosine is produced by the condensation of palmitoyl coenzyme A with serine catalyzed by the enzyme serine palmitoyl coenzyme A transferase to produce 3-ketosphinganine (2-amino-3-keto-octadecanol; Brady and Koval, 1957). The carbonyl oxygen of 3-ketosphinganine is enzymatically reduced to produce the secondary alcohol sphinganine. Most of this substance becomes acylated on the nitrogen atom on carbon 2 of sphinganine to form dihydroceramide. The sphinganine moiety of dihydroceramide undergoes desaturation between carbon atoms 4 and 5 resulting in the formation of N-fatty acyl sphingosine (ceramide; Figure 2).

[1] Developmental and Metabolic Neurology Branch, National Institute of Neurological Disorders and Stroke, Building 10 Room 3D04, National Institutes of Health, Bethesda, Maryland 20892–1260. Tel (301) 496–3285; Fax (301) 496–9480; Email: rb57v@nih.gov.

[2] Present address: Department of Psychiatry, Massachusetts General Hospital, Wang ACC 807, 55 Fruit Street, Boston, MA 02114.

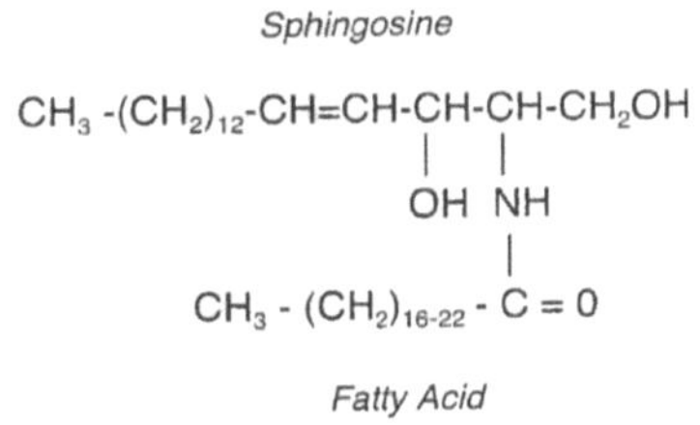

Figure 2. Ceramide.

Ceramide is the fundamental constituent of all of the substances that accumulate in the organs and tissues of patients with sphingolipid storage disorders.

Lipid storage disorders are caused by insufficient activity of specific hydrolytic enzymes required for the biodegradation of sphingolipids. In these disorders, genetic mutations cause specific lipid hydrolases to be less than normally active, impairing catabolism of complex lipids, and resulting in accumulation of the substrate of that hydrolase. For example, the accumulation of glucocerebroside in organs of patients with Gaucher disease is caused by a deficiency of glucocerebrosidase (Brady et al., 1965, 1966a; Figure 3, Reaction 7).

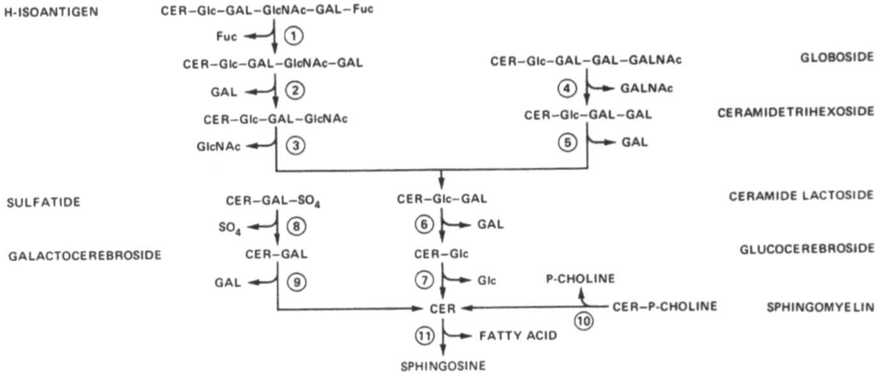

Figure 3. Catabolic pathways of neutral sphingoglycolipids, sulfatide and sphingomyelin. Abbreviations: CER= ceramide; Glc = glucose; GAL = galactose; GlcNAc = N-acetylglucosamine; Fuc = fucose; GALNAc = N-acetylgalactosamine; P-Choline = phosphocholine. Sites of enzymes involved in neutral sphingolipid storage disorders: **1**. Fucosidosis (Durand et al., 1968); **4**. Sandhoff disease (Sandhoff, Andreae, and Jatzkewitz 1968; see Chapter 16, this book); **5**. Fabry disease (Brady et al., 1967a; see Chapter 20, this book); **7**. Gaucher disease (Brady, Kanfer, and Shapiro, 1967; see Chapter 21); **8**. Metachromatic leukodystrophy (Austin et al., 1963; Mehl and Jatzkewitz 1965); **9**. Krabbe disease (Suzuki and Suzuki, 1970; see Chapter 18); **10**. Niemann–Pick disease (Brady et al., 1966b; see Chapter 17, this book); **11**. Farber's disease (Sugita, Dulaney, and Moser, 1972).

Much, if not most of the glucocerebroside in the spleen, and probably the liver as well, in patients with Gaucher disease appears to originate from the biodegradation of components of membranes of senescent white blood cells of which ceramidelactoside is the principal sphingoglycolipid (Kattlove et al., 1969). Lesser amounts of glucocerebroside appear to arise from the catabolism of globoside, the principal neutral glycosphingolipid in red blood cells (Yamakawa, Yokoyama, and Handa, 1963; Ohshima, et al., 1999). Considerably smaller quantities of globotriaosylceramide, ceramidelactoside, and glucocerebroside itself are present in the erythrocyte stroma and are considered to be minor sources of the accumulating glucocerebroside in the tissues of patients with Gaucher disease. Similarly, much of the globotriaosylceramide (ceramidetrihexoside) that accumulates in the tissues of patients with Fabry disease (Sweeley and Klionsky 1963) appears to arise from globoside because of insufficient activity of ceramidetrihexosidase (α-galactosidase A; Brady et al., 1967b). Glucocerebroside that accumulates in the brain of patients with the neuronopathic forms of Gaucher disease appears to arise primarily from the catabolism of glycosphingolipids called gangliosides (Nilsson and Svennerholm, 1982). Gangliosides are acidic glycosphingolipids that contain one or more molecules of N-acetylneuraminic acid (sialic acid; cf. Figure 4).

In the above examples, the rapid turnover of certain cell populations results in the accumulation of the lipid substrate whose catabolism is impaired. Deposition of this material in various tissues causes the morbidity associated with these disorders. However, the source of some of the accumulating materials in lipid storage disorders is the constitutive presence of that particular lipid among the array of sphingolipid components of various tissues. An example of this situation is the accumulation of sphingomyelin in patients with Niemann–Pick disease. Sphingomyelin is a major component of virtually all cell membranes. The catabolism of sphingomyelin is initiated by the hydrolytic cleavage of phosphocholine that is catalyzed by the enzyme sphingomyelinase (Figure 3, Reaction 10; Kanfer et al., 1966). The activity of this enzyme is dramatically reduced in patients with Niemann–Pick disease (Brady et al., 1966b; see Chapter 17 in this book).

Another example of an accumulating constitutive sphingolipid is the accumulation of galactocerebroside in Krabbe disease. Galactocerebroside has been known for more than a century to be the major sphingolipid in the brain (Thudicum 1901). Its degradation is impaired in patients with Krabbe disease due to reduced activity of the enzyme galactocerebrosidase that catalyzes the hydrolytic cleavage of galactose from this lipid (Figure 3, Reaction 9; Suzuki and Suzuki, 1970; see Chapter 18 in this book). A similar situation appears to occur in metachromatic leukodystrophy where sulfatide (galactocerebroside-3-sulfate) accumulates because of a deficiency of arylsulfatase A (Figure 3, Reaction 8; Austin et al., 1963; Mehl and Jatzkewitz 1965). The pathways and enzymatic steps involved in the catabolism of neutral sphingoglycolipids, sphingomyelin, and sulfatide are shown in Figure 3. The sites of enzyme deficiencies in humans with sphingolipid storage disorders in which these lipids accumulate are also indicated in Figure 3.

Neurogenetic conditions caused by the impairment of the catabolism of acidic sphingoglycolipids constitute a major group of lipid storage disorders known as the gangliosidoses. Soon after the enzymatic defects in Gaucher disease and Niemann–Pick disease were discovered, the metabolic defects in generalized (G_{M1}) gangliosidosis and Tay–Sachs disease were considered (Brady, 1966; see chapter 15 on this book). Because ganglioside G_{M2} that accumulates in patients with Tay–Sachs disease is branched in the terminal portion of the molecule, it was conceivable that the missing catabolic enzyme could be either a hexosaminidase that catalyzed the hydrolysis of N-acetylgalactosamine from G_{M2} (Figure 4, Reaction 11) or a neuraminidase (sialidase) that catalyzed the hydrolytic cleavage of N-acetylneuraminic acid (sialic acid) from this sphingolipid (Figure 4, Reaction 10).

An investigation was performed in which hexosaminidase activity in brain tissue from patients with Tay–Sachs disease was assayed with the artificial substrate para-nitophenyl-N-acetylgalactosamine. This experiment indicated that not only was hexosaminidase present in the brain of Tay–Sachs patients, but total hexosaminidase activity was actually several fold greater than that in normal human brain. This discovery caused a serious misdirection in the pursuit of the etiology of Tay–Sachs disease.

Because hexosaminidase was clearly present and more than normally active, it was felt necessary to learn whether human brain contained an enzyme that catalyzed the cleavage of N-acetylneuraminic acid from ganglioside G_{M2}. Because a search for this reaction with unlabeled G_{M2} was unrevealing, it was decided to radioactively label G_{M2}. The chemical synthesis of ganglioside G_{M2} had not been accomplished at that time, therefore the first attempt involved labeling it with tritium (^3H), the radioactive isotope of hydrogen. $G_{M2}$was exposed to a high concentration of tritium gas in a sealed tube for a week, the Wilzbach procedure. A portion of the unlabeled hydrogen atoms in ganglioside G_{M2} became replaced with ^3H producing G_{M2} uniformly labeled with ^3H throughout the molecule. Although this procedure had been used successfully to determine the metabolic defect in Fabry disease (Brady et al., 1967a,b), G_{M2} labeled in this fashion was completely unsuitable for use as a substrate for enzymatic assays. It was therefore decided to try to label G_{M2} biosynthetically.

We injected ^3H-labeled N-acetylmannosamine, a specific precursor of N-acteylneuraminic acid, into the brains of neonatal rats. In this manner, we obtained ganglioside G_{M2} labeled with ^3H in the sialic acid portion of the molecule. We quickly learned that human tissues contained an enzyme that catalyzed the hydrolytic cleavage of N-acetylneuraminic acid from ganglioside G_{M2} and that the catabolism of this material in tissues obtained from patients with Tay–Sachs disease was similar to that in normal individuals (Figure, 4 Reaction 10; Kolodny et al., 1969a). Based on this finding, we felt it was necessary to label ganglioside G_{M2} in the N-acetylgalactosaminyl moiety. The desired compound was produced by injecting ^3H-N-acetylgalactosamine into the brain of neonatal rats. When the catabolism of ganglioside G_{M2} labeled in the N-acetylgalactosaminyl moiety was examined, it was found that the hydrolysis of N-acetylgalactosamine from G_{M2} was completely undetectable in tissues from patients with Tay–Sachs disease and it is the metabolic defect in this condition (Figure 4, Reaction 11; Kolodny et al., 1969b).

Substantiation of this finding was provided in a more extensive subsequent investigation (Tallman, Johnson, and Brady, 1972). One month before the demonstration of the deficiency of the hydrolytic cleavage of N-acetylgalactosamine from ganglioside G_{M2}, Okada and O'Brien (1969) published their paper in which the lack of a specific hexosaminidase isozyme called hexosaminidase A (Robinson and Stirling 1967) was identified. The presence of two hexosaminidases in the human brain with discretely different biochemical properties had been noted a year earlier (Brady 1966). The enzymatic steps involved in the catabolism of gangliosides and asialo-gangliosides are indicated in Figure 4.

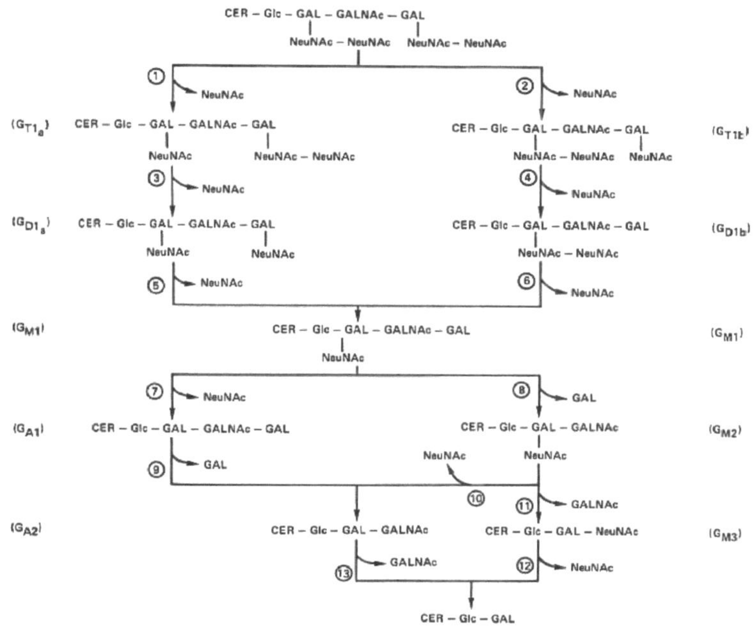

Figure 4. Catabolic pathways of gangliosides. Pathways for the biodegradation of gangliosides and neutral sphingoglycolipids (Figure 3) converge at CER-Glc-Gal (ceramidelactoside). Abbreviations: CER = ceramide; Glc = glucose; GAL = galactose; NeuNAc = N-acetylneuraminic acid (sialic acid); GALNAc = N-acetylgalactosamine. Sites of enzymes involved in storage disorders: **7**. Galactosialidosis (Wenger, Tarby, and Wharton 1978); **8**. Generalized (G_{M1}) gangliosidosis (Okada and O'Brien 1968; see Chapter 17); **11**. Tay–Sachs disease (Okada and O'Brien 1969; Kolodny, Brady, and Volk 1969; Sandhoff disease (Sandhoff et al., 1971; see Chapter 16 in this book); **13**. Sandhoff disease (Sandhoff et al., 1968).

In addition to specific enzymes, a number of sphingolipid cleavage reactions require accessory factors for catalytic activity. Glucocerebrosidase is stimulated by phosphatidylserine and other negatively charged phospholipids (Dale, Villacorte, and Beutler 1976). In addition, a low molecular weight activator protein called saposin C is also required (Ho and O'Brien 1971; Sandhoff, Kolter, and Harzer, 2001). Absence of saposin has been demonstrated in several patients with sphingolipid storage disorders whose clinical presentation was quite complex compared with reduced activity of a single enzyme (Harzer et al., 1989). The catabolism of ceramidetrihexoside (Figure 3, Reaction 5), sulfatide (Figure 3, Reaction 8), and sphingomyelin (Figure 3, Reaction 10) also require the presence of a saposin. A somewhat different activator protein is required for the catabolism of ganglioside G_{M2}. Lack of this activating protein is the basis for the AB variant of Tay–Sachs disease (Sandhoff et al., 1968; Conzelmann and Sandhoff 1978).

We have indicated in this overview seminal discoveries of the enzymatic defects in hereditary sphingolipid storage disorders. Detailed descriptions of phenotypes, genotypes, diagnosis, and treatment strategies for specific disorders are presented in subsequent chapters dedicated to specific metabolic disorders.

REFERENCES

Austin, J.H. et al. (1963) A controlled study of enzymatic activities in three human disorders of glycolipid metabolism. *J. Neurochem.* 10, 805–816.

Brady, R.O. (1966) The sphingolipidoses. *N. Engl. J. Med.* 275, 312–318.

Brady, R.O., and Koval, G.J. (1957) Biosynthesis of sphingosine in vitro. *J. Am. Chem. Soc.* 79, 2648–2649.

Brady, R.O. et al. (1966a) Demonstration of a deficiency of glucocerebroside-cleaving enzyme in Gaucher's disease. *J. Clin. Invest.* 45, 1112–1115.

Brady, R.O. et al. (1966b) The metabolism of sphingomyelin. II. Evidence of an enzymatic deficiency in Niemann–Pick disease. *Proc. Natl. Acad. Sci. USA* 55, 366–369.

Brady, R.O. et al. (1967a) The metabolism of ceramidetrihexosides. I. Purification and properties of an enzyme which cleaves the terminal galactose molecule of galactosylgalactosylglucosylceramide. *J. Biol. Chem.* 242, 1021–1026.

Brady, R.O. et al. (1967b) Enzymatic defect in Fabry's disease. Ceramidetrihexosidase deficency. *N. Engl. J. Med.*, 276, 1163–1167.

Brady, R.O., Kanfer, J.N., and Shapiro, D. (1965) Metabolism of glucocerebrosides. II. Evidence of an enzymatic deficiency in Gaucher's disease. *Biochem. Biophys. Res. Commun.* 18, 221–225.

Conzelmann, E., and Sandhoff, K. (1978) AB variant of infantile GM2 gangliosidosis: Deficiency of a factor necessary for stimulation of hexosaminidase A-catalyzed degradation of ganglioside GM2 and glycolipid GA2. *Proc. Natl. Acad. Sci. USA* 75, 3979–3983.

Dale, G.E., Villacorte, D., and Beutler, E. (1976) Solubilization of glucocerebrosidase from human placenta and demonstration of a phospholipid requirement for its catalytic activity. *Biochem. Biophys. Res. Commun.* 71, 1048–1053.

Durand, F. et al. (1968) Fucosidosis. *Lancet* 1, 1198.

Harzer, K. et al. (1989) Sphingolipid activator protein deficiency in a 16-week-old atypical Gaucher disease patient and his fetal sibling: Biochemical signs of combined sphingolipidoses. *Eur. J. Pediatr.* 149, 31–39.

Ho, M.W., and O'Brien, J.S. (1971), Gaucher's disease; Deficiency of 'acid' β-glucosidase and reconstitution of enzyme activity *in vitro*. *Proc. Natl. Acad. Sci. USA,* 68, 2810–2813.

Kanfer, J.N., Young, O.M., Shapiro, D., and Brady, R.O. (1966) The metabolism of sphingomyelin. I. Purification and properties of a sphingomyelin-cleaving enzyme from rat liver tissue. *J. Biol. Chem.* 241, 1081–1084.

Kattlove, H.E. et al. (1969) Gaucher cells in chronic myelocytic leukemia: An acquired abnormality. *Blood* 33, 379–390.

Kolodny, E.H., Brady, R.O., and Volk, B.W. (1969b) Demonstration of an alteration of ganglioside metabolism in Tay–Sachs disease. Biochem. *Biophys. Res. Commun.* 37: 526–531.

Kolodny, E.H., Brady, R.O., Quirk, J.M., and Kanfer, J.N. (1969a) Studies on the metabolism of Tay–Sachs ganglioside. *Fed. Proc.* 28, 596.

Lieb, H. (1924) Cerebrosidespeicherung bei Splenomegalie Typus Gaucher. *Ztschr. Physiol. Chem.* 140, 305–313.

Mehl, E., and Jatzkewitz, H. (1965) Evidence for the genetic block in metachromatic leukodystrophy (ML). *Biochem. Biophys. Res. Commun.* 19, 407–411.

Nilsson, O., and Svennerholm, L. (1982) Characterization and quantitative determination of gangliosides and neutral glycosphingolipids in human liver. *J. Lipid Res.* 23, 327–334.

Ohshima, T., et al. (1999) Aging accentuates and bone marrow transplantation ameliorates metabolic defects in Fabry disease mice. *Proc. Natl. Acad. Sci. USA,* 96, 6423–6427.

Okada, S., and O'Brien, J.S. (1968) Generalized gangliosidosis: Beta-galactosidase deficiency. *Science* 160, 1002–1004.

Okada, S., and O'Brien, J. (1969) Tay–Sachs disease: Generalized absence of a beta-d-N-acetylhexosaminidase component. *Science*, 165, 698–700.

Robinson, D., and Stirling, J. (1967) N-acetyl-β-glucosaminidases in human spleen. *Biochem. J.* 107, 321–327.

Sandhoff, K., Andreae, U., and Jatzkewitz, H. (1968) Deficient hexosaminidase activity in an exceptional case of Tay–Sachs disease with additional storage of kidney globoside in visceral organs. *Pathol. Eur,* 3, 278–285.

Sandhoff, K., et al. (1971) Enzyme alterations and lipid storage in three variants of Tay–Sachs disease. *J. Neurochem.* 18, 2469–2489.

Sandhoff, K., Kolter, T., and Harzer, K. (2001) Sphingolipid activator proteins. In: *The Metabolic & Molecular Bases of Inherited Disease*, C.R. Scriver, A.L Beaudet, W.S. Sly, and D. Valle, Eds., New York: McGraw-Hill, pp. 3371–3388.

Sugita, M., Dulaney, J.T., and Moser, H.W. (1972) Ceramidase deficiency in Farber's disease (lipogranulomatosis). *Science* 178,1100–1102.

Suzuki, K., and Suzuki, Y. (1970) Globoid cell leukodystrophy (Krabbe's disease): Deficiency of galactocerebroside β-galactosidase. *Proc. Natl. Acad. Sci USA* 66, 302–309.

Sweeley, C.C., and Klionsky, B. (1963) Fabry's disease: Classification as a sphingolipidosis and partial characterization of a novel glycolipid. *J. Biol. Chem.* 238, PC3148–PC3150.

Tallman, J.F., Johnson, W.G., and Brady, R.O. (1972) The metabolism of Tay–Sachs ganglioside: Catabolic studies with lysosomal enzymes from normal and Tay–Sachs brain tissue. *J. Clin. Invest.* 51: 2339–2345.

Thudicum, J.L.W. (1901) *Die chemische Konstitution des Gehirns des Menschen und der Tiere*. Tübingen: Pietziker.

Trams, E.G., and Brady, R.O. (1960) Cerebroside synthesis in Gaucher's disease. *J. Clin. Invest.* 39, 1546–1550.

Van Hoof, F., and Hers, H.G. (1968) Mucopolysaccharidosis by absence of α-fucosidase. *Lancet*, 1, 1198.

Wenger, D.A., Tarby, T.J., and Wharton, C. (1978) Macular cherry-red spots and myoclonus with dementia: Coexistent neuraminidase and β-galactosidase deficiencies. *Biochem. Biophys. Res. Commun.* 82, 589–595.

Yamakawa, T., Yokoyama. S., and Handa, N. (1963) Chemistry of lipids of posthemolytic residue or stroma of erythrocytes. XI. Structure of globoside, the main mucolipid of human erythrocytes. *J. Biochem.* (Tokyo), 53, 28–36.

RETROVIRAL VECTORS FOR GENE THERAPY

Seon-Hee Kim and Paul D Robbins[1]

1 INTRODUCTION

Historically, retroviral vectors have been the most frequently used type of gene delivery vectors for clinical gene therapy. In particular, vectors based on murine leukemia virus (MLV) and related retroviruses have been employed in almost half of the current gene therapy clinical protocols (http://www.wiley.co.uk/genetherapy). The major advantages of retroviral vectors are (1) ease of manipulation for insertion of the therapeutic gene; (2) ability to stably integrate into the target cell genome; (3) relatively high titer of the recombinant retroviruses; (4) a wide range of target species and cells that can be infected without any apparent adverse pathology; and (5) relatively simple procedure for preparation of the recombinant virus. However, the current retroviral vector have potential disadvantages as well, such as (1) requirement for cell division for integration, limiting their in vivo applications; and (2) random integration into host chromosome, resulting in possible insertional mutagenesis or oncogene activation. However, recent developments in virus packaging systems, use of modified or different envelope proteins for packaging, and modifications with the *cis*-acting regulatory elements to regulate transgene expression have allowed for the safe and efficient clinical application of retroviral vectors. This chapter provides a background on retrovirus-based vector systems as well as provides an update regarding improvements in retroviral vector for gene transfer.

2 RETROVIRAL STRUCTURE

Retroviruses are single-stranded RNA viruses (Figure 1). Each retrovirus particle, termed virion, contains two copies of single-stranded positive sense RNA within a protein core surrounded by a lipid envelope. All retroviruses have at least three structural genes, gag, pol, and env coding proteins that are processed post-translationally into multiple polypeptides required for viral replication and packaging. A full-length gag (group antigen) polyprotein

[1] Department of Molecular Genetics and Biochemistry and Molecular Medicine Institute, University of Pittsburgh School of Medicine.
Address correspondence to Seon-Hee Kim PhD, W1246 Biomedical Science Tower, Pittsburgh, PA, 15261. USA

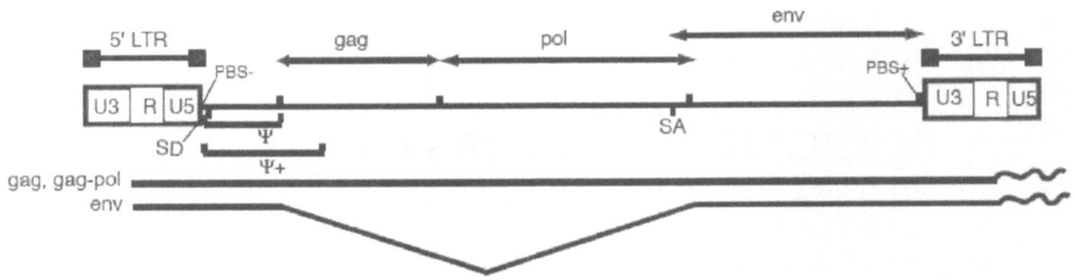

Figure 1. Genome structure of murine leukemia virus. U3, unique 3' end region; R, terminal redundancy; U5, unique 5' end region; PBS, primer binding site; Ψ, packaging signal; SD, splice donor site; SA, splice acceptor site.

is processed to yield four proteins: p15 (matrix, MA), p12, p30 (capsid, CA), and p10 (nucleocapsid, NC). The pol (polymerase) gene is organized into protease (PR), reverse transcriptase/RNase H (RT), and the integrase (IN) coding domains. The env (envelope) gene is synthesized as a polyprotein that is cleaved by a cellular protease to yield the gp70 (surface protein, SU) and the p15E (transmembrane protein, TM).

In addition, certain classes of retroviruses such as lentiviruses and spumaviruses carry additional regulatory genes involved in viral replication and transcription. The retrovirus used most widely in the gene transfer system is the murine leukemia virus (MLV), a mammalian C-type virus. MLV is organized with its gag, pol, and env protein coding domains flanked by *cis*-acting nucleic acid sequences. Duplicated at either end of the provirus are long terminal repeat (LTR), which are the sequences necessary for integration, replication, and the regulation of transcription. The LTR contains a duplicated transcription enhancer element followed by a promoter region, and a prototypical polyadenylation (poly A) signal. At the 3' boundary of the 5' LTR is the primer binding site where the cellular tRNA binds and reverse transcription is initiated. Next is the splicing donor (SD) sequence utilized in the production of the subgenomic envelope messenger RNA. The region between SD and the start of the first protein coding domain contains the packaging signal (Ψ), which is necessary for encapsidation of genomic RNA and dimerization of the virion RNA.

3 RETROVIRAL LIFECYCLE

The retroviral lifecycle is shown in Figure 2. The first step in the retroviral lifecycle is mediated by binding of a retrovirus envelope protein to a specific receptor on the cell surface, followed by endocytosis and uncoating at low pH by fusion of the viral envelope with the lysosomal membrane (McClure et al., 1990). After entry, the viral core complex passes through cytoplasm and eventually enters the nucleus. The process of reverse transcription of the RNA genome into double-strand DNA occurs, using the structure and enzymatic activities that enter the cell in the virion. The resulting double-stranded viral DNA is then transported to the nucleus where it integrates into the host DNA using a virally encoded integrase. The integrated provirus is then transcribed into a single-stranded viral RNA, which is subsequently processed by a polyadenylation at the 3' end

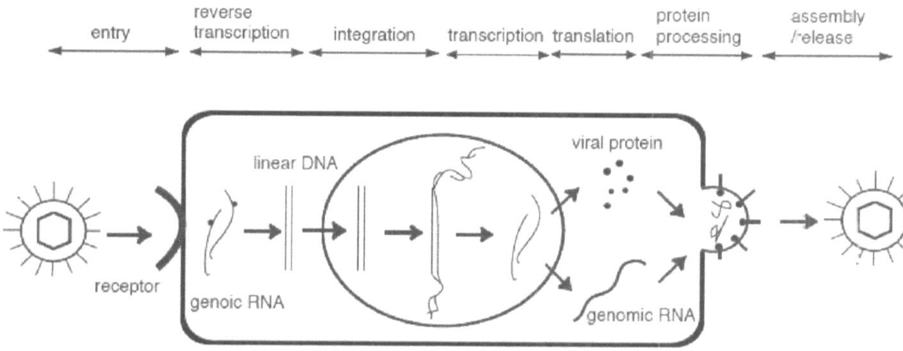

Figure 2. The retroviral lifecycle from entry to release of virions. The virus infects cells by binding to a specific cellular receptor followed by penetration of the virus core into the cell. The viral RNA is then reverse-transcribed in the cytoplasm into a double-stranded proviral DNA. The linear proviral DNA is transported into the nucleus where it integrates into host chromosomal DNA using integrase. Cellular RNA polymerase II then synthesizes the viral RNA and transcripts are processed to genomic RNA and mRNA encoding gag, pol, and env proteins, which allow for packaging of the full length of viral RNA containing the packaging signal, Ψ. The assembled virus particles are released by budding from the env-coated cell membrane.

of the R region in the 3' LTR to yield a genome-length molecule and by splicing of a fraction of the transcripts to generate subgenomic mRNA species. The produced full-length genomic RNA contains the packaging site Ψ, which is then inserted into capsids to create new infectious particles. The infectious virus is then released from the cell by budding from the env-coated cell membrane without affecting the growth properties of the host cell.

4 DEVELOPMENT OF RETROVIRAL VECTORS AND APPLICATION FOR GENE THERAPY

The first generation of retroviral vectors was produced by inserting the gene of interest into the retrovirus genome deleted for the viral proteins. These defective viruses were rescued with replication competent helper virus (Shimotohno and Temin, 1981; Wei et al., 1981; Tabin et al., 1982; Joyner and Bernstein, 1983). Because these vectors require helper virus for infection, their use was limited. Subsequently, Mann, Mulligan, and Baltimore (1983) demonstrated that a specific region between splice donor and the start of the gag gene contained a *cis*-acting element, termed Psi (Ψ), is important for packaging. In addition to LTRs, the packaging site, and tRNA binding site, the presence of the 5' region of gag in *cis* is important to increase packaging efficiency of the viral RNA by stabilizing the RNA and/or affecting the secondary structure (Bender et al., 1987; Adam and Miller, 1988).

A retroviral vector can be generated with a simple conception, deleting the viral genome encoding viral proteins, leaving *cis*-acting elements, and inserting the therapeutic gene(s) (Figure 3). The infectious replication-deficient retrovirus is produced in packaging

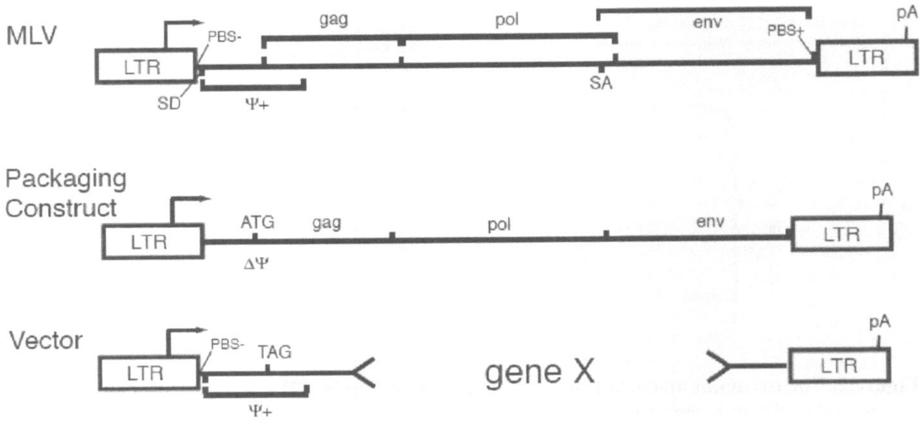

Figure 3. Schematic representation of retroviral vector and packaging construct. Genomic structure of retroviral vector and packaging construct are derived from the MLV genome. LTR. Long terminal repeat; Ψ, packaging signal; pA, polyadenylation signal.

cell-lines that contain the gag, pol, and env that provide in trans all the necessary proteins for packaging of the viral RNA. The vector with its single-stranded RNA carrying the gene of interest cassette enters the target via a specific receptor. In the cytoplasm, the reverse transcriptase carried by the vector converts the vector RNA into proviral DNA that is randomly integrated into the target cell genome, where the expression cassette makes its product for the lifespan on the cell.

5 TYPES OF RETROVIRAL VECTORS

There are many different types of retroviral vectors, which have been developed for gene transfer. The majority of retroviral vectors currently being used are derived from MLV. Two basic retroviral vectors, LN and MFG, are the most frequently used vectors in gene therapy (Figure 4). LN vectors contain an altered 5' gag region, shown to improve viral titer, and the Ψ packaging site. In this vector and in other retroviral vectors carrying the 5' region of gag, the gag translational start codon is mutated to stop codon so no viral gag protein is produced. In addition, the LTR and 5' untranslated region from MLV are replaced with the sequences from MSV (murine sarcoma virus) and all the sequences derived from MLV env gene are deleted (Miller and Rosman, 1989).

The MFG vector was developed by Mulligan and colleagues that mimics the normal pattern of mRNA splicing in the wild-type MLV (Ohashi, et al., 1992). In this vector, a gene of interest is inserted at the normal position of env start codon and is expressed from a sliced transcript whose 5' leader sequence is identical to the normal env message in MLV-infected cells. Although LN and MFG vectors are the most common vectors for gene transfer, modifications of these base vectors have been made to improve their clinical use. Recently, retroviral vectors has been more developed and tuned to a specific manner and purpose for gene therapy trials.

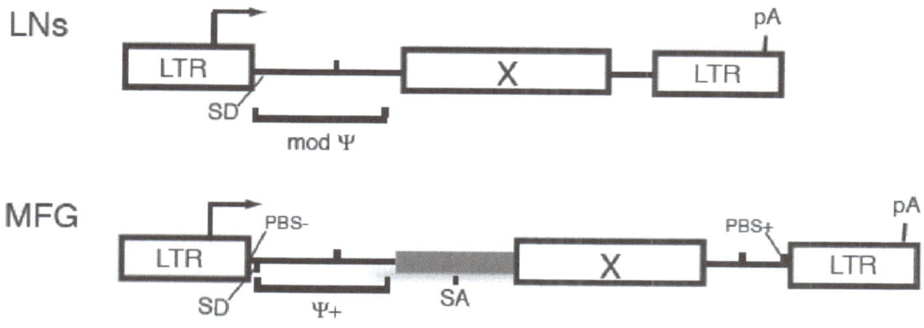

Figure 4. Diagram of LN and MFG retroviral vectors. LN vectors has been modified to prevent translation initiation at the gag gene start codon, and have a substitution of murine sarcoma virus (MSV) sequences instead of MLV. Although LN vectors have deleted 3' sequences derived from MLV env region, the vector contains 420 bp of gag coding sequences. The MFG vector is an example of a splicing vector using the native splicing donor and acceptor sites from MLV to express an inserted gene in a similar manner to the natural RNA coding for the env gene. Thus, MFG contains 420 bp for gag, 377 for pol, and 99 for env sequences in the vector construct.

5.1 Multigene Expression

The requirement in the gene therapy for expression of both marker/reporter genes and therapeutic genes necessitates the use of either multiple transcriptional units or alternative strategies to express two proteins from a single transcript. Approaches for expressing two or more genes from the same vector include the use of internal promoters, alternative splicing, internal ribosome entry site (IRES), and insertion of a copy of a second transcription unit within the retroviral LTR.

In the vector containing an internal promoter (Figure 5a), the primary transcription usually initiates from the LTR and the second promoter is a nonretroviral promoter, either one from a cellular gene such as phosphoglycerate kinase promoter, or a strong viral promoter such as the human cytomegalovirus immediate early gene promoter, SV40 early gene promoter, or the Rous sarcoma virus LTR (Lim, Williams, and Orkin, 1987; Miller and Rosman, 1989).

In the alternative splicing vector (Figure 5b), the retroviral LTR directly promotes the expression of a gene of interest and a selectable marker or reporter gene is produced via splicing (Cepko, Roberts, and Mulligan, 1984). This is very similar to the transcripts produced by wild-type MLV utilizing spliced and unspliced forms to produce gag–pol and env gene products from the same LTR promoter. However, it has proven difficult to predict the efficiency of splicing in these vectors and activation of the cryptic splice donor site can result in deletion of vector sequences or inhibition of expression of the downstream gene (McIvor, 1990).

Picornaviruses such as encephalomyocarditis virus (EMCV) and poliovirus have evolved a unique mechanism for translating two genes from a single messenger RNA molecule. This poly-*cis*tronic messenger RNA is made possible by its incorporation of a region of high conserved secondary structure, called the internal ribosome entry site (IRES), that bypasses the ribosome scanning mechanism and permits direct entry of ribosome downstream from the Kozak ATG site (Davies and Kaufman, 1992). This IRES mechanism has been incorporated into retroviral vectors and is highly effective in permitting expression of two genes without the need for an additional promoter (Figure 5c; Morgan et al., 1992). In addition to the bi-*cis*tronic vector, tri-*cis*tronic or tetra-*cis*tronic vectors have been developed using internal promoters and multiple IRES, derived from different picornaviruses in the same vector (Ciafre et al., 2002) or insertion of two IRES cassettes from different origins (De Felipe and Izquierdro, 2000; Fussenegger, Mazur, and Bailey 1998; Douin et al., 2004). Although this IRES strategy was commonly used for multiple gene expression from a single vector, there are limitations for the large size (~500 bp) and imbalance of co-expression (Mizuguchi et al., 2000). Alternatively, a small peptide of 18 amino acids (2A) from picornaviruses has been used as a linker between two proteins to allow autonomous intraribosomal self-processing of polyproteins (Szymczak et al., 2004; De Felipe, 2004).

5.2 Robust and Long-Term Gene Expression

Robust and long-term gene expression is a requirement for therapeutic effects following gene transfer for therapeutic applications. To try to improve gene expression from retroviral vectors, the *cis*-acting elements in an LTR-based vector or internal promoter vector can be altered to increase expression. For example, the enhancer/promoter element within the U3 region of myeloid proliferative sarcoma virus (MPSV), murine embryonic stem cell virus (MESV), or murine stem cell virus (MSCV) functions more efficiently in hematopoietic progenitor cells than MLV LTR (Grez et al., 1990; Hawley et al., 1994; Challita et al., 1995). It is also possible to insert an enhancer/promoter from an unrelated virus such as HCMV or a cellular gene promoter such as phosphoglycerate kinase (PGK) into the 3'LTR, resulting in enhanced transcription (Figure 5d; Kim et al., 1998; Li et al., 1992). Furthermore, enhancers from cellular promoters such as the immunoglobulin heavy chain enhancer can be inserted in the 3' LTR. To avoid any negative influence of the LTR promoter on the expression of an internal promoter, the U3 region of the 3' LTR was deleted, resulting in inactivating the retroviral LTR called self-inactivating (SIN) vectors (Figure 5e) (Yu et al., 1986; Cone et al., 1987; Guild et al., 1988).

Other approaches to increase expression from retrovirus-transduced gene include the use of locus control regions, insulator elements, and other regulatory sequences. For example, insertion of the ß-globin locus control region (LCR) allows position-independent, copy number-dependent expression and is therefore a potent activator of transcription (Sadelain et al., 1995; Leboulch et al., 1994). It is known that the first 28 nucleotides of the R region of MLV LTR form a stem loop (RSL) supporting the cytoplasmic accumulation of unspliced retroviral transcripts (Trubetskoy, Okenquist, and Lenz, 1999). Removal of the aberrant translational start codons was able to promote correct translation of the unspliced RNA in the retroviral vector (Hildinger et al., 1999). Moreover, RNA export and translation of retroviral vectors may also be improved by incorporating the post-transcriptional regulatory element (PRE) of the woodchuck hepatitis virus into the upstream of 3' LTR (Zufferey et al., 1999; Schambach et al., 2000; Kraunus et al., 2004). Recently, various types of strong and stable promoters and positive regulatory elements have been defined and applied in the retroviral vector system to increase the efficiency of transgene expression.

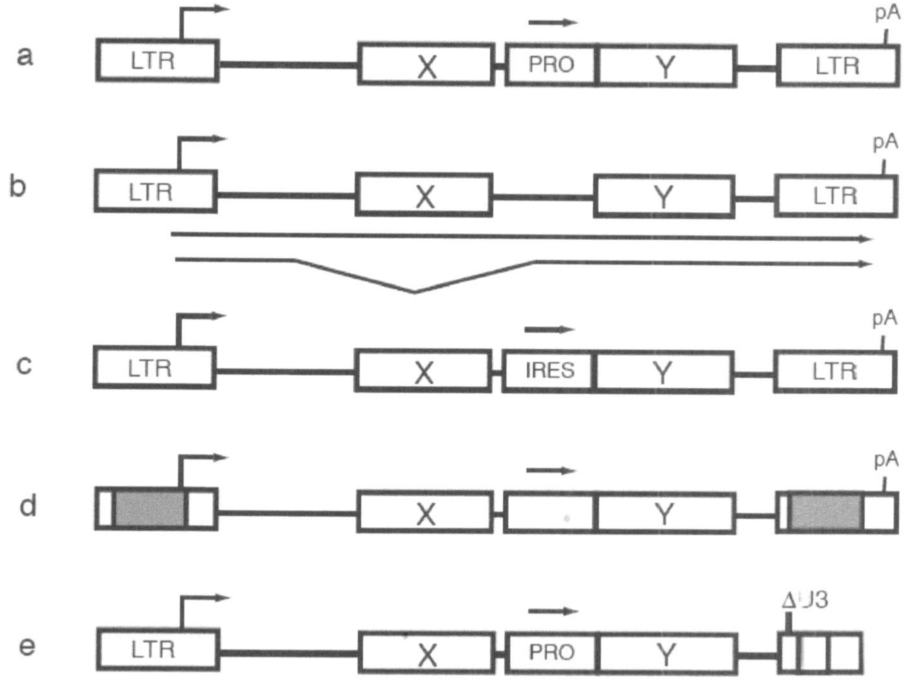

Figure 5. Diagram of various type of retroviral vectors. (a) Internal promoter vectors use two (or more) promoters to express independent genes. (b) Splicing vectors allow expression of the second gene by utilizing a splice acceptor site (SA) upstream of the gene to generate a spliced RNA. (c) The internal ribosomal entry site (IRES) vectors use the LTR to drive the expression of a single gene with the downstream gene translated via an internal ribosomal binding. (d) An alternative promoter is used to replace the viral U3 promoter in the LTR. (e) Self-inactivating vector (SIN) contains deletion of U3 in the 3' LTR that are translated to 5' LTR during reverse transcription.

5.3 Targeted Infection and Gene Expression

Retrovirus has a broad host range and is able to integrate randomly into the chromosome of the infected host cell. One other concern for gene transfer is the specificity of gene expression in the specific cell type. For the targeting to a specific cell, the strategies have been developed to target retrovirus delivery into a certain cell type by modifying the viral env. One of the env subunits, the SU portion, is mainly responsible for the interaction with the cellular receptor, such as recognition, binding, and fusion. The modification of the SU portion is achieved by deletion of a part of SU and replacing it with other ligands including erythropoietin, heregulin, insulinlike growth factor (IGF)-1, or single-chain variable fragment antibodies against various cell membrane proteins (Kasahara, Dozy, and Kan, 1994; Chu and Dornburg, 1995; Han, Kasahara, and Kan, 1995; Somia, Zoppe, and Verma, 1995; Jiang et al., 1998; Konishi et al., 1998; Chadwick et al., 1999). In order to avoid the gross structural changes in env resulting from hybrid env–ligand constructs, simple substitution of random peptides into the cell-targeting region was developed (Bupp and Roth, 2003; Bahrami, Duch, and Pedersen, 2004). Libraries of these substituted env changed tropism of viruses and infected efficiently into a specific cell type.

Transcriptional targeting is also an important aspect of developing vectors for gene therapy to restrict transgene expression to selected target cells. Promoters that are active in certain tissues can be inserted into a retroviral vector as an internal promoter or inserted in place of the viral enhancer within the U3 region of LTR. Tissue-specific expression of the internal promoter can be achieved with the weakening LTR activity by deletion (Cone et al., 1987). Alternatively, hybrid LTR that contains replaced U3 region of 3' LTR induces cell-type specific transgene expression. The insertion of muscle creatine kinase enhancer between, or replacing, the enhancer and proximal promoter in the 3' LTR has been used to direct expression to differentiated myogenic cells (Ferrari et al., 1995; Fassati et al., 1998). Replacement of the viral enhancer with an autoregulatory enhancer of GATA-1 has resulted in targeting expression to the cells of erythroid lineage (Grande et al., 1999) whereas insertion of a silencer into the LTR has been used to target CD4-positive cells (Indraccolo et al., 2000). Because the hybrid LTRs showed tissue-specific gene expression (Richardson, Kaspers, and Porter, 2004), design of retroviral vectors especially manipulation of LTR, should be adequately controlled.

5.4 Inducible Gene Expression

The ability to regulate therapeutic gene expression has become increasingly important for various gene therapy applications. The most commonly studied inducible gene expression system is based on use of a chemical-controlled transactivator and its corresponding synthetic promoter (Gossen and Bujard, 1992; Furth et al., 1994; Rivera et al., 1996; Pollock et al., 2000). Tetracycline is the most widely used small molecule to control *trans*-activation, which either can turn on or turn off the gene expression, depending upon the *trans*-activator. In the tet-off system, the expression of the gene of interest can be induced by withdrawing or suppressed by adding the inducer. In contrast, the tet-on system uses a reversed tetracycline-controlled trans-activator, to activate gene expression upon addition of the tetracyline inducer and suppressed upon withdrawal of the small molecule. In addition to tetracycline-controlled transactivation, rapamycin-inducible and mifepristone-inducible systems have been developed for the transcriptional regulation of the gene of interest (Crittenden et al., 2003; Sirin and Park, 2003).

5.5 Vectors with Improved Safety

Even though a retroviral vector can be generated that does not express any viral proteins, containing only *cis*-acting elements, the retroviral vectors that are currently used for gene delivery still contain extensive viral sequences including regions of gag and env coding sequences. Because these sequences are also present in the packaging genome, homologous recombination can occur, albeit at a low frequency, resulting in the production of replication-competent retrovirus (RCR). Recently, a series of retroviral vectors that are completely lacking any retroviral coding sequences has been developed (Kim et al., 1998; Yu, Kim, and Kim, 2000). In these vectors, all retroviral coding sequences were removed, and heterologous intron/exon sequences containing splicing acceptor were inserted upstream from the start codon of the gene of interest. In addition, vectors containing cre/loxp sequences have been developed for the deletion of the extra viral sequences. The Cre recombinase of bacteriophage P1 recognizes a 32 bp-specific sequence, loxP, and mediates site-specific recombination including deletion, insertion, and inversion of the sequences between the loxP sites (Hoess and Abremski, 1984). Using the cre/loxP system, a portion of the helper construct in the retroviral vector can be deleted after infection into the target cells (Choulika, Guyot, and Nicolas, 1996; Russ et al., 1996). Moreover, the selectable marker or reporter genes also can be deleted using the cre/loxp recombination after the viral DNA integrates into the host genome (Fernex et al., 1997; Loew et al., 2004).

6 GENERATION OF RECOMBINANT RETROVIRUS

Because the viral proteins are only required in *trans* for packaging of RNA containing a Ψ site, it is possible to make a packaging system by providing gag, pol, and env proteins (Figure 6). The initial version of the packaging construct for retroviral vectors was generated simply by deleting the Ψ packaging sequence from the MLV provirus. However, these packaging systems had a high frequency of recombination between the Ψ-site deleted virus and retroviral vector with a functional Ψ site, resulting in the generation of replication-competent virus. The frequency of RCR was reduced significantly by deleting the regulatory elements of the viral genome and by the design of the split-helper constructs (Danos and Mulligan, 1988; Dougherty et al., 1989). Splitting the genome packaging system into two vectors, one expressing gag–pol and the other env reduced the chances of RCR.

The recombinant retrovirus can be produced either by introduction of a retroviral vector into the stable packaging cell-line that already contains a packaging construct or by transient cotransfection of retroviral vector and packaging construct into the cell (Finer et al., 1994; Soneoka et al., 1995). In addition to mouse cells such as NIH3T3, other mammalian cells or avian cells have been used as packaging cells. In particular, human origin 293 cells-based has allowed for the rapid production of high titers by transient transfection (Pear et al., 1993).

Another approach for improving the virus packaging is to use pseudotyped envelopes. The most common example is using the envelope protein from another virus such as gibbon ape leukemia virus (Miller et al., 1991). This envelope allowed for more efficient infection of certain cell types such as CD34+ cells or T cells. The pseudotyping of retrovirus is able to alternate the tropism as well as infectivity for the target cells. Viruses can be generated with an expanded host range through the use of the G protein of vesicular stomatitis virus (Burns, Collignon, and Desrosiers, 1993; Yee, Friedmann, and Burns, 1994), which has a broad host range and is physically stable during speed centrifugation allowing concentration of the viral supernatant.

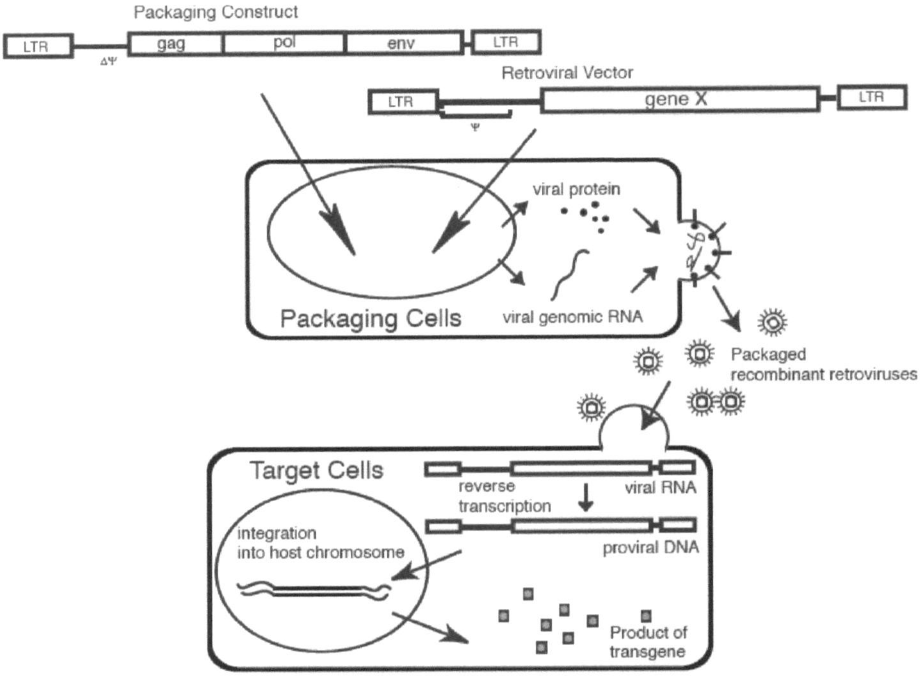

Figure 6. Recombinant retrovirus production and gene transfer. The retroviral vector containing the expression cassette is introduced into the packaging cell that coexpresses the gag, pol, and env proteins necessary to package infectious virus. In the cytoplasm, the reverse transcriptase carried by the virus converts the vector RNA into the proviral DNA that is randomly integrated into the target cell genome, where transcription of the expression cassette occurs for the life of the cell.

7 CLINICAL STUDY SAFETY

There are potential risks associated with the use of retroviral vectors, including insertional mutagenesis and outbreak of retroviral infection resulting from RCR.

Although it is now generally accepted that retroviral vector-mediated gene transfer is an appropriate procedure for clinical trials, possible problems with RCR production must be considered. Rhesus monkeys undergoing retroviral vector-mediated gene transfer via bone marrow transplantation were exposed to RCR, developing T cell lymphoma in three out of eight animals (Donahue et al., 1992). However, the use of newer packaging cells with split genomes, manipulation of codon usage in helper plasmids, deletion of unnecessary sequences in the vectors, and development of SIN vector systems has reduced the risk of RCR. The ability of retroviruses to integrate a viral genome into the host cell chromosome is able to cause oncogene activation via insertional mutagenesis. The recent gene therapy clinical trial for X-linked severe combined immunodeficiency has illustrated the potential disadvantages involved, with two out of ten patients developing T cell leukemia as a consequence of the treatment. However, the possibility of such events can be greatly reduced with simple modification and functional improvement of vectors. Because proviral LTRs contain transcriptional control elements, which are known to activate the host oncogene, careful development of SIN vector and cautious selection of alternative promoters and insulator elements will minimize the possibility of oncogene activation. Even with these potential risks, improved retroviral vectors will continue to be safely utilized in the gene transfer for clinical trials.

8 SUMMARY

MLV-based retroviral vectors are the most frequently used gene delivery vehicle for preclinical and clinical applications. Based on the ability to stably integrate into the host genome, retroviral vectors are well suited for genetic modification of the cells and ex vivo gene delivery. These approaches have been used for the treatment of genetic diseases such as Gaucher's disease, Hunter's syndrome, and ADA deficiency as well as for acquired diseases such as cancer and arthritis. Even though safety concerns occurred for potential dangers in insertional mutagenesis and breakout of RCR, these limitations have been reduced with speedy improvement of the retroviral vector system. The careful design and engineering of the retroviral vectors has been addressed, resulting in vectors with strong and regulative transgene expression as well as safe gene delivery into a target cell. Clearly retroviral vectors have been a useful gene delivery vehicle for preclinical and clinical gene therapy and it is likely that the retroviral vectors will continue to be improved for use for the treatment of a variety of diseases in the future.

REFERENCES

Adam, M.A., and Miller, A.D. 1988, Identification of a signal in a murine retrovirus that is sufficient for packaging of nonretroviral RNA into virions. *J Virol.* **62**: 3802.

Bahrami, S., Duch, M., and Pedersen, F.S. 2004, Change of tropism of SL3-2 murine leukemia virus, using random mutational libraries. *J Virol.* **78**: 9343.

Bender, M.A., Palmer, T.D., Gelinas, R.E., and Miller, A.D. 1987, Evidence that the packaging signal of Moloney murine leukemia virus extends into the gag region. *J Virol.* **61**: 1639.

Bupp, K., and Roth, M.J. 2003, Targeting a retroviral vector in the absence of a known cell-targeting ligand. *Hum Gene Ther.* **14**: 1557.

Burns, D.P., Collignon, C., and Desrosiers, R.C. 1993, Simian immunodeficiency virus mutants resistant to serum neutralization arise during persistent infection of rhesus monkeys. *J Virol.* **67**: 4104.

Cepko, C.L., Roberts, B.E., and Mulligan, R.C. 1984, Construction and applications of a highly transmissible murine retrovirus shuttle vector. *Cell.* **37**: 1053.

Chadwick, M.P., Morling, F.J., Cosset, F.L., and Russell, S.J. 1999, Modification of retroviral tropism by display of IGF-I. *J Mol Biol.* **285**: 485.

Challita, P.M., Skelton, D., el-Khoueiry, A., Yu, X.J., Weinberg, K., and Kohn, D.B. 1995, Multiple modifications in *cis* elements of the long terminal repeat of retroviral vectors lead to increased expression and decreased DNA methylation in embryonic carcinoma cells. *J Virol.* **69**: 748.

Choulika, A., Guyot, V., and Nicolas, J.F. 1996, Transfer of single gene-containing long terminal repeats into the genome of mammalian cells by a retroviral vector carrying the cre gene and the loxP site. *J Virol.* **70**: 1792.

Chu, T.H., and Dornburg, R. 1995, Retroviral vector particles displaying the antigen-binding site of an antibody enable cell-type-specific gene transfer. *J Virol.* **69**: 2659.

Ciafre, S.A., Barillari, G., Bongiorno-Borbone, L., Wannenes, F., Izquierdo, M., and Farace, M.G. 2002, A tricistronic retroviral vector expressing natural antiangiogenic factors inhibits angiogenesis in vitro, but is not able to block tumor progression in vivo. *Gene Ther.* **9**: 297.

Cone, R.D., Weber-Benarous, A., Baorto, D., and Mulligan, R.C. 1987, Regulated expression of a complete human beta-globin gene encoded by a transmissible retrovirus vector. *Mol Cell Biol.* **7**: 887.

Crittenden, M., Gough, M., Chester, J., Kottke, T., Thompson, J., Ruchatz, A., Clackson, T., Cosset, F.L., Chong, H., Diaz, R.M., Harrington, K., Alvarez Vallina, L., and Vile, R. 2003, Pharmacologically regulated production of targeted retrovirus from T cells for systemic antitumor gene therapy. *Cancer Res.* **63**: 3173.

Danos, O. and Mulligan, R.C. 1988, Safe and efficient generation of recombinant retroviruses with amphotropic and ecotropic host ranges. *Proc Natl Acad Sci U S A.* **85**: 6460.

Davies, M.V. and Kaufman, R.J. 1992, The sequence context of the initiation codon in the encephalomyocarditis virus leader modulates efficiency of internal translation initiation. *J Virol.* **66**: 1924.

De Felipe, P. 2004, Skipping the co-expression problem: the new 2A "CHYSEL" technology. *Genet Vaccines Ther.* **2**: 13

De Felipe, P., and Izquierdo, M. 2000, Tricistronic and tetracistronic retroviral vectors for gene transfer. *Hum Gene Ther.* **11**: 1921.

Donahue, R.E., Kessler, S.W., Bodine, D., McDonagh, K., Dunbar, C., Goodman, S., Agricola, B., Byrne, E., Raffeld, M., Moen, R., et al. 1992, Helper virus induced T cell lymphoma in nonhuman primates after retroviral mediated gene transfer. *J Exp Med.* **176**: 1125.

Dougherty, J.P., Wisniewski, R., Yang, S.L., Rhode, B.W., Temin, H.M. 1989, New retrovirus helper cells with almost no nucleotide sequence homology to retrovirus vectors. *J Virol.* **63**: 3209.

Douin, V., Bornes, S., Creancier, L., Rochaix, P., Favre, G., Prats, A.C., and Couderc, B. 2004, Use and comparison of different internal ribosomal entry sites (IRES) in tricistronic retroviral vectors. *BMC Biotechnol.* **4**: 16.

Fassati, A., Bardoni, A., Sironi, M., Wells, D.J., Bresolin, N., Scarlato, G., Hatanaka, M., Yamaoka, S., and Dickson, G. 1998, Insertion of two independent enhancers in the long terminal repeat of a self-inactivating vector results in high-titer retroviral vectors with tissue-specific expression. *Hum Gene Ther.* **9**: 2459.

Fernex, C., Dubreuil, P., Mannoni, P., and Bagnis, C. 1997, Cre/loxP-mediated excision of a neomycin resistance expression unit from an integrated retroviral vector increases long terminal repeat-driven transcription in human hematopoietic cells. *J Virol.* **71**: 7533.

Ferrari, G., Salvatori, G., Rossi, C., Cossu, G., and Mavilio, F. 1995, A retroviral vector containing a muscle-specific enhancer drives gene expression only in differentiated muscle fibers. *Hum Gene Ther.* **6**: 733.

Finer, M.H., Dull, T.J., Qin, L., Farson, D., and Roberts, M.R. 1994, Kat: A high-efficiency retroviral transduction system for primary human T lymphocytes. *Blood.* **83**: 43.

Furth, P.A., St Onge, L., Boger, H., Gruss, P., Gossen, M., Kistner, A., Bujard, H., and Hennighausen, L. 1994, Temporal control of gene expression in transgenic mice by a tetracycline-responsive promoter. *Proc Natl Acad Sci U S A.* **91**: 9302.

Fussenegger, M., Mazur, X., and Bailey, J.E. 1998, pTRIDENT, a novel vector family for tricistronic gene expression in mammalian cells. *Biotechnol Bioeng*. **57**: 1.
Gossen, M., and Bujard, H. 1992, Tight control of gene expression in mammalian cells by tetracycline-responsive promoters. *Proc Natl Acad Sci U S A*. **89**: 5547.
Grande, A., Piovani, B., Aiuti, A., Ottolenghi, S., Mavilio, F., and Ferrari, G. 1999, Transcriptional targeting of retroviral vectors to the erythroblastic progeny of transduced hematopoietic stem cells. *Blood*. **93**: 3276.
Grez, M., Akgun, E., Hilberg, F., and Ostertag, W. 1990, Embryonic stem cell virus, a recombinant murine retrovirus with expression in embryonic stem cells. *Proc Natl Acad Sci U S A*. **87**: 9202.
Guild, B.C., Finer, M.H., Housman, D.E., and Mulligan, R.C. 1988, Development of retrovirus vectors useful for expressing genes in cultured murine embryonal cells and hematopoietic cells in vivo. *J Virol*. **62**: 3795.
Han, X., Kasahara, N., and Kan, Y.W. 1995, Ligand-directed retroviral targeting of human breast cancer cells. *Proc Natl Acad Sci U S A*. **92**: 9747.
Hawley, R.G., Lieu, F.H., Fong, A.Z., and Hawley, T.S. 1994, Versatile retroviral vectors for potential use in gene therapy. *Gene Ther*. **1**: 136.
Hildinger, M., Abel, K.L., Ostertag, W., and Baum, C. 1999, Design of 5' untranslated sequences in retroviral vectors developed for medical use. *J Virol*. **73**: 4083.
Hoess, R.H. and Abremski, K. 1984, Interaction of the bacteriophage P1 recombinase Cre with the recombining site loxP. *Proc Natl Acad Sci U S A*. **81**: 1026.
Indraccolo, S., Minuzzo, S., Habeler, W., Zamarchi, R., Fregonese, A., Gunzburg, W.H., Salmons, B., Uckert, W., Chieco-Bianchi, L., and Amadori, A. 2000, Modulation of Moloney leukemia virus long terminal repeat transcriptional activity by the murine CD4 silencer in retroviral vectors. *Virology*. **276**: 83.
Jiang, A., Chu, T.H., Nocken, F., Cichutek, K., and Dornburg, R. 1998, Cell-type-specific gene transfer into human cells with retroviral vectors that display single-chain antibodies. *J Virol*. **72**: 10148.
Joyner, A.L. and Bernstein, A. 1983, Retrovirus transduction: generation of infectious retroviruses expressing dominant and selectable genes is associated with in vivo recombination and deletion events. *Mol Cell Biol*. **3**: 2180.
Kasahara, N., Dozy, A.M., and Kan, Y.W. 1994, Tissue-specific targeting of retroviral vectors through ligand-receptor interactions. *Science*. **266**: 1373.
Kim, S.H., Yu, S.S., Park, J.S., Robbins, P.D., An, C.S., and Kim, S. 1998, Construction of retroviral vectors with improved safety, gene expression, and versatility. *J Virol*. **72**: 994.
Konishi, H., Ochiya, T., Chester, K.A., Begent, R.H., Muto, T., Sugimura, T., Terada, M., and Begent, R.H. 1998, Targeting strategy for gene delivery to carcinoembryonic antigen-producing cancer cells by retrovirus displaying a single-chain variable fragment antibody. *Hum Gene Ther*. **9**: 235.
Kraunus, J., Schaumann, D.H., Meyer, J., Modlich, U., Fehse, B., Brandenburg, G., von Laer, D., Klump, H., Schambach, A., Bohne, J., and Baum, C. 2000, Self-inactivating retroviral vectors with improved RNA processing. *Gene Ther*. **11**: 1568.
Leboulch, P., Huang, G.M., Humphries, R.K., Oh, Y.H., Eaves, C.J., Tuan, D.Y., and London, I.M. 1994, Mutagenesis of retroviral vectors transducing human beta-globin gene and beta-globin locus control region derivatives results in stable transmission of an active transcriptional structure. *EMBO J*. **13**: 3065.

Li, M., Hantzopoulos, P.A., Banerjee, D., Zhao, S.C., Schweitzer, B.I., Gilboa, E., and Bertino, J.R. 1992, Comparison of the expression of a mutant dihydrofolate reductase under control of different internal promoters in retroviral vectors. *Hum Gene Ther.* **3**: 381.

Lim, B., Williams, D.A., and Orkin, S.H. 1987, Retrovirus-mediated gene transfer of human adenosine deaminase: Expression of functional enzyme in murine hematopoietic stem cells in vivo. *Mol Cell Biol.* **7**: 3459.

Loew, R., Selevsek, N., Fehse, B., von Laer, D., Baum, C., Fauser, A., and Kuehlcke, K. 2004, Simplified generation of high-titer retrovirus producer cells for clinically relevant retroviral vectors by reversible inclusion of a lox-P-flanked marker gene. *Mol Ther.* **9**: 738.

Mann, R., Mulligan, R.C., and Baltimore, D. 1983, Construction of a retrovirus packaging mutant and its use to produce helper-free defective retrovirus. *Cell.* **33**: 153.

McClure, M.O., Sommerfelt, M.A., Marsh, M., and Weiss, R.A. 1990, The pH independence of mammalian retrovirus infection. *J Gen Virol.* **71**: 767-73.

McIvor, R.S. 1990, Deletion in a recombinant retroviral vector resulting from a cryptic splice donor signal in the Moloney leukemia virus envelope gene. *Virology.* **176**: 652.

Miller, A.D. and Rosman, G.J. 1989, Improved retroviral vectors for gene transfer and expression. *Biotechniques.* **7**: 980.

Miller, A.D., Garcia, J.V., von Suhr, N., Lynch, C.M., Wilson, C., and Eiden, M.V. 1991, Construction and properties of retrovirus packaging cells based on gibbon ape leukemia virus. *J Virol.* **65**: 2220.

Mizuguchi, H., Xu, Z., Ishii-Watabe, A., Uchida, E., and Hayakawa, T. 2000, IRES-dependent second gene expression is significantly lower than cap-dependent first gene expression in a bicistronic vector. *Mol Ther.* **1**: 376.

Morgan, R.A., Couture, L., Elroy-Stein, O., Ragheb, J., Moss, B., and Anderson, W.F. 1992, Retroviral vectors containing putative internal ribosome entry sites: Development of a polycistronic gene transfer system and applications to human gene therapy. *Nucleic Acids Res.* **20**: 1293.

Ohashi, T., Boggs, S., Robbins, P.D., Bahnson, A., Patrene, K., Wei, F.S., Wei, J.F., Li, J., Lucht, L., Fei, Y., et al. 1992, Efficient transfer and sustained high expression of the human glucocerebrosidase gene in mice and their functional macrophages following transplantation of bone marrow transduced by a retroviral vector. *Proc Natl Acad Sci U S A.* **89**: 11332.

Pear, W.S., Nolan, G.P., Scott, M.L., and Baltimore, D. 1993, Production of high-titer helper-free retroviruses by transient transfection. *Proc Natl Acad Sci U S A.* **90**: 8392.

Pollock, R., Issner, R., Zoller, K., Natesan, S., Rivera, V.M., and Clackson, T. 2000, Delivery of a stringent dimerizer-regulated gene expression system in a single retroviral vector. *Proc Natl Acad Sci U S A.* **97**: 13221.

Richardson, T.B., Kaspers, J., and Porter, C.D. 2004, Retroviral hybrid LTR vector strategy: functional analysis of LTR elements and generation of endothelial cell specificity. *Gene Ther.* **11**: 775.

Rivera, V.M., Clackson, T., Natesan, S., Pollock, R., Amara, J.F., Keenan, T., Magari, S.R, Phillips, T., Courage, N.L., Cerasoli, F. , Jr, Holt, D.A., and Gilman, M. 1996,

A humanized system for pharmacologic control of gene expression. *Nat Med.* **2**: 1028.

Russ, A.P., Friedel, C., Grez, M., and von Melchner, H. 1996, Self-deleting retrovirus vectors for gene therapy. *J Virol.* **70**: 4927.

Sadelain, M., Wang, C.H., Antoniou, M., Grosveld, F., and Mulligan, R.C. 1995, Generation of a high-titer retroviral vector capable of expressing high levels of the human beta-globin gene. *Proc Natl Acad Sci U S A.* **92**: 6728.

Schambach, A., Wodrich, H., Hildinger, M., Bohne, J., Krausslich, H.G., and Baum, C. 2000, Context dependence of different modules for posttranscriptional enhancement of gene expression from retroviral vectors. *Mol Ther.* **2**: 435.

Shimotohno, K. and Temin, H.M. 1981, Formation of infectious progeny virus after insertion of herpes simplex thymidine kinase gene into DNA of an avian retrovirus. *Cell.* **2**: 67.

Sirin, O. and Park, F. 2003, Regulating gene expression using self-inactivating lentiviral vectors containing the mifepristone-inducible system. *Gene.* **323**: 67.

Somia, N.V., Zoppe, M., and Verma, I.M. 1995, Generation of targeted retroviral vectors by using single-chain variable fragment: An approach to in vivo gene delivery. *Proc Natl Acad Sci U S A.* **92**: 7570.

Soneoka, Y., Cannon, P.M., Ramsdale, E.E., Griffiths, J.C., Romano, G., Kingsman, S.M., and Kingsman, A.J. 1995. A transient three-plasmid expression system for the production of high titer retroviral vectors. *Nucleic Acids Res.* **23**: 628.

Szymczak, A.L., Workman, C.J., Wang, Y., Vignali, K.M., Dilioglou, S., Vanin, E.F., and Vignali, D.A. 2004, Correction of multi-gene deficiency in vivo using a single 'self-cleaving' 2A peptide-based retroviral vector. *Nat Biotechnol.* **22**: 589.

Tabin, C.J., Hoffmann, J.W., Goff, S.P., and Weinberg, R.A. 1982, Adaptation of a retrovirus as a eucaryotic vector transmitting the herpes simplex virus thymidine kinase gene. *Mol Cell Biol.* **2**: 426.

Trubetskoy, A.M., Okenquist, S.A., and Lenz, J. 1999, R region sequences in the long terminal repeat of a murine retrovirus specifically increase expression of unspliced RNAs. *J Virol.* **73**: 3477.

Wei, C.M., Gibson, M., Spear, P.G., and Scolnick, E.M. 1981, Construction and isolation of a transmissible retrovirus containing the src gene of Harvey murine sarcoma virus and the thymidine kinase gene of herpes simplex virus type 1. *J Virol.* **39**: 935.

Yee, J.K., Friedmann, T., and Burns, J.C. 1994, Generation of high-titer pseudotyped retroviral vectors with very broad host range. *Methods Cell Biol.* **43** Pt A: 99.

Yu, S.F., von Ruden, T., Kantoff, P.W., Garber, C., Seiberg, M., Ruther, U., Anderson, W.F., Wagner, E.F., and Gilboa, E. 1986, Self-inactivating retroviral vectors designed for transfer of whole genes into mammalian cells. *Proc Natl Acad Sci U S A.* **83**: 3194.

Yu, S.S., Kim, J.M., and Kim, S. 2000, High efficiency retroviral vectors that contain no viral coding sequences. *Gene Ther.* **7**: 797.

Zufferey, R., Donello, J.E., Trono, D., and Hope, T.J. 1999, Woodchuck hepatitis virus posttranscriptional regulatory element enhances expression of transgenes delivered by retroviral vectors. *J Virol.* **73**: 2886.

ADENOVIRUS IN GENE THERAPY

Angela Montecalvo, [1] **Andrea Gambotto,**[2] **and Leonardo D'Aiuto**[3]

Gene therapy—the transference of genetic material into an individual—was first conceived as an approach to hereditary single-gene disease. Today the subject of gene therapy comprises multifactorial disorders such as cancer, cardiovascular disease, neurodegenerative disorders, and infectious disease. Ideally, a vector system for gene therapy would have the following attributes: (1) allow efficient transduction of the transgene into the target cells, (2) be safe (i.e., toxicity associated with the vector would be minimal or absent), (3) target only the desired cells within the target tissue, (4) express a therapeutic regulatable amount of the transgene, and (5) not integrate into the host genome. The ideal vector system, however, does not currently exist.

Viral-based vectors are the most common gene delivery systems employed for preclinical or clinical applications and adenoviral (Ad) vectors closely follow retroviruses as the most frequently used vectors for gene therapy. Although Ad vectors are not suitable for all applications, they are very efficient in delivering the therapeutic transgene to the cell nucleus. Other advantages of using Ad vectors include the simplicity of vector construction methods, efficient production, high yields and high stability, and reliable transduction of both proliferating and quiescent cell types.

1 ADENOVIRUS STRUCTURE AND BIOLOGY

Ad is a nonenveloped, icosahedral virus with a linear, double-stranded DNA genome of 30–40 Kb. The capsid is 90 nm in diameter with 240 hexon capsomeres and 12 penton capsomeres, and a trimeric fiber protein protruding outward from each of 12 icosahedral vertices. Each penton is located at a vertex of the icosahedral particle and houses the amino termini of the trimeric fiber protein. The carboxy termini of this trimer form a knob that mediates virus attachment to a host cell receptor (Figure 1).

[1] Department of Surgery and Medicine, Starlz Transplantation Institute, University of Pittsburgh School of Medicine, Pittsburgh, Pennsylvania.
[2] Department of Surgery, Division of infectious Diseases, University of Pittsburgh School of Medicine, Pittsburgh, Pennsylvania.
[3] Department of Molecular Genetics and Biochemistry, University of Pittsburgh, Pittsburgh, Pennsylvania

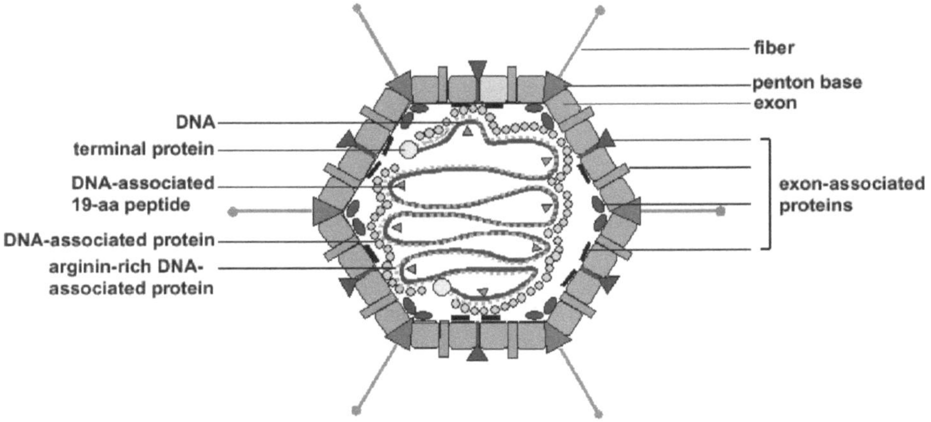

Figure 1. Structure of the adenovirus particle (modified from Shenk).

The Ad genome contains five early transcription regions (E1A, E1B, E2, E3, and E4), two delayed early regions (IX and 1Va2), and one late transcription region that codes for five families of late mRNAs, L1 to L5 (Figure 2).

The Ad infectious cycle can be clearly defined as having early and late phases of gene transcription, with an intervening phase of viral DNA replication. During the early phase, the viral DNA is transported into the nucleus of the host cell where transcription of the early genes begins. The first Ad gene to be expressed is the immediate early E1A gene encoding a transactivator for the transcription of the early genes E1B, E2A, E2B, E3, and E4 as well as genes coding for proteins involved in cell transformation. E2, which encodes the DNA polymerase, the DNA-binding protein, and the precursor of the terminal protein, is the most important region for viral DNA replication. The origins of DNA replication are located at the genome termini in the inverted terminal repeats (ITRs). The Ad DNA polymerase initiates the replication of the viral genome with the covalent coupling of a dCTP to the terminal protein (TP) covalently attached to each 5' end. Late region transcription determines the production of mRNA molecules that encode viral structural proteins as well as the proteins necessary for either protein processing or assembly. After viral proteins are synthesized, structural proteins are assembled into both hexons and pentons. Assembly of the virions occurs in the nucleus where the viral genome is packaged into a preformed empty capsid.

The lifecycle of Ad does not normally involve integration into the host genome. Rather, the viruses replicate as episomal elements in the nuclei of host cells; consequently there is no risk of insertional mutagenesis.

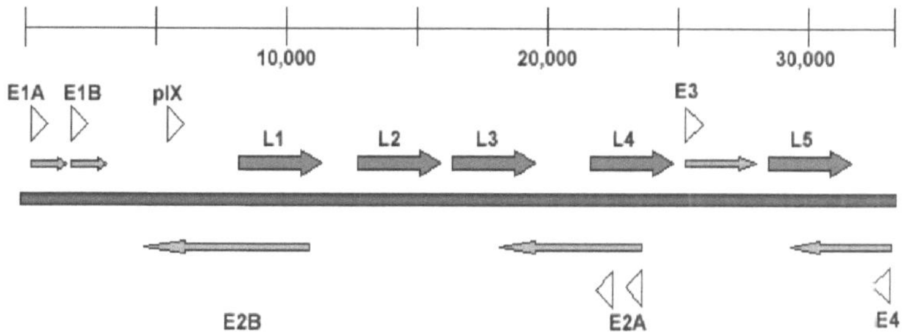

Figure 2. Map of the adenovirus genome. Promoters are indicated by arrowheads. Early (E) andlate (L) mRNA are indicated by thin and heavy arrows, respectively.

2 ADENOVIRUS TROPISM AND IMMUNOLOGY

The adsorption of virus to target cells involves the interaction of the terminal globular domain region of the Ad capsid with both the cell receptors, the coxsackie Ad receptor (CAR; Roelvink et al., 1998; Philipson and Pettersson, 2004) and CD46 (Gaggar, Shayakhmetov, and Lieber, 2003). Following the initial attachment, the interaction of a conserved RGD (Arg–Gly–Asp) motif, which is present on an extended loop in the pentone base with cell surface integrin molecules (avb3 or avb5), determines the internalization of the adenovirus (Wickham et al., 1993). The virus is endocytosed via clathrin-coated vesicles (Meier and Greber, 2004). In the cytoplasm, endosomal disruptions lead to the degradation of the capsid. The partially uncoated virion travels along microtubules in the cytoplasm to the nuclear pore complex (Greber et al., 1993). The viral genome then rapidly associates with the nuclear matrix to allow initiation of the primary transcription events (Nemerrow, 2000).

Widespread application of Ad-based gene therapy has been limited by immune defense mechanisms. The immune response against Ad vectors consists of both innate and adaptive pathways that represent, respectively, the first and second line of immune system defense mechanisms against viral particles. The innate response is nonspecific, depends on the dose of viral particles injected into the recipient, and, it is important to note, is highly variable among recipients.

Once an Ad vector is administered, it is taken up by resident macrophages via a nonspecific mechanism (Zsengeller, 2000; Liu and Muruve, 2003; Schagen et al., 2004); the interaction between protein capsid and cell receptors is nonspecific and leads to downstream activation of genes that encode for inflammatory cytokines and chemokines (Bessis, GarciaCozar, and Boissier, 2004). During this early phase it appears that viral proteins do not play any role in inducing the immune response. Signal transduction related to CAR binding has yet to be demonstrated, but it is known that integrins are involved in a wide variety of signaling events regulating protein kinases, growth factor receptors, and organization of the actin cytoskeleton. The RGD motifs of the penton bases recognize the specific integrin-linked signaling pathways in target cells (Marshall, 2004; Nemerrow, 2000). These interactions translate into activation of a specific intracellular signal transduction pathway that comprises a large number of intracellular proteins.

The cellular immune response towards Ad antigens is activated by antigen-presenting cells (APCs). After uptake of the Ad particle, viral proteins and transgene products are processed into small oligopeptides, which are presented by major histocompatibility complex (MHC) class I molecules at the cell surface. The binding of $CD8^+$ T cells to the peptide-MHC class I complex leads to production of antibodies (Abs) specific for viral or transgene peptides or to the proliferation of specific cytotoxic T lymphocytes (CTLs). The interaction between CD28 and B7 co-stimulatory molecules is important for these reactions (Linsley et al., 1991).

$CD4^+$ T helper cells belonging primarily to the T_H1 subset further stimulate the cellular immune response. In contrast to $CD8^+$ T cells, $CD4^+$ T helper cells are activated by epitopes from the input virion that are presented by MHC class II molecules on the surface of APCs. This activation triggers the T_H1 cells to secrete interleukin-2 (IL-2) and interferon-γ (IFN-γ). These cytokines, in turn, induce the differentiation of $CD8^+$ T cells into CTLs (Maraskovsky, Chen, and Shortman, 1989; Wille et al., 1989), cause the upregulation of MHC class I expression in Ad-transduced cells, and consequently facilitate their recognition by CTLs (Yang et al., 1996, 1995b). Activated $CD4^+$ T_H1 cells have been implicated in the destruction of Ad-transduced cells, similar to primary CTL (Yang and Wilson 1995).

The humoral immune response, which is mediated by B cells, represents a second challenge to transgene expression. After internalization and processing by B cells, Ad-derived epitopes are presented by MHC class II molecules. The resulting antigen-MHC class II complexes can be recognized by activated $CD4^+$ helper cells, which release IL-4, IL-5, IL-6, and IL-10. These cytokines provide important signals for the differentiation of the B cells into plasma cells. As a result, plasma cells secrete antibodies directed against the Ad capsid. The binding of these specific Abs to the Ad vector prevents entry into the cell and promotes opsonization and phagocytosis by macrophages. Consequently, circulating Ad-specific Abs dramatically reduce the efficacy of repeated Ad vector administration (Barr et al., 1995; Chirmule et al., 1999).

Another challenge to the use of adenoviral vector-delivered transgenic material is the existence of neutralizing antibodies against various Ad serotypes (Ritter, Lehmann, and Volk, 2002). In recipients who have such specific neutralizing antibodies, Ad vector administration acts as a second encounter between antigen and the host immune system, enhancing a strong specific immune response and leading to the clearance of almost the 90% of injected Ad vector.

3 ADENOVIRAL VECTORS

Genetic analysis of the Ad genome has highlighted the necessity of deleting the early E1 region during Ad vector construction. This deletion serves two purposes: it renders the virus replication-defective and it provides a location into which foreign genes may be inserted, thereby increasing cloning capacity. In first-generation vectors, deletion of the E1 region enables the insertion of a transgene up to 4.9 Kb in size and prevents the *trans*-activation of viral genes required for viral DNA replication (Danthinne and Imperiale, 2000). The loss of these essential genes prevents propagation of the transgenic vector; however, the use of a packaging cell-line (293) that contains a stable E1a expression cassette overcomes this difficulty by allowing replication and packaging of the E1-deleted transgenic vector (Graham et al., 1977).

Even in the absence of E1 gene products, low-level transcription of the remaining viral genes occurs, resulting in an early innate cytokine response that is followed by an antigen-dependent immune response and cell-mediated destruction of the transduced cells (Yang et al., 1994, 1995a). The final result is a shorter period of transgene expression. To augment viral replication in cell culture and to increase total transgene vector capacity, vectors also contain E3 deletions. Because genes of both E1 and E3 regions are essential for replication of the viral chromosome, their functions must be provided *in trans* by the packaging cell-line. It is thought that the E3 genes may offer protection against the vector. First-generation E1/E3-deleted Ad vectors have the capacity to carry trangenes up to 8 Kb in size and can be produced in large quantities (up to 10^{12}–10^{13} vector particles per ml; Volpers and Kochanek, 2004).

To further reduce toxicity and inflammatory responses associated with first generation Ad vectors, a second generation of vectors was developed. These vectors lack, in addition, E1 and/or E3 regions, the genes coding for E2 and/or E4 (Engelhardt et al., 1994). Although deletions increase the cloning capacity of Ad vectors, some reports indicate that the E4 region may exert a positive effect on long-term expression (Pfeifer and Verma, 2001), and thus, the loss of E4 may be detrimental in some cases.

E1-deleted Ad vectors are powerful gene delivery vehicles, but their therapeutic uses are restricted to applications where only transient expression is needed or to those where stimulation of immunity is beneficial. To overcome the immunological barrier to long-term expression of the gene of interest, a new generation of Ad vectors called helper-dependent (HD) "gutless" vectors has been developed (Morsy and Caskey, 1999; Kochanek, 1999; Parks, 2000). These vectors are devoid of all viral coding sequences, except for the termini consisting of the ITRs and the packaging signals. Deletion of the E1/E2/E3/E4 viral genes expands the capacity of the vector for foreign DNA of ~36 Kb (Kochanek, Schiedner, and Volpers, 2002). The construction of helper-dependent Ad vectors is accomplished through a special vector system in which the helper vector contains all of the viral genes required for replication but has a conditional gene defect in the packaging domain that makes it less likely to be packaged into a virion. The second vector (HD vector) contains only the end of the viral genome, the gene of interest, and the normal packaging signal that allows this genome to be selectively packaged and released from cells (Vorburger and Hunt, 2002).

The generation of gutless vectors is labor-intensive and time-consuming, mainly because of the challenge in purifying the vector from the contaminating helper virus, but also because of the difficulty in scaling up production to pharmaceutical levels. To overcome these limitations a more efficient system for producing gutless vectors has been developed. This system is based on recombinase-mediated excision of the packaging signal, which is flanked by loxP recognition sites, in the helper virus. Initially Cre recombinase 293 cells are transfected with gutless virus carrying the transgene, ITRs as origins of replication, and the packaging signal (Ng et al., 1999). The same cells are infected with helper virus, which cannot be packaged because of excision of the loxP-flanked packaging signal in its genome, but aids in providing the regulatory and structural proteins required for replication and packaging of the vector transgenic DNA (Figure 3). After several rounds of coinfection with vector lysate and helper virus for amplification, vector particles are purified from residual contaminating helper virus by CsCl equilibrium centrifugation. Recently, HPLC has been employed to increase the purity of Ad vector (Green et al., 2002).

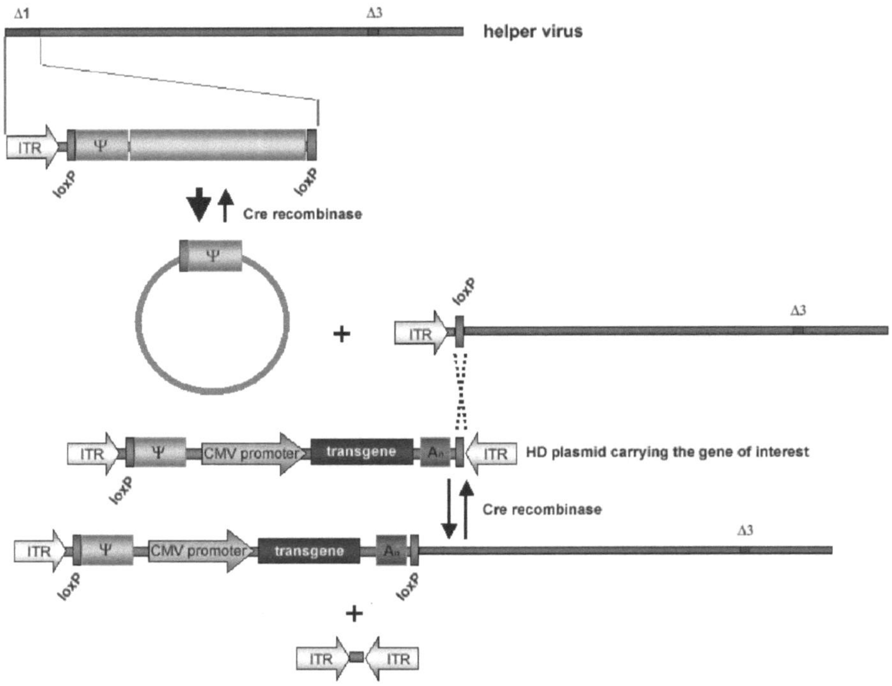

Figure 3. Generation of the helper virus-dependent adenoviral vector. The adenoviral genes that are deleted from the "high density" (HD) vector are provided in *trans* by a helper virus and a complementing cell-line. The propagation of the helper virus is limited by flanking the packaging signal (Ψ) with loxP sites. Expression of the Cre-recombinase in the complementing cell-line allows the removing of the packaging site by intramolecular recombination. An intermolecular recombination between the HD plasmid carrying the transgene and the helper virus creates a recombinant adenoviral HD vector.

Ideally, the risk of immunoresponse to these Ad vectors should be negligible. This is an important goal for many Ad vectors-based clinical trials, given the need to administer treatment frequently (because the inability of Ad to integrate into chromosomal DNA). Unfortunately, deletion of Ad genes can also be counterproductive. Removal of the E3 region abolishes the ability of the virus to encode a protein that protects the virus from immune surveillance mechanisms in the host. The E3 19 KD glycoprotein markedly reduces the capacity of MHC class I molecules for transporting viral antigens to the surface of infected cells, and the E3 14.7 KD protein significantly inhibits the production of TNF-α and, therefore, reduces the polymorphonuclear response (Ginsberg, 1996).

Despite the genetic elegance of the Cre-loxP system, mutant helper virus resistant to Cre-mediated packaging excision can be generated during the propagation of HD Ad, and the amplification of such mutants leads to high helper virus contamination levels. The origin of such mutants may depend on the loss of the 5' loxP site from the helper viral genome because of homologous recombination between the packaging sequences of the

HD Ad and the helper virus (Hardy et al., 1997). To address this problem, nonhomologous packaging signals (Sandig et al., 1999) or packaging signals that are homologous but in opposite orientation in the HD Ad and the helper virus have been used (Hillgenberg et al., 2001). Although it is desirable to avoid or minimize the contamination of the HD Ad preparation with helper virus, the real influence of the currently low levels (0.1 to 0.5%) of helper virus contamination on either the duration of transgene expression or an increase in toxicity has not been determined.

Several strategies have been developed to circumvent the neutralization of Ad vectors by antibodies. One such approach, known as "PEGylation," involves covalent attachment of the polymer polyethylene glycol (PEG) to the surface of the Ad virions (O'Riordan et al., 1999). PEGylated Ad can transduce the transgene into cells even in the presence of Ad-neutralizing antibodies. Recently, an alternative strategy, consisting of the encapsulation of the viral capsid using DOTAP: chol liposomes (Yotnda et al., 2002) have been utilized. Ad encapsidated within such bilamellar cationic liposomes binds to the negatively charged cell membrane independently of any CAR receptor–integrin interaction (Fasbender et al., 1997). Once bound, Ad-liposome complexes cross the membrane by endocytosis, a process whose efficiency depends on the charge density of the complex, the time of incubation with the cells, and the complex size (Zabner et al., 1995). The coated Ad particles, although immunogenic, are nonetheless protected from neutralizing antibodies ex vivo and in vivo and, thus, can be readily readministered.

4 APPLICATIONS

The first clinical trial using recombinant adenoviruses (rAd) was performed for the treatment of cystic fibrosis, an inherited disease that is caused by mutations in the cystic fibrosis transmembrane conductance regulator (CFTR) gene. The use of rAd encoding for the functional *cftr* product to correct the defective gene was not successful, probably because the expression of transgene was transient, whereas long-term expression was required (Ritter, Lehmann, and Volk, 2002).

Successful somatic gene therapy depends on several factors, including efficient delivery of DNA, sustained gene expression, and the presence and activity of a therapeutic protein in the target cells at physiological levels. Although viral vectors are the most efficient vehicles for gene therapy, they are lacking in some properties that are believed necessary in an ideal gene therapy system. Retroviral vectors stably integrate in the host genome, providing long-term expression; however, random integration in the host genome may lead to insertional mutagenesis and to silencing of the transgene (Haviernik and Bunting, 2004). Adeno-associated vectors do not induce an immune response but are limited by the capacity for packaging foreign DNA (Tal, 2000). The ability to accommodate large DNA fragments is important because full controlled expression at physiological levels can often be achieved after the transfer of an intact gene and its regulatory regions. Herpes simplex virus type I-based vectors (Glorioso et al., 1995) and Ad vectors exist as nonreplicating extrachromosomal elements in the nuclei of transduced cells and consequently do not support long-term expression.

It is clear, then, that the ideal vector for somatic gene therapy should have high carrying capacity. Also, it should be retained in the target cells for many cell divisions without integration into the genome (D'Aiuto et al., 2003). This condition could be achieved if a vector contains a human origin of replication and a centromere, necessary for replication and correct segregation of the vector into the daughter cells after cell division, respectively. Although human origin of replications have been characterized (Todorovic, Falaschi, and Giacca, 1999; Paixao et al., 2004) the nature of *cis*-acting sequences necessary for centromere formation remains poorly understood. The absence of the centromere makes all nonintegrating viral vectors unsuitable as vehicles for metabolic diseases that require long-term expression of the therapeutic gene.

Ad vectors, however, have great potential for therapies where short-term expression is sufficient, such as in coronary artery and peripheral vascular diseases. For example, the transient expression of vascular endothelial growth factor (VEGF), a growth factor that induces angiogenesis, can lead to revascularization and reperfusion of affected areas (Rosengart et al., 1999; Khan, Sellke, and Laham, 2003). Also, Ad vectors are attractive vaccines because they induce both innate and adaptive responses in a mammalian host. On the basis of their strong immunogenic and inflammatory properties, Ad vectors act as a powerful adjuvant for a variety of antigens, and are currently being tested as vaccine carriers against numerous infectious agents, ranging from malaria to HIV (Tatsis and Ertl, 2004; Gomez-Roman and Robert-Guroff, 2003). Ad-based live vaccines are able to induce not only systemic immunity, but also a good mucosal response that is essential for the prevention of mucosally and sexually transmitted infection. Another advantage of Ad vaccine vectors is their easy delivery by oral, intranasal, intratracheal, intraperitoneal, intravenous subcutaneous, or intramuscular routes (Jewtoukott and Perricaudet, 1995).

Ad vectors represent a novel approach for treating cancers that are resistant to currently available therapy, mainly because of their unparalleled capacity for gene transfer, their stability in vivo, and the ease of production in high titers (Kanerva and Hemminki, 2004). Various techniques have been adopted to suppress or eliminate tumor cells, the approach in each case depending largely on the type and location of the tumor (Russell, 2000). Ad vectors are used in suicide gene therapy and in approaches that combine gene therapy with chemotherapy. Suicide gene therapy involves delivery of "suicide genes" that encode enzymes capable of converting a prodrug into a cytotoxic compound. This approach is accomplished by injecting a viral vector containing the suicide gene directly into the tumor mass (Wadhwa et al., 2002). The herpes simplex virus thymidine kinase gene (HSV-TK) is the most well-characterized suicide gene; its product activates the prodrug ganciclovir 1000 times more efficiently than its mammalian counterpart (Springer and Niculescu-Duvaz, 2000). Several clinical trials have combined suicide gene therapy with radiotherapy (Teh et al., 2001) or chemotherapy (Hasenburg et al., 2001; Schuler et al., 2001).

Another gene-based immunotherapy approach uses Ad vectors to deliver cytokine genes into solid tumors, resulting in localized high doses of cytokines. There is much reduced toxicity known to be associated with systemic delivery of the recombinant cytokine protein (Liu et al., 2004). Cancer gene therapy with adenoviral vectors is safe and, although there is variability in the gene transfer in tumor cells, in some cases efficacious (Hemminki and Alvarez, 2002).

Although adenoviruses do not allow long-term expression of transgenes, they represent extremely useful tools for genetic diseases and for testing the feasibility of enzyme replacement by gene therapy in animal models. For example, Ad vectors have been used for a large number of lysosomal storage diseases (LDSs) such as Gaucher, Fabry, Pompe, and Tay–Sachs diseases, Mucopolysaccharidosis type II (MPS II, Hunter syndrome), MPS VII-Sly syndrome, Wolman disease, and Aspartyglucosaminuria (Xu et al., 2005; Harkke, Laine, and Jalanko, 2003; Marshall et al., 2002; Ziegler et al., 2002; Daniele et al., 2002; Du et al., 2002; Verdugo et al., 2001; Guidotti et al., 1999). The interest in Ad vectors is due to efficient transduction of the liver and high-level expression and secretion of lysosomal enzymes obtained after systemic delivery of recombinant Ad vectors. Depending on the dose of virus used, these levels can range from 10- to 1000-fold higher than normal levels. Even when the enzyme levels are sufficient to reproduce the basal conditions, the expression is transient, decreasing within few weeks (Cheng and Smith 2003).

Studies on mice models have been conducted to evaluate the effect of administration of modified Ad vectors, such as intravenous administration of vectors encoding human acid α-glucosidase (GAA), which results in efficient hepatic transduction and secretion of high levels of the precursor GAA proenzyme into the plasma of treated animals. These models can potentially be expanded to include the treatment of the other lysosomal enzyme disorders (Amalfitano et al., 1999).

REFERENCES

Amalfitano, A., McVie-Wylie, A.J., Hu, H., Dawson, T.L., Raben, N., Plotz, P., and Chen, Y.T. (1999) Systemic correction of the muscle disorder glycogen storage disease type II after hepatic targeting of a modified adenovirus vector encoding human acid-alpha-glucosidase. *Proc Natl Acad Sci U S A*. **96**, 8861–8866.

Barr, D., Tubb, J., Ferguson, D., Scaria, A., Lieber, A., Wilson C., Perkins, J., and Kay, M.A. (1995) Strain related variations in adenovirally mediated transgene expression from mouse hepatocytes in vivo: Comparisons between immunocompetent and immunodeficient inbred strains. *Gene Ther*. **2**, 151–155.

Bessis, N., GarciaCozar, F.J., and Boissier, M.-C. (2004) Immune response to gene therapy vectors: influence on vector function and effectors mechanisms. *Gene Ther*. **11**, 10–17.

Cheng, S.H. and Smith, A.E. (2003) Gene therapy progress and prospects: Gene delivery of lysosomal storage disorders. *Gene Ther*. **10**, 1275–1281.

Chirmule, N., Propert, K., Magosin, S., Qian, Y., Qian, R. and Wilson, J. (1999) Immune responses to adenovirus and adeno-associated virus in humans. *Gene Ther*. **6**, 1574–1583.

D'Aiuto, L., de las Heras, J.I., Ross, A., Shen, M.H., and Cooke, H. (2003) Generation of a telomere-based episomal vector. *Biotechnol Prog*. **19**, 1775–1780.

Daniele, A., Tomanin, R., Villani, G.R., Zacchello, F., Scarpa, M., and Di Natale, P. (2002) Uptake of recombinant iduronate-2-sulfatase into neuronal and glial cells in vitro. *Biochim Biophys Acta*. **1588**, 203–209.

Danthinne, X. and Imperiale M.J. (2000) Production of first generation adenovirus vector: A review. *Gene Ther*. **7**, 1707–1714.

de Jong, R.N., Meijer, L.A.T., and van der Vliet, P.C. (2003) DNA binding properties of the adenovirus DNA replication priming protein pTP. *Nucleic Acid Res*. **31**, 3274–3286.

Du, H., Heur, M., Witte, D.P., Ameis, D., and Grabowski, G.A. (2002) Lysosomal acid lipase deficiency: Correction of lipid storage by adenovirus-mediated gene transfer in mice. *Hum Gene Ther*. **13**, 1361–1372.

Engelhardt, J.F., Ye, X., Doranz, B., and Wilson, J.M. (1994) Ablation of E2a in recombinant adenoviruses improves transgene persistence and decreases inflammatory response in mouse liver. *Proc Natl Acad Sci USA* **91**, 6196–6200.

Fasbender, A., Zabner, J., Chillon, M., Moninger, T.O., Puga, A.P., Davidson, B.L., and Welsh, M.J. (1997) Complexes of adenovirus with polycationic polymers and cationic lipids increase the efficiency of gene transfer in vitro and in vivo. *J Biol Chem*. **272**, 6479–6489.

Gaggar, A., Shayakhmetov, D.M., and Lieber, A. (2003) CD46 is a cellular receptor for group B adenoviruses. *Nature Med.* **9**, 1408–1412.

Ginsberg, H.S. (1996) The ups and downs of adenovirus vectors. *Bull N Y Acad Med.* **73**, 53–8.

Glorioso, J., De Luca, N.A., and Fink, D.J. (1995) Development and application of herpes simplexmvirus vectors for human gene therapy. *Annu Rev Microbiol.* **49**, 675–710.

Gomez-Roman V.R. and Robert-Guroff, M. (2003) Adenoviruses as vectors for HIV vaccines. *AIDS Reviews.* **5**, 178–185.

Graham, F.L., Smiley, J., Russel, W.C., and Nairn, R. (1977) Characteristics of a human cell line transformed by DNA from human adenovirus type 5. *J Gen Virol.* **36**, 59–74.

Greber, U.F., Willetts, M., Webster, P., and Helenius, A. (1993) Stepwise dismantling of adenovirus 2 during entry into cells. *Cell* **75**, 477–486.

Green, A.P., Huang, J.J., Scott, M.O., Kierstead, T.D., Beaupre, I., Gao G.P., and Wilson, J.M. (2002) A new scalable method for the purification of recombinant adenovirus vectors. *Hum Gene Ther.* **13**, 1921–1934.

Guidotti, J.E., Mignon, A., Haase, G., Caillaud, C., McDonell, N., Kahn, A., and Poenaru, L. (1999) Adenoviral gene therapy of the Tay-Sachs disease in hexosaminidase A-deficient knock-out mice. *Hum Mol Genet.* **8**, 831–838.

Hardy, S., Kitamura, M., Harris-Stansil, T., Dai, Y., and Phipps, M.L. (1997) Construction of adenovirus vectors through Cre-lox recombination. *J Virol.* **71**, 1842–1849.

Harkke, S., Laine, M., and Jalanko, A. (2003) Aspartylglucosaminidase (AGA) is efficiently produced and endocytosed by glial cells: Implication for therapy of a lysosomal storage disorder. *J Gene Med.* **5**, 472–482.

Hasenburg, A. Tong, X.W., Fischer, D.C., Rojas-MArtinez, A.,, Nyberg-Hoffman, C., Kaplan, A.L., Kaufman, R.H., Ramzy, I., Aguilar-Cordova, E., and Kieback, D.G. (2001) Adenovirus-mediated thymidine kinase gene therapy in combination with topotecan for patients with recurrent ovarian cancer: 2.5-Year follow-up. *Gynecol Oncol.* **83**, 549–554,.

Haviernik, P. and Bunting, K.D. (2004) Safety concerns related to hematopoietic stem cell gene transfer using retroviral vectors. *Curr Gene Ther.* **4**, 263–276.

Hemminki, A. and Alvarez, R.D. (2002) Adenoviruses in oncology: A viable option? *BioDrugs.* **16**, 77–87.

Hillgenberg, M., Schnieders, F., Loser, P., and Strauss, M. (2001) System for efficient helper-dependent minimal adenovirus construction and rescue. *Hum Gene Ther.* **12**, 643–657.

Jewtoukott, V.E. and Perricaudet, M. Recombinant adenoviruses as a vaccine. *Biologicals* **23**, 145–157 (1995).

Kanerva, A. and Hemminki, A. (2004) Modified adenoviruses for cancer gene therapy. *Int J Cancer* **11**, 475–480.

Khan, T.A., Sellke F.W. and Laham, R.J. (2003) Gene therapy progress and prospects: therapeutic angiogenesis for limb and myocardial ischemia. *Gene Ther.* **10**, 285–291.

Kochanek, S. (1999) High-capacity adenoviral vectors for gene transfer and somatic gene therapy. *Hum Gene Ther.* **10**, 245–249.

Kochanek, S., Schiedner, G., and Volpers, C. (2002) High-capacity 'gutless' adenoviral vectors. *Curr Opin Mol Ther.* **3**, 454–463.

Linsley, P.S., Brady, W., Grosmaire, L., Aruffo, A., Damle, N.K. and Ledbetter, J.A. (1991) Binding of the B cell activation antigen B7 to CD28 costimulates T cell proliferation and interleukin 2 mRNA accumulation. *J Exp Med.* **173**, 721–730.

Liu, M., Acres, B., Balloul, J.M., Bizouarne, N., Paul, S., Slos, P., and Squiban, P. (2004) Gene-based vaccines and immunotherapeutics. *Proc Natl Acad Sci USA.* **101** Suppl **2**, 14567–14571.

Liu, Q. and Muruve, D.A. (2003) Molecular basis of the inflammatory response to adenoviruses vectors. *Gene Ther.* **10**, 935–940.

Maraskovsky, E., Chen, W.F., and Shortman, K. (1989) IL-2 and IFN-gamma are two necessary lymphokines in the development of cytolytic T cells. *J Immunol.* **143**, 1210–1214.

Marshall, J., McEachern, K.A., Kyros, J.A., Nietupski, J.B., Budzinski, T., Ziegle, R.J., Yew, N.S., Sullivan, J., Scaria, A., van Rooijen, N., Barranger, J.A., and Cheng, S.H. (2002)

Demonstration of feasibility of in vivo gene therapy for Gaucher disease using a chemically induced mouse model. *Mol Ther.* **6**, 179–189.

Marshall, M.S. (2004) Function of adenovirus E3 proteins and their interactions with immunoregulatory cell proteins. *J Gene Med.* **6**, 172–183.

Meier, O. and Greber, U.F. (2004) Adenovirus endocytosed. *J Gene Med.* **6**, 152–163.

Morsy, M.A. and Caskey, C.T. (1999) Expanded-capacity adenoviral vectors—The helper-dependent vectors. *Mol Med Today.* **5**, 18–24.

Nemerrow, G.R. (2000) Cell receptors involved in adenovirus entry. *Virology.* **274**, 1–4.

Ng, P., Parks, R.J., Cummings, D.T., Evelegh, C.M., Sankar, U., and Graham, F.L. (1999) A high-efficiency Cre/loxP-based system for construction of adenoviral vectors. *Hum Gene Ther.* **10**, 2667–2672.

O'Riordan, C.R., Lachapelle, A., Delgado, C., Parkes, V., Wadsworth, S.C., Smith, A.E., and Francis, G.E. (1999) PEGylation of adenovirus with retention of infectivity and protection from neutralizing antibody in vitro and in vivo. *Hum Gene Ther.* **10**, 1349–1358.

Paixao, S., Colaluca, I.N., Cubells, M., Peverali, F.A., Destro, A., Giadrossi, S., Giacca, M., Falaschi, A., Riva, S., and Biamonti, G. (2004) Modular structure of the human lamin B2 replicator. *Mol Cell Biol.* **24**, 2958–2967.

Parks, R.J (2000). Improvements in adenoviral vector technology: overcoming barriers for gene therapy. *Clin Genet*, **58**, 1–11.

Pfeifer, A. and Verma, I.M. (2001) Gene therapy: Promises and problems. *Annu Rev Genomics Hum Genet.* **2**, 177–211.

Philipson, L. and Pettersson, R.F. (2004) The coxsackie-adenovirus receptor: A new receptor in the immunoglobulin family involved in cell adhesion. *Curr Top Microbiol Immunol.* **273**, 87–111.

Ritter, T., Lehmann, M., and Volk, H.-D. (2002) Improvements in gene therapy averting the immune response to adenoviral vectors. *Biodrugs* **16**, 3–10.

Roelvink P.W., Lizonova A., Lee J.G., Li Y., Bergelson J.M., Finberg R.W., Brough, D.E., Kovesdi, I., and Wickham, T.J. (1998) The coxsackievirus adenovirus receptor protein can function as a cellular attachment for subgroups C adenoviruses serotype from subgroups A, C, D, E, and F. *J Virol.* **72**, 7909–7915.

Rosengart, T.K., Lee, L., Patel, SR, Sanborn, T.A., Parikh, M., Bergman, G.W., Hachamovitc, R., Szulk, M., Kligfield, P.D., Okin, P.M., Hanh, R.T., Devereux, R.B., Post, M.R., Hackett, N.R., Foster, T., Grasso, T.M., Lessere, M..L., Isom, O.W., and Crystal, R.G. (1999) Angiogenesis gene therapy: Phase I assessment of direct intramyocardial administration of an adenovirus vector expressing VEGF121 cDNA to individuals with clinically significant severe coronary artery disease. *Circulation.* **100**, 468–474.

Russell, W.C. (2000) Update on adenovirus and its vectors. *J Gen Virol.* **81**, 2573–2604.

Sandig, V., Youil, R., Bett, A.J., Franlin, L.L., Oshima, M., Maione, D., Wang, F., Metzker, M.L., Savino, R., Caskey, C.T. (1999) Optimization of the helper-dependent adenovirus system for production and potency in vivo. *Proc Natl Acad Sci USA.* **97**, 1002–1007.

Schagen, F.H., Ossevoort, M., Toes, R.E., and Hoeben, R.C. (2004) Immune responses against adenoviral vectors and their transgene products: A review of strategies for evasion. *Crit Rev Oncol Hematol.* **50**, 51–70.

Schuler, M., Herrmann, R., De Greve, J.L., Stewart, A.K., Gatzemeier, U., Stewart, D.J., Laufman, L., Gralla, R., Kuball, J., Buhl, R., Heussel, C.P., Kommoss, F., Perruchoud, A.P., Shepherd, F.A., Fritz, M.A., Horowitz, J.A., Huber, C., and Rochlitz, C. (2001) Adenovirus-mediated wild-type p53 gene transfer in patients receiving chemotherapy for advanced non-small-cell lung cancer: results of a multicenter phase II study. *J Clin Oncol.* **19**, 1750–1758.

Springer, C.J. and Niculescu-Duvaz, I. (2000) Prodrug-activating systems in suicide gene therapy. *J Clin Invest.* **105**, 1161–1167.

Tal, J. (2000) Adeno-associated virus-based vectors in gene therapy. *J Biomed Sci.* **7**, 279–91.

Tatsis, N. and Ertl, H.C. (2004) Adenoviruses as vaccine vectors. *Mol Ther.* **10**, 616–29.

Teh, B.S., Aguilar-Cordova, E., Kernen, K., Chou, C.C., Shalev, M., Vlachaki, M.T., MilesD, B., Kadmon, D., Mai, W.Y., Caillouet, J., Davis, M., Ayala, G., Wheeler, T., Brady, J., Carpenter,

L.S., Lu, H.H., Chiu, J.K., Woo, S.Y., Thompson, T., and Butler, E.B. (2001) Phase I/II trial evaluating combined radiotherapy and in situ gene therapy with or without hormonal therapy in the treatment of prostate cancer—A preliminary report. *Int J Radiat Oncol Biol Phys.* **51**, 605–613.

Todorovic, V., Falaschi, A., and Giacca, M. (1999) Replication origins of mammalian chromosomes: The happy few. *Front Biosci.* **4**, D859–868.

Verdugo, M.E., Scarpino, V., Moullier, P., Haskins, M.E., Aguirre G.D., and Ray, J. (2001) Adenoviral vector-mediated beta-glucuronidase cDNA transfer to treat MPS VII RPE in vitro. *Curr Eye Res.* **23**, 357–367.

Volpers, C. and Kochanek, S. (2004) Adenoviral vectors for gene transfer and therapy. *J Gene Med.* **6**, S164–S171.

Vorburger, S.A. and Hunt, K.K. (2002) Adenoviral gene therapy. *Oncologist.* **7**, 46–59.

Wadhwa, P.D., Zielske, S.P., Roth, J.C., Ballas, C.B., Bowman, J.E., and Gerson, S.L. (2002) Cancer gene therapy: scientific basis. *Annu Rev Med.* **53**, 437–52.

Wickham, T.J., Mathias, P., Cheresh, D.A., and Nemerow, G.R. (1993) Integrins $alpha_v beta_3$ and $alpha_v beta_5$ promote adenovirus internalization but not virus attachment. *Cells* **73**, 309–319.

Wille, A., Gessner, A., Lother, H. and Lehmann-Grube, F. (1989) Mechanism of recovery from acute virus infection. VIII. Treatment of lymphocytic choriomeningitis virus-infected mice with anti-interferon-gamma monoclonal antibody blocks generation of virus-specific cytotoxic T lymphocytes and virus elimination. *Eur J Immunol.* **19**, 1283–1288.

Xu, F., Ding, E., Migone, F., Serra, D., Schneider, A., Chen, Y.T., and Amalfitano, A. (2005) Glycogen storage in multiple muscles of old GSD-II mice can be rapidly cleared after a single intravenous injection with a modified adenoviral vector expressing hGAA. *J Gene Med.* **7**, 171–178.

Yang, Y. and Wilson J.M. (1995) Clearance of adenovirus-infected hepatocytes by MHC class I-restricted CD4+ CTLs in vivo. *J Immunol.* **155**, 2564–2570.

Yang, Y., Haecker, S.E., Su, Q., and Wilson J.M. (1996) Immunology of gene therapy with adenoviral vectors in mouse skeletal muscle. *Hum Mol Genet.* **5**, 1703–1712.

Yang, Y., Li, Q., Ertl, H.C., and Wilson J.M. (1995a) Cellular and humoral immune responses to viral antigens create barriers to luna-directed gene therapy with recombinant adenoviruses. *J Virol.* **69**, 2004–2015.

Yang, Y., Nunes, F.A., Berencsi, K., Furth, E.E., Gonczol, E., and Wilson J.M. (1994) Cellular immunity to viral antigens limits E1-deleted adenoviruses for gene therapy. *Proc Natl Acad Sci USA.* **91**, 4407–4411.

Yang, Y., Xiang, Z., Ertl H.C., and Wilson J.M. (1995b) Upregulation of class I major histocompatibility complex antigens by interferon gamma is necessary for T-cell-mediated elimination of recombinant adenovirus-infected hepatocytes in vivo. *Proc Natl Acad Sci USA.* **92**, 7257–7261.

Yotnda, P., Chen, D.H., Chiu, W., Piedra, P.A., Davis, A., Templeton, N.S., and Brenner, M.K. (2002) Bilamellar cationic liposomes protect adenovectors from preexisting humoral immune responses. *Mol Ther.* **5**, 233–241.

Zabner, J., Fasbender, A.J., Moninger, T., Poellinger, K.A., and Welsh, M.J. (1995) Cellular and molecular barriers to gene transfer by a cationic lipid. *J Biol Chem.* **270**, 8997–9007.

Ziegler, R.J., Li, C., Cherry, M., Zhu, Y., Hempel, D., van Rooijen, N., Ioannou, Y.A., Desnick, R.J., Goldberg, M.A., Yew, N.S., and Cheng, S.H. (2002) Correction of the nonlinear dose response improves the viability of adenoviral vectors for gene therapy of Fabry disease. *Hum Gene Ther.* **13**, 935–945.

Zsengeller, Z., Otake, K., Hossain, S.A., Berclaz, P.Y., and Trapnell, B.C. (2000) Internalization of adenovirus by alveolar macrophages initiates early proinflammatory signaling during acute respiratory tract infection. *J Virol.* **4**, 9655–9667.

SETTING BACK THE CLOCK: ADENOVIRAL-MEDIATED GENE THERAPY FOR LYSOSOMAL STORAGE DISORDERS

Dolan Sondhi, Neil R. Hackett, Stephen M. Kaminksy, and Ronald G. Crystal

1 INTRODUCTION

Lysosomal storage diseases (LSD) arise from mutations in the genes for lysosomal proteins that degrade and recycle macromolecules (Futerman and van Meer, 2004; Mach, 2002; Vellodi, 2005; Walkley, 2001). The undegraded waste products accumulate over time, resulting in derangement of cell physiology and eventually cell death. It follows, therefore, that delivery and expression of a wild-type copy of the defective gene to the affected cells should be preventive or therapeutic. For the lysosomal storage disorders, the challenges for gene therapy are to deliver the gene to the target tissue and to achieve reduction of "lysosomal storage," thus preserving cellular function.

Of the various strategies to achieve that goal, replication-deficient adenovirus-derived vectors (Ad) are generally thought to be inappropriate because although Ad vectors mediate high levels of production of their transgene, expression is transient over a period of only a few weeks (Hackett and Crystal, 2003; Trapnell and Gorziglia, 1994; Wilson, 1996). This chapter provides a contrary and counterintuitive view, making the case for using Ad vectors to treat the lysosomal storage disorders. We do so by first providing an overview of the production and properties of Ad vectors and then discussing studies in which Ad has been used to treat animal models of lysosomal storage diseases. The intent of this analysis is to critically evaluate the applicability of Ad for the challenges provided by LSD, including the required spatial and temporal pattern of gene expression. We then discuss a novel hypothesis, which we call "setting back the clock," which holds that, for some lysosomal storage diseases, transient overexpression of the deficient gene at high levels may be sufficient to completely reverse the storage defect and may give a substantial benefit, especially in diseases where the accumulation of the storage defect is slow.

2 GENERAL CHARACTERISTICS OF ADENOVIRUS

This section provides a brief description of the construction and use of Ad for gene transfer. A number of detailed reviews on the biology of adenoviruses and how they are

---

Department of Genetic Medicine, Weil Medical College, Cornell University, U.S.A.

modified as gene transfer vectors are available (Hackett and Crystal, 2003; Trapnell and Gorziglia, 1994; Wilson, 1996).

In humans, adenoviruses generally cause transient mild infections of the upper respiratory tract which are rapidly cleared by the immune system, providing long-term protection against reinfection by the same serotype (Horwitz, 1996). There are at least 51 distinct serotypes of human adenovirus, classified into six groups labeled A to F based on sequence homology and their ability to agglutinate red blood cells (Horwitz, 1996). Most gene therapy related studies have been carried out with derivatives of adenovirus serotype 5 although other serotypes including Ad35 and Ad11 are being evaluated, as are vectors made from nonhuman primate-derived adenoviruses (Cohen et al., 2002; Seshidhar et al., 2003; Stone et al., 2005).

Adenoviruses are nonenveloped double-stranded DNA viruses with an outer protein shell surrounding an inner nucleoprotein (Shenk, 1996). The 20 triangular faces of the capsid are each made up of 60 trimers of the hexon protein, which makes contact with six adjacent pseudo-equivalent neighbors. The 12 vertices are made up of a complex of the five copies of penton base protein and three of the fiber protein (Shenk, 1996). Within the 90 nm capsid, the 36 kilobase (kb) genome is packaged with the core protein and with the terminal protein attached to each end (Shenk, 1996; Figure 1A).

The adenovirus infection pathway is initiated by interaction of Ad with cell surface proteins. For the group C Ad, these include the primary coxsackie-adenovirus receptor (CAR) and coreceptors from the integrin family ($\alpha_V\beta_3$, $\alpha_V\beta_5$, $\alpha_V\beta_1$, $\alpha_M\beta_2$, and $\alpha_5\beta_1$) (Nemerow, 2000; Wickham, 2000). Other serotypes of human adenovirus use different receptors (Sirena et al., 2004). The knob region of fiber contains a binding site for CAR and the penton base protein binds to integrins via a consensus integrin binding domain containing an Arg–Gly–Asp motif (RGD) (Nemerow, 2000; Wickham, 2000). Adenovirus enters cells via clathrin-mediated endocytosis followed by a pH-dependent modification of the capsid and release of the virion to the cytosol (Meier and Greber, 2003). Like many viruses, adenovirus takes advantage of microtubule-associated transport mechanisms to reach the nucleus (Leopold et al., 2000). The capsids bind to the nuclear envelope in the vicinity of nuclear pores and undergo a final round of uncoating during which the viral DNA genome and DNA-binding protein leave the capsid and enter the nucleus (Figure 1B).

The biology of human adenoviruses has been extensively studied and their adaptation to the role of gene transfer vectors was relatively simple. Among the features that make Ad vectors attractive for gene therapy are the mild pathology of wild-type adenovirus, the lack of oncogenic potential, and the ability to infect a wide variety of cells (Hackett and Crystal, 2003; Trapnell and Gorziglia, 1994; Wilson, 1996). These qualities, as well as the ease of manipulating the viral genome, have made Ad a popular gene therapy vector as can be evidenced by the fact that nearly one quarter of all human gene therapy clinical trials in the United States use recombinant adenoviral vectors (http://www4.od.nih.gov/oba/rac/PROTOCOL.pdf). In general Ad are easy to produce in large amounts, achieve efficient infection of many cell types (quiescent or dividing) and rapidly result in high levels of expression of the therapeutic gene (Hackett and Crystal, 2003; Trapnell and Gorziglia, 1994; Wilson, 1996).

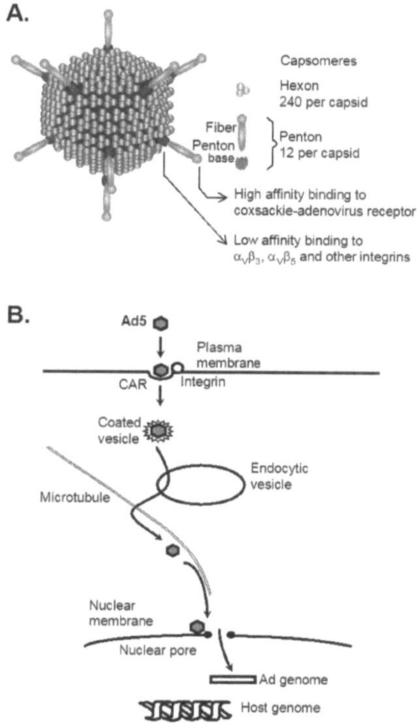

Figure 1. Adenovirus structure and intracellular trafficking. (A) The adenovirus capsid surface is composed primarily of two structures, hexon and penton. The major component is the hexon, but many of the critical functions involved in infection are incorporated in the penton that resides at each vartex. The penton has two parts, the penton base, which binds to integrins, and the high affinity fiber, which binds to the coxsackie adenovirus receptor (CAR). (B) The adenovirus trafficking pathway.

Laboratory investigations of adenovirus serotype 5 have lead to a detailed understanding of its infection and replication cycles and the roles played by the early (E1, E2, E3, and E4) and late (L1, L2, L3, L4, L5) genes (Shenk, 1996; Figure 2). Each gene comprises a complex transcription unit with alternative sites for transcription initiation, termination, and splicing, and each gene expresses multiple proteins (Shenk, 1996). The E1 gene products, E1A and E1B, are expressed immediately upon infection and are essential for expression of all other adenoviral genes (Shenk, 1996). The E1 genes play a critical role in forcing the host cell to enter the replicative state needed by the virus for DNA replication and in preventing cellular apoptosis (Shenk, 1996). Among the genes activated by the E1 proteins are the E2 (encoding proteins required for DNA replication), the E3 proteins (encoding nonessential proteins that help evade host response) and the E4 proteins (encoding proteins that coordinate late gene expression). The late genes encode

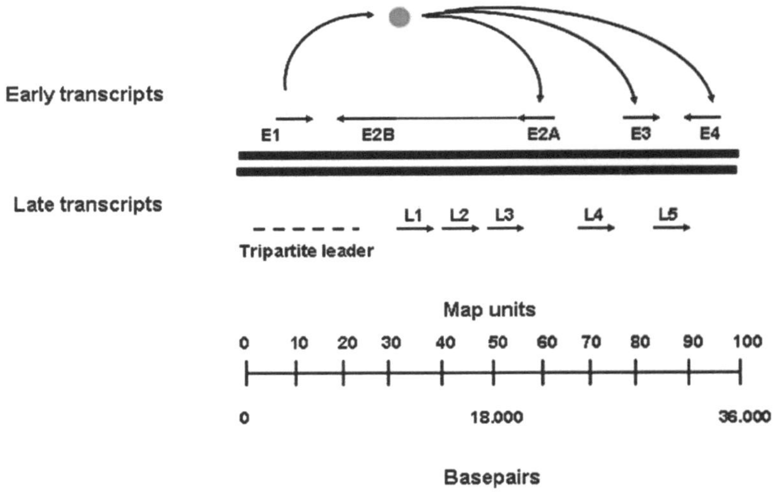

Figure 2. Structure and transcription of the major genes of the adenovirus type 5 genome. The genome is represented as two parallel lines and is divided by the scale shown on top into 100 map units (1 map unit = 360 bp). There are nine major complex transcription units divided into early (above the genome) and late transcripts (below). The four early transcripts are produced before the commencement of DNA replication and specify regulatory proteins and proteins required for DNA replication. Upon initial infection of a cell, the E1A protein is produced from transcripts in the E1 region. E1A is a major regulatory factor required for transcription of E1B, E2, E3, and E4. In replication-deficient adenovirus vectors, the E1 region is deleted. Proteins coded by the E2 and E4 regions are required for late gene transcription. The E3 region codes for proteins that help the virus evade host defenses. All late transcripts originate at the same point and are produced by alternate splicing. The tripartite leader sequence is present at the 5' end of all late transcripts. The L3 region specifies hexon, the L5 specifies fiber and the L2 specifies penton. For conventional adenovirus gene transfer vectors, most of E1 and E3 are deleted, and the expression cassette is inserted into the E1 region.

the components of the viral capsid and are expressed in abundance after DNA replication begins. In the wild-type adenovirus, DNA replication and packaging are coordinated and about 10,000 progeny virus are produced by each infected cell (Hackett and Crystal, 2003; Horwitz, 1996; Shenk, 1996).

3 GENE TRANSFER BY REPLICATION-DEFICIENT ADENOVIRUSES

Because the E1 gene is essential for all subsequent steps of the productive infection cycle, it follows that adenoviruses with E1 deletions are replication-deficient (Shenk, 1996). Gene transfer vectors are made by replacing the E1 gene with an expression cassette for the therapeutic gene. In practice, the nonessential E3 gene is also deleted. These two modifications render the vector incapable of completing the infectious cycle and thus infection is followed by expression of only the therapeutic gene (Hackett and

Crystal, 2003; Trapnell and Gorziglia, 1994). Production of such $E1^{-}E3^{-}$ Ad requires producer cell-lines that provide the deleted viral functions in *trans* (Graham and Prevec, 1995). The 293 cell-line, derived from human embryonic kidney and transformed by the left 11% of the adenovirus genome is generally used (Graham et al., 1977; Figure 3). Recombinant $E1^{-}$ Ad vectors containing therapeutic genes can be expanded on 293 cells and are readily purified in large amounts. These vectors can infect many cell types in vitro and in vivo and result in high-level expression of the therapeutic gene (Hackett and Crystal, 2003; Trapnell and Gorziglia, 1994; Wilson, 1996).

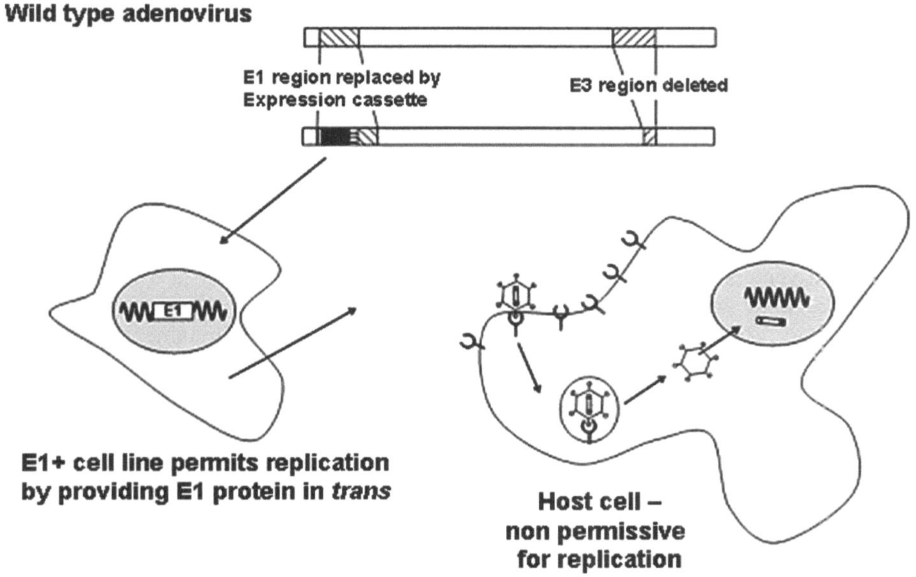

Figure 3. Gene transfer by adenovirus vectors. (Top) The 36,000 basepair genome of wild-type adenovirus serotype 5 with the locations of the E1 and E3 genes cross-hatched. Below this is a schematic of a first-generation adenovirus gene transfer vector. The E3 gene is deleted to make room for the expression cassette for the therapeutic gene. The E1 gene is removed and replaced by an expression cassette, typically consisting of a strong promoter (e.g., cytomegalovirus immediate/early promoter/ enhancer), an artificial splice site to enhance expression, and the reporter or therapeutic gene followed by a transcription stop/polyadenylation site. (Below, left) The genome of the replication deficient Ad vector is constructed in *E. coli* and transfected into human embryonic kidney 293 cells. Because 293 cells have the adenovirus E1 gene expressed from a chromosomal location, they are permissive for the replication of the E1 deleted vector. Serial infection of increasing numbers of 293 cells leads to large amounts of vector which is purified before in vitro or in vivo use. (Below, right) The vector binds to host cells through interaction of the fiber gene, which projects from each vertex of the virion, and the cell surface coxsackie adenovirus receptor integrins. The vector is internalized by endocytosis. It escapes from endosomes and traffics rapidly to the nucleus. The vector DNA, but not the capsid, enters the nucleus. There is no integration into host DNA so the transgene is expressed from an episome which does not replicate and is diluted by cell division.

Extensive use of recombinant Ad in experimental animals and humans has led to a number of general conclusions about their use as vectors. Adenoviruses can be used for gene transfer to most tissues by direct administration to that tissue. When administered by the intravenous route, gene expression is primarily in the liver in rodents, but also in the lung in species such as the pig (Hackett et al., 2000). Due to the packaging capacity of the virus, the expression cassette can be up to ~8 kb and therefore most cDNAs can be expressed using $E1^{-}E3^{-}$ Ad. Ad vectors generally mediate a high level of gene expression, peaking at two to seven days, but diminishing to <5% by about two weeks following gene transfer (Hackett and Crystal, 2003; Trapnell and Gorziglia, 1994; Wilson, 1996). The reason for the decrease is a combination of two factors. First, the Ad genome does not replicate or integrate into the host genome and therefore expression becomes diluted when cells divide. Second, and more important, there are robust cellular and humoral responses against the vector (Yang et al., 1996a,c). Furthermore, if the transgene is foreign to the host, there is also a humoral and cellular response against the transgene. At least part of the immune response is due to efficient infection of dendritic cells and macrophages by Ad, which results in efficient antigen presentation to T and B cells (Yang et al., 1996a,c).

There is extensive experience of administration of Ad to humans (Crystal, 1995). In general Ad vectors are safe and well tolerated when injected locally at moderate doses, but can be toxic when administered by the intravascular route (Crystal et al., 2002; Harvey et al., 2002; Raper et al., 2003). Local inflammatory/immune reactions are sometimes observed. When gene expression has been observed in humans, it is transient, and thus the primary focus of Ad gene therapy has been on scenarios such as angiogenesis and cancer where a therapeutic benefit can be gained from transient transgene expression.

4 SETTING BACK THE CLOCK

The general perception is that Ads are not applicable to treatment of genetic diseases inasmuch as they provide only transient transgene expression. For most hereditary disorders, this seems a rational assumption. For example, diseases such as hemophilia would require a continuous supply of the deficient coagulation proteins for the whole lifetime of the patient. But in the case of lysosomal storage diseases, the validity of this contention is not clear. Contrary to conventional concepts, we propose a testable hypothesis with Ad vectors that we call "setting back the clock" (Figure 4). Waste products accumulate slowly over time in the lysosomal storage disorders, and only when a threshold is reached do they cause cellular pathology (Futerman and van Meer, 2004; Mach, 2002; Vellodi, 2005; Walkley, 2001).

Although the kinetics by which storage products accumulate is complex, and likely differs in different lysosomal storage disorders, we suggest that if gene therapy with transient expression is used, it is theoretically possible that the stage of progression of the cellular pathology can be set back to the wild-type baseline level. As the expression wanes, the abnormal storage will start to reaccumulate, but this should provide a significant window of therapeutic benefit, particularly for the lysosomal storage disorders where the natural history of accumulation is slow. The "setting back the clock" hypothesis predicts that if the storage defect is fully, or even partially, reversed in the recipients, the impact of cessation of therapy (e.g., loss of expression), will be a very slow re-emergence of the disease.

The utility of setting back the clock with an Ad vector is enhanced by another property of many lysosomal storage diseases, referred to as "cross-correction" (Desnick and Schuchman, 2002; Kornfeld, 1986; Sleat et al., 2005). Upon biosynthesis, lysosomal enzymes are glycosylated with carbohydrates that contain mannose-5-phosphate (Kornfeld, 1986). In sorting the newly synthesized enzyme to the lysosome, a fraction of the newly synthesized enzyme is misdirected, secreted from the cell, and can be taken up by mannose-6-phosphate receptor-mediated endocytosis. Thus, a cell that is transduced by an Ad vector may be a depot to produce enzymes for other cells. A good example of the cross-correction concept is the mucopolysaccharidosis VII animal model, which lacks the protein β-glucuronidase (Kosuga et al., 2000). When serum from mucopolysaccharidosis VII mice treated with an adenovirus coding for the b-glucoronidase gene was transfused into untreated mucopolysaccharidosis VII knockout mice, the recipients showed b-glucuronidase activity in liver, spleen, kidney, lung, and heart (Kosuga et al., 2000). In a study involving the mouse model of glycogen storage disease II, administration of an adenovirus coding for acid-α-glucosidase into the gastrocnenius muscle of acid-a-glucosidase knockout mice resulted in secretion and uptake of the transgene product by other muscle groups with amelioration of the storage defect at these distant sites (Martin-Touaux et al., 2002). Cross-correction has also been implicated after use of Ad for treatment of other lysosomal storage diseases, including Fabry (Ziegler et al., 1999) and Tay–Sachs disease (Guidotti et al., 1999). Thus, a single administration of an Ad expressing the missing enzyme has the potential to eliminate the storage defect in cells beyond those transduced and return the pathology to baseline levels.

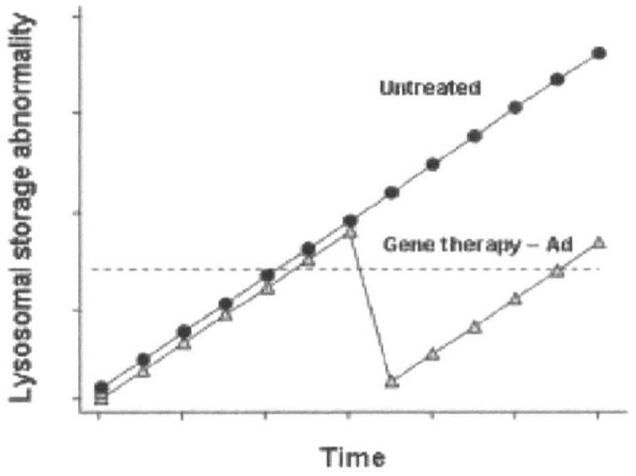

Figure 4. Setting back the clock. The concept is demonstrated graphically with the accumulation of storage defect as a function of time and the impact of intervention. As an example, there is linear accumulation of storage defect that reaches the threshold shown by the dotted line at which point pathology results. Gene therapy with an Ad vector should result in transient expression of the deficient protein, resulting in transient decrease in storage defect which then reaccumulates. This would provide a window of therapeutic benefit.

There are several examples of experimental animal models for which transient expression resulting from a single administration of an Ad vector expressing a lysosomal protein resulted in the correction of storage defect for a much longer duration than that of detectable transgene expression. In all cases, cross-correction was a significant factor enhancing the efficacy of gene transfer. In the mouse model of Fabry disease (α-galactosidase deficiency), a single intravenous administration of an Ad vector expressing the α-galactosidase gene was sufficient to reverse the storage defect (Ziegler et al., 1999).

The deposition of glycosphingolipid globotriaosylceramide that occurs in the mutant Fabry mouse was reversed for up to six months, with only a low level of reaccumulation. Although transgene expression was primarily in the liver, all tissues showed reversal of storage defect (Ziegler et al., 1999). Similarly, when an Ad expressing the acid-α glucosidase gene, was injected intravenously into the acid-α-glucosidase model of the glycogen storage disease type II (Pompe disease) knockout mice, the expression level as assessed by enzymatic activity in the serum peaked on days 2 to 3 post-administration but declined rapidly to background by 10 days. However, this pattern of transgene expression was sufficient to reduce the glycogen storage defect in various muscle groups for 100 to 200 days postvector administration (Ding et al., 2001). In a study where old mice with glycogen storage disease type II were treated by intravenous administration of the same vector, high levels of human acid-aglucosidase were produced and, based on reductions in glycogen storage in heart, quadriceps, and diaphragm, apparently taken up by muscles (Xu et al., 2005).

Surprisingly, in spite of the antivector and antitransgene immune response, long-term correction of the defect was observed (Ding et al., 2001). In another demonstration of this concept, in a lysosomal acid lipase knockout mouse, a model for human Wolman disease and cholesteryl ester storage disease, intravenous administration of Ad expressing lysosomal acid lipase was associated with increased hepatic liposomal acid lipase activity, decreased hepatomegaly, normalization of histopathology, and reductions in trigylceride and cholesterol levels in liver, spleen, and the small intestine for up to 47 days (Du et al., 2002).

Together, these models provide an important element of our hypothesis, that lysosomal storage defects are reversible and not merely preventable by adenovirus-mediated gene transfer. This is important when translating this concept to humans, because the lysosomal storage disorders are generally diagnosed at an advanced stage of storage defect accumulation. These examples verify the overall concept of setting back the clock with storage defect being cleared for a duration that exceeds the duration of transgene expression. None of these examples include behavioral or survival analysis and therefore full assessment of the therapeutic benefit is not known. However, for the experimental animal data available, although setting back the clock may not offer a permanent cure for lysosomal storage diseases, the impact on patients with these diseases may be substantial.

4.1 Enhancing the Efficacy of Ad in the Treatment of Lysosomal Storage Disease

If adenoviral gene transfer can reverse storage defect in lysosomal storage diseases, then the optimal result would be to return the storage defect to baseline levels of cellular pathology. It follows that a small dose and a short duration of expression of the therapeutic gene after Ad gene transfer may give only partial relief. Therefore manipulations that enhance adenoviral gene transfer or prolong transgene duration may be advantageous to "setting back the clock" as far as possible. Studies with Ad in experimental models of lysosomal storage diseases have used various strategies to overcome the short-term transgene expression resulting from the immune responses against the Ad vector. Examples of different approaches to this challenge as applied to lysosomal storage disease are as follows.

4.1.1 Vectors with Additional Early Gene Deletions

Gene transfer with first generation Ad ($E1^{-}$ $E3^{-}$ Ad), results in gene expression that is limited to a few weeks. This is, in part, attributed to the presence of E1-like activities in cells that potentiate expression of other adenovirus genes. The result is that Ad infected cells can be detected by the cellular immune system (Yang et al., 1996a,c). This problem can be reduced, at least in part, by making additional deletions of essential Ad genes such as E2 or E4 (Brough et al., 1997; Ding et al., 2001; Engelhardt et al., 1994; Gao, Yang, and Wilson, 1996; Yang et al., 1994). Propagation of these vectors requires specially constructed cell-lines that complement both missing adenoviral genes. There has been no direct comparison of such vectors to conventional E1ΓE3Γ Ad vectors in experimental models of lysosomal storage disease, but a double mutant Ad with both E1 and E2B (DNA polymerase) deletions expressing the acid-α-glucosidase gene (the gene that is defective in glycogen storage disease type II, Pompe disease) did lead to long-term clearance of glycogen storage defect (Ding et al., 2001).

4.1.2 Immuno-Supression Strategies

The most important factor limiting the duration of transgene expression mediated by Ad vectors is the acquired immune response to Ad gene products. As a dramatic demonstration of this concept relevant to the lysosomal storage diseases, an immune-deficient mouse model of glycogen storage disease type II was developed by crossing acid α-glucosidase knockout mice with severe combined immune-deficient mice. In these recipients, long-term high-level expression of acid-α-glucosidase was achieved for at least six months following intravenous administration of an Ad-vector coding for human-acid-a-glucosidase compared to a limited duration of approximately ten days in immune competent recipients (Xu et al., 2004).

Because there is such a vigorous immune response to Ad and to nonautologous transgenes, a number of groups have used immunosuppression to blunt this response in an attempt to prolong transgene expression (Guibinga et al., 1998; Scaria et al., 1997; Yang et al., 1996b). For example, intravenous administration of an Ad vector coding for b-glucuronidase led to detectable serum levels of b-glucuronidase that declined to undetectable by 50 days (Kosuga et al., 2000). However, simultaneous injection of another Ad expressing a decoy for immune costimulatory molecules, Ad-CLTLA4Ig, resulted in expression of detectable levels of glucuronidase for greater than 300 days.

In the mouse model of Fabry disease (α-galactosidase deficiency), high doses of Ad vector expressing the α-galactosidase transgene were needed to provide the required level of enzyme to cause clearance of storage defect, but as a result of the high dose, a strong immune response occurred, thereby limiting the duration of gene expression (Ziegler et al., 2002). Pretreatment with clodronate liposomes to eliminate Kupfer cells and other antigen-presenting cells (thereby blunting both the innate and acquired immune responses to the vector) permitted complete clearance of the storage defect with an Ad vector 100-fold lower than required in control mice (Ziegler et al., 2002). Similarly, treatment of mice by gamma globulin prior to Ad infusion resulted in a transient blockade of cells expressing IgG receptors (macrophage and dendritic cells) allowing 10-fold higher transgene expression levels at the same dose (Ziegler et al., 2002).

Another strategy to circumvent the immune system is to use newborn animals in which the immature immune system is unable to detect the Ad (Chou, Zingone, and Pan, 2002; Kamata et al., 2003; Martin-Touaux et al., 2002). Using the mucopolysaccharidosis type VII (MPSVII) model, mice were administered within 24 hours of birth an Ad expressing the wild-type form of the defective gene, β-glucuronidase (Kamata et al., 2003). More than 20% of normal β-glucuronidase activity was maintained for at least 20 weeks after birth in the brains of the Ad-injected mice. Histopathological analysis showed no obvious lysosomal storage defect in any of the visceral organs or the brain. The treated neonatal mice also failed to develop the facial skeletal deformities that characterize MPS VII mice. Similarly, the feasibility of adenovirus-mediated gene transfer was assessed in newborn glucose-6-phosphatase (G6Pase) deficient mice (a model of Type 1A glycogen storage disease) by infusion of a recombinant adenovirus containing the murine G6Pase gene. Whereas only 15% of untreated mutants survived weaning, 100% of the Ad vector-treated mice lived to three months of age, and liver and kidney enlargement was less pronounced with near normal levels of glycogen depositions in both organs (Chou, Zingone, and Pan, 2002). The concept of early therapy with Ad vectors has been extended to the period before birth, based on the observation that long term lacZ expression can be achieved by *in utero* gene transfer using Ad vectors (Shen et al., 2004).

4.1.3 Targeting Immuno-Privileged Tissues

Ad yields short duration expression in most tissues, however, the brain and eye are exceptions due to their partially immunoprivileged nature (Chaum and Hatton, 2002; Lowenstein and Castro, 2003). It follows that complete reversal of the storage defect using Ad to set back the clock is most likely in these organs. Inasmuch as several lysosomal storage diseases manifest in these organs and methods exist for direct administration to these sites, there is a possibility of treating the local symptoms of such lysosomal storage diseases with Ad. For example, when MPS VII knockout mice were treated by direct administration of Ad-β-glucuronidase into the brain, long-term (>16 wk) expression of β-glucuronidase protein was observed (Stein et al., 1999). Interestingly, there was correction of the storage defect in many parts of the brain, including areas where there was no β-glucuronidase activity detectable by immunohistochemistry. Animal models of other lysosomal storage diseases with neurological symptoms such as aspartylglucosaminuria (Peltola et al., 1998) have also been treated by direct CNS administration of Ad with consequent clearance of the storage defect.

Due to the limited size of the eye, it is easy to deliver Ad expressing the deficient gene to a large proportion of this target tissue, and because the eye is partially immune-protected, long-term transgene expression may result (Bennett et al., 1994; Mashhour et al., 1994). For example, phenotypic correction of storage defect in the retinal pigment epithelium of mucopolysaccharidosis VII mice has been demonstrated after Ad-β-glucuronidase injection by subretinal and intravitreous routes (Li and Davidson, 1995).

The blood–brain barrier precludes targeting and delivery of Ad to the brain by the vascular route and necessitates gene transfer directly to the CNS. In this regard, intravenous injection of an Ad expressing β-glucuronidase into MPS VII knockout mice resulted in no detectable β-glucuronidase activity in the brain (Ohashi et al., 1997). In an attempt to enhance delivery of Ad vectors to the brain, methods of delivery have been assessed in which the brain barriers are transiently compromised to allow Ad expressing the therapeutic gene to reach the brain parenchyma. For example, when intraperitoneal mannitol was administered at the time of the intraventricular injection of Ad-β-glucuronidase into MPS VII knockout mice, there was penetration of vector across the ependymal cell layer, with infection of cells in the subependymal region (Ghodsi et al., 1999).

5 CONCLUSIONS

The argument made in this chapter is that transient, adenovirus-mediated expression of a therapeutic enzyme should be sufficient to provide a prolonged therapeutic effect for lysosomal storage diseases due to the high levels of therapeutic protein, cross-correction, and long residence time of degradative enzymes in the lysosome. This effect may be further enhanced by some of the innovations that interrupt the immune response against first-generation Ad vectors thereby providing a longer duration of expression and a greater therapeutic benefit. The concept of setting back the clock merits a full experimental investigation in animals and consideration as an approach to human therapy for lysosomal storage diseases.

ACKNOWLEDGMENTS

We thank N. Mohamed for help in preparing this manuscript. These studies were supported, in part, by NIH U01 NS047458 and M01RR00047; Will Rogers Memorial Fund, Los Angeles, CA; and Nathan's Battle Foundation, Greenwood, IN.

REFERENCES

Bennett, J., Wilson, J., Sun, D., Forbes, B., and Maguire, A., 1994, Adenovirus vector-mediated in vivo gene transfer into adult murine retina, *Invest Ophthalmol. Vis. Sci.* **35:** 2535.

Brough, D.E., Hsu, C., Kulesa, V.A., Lee, G.M., Cantolupo, L.J., Lizonova, A., and Kovesdi, I., 1997, Activation of transgene expression by early region 4 is responsible for a high level of persistent transgene expression from adenovirus vectors in vivo, *J. Virol.* **71:** 9206.

Chaum, E. and Hatton, M.P., 2002, Gene therapy for genetic and acquired retinal diseases, *Surv. Ophthalmol.* **47:** 449.

Chou, J.Y., Zingone, A., and Pan, C.J., 2002, Adenovirus-mediated gene therapy in a mouse model of glycogen storage disease type 1a, *Eur. J. Pediatr.* **161 Suppl 1:** S56.

Cohen, C.J., Xiang, Z.Q., Gao, G.P., Ertl, H.C., Wilson, J.M., and Bergelson, J.M., 2002, Chimpanzee adenovirus CV-68 adapted as a gene delivery vector interacts with the coxsackievirus and adenovirus receptor, *J. Gen. Virol.* **83:** 151.

Crystal, R.G., 1995, Transfer of genes to humans: Early lessons and obstacles to success, *Science* **270:** 404.

Crystal, R.G., Harvey, B.G., Wisnivesky, J.P., O'Donoghue, K.A., Chu, K.W., Maroni, J., Muscat, J.C., Pippo, A.L., Wright, C.E., Kaner, R.J., Leopold, P.L., Kessler, P.D., Rasmussen, H.S., Rosengart, T.K., and Hollmann, C., 2002, Analysis of risk factors for local delivery of low- and intermediate-dose adenovirus gene transfer vectors to individuals with a spectrum of comorbid conditions, *Hum. Gene Ther.* **13:** 65.

Desnick, R.J. and Schuchman, E.H., 2002, Enzyme replacement and enhancement therapies: lessons from lysosomal disorders, *Nat. Rev. Genet.* **3:** 954.

Ding, E.Y., Hodges, B.L., Hu, H., McVie-Wylie, A.J., Serra, D., Migone, F.K., Pressley, D., Chen, Y.T., and Amalfitano, A., 2001, Long-term efficacy after [E1-, polymerase-] adenovirus-mediated transfer of human acid-alpha-glucosidase gene into glycogen storage disease type II knockout mice, *Hum. Gene Ther* **12:** 955.

Du, H., Heur, M., Witte, D.P., Ameis, D., and Grabowski, G.A., 2002, Lysosomal acid lipase deficiency: Correction of lipid storage by adenovirus-mediated gene transfer in mice, *Hum. Gene Ther* **13:** 1361.

Engelhardt, J.F., Ye, X., Doranz, B., and Wilson, J.M., 1994, Ablation of E2A in recombinant adenoviruses improves transgene persistence and decreases inflammatory response in mouse liver, *Proc. Natl. Acad Sci USA.* **91:** 6196.

Futerman, A.H. and van Meer, G., 2004, The cell biology of lysosomal storage disorders, *Nat. Rev. Mol. Cell Biol.* **5:** 554.

Gao, G.P., Yang, Y., and Wilson, J.M., 1996, Biology of adenovirus vectors with E1 and E4 deletions for liver-directed gene therapy, *J. Virol* **70:** 8934.

Ghodsi, A., Stein, C., Derksen, T., Martins, I., Anderson, R.D., and Davidson, B.L., 1999, Systemic hyperosmolality improves beta-glucuronidase distribution and pathology in murine MPS VII brain following intraventricular gene transfer, *Exp. Neurol.* **160:** 109.

Graham, F.L. and Prevec, L., 1995, Methods for construction of adenovirus vectors, *Mol. Biotechnol.* Jun, **3:** 207.

Graham, F.L., Smiley, J., Russell, W.C., and Nairn, R., 1977, Characteristics of a human cell line transformed by DNA from human adenovirus type 5, *J. Gen. Virol.* **36:** 59.

Guibinga, G.H., Lochmuller, H., Massie, B., Nalbantoglu, J., Karpati, G., and Petrof, B.J., 1998, Combinatorial blockade of calcineurin and CD28 signaling facilitates primary and secondary therapeutic gene transfer by adenovirus vectors in dystrophic (mdx) mouse muscles, *J. Virol* **72:** 4601.

Guidotti, J.E., Mignon, A., Haase, G., Caillaud, C., McDonell, N., Kahn, A., and Poenaru, L., 1999, Adenoviral gene therapy of the Tay-Sachs disease in hexosaminidase A-deficient knock-out mice, *Hum. Mol. Genet.* **8:** 831.

Hackett, N.R. and Crystal, R.G., 2003, Adenovirus vectors for gene therapy. In *Gene and Cell Therapy: Therapeutic Mechanisms and Strategies* (Ed. N. Smyth Templeton). Marcel Dekker, New York, pp. 17–42.

Hackett, N.R., El Sawy, T., Lee, L.Y., Silva, I., O'Leary, J., Rosengart, T.K., and Crystal, R.G., 2000, Use of quantitative TaqMan real-time PCR to track the time-dependent distribution of gene transfer vectors in vivo, *Mol. Ther.* **2:** 649.

Harvey, B.G., Maroni, J., O'Donoghue, K.A., Chu, K.W., Muscat, J.C., Pippo, A.L., Wright, C.E., Hollmann, C., Wisnivesky, J.P., Kessler, P.D., Rasmussen, H.S., Rosengart, T.K., and Crystal, R.G., 2002, Safety of local delivery of low- and intermediate-dose adenovirus gene transfer vectors to individuals with a spectrum of morbid conditions, *Hum. Gene Ther.* **13:** 15.

Horwitz, M.S., 1996, Adenoviruses. In *Fields Virology* (Eds. B.N. Fields, D.M. Knipe, and P.M. Howley), Philadelphia: Lippincott-Raven, pp. 2149–2171.

Kamata, Y., Tanabe, A., Kanaji, A., Kosuga, M., Fukuhara, Y., Li, X.K., Suzuki, S., Yamada, M., Azuma, N., and Okuyama, T., 2003, Long-term normalization in the central nervous system, ocular manifestations, and skeletal deformities by a single systemic adenovirus injection into neonatal mice with mucopolysaccharidosis VII, *Gene Ther.* **10:** 406.

Kornfeld, S., 1986, Trafficking of lysosomal enzymes in normal and disease states, *J. Clin. Invest* **77:** 1.

Kosuga, M., Takahashi, S., Sasaki, K., Li, X.K., Fujino, M., Hamada, H., Suzuki, S., Yamada, M., Matsuo, N., and Okuyama, T., 2000, Adenovirus-mediated gene therapy for mucopolysaccharidosis VII: Involvement of cross-correction in widespread distribution of the gene products and long-term effects of CTLA-4Ig coexpression, *Mol. Ther.* **1:** 406.

Leopold, P.L., Kreitzer, G., Miyazawa, N., Rempel, S., Pfister, K.K., Rodriguez-Boulan, E., and Crystal, R.G., 2000, Dynein- and microtubule-mediated translocation of adenovirus serotype 5 occurs after endosomal lysis, *Hum. Gene Ther* **11:** 151.

Li, T. and Davidson, B.L., 1995, Phenotype correction in retinal pigment epithelium in murine mucopolysaccharidosis VII by adenovirus-mediated gene transfer, *Proc. Natl. Acad. Sci. USA* **92:** 7700.

Lowenstein, P.R. and Castro, M.G., 2003, Inflammation and adaptive immune responses to adenoviral vectors injected into the brain: Peculiarities, mechanisms, and consequences, *Gene Ther.* **10:** 946.

Mach, L., 2002, Biosynthesis of lysosomal proteinases in health and disease, *Biol. Chem.* **383:** 751.

Martin-Touaux, E., Puech, J.P., Chateau, D., Emiliani, C., Kremer, E.J., Raben, N., Tancini, B., Orlacchio, A., Kahn, A., and Poenaru, L., 2002, Muscle as a putative producer of acid alpha-glucosidase for glycogenosis type II gene therapy, *Hum. Mol. Genet.* **11:** 1637.

Mashhour, B., Couton, D., Perricaudet, M., and Briand, P., 1994, In vivo adenovirus-mediated gene transfer into ocular tissues, *Gene Ther.* **1:** 122.

Meier, O. and Greber, U.F., 2003, Adenovirus endocytosis, *J. Gene Med.* **5:** 451.

Nemerow, G.R., 2000, Cell receptors involved in adenovirus entry, *Virology* **274:** 1.

Ohashi, T., Watabe, K., Uehara, K., Sly, W.S., Vogler, C., and Eto, Y., 1997, Adenovirus-mediated gene transfer and expression of human beta-glucuronidase gene in the liver, spleen, and central nervous system in mucopolysaccharidosis type VII mice, *Proc. Natl. Acad. Sci. USA* **94:** 1287.

Peltola, M., Kyttala, A., Heinonen, O., Rapola, J., Paunio, T., Revah, F., Peltonen, L., and Jalanko, A., 1998, Adenovirus-mediated gene transfer results in decreased lysosomal storage in brain and total correction in liver of aspartylglucosaminuria (AGU) mouse, *Gene Ther.* **5:** 1314.

Raper, S.E., Chirmule, N., Lee, F.S., Wivel, N.A., Bagg, A., Gao, G.P., Wilson, J.M., and Batshaw, M.L., 2003, Fatal systemic inflammatory response syndrome in a ornithine transcarbamylase deficient patient following adenoviral gene transfer, *Mol. Genet. Metab.* **80:** 148.

Scaria, A., St George, J.A., Gregory, R.J., Noelle, R.J., Wadsworth, S.C., Smith, A.E., and Kaplan, J.M., 1997, Antibody to CD40 ligand inhibits both humoral and cellular immune responses to adenoviral vectors and facilitates repeated administration to mouse airway, *Gene Ther.* **4:** 611.

Seshidhar, R.P., Ganesh, S., Limbach, M.P., Brann, T., Pinkstaff, A., Kaloss, M., Kaleko, M., and Connelly, S., 2003, Development of adenovirus serotype 35 as a gene transfer vector, *Virology* **311:** 384.

Shen, J.S., Meng, X.L., Maeda, H., Ohashi, T., and Eto, Y., 2004, Widespread gene transduction to the central nervous system by adenovirus in utero: implication for prenatal gene therapy to brain involvement of lysosomal storage disease, *J. Gene Med.* **6:** 1206.

Shenk, T., 1996, Adenoviridae: The viruses and their replication. In *Fields Virology* (Eds. B.N. Fields, D.M. Knipe, and P.M. Howley), Philadelphia: Lippincott-Raven, pp. 2111–2148.

Sirena, D., Lilienfeld, B., Eisenhut, M., Kalin, S., Boucke, K., Beerli, R.R., Vogt, L., Ruedl, C., Bachmann, M.F., Greber, U.F., and Hemmi, S., 2004, The human membrane cofactor CD46 is a receptor for species B adenovirus serotype 3, *J. Virol.* **78:** 4454.

Sleat, D.E., Lackland, H., Wang, Y., Sohar, I., Xiao, G., Li, H., and Lobel, P., 2005, The human brain mannose 6-phosphate glycoproteome: a complex mixture composed of multiple isoforms of many soluble lysosomal proteins, *Proteomics.* **5:** 1520.

Stein, C.S., Ghodsi, A., Derksen, T., and Davidson, B.L., 1999, Systemic and central nervous system correction of lysosomal storage in mucopolysaccharidosis type VII mice, *J. Virol.* **73:** 3424.

Stone, D., Ni, S., Li, Z.Y., Gaggar, A., Di Paolo, N., Feng, Q., Sandig, V., and Lieber, A., 2005, Development and assessment of human adenovirus type 11 as a gene transfer vector, *J. Virol.* **79:** 5090.

Trapnell, B.C. and Gorziglia, M., 1994, Gene therapy using adenoviral vectors, *Curr. Opin. Biotechnol.* **5:** 617.

Vellodi, A., 2005, Lysosomal storage disorders, *Br. J. Haematol.* **128:** 413.

Walkley, S.U., 2001, New proteins from old diseases provide novel insights in cell biology, *Curr. Opin. Neurol.* **14:** 805.

Wickham, T.J., 2000, Targeting adenovirus, *Gene Ther.* **7:** 110.

Wilson, J.M., 1996, Adenoviruses as gene-delivery vehicles, *N. Engl. J. Med.* **334:** 1185.

Xu, F., Ding, E., Liao, S.X., Migone, F., Dai, J., Schneider, A., Serra, D., Chen, Y.T., and Amalfitano, A., 2004, Improved efficacy of gene therapy approaches for Pompe disease using a new, immune-deficient GSD-II mouse model, *Gene Ther.* **11:** 1590.

Xu, F., Ding, E., Migone, F., Serra, D., Schneider, A., Chen, Y.T., and Amalfitano, A., 2005, Glycogen storage in multiple muscles of old GSD-II mice can be rapidly cleared after a single intravenous injection with a modified adenoviral vector expressing hGAA, *J. Gene Med.* **7:** 171.

Yang, Y., Jooss, K.U., Su, Q., Ertl, H.C., and Wilson, J.M., 1996a, Immune responses to viral antigens versus transgene product in the elimination of recombinant adenovirus-infected hepatocytes in vivo, *Gene Ther.* **3:** 137.

Yang, Y., Nunes, F.A., Berencsi, K., Gonczol, E., Engelhardt, J.F., and Wilson, J.M., 1994, Inactivation of E2a in recombinant adenoviruses improves the prospect for gene therapy in cystic fibrosis, *Nat. Genet.* **7:** 362.

Yang, Y., Su, Q., and Wilson, J.M., 1996c, Role of viral antigens in destructive cellular immune responses to adenovirus vector-transduced cells in mouse lungs, *J. Virol.* **70:** 7209.

Yang, Y., Su, Q., Grewal, I.S., Schilz, R., Flavell, R.A., and Wilson, J.M., 1996b, Transient subversion of CD40 ligand function diminishes immune responses to adenovirus vectors in mouse liver and lung tissues, *J. Virol.* **70:** 6370.

Ziegler, R.J., Li, C., Cherry, M., Zhu, Y., Hempel, D., van Rooijen, N., Ioannou, Y.A., Desnick, R.J., Goldberg, M.A., Yew, N.S., and Cheng, S.H., 2002, Correction of the nonlinear dose response improves the viability of adenoviral vectors for gene therapy of Fabry disease, *Hum. Gene Ther.* **13:** 935.

Ziegler, R.J., Yew, N.S., Li, C., Cherry, M., Berthelette, P., Romanczuk, H., Ioannou, Y.A., Zeidner, K.M., Desnick, R.J., and Cheng, S.H., 1999, Correction of enzymatic and lysosomal storage defects in Fabry mice by adenovirus-mediated gene transfer, *Hum. Gene Ther.* **10:** 1667.

ADENO-ASSOCIATED VIRAL-MEDIATED GENE THERAPY OF LYSOSOMAL STORAGE DISORDERS

M.A. Cabrera-Salazar and Seng H. Cheng*

1 INTRODUCTION

Lysosomal storage disorders (LSD) are a group of inherited diseases that result as a consequence of the loss of one or more of the several lysosomal enzymes that are responsible for the catabolism of a variety of macromolecules. The LSD as a group, present with an incidence of approximately 1 in 7500 live births making it one of the more prevalent groups of genetic diseases. The description of the pathophysiology of LSD, as well as their biochemical basis has facilitated the development of several therapeutic approaches for the treatment of this group of disorders. These include the use of enzyme replacement (see Chapter 3, this book), substrate inhibition (see Chapter 11, this book), and cell and gene-based therapies (see Chapter 14, this book). Enzyme and small molecule therapies have been successfully developed for a small number of these diseases. However, for a large number of the LSD, especially those affecting the central nervous system, no therapies are available as yet. Moreover, improvements to the current treatments are desirable and are under active consideration by a number of investigators. This review focuses specifically on the potential of gene therapy and in particular the use of recombinant adeno-associated viral (AAV) vectors as an alternate approach to treat the visceral and CNS manifestations shown associated with these enzymatic deficiencies.

Several features of LSD make them attractive candidates for treatment by gene therapy. Foremost is the demonstration that a proportion of the lysosomal enzymes are normally secreted into systemic circulation and that these secreted enzymes can be recaptured by adjacent and distal cells through the cation-independent mannose-6-phosphate receptor that is present ubiquitously on the cell surface of most cells (Fratantoni, Hall, and Neufeld, 1968; Hickman, Shapiro, and Neufeld, 1974; Furbish et al., 1978). This ability of the secreted enzymes to facilitate metabolic cooperativity suggests that genetic modification of a depot organ such as the liver, lung, or muscle may allow for production and secretion of therapeutic levels of the deficient enzymes. Secondly, it is likely that the levels of enzyme that need to be reconstituted in the affected subjects may not be very high based on enzyme levels in heterozygote individuals. Although the levels will differ for each of the LSD, heterozygotes with approximately 5–15% of normal levels of enzyme would appear to be healthy. Hence gene transduction of a small number of cells

* Applied Discovery Research, Genzyme Corporation, Framingham, MA, USA. Address correspondence to Seng H. Cheng. 31 New York Avenue, Framingham, MA 01701; e-mail: Seng.Cheng@genzyme.com

may suffice for therapeutic efficacy for several of these LSD. It is also unlikely that tight gene regulation will be necessary for this group of diseases as shown by the experience with enzyme replacement therapy, where relatively large boluses of the enzyme have been administered without any obvious adverse clinical consequences. Also, because the optimal pH profile for these enzymes is within the acidic range, they are unlikely to have untoward effects if high levels are inadvertently present within the neutral pH milieu of the circulation. Thirdly, most of the genes for the lysosomal enzymes have been isolated and the pathophysiology of the different diseases is fairly well understood. Finally, several animal models of LSD have been generated that share some of the disease characteristics observed in man that should facilitate preclinical research and evaluation of different gene therapeutic strategies (for a complete review of animal models of lysosomal storage disorders, see Ellinwood, Vite, and Haskins, 2004).

Early development efforts to treat the LSD had focused on those without significant CNS manifestations such as Gaucher, Fabry, Niemann–Pick B, MPS I, and Pompe disease. This was influenced by the paucity of facile strategies to facilitate the transfer of the lysosomal enzymes or any large macromolecular therapeutics across the blood–brain barrier. However, this trend is changing with the realization that long-term and in some instances, global therapy may be attained through the direct administration of cell and gene-based therapeutics into the CNS (see Chapter 14, this book). In addition, bone marrow transplantation continues to be a consideration although this procedure is still associated with some frequency of morbidity and mortality and does not appear to be equally beneficial to all LSD (Barranger, 1984).

Several gene delivery vectors and systems have been evaluated for the treatment of a number of different LSD. These include both viral (adenoviral, retroviral, herpes simplex virus, and adeno-associated viral) and nonviral (cationic lipids, polymers, and molecular conjugates) based gene transfer vectors. Of these, the adeno-associated viral (AAV) vector is emerging as the gene delivery vector of choice for treating both the visceral and CNS diseases associated with this group of disorders. Several features of this vector make it attractive for treating chronic genetic diseases such as the LSD. They are efficient at transducing a variety of cell types, are reportedly only mildly inflammatory in vivo, and are capable of supporting long-term transgene expression. Moreover, recently, several new viral serotypes with different tissue tropisms and significantly greater gene transduction activity than the prototypical AAV2 serotype have been isolated.

2 BIOLOGY OF AAV

AAV is a human parvovirus containing a single-stranded molecule of DNA with either a positive or negative polarity surrounded by a protein capsule. Both are equally infectious and are packaged with equal frequencies. AAV constitutes a particle of icosahedral symmetry with an approximate diameter of 20 nm. It was first discovered as a contaminant of adenoviral preparations (Hoggan, Blackow, and Rowe, 1966) but was later shown to require the presence of adenoviral or herpes viral genes for replication and function (Atchison, Casto, and Hammon, 1965). It has also been shown that human papilomaviruses (HPV 16) and vaccinia viruses can provide the proteins necessary for the replication of adeno-associated virus (Walz et al., 1997). The genome of all AAV serotypes is comprised of two, 145 base ITR flanking sequences, which provide the origin of replication, signals for encapsidation, integration into the cell genome, and rescue of the genome in latent cells (Carter, Burstein, and Peluso, 2004). Between the two ITRs the AAV genome is divided into two open reading frames, rep and cap. In the presence of a helper virus the single-stranded genome of AAV is converted into a duplex form; this conversion is known as metabolic activation and requires the expression of several proteins provided by helper viruses. (For a more detailed review on AAV structure and biology please see Carter, Burstein, and Peluso, 2004.)

3 AAV-MEDIATED GENE THERAPY OF LSD

The observation that lysosomal enzymes secreted into the systemic circulation are capable of supporting metabolic cross-correction of distal cells suggests that the genetic modification of a small number of depot cells engineered to produce the deficient enzyme should allow for broad therapeutic correction of affected cells. Although this concept has obvious implications for treating the visceral manifestations associated with the LSD, gratifyingly, this has also been shown to apply, at least in part, to the treatment of the CNS disease as well. Several of the lysosomal enzymes produced in the brain parenchyma following stereotaxic injections of various viral vectors exhibited surprisingly good diffusion capacity to surrounding cells. In addition, some AAV serotypes are capable of undergoing retrograde axonal transport providing yet another route to broader distribution of therapy in the CNS. This mechanism of viral uptake by terminal axons, and retrograde transport was first evidenced by Kaspar et al. (2002) following injection of axon terminal fields in the hippocampus and striatum with AAV vectors. They observed transgene expression not only at the site of injection but also at distal sites such as the entorhinal cortex and substantia nigra. Examples of the effectiveness of this approach are given when discussing treatment of LSD that affect the CNS. The possibility of using a common gene-delivery platform such as AAV to treat both the neurological and nonneurological aspects of LSD is an attractive consideration.

3.1 Systemic Delivery of Recombinant AAV Vectors for Treating the Visceral Disease

Recombinant AAV2 serotype vectors encoding lysosomal enzymes have been constructed and evaluated for their ability to correct the visceral pathology of a number of LSD including Fabry, Pompe, MPS I, and MPS VII. Following systemic administration, AAV vectors can be detected in a number of tissues but in greatest abundance in the spleen and liver. Hence, expression of the transgene product can be realized in multiple organs but frequently at the highest levels from the liver. This is likely influenced by a combination of the tropism of the viral vector for the organ and the selection of promoters used in the studies.

Systemic delivery of a recombinant AAV2 vector encoding α-galactosidase A into Fabry mice (Jung et al., 2001) or encoding β-glucuronidase (Daly et al., 2001) into neonatal MPS VII mice, resulted in the reconstitution of the respective enzymes in several tissues to 10–80% of normal levels. These levels are 100- to 1000-fold lower than those attained using recombinant adenoviral vectors, despite the use of much higher doses of AAV2. The kinetics of expression were consistent with those reported for AAV2 vectors, with peak expression levels generally attained between two and four weeks, and with sustained expression for several months posttreatment.

Despite the modest increase in enzyme levels observed in the different disease-affected tissues, they were sufficient to reduce measurably the amount of the substrates that accumulated in the different animal models. These observations support the notion that continuous expression of low levels of enzyme activity may be sufficient to reduce the storage pathology in the lysosomes. However, the kinetics for the reduction of the storage materials were relatively slow requiring several weeks to effect complete clearance. Coupled with its favorable safety profile, these results are supportive of the continued evaluation of recombinant AAV2 vectors for treating LSDs. However, because AAV2-mediated expression levels were relatively low and close to the threshold for therapeutic efficacy in some of the affected tissues, an improvement in transduction activity is clearly desirable.

In this regard, other AAV serotypes such as AAV1, and in particular AAV8, have recently been shown to have substantially greater liver transduction activity than AAV2 (Gao et al., 2002) The 10- to 100-fold higher expression levels attained with recombinant AAV8 vectors compared with AAV2 vectors were correlated with a higher number of transduced hepatocytes and greater persistence of vector DNA. Moreover, AAV8 vectors would appear to have a lower reactivity to neutralizing antibodies directed to human AAVs. This relative lack of pre-existing immunity to AAV8 coupled with its higher hepatic transduction activity support the selection of this serotype vector for further consideration in gene therapy of LSD. The use of a more potent vector will also allow for a lower amount of viral particles required for therapeutic efficacy. This in turn may further improve the safety profile and lessen the burden of vector production.

A similar strategy was also used to compare the therapeutic efficacy of AAV8- and AAV1-mediated gene therapy of Niemann–Pick type B disease. Higher levels of acid sphingomyelinase were attained with a recombinant AAV8 vector, which led to near complete clearance of sphingomyelin in the liver, spleen, kidney, and the lung of the acid sphingomyelinase knockout mice (ASMKO). A decrease in inflammatory proteins, an increase in the phagocytic activity of the alveolar macrophages and a reduction in the percentage of neutrophils in broncho-alveolar lavage fluid were also observed in this study (Barbon et al., 2005). As noted in Chapter 17, a primary concern in the development of therapies for Niemann–Pick B disease is avoidance of the development of respiratory infections. As with the Fabry studies, expression was sustained for the period of the study when a liver-restricted promoter was used to mediate the production of the enzyme.

3.2 Liver-Directed Gene Transfer Using Adeno-Associated Virus

Expression of lysosomal hydrolases following viral-mediated gene transfer to immunocompetent mouse models of LSD was invariably associated with the generation of a robust humoral response against the enzymes. This had the effect of extinguishing transgene expression and thereby limiting the duration of therapy. This problem is likely to be particularly pertinent in LSD subjects that harbor null mutations. The proportion of patients among the different LSD carrying null mutations is varied but can be as high as 70% as in the case of MPS I. However, it has been shown that this immune response could be circumvented in AAV-treated animals provided a tissue-restricted promoter was used to direct transgene expression (Wang et al., 2000). Systemic injection of recombinant AAV vectors in which the transgenes were placed under the transcriptional control of liver-restricted promoters reduced the extent of the antibody response and increased the longevity of transgene expression in immunocompetent mice. The reduced tendency to provoke an immune response was thought to be related to the reduced expression of the transgenes in antigen presenting cells (Ziegler et al., 2004; Franco et al.,

2005). This effect was independent of the AAV serotype used and was achieved with different liver-restricted promoters and for different lysosomal enzymes.

In Fabry mice, injection of a recombinant AAV2 vector encoding α-galactosidase A under the transcriptional control of a chimeric liver-restricted promoter consisting of two copies of the prothrombin enhancer, linked to a human serum albumin promoter, resulted in an undiminished expression for up to one year (Ziegler et al., 2004). In contrast to Fabry mice treated with an AAV vector where transcription of the enzyme was placed under the control of the ubiquitous CMV promoter, no antibodies to the transgene product were detected in the animals treated with the liver-restricted promoter cassette. The ability of this chimeric promoter to sustain prolonged expression in larger animals such as in a nonhuman primate has also been demonstrated (Sondhi et al., 2005).

3.3 Skeletal Muscle as an Alternate Depot Organ for Production of Lysosomal Enzymes

The large mass of muscle tissue in the human body makes this organ an attractive depot for the production of lysosomal enzymes. The use of intramuscular AAV injection to produce lysosomal enzymes has been reported for MPS VII (Daly et al., 2001) Pompe disease (Fraites et al., 2002; Ding et al., 2002; Martin-Touaux et al., 2002) and Fabry disease (Takahashi et al., 2002). Despite high levels of localized expression of the enzymes in the muscle, very low levels of the enzymes were detected in the circulation. Although the levels attained locally in the muscle were generally corrective, the levels secreted into circulation were too low to completely revert the pathology in the affected visceral organs of Pompe and MPS VII mice. These findings were consistent with the report by Raben et al. (2001), showing that the muscle is significantly less efficient than the liver at secreting the lysosomal enzyme α-glucosidase for Pompe disease. However, these observations did not extend to Fabry disease where a low-level secretion of α-galactosidase A into the circulation following intramuscular injection was sufficient to correct the storage defect locally and globally. Therefore, it would appear that the selection of muscle as a depot organ might be applicable for use in some but not all LSDs. The use of different AAV serotypes with higher transduction efficiency in muscle such as AAV1 and AAV7, may improve the utility of this organ for the production of lysosomal enzymes (Gao et al., 2002).

Skeletal muscle is severely affected in some LSD such as Pompe disease and consequently may not be as amenable to gene transfer as healthy muscle. To address the disease manifestations in the Pompe-affected muscles, the ability to deliver therapeutic quantities of the lysosomal enzyme to this target organ is paramount (see Reuser, Chapter 30 in this book). In this regard, the recently reported property of AAV8-vectors to deliver a mini-agrin gene construct to striated and cardiac muscle tissue of laminin-2-deficient neonatal mice (model for congenital muscular dystrophy) would suggest that is another vector and route of viral delivery that may be considered for treating Pompe disease. (Qiao et al., 2005). Similar observations were also reported in Duchenne Muscular Dystrophy (DMD) mice, where systemic administration of a recombinant AAV8 vector encoding a mini-dystrophin construct led to widespread transduction of muscle (Blankinship, Gregorevic, and Chamberlain, 2006).

3.4 AAV-Mediated Delivery of Lysosomal Enzymes to the Central Nervous System

The inability of lysosomal enzymes secreted into the circulation following systemic gene transfer to traverse the blood–brain barrier prompted the evaluation of intracranial injection as an alternate route of delivery of the lysosomal enzymes. As the CNS diseases associated with a number of LSDs are rapidly progressive and because intracranial injections are necessarily invasive, preference has been afforded to vectors that support long-term transgene expression. Integrating vectors such as retroviral- and lentiviral-based vectors have been used with some success (see Naldini, Chapter 10, this book). However, a large number of LSD-related studies have also been performed using AAV-based vectors.

Use of recombinant AAV vectors to treat CNS diseases, in particular those associated with LSD, is gaining increasing popularity. This is borne of observations that AAV vectors are capable of transducing a variety of cells in the brain and supporting long-term transgene expression (Kaplitt et al., 1994). Moreover, several AAV serotypes have been demonstrated to be capable of undergoing diffusion from the site of injection and also axonal transport to distal sites, thereby providing broader distribution of the vector than may be expected from localized parenchymal injections. In addition, this viral vector platform also displays a favorable safety profile. These features are desirable because the CNS manifestations associated with LSD are global and progressive in nature and because the procedure will necessarily involve invasive surgery. Hence vectors such as AAV, that are relatively innocuous and that facilitate extended and global therapy of the CNS following a minimal number of surgical interventions are good candidates for consideration.

Different AAV serotypes are able to transduce the various cell types in the brain with varying efficiencies. Presumably, this is dictated in part by the relative propensity of the appropriate receptors on the surface of the cells for the different AAV capsids. Recombinant AAV2 vectors have been shown to be able to transduce the hippocampus, inferior colliculus, piriform cortex, olfactory tubercle, and striatum (McCown et al., 1996). The neuronal and glial components of these regions are rich in heparan sulphate proteoglycans (HSPG; Fuxe et al., 1994), an extracellular component that mediates viral attachment and entry (Summerford and Samulski, 1988). Different serotypes of AAV present alternate transduction patterns and different cellular tropisms. For example, AAV4 does not require the expression of HSPG for infection and can transduce ependymal cells (Davidson et al., 2000). Recent studies suggest that AAV1 and 5 serotype vectors are more efficient than AAV2 at transducing different regions of the brain.

Early gene therapy studies in the CNS have focused primarily on the use of AAV2 serotype vectors with the transgenes placed under the transcriptional control of the CMV or β-actin promoters. Transgene expression from these vectors reportedly persisted for as long as one year postinjection (Tenenbaum et al., 2000). However, the CMV promoter can be subject to methylation and subsequent reduction in transgene expression levels over time. To address this limitation, alternate promoters that are not subject to this modification such as the neuron-specific enolase promoter have been used with good success (Xu et al., 2001). A common enhancer element for improving gene expression is the Woodchuck hepatitis virus element posttranscriptional regulator (WPRE), which reportedly increases the steady-state level of messenger RNA and the efficiency of translation and therefore gene expression (Loeb et al., 1999; Paterna et al., 2000).

3.4.1 Preclinical Studies in Animal Models of LSD

Early studies demonstrating the feasibility of AAV vectors at correcting the pathology in the CNS of mouse models of LSD were achieved by localized stereotaxic delivery of the vectors into the brain parenchyma. Localized transduction and expression of high levels of lysosomal enzymes in a variety of cell types at the sites of injections could be discerned using immunohistochemical and chemical staining. However, it was noted that some of the enzymes such as β-glucuronidase (enzyme-deficient in MPS VII) were also present in adjacent nontransduced cells suggesting that enzyme secretion and uptake also occurred in the CNS (Daly et al., 2001; Passini and Wolfe, 2001; Haskell et al., 2003). Moreover, the viral vectors were also shown to be capable of undergoing retrograde transport to distal sites resulting in secondary areas of transduction and expression of the enzymes (Kaspar et al., 2002).

These observations of enzyme diffusion and retrograde transport of the viral vectors likely accounted for the larger areas of correction of pathology than would have been expected from the localized intraparenchymal injections. These findings in the MPS VII mice have also been reported in other models of LSD such as Niemann–Pick A (Dodge et al., 2005; Passini et al., 2005), MPS I (Hartung et al., 2004), and MPS III (Fu et al., 2002; Cressant et al., 2004). Together, they suggest that global correction of the CNS pathology in a number of LSD may be attainable with a relatively small number of parenchymal injections into select regions of the brain with high neuronal circuitry.

Research on AAV-mediated intracranial gene therapy has also focused on finding the AAV serotypes that support optimal transduction and expression of the lysosomal enzymes. The transduction property of AAV2 and AAV5 has been compared in the mouse model of MPS I (Passini et al., 2006). Both vectors demonstrated widespread reduction of the storage materials in the brain, however, AAV5 generated higher levels of enzyme and supported greater vector distribution than AAV2 when administered at similar doses. Recombinant AAV1 vectors have also been tested and shown to result in greater transduction than AAV2 vectors in a number of brain structures. Different regions were preferentially transduced by the two AAV serotypes suggesting that there may be differential expression of the receptors for the different viral serotypes on neuronal surfaces leading to complementary patterns of neuronal transduction that could be used for broader transduction strategies when using AAV (Passini et al., 2003).

Intravitreal injection of recombinant AAV vectors represents another route to deliver genes into the CNS. The administration of such vectors into MPS VII mice resulted in detection of enzymatic activity not only at the site of injection but also in regions receiving inputs from the injected eye. The presence of the enzyme at the distal sites was not due to secondary viral transduction but from transit of the enzyme, presumably along the circuitory pathways. Although this approach could correct the pathology in the eyes shown associated with a number of LSDs, it is unlikely to provide broad significant benefit to the CNS. Delivery of AAV vectors to the CNS via the intraventricular route has also been attempted. This approach exploits the potential ability of the cerebrospinal fluid to distribute the therapy throughout the CNS. However, studies to date with recombinant AAV2 vectors suggest that this approach is less efficient as evidenced by significantly lower levels of transduction and expression of the enzymes (Hennig et al., 2004).

3.4.2 Gene Transfer to Neonates

Some lysosomal storage diseases exhibit storage pathology and present clinical manifesttations at or soon after birth. Consequently, it may be important to consider strategies that allow for early intervention to prevent onset of irreversible developmental damage. This is particularly important in those LSDs where the CNS and the disease are often rapidly progressive in nature. In this regard, investigators have explored the viability of treating neonatal brains using AAV-mediated gene therapy. Studies of neonatal transduction of CNS in animal models of LSD have shown widespread activity and persistent expression of the encoded transgenes. Treatment of neonatal animals prevented onset of pathology and progression of disease that persisted to adulthood. For example, intraventricular or intracerebral injections of AAV2 vectors encoding galactocerebrosidase into neonatal Krabbe mice decreased storage pathology and increased lifespan. A significant increase in galactocerebrosidase levels could be measured throughout the brain (hippocampus, cerebral cortex, and periventricular areas) following either route of injection. Viral genomes could also be amplified in several areas of the brain confirming that widespread transduction had occurred (Rafi et al., 2005).

3.4.3 Clinical Studies

There are almost ten years of accumulated clinical experience with recombinant AAV vectors in man. Presently, at least 11 different AAV vectors have been administered to over 200 human subjects, some repeatedly without any serious adverse events (Carter, 2005). Two of these studies are in subjects with lysosomal storage disorders: Canavan disease (deficiency of aspartoacylase; Leone et al., 2000) and late-infantile Batten's disease (deficiency of tripeptidyl peptidase). Based on preclinical studies that showed a decrease in the substrate (N-acetyl aspartate) that accumulates in Krabbe disease following AAV2-mediated gene delivery (Kitada et al., 2000), a study is ongoing with the plan to enroll 21 patients. The study will utilize an AAV2 vector encoding aspartoacylase under the control of a neuron-specific enolase promoter. A dose of nine $X10^{11}$ viral particles will be administered intraparenchymally to different sites of the brain including the frontal, parietal and occipital regions. Treatment has been well tolerated in the 10 patients enrolled thus far (Carter, 2005).

In the case of late infantile Battens disease, a clinical trial using an AAV2 vector encoding tripeptidyl peptidase under the transcription control of a chicken β-actin promoter has been initiated. The vector (a total of 3.6×10^{12} viral particles) will be administered at six different subcortical sites (Crystal et al., 2004). Up to ten patients will be enrolled in the trial, four of them presenting severe disease and six with moderate disease. The study will evaluate clinical findings using the LICNL scale and imaging studies will be used as secondary endpoints. So far one death has been observed in this trial, which appears not to be related to treatment but rather to the advanced nature of clinical disease in this patient at commencement.

4 CHALLENGES TO AAV-MEDIATED GENE TRANSFER

Exposure to AAV vectors has been shown to elicit an immune response that begins with the production of proinflammatory cytokines, antigen presentation, and followed by specific cellular or humoral immune responses (Bessis, Garcia-Cozar, and Boissier, 2004).

The responses can either be innate or acquired although with AAV vectors; it has been suggested that the innate response is minimal relative to other viral-based vectors (Zaiss et al., 2002, Bessis, Garcia-Cozar, and Boissier, 2004). Adaptative cell-mediated responses can also be triggered by recombinant AAV vectors. However, again the reaction is reportedly weaker than with other viral vectors such as recombinant adenoviral vectors. It has been suggested that this may be due to the relatively low efficiency of AAV vectors at transducing mature dendritic cells and macrophages (Zhang et al., 2000; Bessis, Garcia-Cozar, and Boissier, 2004).

Humoral immune responses to recombinant AAV vectors are triggered presumably following presentation of the viral capsid antigens. The production of antibodies against the viral capsids prevents subsequent readministrations of the vector. It should be noted that approximately 80% of the population have detectable antibodies against AAV2 vectors as a result of naturally occurring infections of which approximately 12% have been determined to be neutralizing. This is of significant concern because the presence of pre-existing immunity has been shown to abrogate viral transduction (Moskalenko et al., 2000). Additionally, antibodies can also be generated against the expressed transgene product. However, as indicated earlier, this could be minimized by restricting the expression of the transgene to the liver.

Neonatal gene transfer approaches also present immunological concerns, especially because maternal and fetal antibodies against wild-type vectors (AAV2) can be found in 63% of female adults and in 85% of pregnant females (Erles, Sebokov, and Schlehofer, 1999). In utero AAV infections can occur as the virus has been detected in uterine samples, curettage samples from early miscarriage, and amniotic fluid samples, and is a latent virus in the genital tract. (Burguete et al., 1999). The detection of IgM antibodies against AAV in fetuses and the observation of placental transfer of specific immune globulins (Bona, 2005) may facilitate the endocytosis and neutralization of the therapeutic vector by the reticuloendothelial system if gene transfer is performed in the first days or weeks of life.

Most efficacy studies performed to date in adult and neonatal mice, although encouraging, have been done in naïve animals, thus obviating these immune-related hurdles that may hamper the efficacy of this therapeutic approach. Pre-existing immunity or the induction of immune responses to the viruses and/or the expressed transgene products can affect the safety profile of the vectors and ultimately the efficacy of treatment.

5 CONCLUSIONS

The prospects offered by AAV-mediated gene transfer to alleviate the visceral and CNS manifestations shown associated with LSD appear to be promising. It is the hope that with further improvements in vector design and strategies to address the limitations associated with this vector platform that this therapeutic approach be brought to fruition in the not too distant future.

REFERENCES

Atchison RW, Casto BC, Hammon WM., 1965, Adenovirus associated defective virus particles. *Science* **49**:754.

Barbon CM, Ziegler RJ, Li C, Armentano D, Cherry M, Desnick RJ, Schuchman EH, Cheng SH., 2005, AAV8-mediated hepatic expression of acid sphingomyelinase corrects the metabolic defect in the visceral organs of a mouse model of Niemann-Pick disease. *Mol Ther.* **3**:431.

Barranger JA., 1984, Marrow transplantation in genetic disease. *N Engl J Med.* **311**:1629.

Bessis N, Garcia-Cozar FJ, Boissier MC., 2004, Immune responses to gene therapy vectors: influence on vector function and effector mechanisms. *Gene Ther.* Suppl **1**:S10–7.

Blankinship MJ, Gregorevic P, Chamberlain JS., 2006, Gene therapy strategies for Duchenne muscular dystrophy utilizing recombinant adeno-associated virus vectors. *Mol Ther.* **2**:241–9.

Bona C., 2005, Effect of maternal antibodies on neonatal B cell response. In Bona C (Ed.) *Neonatal Immunity*. Totowa, NJ: Humana Press, 159–178.

Burguete T, Rabreau M, Fontanges-Darriet M, Roset E, Hager HD, Koppel A, Bischof P, Schlehofer JR., 1999, Evidence for infection of the human embryo with adeno-associated virus in pregnancy. *Hum Reprod.* **14**:2396.

Carter BJ., 2005, Adeno-associated virus vectors in clinical trials. *Hum Gene Ther.* **16**:541.

Carter BJ, Burstein H, Peluso RW., 2004, Adeno-associated virus and AAV vectors for gene delivery. In: Templeton NS (Ed.) *Gene and Cell Therapy Therapeutic Mechanisms and Strategies*, second edition. Switzerland: Marcel Dekker, pp. 71–101.

Cressant A, Desmaris N, Verot L, Brejot T, Froissart R, Vanier MT, Maire I, Heard JM., 2004, Improved behavior and neuropathology in the mouse model of Sanfilippo type IIIB disease after adeno-associated virus-mediated gene transfer in the striatum. *J Neurosci.* **24**:10229.

Crystal RG, Sondhi D, Hackett NR, Kaminsky SM, Worgall S, Stieg P, Souweidane M, Hosain S, Heier L, Ballon D, Dinner M, Wisniewski K, Kaplitt M, Greenwald BM, Howell JD, Strybing K, Dyke J, Voss H., 2004, Clinical protocol. Administration of a replication-deficient adeno-associated virus gene transfer vector expressing the human CLN2 cDNA to the brain of children with late infantile neuronal ceroid lipofuscinosis. *Hum Gene Ther.* **15**:1131.

Daly TM, Ohlemiller KK, Roberts MS, Vogler CA, Sands MS., 2001, Prevention of systemic clinical disease in MPS VII mice following AAV-mediated neonatal gene transfer. *Gene Ther.* **8**:1291.

Davidson BL, Stein CS, Heth JA, Martins I, Kotin RM, Derksen TA, Zabner J, Ghodsi A, Chiorini JA., 2000, Recombinant adeno-associated virus type 2, 4, and 5 vectors: Transduction of variant cell types and regions in the mammalian central nervous system. *Proc Natl Acad Sci USA*. **97**:3428.

Ding E, Hu H, Hodges BL, Migone F, Serra D, Xu F, Chen YT, Amalfitano A., 2002, Efficacy of gene therapy for a prototypical lysosomal storage disease (GSD-II) is critically dependent on vector dose, transgene promoter, and the tissues targeted for vector transduction. *Mol Ther.* **5**:436.

Dodge JC, Clarke J, Song A, Bu J, Yang W, Taksir TV, Griffiths D, Zhao MA, Schuchman EH, Cheng SH, O'Riordan CR, Shihabuddin LS, Passini MA, Stewart GR., 2005, Gene transfer of human acid sphingomyelinase corrects neuropathology and motor deficits in a mouse model of Niemann–Pick type A disease. *Proc Natl Acad Sci USA*. **102**:17822.

Ellinwood NM, Vite CH, Haskins ME., 2004, Gene therapy for lysosomal storage diseases: the lessons and promise of animal models. *J Gene Med*. **6**:481.

Erles K, Sebokova P, Schlehofer JR., 1999, Update on the prevalence of serum antibodies (IgG and IgM) to adeno-associated virus (AAV). *J Med Virol*. **59**:406.

Fraites TJ Jr, Schleissing MR, Shanely RA, Walter GA, Cloutier DA, Zolotukhin I, Pauly DF, Raben N, Plotz PH, Powers SK, Kessler PD, Byrne BJ., 2002, Correction of the enzymatic and functional deficits in a model of Pompe disease using adeno-associated virus vectors. *Mol Ther*. **5**:571.

Franco LM, Sun B, Yang X, Bird A, Zhang H, Schneider A, Brown T, Young SP, Clay TM, Amalfitano A, Chen YT, Koeberl DD., 2005, Evasion of immune responses to introduced human acid alpha-glucosidase by liver-restricted expression in glycogen storage disease Type II. *Mol Ther*. **12**:876.

Fratantoni JC, Hall CW, Neufeld EF., 1968, Hurler and Hunter syndromes: Mutual correction of the defect in cultured fibroblasts. *Science* **162**:570–2.

Fu H, Samulski RJ, McCown TJ, Picornell YJ, Fletcher D, Muenzer J., 2002, Neurological correction of lysosomal storage in a mucopolysaccharidosis IIIB mouse model by adenoassociated virus-mediated gene delivery. *Mol Ther*. **5**:42.

Furbish FS, Steer CJ, Barranger JA, Jones EA, Brady RO., 1978, The uptake of native and desialylated glucocerebrosidase by rat hepatocytes and Kupffer cells. *Biochem Biophys Res Commun*. **8**:1047.

Fuxe K, Tinner B, Chadi G, Harfstrand A, Agnati LF., 1994, Evidence for a regional distribution of hyaluronic acid in the rat brain using a highly specific hyaluronic acid recognizing protein. *Neurosci Lett*. **169**:25.

Gao GP, Alvira MR, Wang L, Calcedo R, Johnston J, Wilson JM., 2002, Novel adeno-associated viruses from rhesus monkeys as vectors for human gene therapy. *Proc Natl Acad Sci USA*. **99**:11854.

Hartung SD, Frandsen JL, Pan D, Koniar BL, Graupman P, Gunther R, Low WC, Whitley CB, McIvor RS., 2004, Correction of metabolic, craniofacial, and neurologic abnormalities in MPS I mice treated at birth with adeno-associated virus vector transducing the human alpha-L-iduronidase gene. *Mol Ther*. **9**:866.

Haskell RE, Hughes SM, Chiorini JA, Alisky JM, Davidson BL., 2003, Viral-mediated delivery of the late-infantile neuronal ceroid lipofuscinosis gene, TPP-I to the mouse central nervous system. *Gene Ther*. **10**:34.

Hennig AK, Ogilvie JM, Ohlemiller KK, Timmers AM, Hauswirth WW, Sands MS., 2004, AAV-mediated intravitreal gene therapy reduces lysosomal storage in the retinal pigmented epithelium and improves retinal function in adult MPS VII mice. *Mol Ther*. **10**:106.

Hickman S, Shapiro LJ, Neufeld EF., 1974, A recognition marker required for uptake of a lysosomal enzyme by cultured fibroblasts. *Biochem Biophys Res Commun*. **57**:55.

Hoggan MD, Blackow NR, Rowe WP., 1966, Studies of small DNA viruses found in different adenovirus preparations: Physical, biological and immunological characteristics *Proc Natl Acad Sci USA*. **55**:1467.

Jung SC, Han IP, Limaye A, Xu R, Gelderman MP, Zerfas P, Tirumalai K, Murray GJ, During MJ, Brady RO, Qasba P., 2001, Adeno-associated viral vector-mediated gene transfer results in long-term enzymatic and functional correction in multiple organs of Fabry mice. *Proc Natl Acad Sci USA*. **98**:2676.

Kaplitt MG, Leone P, Samulski RJ, Xiao X, Pfaff DW, O'Malley KL, During MJ., 1994, Long-term gene expression and phenotypic correction using adeno-associated virus vectors in the mammalian brain. *Nat Genet.* **8**:148.

Kaspar BK, Erickson D, Schaffer D, Hinh L, Gage FH, Peterson DA., 2002, Targeted retrograde gene delivery for neuronal protection. *Mol Ther.* **5**:50.

Kitada K, Akimitsu T, Shigematsu Y, Kondo A, Maihara T, Yokoi N, Kuramoto T. Sasa M, Serikawa T., 2000, Accumulation of N-acetyl-L-aspartate in the brain of the tremor rat, a mutant exhibiting absence-like seizure and spongiform degeneration in the central nervous system. *J Neurochem.* **74**:2512.

Leone P, Janson CG, Bilaniuk L, Wang Z, Sorgi F, Huang L, Matalon R, Kaul R, Zeng Z, Freese A, McPhee SW, Mee E, During MJ., 2000, Aspartoacylase gene transfer to the mammalian central nervous system with therapeutic implications for Canavan disease. *Ann Neurol.* **48**:27.

Loeb JE, Cordier WS, Harris ME, Weitzman MD, Hope TJ., 1999, Enhanced expression of transgenes from adeno-associated virus vectors with the woodchuck hepatitis virus posttranscriptional regulatory element: implications for gene therapy. *Hum Gene Ther.* **10**:2295.

Martin-Touaux E, Puech JP, Chateau D, Emiliani C, Kremer EJ, Raben N, Tancini B, Orlacchio A, Kahn A, Poenaru L., 2002, Muscle as a putative producer of acid alpha-glucosidase for glycogenosis type II gene therapy. *Hum Mol Genet.* **11**:1637.

McCown TJ, Xiao X, Li J, Breese GR, Samulski RJ., 1996, Differential and persistent expression patterns of CNS gene transfer by an adeno-associated virus (AAV) vector. *Brain Res.* **713**:99.

Moskalenko M, Chen L, van Roey M, Donahue BA, Snyder RO, McArthur JG, Patel SD., 2000, Epitope mapping of human anti-adeno-associated virus type 2 neutralizing antibodies: implications for gene therapy and virus structure. *J Virol.* **74**:1761.

Passini MA, Dodge JC, Bu J, Yang W, Zhao Q, Sondhi D, Hackett N, Kaminsky S, Mao Q, Shihabuddin LS, Cheng SH, Sleat D, Stewart GR, Lobel P, Crystal RG., 2006, Intracranial delivery of CLN2 reduces brain pathology in a mouse model of classical late infantile neuronal ceroid lipofuscinosis. *J Neurosci.* **26**:1334.

Passini MA, Macauley SL, Huff MR, Taksir TV, Bu J, Wu IH, Piepenhagen PA, Dodge JC, Shihabuddin LS, O'Riordan CR, Schuchman EH, Stewart GR., 2005, AAV vector-mediated correction of brain pathology in a mouse model of Niemann–Pick A disease. *Mol Ther.* **11**:754.

Passini MA, Watson DJ, Vite CH, Landsburg DJ, Feigenbaum AL, Wolfe JH., 2003, Intraventricular brain injection of adeno-associated virus type 1 (AAV1) in neonatal mice results in complementary patterns of neuronal transduction to AAV2 and total long-term correction of storage lesions in the brains of beta-glucuronidase-deficient mice. *J Virol.* **77**:7034.

Passini MA, Wolfe JH., 2001, Widespread gene delivery and structure-specific patterns of expression in the brain after intraventricular injections of neonatal mice with an adeno-associated virus vector. *J Virol.* **75**:12382.

Paterna JC, Moccetti T, Mura A, Feldon J, Bueler H., 2000, Influence of promoter and WHV post-transcriptional regulatory element on AAV-mediated transgene expre-ssion in the rat brain. *Gene Ther.* **7**:1304.

Qiao C, Li J, Zhu T, Draviam R, Watkins S, Ye X, Chen C, Li J, Xiao X., 2005, Amelioration of laminin-{alpha}2-deficient congenital muscular dystrophy by somatic gene transfer of miniagrin. *Proc Natl Acad Sci USA.* **102**:11999.

Raben N, Lu N, Nagaraju K, Rivera Y, Lee A, Yan B, Byrne B, Meikle PJ, Umapathysivam K, Hopwood JJ, Plotz PH., 2001, Conditional tissue-specific expression of the acid alpha-glucosidase (GAA) gene in the GAA knockout mice: Implications for therapy. *Hum Mol Genet.* **10**:2039.

Rafi MA, Zhi Rao H, Passini MA, Curtis M, Vanier MT, Zaka M, Luzi P, Wolfe JH, Wenger DA., 2005, AAV-mediated expression of galactocerebrosidase in brain results in attenuated symptoms and extended life span in murine models of globoid cell leukodystrophy. *Mol Ther.* **11**:734.

Sondhi D, Peterson DA, Giannaris EL, Sanders CT, Mendez BS, De B, Rostkowski AB, Blanchard B, Bjugstad K, Sladek JR Jr, Redmond DE Jr, Leopold PL, Kaminsky SM, Hackett NR, Crystal RG., 2005, AAV2-mediated CLN2 gene transfer to rodent and non-human primate brain results in long-term TPP-I expression compatible with therapy for LINCL. *Gene Ther.* **12**:1618.

Summerford C, Samulski RJ., 1988, Membrane-associated heparan sulfate proteoglycan is a receptor for adeno-associated virus type 2 virions. *J Virol.* **72**:1438.

Takahashi H, Hirai Y, Migita M, Seino Y, Fukuda Y, Sakuraba H, Kase R, Kobayashi T, Hashimoto Y, Shimada T., 2002, Long-term systemic therapy of Fabry disease in a knockout mouse by adeno-associated virus-mediated muscle-directed gene transfer. *Proc Natl Acad Sci USA.* **99**:13777

Tenenbaum L, Jurysta F, Stathopoulos A, Puschban Z, Melas C, Hermens WT, Verhaagen J, Pichon B, Velu T, Levivier M., 2000, Tropism of AAV-2 vectors for neurons of the globus pallidus. *Neuroreport.* **11**:2277.

Wadsworth SC, Cheng SH., 2004, AAV2 vector harboring a liver-restricted promoter facilitates sustained expression of therapeutic levels of alpha-galactosidase A and the induction of immune tolerance in Fabry mice. *Mol Ther.* **9**:231.

Walz C, Deprez A, Dupressoir T, Durst M, Rabreau M, Schlehofer JR. 1997, Interaction of human papillomavirus type 16 and adeno-associated virus type 2 co-infecting human cervical epithelium. *J Gen Virol.* **78**:1441.

Wang L, Nichols TC, Read MS, Bellinger DA, Verma IM., 2000, Sustained expression of therapeutic level of factor IX in hemophilia B dogs by AAV-mediated gene therapy in liver. *Mol Ther.* **1**:154.

Xu R, Janson CG, Mastakov M, Lawlor P, Young D, Mouravlev A, Fitzsimons H, Choi KL, Ma H, Dragunow M, Leone P, Chen Q, Dicker B, During MJ., 2001, Quantitative comparison of expression with adeno-associated virus (AAV-2) brain-specific gene cassettes. *Gene Ther.* **8**:1323.

Zaiss AK, Liu Q, Bowen GP, Wong NC, Bartlett JS, Muruve DA., 2002, Differential activation of innate immune responses by adenovirus and adeno-associated virus vectors. *J Virol.* **76**:4580.

Zhang Y, Chirmule N, Gao G, Wilson J., 2000, CD40 ligand-dependent activation of cytotoxic T lymphocytes by adeno-associated virus vectors in vivo: Role of immature dendritic cells. *J Virol.* **74**:8003.

Ziegler RJ, Lonning SM, Armentano D, Li C, Souza DW, Cherry M, Ford C, Barbon CM, Desnick RJ, Gao G, Wilson JM, Peluso R, Godwin S, Carter BJ, Gregory RJ, Wadsworth SC, Cheng SH., 2004, AAV2 vector harboring a liver-restricted promoter facilitates sustained expression of therapeutic levels of alpha-galactosidase A and the induction of immune tolerance in Fabry mice. *Mol Ther.* **9**:231–40.

HERPES SIMPLEX VIRUS VECTORS FOR GENE THERAPY OF LYSOSOMAL STORAGE DISORDERS

Edward A. Burton[1] and Joseph C. Glorioso[2]

1 INTRODUCTION

Lysosomal storage diseases (LSDs) are a genetically heterogeneous group of conditions in which loss of specific lysosomal enzymes results in progressive accumulation of undegraded substrate, which results in cytotoxicity. LSDs represent an attractive target for gene therapy for several reasons. First, they are monogenic diseases, and, in the vast majority of cases, the causative genetic mutations are well characterised. Second, the diseases are recessive and due to genetic loss-of-function mutations, so that transfer of a single transgene would be expected to effect biochemical complementation. Third, experimental studies show that in most cases, low-level unregulated expression of the missing lysosomal enzyme can result in phenotypic correction. This is important, because current gene delivery technology is not capable of restoring precisely physiological amounts of the gene product to the cell. Finally, many lysosomal enzymes are released into the extracellular space and taken up by adjacent cells, so that protection of a broad area of tissue, or even cells at remote sites, may be possible through transduction of only a proportion of cells at a specific anatomical location.

Various methods have been used to successfully deliver the genes encoding lysosomal enzymes to skeletal muscle and the liver in experimental models of LSDs (Cheng and Smith, 2003). However, many of the diseases have central nervous system manifestations, which are progressive and not amenable to therapies based on systemic enzyme replacement or extracranial gene transfer. The CNS presents a number of formidable challenges for gene delivery, including the inaccessibility of the tissue compartment from the circulation and the delicate and essentially nonregenerating nature of the brain. To compound matters for the aspiring LSD gene therapist, CNS gene replacement would need to be lifelong in order to prevent neurodegeneration. Initial reports of β-glucuronidase gene transfer in murine models of MPSVII, using AAV (Elliger et al., 1999; Passini et al., 2003) and FIV (Bosch et al., 2000; Brooks et al., 2002) based viral vectors, have shown encouraging results, indicating that gene delivery may be a viable approach to tackling this complex problem. However, there are difficulties with these vectors, including the potential for toxicity (in particular insertional mutagenesis following genomic integration, see Kohn, Sadelain, and Glorioso, 2003), inability to repeat dose due to the presence of neutralising antibodies, and limitations in vector production technology resulting in low

1. Department of Neurology, University of Pittsburgh School of Medicine W957 Biomedical Science Tower, Pittsburg, PA, 15261. e-mail: eab25@pitt.edu
2. Department of Molecular Genetics and Biochemistry, University of Pittsburgh School of Medicine E1246 Biomedical Science Tower, Pittsburgh, PA, 15261. e-mail: glorioso@pitt.edu

manufacturing yields. These issues suggest that other vector systems are worthy of evaluation for this application.

In this chapter, we discuss neurotropic gene therapy vectors based on the herpes simplex virus and their potential application to the development of gene therapy for the CNS manifestations of LSDs. We present a brief overview of the basic biology of the herpes simplex virus in order to explain aspects of the wild-type viral lifecycle that are exploited in the generation of gene transfer vectors. We then discuss HSV vector engineering and summarise the results of CNS gene delivery experiments carried out in animal models of degenerative diseases, using HSV vectors.

2 BASIC BIOLOGY OF HSV-1

Herpes simplex virus is an enveloped double-stranded DNA virus (Roizman and Sears, 1996). The mature virion represents a highly ordered and stereotyped structure (Figure 1). The external face of the virion consists of a lipid envelope, embedded in which are viral glycoproteins. These are responsible for several functions including receptor-mediated cellular entry (Rajcani and Vojvodova, 1998; Spear, 1993; Stevens and Spear, 1997). A layer of proteins, the tegument, is intercalated between the envelope and the underlying capsid. Tegument proteins are responsible for induction of viral gene expression (Batterson and Roizman, 1983; Campbell, Palfreyman, and Preston, 1984) and shutoff of host protein synthesis immediately following infection (Kwong and Frenkel, 1987; Kwong and Frenkel, 1989; Kwong, Kruper, and Frenkel, 1988; Read and Frenkel, 1983), in addition to virion assembly functions. The viral capsid is a regular icosadeltahedron typical of herpes virus family members, (Newcomb et al., 1999). The capsid contains the viral genome in the form of a core of toroidal dsDNA (Puvion Dutilleul, Pichard, and Leduc, 1985).

2.1 The Viral Genome

Viral genes encode the structural components of the mature virion in addition to various regulatory proteins and synthetic enzymes. The HSV genome consists of 152 kb of dsDNA, arranged as long and short unique segments (UL and US) flanked by repeated sequences (McGeoch et al., 1988, 1986, 1985; Perry and McGeoch, 1988; Figure 2). Eighty-four viral genes are encoded, and these may be divided into essential and nonessential genes, according to whether their expression is necessary for viral replication in a permissive tissue culture environment. Nonessential genes often encode functions that are important for specific virus–host interactions in vivo, for example, immune evasion, replication in

Figure 1. The structure of herpes simplex virus. A schematic depiction of a mature HSV-1 virion is shown to illustrate the key structural components discussed in the text.

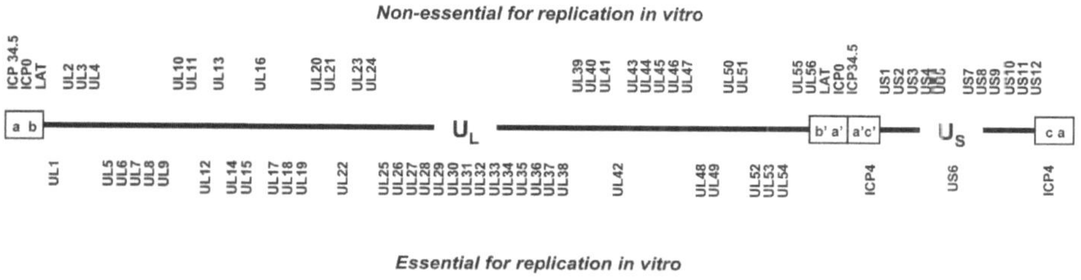

Figure 2. The HSV-1 genome. The 152 kb HSV-1 genome is illustrated diagrammatically (not to scale) to illustrate genomic locations of the various viral genes discussed in the text. Genes that are essential and nonessential for viral replication in vitro are indicated. Of note, most of the genes in the unique short segment of the genome are not essential for replication in vitro. Consequently, these may be deleted in replication-defective gene transfer vectors, without eliminating the ability to produce them in tissue culture. This manoeuvre alone could generate approximately 30 kb of capacity for insertion of transgenes.

nondividing cells, or shutdown of host protein synthesis. These genes may be deleted in the generation of gene therapy vectors, allowing the vector to replicate in culture, but attenuating pathogencity in vivo and generating capacity for the insertion of exogenous genetic material (Krisky et al., 1998a,b).

2.2 Gene Expression During Lytic Infection

During lytic infection, viral genes are expressed in a tightly regulated, interdependent temporal sequence (Honess and Roizman, 1974, 1975, reviewed in Roizman and Sears, 1996; Figure 3). Transcription of the five immediate–early (IE) genes, ICP0, ICP4, ICP22, ICP27, and ICP47 commences on viral DNA entry to the nucleus. Expression of these genes is regulated by viral promoters that contain a cognate DNA binding motif for the viral tegument protein, VP16. This protein is transported to the host cell nucleus with viral DNA following cell entry and associates with cellular transcription factors to promote transcriptional activation of the IE genes (Campbell, Palfreyman, and Preston, 1984; Mackem and Roizman, 1982; Preston, Frame, and Campbell, 1988). Expression of IE genes initiates a cascade of viral gene expression.

Transcription of early (E) genes, which primarily encode enzymes involved in DNA replication, is followed by expression of late (L) genes mainly encoding structural components of the virion (Honess and Roizman, 1974, 1975; Roizman and Sears, 1996). Of the IE gene products, only ICP4 and ICP27 are essential for expression of E and L genes, and hence viral replication (DeLuca, McCarthy, and Schaffer, 1985; Dixon and Schaffer, 1980; Sacks et al., 1985). The molecular events underlying HSV infection in vivo are more complex than the lytic lifecycle that occurs in tissue culture. Prior to cellular entry, the viral glycoproteins afford protection of the virion from complement-mediated lysis and antibody-mediated neutralisation. Following cellular entry, the viral immediate early gene ICP47 inhibits the presentation of viral antigens on the cell surface, contributing to immune evasion. Virus-encoded enzymes such as ribonucleotide reductase and thymidine kinase allow synthesis of dNTP precursors in nondividing cells to allow viral genome replication.

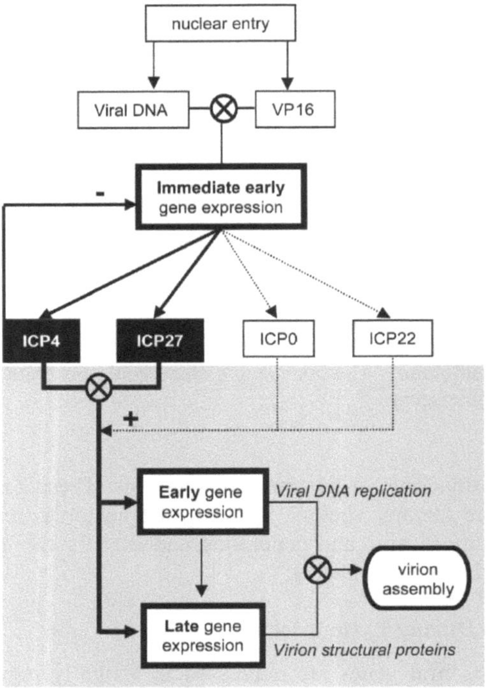

Figure 3. Regulation of HSV-1 gene expression during lytic infection. This flowchart illustrates the important regulatory events in gene expression occurring during lytic HSV-1 infection. In order to proceed to the later stages of infection (contained within the grey box), during which the viral genome is replicated and new virions assembled, both of the essential immediate early genes ICP4 and ICP27 must be expressed (black boxes). Deletion of one or other essential IE gene results in a replication-defective virus that is functionally null for all of the remaining viral genes unless the missing IE function is complemented in vitro. In practice, it is necessary to delete at least three of the IE genes to eliminate cytotoxicity in the absence of replication. This results in gene transfer vectors that may be produced to high titre in vitro in cell lines that complement the three missing IE genes.

2.3 HSV Lifecycle in vivo

The in vivo lifecycle of HSV (depicted schematically in Figure 4) involves both lytic replication in various cell types and a latent infection in neurons, which may persist for the lifetime of the host, apparently without causing significant metabolic derangement. Consequently, the biochemical events underlying latency are of considerable interest to the neurological gene therapist (see below). Following primary cutaneous or mucosal inoculation, the infected epithelia support a cycle of viral replication resulting in cell lysis and release of infectious particles into the subcutaneous tissues. These particles may enter sensory neurons whose axon terminals innervate the affected area. The nucleocapsid and tegument are carried by retrograde axonal transport from the site of entry to the neuronal soma in the dorsal root ganglia or trigeminal ganglia, where the viral genome and VP16 enter the nucleus (Bak et al., 1977; Bearer et al., 2000; Cook and Stevens, 1973). At this point, one of two chains of events may ensue. First, the lytic replication cycle described above may take place. This pathway results in neuronal cell death and egress of infectious particles.

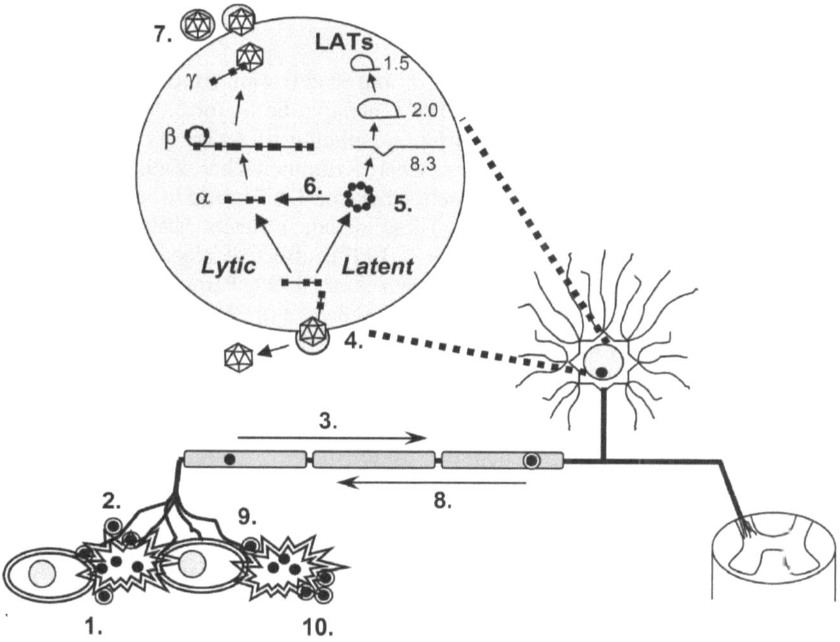

Figure 4. The lifecycle of HSV-1 in vivo. The key events occurring during infection of a human host are depicted schematically. A bipolar sensory neuron, located within a dorsal root ganglion, innervating the skin is shown for illustration: (1) Lytic cycle of replication at epithelial port of entry. (2) Virions released from epithelia enter sensory nerve terminals. (3) Nucleocapsid and tegument undergoes retrograde axonal transport to soma. (4) Viral DNA enters neuronal nucleus and either initiates lytic cascade of gene expression or becomes latent. (5) During latency, viral genome remains episomal and nuclear. Only the LAT genes are expressed. (6) Immunosuppression, intercurrent illness or other stimulus 'reactivates' lytic infection. (7) Virions formed by budding from nuclear membrane. (8) Nucleocapsid and glycoproteins transported separately by anterograde axonal transport. (9) Virion assembly and egress from nerve terminal. (10) Recurrent epithelial infection at or near site of primary lesion.

Alternatively, the viral DNA can enter the latent state. During latency, the viral genome persists as a stable episomal element (Mellerick and Fraser, 1987). The DNA adopts a chromatinlike structure; it is probably not extensively methylated (Deshmane and Fraser, 1989; Dressler, Rock, and Fraser, 1987). No IE, E, or lytic L genes are expressed during latency, but a set of nontranslated RNA species, the latency-associated transcripts (LATs), is produced and detectable in the nuclei of latently infected neurons (Croen et al., 1987; Gordon et al., 1988; Rock et al., 1987; Spivack and Fraser, 1987; Stevens et al., 1987; and see below). At a time point that may be remote from the establishment of latency, alterations in the host–virus interaction, for example, intercurrent illness or immunosuppression, may cause 'reactivation' of the viral infection. IE genes are expressed and the lytic cascade of gene expression follows, resulting in the production of mature virions. The nucleocapsid and glycoproteins are transported by separate anterograde axonal transport pathways to the peripheral nerve terminals, where they are assembled and released (Miranda Saksena et al., 2000; Rivera, Beuerman, and Hill, 1988).

2.4 Latency

The processes regulating the establishment of and reactivation from latency are not well understood. The LATs are a hallmark of HSV latency; the major 2.0 and 1.5 kb species are abundant, stable lariat introns that arise by splicing of a primary transcript (Alvira et al., 1999; Farrell, Dobson, and Feldman, 1991; Krummenacher, Zabolotny, and Fraser, 1997; Rodahl and Haarr, 1997). The functions of the LATs remain unknown, although several putative roles have been suggested. These include: efficient establishment of latency (Perng et al., 2000b; Thompson and Sawtell, 1997); effective reactivation from latency (Block et al., 1993; Drolet et al., 1999; Loutsch et al., 1999; Perng et al., 1994, 1996a,b, 1999); antisense regulation of IE gene transcripts (Chen et al., 1997; Garber, Schaffer, and Knipe, 1997; Mador et al., 1998); prevention of apoptosis in infected neurons (Perng et al., 2000a); expression of proteins (Goldenberg et al., 1997) that may compensate for the absence of IE gene expression during latency (Thomas et al., 1999); and functions relating to RNA-mediated catalysis (Hui and Lo, 1998). However, it is clear that the LAT genes are not an absolute requirement for establishment, maintenance, or reactivation from latency (Ho and Mocarski, 1989; Javier et al., 1988; Sedarati et al., 1989; Steiner et al., 1989) and they do not appear to have a definite role in viral IE gene regulation (Burton, Hong, and Glorioso, 2003). This has important implications for vector construction, as it is possible to insert transgenes within the LAT loci, disrupting the LAT genes and using the LAT *cis*-acting regulatory sequences to drive transgene expression (see below).

3 USING HSV-1 TO MAKE GENE THERAPY VECTORS

Various aspects of the basic biology of HSV-1 are attractive when considering the design of gene therapy vectors. The virus is highly infectious and has a broad host cell range, so that a wide variety of applications may be contemplated (Burton et al., 2001). Nondividing cells may be efficiently transduced, which is important in postmitotic tissues such as the CNS, and the latent behaviour of the virus may be exploited for the stable long-term expression of therapeutic transgenes in neurons (Goins et al., 1999; Wolfe et al., 2001). The vectors have a large capacity for exogenous transgene insertion; approximately half of the 84 viral genes are nonessential for growth in tissue culture and can be replaced by therapeutic transgene cassettes (Krisky et al., 1998a), allowing accommodation of large or multiple transgenes. Recombinant replication-defective HSV-1 may be prepared to high titre and purity without contamination from wild-type recombinants. These manufacturing considerations are of paramount importance when considering the development of a clinical product.

3.1 Preventing Viral Replication

Wild-type HSV infection is toxic and results in lysis of many cell types. Blocking viral replication after cellular entry arrests the HSV lifecycle and prevents lytic infection.

Strategies for generating nontoxic vectors from the HSV genome all depend on generating particles that lack essential genes or structural components that allow viral replication. There are two basic ways that this may be accomplished. The first depends on adding the minimal packaging and DNA replication signals to plasmid or BAC DNA, and then supplying all viral functions necessary for particle production in *trans* through either a series of BACs or cosmids, or using a helper virus. The resulting defective particles are called amplicons; they are very difficult to generate at high titre without contamination from replication-competent virus, which could have disastrous consequences in the brain.

An alternative approach relies on deleting the minimal number of viral genes that will arrest the lifecycle and prevent cytotoxicity. As E and L gene expression, and therefore replication, are fully dependent upon the expression of IE genes, generation of replication-incompetent vectors can be accomplished by disruption of one or other essential IE gene, ICP4 or ICP27 (DeLuca, McCarthy, and Schaffer, 1985). An ICP4 null mutant, for example, is unable to replicate in noncomplementing cells in culture (DeLuca, McCarthy, and Schaffer, 1985). It is important to note that, as the viral gene expression cascade is dependent on the essential IE genes, vectors deleted for ICP4 or ICP27 are functionally null for all viral E and L genes in vitro in noncomplementing cells, and in vivo. Complementing cell-lines can be made that will support replication of the vectors in vitro to high titre, without the need for cotransfections or helper virus. Furthermore, transgene expression occurs from the intact viral backbone, which adopts a state very similar to viral latency, allowing use of the viral latency *cis*-acting sequences to drive transgene expression.

3.2 Minimising Cytotoxicity

Further manipulations are necessary to prevent cytotoxicity resulting from expression of intact IE genes. ICP22, ICP27, and ICP0 are toxic to host cells in many situations (Johnson, Wang, and Friedmann, 1994; Samaniego, Neiderhiser, and DeLuca, 1998; Wu et al., 1996), and are negatively regulated by ICP4, such that infection with an ICP4 null mutant results in their overexpression, resulting in cell death in the absence of viral replication (DeLuca, McCarthy, and Schaffer, 1985; Krisky et al., 1998b; Moriuchi et al., 2000). To address this problem, a series of vectors deleted for multiple IE genes has been generated (Krisky et al., 1998b; Samaniego, Neiderhiser, and DeLuca, 1998; Samaniego, Wu, and DeLuca, 1997). Characterisation of these vectors shows that deletion of multiple IE genes improves the vector cytotoxicity profile (Krisky et al., 1998b). Deletion of all IE genes results in a vector that is nontoxic to cells and the vector genome correspondingly persists for long periods. However, loss of ICP0 results in poor growth in culture and low-level transgene expression.

ICP0 is a potent activator of transcription from the viral backbone, and appears to prevent shut-off of gene expression from HSV vectors. Retention of ICP0 allows efficient expression of viral genes and transgenes, and allows the virus to be prepared to higher titre. Although ICP0 is toxic in many cell types, neurons are not adversely affected by ICP0-expressing vectors on account of neuron-specific proteolytic degradation of ICP0 (Chen et al., 2000). This means that the advantage of the presence of ICP0 in the vector genome with regard to vector production issues can be exploited with relative impunity in the design of neurological gene transfer vectors. For many applications vectors deleted for ICP4, ICP27, and ICP22 have been used. These vectors show minimal cytotoxicity in vitro and in vivo, are efficient vehicles for transgene delivery, and can be grown efficiently in cells that stably supply ICP4 and ICP27 in *trans* (Krisky et al., 1998b; Wu et al., 1996).

3.3 Promoter Elements for Long-Term Gene Expression

Neurological LSDs are chronic pathological disturbances. Effective in situ enzyme replacement through gene transfer will depend upon long-term transgene expression. Exploitation of aspects of HSV latency is a promising means to achieving this goal. The latency loci of HSV lie in the repeats flanking the unique long segment of the genome. There are two promoters, latency active promoters 1 and 2 (LAP1 (Chen et al., 1995; Dobson et al., 1989; Goins et al., 1994; Leib et al., 1991; Soares et al., 1996; Zwaagstra et al., 1989, 1991, 1990), LAP2 (Chen et al., 1995; French, Schmidt, and Glorioso, 1996; Goins et al., 1994; Nicosia et al., 1993).

LAP1 is a typical RNA polymerase II promoter with a TATA element, which directs transcription starting at position –736 with respect to the 5' end of the 2.0kb LAT intron (Dobson et al., 1989; Zwaagstra et al., 1989, 1991, 1990). LAP1 is primarily responsible for LAT expression during latent infection, as determined by deletion analysis (Chen et al., 1995). LAP2 is a GC-rich promoter, typical of eukaryotic 'housekeeping' promoters (Chen et al., 1995; French, Schmidt, and Glorioso, 1996; Goins et al., 1994). It is situated 3' to LAP1 and deletion analysis suggests that it is primarily responsible for LAT expression during lytic infection (Chen et al., 1995). The situation arising in the intact virus is, however, complex. Several studies suggest that sequences contained within LAP2 facilitate the sustained transcription from LAP1 that occurs during latency (Berthomme et al., 2000; Goins et al., 1999, 1994; Lokensgard, Berthomme, and Feldman, 1997; Lokensgard et al., 1994). Furthermore, it has been possible to drive latent phase transcription of a gene placed at an ectopic locus within the HSV genome from LAP2, but not LAP1 (Goins et al., 1999, 1994; virus SLZ, Figure 4).

Finally, sequences contained within LAP2 seem able to direct high-level and sustained latent-phase transcription from some heterologous promoter elements (Berthomme et al., 2000; Palmer et al., 2000; Goins, unpublished). As the LATs are not an absolute requirement for the establishment of latency, it has been possible to insert transgenes into the LAT loci to utilise all of the relevant *cis*-acting sequences (Lachmann and Efstathiou, 1997; Marshall et al., 2000). Studies aimed at identifying the best ways in which to use the latency promoter system for neurological gene therapy are ongoing. However, the simplest mechanism may be to utilise the ability of LAP2 to drive sustained expression of a transgene from an ectopic locus within the viral genome. Recent work has demonstrated that a biologically relevant level of glial cell-line derived neurotrophic factor (GDNF) was expressed for over six months in vivo using the LAP2 promoter (Puskovic et al., 2004).

3.4 Vector Manufacturing

Vector production issues are of great importance when clinical applications are envisaged, because the supply of sufficient material of high enough purity and quality for clinical trials or therapy is crucial for successful outcomes. Replication-defective HSV vectors present many advantages in this regard, including the high titres that may be produced on complementing cells and the lack of helper virus. Recent work has examined in detail the factors dictating optimal vector production. First, alterations in culture conditions, infection parameters, and harvest time point have a profound effect on yield; optimisation of each of these variables leads to a highly significant improvement in production efficiency (Ozuer et al., 2002b). Second, the most disabled vector backbones yield the lowest numbers of infectious particles under controlled conditions, underlining the importance of optimising production processes for these vectors, which are least cytotoxic and thus most likely to find clinical application (Ozuer et al., 2002a). Finally, it has been possible to design methods for viral purification that do not rely on centrifugation, a process that can damage the delicate viral envelope.

Thus, in one study, a binding site for divalent metal cations was inserted into one of the viral glycoproteins, allowing selective retention of mature vector particles on a cobalt affinity column, and consequent separation of the viral particles from contaminants without disrupting essential viral structures (Jiang et al., 2004). By combining optimised techniques, it is now possible to generate clinically relevant yields of vector particles at appropriate levels of purity under GMP conditions.

3.5 Preclinical Safety Studies

Safety issues concerning the use of HSV vectors in the CNS have been examined in a recent study in nonhuman primates. The vector was a replication-defective HSV vector bearing a number of transgenes for the treatment of malignant glioma, which is poised to enter clinical trials (Wolfe et al., 2004). At the highest dose, 1 × 109 pfu inoculated directly into the cortex, transient inflammatory changes were found (this vector encodes TNFα and a suicide gene therapy substrate). It is important to note, however, the vector did not replicate, remained localised to the site of injection, and did not provoke the formation of neutralising antibodies. These data bode well for the development of vectors for cerebral inoculation to treat degenerative disease.

4 DATA FROM IN VIVO MODELS OF CNS DISEASE

Much of the work establishing proof of principle that HSV vectors will be useful in the development of neurological gene therapy has been carried out in the peripheral nervous system, which is the natural host tissue for latent infection with the wild-type virus (Chattopadhyay et al., 2003, 2004, 2002; Goss et al., 2002a,b; Hao et al., 2003a,b; Liu et al., 2004; Natsume et al., 2003; Sasaki et al., 2004). This work is not discussed further here, except to say that the vectors are able to efficiently infect peripheral nerve sensory cells with minimal toxicity and no demonstrable effect in neuronal electrophysiology (Howard et al., 1998; Krisky et al., 1998b). In contrast to viral latency observed in the PNS, infection of the CNS with wild-type HSV-1 results in rapidly fatal haemorrhagic encephalitis. This is dependent upon viral replication. Consequently, replication-defective HSV vectors do not cause this dramatic effect. However, further vector engineering is necessary to allow stable nontoxic transduction of CNS tissue in vivo.

4.1 Vector Effects

Eliminating expression of multiple IE genes appears crucial in minimising CNS neuronal toxicity. Thus, a single IE mutant (ICP4-) virus was toxic to cultured cortical neurons (Johnson et al., 1992), which showed minimal evidence of toxicity or metabolic disturbance when infected with a triple IE mutant vector (Krisky et al., 1998b). The same appears true of the brain in vivo. Thus, a single IE mutant gave rise to cell death and an inflammatory response following intraparenchymal injection (Howard et al., 1998), whereas a triple IE mutant (Krisky et al., 1998b) caused a small degree of tissue damage that was similar to that seen with saline injection, and was presumably partially mechanical in origin (Marconi et al., 1999). It is important to note, there is no evidence that direct introduction of disabled HSV into the cerebral parenchyma can effect reactivation of latent wild-type virus (Wang, Guo, and Jia, 1997).

4.2 Distribution of Vector

Following direct inoculation of the CNS parenchyma with replication-defective HSV-1, local transduction occurs only for a few millimetres around the needle track (Bloom et al., 1995; Howard et al., 1998; Marconi et al., 1999). This is not enhanced significantly by increasing injection volume, although the utility of convection-enhanced delivery is currently being tested. The axonal transport of HSV in transduced neurons gives rise to transduction of neurons at locations remote from the injection site, for example, within the substantia nigra following striatal injection (Maidment et al., 1996), which might be exploited to enhance vector distribution into anatomically removed, but functionally related, brain areas. Introduction of the virus into the cerebrospinal fluid by cisternal puncture (Martino et al., 2000a,b) allows transduction of the pia and arachnoid mater overlying the brain; meningeal cells are transduced over a wide area after a single injection, but there is little expression in the underlying neuropil. These results seem similar to those obtained using other vector systems. However, as explained in the introduction, widespread vector distribution may not be an absolute requirement of LSD gene therapy owing to the ability of the enzyme to pass between cells.

4.3 Long-Term Gene Expression

Viral DNA persists long-term following intracerebral inoculation with a replication-defective HSV vector (Bloom et al., 1995). Use of viral promoters other than the latency promoters gives rise to short-term transgene expression in CNS neurons, as might be expected (Bloom et al., 1995; Howard et al., 1998; Marconi et al., 1999; Yamada et al., 1999). Long-term expression using the latency promoter system has been demonstrated in the context of a replication-competent attenuated vector (Smith, Lachmann, and Efstathiou, 2000), and it is known that, following acute infection, replication-competent and neuro-attenuated vectors persist in CNS neurons where they transcribe the LAT genes (Drummond, Eglin, and Esiri, 1994; Kesari et al., 1996; Smith, Lachmann, and Efstathiou, 2000). Stable CNS gene expression using the LAP2 promoter in a replication-defective background was recently reported (Puskovic et al., 2004). In this study, detectable GDNF was expressed in abiologically relevant levels for at least six months following direct intraparenchymal vector inoculation.

4.4 Neuroprotection

It is now established that CNS neuroprotection may be achieved by HSV-mediated expression of appropriate transgenes. Expression of bcl-2 using an HSV vector in the substantia nigra prevented 6-hydroxydopamine mediated cell death in a mouse model of Parkinson's disease (Yamada et al., 1999). In a similar model, GDNF was also protective after expression from a replication-defective HSV vector (Puskovic et al., 2004). Ii is important to note that this occurred many months after vector inoculation, using the LAP2 promoter to drive transgene expression. This provides proof of principle that long-term transgene expression is possible and protective in the appropriate setting. This study is illustrated schematically in Figure 5.

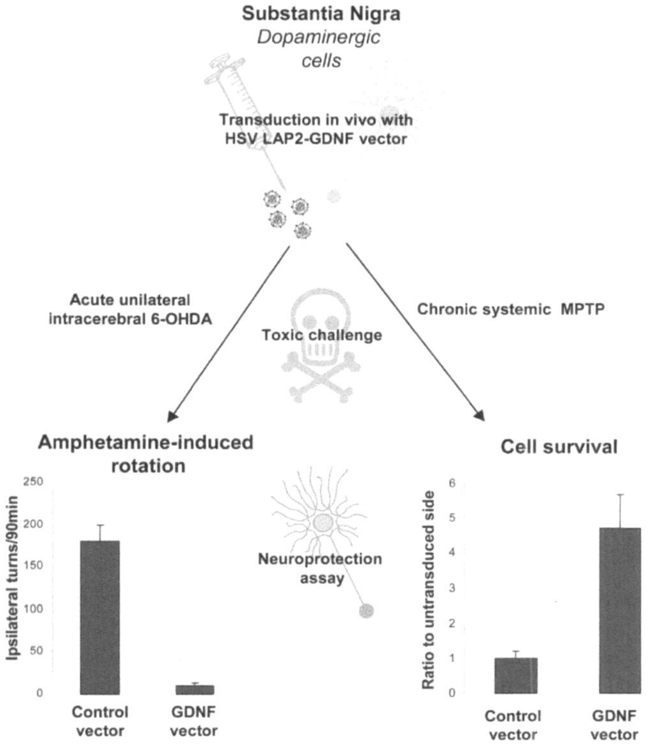

Figure 5. CNS neuroprotection using replication-defective HSV vectors. A replication-defective HSV vector expressing the neurotrophic agent GDNF was used to demonstrate long-term delivery of a biologically active neuroprotective molecule to the CNS, in vivo. GDNF is a powerful neurotrophic agent that is able to protect dopaminergic neurons from a variety of insults. Vector inoculation into the substantia nigra of the brain was carried out six months before a dopaminergic neuron-specific toxic challenge was presented. Two different experimental paradigms were used. First, unilateral injection of a dopamine neurotoxin, 6-OHDA was made directly into the substantia nigra, resulting in unilateral loss of the nigrostriatal projection and circling behaviour in response to amphetamine. This effect was substantially mitigated by prior inoculation with the GDNF vector. Second, chronic systemic MPTP administration resulted in bilateral loss of dopaminergic neurons. In animals injected unilaterally with the GDNF vector, cell loss was distinctly asymmetric due to protection of cells on the side of the vector inoculation. Together, these data show that robust GDNF-related neuroprotection occurred six months after the vector was introduced into the brain. This provides proof of principle that the vector system may be used to deliver biologically active molecules to protect neurons from toxicity in the long term (see Puskovic et al., 2004).

5 EXTRACEREBRAL APPLICATION OF HSV VECTORS IN LSDs

The utility of HSV vectors for treating LSDs might not be limited to treating the serious CNS effects of these diseases.

5.1 Stem Cells

Bone marrow transplantation has been used with variable results in the treatment of LSDs (Malatack, Consolini, and Bayever, 2003; Malm et al., 2004). It is conceivable that autologous bone marrow or neural stem cells might be used instead of donated marrow, if it were possible to genetically modify these stem cells to express the missing enzymes (Snyder, Taylor, and Wolfe, 1995). Infection of stem cell populations has been demonstrated using replication-defective HSV vectors. Almost all CD34+ human mobilized peripheral blood and monkey bone marrow cells infected with a replication-defective HSV-1 vector showed reporter gene expression after 12 hours (Gomez Navarro et al., 2000). The infected culture showed similar cell viability to an uninfected sample at days 2, 4, and 7 postinfection, consistent with previous experience using this vector in other cell populations (Krisky et al., 1998b).

Transduced monkey CD34+ cells were detected in the walls of nascent vasculature in a skin graft model, and mononuclear cells from the peripheral blood and bone marrow showed reporter gene expression for over three weeks following transplantation (Gomez Navarro et al., 2000). This indicates that a functional and potentially therapeutic gene product may be introduced into stem cells using an HSV vector, although it is worth noting that the nonintegrating nature of HSV-1 vectors will preclude their use in unmodified form for delivering genes to cells whose final destinations are many cell divisions away from the transduction event, because it is likely that the vector genome will be lost during mitotic segregation.

5.2 Adipocytes

Adipose tissue represents an attractive target for vector transduction and delivery of transgene products; fat is accessible, abundant, and well vascularised enabling secreted factors to gain access to the intravascular compartment. Thus, adipose tissue may be a suitable depot site for HSV transduction and secretion of circulating enzyme, which may be of interest in developing treatment for some LSDs (Jung et al., 2001). Using an in vitro model of human adipose differentiation, it was demonstrated that mature adipocytes and their precursor cells expressed the two principal HSV viral entry receptors and were efficiently transduced at a low multiplicity of infection (Fradette et al., 2004). Extended expression of beta-galactosidase and secretion of GDNF occurred in transduced fat tissue explants from rabbits. In vivo gene transfer to rabbit subcutaneous adipose tissue resulted in local gene expression for at least two months. It appears, therefore, that replication-defective HSV vectors are able to persist in adipose and express transgenes long-term, raising the possibility that systemic expression of lysosomal enzymes could be effected through this route.

6 CONCLUSIONS

Replication-defective HSV vectors present a flexible and efficient tool for transduction and prolonged gene expression in the CNS. The system has yet to be tested for its ability to deliver lysosomal enzymes to the brain and elsewhere, probably on account of the relatively few laboratories working on this vector system, which is comparatively complex in the context of more popular systems. However, the numerous advantages of HSV vectors suggest that, at least, they are worthy of consideration in the development of gene therapy-based treatments for the CNS complications of these diseases. Consequently, the first reports of LSD gene therapy using HSV vectors are eagerly awaited.

REFERENCES

Alvira, M. R., Goins, W. F., Cohen, J. B., and Glorioso, J. C. (1999). Genetic studies exposing the splicing events involved in herpes simplex virus type 1 latency-associated transcript production during lytic and latent infection. J Virol *73*, 3866–3876.

Bak, I. J., Markham, C. H., Cook, M. L., and Stevens, J. G. (1977). Intraaxonal transport of Herpes simplex virus in the rat central nervous system. Brain Res *136*, 415–429.

Batterson, W., and Roizman, B. (1983). Characterization of the herpes simplex virion-associated factor responsible for the induction of alpha genes. J Virol *46*, 371–377.

Bearer, E. L., Breakefield, X. O., Schuback, D., Reese, T. S., and LaVail, J. H. (2000). Retrograde axonal transport of herpes simplex virus: Evidence for a single mechanism and a role for tegument. Proc Natl Acad Sci USA *97*, 8146–8150.

Berthomme, H., Lokensgard, J., Yang, L., Margolis, T., and Feldman, L. T. (2000). Evidence for a bidirectional element located downstream from the herpes simplex virus type 1 latency-associated promoter that increases its activity during latency. J Virol *74*, 3613–3622.

Block, T. M., Deshmane, S., Masonis, J., Maggioncalda, J., Valyi Nagi, T., and Fraser, N. W. (1993). An HSV LAT null mutant reactivates slowly from latent infection and makes small plaques on CV-1 monolayers. Virology *192*, 618–630.

Bloom, D. C., Maidment, N. T., Tan, A., Dissette, V. B., Feldman, L. T., and Stevens, J. G. (1995). Long-term expression of a reporter gene from latent herpes simplex virus in the rat hippocampus. Brain Res Mol Brain Res *31*, 48–60.

Bosch, A., Perret, E., Desmaris, N., and Heard, J. M. (2000). Long-term and significant correction of brain lesions in adult mucopolysaccharidosis type VII mice using recombinant AAV vectors. Mol Ther *1*, 63–70.

Brooks, A. I., Stein, C. S., Hughes, S. M., Heth, J., McCray, P. M., Jr., Sauter, S. L., Johnston, J. C., Cory-Slechta, D. A., Federoff, H. J., and Davidson, B. L. (2002). Functional correction of established central nervous system deficits in an animal model of lysosomal storage disease with feline immunodeficiency virus-based vectors. Proc Natl Acad Sci USA *99*, 6216–6221.

Burton, E. A., Hong, C. S., and Glorioso, J. C. (2003). The stable 2.0-kilobase intron of the herpes simplex virus Type 1 latency-associated transcript does not function as an antisense repressor of ICP0 in nonneuronal cells. J Virol *77*, 3516–3530.

Burton, E. A., Wechuck, J. B., Wendell, S. K., Goins, W. F., Fink, D. J., and Glorioso, J. C. (2001). Multiple applications for replication-defective herpes simplex virus vectors. Stem Cells *19*, 358–377.

Campbell, M. E., Palfreyman, J. W., and Preston, C. M. (1984). Identification of herpes simplex virus DNA sequences which encode a trans-acting polypeptide responsible for stimulation of immediate early transcription. J Mol Biol *180*, 1–19.

Chattopadhyay, M., Goss, J., Lacomis, D., Goins, W. C., Glorioso, J. C., Mata, M., and Fink, D. J. (2003). Protective effect of HSV-mediated gene transfer of nerve growth factor in pyridoxine neuropathy demonstrates functional activity of trkA receptors in large sensory neurons of adult animals. Eur J Neurosci *17*, 732–740.

Chattopadhyay, M., Goss, J., Wolfe, D., Goins, W. C., Huang, S., Glorioso, J. C., Mata, M., and Fink, D. J. (2004). Protective effect of herpes simplex virus-mediated neurotrophin gene transfer in cisplatin neuropathy. Brain *127*, 929–939.

Chattopadhyay, M., Wolfe, D., Huang, S., Goss, J., Glorioso, J. C., Mata, M., and Fink, D. J. (2002). *In vivo* gene therapy for pyridoxine-induced neuropathy by herpes simplex virus-mediated gene transfer of neurotrophin-3. Ann Neurol *51*, 19–27.

Chen, S. H., Kramer, M. F., Schaffer, P. A., and Coen, D. M. (1997). A viral function represses accumulation of transcripts from productive-cycle genes in mouse ganglia latently infected with herpes simplex virus. J Virol *71*, 5878–5884.

Chen, X., Li, J., Mata, M., Goss, J., Wolfe, D., Glorioso, J. C., and Fink, D. J. (2000). Herpes simplex virus type 1 ICP0 protein does not accumulate in the nucleus of primary neurons in culture. J Virol *74*, 10132–10141.

Chen, X., Schmidt, M. C., Goins, W. F., and Glorioso, J. C. (1995). Two herpes simplex virus type 1 latency-active promoters differ in their contributions to latency-associated transcript expression during lytic and latent infections. J Virol *69*, 7899–7908.

Cheng, S. H., and Smith, A. E. (2003). Gene therapy progress and prospects: gene therapy of lysosomal storage disorders. Gene Ther *10*, 1275–1281.

Cook, M. L., and Stevens, J. G. (1973). Pathogenesis of herpetic neuritis and ganglionitis in mice: evidence for intra-axonal transport of infection. Infect Immun *7*, 272–288.

Croen, K. D., Ostrove, J. M., Dragovic, L. J., Smialek, J. E., and Straus, S. E. (1987). Latent herpes simplex virus in human trigeminal ganglia. Detection of an immediate early gene "anti-sense" transcript by in situ hybridization. N Engl J Med *317*, 1427–1432.

DeLuca, N. A., McCarthy, A. M., and Schaffer, P. A. (1985). Isolation and characterization of deletion mutants of herpes simplex virus type 1 in the gene encoding immediate-early regulatory protein ICP4. J Virol *56*, 558–570.

Deshmane, S. L., and Fraser, N. W. (1989). During latency, herpes simplex virus type 1 DNA is associated with nucleosomes in a chromatin structure. J Virol *63*, 943–947.

Dixon, R. A., and Schaffer, P. A. (1980). Fine-structure mapping and functional analysis of temperature-sensitive mutants in the gene encoding the herpes simplex virus type 1 immediate early protein VP175. J Virol *36*, 189–203.

Dobson, A. T., Sederati, F., Devi Rao, G., Flanagan, W. M., Farrell, M. J., Stevens, J. G., Wagner, E. K., and Feldman, L. T. (1989). Identification of the latency-associated transcript promoter by expression of rabbit beta-globin mRNA in mouse sensory nerve ganglia latently infected with a recombinant herpes simplex virus. J Virol *63*, 3844–3851.

Dressler, G. R., Rock, D. L., and Fraser, N. W. (1987). Latent herpes simplex virus type 1 DNA is not extensively methylated *in vivo*. J Gen Virol *68*, 1761–1765.

Drolet, B. S., Perng, G. C., Villosis, R. J., Slanina, S. M., Nesburn, A. B., and Wechsler, S. L. (1999). Expression of the first 811 nucleotides of the herpes simplex virus type 1 latency-associated transcript (LAT) partially restores wild-type spontaneous reactivation to a LAT-null mutant. Virology *253*, 96–106.

Drummond, C. W., Eglin, R. P., and Esiri, M. M. (1994). Herpes simplex virus encephalitis in a mouse model: PCR evidence for CNS latency following acute infection. J Neurol Sci *127*, 159–163.

Elliger, S. S., Elliger, C. A., Aguilar, C. P., Raju, N. R., and Watson, G. L. (1999). Elimination of lysosomal storage in brains of MPS VII mice treated by intrathecal administration of an adeno-associated virus vector. Gene Ther *6*, 1175–1178.

Farrell, M. J., Dobson, A. T., and Feldman, L. T. (1991). Herpes simplex virus latency-associated transcript is a stable intron. Proc Natl Acad Sci USA *88*, 790–794.

Fradette, J., Wolfe, D., Goins, W. F., Huang, S., Flanigan, R. M., and Glorioso, J. C. (2005). HSV vector-mediated transduction and GDNF secretion from adipose cells. Gene Ther. *12*, 48–58.

French, S. W., Schmidt, M. C., and Glorioso, J. C. (1996). Involvement of a high-mobility-group protein in the transcriptional activity of herpes simplex virus latency-active promoter 2. Mol Cell Biol *16*, 5393–5399.

Garber, D. A., Schaffer, P. A., and Knipe, D. M. (1997). A LAT-associated function reduces productive-cycle gene expression during acute infection of murine sensory neurons with herpes simplex virus type 1. J Virol *71*, 5885–5893.

Goins, W. F., Lee, K. A., Cavalcoli, J. D., O'Malley, M. E., DeKosky, S. T., Fink, D. J., and Glorioso, J. C. (1999). Herpes simplex virus type 1 vector-mediated expression of nerve growth factor protects dorsal root ganglion neurons from peroxide toxicity. J Virol *73*, 519–532.

Goins, W. F., Sternberg, L. R., Croen, K. D., Krause, P. R., Hendricks, R. L., Fink, D. J., Straus, S. E., Levine, M., and Glorioso, J. C. (1994). A novel latency-active promoter is contained within the herpes simplex virus type 1 UL flanking repeats. J Virol *68*, 2239–2252.

Goldenberg, D., Mador, N., Ball, M. J., Panet, A., and Steiner, I. (1997). The abundant latency-associated transcripts of herpes simplex virus type 1 are bound to polyribosomes in cultured neuronal cells and during latent infection in mouse trigeminal ganglia. J Virol *71*, 2897–2904.

Gomez Navarro, J., Contreras, J. L., Arafat, W., Jiang, X. L., Krisky, D., Oligino, T., Marconi, P., Hubbard, B., Glorioso, J. C., Curiel, D. T., and Thomas, J. M. (2000). Genetically modified CD34+ cells as cellular vehicles for gene delivery into areas of angiogenesis in a rhesus model. Gene Ther *7*, 43–52.

Gordon, Y. J., Johnson, B., Romanowski, E., and Araullo Cruz, T. (1988). RNA complementary to herpes simplex virus type 1 ICP0 gene demonstrated in neurons of human trigeminal ganglia. J Virol *62*, 1832–1835.

Goss, J. R., Goins, W. F., Lacomis, D., Mata, M., Glorioso, J. C., and Fink, D. J. (2002a). Herpes simplex-mediated gene transfer of nerve growth factor protects against peripheral neuropathy in streptozotocin-induced diabetes in the mouse. Diabetes *51*, 2227–2232.

Goss, J. R., Harley, C. F., Mata, M., O'Malley, M. E., Goins, W. F., Hu, X., Glorioso, J. C., and Fink, D. J. (2002b). Herpes vector-mediated expression of proenkephalin reduces bone cancer pain. Ann Neurol *52*, 662–665.

Hao, S., Mata, M., Goins, W., Glorioso, J. C., and Fink, D. J. (2003a). Transgene-mediated enkephalin release enhances the effect of morphine and evades tolerance to produce a sustained antiallodynic effect in neuropathic pain. Pain *102*, 135–142.

Hao, S., Mata, M., Wolfe, D., Huang, S., Glorioso, J. C., and Fink, D. J. (2003b). HSV-mediated gene transfer of the glial cell-derived neurotrophic factor provides an antiallodynic effect on neuropathic pain. Mol Ther *8*, 367–375.

Ho, D. Y., and Mocarski, E. S. (1989). Herpes simplex virus latent RNA (LAT) is not required for latent infection in the mouse. Proc Natl Acad Sci U S A *86*, 7596–7600.

Honess, R. W., and Roizman, B. (1974). Regulation of herpesvirus macromolecular synthesis. I. Cascade regulation of the synthesis of three groups of viral proteins. J Virol *14*, 8–19.

Honess, R. W., and Roizman, B. (1975). Regulation of herpesvirus macromolecular synthesis: sequential transition of polypeptide synthesis requires functional viral polypeptides. Proc Natl Acad Sci USA *72*, 1276–1280.

Howard, M. K., Kershaw, T., Gibb, B., Storey, N., MacLean, A. R., Zeng, B. Y., Tel, B. C., Jenner, P., Brown, S. M., Woolf, C. J., et al. (1998). High efficiency gene transfer to the central nervous system of rodents and primates using herpes virus vectors lacking functional ICP27 and ICP34.5. Gene Ther *5*, 1137–1147.

Hui, E. K., and Lo, S. J. (1998). Does the latency associated transcript (LAT) of herpes simplex virus (HSV) function as a ribozyme during viral reactivation? Virus Genes *16*, 147–148.

Javier, R. T., Stevens, J. G., Dissette, V. B., and Wagner, E. K. (1988). A herpes simplex virus transcript abundant in latently infected neurons is dispensable for establishment of the latent state. Virology *166*, 254–257.

Jiang, C., Wechuck, J. B., Goins, W. F., Krisky, D. M., Wolfe, D., Ataai, M. M., and Glorioso, J. C. (2004). Immobilized cobalt affinity chromatography provides a novel, efficient method for herpes simplex virus type 1 gene vector purification. J Virol *78*, 8994–9006.

Johnson, P. A., Wang, M. J., and Friedmann, T. (1994). Improved cell survival by the reduction of immediate-early gene expression in replication-defective mutants of herpes simplex virus type 1 but not by mutation of the virion host shutoff function. J Virol *68*, 6347–6362.

Johnson, P. A., Yoshida, K., Gage, F. H., and Friedmann, T. (1992). Effects of gene transfer into cultured CNS neurons with a replication-defective herpes simplex virus type 1 vector. Brain Res Mol Brain Res *12*, 95–102.

Jung, S. C., Han, I. P., Limaye, A., Xu, R., Gelderman, M. P., Zerfas, P., Tirumalai, K., Murray, G. J., During, M. J., Brady, R. O., and Qasba, P. (2001). Adeno-associated viral vector-mediated gene transfer results in long-term enzymatic and functional correction in multiple organs of Fabry mice. Proc Natl Acad Sci USA *98*, 2676–2681.

Kesari, S., Lee, V. M., Brown, S. M., Trojanowski, J. Q., and Fraser, N. W. (1996). Selective vulnerability of mouse CNS neurons to latent infection with a neuro-attenuated herpes simplex virus-1. J Neurosci *16*, 5644–5653.

Kohn, D. B., Sadelain, M., and Glorioso, J. C. (2003). Occurrence of leukaemia following gene therapy of X-linked SCID. Nat Rev Cancer *3*, 477–488.

Krisky, D. M., Marconi, P. C., Oligino, T. J., Rouse, R. J., Fink, D. J., Cohen, J. B., Watkins, S. C., and Glorioso, J. C. (1998a). Development of herpes simplex virus replication-defective multigene vectors for combination gene therapy applications. Gene Ther *5*, 1517–1530.

Krisky, D. M., Wolfe, D., Goins, W. F., Marconi, P. C., Ramakrishnan, R., Mata, M., Rouse, R. J., Fink, D. J., and Glorioso, J. C. (1998b). Deletion of multiple immediate-early genes from herpes simplex virus reduces cytotoxicity and permits long-term gene expression in neurons. Gene Ther *5*, 1593–1603.

Krummenacher, C., Zabolotny, J. M., and Fraser, N. W. (1997). Selection of a noncomsensus branch point is influenced by an RNA stem-loop structure and is important to confer stability to the herpes simplex virus 2-kilobase latency-associated transcript. J Virol *71*, 5849–5860.

Kwong, A. D., and Frenkel, N. (1987). Herpes simplex virus-infected cells contain a function(s) that destabilizes both host and viral mRNAs. Proc Natl Acad Sci USA *84*, 1926–1930.

Kwong, A. D., and Frenkel, N. (1989). The herpes simplex virus virion host shutoff function. J Virol *63*, 4834–4839.

Kwong, A. D., Kruper, J. A., and Frenkel, N. (1988). Herpes simplex virus virion host shutoff function. J Virol *62*, 912–921.

Lachmann, R. H., and Efstathiou, S. (1997). Utilization of the herpes simplex virus type 1 latency-associated regulatory region to drive stable reporter gene expression in the nervous system. J Virol *71*, 3197–3207.

Leib, D. A., Nadeau, K. C., Rundle, S. A., and Schaffer, P. A. (1991). The promoter of the latency-associated transcripts of herpes simplex virus type 1 contains a functional cAMP-response element: role of the latency-associated transcripts and cAMP in reactivation of viral latency. Proc Natl Acad Sci USA *88*, 48–52.

Liu, J., Wolfe, D., Hao, S., Huang, S., Glorioso, J. C., Mata, M., and Fink, D. J. (2004). Peripherally delivered glutamic acid decarboxylase gene therapy for spinal cord injury pain. Mol Ther *10*, 57–66.

Lokensgard, J. R., Berthomme, H., and Feldman, L. T. (1997). The latency-associated promoter of herpes simplex virus type 1 requires a region downstream of the transcription start site for long-term expression during latency. J Virol *71*, 6714–6719.

Lokensgard, J. R., Bloom, D. C., Dobson, A. T., and Feldman, L. T. (1994). Long-term promoter activity during herpes simplex virus latency. J Virol *68*, 7148–7158.

Loutsch, J. M., Perng, G. C., Hill, J. M., Zheng, X., Marquart, M. E., Block, T. M., Ghiasi, H., Nesburn, A. B., and Wechsler, S. L. (1999). Identical 371-base-pair deletion mutations in the LAT genes of herpes simplex virus type 1 McKrae and 17syn+ result in different in vivo reactivation phenotypes. J Virol *73*, 767–771.

Mackem, S., and Roizman, B. (1982). Structural features of the herpes simplex virus alpha gene 4, 0, and 27 promoter-regulatory sequences which confer alpha regulation on chimeric thymidine kinase genes. J Virol *44*, 939–949.

Mador, N., Goldenberg, D., Cohen, O., Panet, A., and Steiner, I. (1998). Herpes simplex virus type 1 latency-associated transcripts suppress viral replication and reduce immediate-early gene mRNA levels in a neuronal cell line. J Virol *72*, 5067–5075.

Maidment, N. T., Tan, A. M., Bloom, D. C., Anton, B., Feldman, L. T., and Stevens, J. G. (1996). Expression of the lacZ reporter gene in the rat basal forebrain, hippocampus, and nigrostriatal pathway using a nonreplicating herpes simplex vector. Exp Neurol *139*, 107–114.

Malatack, J. J., Consolini, D. M., and Bayever, E. (2003). The status of hematopoietic stem cell transplantation in lysosomal storage disease. Pediatr Neurol *29*, 391–403.

Malm, G., Mansson, J. E., Winiarski, J., Mosskin, M., and Ringden, O. (2004). Five-year follow-up of two siblings with aspartylglucosaminuria undergoing allogeneic stem-cell transplantation from unrelated donors. Transplantation *78*, 415–419.

Marconi, P., Simonato, M., Zucchini, S., Bregola, G., Argnani, R., Krisky, D., Glorioso, J. C., and Manservigi, R. (1999). Replication-defective herpes simplex virus vectors for neurotrophic factor gene transfer *in vitro* and *in vivo*. Gene Ther *6*, 904–912.

Marshall, K. R., Lachmann, R. H., Efstathiou, S., Rinaldi, A., and Preston, C. M. (2000). Long-term transgene expression in mice infected with a herpes simplex virus type 1 mutant severely impaired for immediate-early gene expression. J Virol *74*, 956–964.

Martino, G., Poliani, P. L., Furlan, R., Marconi, P., Glorioso, J. C., Adorini, L., and Comi, G. (2000a). Cytokine therapy in immune-mediated demyelinating diseases of the central nervous system: A novel gene therapy approach. J Neuroimmunol *107*, 184–190.

Martino, G., Poliani, P. L., Marconi, P. C., Comi, G., and Furlan, R. (2000b). Cytokine gene therapy of autoimmune demyelination revisited using herpes simplex virus type-1-derived vectors. Gene Ther *7*, 1087–1093.

McGeoch, D. J., Dalrymple, M. A., Davison, A. J., Dolan, A., Frame, M. C., McNab, D., Perry, L. J., Scott, J. E., and Taylor, P. (1988). The complete DNA sequence of the long unique region in the genome of herpes simplex virus type 1. J Gen Virol *69*, 1531–1574.

McGeoch, D. J., Dolan, A., Donald, S., and Brauer, D. H. (1986). Complete DNA sequence of the short repeat region in the genome of herpes simplex virus type 1. Nucleic Acids Res *14*, 1727–1745.

McGeoch, D. J., Dolan, A., Donald, S., and Rixon, F. J. (1985). Sequence determination and genetic content of the short unique region in the genome of herpes simplex virus type 1. J Mol Biol *181*, 1–13.

Mellerick, D. M., and Fraser, N. W. (1987). Physical state of the latent herpes simplex virus genome in a mouse model system: Evidence suggesting an episomal state. Virology *158*, 265–275.

Miranda Saksena, M., Armati, P., Boadle, R. A., Holland, D. J., and Cunningham, A. L. (2000). Anterograde transport of herpes simplex virus type 1 in cultured, dissociated human and rat dorsal root ganglion neurons. J Virol *74*, 1827–1839.

Moriuchi, S., Krisky, D. M., Marconi, P. C., Tamura, M., Shimizu, K., Yoshimine, T., Cohen, J. B., and Glorioso, J. C. (2000). HSV vector cytotoxicity is inversely correlated with effective TK/GCV suicide gene therapy of rat gliosarcoma. Gene Ther *7*, 1483–1490.

Natsume, A., Wolfe, D., Hu, J., Huang, S., Puskovic, V., Glorioso, J. C., Fink, D. J., and Mata, M. (2003). Enhanced functional recovery after proximal nerve root injury by vector-mediated gene transfer. Exp Neurol *184*, 878–886.

Newcomb, W. W., Homa, F. L., Thomsen, D. R., Trus, B. L., Cheng, N., Steven, A., Booy, F., and Brown, J. C. (1999). Assembly of the herpes simplex virus procapsid from purified components and identification of small complexes containing the major capsid and scaffolding proteins. J Virol *73*, 4239–4250.

Nicosia, M., Deshmane, S. L., Zabolotny, J. M., Valyi Nagy, T., and Fraser, N. W. (1993). Herpes simplex virus type 1 latency-associated transcript (LAT) promoter deletion mutants can express a 2-kilobase transcript mapping to the LAT region. J Virol *67*, 7276–7283.

Ozuer, A., Wechuck, J. B., Goins, W. F., Wolfe, D., Glorioso, J. C., and Ataai, M. M. (2002a). Effect of genetic background and culture conditions on the production of herpesvirus-based gene therapy vectors. Biotechnol Bioeng *77*, 685–692.

Ozuer, A., Wechuck, J. B., Russell, B., Wolfe, D., Goins, W. F., Glorioso, J. C., and Ataai, M. M. (2002b). Evaluation of infection parameters in the production of replication-defective HSV-1 viral vectors. Biotechnol Prog *18*, 476–482.

Palmer, J. A., Branston, R. H., Lilley, C. E., Robinson, M. J., Groutsi, F., Smith, J., Latchman, D. S., and Coffin, R. S. (2000). Development and optimization of herpes simplex virus vectors for multiple long-term gene delivery to the peripheral nervous system. J Virol *74*, 5604–5618.

Passini, M. A., Watson, D. J., Vite, C. H., Landsburg, D. J., Feigenbaum, A. L., and Wolfe, J. H. (2003). Intraventricular brain injection of adeno-associated virus type 1 (AAV1) in neonatal mice results in complementary patterns of neuronal transduction to AAV2 and total long-term correction of storage lesions in the brains of beta-glucuronidase-deficient mice. J Virol *77*, 7034–7040.

Perng, G. C., Dunkel, E. C., Geary, P. A., Slanina, S. M., Ghiasi, H., Kaiwar, R., Nesburn, A. B., and Wechsler, S. L. (1994). The latency-associated transcript gene of herpes simplex virus type 1 (HSV-1) is required for efficient in vivo spontaneous reactivation of HSV-1 from latency. J Virol *68*, 8045–8055.

Perng, G. C., Ghiasi, H., Slanina, S. M., Nesburn, A. B., and Wechsler, S. L. (1996a). The spontaneous reactivation function of the herpes simplex virus type 1 LAT gene resides completely within the first 1.5 kilobases of the 8.3-kilobase primary transcript. J Virol *70*, 976–984.

Perng, G. C., Jones, C., Ciacci Zanella, J., Stone, M., Henderson, G., Yukht, A., Slanina, S. M., Hofman, F. M., Ghiasi, H., Nesburn, A. B., and Wechsler, S. L. (2000a). Virus-induced neuronal apoptosis blocked by the herpes simplex virus latency-associated transcript. Science *287*, 1500–1503.

Perng, G. C., Slanina, S. M., Ghiasi, H., Nesburn, A. B., and Wechsler, S. L. (1996b). A 371-nucleotide region between the herpes simplex virus type 1 (HSV-1) LAT promoter and the 2-kilobase LAT is not essential for efficient spontaneous reactivation of latent HSV-1. J Virol *70*, 2014–2018.

Perng, G. C., Slanina, S. M., Yukht, A., Drolet, B. S., Keleher, W., Jr., Ghiasi, H., Nesburn, A. B., and Wechsler, S. L. (1999). A herpes simplex virus type 1 latency-associated transcript mutant with increased virulence and reduced spontaneous reactivation. J Virol *73*, 920–929.

Perng, G. C., Slanina, S. M., Yukht, A., Ghiasi, H., Nesburn, A. B., and Wechsler, S. L. (2000b). The latency-associated transcript gene enhances establishment of herpes simplex virus type 1 latency in rabbits. J Virol *74*, 1885–1891.

Perry, L. J., and McGeoch, D. J. (1988). The DNA sequences of the long repeat region and adjoining parts of the long unique region in the genome of herpes simplex virus type 1. J Gen Virol *69*, 2831–2846.

Preston, C. M., Frame, M. C., and Campbell, M. E. (1988). A complex formed between cell components and an HSV structural polypeptide binds to a viral immediate early gene regulatory DNA sequence. Cell *52*, 425–434.

Puskovic, V., Wolfe, D., Goss, J., Huang, S., Mata, M., Glorioso, J. C., and Fink, D. J. (2004). Prolonged biologically active transgene expression driven by HSV LAP2 in brain *in vivo*. Mol Ther *10*, 67–75.

Puvion Dutilleul, F., Pichard, E., and Leduc, E. H. (1985). Influence of embedding media on DNA structure in herpes simplex virus type 1. Biol Cell *54*, 195–198.

Rajcani, J., and Vojvodova, A. (1998). The role of herpes simplex virus glycoproteins in the virus replication cycle. Acta Virol *42*, 103–118.

Read, G. S., and Frenkel, N. (1983). Herpes simplex virus mutants defective in the virion-associated shutoff of host polypeptide synthesis and exhibiting abnormal synthesis of alpha (immediate early) viral polypeptides. J Virol *46*, 498–512.

Rivera, L., Beuerman, R. W., and Hill, J. M. (1988). Corneal nerves contain intra-axonal HSV-1 after virus reactivation by epinephrine iontophoresis. Curr Eye Res *7*, 1001–1008.

Rock, D. L., Nesburn, A. B., Ghiasi, H., Ong, J., Lewis, T. L., Lokensgard, J. R., and Wechsler, S. L. (1987). Detection of latency-related viral RNAs in trigeminal ganglia of rabbits latently infected with herpes simplex virus type 1. J Virol *61*, 3820–3826.

Rodahl, E., and Haarr, L. (1997). Analysis of the 2-kilobase latency-associated transcript expressed in PC12 cells productively infected with herpes simplex virus type 1: Evidence for a stable, nonlinear structure. J Virol *71*, 1703–1707.

Roizman, B., and Sears, A. E. (1996). Chapter 72: Herpes simplex viruses and their replication. In Fields Virology, B. N. Fields, D. M. Knipe, and P. M. Howley, eds. (Philadelphia, Lippincott- Raven), pp. 2231–2295.

Sacks, W. R., Greene, C. C., Aschman, D. P., and Schaffer, P. A. (1985). Herpes simplex virus type 1 ICP27 is an essential regulatory protein. J Virol *55*, 796–805.

Samaniego, L. A., Neiderhiser, L., and DeLuca, N. A. (1998). Persistence and expression of the herpes simplex virus genome in the absence of immediate-early proteins. J Virol *72*, 3307–3320.

Samaniego, L. A., Wu, N., and DeLuca, N. A. (1997). The herpes simplex virus immediate-early protein ICP0 affects transcription from the viral genome and infected-cell survival in the absence of ICP4 and ICP27. J Virol *71*, 4614–4625.

Sasaki, K., Chancellor, M. B., Goins, W. F., Phelan, M. W., Glorioso, J. C., de Groat, W. C., and Yoshimura, N. (2004). Gene therapy using replication-defective herpes simplex virus vectors expressing nerve growth factor in a rat model of diabetic cystopathy. Diabetes *53*, 2723–2730.

Sedarati, F., Izumi, K. M., Wagner, E. K., and Stevens, J. G. (1989). Herpes simplex virus type 1 latency-associated transcription plays no role in establishment or maintenance of a latent infection in murine sensory neurons. J Virol *63*, 4455–4458.

Smith, C., Lachmann, R. H., and Efstathiou, S. (2000). Expression from the herpes simplex virus type 1 latency-associated promoter in the murine central nervous system. J Gen Virol *3*, 649–662.

Snyder, E. Y., Taylor, R. M., and Wolfe, J. H. (1995). Neural progenitor cell engraftment corrects lysosomal storage throughout the MPS VII mouse brain. Nature *374*, 367–370.

Soares, K., Hwang, D. Y., Ramakrishnan, R., Schmidt, M. C., Fink, D. J., and Glorioso, J. C. (1996). cis-acting elements involved in transcriptional regulation of the herpes simplex virus type 1 latency-associated promoter 1 (LAP1) *in vitro* and *in vivo*. J Virol *70*, 5384–5394.

Spear, P. G. (1993). Entry of alphaherpesviruses into cells. Seminars in Virology *4*, 167–180.

Spivack, J. G., and Fraser, N. W. (1987). Detection of herpes simplex virus type 1 transcripts during latent infection in mice. J Virol *61*, 3841–3847.

Steiner, I., Spivack, J. G., Lirette, R. P., Brown, S. M., MacLean, A. R., Subak Sharpe, J. H., and Fraser, N. W. (1989). Herpes simplex virus type 1 latency-associated transcripts are evidently not essential for latent infection. Embo J *8*, 505–511.

Stevens, A. C., and Spear, P. G. (1997). Herpesvirus capsid assembly and envelopment. In Structural Biology of Viruses, W. Chiu, R. Burnett, and R. Garcea, Eds. (New York, Oxford University Press).

Stevens, J. G., Wagner, E. K., Devi Rao, G. B., Cook, M. L., and Feldman, L. T. (1987). RNA complementary to a herpesvirus alpha gene mRNA is prominent in latently infected neurons. Science *235*, 1056–1059.

Thomas, S. K., Gough, G., Latchman, D. S., and Coffin, R. S. (1999). Herpes simplex virus latency-associated transcript encodes a protein which greatly enhances virus growth, can compensate for deficiencies in immediate-early gene expression, and is likely to function during reactivation from virus latency. J Virol *73*, 6618–6625.

Thompson, R. L., and Sawtell, N. M. (1997). The herpes simplex virus type 1 latency-associated transcript gene regulates the establishment of latency. J Virol *71*, 5432–5440.

Wang, Q., Guo, J., and Jia, W. (1997). Intracerebral recombinant HSV-1 vector does not reactivate latent HSV-1. Gene Ther *4*, 1300–1304.

Wolfe, D., Goins, W. F., Kaplan, T. J., Capuano, S. V., Fradette, J., Murphey-Corb, M., Robbins, P. D., Cohen, J. B., and Glorioso, J. C. (2001). Herpesvirus-mediated systemic delivery of nerve growth factor. Molecular Therapy *3*, 61–69.

Wolfe, D., Niranjan, A., Trichel, A., Wiley, C., Ozuer, A., Kanal, E., Kondziolka, D., Krisky, D., Goss, J., DeLuca, N., et al. (2004). Safety and biodistribution studies of an HSV multigene vector following intracranial delivery to non-human primates. Gene Ther. *11*, 1675–84.

Wu, N., Watkins, S. C., Schaffer, P. A., and DeLuca, N. A. (1996). Prolonged gene expression and cell survival after infection by a herpes simplex virus mutant defective in the immediate-early genes encoding ICP4, ICP27, and ICP22. J Virol *70*, 6358–6369.

Yamada, M., Oligino, T., Mata, M., Goss, J. R., Glorioso, J. C., and Fink, D. J. (1999). Herpes simplex virus vector-mediated expression of Bcl-2 prevents 6-hydroxydopamine-induced degeneration of neurons in the substantia nigra *in vivo*. Proc Natl Acad Sci USA *96*, 4078–4083.

Zwaagstra, J., Ghiasi, H., Nesburn, A. B., and Wechsler, S. L. (1989). *In vitro* promoter activity associated with the latency-associated transcript gene of herpes simplex virus type 1. J Gen Virol *70*, 2163–2169.

Zwaagstra, J. C., Ghiasi, H., Nesburn, A. B., and Wechsler, S. L. (1991). Identification of a major regulatory sequence in the latency associated transcript (LAT) promoter of herpes simplex virus type 1 (HSV-1). Virology *182*, 287–297.

Zwaagstra, J. C., Ghiasi, H., Slanina, S. M., Nesburn, A. B., Wheatley, S. C., Lillycrop, K., Wood, J., Latchman, D. S., Patel, K., and Wechsler, S. L. (1990). Activity of herpes simplex virus type 1 latency-associated transcript (LAT) promoter in neuron-derived cells: evidence for neuron specificity and for a large LAT transcript. J Virol *64*, 5019–5028.

GENE THERAPY OF LYSOSOMAL STORAGE DISORDERS BY LENTIVIRAL VECTORS

Alessandra Biffi and Luigi Naldini

1 INTRODUCTION

Lysosomal storage disorders (LSDs) comprise a class of inherited diseases characterized by disruption of normal lysosomal function and the consequent accumulation of incompletely degraded substrates. Most LSDs are caused by loss of function of specific lysosomal acid hydrolases, which act to degrade complex substrates that have been targeted for degradation after endocytosis or autophagy. The degradation occurs by a stepwise pathway, and if one step in the process fails, further degradation often ceases and the partially degraded substrate accumulates. The ensuing substrate accumulation in lysosomes affects the architecture and function of cells, tissues, and organs. In some cases, the accumulated substrate itself (as in Galactocerebrosidosis) or the product of an alternative metabolic route, which is upregulated by the accumulated primary substrate (as in the case of psycosine in Globoid Cell Leukodystrophy), is cytotoxic and leads to cell dysfunction or death. In other cases, the actual molecular mechanism triggered by the accumulated metabolite and leading to cellular toxicity and tissue pathology remains elusive.

The primary defect may be further exacerbated by secondary responses. For instance, microglial activation occurring in the central nervous system (CNS) of several LSDs may represent a primary reaction to substrate accumulation within these cells and/or an inflammatory response to a primary neuronal damage. Despite sharing a similar pathogenetic mechanism, the over 40 different LSDs that have been described to date differ for several disease-specific features. The main differences are represented by the different pattern of peripheral organ involvement and by the presence and severity of nervous system (NS) involvement. It has been reported that neurodegeneration in the CNS and/or dismyelination affecting the entire NS are the hallmark of roughly 60% of LSDs. The severity of NS disease varies among different types of LSDs and among different forms of the same LSD, and it ranges from mild mental retardation with minimal alteration of cognitive function to severe and rapidly progressing disability and death.

Most lysosomal enzymes are sorted to the lysosomes as inactive proenzymes from the Golgi apparatus through a specific pathway controlled by the mannose-6-phosphate receptor (M6PR). As originally shown (Neufeld, 1991), a fraction of proenzyme escapes the lysosomal sorting pathway and is secreted into the extracellular space. The secreted proenzyme may then be endocytosed by the producer cell or other neighbouring cells via

San Raffaele Telethon Institute for Gene Therapy, Vita-Salute San Raffaele University, Milan, Italy. e-mail: naldini.luigi@hsr.it

M6PR molecules expressed on the cell membrane and sorted to the lysosomes. Whether this extracellular pathway provides for some specific physiological function or mainly represents salvage of enzyme leaked from the intracellular route remains to be established. Based on this feature, however, most treatment strategies for LSDs are based on providing an exogenous supply of the missing or defective lysosomal enzyme in the appropriate target tissue. The exogenous enzyme can be taken up by the deficient cells leading to cross-correction of the metabolic defect.

Studies in preclinical in vitro systems and animal models have allowed the development of enzyme replacement therapy (ERT) – parenteral injection of purified recombinant proenzyme – as a therapeutic option for some LSDs. As recently reported (Barranger and O'Rourke, 2001; Desnick, Ioannou, and Eng, 2001; Eng et al., 2001; Schiffmann et al., 2001; Kakkis et al., 2001; Sly, 1993) ERT is currently the standard therapy for nonneuropathic Gaucher patients and it was recently approved or is being evaluated for the treatment of Fabry disease, Pompe disease, MPS I, MPS II, and MPS VI. Despite the major therapeutic advances made possible by ERT for the treatment of several LSDs, protein delivery poses serious challenges when sustained administration is required and when the CNS and PNS are the major disease targets. Recurring parenteral administration of an exogenous protein carries the risk of inducing an immune response against the therapeutic enzyme that may interfere with, or even neutralize its activity, as indicated by some recent studies (Van den Hout et al., 2004; Kakkis et al., 2004). This risk may be increased in patients carrying null mutations.

Moreover, for LSDs with NS involvement, the blood–brain and the blood–nerve barriers may severely limit access of systemically administered therapeutic molecules to these tissues. Gene-based delivery may allow establishing a sustained source of therapeutic proteins within the body for peripheral organs correction, or within the NS, overcoming the anatomical barriers which limit enzyme diffusion from the circulation (Kay, Glorioso, and Naldini, 2001; Glorioso, Mata, and Fink, 2003).

Gene therapy, according to the route of administration and the type of target tissue or organ, may thus provide one or more of the following treatment strategies for LSDs: i) direct metabolic correction of specific cell types representing major disease targets, such as oligodendrocytes and neurons upon vector administration into the CNS; ii) establishment of a cellular pump releasing the therapeutic enzyme into the extracellular fluids for the cross-correction of affected cells within the tissue and/or widespread in the body, such as in the case of liver-directed gene transfer which allows release of enzyme into the bloodstream; iii) cell replacement in the affected tissues by genetically corrected cells or progenitors, such as in the transplantation of ex vivo transduced hematopoietic or neural stem cells. Lentiviral vectors (LV), with their capability to efficiently transduce several cell types both in vitro and in vivo and stably integrate into the genome (Follenzi and Naldini, 2002a,b), may represent a profitable tool for these purposes. In addition, LV have proved to be useful tools for the study of cellular biology and disease mechanisms. Regarding their potential clinical use, significant advances in vector design have led to highly improved vector safety, and the concern for the formation of a pathogenic, replication competent virus during vector production or target cell infection has been significantly alleviated (Follenzi and Naldini, 2002a,b).

In this chapter we review the recent and prospective applications of LV-mediated gene transfer to LSDs therapy by discussing the following approaches: i) systemic vector administration; ii) direct CNS gene transfer; iii) hematopoietic stem cell (HSC)-based ex vivo gene therapy; and iv) transplantation of genetically modified neural stem cell (NSC).

2 In Vivo Gene Therapy Applications

2.1 Systemic Delivery of LV

As shown by several studies (Neufeld, 1991; Barranger et al., 2001; Desnick et al., 2001; Eng et al., 2001; Schiffmann et al., 2001; Kakkis et al., 2001; Sly, 1993), enzyme replacement therapy is currently the standard therapy for several LSDs. It is likely that similar efficacy of ERT could be achieved by establishing a tissue source of the therapeutic enzyme by means of gene therapy, thus obviating the need for continuous administration of exogenous enzyme. Intravenous delivery of viral vectors may represent an effective strategy to target gene transfer to tissues which are readily accessible from the bloodstream and may thus become efficient sources for systemic enzyme distribution.

It has been shown that intravenous administration of adenoviral (Reddy et al., 2002), adeno-associated (Daly, 2004), retroviral (in selected conditions; McCormack et al., 2001), and lentiviral (Follenzi et al., 2002a; Nguyen et al., 2002) vectors leads to efficient liver transduction. By these approaches, the liver can produce large amounts of therapeutic enzyme, thus becoming a depot organ. This strategy has been demonstrated to be effective in controlling visceral and skeletal disease manifestations in several LSD models, particularly when applied in the neonatal age. Neonatal intravenous retroviral gene therapy has been evaluated in mice and dogs affected by mucopolysaccharidosis (MPS) VII (Parker Ponder et al., 2002; Mango et al., 2004).

Dogs injected with retroviral vectors expressing the canine β-glucuronidase (cGUSB) displayed stable levels of GUSB in serum, at or above the heterozygous levels, absence of major clinical signs of the disease, such as cardiac abnormalities, and marked amelioration of the skeletal, cartilage, and synovial disease. Moreover, almost complete prevention of corneal clouding was observed in treated dogs likely due to uptake of the enzyme from blood. This study, however, showed little evidence of therapeutic effect on the CNS, despite the fact that mental retardation, a typical manifestation of MPS VII in humans, is difficult to assess in MPS VII dogs. Two RV-treated dogs had low (1.5% of normal) GUSB activity and reductions of cytoplasmic vacuolation in the brains at 6 months. The authors claimed that the enzyme activity detected in the brain may have originated from vector taken up in the CNS before the formation of the blood–brain barrier.

A similar correction of disease manifestations in the periphery coupled to stronger effects in the CNS have been obtained in MPS I mice treated with neonatal systemic administration of AAV vectors (Hartung et al., 2004). This study showed that injection of AAV vectors carrying the α-L-iduronidase (IDUA) enzyme in neonate mice provided a major curative impact on several of the most important parameters of the disease, including neurological abnormalities, both at the hystopathological and behavioral level. The authors, however, were unable to detect vector DNA in brain extracts, suggesting that the vector was present in the CNS to very low levels and below the detection threshold. Whether a similar systemic gene therapy approach could provide therapeutic benefits on CNS disease manifestations when applied in adulthood remains to be clarified.

Preliminary evidence has been recently reported in this regard (Sferra et al., 2004). The authors demonstrated partial correction of CNS biochemical and hystophatological abnormalities in MPS VII mice that received high doses of GUSB-AAV by intrahepatic injection. The reported finding of very low but detectable levels of vector genome in the brain of treated mice indicates that AAV2 vector, if administered in sufficient amounts, may cross the MPS VII blood–brain barrier and contribute to disease correction.

Additional experimental evidence will be required to elucidate the possible therapeutic role of systemic gene therapy for NS disease manifestations in LSDs and whether LV may provide a suitable delivery platform for exploiting this approach. Moreover, careful evaluation of the immune responses against the therapeutic protein and the vector used for in vivo gene transfer has to be performed. Several studies using either reporter or therapeutic genes have shown the occurrence of cellular and/or humoral immune responses following administration of purified enzyme (Sly et al., 2001) and liver-directed in vivo gene therapy, responsible for the clearance of the transduced cells and/or the disappearance of enzyme activity from the circulation (Stein et al., 1999; Gao et al., 2000; Follenzi et al., 2004; Di Domenico et al., 2005).

On the contrary, when in vivo gene therapy was performed in the neonatal age (Parker Ponder et al., 2002; Mango et al., 2004; Hartung et al., 2004), when the immune system is relatively immature, stable, long-term enzyme expression was observed in the tissues and in the circulation, indicating that no antibody or cytotoxic T cell response was induced. Sferra and colleagues (2004), however, showed stable levels of circulating beta-glucuronidase and lack of inflammatory cell infiltrate within the liver following r-AAV-GUSB administration to adult mice, suggesting the absence of inactivating antibodies and of overt cytotoxic immune reaction. To assess whether this phenomenon was due to induction of tolerance by vector-mediated expression of foreign protein within the liver (Chao and Walsh, 2001; Mount et al., 2002; Mingozzi et al., 2003; Follenzi et al., 2004), or to a specific immune defect of MPS VII mice (Daly, Lorenz, and Sands, 2000), or finally to recognition of the normal GUSB protein as a self protein by MPS VII mice immune system (Birkenmeier et al., 1989; Watson et al., 1998), will require further studies.

Some recent studies indicate that under specific experimental conditions, in vivo gene delivery in adult mice may escape immune responses, such as by using hepatocyte-specific promoters that restrict transgene expression to parenchymal cells of the liver and limit direct transgene expression within antigen presenting cells (APC; Chao and Walsh, 2001; Mount et al., 2002; Mingozzi et al., 2003; Follenzi et al., 2004). Such a strategy may alleviate the risk that administration of viral vectors may exert an adjuvant effect and trigger an immune response against the transgene product in recipients that are tolerant to ERT.

Other issues that need to be addressed for a clinical translation of LV-mediated systemic gene therapy include the optimal choice of envelope pseudotype to avoid complement-dependent vector neutralization in the plasma and vector sequestration in blood-filtering organs, and to achieve, if possible, specific targeting to the desired cell population. Moreover, the safety and toxicity of parenteral vector administration and the possible risk of inadvertent gene transfer to the gonads and germ-line transmission of the vector will have to be established.

2.2 Direct CNS Gene Transfer

Currently, a number of vector delivery systems exist that can be used for direct in vivo gene transfer into the CNS and the delivery of therapeutic molecules to brain regions vulnerable to neurodegenerative diseases. Direct injection of gene transfer vectors into the CNS has achieved long-term protein expression and therapeutic benefit in several disease models of LSDs (Bosch et al., 2000; Stein et al., 1999, 2001; Skorupa et al., 1999; Consiglio et al., 2001). Efficient transduction of the neurons of adult rodent brains was observed with all generations of lentiviral vectors (Naldini et al., 1996; Blomer et al.,

1997; Zufferey et al., 1997; Naldini, 1998; Miyoshi et al., 1998), and effective gene transfer was also observed to brain neurons of nonhuman primates (Kordower et al., 1999). The main cellular targets of LV in the CNS are neurons (over 90% of transgene-expressing cells after administration of VSV-pseudotyped LV), whereas glial cells are transduced in vivo to a lower efficiency. Cross-correction from transduced neurons may have the advantage of achieving a widespread enzyme distribution within the CNS thanks to the long-range reach of neural processes.

Bosch and colleagues (2000) showed widespread distribution of the therapeutic GUSB enzyme and reversal of pathology in large sectors of the brain of MPS type VII mice after lentivirus-mediated gene transfer. The relevance of this observation was increased by the fact that mice were treated as adult, an age at which pathology was already well established in the brain. Consiglio and colleagues (2001) showed sustained expression of active enzyme throughout a large portion of the brain, with long-term protection from development of neuropathology and hippocampal-related learning impairments of metachromatic leukodystrophy (MLD) mice treated with LV carrying the Arylsulfatase A (ARSA) cDNA. These initial studies, later reproduced with other vector systems on several disease models (Haskell et al., 2003; Griffey et al., 2004; Desmaris et al., 2004), provided the proof of principle that direct CNS gene delivery by viral vectors is an effective strategy to restore the biologic functions of LSDs' brains.

Moreover, several observations indicate that this strategy may not only halt disease progression but also reverse the disease phenotype once it has been established. Brooks and colleagues (2002) have recently shown functional correction of established CNS deficits in MPS VII mice after feline immunodeficiency viruses (FIV)-mediated direct CNS gene transfer. Recovery of behavioral functions was observed in mice treated when already bearing clear impairments in spatial learning and memory. These data indicate that enzyme replacement in the MPS VII nervous system goes beyond restoration of β-glucuronidase activity in the lysosomes. Moreover, they suggest that enzyme replacement by gene transfer in LSDs' brains has the potential to reverse severe neurological deficit in mice with established brain lysosomal storage disease. The mechanisms by which such correction could be achieved remain to be elucidated.

In this context, the use of LV to obtain sustained expression of therapeutic genes in different neural cell populations in vivo (Naldini et al., 1998; Baekelandt et al., 2002; Georgievska, 2002; Consiglio et al., 2004) is particularly relevant. An intriguing hypothesis is that endogenous neural stem cells are targeted by the vector, and that their progeny may be capable of replacing, at least in part, the damaged tissue. It was recently demonstrated (Consiglio et al., 2004) that LV efficiently target long-term repopulating adult neural stem cells (NSCs) upon direct injection into neurogenic areas of the adult mouse brain. These findings raise the possibility that endogenous progenitors may be genetically modified in vivo to compensate for inherited deficiencies and/or to steer their differentiation towards specific differentiation pathways. It is thus possible that the correction of NS pathology observed in some studies may be due to neuroregeneration from gene-corrected progenitors.

An additional question raised by these findings is the cellular mechanism responsible for the widespread distribution of therapeutic enzyme within the injected brain. This could be diffusion in the extracellular space or, more likely, axonal transport within secretory vesicles that may be released at nerve endings, uptake, and retrograde transport along neural processes. The latter mechanism would allow active transport across long distances and crossing the midline. Recent experimental evidence indicates that axonal transport across the midline is occurring in the brain of mice injected in the hippocampus with LV carrying an ARSA-tagged construct and may effectively contribute to the widespread correction observed in treated mice (Luca et al., 2005). Such a mechanism was previously reported upon AAV-mediated gene transfer in MPS VII mice (Watson et al., 1998).

Regarding the possible clinical translation of this strategy, preliminary studies on nonhuman primates (Kordower et al., 1999) indicate robust transduction of injected brain by LV. Moreover, Kordower and coworkers demonstrated reversal of functional deficits and prevention of nigrostriatal degeneration when they injected LV-expressing glial cell-line-derived neurotrophic factor (GDNF) into the brains of aged monkeys or of monkeys previously treated with 1-methyl-4-phenyl-1,2,3,6-tetrahydropyridine (MPTP) to induce a Parkinson's diseaselike phenotype. However, scaling up such procedures to the large brain of human beings still represents a challenge in view of the need for correcting large portions of the CNS, as required in most LSDs. Moreover, safety issues have to be addressed, not only related to the injection procedure itself, but also to enzyme expression levels. It remains to be determined what expression levels are required for correction of CNS pathology in different LSDs, and whether enzyme expression needs to be regulated. The recent identification of Sulfatase Modifying Factor 1 (SUMF1; Cosma et al., 2003) as a common activator of sulfatases and a rate-limiting factor in the biological activation of these enzymes raises concerns for possible adverse effects of enzyme overexpression within certain cell types. In fact, overexpression of one type of sulfatase may lead to reduced activity of other sulfatases by a competitive interaction with their common activator. These and other issues will have to be addressed before moving forward with direct CNS gene therapy to humans.

3 Ex Vivo Gene Therapy Applications

3.1 Hematopoietic Stem Cells

Hematopoietic stem cell (HSC)-mediated gene therapy has long been considered an attractive option for the treatment of LSDs, and in particular for NS manifestations. This strategy relies on the positive clinical experience with allogeneic bone marrow transplantation (BMT) in several LSDs (Krivit et al., 1999). Over the past two decades, HSC transplantation has been used with increased frequency to treat LSD patients by replacing the intra- and extravascular hematopoietic compartments with a cell population expressing a functional enzyme. The therapeutic impact of bone marrow transplantation depends on the specific enzymatic deficiency and the stage of the disease. Usually the visceral symptoms can be improved, whereas the established skeletal lesions remain relatively unaffected. The effect on the neurologic symptoms varies.

BMT remains a viable treatment option in those LSD for which data indicative of disease stabilization or amelioration are known, such as late onset forms of globoid cell leukodystrophy or metachromatic leukodystrophy (Peters and Steward, 2003). Early transplantation is the goal so that enzyme replacement may occur before extensive CNS injury becomes evident. However, the lack of adequate donors and the high morbidity and mortality related to the allogenic transplant greatly reduce the overall number of candidate patients for BMT. Gene therapy strategies aimed at correcting the genetic defect in patients' HSC could represent a significant improvement as compared to conventional allogeneic BMT. The autologous procedure is associated to a reduced transplant-related morbidity and mortality and avoids the risks of graft versus host disease (GVHD). Moreover, autologous cells may be genetically modified to constitutively express high levels of the therapeutic enzyme and become a quantitatively more effective source of enzyme than wild-type cells, possibly also in the NS.

Hematopoietic stem cells (HSC) are important targets for gene therapy, due to the ease with which they can be manipulated ex vivo and returned to the host, as well as to the broad range of diseases that can be potentially treated in this way. The largely quiescent nature of HSC, combined with the need for vector integration to ensure gene delivery to the HSC progeny, makes them prime candidates for lentiviral vector transduction. Successful gene marking of primitive human cord blood and bone-marrow derived nonobese diabetic/severe combined (NOD/SCID) repopulating cells has been shown with both early (Miyoshi et al., 1999) and late generation (Guenechea et al., 2001; Ailles et al., 2002) HIV-1-derived vectors using a short-term exposure to the vector in the absence of cytokines.

Therapeutic efficacy of HSC-based ex vivo gene therapy approaches in controlling some disease manifestations has been shown in preclinical experiments on LSD models (Walkley et al., 1994; Leiming et al., 2002; Zheng et al., 2003). Ex vivo gene therapy based on oncoretroviral vectors has been proven to be effective in restoring the missing enzyme activity at several visceral organs and providing therapeutic effects in different LSD animal models (Leiming et al., 2002; Zheng et al., 2003). Interestingly, it was recently shown that a similar therapeutic effect could be also achieved using human CD34+ hematopoietic cells, transduced with a lentiviral vector. Hofling and colleagues (2004) demonstrated GUSB activity in several tissues of NOD/SCID/MPS VII mice transplanted with CD34+ cells from an MPS VII patient transduced with LV carrying the GUSB cDNA. These data were associated with improvement of biochemical parameters and reduction of the lysosomal distension in several visceral tissues. These data strengthen the rationale for such an approach to be used in MPS and similar diseases with predominant visceral organ/bone and cartilage involvement.

The proof of the therapeutic potential of this strategy for LSDs characterized by extensive CNS and PNS involvement came more recently. Microglia cells have been implicated in the pathogenesis of a number of neurodegenerative conditions. In the case of LSD pathogenesis, microglia involvement was demonstrated in several mouse models (Hess et al., 1996; Wada, Tifft, and Proia, 2000; German et al., 2002; Ohmi et al., 2003). Microglial cells are considered a primary site of lipid storage, resulting in cell activation and secretion of cytokines and proinflammatory molecules which trigger the focal inflammation, demyelination, and neurodegeneration characteristic of these diseases. Thus, microglia should be considered a primary target cell type in therapeutic strategies for LSDs.

The first formal proof of HSC contribution to the turnover of CNS-resident microglia in adult mice came from gene marking studies (Eglitis and Mezey, 1997; Kennedy and Abkowitz, 1997; Priller et al., 2001; Biffi et al., 2004). The use of green fluorescent protein (GFP) allowed demonstrating progressive engraftment of transgene-expressing microglia both in physiological and pathological conditions upon transplantation of HSC transduced with GFP encoding RV (Priller et al., 2001) or LV (Biffi et al., 2004). This phenomenon led to the reconstitution of up to one third of well-differentiated, resting microglia in a few months after BMT. Interestingly, microglia replacement by HSC-derived cells seems to be independent of the conditioning regimen applied prior to BMT, being reproducible after both total body irradiation of the recipient mice or pharmacological conditioning (Yeager, Shinohara, and Shinn, 1991; Yeager et al., 1993).

It has been recently demonstrated that similar findings can be reproduced using human CD34+ HSC in human-mouse chimeras (Hofling et al., 2004; Asheuer et al., 2004). Asheuer and colleagues showed that cells derived from human CD34+ cells isolated from either cord blood or mobilized peripheral blood of adrenoleukodystrophy (ALD) patients migrated into the brain and differentiated in resident microglia after infusion into NOD/SCID immunodeficient mice (Asheuer et al., 2004). A fraction of CD34-derived, transduced cells expressed the human ALD protein in the brain after peripheral infusion.

These results strengthen the concept that genetically modified HSC can be used to deliver therapeutic proteins to the CNS. Remarkably, a similar phenomenon was demonstrated in the PNS of transplanted mice, in which HSC-derived, transgene-expressing cells progressively replaced endoneurial macrophages. The frequency of transgene-expressing monocyte lineage cells found in the PNS approached that observed in the BM and peripheral blood, indicating a faster turnover of endoneurial macrophages as compared to that of CNS microglia, and possibly reflecting the more permeable blood–nerve barrier. The significant recruitment of BM-derived macrophages provides a new avenue to vehicle gene therapy to this widespread and hardly accessible tissue, which plays a crucial role in the disease evolution of most LSDs.

The therapeutic potential of HSC-based gene therapy might thus rely on the progresssive replacement of endogenous microglia and endoneurial macrophages by gene-corrected HSC derived cells, leading to prevention/correction of microglia activation, reduced inflammation, and neurodegeneration. In most LSDs, this therapeutic effect could be enhanced by enzyme delivery to resident cell types within both the CNS and PNS: microglia derived from genetically corrected HSC could represent a tissue source of therapeutic enzyme, which could be taken up by surrounding cells and cross-correct their deficiency.

One of the first successes in controlling CNS disease manifestations by conventional allogeneic BMT was obtained in the feline model of α-mannosidosis (Walkley et al., 1994). Clinical experience has later on shown significant CNS therapeutic benefit in some LSDs, such as childhood cerebral forms of ALD (which is, however, a peroxysomal deficiency) or Krabbe disease (Krivit et al., 1999; Peters and Steward, 2003). Recently, Leiming and colleagues (2002) demonstrated partial protection from CNS disease manifestations in the mouse model of galactosialidosis following ex vivo gene therapy with a retroviral vector carrying the therapeutic protein catepsin A (PPCA) cDNA. The authors observed PPCA expression in the CNS of transplanted mice, associated with delayed Purkinje cell degeneration and prevention of the ataxia which is usually observed in aging PPCA-/- mice. However, in other LSDs such as MLD or GM2 gangliosidosis, both BMT and ex vivo gene therapy based on conventional oncoretroviral vectors did not halt CNS disease progression.

Zheng and colleagues (2003) have shown a significant reduction of soluble glycosaminoglycan accumulation in liver and spleen, but not in the brain, following HSC-based ex vivo gene therapy with retroviral vectors in the MPS I mouse model. These results were associated with partial recovery of α-L-iduronidase activity in visceral organs, and to the correction of pathology in the kidney, bladder epithelium, and fibrocartilage. On the contrary, only minimal improvement in neuronal phenotype was observed in selected brain areas.

Similarly, Matzner and colleagues (2000, 2001) showed that HSC-based gene therapy with bone marrow cells overexpressing the human ARSA cDNA from a retroviral vector resulted in the expression of high enzyme levels in various tissues. The treatment almost completely prevented sulfatide storage in livers and kidney, whereas only a partial correction of the lipid metabolism was detectable in the brain. This partial correction was accompanied by modest amelioration of neuropathology and behavioral tests suggesting some improvement of neuromotor abilities, even if in the absence of statistical significance. As suggested by the authors, the limited success of ex vivo gene therapy might have been due to the requirement of unexpectedly high levels of ARSA for the correction of the metabolic defect in the CNS. Differences in the enzyme expression levels necessary to obtain therapeutic efficacy and/or in the efficiency of enzyme secretion by gene-corrected cells could explain the different outcomes of ex vivo gene therapy protocols in different LSD models.

In the more challenging conditions, LV may provide additional benefits due to higher transduction efficiency of murine and human HSC, as compared to that obtained by oncoretroviral vectors, as was recently shown in the mouse model of MLD70. By transplanting HSC transduced with third-generation LV carrying the therapeutic ARSA cDNA, we reconstituted enzyme activity in the hematopoietic system of MLD mice at supranormal levels and prevented the development of major CNS and PNS disease manifestations, at the functional, histopathological, and behavioral levels. Remarkably, ex vivo gene therapy had a significantly higher therapeutic impact than wild-type HSC transplantation, indicating a critical role for enzyme overexpression in the HSC progeny.

These results highlighted the concept that therapeutic HSC transplantation actually corrects a nervous system damage-response pathway defective in LSDs. LV-mediated overexpression of the therapeutic enzyme in macrophages/microglia may serve to enhance the corrective potential of this pathway and to dampen its destructive capacity. Transplantation of LV-transduced autologous HSC may ultimately herald an effective treatment of the nervous system manifestations of LSD.

Some relevant issues are raised by these data. Although clear indications about the turnover rate of microglia cells from circulating monocytes have been provided in rodents, few data are available in humans. Relevant information is being obtained from long-term monitoring of LSDs patients who underwent allogeneic BMT. Significant infiltration of donor-derived cells with the antigenic features of microglia cells have been detected in the brain of GLD patients after transplantation of CD34+ cells from normal cord blood (J. Kurtzberg and E. Y. Snyder, personal communication). Lower rates of cell replacement in the CNS by hematopoietic elements have been reported to date in patients who underwent allo-BMT for hematologic malignances (Unger et al., 1993). This discrepancy may suggest that active CNS disease may enhance microglia turnover in humans as it does in rodents.

Another important issue is whether the turnover rate could be adequate to guarantee some protection from disease progression. This parameter could change greatly with the disease considered and the stage of the disease itself. Krabbe disease, as other LSDs, is characterized by a major inflammatory reaction taking place in the CNS, which may favor microglia turnover versus disease progression and allow better therapeutic outcomes of HSC transplantation. Among LSDs with minor inflammation, late onset and slowly progressive forms may benefit more significantly from the treatment due to a longer window of disease-free time allowing reconstitution of microglia. Moreover, whether conditioning regimens could play any role in microglia reconstitution in humans has to be evaluated. In humans CNS and PNS lesions induced by either cytotoxic drugs or total body irradiation used for conditioning prior to BMT have been described (Solaro et al., 2001; Krouwer and Wijdicks, 2003). A conditioning-dependent toxic effect might account for both an enhancement of microglia reconstitution by donor-derived cells, and for a direct negative effect on the disease course.

Finally, the issue of possible HSC transdifferentiation has to be mentioned. To date clear-cut data have been provided only regarding the differentiation of transplanted hematopoietic progenitors into either perivascular or mature ramified microglia. Limited evidence of transdifferentiation of hematopoietic cells into neurons or, more likely, of fusion between the two cell types has been reported in the literature (Mezey et al., 2003; Weimann et al., 2003). Whatever their actual mechanism may be, these events were too rare (2–5 donor derived neurons/Purkinje cells over 10,000 neurons detected in human brains) for bearing relevance to current therapeutic modalities.

Regarding the clinical translation of this strategy, other important issues need to be addressed. As mentioned for the in vivo approaches, it is necessary to define the threshold expression levels necessary to achieve therapeutic efficacy also in the case of gene-corrected HSC transplantation. Moreover, it has to be defined whether such expression levels could be achieved by a high engraftment of cells expressing the enzyme to a low level, or also by HSC expressing the enzyme at high level and engrafted to a lower extent. These points are crucial in consideration of possible toxicity and side effects. High frequency of gene-corrected HSC engraftment requires a severe conditioning regimen, with its potential side effects, both systemic and on the NS.

On the contrary, sparing the patient a severe conditioning regimen may cause low engraftment of gene-corrected HSC, and thus limit the potential benefit of the procedure. Moreover, in order to achieve high enzyme activity and therapeutic efficacy in conditions of mixed hematopoietic chimerism, a high level of transgene expression by HSC and their progeny might be required. High transgene expression relies on multiple vector integration into HSC, a condition that might be associated with more relevant integration-dependent side effects. After the recent report of adverse events occurring in some gene therapy treated X-SCID patients (Fischer et al., 2004), a great deal of attention has been dedicated to RV integration-related mutagenesis and leukemogenesis. In addition, RV and LV have been shown to have a strong bias for integration in the proximity of expressed genes, thus making transcriptional interference between the vector and flanking endogenous genes more likely (Schroder et al., 2002; Wu et al., 2002; De Palma et al., 2004). Such interference may result both in transcriptional deregulation of the endogenous genes flanking the integration site and in silencing of the endogenous gene targeted by the integration.

The advanced design of late-generation LV provides safety features that may limit such interference. LV are self-inactivating (SIN) vectors with transcriptionally inactive LTR upon transduction, and express the transgene from an internal promoter of choice. In addition, several differences in lentiviral and oncoretroviral biology and pathogenesis suggest that the risk of insertional mutagenesis by RV and LV may be different. However, safety studies specifically designed to address this issue are required before moving LV into the clinic in an autologous HSC transplantation setting.

3.2 Neural Stem Cells

The presence of a reservoir of neural stem cells (NSC) in the CNS of adult mammals has recently raised a great deal of interest among investigators not only because of their possible physiological role, but also for their therapeutic potential in degenerative and other neurological disorders (Gage et al., 1995). Their abundance, multipotency, possibility of ex vivo manipulation, and engrafting ability make NSC attractive targets for CNS gene therapy and repair. In comparison to extant approaches, NSC represent a relatively well-defined neural cell population that can be isolated and expanded ex vivo, and, if necessary, genetically manipulated by the means of gene therapy vectors such as LV. The expression of therapeutic genes into NSC could allow efficient import of these genes into the CNS following transplantation of the modified cells into the developing and mature brain.

Genetically engineered human neural progenitors survived long-term and stably expressed foreign genes after transplantation (Flax et al., 1998; Fricker et al., 1999; Rubio et al., 2000; Ourednik et al., 2001; Tamaki et al., 2002; Englund, Bjorklund, and Wictorin, 2002; Wu et al., 2002; Buchet et al., 2002). Moreover, NSC migrated extensively in the host brain and differentiated into neurons and glia in a region-specific manner upon transplantation into the rodent and primate brain in the prenatal or neonatal period (Flax et al., 1998; Ourednik et al., 2001; Tamaki et al., 2002). When transplanted in the brain of adult animals, human NSC migrated out of the injection site in the brain parenchyma to surrounding regions and showed regionally restricted differentiation (Fricker et al., 1999; Rubio et al., 2000; Englund, 2002; Wu et al., 2002). In addition NSC may be recruited to degenerating neural tissue in response to poorly understood signals.

The original work of Snyder and coworkers provided the first evidence that NSC engraftment has the potential to correct lysosomal storage in the mouse CNS (Rubio et al., 2000). More recently, it was shown that human NSC transduced with LV or RV encoding the therapeutic GUSB grafted into the CNS of MPS VII mice differentiated into neurons and astrocytes and expressed high levels of therapeutic enzyme allowing correction of brain lesions (Buchet et al., 2002; Meng et al., 2003).

There are several possible mechanisms for the corrective activity of transplanted NSC in the NS of LSDs models. The progeny of transplanted NSC may act as an intraparenchymal enzyme source even if they fail to properly differentiate and functionally integrate within the brain. The therapeutic potential of this mechanism has been demonstrated using genetically engineered fibroblasts (Taylor and Wolfe, 1997), encapsulated cells (Ross, Ralph, and Chang, 2000; Barsoum et al., 2003) and amniotic epithelial cells (Kosuga et al., 2001).

A second possibility is neuronal and glial cell replacement by the engineered and transplanted NSC. The extent of such replacement may be limited (Pluchino et al., 2003). Moreover, the possibility that newly replaced neurons may faithfully recapitulate the differentiated features of specific neuronal subtypes and re-establish relevant neural networks formed during brain development remains an outstanding challenge for these strategies. To overcome these limitations a better knowledge of the environmental cues that control NSC migration and differentiation in vivo is required.

Thirdly, it has been proposed that NSC might have a homeostatic effect on the damaged brain, providing survival factors and downregulating local inflammation. Such a mechanism, rather than neural cells replacement, might explain their therapeutic effects on the CNS disease in LSDs or other degenerative conditions. In the perspective of clinical translation, the identification of the most appropriate NSC source is crucial. Using established cell-lines or gene-corrected autologous NSC, thus circumventing ethical and immunological concerns, should be considered. Additional strategies currently being explored are to transplant committed oligodendrocyte progenitors (Blakemore, Gilson, and Crang, 2003; Espinosa de los Monteros et al., 2001) and to use embryonic stem (ES) cells as a source of specialized glial and neuronal subtypes for cell replacement therapies (Wichterle et al., 2002; Kim et al., 2002).

4 CONCLUSIONS

Here we reviewed several gene therapy strategies that have been developed taking advantage of the gene transfer properties of lentiviral vectors and have shown promising results in LSDs animal models. Based on the preclinical efficacy data and the feasibility of their translation to patients, some of these strategies are closely approaching the stage of clinical testing. Clinical translation will require stringent demonstration of the efficacy and safety of the new vectors and gene transfer protocols in the most appropriate models, development of clinical-grade large-scale vector manufacturing and quality assays, mobilization of the financial resources required to support such an endeavor, and securing the consensus and/or approval of the scientific and biomedical communities, national and international regulatory bodies, and patients' associations.

Moreover, a comprehensive knowledge of disease manifestations and evolution will be crucial to allow proper assessment of the risk–benefit ratio and selection of the best candidate patients for testing the new gene therapy strategies. In conclusion, the work described in this chapter represents the fruitful outcome of a decade-long series of efforts aiming at the improvement of gene transfer tools and gene delivery strategies for LSDs and other diseases. The results of these studies will, it is hoped, set the stage for reaching a long-sought effective treatment for these rare and devastating disorders.

REFERENCES

Ailles L., Schmidt M., Santoni de Sio F.R., Glimm H., Cavalieri S., Bruno S., Piacibello W., Von Kalle C., and Naldini L. (2002). Molecular evidence of lentiviral vector-mediated gene transfer into human self-renewing, multi-potent, long-term NOD/SCID repopulating hematopoietic cell. *Mol. Ther.* 6, 615–626.

Asheuer M., Pflumio F., Benhamida S., Dubart-Kupperschmitt A., Fouquet F., Imai Y., Aubourg P., and Cartier N. (2004). Human CD34+ cells differentiate into microglia and express recombinant therapeutic protein. *Proc. Natl. Acad. Sci. U.S.A.* 101(10), 3557–3562.

Baekelandt V., Claeys A., Eggermont K., Lauwers E., De Strooper B., Nuttin B., and Debyser Z. (2002). Characterization of lentiviral vector-mediated gene transfer in adult mouse brain. *Hum. Gene Ther.* 13, 841–853.

Barranger J.A. and O'Rourke E. (2001). Lessons learned from the development of enzyme therapy for Gaucher disease. *J. Inherit. Metab. Dis.* 24 Suppl 2, 89–96.

Barsoum S.C., Milgram W., Mackay W., Coblentz C., Delaney K.H., Kwiecien J.M., Kruth S.A., and Chang P.L. (2003). Delivery of recombinant gene product to canine brain with the use of microencapsulation. *J. Lab. Clin. Med.* 142(6), 399–413.

Biffi A., De Palma M., Quattrini A., Del Carro U., Amadio S., Visigalli I., Sessa M., Fasano S., Brambilla R., Marchesini S., Bordignon C., and Naldini L. (2004). Correction of metachromatic leukodystrophy in the mouse model by transplantation of genetically modified hematopoietic stem cells. *J. Clin. Invest.* 113(8), 1118–1129.

Birkenmeier E.H., Davisson M.T., Beamer W.G., Ganschow R.E., Vogler C.A., Gwynn B., Lyford K.A., Maltais L.M., and Wawrzyniak C.J. (1989). Murine mucopolysaccharidosis type VII. Characterization of a mouse with beta-glucuronidase deficiency. *J. Clin. Invest.* 83(4), 1258–1266.

Blakemore W.F., Gilson J.M., and Crang A.J. (2003). The presence of astrocytes in areas of demyelination influences remyelination following transplantation of oligodendrocyte progenitors. *Exp. Neurol.* 184(2), 955–963.

Blomer U., Naldini L., Kafri T., Trono D., Verma I.M., and Gage F.H. (1997). Highly efficient and sustained gene transfer in adult neurons with a lentivirus vector. *J. Virol.* 71(9), 6641–6649.

Bosch A., Perret E., Desmaris N., Trono D., and Heard J.M. (2000). Reversal of pathology in the entire brain of mucopolysaccharidosis type VII mice after lentivirus-mediated gene transfer. *Hum. Gene Ther.* 11, 1139–1150.

Brooks A.I., Stein C.S., Hughes S.M., Heth J., McCray P.M., Sauter S.L., Johnston J.C., Selechta D.A., Federoff H.J., and Davidson B.L. (2002). Functional correction of established central nervous system deficits in an animal model of lysosomal storage disease with feline immunodeficiency virus-based vectors. *Proc. Natl. Acad. Sci.* 99, 6216–6221.

Buchet D., Serguera C., Zennou V., Charneau P., and Mallet J. (2002). Long-term expression of beta-glucuronidase by genetically modified human neural progenitor cells grafted into the mouse central nervous system. *Mol. Cell. Neurosci.* 19(3), 389–401.

Chao H. and Walsh C.E. (2001). Induction of tolerance to human factor VII in mice. *Blood* 97, 3311–3312.

Consiglio A., Gritti A., Dolcetta D., Follenzi A., Bordignon C., Gage F.H., Vescovi A.L., and Naldini L. (2004). Robust in vivo gene transfer into adult mammalian neural stem cells by lentiviral vectors. *Proc. Natl. Acad. Sci. U.S.A.* 101(41), 14835–14840.

Consiglio A., Quattrini A., Martino S., Bensadoun J.C., Dolcetta D., Trojani A., Benaglia G., Marchesini S., Cestari V., Oliverio A., Bordignon C., and Naldini L. (2001). In vivo gene therapy of metachromatic leukodystrophy by lentiviral vectors: Correction of neuropathology and protection against learning impairments in affected mice. *Nat. Med.* 7, 310–316.

Cosma M.P., Pepe S., Annunziata I., Newbold R.F., Grompe M., Parenti G., and Ballabio A. (2003). The multiple sulfatase deficiency gene encodes an essential and limiting factor for the activity of sulfatases. *Cell* 113(4), 445–456.

Daly T.M. (2004). AAV-mediated gene transfer to the liver. *Methods Mol. Biol.* 246, 195–199.

Daly T.M., Lorenz R., and Sands M.S. (2000). Abnormal immune function in vivo in a murine model of lysosomal storage disease. *Pediatr. Res.* 47, 757–762.

De Palma M., Montini E., Santoni de Sio F.R., Benedicenti F., Gentile A., Medico E., and Naldini L. (2004). Promoter trapping reveals significant differences in integration site selection between MLV and HIV vectors in primary hematopoietic cells. *Blood*, 105(6), 2307–2315.

Desmaris N., Verot L., Puech J.P., Caillaud C., Vanier M.T., and Heardt J.M. (2004). Prevention of neuropathology in the mouse model of Hurler syndrome. *Ann. Neurol.* 56, 68–76.

Desnick R.J., Ioannou Y.A., and Eng C.M. (2001). Metabolic and Molecular Bases of Inherited Diseases. Scriver C.R., Beaudet A.L., Sly W.S., Valle D. (Eds.) (New York: McGraw-Hill) pp. 3733–3774.

Di Domenico C., Villani G.R., Di Napoli D., Reyero E.G., Lombardo A., Naldini L., and Di Natale P. (2005). Gene therapy for a mucopolysaccharidosis type I murine model with lentiviral-IDUA vector. *Hum Gene Ther.* 16(1), 81–90.

Eglitis M.A. and Mezey E. (1997). Hematopoietic cells differentiate into both microglia and macroglia in the brains of adult mice. *Proc. Natl. Acad. Sci. U.S.A.* 94, 4080–4085.

Eng C.M., Guffon N., Wilcox W.R., Germain D.P., Lee P., Waldek S., Caplan L., Linthorst G.E., and Desnick R.J. (2001). Safety and efficacy of recombinant human alpha-galactosidase A–replacement therapy in Fabry's disease. *N. Engl. J. Med.* 345, 9–16.

Englund U., Bjorklund A., and Wictorin K. (2002). Migration patterns and phenotypic differentiation of long-term expanded human neural progenitor cells after transplantation into the adult rat brain. *Brain Res. Dev. Brain Res.* 134(1–2), 123–141.

Espinosa de los Monteros A., Baba H., Zhao P.M., Pan T., Chang R., de Vellis J., and Ikenaka K. (2001). Remyelination of the adult demyelinated mouse brain by grafted oligodendrocyte progenitors and the effect of B-104 cografts. *Neurochem. Res.* 26(6), 673–682.

Fischer A., Abina S.H., Thrasher A., von Kalle C., and Cavazzana-Calvo M. (2004). LMO2 and gene therapy for severe combined immunodeficiency. *N. Engl. J. Med.* 350, 2526–2527.

Flax J.D., Aurora S., Yang C., Simonin C., Wills A.M., Billinghurst L.L., Jendoubi M., Sidman R.L., Wolfe J.H., Kim S.U., and Snyder E.Y. (1998). Engraftable human neural stem cells respond to developmental cues, replace neurons, and express foreign genes. *Nat. Biotechnol.* 16(11), 1033–1039.

Follenzi A. and Naldini L. (2002a). HIV-based vectors. Preparation and use. *Methods Mol. Med.* 69, 259–274.

Follenzi A. and Naldini L. (2002b). Generation of HIV-1 derived lentiviral vectors. *Methods Enzymol.* 346, 454–465.

Follenzi A., Battaglia M., Lombardo A., Annoni A., Roncarolo M.G., and Naldini L. (2004). Targeting lentiviral vector expression to hepatocytes limits transgene-specific immune response and establishes long-term expression of human antihemophilic factor IX in mice. *Blood* 103(10), 3700–3709.

Follenzi A., Sabatino G., Lombardo A., Boccaccio C., and Naldini L. (2002). Efficient gene delivery and targeted expression to hepatocytes in vivo by improved lentiviral vectors. *Hum. Gene Ther.* 13(2), 243–260.

Fricker R.A., Carpenter M.K., Winkler C., Greco C., Gates M.A., and Bjorklund A. (1999). Site-specific migration and neuronal differentiation of human neural progenitor cells after transplantation in the adult rat brain. *J. Neurosci.* 19(14), 5990–6005.

Gage F.H., Ray J., and Fisher L.J. (1995). Isolation, characterization, and use of stem cells from the CNS. *Annu. Rev. Neurosci.* 18, 159–192.

Gao C., Sands M.S., Haskins M.E., and Parker Ponder K. (2000). Delivery of retroviral vector expressing human beta-glucuronidase to the liver and spleen decreases lysosomal storage in mucopolysaccharidosis VII mice. *Mol. Ther.* 2, 233–244.

Georgievska B., Kirik D., Rosenblad C., Lundberg C., and Bjorklund A. (2002). Neuroprotection in the rat Parkinson model by intrastriatal GDNF gene transfer using a lentiviral vector. *Neuroreport* 13, 75–82.

German D.C., Liang C.L., Song T., Yazdani U., Xie C., and Dietschy J.M. (2002). Neurodegeneration in the Niemann-Pick C mouse: Glial involvement. *Neuroscience* 109(3), 437–450.

Glorioso J.C., Mata M., and Fink D.J. (2003). Therapeutic gene transfer to the nervous system using viral vectors. *J. Neurovirol.* 9, 165–172.

Griffey M., Bible E., Vogler C., Levy B., Gupta P., Cooper J., and Sands M. (2004). Adeno-associated virus 2-mediated gene therapy decreases autofluorescent storage material and increases brain mass in a murine model of infantiole neuronal ceroid lipofuscinosis. *Neurobiol. Dis.* 16(2), 360–369.

Guenechea G., Gan O.I., Dorrell C., and Dick J.E. (2001). Distinct classes of human stem cells that differ in proliferative and self-renewal potential. *Nat. Immunol.* 2, 75–82.

Hartung S.D., Frandsen J.L., Pan D., Koniar B.L., Graupman P., Gunther R., Low W.C., Whitley C.B., and McIvor R.S. (2004). Correction of metabolic, craniofacial, and neurologic abnormalities in MPSI mice treated at birth with adeno-associated virus vector transducing the human alpha-L-iduronidase gene. *Mol. Ther.* 9(6), 866–875.

Haskell R.E., Hughes S.M., Chiorini J.A., Alisky J.M., and Davidson B.L. (2003). Viral-mediated delivery of the late-infantile neuronal ceroid lipofuscinosis gene, TPP-I to the mouse central nervous system. *Gene Ther.* 10(1), 34–42.

Hess B., Saftig P., Hartmann D., Coenen R., Lullmann-Rauch R., Goebel H.H., Evers M., von Figura K., D'Hooge R., Nagels G., De Deyn P., Peters C., and Gieselmann V. (1996). Phenotype of arylsulfatase A-deficient mice: Relationships to human metachromatic leukodystrophy. *Proc. Natl. Acad. Sci. U.S.A.* 93, 14821–14826.

Hofling A.A., Devine S., Vogler C., and Sands M.S. (2004). Human CD34+ hematopoietic progenitor cell-directed lentiviral-mediated gene therapy in a xenotransplantation model of lysosomal storage disease. *Mol. Ther.* 9(6), 856–865.

Kakkis E., Lester T., Yang R., Tanaka C., Anand V., Lemontt J., Peinovich M., and Passage M. (2004). Successful induction of immune tolerance to enzyme replacement therapy in canine mucopolysaccharidosis I. *Proc. Natl. Acad. Sci. U.S.A.* 101(3), 829–834.

Kakkis E.D., Muenzer J., Tiller G.E., Waber L., Belmont J., Passage M., Izykowski B., Phillips J., Doroshow R., Walot I., Hoft R., and Neufeld E.F. (2001). Enzyme-replacement therapy in mucopolysaccharidosis I. *N. Engl. J. Med.* 344, 182–188.

Kay M.A., Glorioso J.C., and Naldini L. (2001). Viral vectors for gene therapy: the art of turning infectious agents into vehicles of therapeutics. *Nat. Med.* 7, 33–40.

Kennedy D.W. and Abkowitz J.L. (1997). Kinetics of central nervous system microglial and macrophage engraftment. Analysis using a transgenic bone marrow transplanttation model. *Blood* 90, 986–993.

Kim J.H., Auerbach J.M., Rodriguez-Gomez J.A., Velasco I., Gavin D., Lumelsky N., Lee S.H., Nguyen J., Sanchez-Pernaute R., Bankiewicz K., and McKay R. (2002). Dopamine neurons derived from embryonic stem cells function in an animal model of Parkinson's disease. *Nature* 418, 50–56.

Kordower J.H., Bloch J., Ma S.Y., Chu Y., Palfi S., Roitberg B.Z., Emborg M., Hantraye P., Deglon N., and Aebischer P. (1999). Lentiviral gene transfer to the nonhuman primate brain. *Exp. Neurol.* 160, 1–16.

Kosuga M., Takahashi S., Tanabe A., Fujino M., Li X.K., Suzuki S., Yamada M., Kakishita K., Ono F., Sakuragawa N., and Okuyama T. (2001). Widespread distribution of adenovirus-transduced monkey amniotic epithelial cells after local intracerebral injection: Implication for cell-mediated therapy for lysosome storage disorders. *Cell Transplant.* 10(4–5), 435–439.

Krivit W., Aubourg P., Shapiro E., and Peters C. (1999). Bone marrow transplantation for globoid cell leukodystrophy, adrenoleukodystrophy, metachromatic leukodystrophy, and Hurler syndrome. *Curr. Opin. Hematol.* 6, 377–382.

Krouwer H.G. and Wijdicks E.F. (2003). Neurologic complications of bone marrow transplantation. *Neurol. Clin.* 21(1), 319–352.

Leiming T., Mann L., del Pilar Martin M., Bonten E., Persons D., Knowles J., Allay J.A., Cunningham J., Nienhuis A.W., Smeyne R., and d'Azzo A. (2002). Functional amelioration of murine galactosialidosis by genetically modified bone marrow hematopoietic progenitor cells. *Blood* 99(9), 3169–3178.

Luca T., Givogri M.I., Perani L., Galbiati F., Follenzi A., Naldini L., and Bongarzone E.R. (2005). Axons mediate the distribution of arylsulfatase A within the mouse hippocampus upon gene delivery. *Mol. Ther.* 12(4), 669–679.

Mango R.L., Xu L., Sands M.S., Vogler C., Seiler G., Schwarz T., Haskins M.E., and Parker Ponder K. (2004). Neonatal retroviral vector-mediated hepatic gene therapy reduces bone, joint, and cartilage disease in mucopolysaccharidosis VII mice and dogs, *Mol. Genetics and Metab.* 82, 4–19.

Matzner U., Harzer K., Learish R.D., Barranger J.A., and Gieselmann V. (2000). Long-term expression and transfer of arylsulfatase A into brain of arylsulfatase A-deficient mice transplanted with bone marrow expressing the arylsulfatase A cDNA from a retroviral vector. *Gene Ther.* 7, 1250–1257.

Matzner U., Schestag F., Hartmann D., Lullmann-Rauch R., D'Hooge R., De Deyn P.P., and Gieselmann V. (2001). Bone marrow stem cell gene therapy of arylsulfatase A-deficient mice, using an arylsulfatase A mutant that is hypersecreted from retrovirally transduced donor-type cells. *Hum. Gene Ther.* 12, 1021–1033.

McCormack J.E., Edwards W., Sensintaffer J., Lillegren L., Kozloski M., Brumm D., Karavodin L., Jolly D.J., and Greengard J. (2001). Factors affecting long-term expression of a secreted transgene product after intravenous administration of a retroviral vector. *Mol. Ther.* 3(4), 516–525.

McKay R. (1997). Stem cells in the central nervous system. *Science* 276(5309), 66–71.

Meng X.L., Shen J.S., Ohashi T., Maeda H., Kim S.U., and Eto Y. (2003). Brain transplantation of genetically engineered human neural stem cells globally corrects brain lesions in the mucopolysaccharidosis type VII mouse. *J. Neurosi. Res.* 74(2), 266–277.

Mezey E., Key S., Vogelsang G., Szalayova I., Lange G.D., and Crain B. (2003). Transplanted bone marrow generates new neurons in human brains. *Proc. Natl. Acad. Sci. U.S.A.* 100(3), 1364–1369.

Mingozzi F., Liu Y.L., Dobrzynski E., Kaufhold A., Liu J.H., Wang Y., Arruda V.R., High K.A., and Herzog R.W. (2003). Induction of immune tolerance to coagulation factor IX antigen by in vivo hepatic gene transfer. *J. Clin. Invest.* 111, 1347–1356.

Miyoshi H., Blomer U., Takahashi M., Gage F.H., and Verma I.M. (1998). Development of a self-inactivating lentivirus vector. *J. Virol.* 72(10), 8150–8157.

Miyoshi H., Smith K.A., Mosier D.E., Verma I.M., and Torbett B.E. (1999). Transduction of human CD34+ cells that mediate long-term engraftment of NOD/SCID mice by HIV vectors. *Science* 283, 682–686.

Mount J.D., Herzog R.W., Tillson D.M., Goodman S.A., Robinson N., McCleland M.L., Bellinger D., Nichols T.C., Arruda V.R., Lothrop C.D. Jr., and High K.A. (2002). Sustained phenotypic correction of hemophilia B dogs with factor IX null mutation by liver-directed gene therapy. *Blood* 99, 2670–2676.

Naldini L. (1998). Lentiviruses as gene transfer agents for delivery to non-dividing cells. *Curr. Opin. Biotechnol.* 9(5), 457–463.

Naldini L., Blomer U., Gage F.H., Trono D., and Verma I.M. (1996). Efficient transfer, integration, and sustained long-term expression of the transgene in adult rat brains injected with a lentiviral vector. *Proc. Natl. Acad. Sci. U.S.A.* 93, 11382–11388.

Neufeld E.F. (1991). Lysosomal disease. *Annu. Rev. Biochem.* 60, 257–280.

Nguyen T.H., Oberholzer J., Birraux J., Majno P., Morel P., and Trono D. (2002). Highly efficient lentiviral vector-mediated transduction of nondividing, fully reimplantable primary hepatocytes. *Mol. Ther.* 6(2), 199–209.

Ohmi K., Greenberg D.S., Rajavel K.S., Ryazantsev S., Li H.H., and Neufeld E.F. (2003). Activated microglia in cortex of mouse models of mucopolysaccharidoses I and IIIB. *Proc. Natl. Acad. Sci. U.S.A.* 100, 1902–1907.

Ourednik V., Ourednik J., Flax J.D., Zawada W.M., Hutt C., Yang C., Park K.I., Kim S.U., Sidman R.L., Freed C.R., and Snyder E.Y. (2001). Segregation of human neural stem cells in the developing primate forebrain. *Science* 293(5536), 1820–1824.

Parker Ponder K., Melniczek J.R., Xu L., Weil M.A., O'Malley T.M., O'Donnel P.A., Knox V.W., Aguirre G.D., Mazrier H., Ellinwood N.M., Sleeper M., Maguire A.M., Volk S.W., Mango R.L., Zweigle J., Wolfe J.H., and Haskins M.E. (2002). Therapeutical neonatal hepatic gene therapy in mucopolisaccharidosis VII dogs. *Proc. Natl. Acad. Sci. U.S.A.* 99(20), 13102–13107.

Peters C. and Steward C.G. (2003). Hematopoietic cell transplantation for inherited metabolic diseases: An overview of outcomes and practice guidelines. *Bone Marrow Transplant.* 31, 229–239.

Pluchino S., Quattrini A., Brambilla E., Gritti A., Salani G., Dina G., Galli R., Del Carro U., Amadio S., Bergami A., Furlan R., Comi G., Vescovi A.L., and Martino G. (2003). Injection of adult neurospheres induces recovery in a chronic model of multiple sclerosis. *Nature* 422(6933), 688–694.

Priller J., Flugel A., Wehner T., Boentert M., Haas C.A., Prinz M., Fernandez-Klett F., Prass K., Bechmann I., de Boer B.A., Frotscher M., Kreutzberg G.W., Persons D.A., and Dirnagl U. (2001). Targeting gene-modified hematopoietic cells to the central nervous system: Use of green fluorescent protein uncovers microglial engraftment. *Nat. Med.* 7, 1356–1361.

Reddy P.S., Sakhuja K., Ganesh S., Yang L., Kayda D., Brann T., Pattison S., Golightly D., Idamakanti N., Pinkstaff A., Kaloss M., Barjot C., Chamberlain J.S., Kaleko M., and Connelly S. (2002). Sustained human factor VIII expression in hemophilia A mice following systemic delivery of a gutless adenoviral vector. *Mol. Ther.* 5(1), 63–73.

Ross C.J., Ralph M., and Chang P.L. (2000). Somatic gene therapy for a neurodegenerative disease using microencapsulated recombinant cells. *Exp. Neurol.* 166(2), 276–286.

Rubio F.J., Bueno C., Villa A., Navarro B., and Martinez-Serrano A. (2000). Genetically perpetuated human neural stem cells engraft and differentiate into the adult mammalian brain. *Mol. Cell. Neurosci.* 16(1), 1–13.

Schiffmann R., Kopp J.B., Austin H.A., Sabnis S., Moore D.F., Weibel T., Balow J.E., and Brady R.O. (2001). Enzyme replacement therapy in Fabry disease: A randomized controlled trial. *J. Am. Med. Assoc.* 285, 2743–2749.

Schroder A., Shinn P., Chen H., Berry C., Ecker J., and Bushman F. (2002). HIV-1 integration in the human genome favors active genes and local hotspots. *Cell.* 110, 521.

Sferra T.J., Backstrom K., Chuansong W., Rennard R., Miller M., and Hu Y. (2004). Widespread correction of lysosomal storage following intrahepatic injection of a recombinant adeno-associated virus in the adult MPS VII mouse. *Mol. Ther.* 10(3), 478–491.

Skorupa A.F., Fisher K.J., Wilson J.M., Parente M.K., and Wolfe J.H. (1999). Sustained production of beta-glucuronidase from localized sites after AAV vector gene transfer results in widespread distribution of enzyme and reversal of lysosomal storage lesions in a large volume of brain in mucopolysaccharidosis VII mice. *Exp. Neurol.* 160(1), 17–27.

Sly W.S. (1993). Gene therapy on the Sly. *Nat Genet.* 4(2),105–106.

Sly W.S., Vogler C., Grubb J.H., Zhou M., Jiang J., Zhou X.Y., Tomatsu S., Bi Y., and Snella E.M. (2001). Active site mutant transgene confers tolerance to human beta-glucuronidase without affecting the phenotype of MPSVII mice. *Proc. Natl. Acad. Sci. U.S.A.* 98, 3311–3312.

Snyder E.Y. (1995). Immortalized neural stem cells: insights into development; prospects for gene therapy and repair. *Proc. Natl. Acad. Sci. U.S.A.* 107(2), 195–204.

Solaro C., Nurialdo A., Giunti D., Mancardi G., and Uccelli A. (2001). Central and peripheral nervous system complications following allogeneic bone marrow transplantation. *Europ. J. Neurol.* 8, 77–80.

Stein C.S., Ghodsi A., Dersen T., and Davidson B.L. (1999). Systemic and central nervous system correction of lysosomal storage in mucopolysaccharidosis type VII mice. *J. Virol.* 73, 3424–3429.

Stein C.S., Kang Y., Sauter S.L., Townsend K., Staber P., Derksen T.A., Martins I., Qian J., Davidson B.L., and McCray P.B. Jr. (2001). In vivo treatment of hemophilia A and mucopolysaccharidosis type VII using nonprimate lentiviral vectors. *Mol. Ther.* 3(6), 850–856.

Tamaki S., Eckert K., He D., Sutton R., Doshe M., Jain G., Tushinski R., Reitsma M., Harris B., Tsukamoto A., Gage F., Weissman I., and Uchida N. (2002). Engraftment of sorted/expanded human central nervous system stem cells from fetal brain. *J. Neurosci. Res.* 69(6), 976–986.

Taylor R.M. and Wolfe J.H. (1997). Decreased lysosomal storage in the adult MPS VII mouse brain in the vicinity of grafts of retroviral vector-corrected fibroblasts secreting high levels of beta-glucuronidase. *Nat. Med.* 3(7), 771–774.

Unger, E.R., Sung J.H., Manivel J.C., Chenggis M.L., Blazar B.R., and Krivit W. (1993). Male donor-derived cells in the brains of female sex-mismatched bone marrow transplant recipients: a Y-chromosome specific in situ hybridization study. *J. Neuropathol. Exp. Neurol.* 52(5), 460–470.

Van den Hout J.M., Kamphoven J.H., Winkel L.P., Arts W.F., De Klerk J.B., Loonen M.C., Vulto A.G., Cromme-Dijkhuis A., Weisglas-Kuperus N., Hop W., Van Hirtum H., Van Diggelen O.P., Boer M., Kroos M.A., Van Doorn P.A., Van der Voort E., Sibbles B., Van Corven E.J., Brakenhof J.P., Van Hove J., Smeitink J.A., de Jong G., Reuser A.J., and Van der Ploeg A.T. (2004). Long-term intravenous treatment of Pompe disease with recombinant human alpha-glucosidase from milk. *Pediatrics* 113(5), 448–457.

Wada R., Tifft C.J., and Proia R.L. (2000). Microglial activation precedes acute neurodegeneration in Sandhoff disease and is suppressed by bone marrow transplantation. *Proc. Natl. Acad. Sci. U.S.A.* 97(20), 10954–10959.

Walkley S.U., Thrall M.A., Dobreins K., Huang M., March P.A., Siegel D.A., and Wurzelmann S. (1994). Bone marrow transplantation corrects the enzyme defect in neurons of the central nervous system in a lysosomal storage disease. *Proc. Natl. Acad. Sci. U.S.A.* 91, 2970–2974.

Watson G.L., Sayles J.N., Chen C., Elliger S.S., Elliger C.A., Raju N.R., Kurtzman G.J., and Podsakoff G.M. (1998). Treatment of lysosomal storage disease in MPS VII mice using a recombinant adeno-associated virus. *Gene Ther.* 5(12), 1642–1649.

Weimann J.M., Charlton C.A., Brazelton T.R., Hackman R.C., and Blau H.M. (2003). Contribution of transplanted bone marrow cells to Purkinje neurons in human adult brains. *Proc. Natl. Acad. Sci. U.S.A.* 100(4), 2088–2093.

Wichterle H., Lieberam I., Porter J.A., and Jessell T.M. (2002). Directed differentiation of embryonic stem cells into motor neurons. *Cell* 110, 385–397.

Wu P., Tarasenko Y.I., Gu Y., Huang L.Y., Coggeshall R.E., and Yu Y. (2002). Region-specific generation of cholinergic neurons from fetal human neural stem cells grafted in adult rat. *Nat. Neurosci.* 5(12), 1271–1278.

Wu X., Li Y., Crise B., and Burgess S.M. (2003). Transcription start regions in the human genome are favored targets for MLV integration. *Science* 300, 1749–1751.

Yeager A.M., Shinn C., Shinohara M., and Pardoll D.M. (1993). Hematopoietic cell transplantation in the twitcher mouse. The effects of pretransplant conditioning with graded doses of busulfan. *Transplantation* 56(1), 185–189.

Yeager A.M., Shinohara M., and Shinn C. (1991). Hematopoietic cell transplantation after administration of high-dose busulfan in murine globoid cell leukodystrophy (the twitcher mouse). *Pediatr. Res.* 29(3), 302–305.

Zheng Y., Rozengurt N., Ryazantsev S., Kohn D.B., Satyake N., and Neufeld E.F. (2003). Treatment of the mouse model of mucopolysaccharidosis I with retrovirally tyransduced bone marrow. *Mol. Genetics and Metab.* 79, 233–244.

Zufferey R., Dull T., Mandel R.J., Bukovsky A., Quiroz D., Naldini L., and Trono D. (1998). Self-inactivating lentivirus vector for safe and efficient in vivo gene delivery. *J Virol.* 72(12), 9873–9880.

Zufferey R., Nagy D., Mandel R.J., Naldini L., and Trono D. (1997). Multiply attenuated lentiviral vector achieves efficient gene delivery in vivo. *Nat. Biotechnol.* 15(9), 871–875.

SUBSTRATE REDUCTION THERAPY

Frances M. Platt and Terry D. Butters

1 INTRODUCTION

1.1 Glycosphingolipid Storage Diseases

The glycosphingolipidoses are a family of storage diseases that arise due to incomplete catabolism of glycosphingolipids (GSLs) in the lysosome (Wraith, 2002). The majority are autosomal recessive disorders and result from mutations in the genes that encode the catabolic enzymes of the lysosome (Winchester, 2004). Clinically they are highly variable (Beck, 2001) but typically have a neurodegenerative course and commonly present in infancy or early childhood (Wraith, 2004). Adult-onset variants also occur (Rapola, 1994; Wraith, 2004). The age of onset is influenced by the residual enzyme activity present, that in turn reflects the impact a specific mutation has on the properties of the enzyme (Rapola, 1994; Winchester, 2004). Little or no activity leads to rapid storage and early onset of symptoms whereas higher levels of residual activity lead to a slower rate of storage and a longer presymptomatic period. In this chapter, we focus on a drug-based therapy that is relevant to all lysosomal diseases involving the storage of glucosylceramide-derived GSLs, including Gaucher, Fabry, Tay–Sachs, Sandhoff, and GM1 gangliosidosis. In addition, storage diseases involving the secondary storage of GSLs (Walkley, 2004; e.g., Niemann–Pick type C disease, NPC) in which GSLs are implicated in the pathology, may also benefit from this approach (Table 1).

1.2 Therapeutic Intervention: Potential Strategies

Until recently, the main approach being used for clinical intervention in storage diseases was enzyme replacement therapy (ERT; Neufeld, 2004, Chapter 8, this book). This utilizes wild-type enzyme administered intravenously at regular intervals for the lifetime of the patient. This has been highly successful in the treatment of one of these disorders, type 1 Gaucher disease and more recently has also been evaluated in Fabry disease (Brady, 2003). This approach has one major limitation in that glycoprotein enzymes do not cross the blood–brain barrier and therefore this form of therapy has no impact on the majority of these disorders that have pathology in the brain. Enzyme-augmenting approaches are therefore required that can deliver enzyme to the brain and these include bone marrow transplantation (BMT; Krivit, 2002, Ringden et al., 1995), gene therapy (Cabrera-Salazar, Novelli, & Barranger, 2002; Sands, 2004), and neural stem cell therapy (Snyder, Daley, & Goodell, 2004; Chapter 14, this book). BMT is limited by availability of matched

Oxford Glycobiology Institute, Department of Biochemistry, Oxford University, South Parks Road, Oxford, OX1 3QU, UK. e-mail: Frances.platt@pharm.ox.ac.uk

donors and the mortality/morbidity associated with the procedure. It is also a relatively inefficient delivery system due to the limited numbers of donor-derived microglial cells that are present in the brain posttransplantation (Krivit, Sung, Shapiro, & Lockman, 1995). Gene therapy and stem cell therapy remain experimental (Dobrenis, 2004).

Table 1. Diseases with primary or secondary storage of GSL that are candidates for SRT upon preclinical or clinical dataset (see text).

Diseases with Primary GSL Storage
Gaucher types 1,2 and 3
Fabry
GM2 gangiosidoses (Tay–Sachs and Sandhoff)
GM1 gangliosidosis
Disease with Secondary GSL Storage
Niemann–Pick type C

About two decades ago an alternative approach was suggested by Radin (Vunnam & Radin, 1980). He proposed the use of an inhibitor of GSL biosynthesis as a means of treating type 1 Gaucher disease (Radin, 1996). This approach will be referred to in this chapter as substrate reduction therapy (SRT; F. M. Platt & Butters, 2004; F.M. Platt et al., 2003). The principle is that a small molecule drug could be used to partially inhibit GSL biosynthesis reducing the number of GSL molecules requiring catabolism in the lysosome. The aim is to balance the rate of synthesis with the impaired rate of catabolism. This approach has a number of potential advantages relative to biotherapies, including oral availability, the use of a single drug to treat several diseases, and the possibility of being able to affect storage in the brain (F. M. Platt & Butters, 2004). The diseases that could potentially benefit from this approach are listed in Table 1.

2 SUBSTRATE REDUCTION THERAPY (SRT)

In a diseased cell, the mutant enzyme is impaired in its ability to hydrolyse a lysosomal GSL(s) and the constant rate of GSL influx increases the GSL concentration above the optimal catalytic efficiency of the enzyme, thus resulting in storage. By reducing the rate of GSL synthesis, the rate of influx of GSLs into the lysosome is decreased, slowing the rate of storage.

Therapeutic intervention at an early step of the pathway of glycolipid biogenesis has been achieved using small molecule inhibitors (imino sugars) that have enzymatic selectivity for the key step in the biosynthesis of all glucosphingolipids in the Golgi apparatus (Butters, Dwek, & Platt, 2003; Butters, Mellor, Narita, Dwek, & Platt, 2003; Dwek, Butters, Platt, & Zitzmann, 2002; see Figure 1). This step is the conversion of ceramide to glucosylceramide catalysed by ceramide glucosyltransferase (CGT). These imino sugar inhibitors are stereochemical mimics for the 6-membered pyranose sugars where a nitrogen atom replaces the ring oxygen. This structural similarity provides imino sugars with potent inhibitory activity towards glycosidases specific for monosaccharide hydrolysis. The important discovery that imino sugars with glucose and galactose stereochemistry containing an alkyl, alkenyl, or aryl group attached to the nitrogen (see Figure 1), also inhibit ceramide glucosyltransferase (Figure 1) with low micromolar potency (Butters, van den Broek et al., 2000), allowed clinical exploitation of these compounds for treating GSL storage disorders (Butters, Dwek, & Platt, 2000). The mechanism for this inhibitory property has not been fully determined but is best explained by a structural similarity between imino sugars and ceramide (Butters, Mellor et al., 2003).

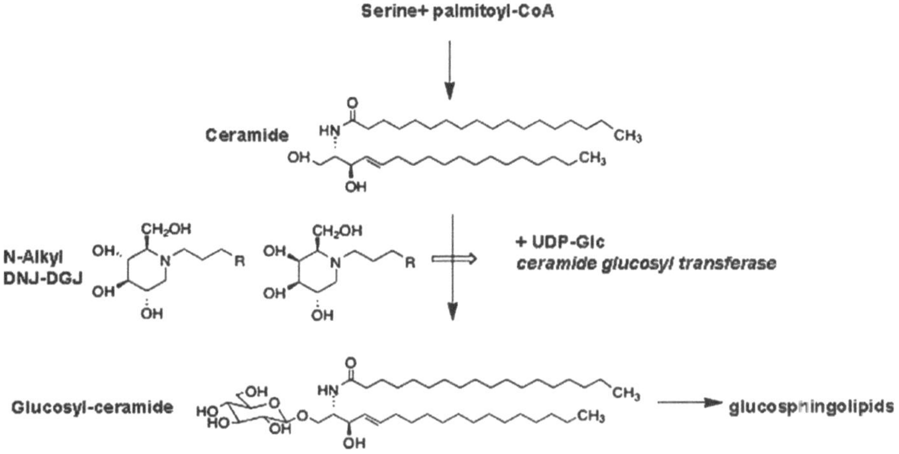

Figure 1. Key steps of the glyscosphingolipid biosynthesis pathway. Ceramide glucosyl transferase (CGT) transfers glucose from UDP glucose to ceramide to form glucosylceramide, the intermediate required for further synthesis of glycosphingolipids. This step is inhibited by N-alkyl imino-sugars and N-alkyd DGJ. Shown above. R = H, N-butyl-DNJ(DGJ); R= C5H11, N-nonyl-DNJ(DGJ).

The other class of compounds that inhibit the same step in the pathway are the morpholino and pyrrolidino derivatives (PDMP series) that structurally resemble ceramide and were developed by Radin and colleagues. In this review, we focus on the imino sugars as *N*-butyldeoxynojirimycin (*N*B-DNJ, miglustat or Zavesca) is approved and in clinical use. The current status of the development of the PDMP series compounds has recently been reviewed in detail elsewhere (F. M. Platt & Butters, 2004).

2.1 Structure/Function Relationships of CGT Inhibitors

The orientation of the hydroxyl groups in both 6- and 5-membered imino sugars critically affects inhibitory potency for CGT. The 6-membered piperidines, deoxynojirimycins (DNJ), and deoxygalactonojirimycin (DGJ) contain 3 chiral centres (C2, 3, and 4) with the correct stereochemistry that is required for inhibition (Butters, van den Broek et al., 2000). Configurational inversion at C5 to generate the idose or altrose isomers of DNJ and DGJ respectively, retain inhibitory potency.

It was predicted from modelling studies of *N*-butyl-DNJ (*N*B-DNJ) and ceramide that better mimicry of the alkyl chain would improve inhibitory potency. This was found to be correct when a series of N-alk(en)yl-DNJ compounds was synthesized and assayed for in vitro inhibitory activity. Extending the N-alkyl chain to 18 carbon atoms increased potency tenfold (Mellor et al., 2002) and a further fivefold increase in efficacy, as assessed by the reduction of GSLs, was demonstrated in cellular experiments (Mellor et al., 2004). A computational model of CGT confirmed the presence of a large hydrophobic groove that would accommodate longer alkyl chain compounds and could mimic the conformation of the N-acyl chain of ceramide bound in the groove as the enzyme directs substrate to aid glycosyl transfer (Butters, Mellor et al., 2003).

2.2 Cell and Tissue Penetration of Imino Sugars

The entry of imino sugars into cells appears to be by passive, nonfacilitated diffusion, or flip-flop across the membrane (Mellor et al., 2002). CGT is associated with Golgi membrane with its active site facing the cytosol. Experiments in tissue culture cells indicate that imino sugars are able to rapidly inhibit glycosyl transfer, in less than 1 min. (Mellor et al., 2004). The rate of entry is independent of N-alkyl chain length but increasing lipophilicity results in more protein and membrane binding (Mellor et al., 2002). It is therefore likely that the improved cellular potency of longer alkyl chain compounds reflects the increased deposition of compound in membranes close to the site of ceramide glucosylation.

In vivo studies have also revealed a correlation between increasing N-alkyl-imino sugar hydrophobicity and tissue access and sequestration. Using radiolabelled compounds (Mellor et al., 2002) administered by gavage to mice, slowed penetration to the liver and brain from the gut was observed as the N-alkyl chain was increased in length. The proportion of N-nonyl-DNJ containing a 9-carbon chain compound was 15-fold greater in liver and brain than the 4-carbon chain compound, *N*B-DNJ 90 min post gavage. Because hydrophobic compounds in general cross the blood–brain barrier more effectively this might be expected, but the relative proportions of imino sugars gaining access from plasma are restricted. Approximately 25% of the serum concentration was found in brain tissue from a mouse model for Sandhoff disease fed a diet of either *N*B-DNJ or N-butyl-DGJ (*N*B-DGJ; Andersson, Smith et al., 2004). The level of *N*B-DNJ found in the brain of Sandhoff mice was 25% higher than the wild-type littermate mice indicating that the blood–brain barrier is compromised in this disease (Jeyakumar et al., 2003).

These factors, if extrapolated to humans offer considerable potential benefit for treating the neuronopathic storage disorders where enzyme replacement therapy is of marginal efficacy due to lack of blood–brain barrier access. However, compound selection to provide a better therapeutic outcome is critical because of the many other properties that N-alkylated imino sugars display (Andersson, Butters, Dwek, & Platt, 2000).

2.3 Toxicity of GSL Synthesis Inhibitors

2.3.1 Reversibility of CGT Inhibition

We understand very little about the long-term effects of reducing GSL levels in cells and tissues because of our less than precise knowledge of the biological roles played by these molecules. N-alkylated imino sugar inhibitors have relatively high affinity for CGT, reach their site of action rapidly, and can accumulate in tissues depending on hydrophobicity. All of these properties are reversible. In vitro administration of *N*B-DNJ (a hydrophilic compound) to cells to inhibit CGT completely, followed by removal of compound, led to a full restoration of GSL levels after 24 h. By contrast, the more hydrophobic compounds, because of greater cellular partitioning, took much longer, an important consideration when selecting compounds for therapy. A small molecule that is rapidly cleared by the body is much easier to regulate than one where residency may be tissue- and time-dependent.

Further support for the lack of long-term effects following administration of N-alkylated imino sugars to rodents is the absence of up-regulation of CGT that might increase GSL synthesis after compound removal (Butters, Mellor et al., 2003) and the return of fertility to male mice after *N*B-DNJ withdrawal (van der Spoel et al., 2002). No overt behavioral changes have been reported in mice treated with these drugs, however, more sophisticated testing will be necessary to determine whether there are any more subtle consequences of long-term GSL depletion in the mouse.

2.3.2 Cytotoxicity of N-Alkylated Imino Sugars

Imino sugars are not metabolized and are excreted intact, primarily by the kidney. Cytotoxicity might be expected to occur given that increased hydrophobicity leads to increased tissue levels. However, cellular studies have revealed that the compounds with greatest therapeutic value at present, *N*B-DNJ and *N*B-DGJ, have extremely high CC_{50} values (concentration at which 50% of cells become nonviable) and are in the mM range (Durantel et al., 2001; Mellor et al., 2002). As predicted, the more hydrophobic compounds are more cytotoxic and have lower CC_{50} values, again an important factor in therapeutic compound design. The toxicity associated with long chain N-alkylated imino sugars in cells is unrelated to inhibition of GSL biosynthesis, the generation of ceramide or any detergentlike effects. The major cause of cellular toxicity with these amphiphilic compounds appears to be cell lysis following increased membrane insertion and pore formation (Mellor, Platt, Dwek, & Butters, 2003).

2.3.3 Side Effects

The inhibitory activity of imino sugars for glycosidases has led to their evaluation as anti-viral drugs (Block et al., 1998; Durantel et al., 2001; Fischer et al., 1995; Zitzmann et al., 1999). This is based on inhibition of glycoprotein processing α-glucosidases in the endoplasmic reticulum (ER) that can lead to misfolding of some proteins. Certain viruses have glycoproteins that are exquisitely sensitive to folding events in the ER and can be rendered noninfective by inhibiting *N*-glycan processing. To obtain global effects on host cell biosynthesis of glycoproteins, concentrations 1000–10,000-fold in excess of that which inhibits CGT is required. In vivo data support the observation that NB-DNJ is a poor α-glucosidase inhibitor; because of its restricted access to the ER (Butters, Mellor et al., 2003).

Other glucosidases, such as the gastrointestinal tract enzyme sucrase/isomaltase and the liver and muscle glycogenolytic enzymes (Andersson, Reinkensmeier, Butters, Dwek, & Platt, 2004) have also been shown to be sensitive to inhibition by *N*B-DNJ. In humans, the former appears to be easily controlled (R. H. Lachmann, 2003) and the latter in mouse is without an apparent physiological consequence (Andersson, Reinkensmeier et al., 2004).

Lack of galactosylceramide synthesis inhibition by both *N*B-DNJ and *N*B-DGJ means these drugs do not compromise myelin stability in neural tissue (Butters, Mellor et al., 2003).

3 EFFICACY OF SRT IN MOUSE MODELS OF GSL STORAGE DISEASES

Having established the potential of these compounds to inhibit the GLS biosynthetic pathway, proof of principle of efficacy in a cell culture model of a GSL storage disease was performed. Imino sugar treatment prevented storage in a cell-based system exposed to the irreversible inhibitor of glucocerebrosidase conduritol β epoxide, a model of Gaucher disease (F. M. Platt, Neises, Dwek, & Butters, 1994; F. M. Platt, Neises, Karlsson, Dwek, & Butters, 1994). There was therefore the prospect that these drugs could show clinical efficacy in humans. Proof of principle was required in animal models of these disorders before clinical trials could be initiated. As the majority of these diseases involve storage and pathology in the central nervous system (CNS; Wraith, 2004) it was essential to determine whether the imino sugars could cross the blood–brain barrier to a great enough extent to reduce the rate of storage. Ultimately, the objective was to establish whether there was clinical benefit resulting from imino sugar treatment.

3.1 SRT in the Tay–Sachs Disease Mouse

There are a number of well-characterised authentic lysosomal storage disease models available (Hopwood, Crawley, & Taylor, 2004) in which experimental therapies can be evaluated. However, several of the GSL storage disease models are in large-animal species that are impractical for evaluating SRT in vivo. Quite fortuitously the discovery that some imino sugar drugs inhibit GSL biosynthesis coincided with the first report of a knockout mouse model of one of these diseases, the Tay–Sachs mouse (Yamanaka et al., 1994). Tay–Sachs disease results from mutations in the hexa gene that encodes the αsubunit of β-hexosaminidase (causing deficiency in the HexA and HexS isoenzymes) leading to the storage of GM2 ganglioside. The mouse exhibits significant levels of storage in the brain, although it never develops the clinical signs of the disease (Yamanaka et al., 1994). This is the result of a bypass pathway particularly active in the mouse relative to human that is mediated by sialidase that converts GM2 into GA2 that can then be catabolised by HexB (ββ), that remains fully functional in this disorder.

This mouse model had the advantage that it has storage in the brain and so the effect of imino sugar administration could be evaluated, even in the absence of a clinical phenotype. The prototypic imino sugar drug for SRT, NB-DNJ, was therefore administered orally to Tay–Sachs mice from 4 weeks of age by incorporating the drug into powdered mouse chow (F. M. Platt et al., 1997). This allowed noninvasive long-term administration. The mice were sacrificed at 12 weeks of age and the degree of storage in the brain of treated mice compared with untreated littermates. Three assessments of GSL storage were made: (a) storage levels in whole brain measured by high-performance thin-layer chromatography, (b) histochemical visualisation of storage material using periodic acid-Schiff staining of brain sections, and (c) electron microscopy of storage neurons of the brain (F. M. Platt et al., 1997). The analyses revealed reduced biochemical storage burden in whole brain, reduced storage material in storage regions of the brain at a histological level, and reduced electron density of membranous cytoplasmic bodies in storage neurons examined by EM. Taken together these data provided proof of principle of SRT in vivo and demonstrated that sufficient drug crossed the blood–brain barrier to affect storage in the CNS.

3.2 SRT in the Sandhoff Disease Mouse

The next engineered mouse model to be reported was the β-subunit knockout, the Sandhoff disease mouse (Sango et al., 1995). The mouse lacked the β-subunit of β-hexosaminidase resulting in loss of the HexA (αβ) and Hex B (ββ) isoenzymes, with only very low level of residual enzyme activity conferred by Hex S (αα). This mouse model had a clinical phenotype that closely resembled the human presentation of infantile/juvenile Tay–Sachs and Sandhoff disease. This therefore served as an authentic model of the human disease in which SRT could be evaluated, in terms of its potential impact on clinical signs. Sandhoff mice were therefore fed on a diet containing *N*B-DNJ and noninvasively monitored for disease progression using behavioural tests of motor coordination and muscle strength (Jeyakumar et al., 1999). In addition, survival was recorded.

It was found that the presymptomatic period was extended in response to SRT, the rate of clinical decline was slowed, and life expectancy increased by approximately 40%. This study therefore suggested that if similar penetrance of the CNS by *N*B-DNJ was achievable in humans, clinical benefit could potentially be achieved. These results provide support for studies in lysosomal diseases with neurological involvement. Clinical trials in CNS diseases are problematic for a number of reasons: (a) small numbers of patients, (b) clinical heterogeneity, (c) lack of consensus on clinically relevant endpoints, and (d) no consensus on how long a trial would be needed to observe efficacy, if achievable. A clinical study in the neuronopathic disorders was therefore discounted in favour of a trial in type 1 Gaucher disease (nonneuronopathic) in which there was an effective therapy (ERT) with which to compare SRT and accepted clinically relevant endpoints.

4 CLINICAL EVALUATION OF SRT

4.1 Clinical Trials with NB-DNJ (Miglustat)

In 1998–1999 patients with nonneuronopathic Gaucher disease were recruited (Cambridge, Amsterdam, Prague, and Jerusalem) into a one-year open-label clinical trial of *N*B-DNJ (miglustat; T. Cox et al., 2000). Type 1 Gaucher disease is a macrophage disorder characterised by hepatosplenomegaly, anaemia, and bone disease (Beutler & Grabowski, 2001). All patients were unable or unwilling to receive ERT. Liver and spleen volumes were measured by MRI or computed tomography and haematological parameters assessed. Biochemical markers were also measured including chitotriosidase (Aerts & Hollak, 1997), cell surface leukocyte GM1 as an indicator of whether GSL levels were depleted in response to treatment, and the plasma levels of GlcCer, the storage lipid. Oral dosing was typically 100 mg OGT-918 three times daily. Individualised dosing in a small number of patients was in response to variation in the pharmacokinetics of the compound, tolerability, and organ volume response after six months of treatment.

4.2 Pharmacokinetics

Pharmacokinetic profiling in a subgroup of the 28 patients enrolled showed that the drug reached maximum plasma concentrations by 2.5 hours with a plasma half-life of 6.3 hours. Steady state concentrations of OGT-918 were achieved after 15 days of dosing and the mean peak level of OGT-918 over the 12-month study was 6.8 μM with trough values of 3.9 μM (T. Cox et al., 2000; Moyses, 2003).

4.3 Reasons for Patient Withdrawals

Of the 28 patients enrolled in the trial, six withdrew. Two were unable to tolerate the GI tract side effects (one suffered from Parkinson's disease and the other travelled extensively). The GI tract side effects are due to inhibition of sucrase/isomaltase by leading to osmotic diarrhoea. Two patients withdrew due to pre-existing medical conditions (hepatocellular carcinoma and pulmonary hypertension), one additional patient left to start a family, and another after one day on study. The remaining 22 patients were assessed at 6 and 12 months. Two further patients withdrew from an extension study because of symptoms of peripheral neuropathy. Currently, two of the original patients enrolled early on in the trial have been receiving continuous therapy for over six years.

4.4 Clinical Efficacy: 1-Year Study

Spleen and liver volumes showed a statistically significant reduction (of 15%, 11.8–18.4, $p < 0.001$ and 7%, 3.4–10.5, $p < 0.001$, respectively) after 6 months of therapy. By 12 months the decrease from baseline was 19% (14.3–23.7, $p < 0.001$) and 12% (7.8–16.4, $p < 0.001$), respectively (T. Cox et al., 2000). This was comparable to the response observed in patients of comparable disease severity at baseline receiving ERT (R.H. Lachmann & Platt, 2001). Chitotriosidase activity showed a time-dependent reduction, reflecting a reduction in the number of Gaucher cells within the patients treated with miglustat (T. Cox et al., 2000). Haemoglobin and platelet counts showed trends towards improvement, with a greater improvement in haemoglobin noted in patients who were anaemic at baseline. A statistically significant improvement in platelet counts was achieved following 12 months of treatment.

4.5 Extension Study: 12–36 Months

Longer-term efficacy and safety were evaluated in patients that had completed 12 months of therapy (Elstein et al., 2004). Eighteen patients of 22 that were eligible entered the extension phase and were followed for a further two years. Continued and increasing efficacy was observed in all clinical parameters measured. In particular, there was a marked improvement in platelet counts and haemoglobin levels relative to the 12-month time point. Bone marrow fat fraction measurements were made in two patients and the results showed an improvement in fat fraction from baseline by 12 months and an even greater improvement was noted in both individuals at 36 months. This indicates a reduction in the number of Gaucher cells in the bone marrow in response to therapy. In keeping with this observation, chitotriosidase levels continued to decline in the extension phase. No serious adverse events were reported and no further cases of peripheral neuropathy emerged in the extension phase. GI tract side effects persisted in these patients but to a lesser extent than in the first 12 months.

4.6 Low-Dose Clinical Study

The efficacy and safety of low-dose SRT was evaluated (Heitner, Elstein, Aerts, Weely, & Zimran, 2002). Eighteen patients with type 1 Gaucher disease were enrolled in two centres in an open-label 6-month study of OGT 918, 50 mg (half the dose used in the first trial) taken three times daily (TID), followed by an optional extended-use phase. Seventeen patients completed 6 months and 13 were evaluated at 12 months. Percentage reduction in liver (–5.9%, $p = 0.007$) and spleen (–4.5%, $p = 0.025$) volumes and in chitotriosidase levels (–4.6%, $p = 0.039$) at 6 months were lower than in the higher dose trial (T. Cox et al., 2000; T. M. Cox et al., 2003). At 12 months there were further mean decreases from baseline in liver volume (–6.2%, $p = 0.037$), spleen volume (–10.1%, $p < 0.05$), and chitotriosidase levels (–15.3%, $p < 0.05$) as well as mean changes of +1.2 and

+14.7% in hemoglobin and platelet concentrations, respectively. There were no serious adverse effects throughout the 6-month study period. There was no improvement in the rate of hematological response and no reduction in side effects at the lower dose. This study demonstrated the dose-dependent nature of the response to miglustat and suggested that 100 mg TID should be the preferred starting dose for patients with symptomatic type I Gaucher disease (Zimran & Elstein, 2003).

4.7 Baseline Neurological Assessment

Clinical studies in Gaucher disease have not previously included comprehensive baseline neurological and neuropsychological assessment of patients. In order to better understand the context of the two cases of peripheral neuropathy observed in the first miglustat clinical study, a 12-month study with an additional 12-month extension phase was conducted, focusing on the neurological and neuropsychological assessment of patients at baseline and throughout the study (Pasteurs, 2003).

Ten patients were enrolled and received miglustat 100 mg tid. For the seven patients who continued into the extension study, the mean percentage change in liver volume was –6% ($p = 0.084$) and in spleen volume was –15% ($p = 0.094$), which were consistent with results of earlier studies. The three discontinuations were due to paresthesiae (1), flatulence (1), and lost to follow-up (1). Neurological assessments, including nerve conduction velocity studies (NCV), accelerometry for the detection of tremor, and the International Cooperative Ataxia Rating Scale for the assessment of ataxia were carried out. Cognitive function and dexterity were tested using the Mini-Mental State Examination (MMSE) and Purdue Peg Board tests. There were no emergent cases of peripheral neuropathy or modification of a pre-existing peripheral neuropathy, and there was an absence of significant adverse effects on neurological and neuropsychological function. A symptom survey of Gaucher type 1 patients receiving ERT has shown that patients had a high prevalence of neurological complaints, including paresthesia, muscle weakness, and muscle stiffness (Pastores, Barnett, Bathan, & Kolodny, 2003). In order to address the limited knowledge regarding neurological adverse events in Gaucher type 1, a clinical study is underway to establish the natural history of disease.

4.8 Regulatory Approval

In 2002 the European regulatory authority (EMEA) approved *N*B-DNJ (miglustat, Zavesca®) for the treatment of type 1 Gaucher disease (mild to moderate disease, unwilling or unable to receive ERT; T. M. Cox et al., 2003; R. H. Lachmann, 2003). Miglustat has subsequently been approved in the United States, Israel, Canada, and Switzerland with similar labelling.

5 SRT and Niemann–Pick Type C (NPC) Disease

5.1 NPC Disease

The classical GSL storage diseases store GSLs because of a primary defect in GSL catabolism, due to an enzyme or cofactor deficiency (Wraith, 2002). However, many storage diseases that have primary defects unrelated to GSL catabolism also store GSLs in diseased neurons, in particular GM2 and GM3 gangliosides (Walkley, 2004). We currently do not understand the mechanism(s) that underlie this phenomenon. However, in the absence of knowledge about the aetiology of this secondary GSL storage it is possible to address the question as to whether these stored GSLs contribute to the disease process.

One of the diseases with secondary GSL accumulation is Niemann–Pick type C (NPC). NPC disease is caused by mutations in either the NPC1 or NPC2 genes (Ikonen & Holtta-Vuori, 2004). The vast majority of cases involve defects in NPC1. Both genes have been cloned and they encode two very different proteins, the precise function of which remains enigmatic. NPC1 is a glycoprotein with 13 putative transmembrane domains that include a sterol-sensing domain between the third and seventh transmembrane domains (Carstea et al., 1997). This protein has recently been shown to bind an analogue of cholesterol via an interaction with the sterol-sensing domain (Ohgami et al., 2004). The NPC 2 protein is quite different and is a soluble lysosomal protein that binds cholesterol with high affinity (Ko, Binkley, Sidow, & Scott, 2003). These proteins appear to act in concert to regulate lipid transport from the lysosome to other parts of the cell (Sleat et al., 2004). One aspect of NPC disease that remains an area of considerable debate is why do so many different lipid species accumulate and which ones contribute to the pathology associated with this disease (Liscum, 2000). The lipids that are known to accumulate in this disorder include unesterified cholesterol, sphingomyelin, sphingosine, and glycosphingolipids (including GM2, GM3, glucosylceramide, and lactosylceramide). NPC1 may therefore transport multiple lipid cargoes in late endosomes/lysosomes (Vanier & Millat, 2003).

Walkley and colleagues tested the hypothesis that the glycosphingolipids that accumulate in NPC disease significantly contribute to the disease process by treating the spontaneous mouse model of NPC1 with *N*B-DNJ (Zervas, Somers, Thrall, & Walkley, 2001). They found that life expectancy of the NPC mouse was significantly extended and the neuropathology significantly delayed. Similar findings were reported when the NPC cat model was treated with NB-DNJ (Zervas et al., 2001). These observations suggest that GSLs are involved in the neuropathology of NPC. The other possibility is that *N*B-DNJ mediates the clinical improvement via a currently unidentified activity of this drug. However, the imino sugar *N*B-DGJ (galactose analogue) which also inhibits GSL biosynthesis, but does not cause any side effects attributable to *N*B-DNJ (Andersson et al., 2000), had the same effect in the NPC mouse making a GSL-based mechanism of clinical improvement likely.

5.2 SRT in a Patient with NPC Disease

The findings from the NPC1 mouse study provided the impetus for the evaluation of miglustat in an NPC patient (R. H. Lachmann et al., 2004). Peripheral blood B cells were isolated from the patient at time intervals spanning 6 months. B cells were selected as they are a homogeneous resting cell population that are readily purified making comparisons over time on a single patient and between patients possible. In NPC disease several cell biological abnormalities are known. These include GSL accumulation (Vanier & Millat, 2003), altered GSL trafficking (Puri et al., 1999; te Vruchte et al., 2004), reduced endosomal uptake of fluid phase markers (Mayran, Parton, & Gruenberg, 2003) and expansion of the late endosomal/lysosomal compartment. In the NPC patient treated with miglustat pathological lipid storage was reduced, the total late endosomal/lysosomal compartment was almost normalised, improved endosomal uptake was observed and lipid trafficking in peripheral blood B lymphocytes was greatly improved (R. H. Lachmann et al., 2004). The demonstration that treatment with miglustat, which has no direct effect on cholesterol metabolism, corrected the abnormal lipid trafficking seen in B lymphocytes in NPC indicates that GSL accumulation is a significant pathogenic mechanism in NPC. Although this study does not reveal the status of storage and pathology in the brain it serves to illustrate those key disease parameters attributable to NPC disease can be improved in peripheral cells. Long-term follow-up and clinical and biochemical monitoring of this patient is ongoing(R. H. Lachmann et al., 2004).

A clinical trial in NPC patients in centres in the UK and USA has subsequently been performed to evaluate clinical efficacy. Interim data published after 12 months of a 2-year trial demonstrated benefit (stabilization or improvement in saccadic eye movement)(Patterson et al., 2006). These findings therefore hold promise for the use of miglustat in the clinical management of NPC patients in the future.

6. SECOND GENERATION COMPOUNDS

In the original screen of imino sugar compounds with inhibitory properties against the ceramide glucosyltransferase, the galactose analogue *N*B-DGJ was identified (F. M. Platt, Neises, Karlsson et al., 1994a) This compound is equivalent to *N*B-DNJ in terms of potency but lacks many of the additional enzyme inhibitory properties associated with *N*B-DNJ (Andersson et al., 2000). Significantly, it does not inhibit the gut disaccharidases, the property of *N*B-DNJ that causes osmotic diarrhoea. Also, *N*B-DGJ does not cause weight loss in mice, which may be an advantage particularly in the treatment of paediatric patients. Recently, *N*B-DGJ was evaluated in the mouse model of Sandhoff disease and dose escalation was possible with increasing clinical benefit (Andersson, Smith et al., 2004). This compound has progressed to phase 1 in healthy volunteers.

7. FUTURE PROSPECTS

The majority of patients with GSL storage diseases have brain disease and become symptomatic soon after birth or in early infancy (Wraith, 2004). The early presentation of symptoms reflects the low levels of residual enzyme activity present, as a result of the disease causing mutation(s). This group of patients is currently untreatable. Although SRT monotherapy may help slow disease in these severely affected individuals it will not prevent the relentless progression of the disease, only slow the rate of progression. The question this raises is whether there are any approaches that could be combined with SRT to provide greater benefit to this currently intractable group of patients. Enzyme augmentation in combination with SRT is the obvious choice, but we currently lack the technology to do this safely and effectively in the human brain. This may change with time as gene delivery, stem cell delivery, or ERT delivery across the blood–brain barrier is developed. However, in the meantime is there any other avenue that can be explored?

Over the past few years much greater emphasis has been placed on trying to understand the downstream consequences of storage of GSLs, with the hope that the insights gained may reveal more about GSL biology and suggest new avenues for clinical intervention. We therefore have the prospect of targeting the disease via at least three independent routes including enzyme augmentation, SRT, and targeting downstream events. To date two main consequences of storage have been reported that are potentially amenable to pharmacological intervention, altered calcium homeostasis (Ginzburg, Kacher, & Futerman, 2004) and macrophage/microglial cell-mediated inflammation in the brain (Jeyakumar et al., 2003; Wada, Tifft, & Proia, 2000; Wu & Proia, 2004). This opens up not only the prospect of improving quality of life for patients but to do so using drugs that may well be currently in clinical use for other indications and therefore could be rapidly translated to clinical use in these diseases. Recently, anti-inflammatory drugs and antioxidants have been evaluated in the mouse model of Sandhoff disease and found to show efficacy as monotherapies and to synergise with SRT (Jeyakumar et al., 2004). We therefore have the prospect that within the next few years even the most intractable group of patients, those with storage in the brain, could begin to benefit from therapies emerging from the studies conducted in mouse models of these disorders.

ACKNOWLEDGEMENTS

We thank Raymond Dwek for his comments on the manuscript. The authors are in receipt of a research grant from Celltech/Oxford GlycoSciences.

REFERENCES

Aerts, J. M., & Hollak, C. E. (1997). Plasma and metabolic abnormalities in Gaucher's disease. *Baillieres Clin Haematol, 10*(4), 691-709.

Andersson, U., Butters, T. D., Dwek, R. A., & Platt, F. M. (2000). N-butyldeoxygalactonojirimycin: a more selective inhibitor of glycosphingolipid biosynthesis than N-butyldeoxynojirimycin, in vitro and in vivo. *Biochem Pharmacol, 59*(7), 821-829.

Andersson, U., Reinkensmeier, G., Butters, T. D., Dwek, R. A., & Platt, F. M. (2004). Inhibition of glycogen breakdown by imino sugars in vitro and in vivo. *Biochem Pharmacol, 67*(4), 697-705.

Andersson, U., Smith, D., Jeyakumar, M., Butters, T. D., Borja, M. C., Dwek, R. A., et al. (2004). Improved outcome of N-butyldeoxygalactonojirimycin-mediated substrate reduction therapy in a mouse model of Sandhoff disease. *Neurobiol Dis, 16*(3), 506-515.

Beck, M. (2001). Variable clinical presentation in lysosomal storage disorders. *J Inherit Metab Dis, 24 Suppl 2*, 47-51; discussion 45-46.

Beutler, E., & Grabowski, G. (2001). Gaucher disease. In C. R. Scriver, A. L. Beadet, D. Valle & W. S. Sly (Eds.), *The metabolic and molecular bases of inherited diseases* (8 ed., Vol. 3, pp. 3636-3668). New York: McGraw Hill.

Block, T. M., Lu, X. Y., Mehta, A. S., Blumberg, B. S., Tennant, B., Ebling, M., et al. (1998). Treatment of chronic hepadnavirus infection in a woodchuck animal model with an inhibitor of protein folding and trafficking. *Nature Med, 4*(5), 610-614.

Brady, R. O. (2003). Gaucher and Fabry diseases: from understanding pathophysiology to rational therapies. *Acta Paediatr Suppl, 92*(443), 19-24.

Butters, T. D., Dwek, R. A., & Platt, F. M. (2000). Inhibition of Glycosphingolipid Biosynthesis: Application to Lysosomal Storage Disorders. *Chem Rev, 100*, 4683-4696.

Butters, T. D., Dwek, R. A., & Platt, F. M. (2003). Therapeutic applications of imino sugars in lysosomal storage disorders. *Curr Top Med Chem, 3*(5), 561-574.

Butters, T. D., Mellor, H. R., Narita, K., Dwek, R. A., & Platt, F. M. (2003). Small-molecule therapeutics for the treatment of glycolipid lysosomal storage disorders. *Philos Trans R Soc Lond B Biol Sci, 358*(1433), 927-945.

Butters, T. D., van den Broek, L. A. G. M., Fleet, G. W. J., Krulle, T. M., Wormald, M. R., Dwek, R. A., et al. (2000). Molecular requirements of imino sugars for the selective control of N-linked glycosylation and glycospingolipid biosynthesis. *Tetrahedron Assymetry, 11*, 113-124.

Cabrera-Salazar, M. A., Novelli, E., & Barranger, J. A. (2002). Gene therapy for the lysosomal storage disorders. *Curr Opin Mol Ther, 4*(4), 349-358.

Carstea, E. D., Morris, J. A., Coleman, K. G., Loftus, S. K., Zhang, D., Cummings, C., et al. (1997). Niemann-Pick C1 disease gene: homology to mediators of cholesterol homeostasis. *Science, 277*(5323), 228-231.

Cox, T., Lachmann, R., Hollak, C., Aerts, J., van Weely, S., Hrebicek, M., et al. (2000). Novel oral treatment of Gaucher's disease with N-butyldeoxynojirimycin (OGT 918) to decrease substrate biosynthesis. *Lancet, 355*(9214), 1481-1485.

Cox, T. M., Aerts, J. M., Andria, G., Beck, M., Belmatoug, N., Bembi, B., et al. (2003). The role of the iminosugar N-butyldeoxynojirimycin (miglustat) in the management of type I (non-neuronopathic) Gaucher disease: a position statement. *J Inherit Metab Dis, 26*(6), 513-526.

Dobrenis, K. (2004). Cell-mediated delivery systems. In F. M. Platt & S. U. Walkley (Eds.), *Lysosomal Disorders of the Brain* (pp. 339-380). Oxford: Oxford University Press.

Durantel, D., Branza-Nichita, N., Carrouee-Durantel, S., Butters, T. D., Dwek, R. A., & Zitzmann, N. (2001). Study of the Mechanism of Antiviral Action of Iminosugar Derivatives against Bovine Viral Diarrhea Virus. *J Virol, 75*(19), 8987-8998.

Dwek, R. A., Butters, T. D., Platt, F. M., & Zitzmann, N. (2002). Targetting glycosylation as a therapeutic approach. *Nature drug discovery, 1*(1), 65-75.

Elstein, D., Hollak, C., Aerts, J. M., Van Weely, S., Maas, M., Cox, T. M., et al. (2004). Sustained therapeutic effects of oral miglustat (Zavesca, N-butyldeoxynojirimycin, OGT-918) in type 1 Gaucher disease. *J Inherit Metab Dis, In Press*.

Fischer, P. B., Collin, M., Karlsson, G. B., James, W., Butters, T. D., Davis, S. J., et al. (1995). The alpha-glucosidase inhibitor N-butyldeoxynojirimycin inhibits human immunodeficiency virus entry at the level of post-CD4 binding. *J Virol, 69*(9), 5791-5797.

Ginzburg, L., Kacher, Y., & Futerman, A. H. (2004). The pathogenesis of glycosphingolipid storage disorders. *Semin Cell Dev Biol, 15*(4), 417-431.

Heitner, R., Elstein, D., Aerts, J., Weely, S., & Zimran, A. (2002). Low-dose N-butyldeoxynojirimycin (OGT 918) for type I Gaucher disease. *Blood Cells Mol Dis, 28*(2), 127-133.

Hopwood, J. J., Crawley, A. C., & Taylor, R. M. (2004). Spontaneous and engineered mammalian storage disease models. In F. M. Platt & S. U. Walkley (Eds.), *Lysosomal disorders of the brain* (Vol. 1, pp. 257-289). Oxford: Oxford University Press.

Ikonen, E., & Holtta-Vuori, M. (2004). Cellular pathology of Niemann-Pick type C disease. *Semin Cell Dev Biol, 15*(4), 445-454.

Jeyakumar, M., Butters, T. D., Cortina-Borja, M., Hunnam, V., Proia, R. L., Perry, V. H., et al. (1999). Delayed symptom onset and increased life expectancy in Sandhoff disease mice treated with N-butyldeoxynojirimycin. *Proc Natl Acad Sci U S A, 96*(11), 6388-6393.

Jeyakumar, M., Smith, D. A., Williams, I. M., Borja, M. C., Neville, D. C., Butters, T. D., et al. (2004). NSAIDs increase survival in the Sandhoff disease mouse: Synergy with N-butyldeoxynojirimycin. *Ann Neurol, 56*(5), 642-649.

Jeyakumar, M., Thomas, R., Elliot-Smith, E., Smith, D. A., van der Spoel, A. C., d'Azzo, A., et al. (2003). Central nervous system inflammation is a hallmark of pathogenesis in mouse models of GM1 and GM2 gangliosidosis. *Brain, 126*(Pt 4), 974-987.

Ko, D. C., Binkley, J., Sidow, A., & Scott, M. P. (2003). The integrity of a cholesterol-binding pocket in Niemann-Pick C2 protein is necessary to control lysosome cholesterol levels. *Proc Natl Acad Sci U S A, 100*(5), 2518-2525.

Krivit, W. (2002). Stem cell bone marrow transplantation in patients with metabolic storage diseases. *Adv Pediatr, 49*, 359-378.

Krivit, W., Sung, J. H., Shapiro, E. G., & Lockman, L. A. (1995). Microglia: the effector cell for reconstitution of the central nervous system following bone marrow transplantation for lysosomal and peroxisomal storage diseases. *Cell-Transplant, 4*(4), 385-392.

Lachmann, R. H. (2003). Miglustat. Oxford GlycoSciences/Actelion. *Curr Opin Investig Drugs, 4*(4), 472-479.

Lachmann, R. H., & Platt, F. M. (2001). Substrate reduction therapy for glycosphingolipid storage disorders. *Exp. Opin. Invest. Drugs, 10*, 455-466.

Lachmann, R. H., te Vruchte, D., Lloyd-Evans, E., Reinkensmeier, G., Sillence, D. J., Fernandez-Guillen, L., et al. (2004). Treatment with miglustat reverses the lipid-trafficking defect in Niemann-Pick disease type C. *Neurobiol Dis, 16*(3), 654-658.

Liscum, L. (2000). Niemann-Pick type C mutations cause lipid traffic jam. *Traffic, 1*(3), 218-225.

Mayran, N., Parton, R. G., & Gruenberg, J. (2003). Annexin II regulates multivesicular endosome biogenesis in the degradation pathway of animal cells. *Embo J, 22*(13), 3242-3253.

Mellor, H. R., Neville, D. C., Harvey, D. J., Platt, F. M., Dwek, R. A., & Butters, T. D. (2004). Cellular effects of deoxynojirimycin analogues: uptake, retention and inhibition of glycosphingolipid biosynthesis. *Biochem J, 381*(Pt 3), 861-866.

Mellor, H. R., Nolan, J., Pickering, L., Wormald, M. R., Platt, F. M., Dwek, R. A., et al. (2002). Preparation, biochemical characterization and biological properties of radio-labelled N-alkylated deoxynojirimycins. *Biochem J, 366*(Pt 1), 225-233.

Mellor, H. R., Platt, F. M., Dwek, R. A., & Butters, T. D. (2003). Membrane disruption and cytotoxicity of hydrophobic N-alkylated imino sugars is independent of the inhibition of protein and lipid glycosylation. *Biochem J, 374*(Pt 2), 307-314.

Moyses, C. (2003). Substrate reduction therapy: clinical evaluation in type 1 Gaucher disease. *Philos Trans R Soc Lond B Biol Sci, 358*(1433), 955-960.

Neufeld, E. F. (2004). Enzyme replacement therapy. In F. M. Platt & S. U. Walkley (Eds.), *Lysosomal disorders of the brain* (pp. 327-338). Oxford: Oxford University Press.

Ohgami, N., Ko, D. C., Thomas, M., Scott, M. P., Chang, C. C., & Chang, T. Y. (2004). Binding between the Niemann-Pick C1 protein and a photoactivatable cholesterol analog requires a functional sterol-sensing domain. *Proc Natl Acad Sci U S A, 101*(34), 12473-12478.

Pastores, G. M., Barnett, N. L., Bathan, P., & Kolodny, E. H. (2003). A neurological symptom survey of patients with type I Gaucher disease. *J Inherit Metab Dis, 26*(7), 641-645.

Patterson, M. C., Vecchio, D., Prady, H., Abel, L., Ait-Alissa, N., & Wraith, J. E. (2006). Oral miglustat in Niemann-Pick type C (NPC) disease. *Rev Neurol (separata) 43*, 8.

Platt, F. M., & Butters, T. D. (2004). Inhibition of substrate synthesis: a pharmacological approach for glycosphingolipid storage disease therapy. In F. M. Platt & S. U. Walkley (Eds.), *Lysosomal Disorders of the Brain* (pp. 381-408). Oxford: Oxford University Press.

Platt, F. M., Jeyakumar, M., Andersson, U., Heare, T., Dwek, R. A., & Butters, T. D. (2003). Substrate reduction therapy in mouse models of the glycosphingolipidoses. *Phil. Trans. R. Soc. Lond. B, 358*, 947-954.

Platt, F. M., Neises, G. R., Dwek, R. A., & Butters, T. D. (1994). N-butyldeoxy-nojirimycin is a novel inhibitor of glycolipid biosynthesis. *J Biol Chem, 269*(11), 8362-8365.

Platt, F. M., Neises, G. R., Karlsson, G. B., Dwek, R. A., & Butters, T. D. (1994a). N-butyldeoxygalactonojirimycin inhibits glycolipid biosynthesis but does not affect N-linked oligosaccharide processing. *J Biol Chem, 269*(43), 27108-27114.

Platt, F. M., Neises, G. R., Reinkensmeier, G., Townsend, M. J., Perry, V. H., Proia, R. L., et al. (1997). Prevention of lysosomal storage in Tay-Sachs mice treated with N-butyldeoxynojirimycin. *Science, 276*(5311), 428-431.

Puri, V., Watanabe, R., Dominguez, M., Sun, X., Wheatley, C. L., Marks, D. L., et al. (1999). Cholesterol modulates membrane traffic along the endocytic pathway in sphingolipid-storage diseases. *Nat Cell Biol, 1*(6), 386-388.

Radin, N. S. (1996). Treatment of Gaucher disease with an enzyme inhibitor. *Glycoconj J, 13*(2), 153-157.

Rapola, J. (1994). Lysosomal storage diseases in adults. *Pathol Res Pract, 190*(8), 759-766.

Ringden, O., Groth, C. G., Erikson, A., Granqvist, S., Mansson, J. E., & Sparrelid, E. (1995). Ten years' experience of bone marrow transplantation for Gaucher disease. *Transplantation, 59*(6), 864-870.

Sands, M. S. (2004). Gene therapy. In F. M. Platt & S. U. Walkley (Eds.), *Lysosomal Disorders of the Brain* (pp. 409-430). Oxford: Oxford University Press.

Sango, K., Yamanaka, S., Hoffmann, A., Okuda, Y., Grinberg, A., Westphal, H., et al. (1995). Mouse models of Tay-Sachs and Sandhoff diseases differ in neurologic phenotype and ganglioside metabolism. *Nat Genet, 11*(2), 170-176.

Sleat, D. E., Wiseman, J. A., El-Banna, M., Price, S. M., Verot, L., Shen, M. M., et al. (2004). Genetic evidence for nonredundant functional cooperativity between NPC1 and NPC2 in lipid transport. *Proc Natl Acad Sci U S A, 101*(16), 5886-5891.

Snyder, E. Y., Daley, G. Q., & Goodell, M. (2004). Taking stock and planning for the next decade: realistic prospects for stem cell therapies for the nervous system. *J Neurosci Res, 76*(2), 157-168.

te Vruchte, D., Lloyd-Evans, E., Veldman, R. J., Neville, D. C., Dwek, R. A., Platt, F. M., et al. (2004). Accumulation of glycosphingolipids in Niemann-Pick C disease disrupts endosomal transport. *J Biol Chem, 279*(25), 26167-26175.

van der Spoel, A. C., Jeyakumar, M., Butters, T. D., Charleton, H. M., Moore, H. D., Dwek, R. A., et al. (2002). Reversible infertility in male mice following oral administration of alkylated imino sugars: a non-hormonal approach to male contraception. *Proc. Natl. Acad. Sci. U.S.A., In Press.*

Vanier, M. T., & Millat, G. (2003). Niemann-Pick disease type C. *Clin Genet, 64*(4), 269-281.

Vunnam, R. R., & Radin, N. S. (1980). Analogs of ceramide that inhibit glucocerebroside synthetase in mouse brain. *Chem-Phys-Lipids, 26*(3), 265-278 issn: 0009-3084.

Wada, R., Tifft, C. J., & Proia, R. L. (2000). Microglial activation precedes acute neurodegeneration in Sandhoff disease and is supressed by bone marrow transplanttation. *Proc Natl Acad Sci USA, 97*(20), 10954-10959.

Walkley, S. U. (2004). Secondary accumulation of gangliosides in lysosomal storage disorders. *Semin Cell Dev Biol, 15*(4), 433-444.

Winchester, B. (2004). Primary defects in lysosomal enzymes. In F. M. Platt & S. U. Walkley (Eds.), *Lysosomal Disorders of the Brain* (pp. 81-130). Oxford: Oxford University Press.

Wraith, J. E. (2002). Lysosomal disorders. *Semin Neonatol, 7*(1), 75-83.

Wraith, J. E. (2004). Clinical aspects and diagnosis. In F. M. Platt & S. U. Walkley (Eds.), *Lysosomal disorders of the brain* (pp. 50-77). Oxford: Oxford University Press.

Wu, Y. P., & Proia, R. L. (2004). Deletion of macrophage-inflammatory protein 1 alpha retards neurodegeneration in Sandhoff disease mice. *Proc Natl Acad Sci U S A, 101*(22), 8425-8430.

Yamanaka, S., Johnson, M. D., Grinberg, A., Westphal, H., Crawley, J. N., Taniike, M., et al. (1994). Targeted disruption of the Hexa gene results in mice with biochemical and pathologic features of Tay-Sachs disease. *Proc Natl Acad Sci USA, 91*(21), 9975-9979.

Zervas, M., Somers, K. L., Thrall, M. A., & Walkley, S. U. (2001). Critical role for glycosphingolipids in Niemann-Pick disease type C. *Curr Biol, 11*(16), 1283-1287.

Zimran, A., & Elstein, D. (2003). Gaucher disease and the clinical experience with substrate reduction therapy. *Philos Trans R Soc Lond B Biol Sci, 358*(1433), 961-966.

Zitzmann, N., Mehta, A. S., Carrouee, S., Butters, T. D., Platt, F. M., McCauley, J., et al. (1999). Imino sugars inhibit the formation and secretion of bovine viral diarrhea virus, a pestivirus model of hepatitis C virus: implications for the development of broad spectrum anti-hepatitis virus agents. *Proc Natl Acad Sci U S A, 96*(21), 11878-11882.

NEWBORN SCREENING FOR LYSOSOMAL STORAGE DISORDERS

C. Ronald Scott, M.D.,[1] Frantisek Turecek, Ph.D.,[2] Michael H. Gelb, Ph.D.[2,3]

Screening newborn infants for treatable metabolic diseases has been in existence for over four decades. It was initiated by Dr. Robert McCready in 1962 in the Massachusetts Department of Public Health and was based on the work of Dr. Robert Guthrie for the detection of phenylketonuria. The initiation of screening for phenylketonuria and its subsequent expansion to include congenital hypothyroidism, galactosemia, congenital adrenal hyperplasia, biotinidase deficiency, and hemoglobin abnormalities has had a profound effect in preventing the serious consequences from these disorders in susceptible individuals (for review, see Levy and Albers, 2000).

The basis of newborn screening is founded on the availability of a simple test that can be performed on a drop of blood obtained from an infant. In practice, several drops of blood are obtained from the infant near the time of birth, placed on specific filter papers, and submitted to the laboratory for analysis. Thus, the simple availability of a blood spot is both the power and limitation behind newborn screening. The limitation is devising techniques for the detection of some disorders that may more easily be detected by using a different biologic specimen. Wilson and Junger (1968) outlined the principles that needed to be met to have a successful candidate for newborn screening. This was prepared as a statement of the World Health Organization (WHO). The criteria are as shown in Table 1.

Table 1. Criteria for newborn screening approved by the WHO.

1.	Condition should be an important health problem
2.	There exists an accepted treatment for the patient
3.	Facilities for diagnosis and treatment should be available
4.	The detected disorder should be in a latent or early symptomatic stage
5.	There should exist a suitable screening test
6.	The test should be acceptable to the public
7.	The natural history of the condition should be adequately understood
8.	There should be an agreed-upon policy on whom to treat
9.	The cost of case finding, including diagnosis and treatment, should be economically balanced in relation to the expenditures of medical care as a whole
10.	Case finding should be a continuing process

Wilson and Junger, 1968

Depts. of Pediatrics,[1] Chemistry,[2] and Biochemistry,[3] University of Washington, Seattle, WA 98195 Address correspondence to C. Ronald Scott, M.D. Dept. of Pediatrics, 56320 1959 NE Pacific Street, RR-310. University of Washington, Seattle, WA 98195-6320. tel: (206) 543-3370; fax: (206) 543-3379; e-mail: crscott@u.washington.edu

In today's practice of genetic medicine, it is unlikely that the disease candidates advocated for newborn screening will be able to meet all of the criteria proposed by Wilson and Junger. For lysosomal storage diseases, these criteria present particular problems because of the rapid changes that are occurring in understanding their biological basis, the availability of more modern detection methods, and the changing face of therapeutic intervention with little information yet available on long-term therapeutic outcomes.

Traditional newborn screening has relied on the quantitation of specific analytes present in the blood spot as an indicator of the presence of disease. Each of the classic disorders requires a different methodology for its detection. The bacterial inhibition assay was used for the detection of increased concentration of phenylalanine observed in phenylketonuria, immunological binding assays were required for quantitation of T4 or TSH for congenital hypothyroidism, and fluorogenic enzyme assays were developed for the detection of deficiencies of biotinidase or galactose-1-phosphate uridyltransferase. The common denominator in all of these was the use of a blood spot as the biological sample.

In recent years, the introduction of tandem mass spectrometry has begun to revolutionize the practice of newborn screening. Its strength lies in the ability to differentiate multiple components from a single sample and to accurately quantitate the analyte of interest. The tandem mass spectrometer has been adapted for the identification and quantitation of amino acids that are present in blood samples and can be utilized for the detection of phenylketonuria, tyrosinemia, maple syrup urine disease, and homocystinuria, all diseases that require early detection and therapeutic intervention to prevent chronic disabilities or early mortality. In addition to the detection of amino acids, the mass spectrometer can be adapted to measure the acyl-carnitine profile in a single blood spot. When properly used, the technique can detect as many as 30 disorders of fatty acid oxidation and organic acid metabolism (Wilcken et al., 2003; Zytkovicz et al., 2001). The more critical are the identification of medium chain acyl-Co-A dehydrogenase (MCADD) and the long-chain disorders of fatty acid oxidation, long-chain dehydrogenase deficiency, and long-chain hydroxy dehydrogenase deficiency (LCAD and LCHD, respectively). It is currently estimated that the tandem mass spectrometer, when used as a comprehensive tool for newborn screening, will detect approximately 1 in 4000 infants with a metabolic disorder.

The lysosomal storage diseases now present the possibility for early detection using the mass spectrometers that are currently in place in newborn screening laboratories; but which disorders should be considered for newborn screening?

There exist several lysosomal disorders that are potential candidates for early detection because there may exist effective therapeutic intervention (Table 2). The introduction of enzyme replacement therapy for Gaucher (Barton et al., 1991) and Fabry (Eng et al., 2001) disease has been demonstrated to be effective in alleviating the major clinical symptoms and anatomic alterations leading to morbidity and mortality. Although there is no convincing evidence that enzyme replacement therapy can alter the neurologic components in Gaucher disease, type II and III, enzyme replacement therapy can certainly improve the somatic alterations caused by the disease. This is especially true for Gaucher disease, type I, where there is reduction in liver and spleen volume and prevention of bony degeneration once enzyme replacement therapy has been initiated. Similarly for Fabry disease, there is good evidence for resolution of peripheral pain, improvement of cardiac function, and removal of the lipid storage from skin and kidney (Wilcox et al., 2004; Eng et al., 2001).

Although Gaucher and Fabry disease typically do not present in early childhood with irreversible changes, an argument can be made that their early detection would allow for appropriate timing of enzyme replacement therapy.

Other disorders for which enzyme therapy is currently available or being evaluated are mucopolysaccharidosis-I (Hurler), mucopolysaccharidosis-II (Hunter syndrome), mucopolysaccharidosis-VI (Maroteaux–Lamy syndrome), and glycogenosis, type II (Pompe disease). Early evidence would suggest that MPS-I benefits from enzyme replacement therapy by improvement in the somatic features of the disorder. There is some evidence that early intervention for some patients with Pompe disease improves muscle strength and allows for improved developmental landmarks.

The other therapeutic modality that may alter the natural history of Krabbe disease, metachromatic leukodystrophy, and Niemann–Pick B is bone marrow transplantation. For these disorders bone marrow transplantation prior to the onset of neurological symptoms has been reported to dramatically alter the slowly progressive neurological involvement (Krivit, Peters, and Shapiro, 1999). Because of the above observations from enzyme replacement therapy and bone marrow transplantation, there have been initial efforts to devise technologies for their early recognition that would allow for presymptomatic treatment. Several gene delivery vectors and systems have been evaluated for the treatment of a number of different LSD. These include both viral (adenoviral, retroviral, herpes simplex virus, and adeno-associated viral) and nonviral (cationic lipids, polymers, and molecular conjugates) based gene transfer vectors.

Of these, the adeno-associated viral (AAV) vector is emerging as the gene delivery vector of choice for treating both the visceral and CNS diseases associated with this group of disorders. Several features of this vector make it attractive for treating chronic genetic diseases such as the LSD. They are efficient at transducing a variety of cell types, are reportedly only mildly inflammatory in vivo, and are capable of supporting long-term transgene expression. Moreover, recently, several new viral serotypes with different tissue tropisms and significantly greater gene transduction activity than the prototypical AAV2 serotype have been isolated.

Table 2. Potential candidates for newborn screening of lysosomal storage disorders.

Disorders	Effective Rx	Meets Criteria for NB Screening?
Gaucher	ERT	Possibly
Fabry	ERT	Possibly
Mucopolysaccharidosis-1	ERT/BMT	Yes
Pompe	ERT	Yes
Krabbe	BMT	Yes
Metachromatic leukodystrophy	BMT	Yes
Niemann–Pick B	BMT	Possibly
Maroteaux–Lamy	ERT	Possibly

ERT = enzyme replacement therapy
BMT = bone marrow transplantation

Interest in screening newborns for lysosomal storage disorders has been prevalent for the last decade (Meikle and Hopwood, 2003). The interest was stimulated by the results of clinical studies on the effectiveness of enzyme replacement therapy or bone marrow transplantation in modifying the natural history of the disorders. A leader in this approach has been the lysosomal storage disease unit in North Adelaide, Australia, under the direction of Dr. John Hopwood. Their program has focused on identifying analytes that are increased in individuals affected with these disorders. They have identified two proteins that are associated with the lysosomal membrane and have developed techniques for quantitating the proteins in blood. These proteins are called lysosome-associated membrane protein-1 (LAMP-1) and lysosome-associated membrane protein-2 (LAMP-2). The method involves a sensitive fluorescence immunoassay that can be performed on extracts obtained from blood spots. They have obtained quantitative measurements from individuals affected with lysosomal storage disorders and have compared the results to normal newborns (Meikle et al., 1997; Hua et al., 1998). Their published data indicate that for LAMP-1, if they use a cutoff that is above the 95th percentile for the normal population, they are able to achieve a sensitivity of 72% for patients with known LSDs. This included 320 affected individuals representing 25 different LSDs. A selection of disorders for which LAMP-1 was greater than the 95th percentile as compared to controls, is listed in Table 3.

Table 3. Detection of LSDs by increased concentration of LAMP-1 or LAMP-2 in plasma.

Disorder	**% > 95 Percentile of Normal Newborns**	
	LAMP-1	**LAMP-2**
Galactosialidosis	100%	100%
I-cell	100	100
MPS-1	100	100
MPS-II & III	100	100
Sandoff	100	0
Tay–Sachs	100	0
Gaucher	92	92
Fabry	5	0
Niemann–Pick A &B	33	20
Krabbe	17	8
Pompe	25	0
MLD	19	16

From: Meikle, P. et al., 1997
Hua, C. et al., 1998

A second protein, LAMP-2 (Hua et al., 1998), was separately evaluated using an immunoquantitation assay. When measured, a wider quantitative range was observed in patients affected with lysosomal storage disorders, as compared to controls from normal infants. LAMP-2, however, turned out to be less sensitive than LAMP-1 in discriminating between affected and unaffected individuals, and was elevated beyond the 95th percentile in only 66% of patients with LSDs (Table 3).

The measurement of LAMP-1 and 2, however, shows promise as an initial screen for newborns unsuspected of having an LSD. The Adelaide program continues to work on technology that would select samples above the 95–99th percentile and follow up the procedure by utilizing tandem mass spectroscopy that can identify specific abnormal lipids or oligosaccharides present in the newborn blood spot to more accurately identify affected patients. This approach depends on a two-tiered system of evaluation, similar to that used by some programs that are screening for cystic fibrosis.

Another analyte used for detection of a lysosomal storage disease has been the evaluation of a glucose tetrasacharide as a putative biomarker for the diagnosis of Pompe disease (Table 4). It was noted by the group at Duke University that patients affected with Pompe disease excreted in urine a tetraglucoseoligimer (Glc-Glc-Glc-Glc, or Glc_4) in increased quantities (An et al., 2000). A high-pressure liquid chromatography method was developed for identification and quantitation of this Glc_4 molecule. It was confirmed that all patients affected with Pompe disease, irrespective of age, excreted an increased quantity of Glc_4 compared to normal controls.

Table 4. Data on the assay of Glc_4 in urine and blood.

Glc_4 in urine:

Age (yr)	(mmole/molCr)	
	Affected Pompe	Normal Persons
<1	34.6	8.9 ± 8.2
2–5	60	~3.6 ± 3.8
11–31	29.6	~0.9 ± 1.0
40–61	5.5	~0.4 ± 0.3

An et al., 2000

Glc_4 in plasma:

	(mmole/molCr)	
< 1	3.0	0.3 ± 0.39

Young et al., 2000

Glc_4 in blood spots:

	(μmole/mL)	
	5.3 (2.7 – 10)	7.1 (3–30)

Rozaklis et al., 2002

The method was improved by adaptation to electrospray ionization tandem mass spectroscopy (Young et al., 2003). The technique was validated on urine samples from affected patients and extended to the documentation that the Glc_4 was also elevated in plasma. There was good agreement between the urine samples, with an $r^2 = 0.94$, but less agreement between the HPLC method and the tandem mass spectrometry method for plasma samples, $r^2 = 0.53$. This raised a question regarding the immediate application of the method to newborn blood spots for the detection of Pompe disease.

In evaluating this method in Adelaide, the Australian program confirmed that an increase in Glc_4 levels could be detected in urine from infantile onset Pompe disease, when measured by electrospray ionization tandem mass spectroscopy, and that Glc_4 levels could be monitored in plasma samples (Umapathysivam, Hopwood, and Meikle, 2001). Although the method was sensitive enough to determine Glc_4 levels in a single 3 mm blood spot, they could not confirm the difference between individuals affected with Pompe disease and normal controls (Rozaklis et al., 2002). Thus, this particular method appears limited in its application for the detection of Pompe disease from blood spots routinely submitted for newborn screening.

A more direct method for evaluating newborns at risk for a lysosomal storage disorder would be to measure the gene product responsible for each of the disorders. This involves the assay of lysosomal enzymes from blood samples submitted on newborn screening cards. Chamoles et al. (2001, 2002a,b, 2004) have documented that in retrospective sampling of newborn blood spots from individuals later diagnosed with lysosomal storage disorders, the enzyme deficiency can be detected using conventional artificial substrates, the 4-methylumbelliferyl substrates that are traditionally used in fluorescent enzyme assays of white cells or fibroblasts. These disorders included Gaucher, Fabry, Hurler, Niemann–Pick, and the GM-2 gangliosidoses. Samples of these results are shown in Table 5.

Table 5. Lysosomal enzyme activity using blood spots as the biologic sample.

Disorder (n)	**(4-MU-substrates) umole/hr/L**	**Normal newborns**
Gaucher (54)	0.48	5.27
Fabry (10)	0–0.08	4.3 ± 1.32
Mps-1 & 1S (10)	0–5.2	91 ± 27.2
Niemann–Pick (8)*	125	4483

*^{14}C assay

Chamoles, N. et al., 2001, 2002

There was clear separation between affected individuals and normal controls. There was no overlap in enzyme activity reported between individuals affected with the lysosomal storage disease and normal controls or obligate carriers. This particular method involves 3-mm punches taken from newborn screening cards and individually incubated with the appropriate buffer and substrate and the measurement of fluorescence after a specified time period. Each assay must be performed individually. In the case of Niemann–Pick disease, rather than a fluorescent substrate being utilized, a radioactive substrate was utilized. These publications clearly demonstrate that lysosomal enzymes are reasonably stable in dried blood samples and can be used to measure selected enzymes. This approach offers an alternative to evaluating analytes that may be elevated on the basis of an abnormality in lysosomal integrity secondary to the accumulation of a complex molecule.

Pompe disease represents a special case within the family of storage diseases. The α-glucosidase responsible for Pompe disease is typically measured in muscle tissue or cultured skin fibroblasts. There exists a second enzyme, a neutral α-glucosidase (or renal glucosidase) that is primarily expressed in renal tissue, but also present in white cells and interferes with the accurate quantitation of the acid α-glucosidase in lysosomes. The pH optimum of the neutral enzyme overlaps with the optimum pH of the acid glucosidase, and neutrophils have a higher proportion of this neutral enzyme than do lymphocytes. The neutral acid α-glucosidase is significantly less abundant in muscle or cultured fibroblasts, making these tissues a better biological source for the evaluation of Pompe disease. In an attempt to overcome this problem, Umapathysivam et al. (2000) developed a monoclonal antibody that was specific for the acid form of the enzyme and showed that in the majority of patients affected with Pompe disease, the mature form of the acid α-glucosidase was either absent or present in very low concentrations. To directly measure the enzyme using the aid of the antibody, the antibody was coated onto plastic wells and used to trap the enzyme from blood. This proved to be effective. They demonstrated that enzyme activity could not be detected in blood samples obtained from infantile-, juvenile-, or the majority of adult-onset patients (Table 6) (Umapathysivam, Hopwood, and Meikle, 2001). There was no overlap between values obtained from affected patients compared to normal controls or obligate heterozygotes for Pompe disease.

Table 6. Acid α-glucosidase activity in blood spots by immunocapture assay.

	umole/hr/L Mean and (Range)
Normal newborns (96)	3.5 (0.3–10)
Normal adults (215)	1.3 (0.3–3.0)
Infantile Pompe (1)	0
Juvenile Pompe (3)	0
Adult Pompe (13)	0 (0–0.04)

Umapathysivam et al., 2001

An expansion of the idea of using enzyme activity from dried blood spots as a means of assaying for lysosomal storage diseases has been developed by the group at the University of Washington. They have used tandem mass spectroscopy as a common platform for measuring enzyme reactions (Gerber et al., 2001). Artificial substrates are synthesized for each of the lysosomal enzymes of interest, along with appropriate internal standards. An example of one of the synthesized artificial substrates and the resultant product for Fabry disease is shown in Figure 1. Each of the substrates and products has a different mass (m/z) that can be uniquely identified by the tandem mass spectrometer. The method consists of taking a single, 5-mm punch, using a common extraction buffer of sodium phosphate solution at pH 7.0 to solubilize the bound enzymes into a liquid phase. After extraction, 10 μL of the extracted enzyme solution can be used for each assay. Following an overnight incubation with the appropriate substrate, the internal standards are added, and the multiple reaction mixtures can be combined for further processing and injection into a mass spectrometer. Because of the mass differences of the products being assayed, a single injection is used to quantitate multiple enzymes simultaneously (multiplexing).

In a 2004 paper (Li et al., 2004), the authors report the results of five lysosomal enzymes that can be measured simultaneously. These are the enzymes responsible for Fabry, Gaucher, Krabbe, Niemann–Pick A/B, and Pompe disease. The assays were documented to be linear with regard to protein concentration of each assay, and linear over the

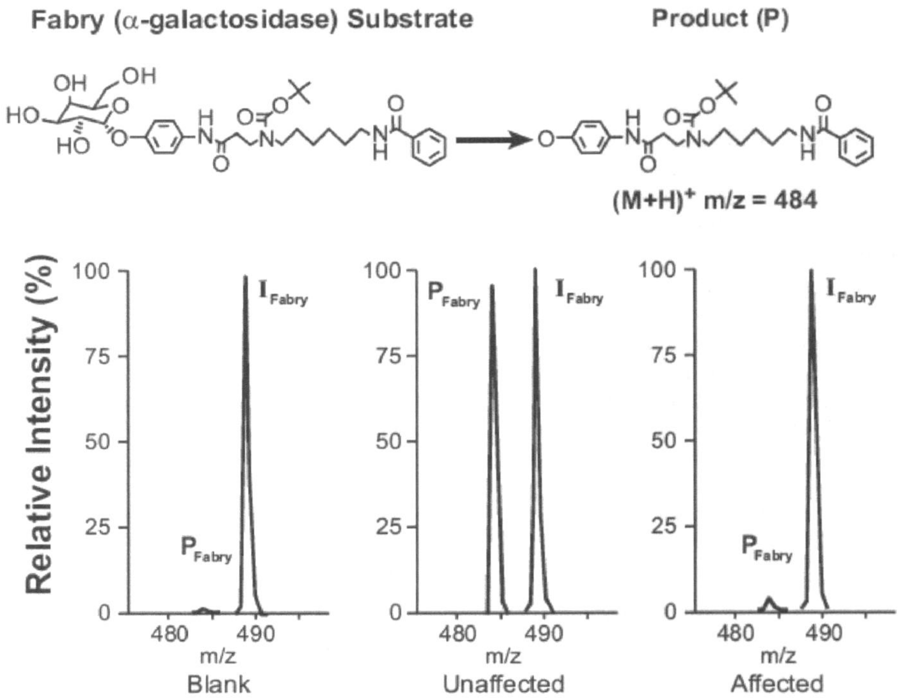

Figure 1. Results from an enzyme assay for Fabry disease using the tandem mass spectrometer. The α-linked galactose is cleaved from the artificial substrate forming a product (P) of m/z: 484. A deuterated product of m/z: 489 serves as an internal (I) standard. The affected patient has ~3% of normal activity.

time period of the incubations. An example of the recording obtained from the mass spectrometer for Fabry disease is shown in Figure 1. An affected patient has approximately 4% residual activity, compared to a nonaffected person. A summary of the results from this method is shown in Table 7.

Table 7. Lysosomal enzyme activity from a single 5 mm blood spot. The maximum observed activity is less than the minimum activity observed in random samples obtained from healthy newborns (Li et al., 2004).

Disorder	(n)	Enzyme Activity (µmole/hr/L)	
		Affected Persons (Maximum Activity)	Normal Infants Median (Range)
Gaucher (ABG)	(6)	0.18	3.31 (0.89– 9.6)
Fabry (GLA)	(5)	0.17	2.23 (0.77–5.65)
Krabbe (GALC)	(9)	0.20	1.01 (0.42–1.53)
Niemann–Pick A/B (ASM)	(5)	0.32	4.41 (0.92–11.3)
Pompe (GAA)	(5)	0.33	3.12 (0.93–7.33)

The authors were able to obtain accurate quantitation of the α-glucosidase for the detection of Pompe disease because of the innovative use of an inhibitor, acarbose, that inhibits the neutral α-glucosidase while retaining the activity of the acid α-glucosidase. The α-glucosidase assay performed on blood spots clearly distinguishes between individuals affected with Pompe and healthy infants.

With the emerging technologies of enzyme replacement therapy, the use of bone marrow transplantation, and, it is hoped, the eventual success of gene therapy, selected lysosomal storage diseases can now be considered potential candidates for newborn screening programs. Enough preliminary evidence exists that the direct analysis of analytes or enzymes for selected lysosomal disorders may be an approach for their early detection. The LAMP-1 protein continues to hold promise for disorders involved with mucopolysaccharide metabolism and potentially the GM-2 gangliosidoses. The direct measurement of enzyme activity from newborn blood spots appears to have greater versatility, either by immunocapture assays or by tandem mass spectroscopy. The development of tandem mass spectrometry as a platform for measuring enzyme activity directly from blood spots, may offer an opportunity for easy introduction into existing newborn screening programs.

The screening for lysosomal storage diseases will not be without its tribulations. As we have learned from previous screening programs, the introduction of new technology or new disorders to screening programs will unveil unanticipated surprises. We know there exist pseudo-deficiency alleles that are likely to be detected by measuring enzymes directly, and extensive data are not yet available on the potential overlap of enzyme activity between some heterozygotes and affected infants. It is also possible that new disorders will become "unmasked" through newborn screening programs, similar to what happened with the detection of biopterin defects when screening for phenylketonuria. Because of these cautionary notes, any initiation of newborn screening programs for lysosomal diseases will need to be carefully evaluated and will require careful clinical confirmation.

REFERENCES

An, Y., Young, S.P., Hillman, S.L., Van Hove, J.L., Chen, Y.T., and Millington, D.S., 2000, Liquid chromatographic assay for a glucose tetrasaccharide, a putative biomarker for the diagnosis of Pompe disease, *Anal Biochem.* **287**:136.

Barton, N.W., Brady, R.O., Dambrosia, J.M., Di Bisceglie, A.M., Doppelt, S.H., Hill, S.C., Mankin, H.J., Murray, G.J., Parker, R.I., Argoff, C.E., et al., 1991, Replacement therapy for inherited enzyme deficiency—Macrophage-targeted glucocerebrosidase for Gaucher's disease, *N Engl J Med.* **324**:1464.

Chamoles, N.A., Blanco, M.B., Gaggioli, D., and Casentini, C., 2001, Hurler-like phenotype: Enzymatic diagnosis in dried blood spots on filter paper, *Clin Chem.* **47**:2098.

Chamoles, N.A., Blanco, M., Gaggioli, D., and Casentini, C., 2002a, Gaucher and Niemann-Pick diseases—Enzymatic diagnosis in dried blood spots on filter paper: Retrospective diagnoses in newborn-screening cards, *Clin Chim Acta.* **317**:191.

Chamoles, N.A., Blanco, M., Gaggioli, D., and Casentini, C., 2002b, Tay-Sachs and Sandhoff diseases: Enzymatic diagnosis in dried blood spots on filter paper: Retrospective diagnoses in newborn-screening cards, *Clin Chim Acta.* **318**:133.

Chamoles, N.A., Niizawa, G., Blanco, M., Gaggioli, D., and Casentini, C., 2004, Glycogen storage disease type II: Enzymatic screening in dried blood spots on filter paper, *Clin Chim Acta.* **347**:97.

Eng, C.M., Guffon, N., Wilcox, W.R., Germain, D.P., Lee, P., Waldek, S., Caplan, L., Linthorst, G.E., and Desnick, R.J., 2001, Safety and efficacy of recombinant human alpha-galactosidase A—Replacement therapy in Fabry's disease, *N Engl J Med.* **345**:9.

Gerber, S.A., Scott, C.R., Turecek, F., and Gelb, M.H., 2001, Direct profiling of multiple enzyme activities in human cell lysates by affinity chromatography/electrospray ionization mass spectrometry: Application to clinical enzymology, *Anal Chem.* **73**:1651.

Hua, C.T., Hopwood, J.J., Carlsson, S.R., Harris, R.J., and Meikle, P.J., 1998, Evaluation of the lysosome-associated membrane protein LAMP-2 as a marker for lysosomal storage disorders, *Clin Chem.* **44**:2094.

Krivit, W., Peters, C., and Shapiro, E.G., 1999, Bone marrow transplantation as effective treatment of central nervous system disease in globoid cell leukodystrophy, metachromatic leukodystrophy, adrenoleukodystrophy, mannosidosis, fucosidosis, aspartylglucosaminuria, Hurler, Maroteaux-Lamy, and Sly syndromes, and Gaucher disease type III, 1999, *Curr Opinion Neurol.* **12**:167.

Levy, H.L., and Albers, S., 2000, Genetic screening of newborns, *Annu Rev Genomics Hum Genet.* **1**:139.

Li, Y., Scott, C.R., Chamoles, N.A., Ghavami, A., Pinto, B.M., Turecek, F., and Gelb, M.H., 2004, Direct multiplex assay of lysosomal enzymes in dried blood spots for newborn screening, *Clin Chem.* doi:10.1373/clinchem.2004.035907.

Meikle, P.J., and Hopwood, J.J., 2003, Lysosomal storage disorders: Emerging therapeutic options require early diagnosis, *Eur J Pediatr.* **162 Suppl 1**:S34.

Meikle, P.J., Brooks, D.A., Ravenscroft, E.M., Yan, M., Williams, R.E., Jaunzems, A.E., Chataway, T.K., Karageorgos, L.E., Davey, R.C., Boulter, C.D., Carlsson, S.R., and Hopwood, J.J., 1997, Diagnosis of lysosomal storage disorders: Evaluation of lysosome-associated membrane protein LAMP-1 as a diagnostic marker, *Clin Chem.* **43**:1325.

Rozaklis, T., Ramsay, S.L., Whitfield, P.D., Ranieri, E., Hopwood, J.J., and Meikle, P.J., 2002, Determination of oligosaccharides in Pompe disease by electrospray ionization tandem mass spectrometry, *Clin Chem.* **48**:131.

Umapathysivam, K., Hopwood, J.J., and Meikle, P.J., 2001, Determination of acid alpha-glucosidase activity in blood spots as a diagnostic test for Pompe disease, *Clin Chem.* **47**:1378.

Umapathysivam, K., Whittle, A.M., Ranieri, E., Bindloss, C., Ravenscroft, E.M., van Diggelen, O.P., Hopwood, J.J., and Meikle, P.J., 2000, Determination of acid alpha-glucosidase protein: Evaluation as a screening marker for Pompe disease and other lysosomal storage disorders, *Clin Chem.* **46**:1318.

Wilcken, B., Wiley, V., Hammond, J., and Carpenter, K., 2003, Screening newborns for inborn errors of metabolism by tandem mass spectrometry, *N Engl J Med.* **348**:2304.

Wilcox, W.R., Banikazemi, M., Guffon, N., Waldek, S., Lee, P., Linthorst, G.E., Desnick, R.J., and Germain, D.P., 2004, Long-term safety and efficacy of enzyme replacement therapy for Fabry disease, *Am J Hum Genet.* **75**:65.

Wilson, J., and Junger, G., 1968, *The Principles and Practice of Screening for Disease*, World Health Organization, Geneva.

Young, S.P., Stevens, R.D., An, Y., Chen, Y.T., and Millington, D.S., 2003, Analysis of a glucose tetrasaccharide elevated in Pompe disease by stable isotope dilution-electrospray ionization tandem mass spectrometry, *Anal Biochem.* **316**:175.

Zytkovicz, T.H., Fitzgerald, E.F., Marsden, D., et al., 2001, Tandem mass spectrometric analysis for amino, organic, and fatty acid disorders in newborn dried blood spots: A two-year summary from the New England Newborn Screening Program, *Clin Chem.* **47**:1945.

GENETIC COUNSELING FOR LYSOSOMAL STORAGE DISEASES

Erin O'Rourke,[1] Dawn Laney,[2] Cindy Morgan,[3] Kim Mooney,[4] Jennifer Sullivan[5]

In many ways, the role of the genetic counselor working with patients and families with a lysosomal storage disease is similar to a counselor in other pediatric and adult counseling situations. The goals of counseling; education, access to health care, and supportive counseling are the same. Although the goals of counseling are simply stated, an effective counseling session is always a complex interaction. It is an empathic exchange where the counselor promotes understanding of the particular disease, the inheritance pattern, takes a family history, provides recurrence risks, and reviews diagnostic testing, disease management issues, and supportive resources, while simultaneously learning about family structure, dynamics, and the impact of the diagnosis. The counselor assesses the level of understanding about the disease and the appropriate resources to assist the patient and family facing an inherited disease. A balance of teaching and counseling and the ability to be flexible and responsive to the counselee is key for achieving an effective interaction.

There are a few important differences in the genetic counseling for this group of genetic diseases. Because these are individually rare diseases, the initial counseling session is often the first time the patient and family have seen a health care professional who is familiar with the lysosomal disease and therefore a basic description of the disease and the basis for the disease is usually an appropriate and very much appreciated starting point. Because there is treatment available for most of the lysosomal diseases discussed in this chapter, the related counseling includes discussions about treatment options and expectations regarding treatment outcomes. The requirement for lifelong treatments at regular intervals provides the opportunity for continued interaction and a long-term relationship between the counselor and the patient/family. This can be mutually rewarding as there are numerous opportunities to educate and counsel as the patient and family adjust to the diagnosis, and deal with emerging issues such as disease management, family planning, and living with a lysosomal storage disease.

The chapter is designed as a resource for counselors or other health care providers interested in an update on counseling issues for this group of treatable lysosomal storage diseases. Six genetic counselors who have been involved with the lysosomal storage disease community for the last decade draw on their counseling expertise to provide a disease summary and clinical counseling pearls for Gaucher disease, Fabry disease, Mucopolysaccharidosis (MPS) type 1, MPS II, MPS VI, and Pompe disease.

---

1. University of Pittsburgh, Department of Human Genetics/ Genzyme corporation. E-mail: erin.orourke@hgen.pitt.edu
2. Emory University, Atlanta, GA
3. University of California, San Francisco
4. Biomarin Pharmaceuticals
5. Duke University Medical Center

1 GAUCHER DISEASE

Erin O'Rourke; University of Pittsburgh/Genzyme Corporation

Gaucher disease is an autosomal recessive lysosomal storage disease caused by a deficiency of the enzyme glucocerebrosidase, the enzyme important in the catabolism of glucosylceramide into ceramide and glucose. When this enzyme activity is deficient, glucosylceramide accumulates in the lysosomes of macrophages in a variety of tissues causing hematologic, visceral, skeletal, and rarely neurologic disease symptoms. The glucosylceramide storage and resultant toxic process in surrounding tissues leads to the symptoms of Gaucher disease which include anemia, thrombocytopenia, splenomegaly, hepatomegaly, growth retardation, osteopenia, pathologic fractures, pulmonary hypertension, and rarely central nervous system disease (Barranger & Ginns 1989; Grabowski & Beutler 2001).

There are three types of Gaucher disease that are differentiated based on the presence of neurologic involvement. Type 1 disease does not involve the central nervous system (CNS) and occurs more frequently in the Ashkenazi Jewish (AJ) population where the carrier frequency is 1 in 10 and the incidence is approximately 1 in 450 AJ persons. Types 2 and 3 diseases involve the CNS and occur less frequently than type 1 disease with incidences of 1 in 100,000 and 1 in 50,000, respectively. Type 2 Gaucher disease has an acute neurologic course whereas type 3 disease is the chronic neurologic form. Types 2 and 3 diseases have no ethnic predilection; however, there is a genetic isolate of type 3 disease in Sweden (Barranger & Ginns 1989; Grabowski & Beutler 2001).

The age of onset in type 1 Gaucher disease is variable; children most often present with splenomegaly whereas many adults are diagnosed secondary to unexplained anemia, thrombocytopenia, or bone pain. Infants with acute neurologic disease usually present at 3 to 6 months with failure to thrive, hyperreflexia, splenomegaly, and strabismus. Many of these infants never sit alone, become opisthotonic, and die due to aspiration by 24 months of age. Children with type 3 disease most often present with an eye movement disorder and compensatory head thrusting; splenomegaly is present but variable in severity (Barranger & Ginns 1989; Grabowski & Beutler 2001).

The diagnosis of Gaucher disease is made by measuring the activity of glucocerebrosidase in peripheral blood leukocytes, amniocytes, chorionic villi, or fibroblasts (Grabowski & Beutler 2001). All individuals with Gaucher disease have decreased glucocerebrosidase activity. The preferred method for determining Gaucher disease carrier status is mutation analysis. There are four common mutations in the glucocerebrosidase gene, N370S, L444P, 84gg, and IVS2+1, that account for approximately 97% of the mutations in the AJ population and 50–65% of the mutations in the non-AJ population (Grabowski & Beutler 2001). The genotype phenotype correlation is limited. The N370S mutation is associated with type 1 (nonneurologic) disease (Barranger & O'Rourke 2001).

Prior to enzyme replacement therapy, the management of Gaucher disease was symptomatic. Splenectomy temporarily improved anemia and thrombocytopenia; analgesics and narcotics were and continue to be used to manage bone pain and bone crises; and joint replacement is used for knees, hips, and shoulders that are irreversibly damaged from Gaucher-related avascular necrosis.

Treatment in the form of enzyme replacement therapy (Cerezyme®) has been available since 1991 and is recommended for children and adults with types 1 and 3 Gaucher disease. Enzyme replacement therapy (ERT) is a safe and effective treatment.

Most treated patients experience a reversal of anemia, thrombocytopenia, and organomegaly in 6 to 12 months, and an improvement in bone disease in 12 to 24 months.

Other approaches to therapy in Gaucher disease include substrate synthesis inhibition (SSI). SSI has been studied in type 1 Gaucher disease and does not compete with ERT in either efficacy or safety (Cox et al., 2000; Elstein et al., 2004). Substrate synthesis inhibittion may be appropriate in the neurological form of Gaucher disease as these molecules may cross the blood–brain barrier. Clinical trials are underway to evaluate the utility of this approach. There is considerable effort at present being directed towards developing small molecules with more therapeutic and fewer side effects.

There is a well-established standard of care for monitoring patients with Gaucher disease and a registry that has been available to physicians and their patients for more than a decade. The schedule of monitoring assessments is available on the registry Web site at www.gaucherregistry.com (Charrow et al., 2004; Damiano, Pastores, & Ware 1998; Pastores, Weinreb, & Aerts 2004).

It is known that treatment with ERT does not prevent neurological disease progression and death in infants with the acute neurological form of Gaucher disease and therefore it is not recommended for infants diagnosed with type 2 disease. However, neurological Gaucher disease is thought of as a continuum of disease, therefore it is important to allow infants with significant but not classic acute neurological disease to declare themselves. With careful and frank counseling about the uncertainty of the child's long-term prognosis and knowledge that treatment does not prevent progression of acute disease, some families may choose a trial of ERT with careful monitoring for neurological progression and a plan to withdraw ERT if neurological progression is evident.

Counseling for families with type 2 disease includes discussions about the implications of the diagnosis and the prognosis for their child, options in future pregnancies (egg donation, sperm donation, IVF with gamete testing, adoption, prenatal diagnosis), supportive counseling related to facing death and loss, and support group resources that are often helpful both while a child with acute neurological disease is alive and afterwards.

Families with nonneurological and chronic neurological Gaucher disease must adjust to incorporating every other week infusions into their lifestyle. Compliance is sometimes an issue particularly in teenagers and young adults inasmuch as the therapy is lifelong and treated patients, especially those identified and adequately treated since childhood, usually feel very well. Older patients with disease that had progressed prior to the availability of ERT may have significant burden of disease such as bone pain and/or disability which can lead to depression. Insurance concerns are a reality for patients faced with the expense of lifelong treatment. Quality of life in treated Gaucher patients is improved both by individual report and collectively using standardized measures such as the SF-36 completed at regular intervals.

Resources for the Gaucher community include:
National Gaucher Foundation www.gaucherdisease.org
National Society of Genetic Counselors www.nsgc.org
Genzyme Treatment Support 800-745-4447.

2 FABRY DISEASE

Dawn Laney; Emory University

Fabry disease (FD) is an X-linked genetic condition caused by a deficiency of the enzyme α-galactosidase A (α-Gal A) resulting from a mutation in the α-galactosidase A gene (GLA). The lack of α-Gal A leads to a progressive accumulation of glycosphingolipids in vascular endothelial cells and other cell types. Symptoms of FD often present in childhood in males and females as severe pain in the hands and feet (acroparesthesias), characteristic skin rash (angiokeratomas), gastrointestinal complications, headaches, decreased sweating (hypohidrosis), decreased urine concentration (isosthenuria), fatigue, and protein in the urine (proteinuria). The progression of FD typically leads to renal, cardiovascular, and cerebrovascular disease in the third to fifth decade of life. Variant forms of FD that predominantly affect the heart or kidneys are also part of the FD spectrum (Desnick et al., 2005; Brady, Grabowski, and Thadhani 2000; Mac Dermot and Holmes 2001).

The least invasive and most cost-effective method of diagnosing Fabry disease in males is by measuring α-Gal A levels in plasma, leukocytes, and/or cultured cells. The enzyme analysis of α-Gal A is not as accurate in women due to their variable patterns of cellular enzyme activity. Accordingly, diagnosis of at-risk females is achieved through a combination of enzyme analysis of α-Gal A and molecular testing of the GLA gene on Xq22. If a known mutation is present in the family, targeted mutation analysis is available. Therefore, molecular testing of affected males is important to help identify affected female relatives. If no mutation is known in the family of the at-risk female, sequence analysis of the alpha-galactosidase A gene is available. The condition is present in all ethnicities and most families have private mutations. Prenatal testing is available.

Enzyme replacement therapy (ERT) is available for the treatment of Fabry disease. ERT is given intravenously every other week as an exogenous source of α-Gal A. ERT is not a cure for FD, but it does reduce globotriaosylceramide (GL3) deposition in the capillary endothelium of the kidney, heart, and skin (Desnick et al., 2005; Brady, Grabowski, and Thadhani, 2001; Eng et al., 2001).

Several FD-related symptoms can be effectively treated by ERT in combination with standard symptomatic treatments. As examples, antiepileptic medications such as carbamazepine and phenytoin can offer relief of pain from acroparesthesias and Fabry pain crises, and Angiotensin-converting enzyme (ACE) inhibitors are often used to treat proteinuria (Desnick et al., 2005).

In addition to ERT, most FD patients will still need monitoring and treatment of existing FD-related health issues by a team of specialists. Many FD patients are identified later in the course of their disease progression and will have irreversible damage to their kidneys and heart. Patients affected by FD should be monitored whether or not they are undergoing ERT. Recommended assessments include monitoring quality of life and pain by questionnaire and annual follow-up of renal, cardiac, audiological, neurological, and ophthalmological parameters. Biomarker (GL3) and antibody studies should be added to assessments if the patient is on ERT. The Fabry registry board has developed a series of recommended assessments for symptomatic FD patients. Recommendations for the testing and treatment of asymptomatic pediatric patients are under development (Desnick et al., 2000; Bennett et al., 2002).

FD is a chronic, progressive genetic disease. The psychosocial themes of FD revolve around the individual disease management issues, family dynamics, guilt, and denial. The monetary and time commitment of ERT, the assessments, and specialist visits can be overwhelming to FD patients. It is important to provide and prioritize care, connect them with strong support systems, and provide resources to decrease FD's impact on their lives as much as possible. It is also important to remember to give patients a realistic expectation for ERT: it is a new therapy that may not be effective in reversing damage already done. In addition, because depression and anxiety are part of FD, treatment for both conditions must be addressed for successful patient management.

Given the X-linked inheritance pattern of the disease, management includes multiple affected family members. Often, the identification of one proband will result in an average of six other affected family members. Learning the family dynamics is key for successful diagnosis and treatment. Each family member will have a different perspective on the disease. Did they watch their relatives suffer and die from FD? Are they still grieving the loss of a relative to FD? Are they angry at the medical community for past medical care for themselves or other family members? Do they have "survivor guilt" because they are unaffected by Fabry disease? Are they mothers who refuse any evaluation or treatment until their sons are on ERT and improving? Were they, as women, told by their physicians and family members that they cannot be affected by FD?

Most counseling challenges in FD can be addressed through discussion of psychosocial issues directly. Guided discussion with patients and their families provides knowledge about family dynamics. During discussion genetic counselors should be prepared to provide resources, develop open lines of communication, and determine the issues and impediments to medical care for each patient.

Below is a list of challenges to consider when developing a counseling plan for a FD patient and family:

- FD patients often do not appear ill to the casual observer may be thought of as lazy, unmotivated, and manipulative.

- Late diagnosis and ERT treatment may delay, but not prevent, serious FD-related complications.

- The patient may not fully understand FD and may need a genetic refresher course.

- Depression and anxiety can make a patient appear noncompliant.

- FD patients may be addicted to opioids or be self-medicating for pain therapy.

- ERT is a biweekly lifetime treatment that interrupts normal life activity and work.

- Lifetime ERT is expensive and financial and emotional burdens should be addressed directly.

- Transitioning teenagers from parent-directed care to independence is crucial to continued care.
- FD treatment is focused on the individual, but involves the whole family.

Diagnosis and treatment of FD are still in their infancy. Ongoing research will provide more information about FD and ERT treatment. The development of quick, inexpensive blood spot testing may soon result in FD being added to the federally recommended newborn screening panel. Gene therapy may provide a cure for FD, but this technology is still in development.

3 MUCOPOLYSACCHARIDOSIS

3.1 Mucopolysaccharidosis Type 1

Cindy Morgan; University of California, San Francisco

Mucopolysaccharidoses (MPS) are a group of lysosomal storage disorders caused by the deficiency of enzymes that degrade glycosaminoglycans. Mucopolysaccharidosis type I (MPS I) is a rare disorder caused by the deficiency of the lysosomal enzyme α-L-iduronidase. This enzyme catalyzes the second reaction in the stepwise degredation of the glycosaminoglycans (GAGs) dermatan and heparan sulfate. Deficiency of α-L-iduronidase causes progressive accumulation of these GAGs in the lysosome and leads to tissue and organ dysfunction (Terlato and Cox, 2003).

MPS I displays a wide clinical spectrum and historically there have been three recognized subtypes: Hurler, Hurler–Scheie, and Scheie. Hurler syndrome is the most severe, Scheie is the mildest, and Hurler–Scheie an intermediate phenotype. However, not all patients fit into one of these distinct phenotypes and MPS I is often viewed as a clinical continuum. Assignment along the continuum is made by clinical findings as all three are relatively indistinguishable biochemically. All forms have excessive urinary excretion of dermatan and heparan sulfate, deficient α-L-iduronidase, and storage of GAGs in cultured fibroblasts (Terlato and Cox, 2003). Mutation analysis can be helpful in diagnosing Hurler syndrome, but not the others. Most families have private mutations and nonsense, missense, insertion, deletion, and splice site mutations have all been identified. Patients with two nonsense mutations develop Hurler syndrome, but the phenotypes of the other mutations are much more variable. Consequently genotype–phenotype correlations can not be made unless two nonsense mutations are identified (Peters et al., 1998).

MPS I patients have a normal appearance at birth. Individuals with Hurler syndrome are generally diagnosed between 4 and 18 months of age and most present with any of the following features: skeletal deformities, recurrent ear and nose infections, inguinal and/or umbilical hernias, coarse facial features, hepatosplenomegaly, and macroglossia. MPS I is a progressive disorder and patients will also develop corneal clouding, glaucoma, joint contractures, dysostosis multiplex, short stature, cardiac valve disease, hearing loss, and obstructive airway disease. Hurler patients also have developmental delay, which usually is apparent by 12–24 months age. Maximum functional capacity is about 2–4 years of age followed by deterioration in intellectual function. Hurler–Scheie and Scheie syndromes spare cognitive abilities although patients can suffer from neurologic morbidity due to spinal cord compression from cirumferential meningeal storage. Joint stiffness, aortic valve disease, corneal clouding, coarse facial features, and possibly some other findings generally characterize Scheie syndrome. Hurler–Scheie patients have physical features in between Hurler and Hurler–Scheie. Obstructive airway disease, respiratory infection, and cardiac disease are generally the causes of death in all three forms. Hurler patients often die by five years of age whereas Scheie patients may have a relatively normal lifespan (Peters et al., 1998, Wraith et al., 2004).

Hematopoetic stem cell transplant (HSCT) has been successfully used to treat Hurler patients since 1980. Successfully engrafted HSCT corrects the enzyme deficiency and GAG storage, improves many of the systemic abnormalities, helps prevent physical progression, and increases lifespan. If Hurler patients are treated before age two and before developmental regression occurs, HSCT can prevent intellectual decline and permit normal or near normal intellectual development. Microglial cells in the brain originate in the bone marrow and are believed to be the source of α-L-iduronidase in the brains of successfully engrafted patients. HSCT does not treat the skeletal or ocular disease so natural history of these features is virtually the same as nontransplanted patients.

Enzyme replacement therapy (ERT) became available for MPS I in 2003 and is marketed under the trade name Aldurazyme®. In clinical studies Aldurazyme has been shown to ameliorate clinical disease by improving pulmonary function and endurance, and decreasing hepatosplenomegaly and GAGs in peripheral tissues. Additionally, recipients reported an increased sense of well-being and increased range of motion in large joints. Aldurazyme can not cross the blood–brain barrier and therefore has not proven to have any effect on the central nervous system. Long-term data on the effects of Aldurazyme are not available. Adurazyme therapy is currently indicated for individuals with Hurler, Hurler–Scheie, and Scheie patients with moderate to severe symptoms. Aldurazyme is given as a weekly 4–6 hour intravenous infusion at 0.58 mg/kg. The greatest risk of Aldurazyme ERT is anaphylactic reaction (Kakkis et al., 2004).

MPS I is an autosomal recessive disorder with an estimated incidence of 1/100,000 births. Prenatal diagnosis is available by enzyme or mutation analysis on CVS tissue or amniocytes. Carrier testing is available but is only really reliable when done via mutation analysis.

MPS I is a progressive and fatal disease that is a devastating diagnosis for families to receive. It is a rare condition and consequently most families feel isolated and have difficulty finding adequate emotional support. The MPS Society offers several resources including layman's literature, national conferences, and family contacts.

One of the biggest remaining challenges in MPS I is treating the central nervous system manifestations. Studies have recently been conducted using intrathecal administration of recombinant α-L-iduronidase in MPS I affected dogs (Kakkis et al., 2004). These studies found that mean total GAG levels in the brain returned to normal, the meninges had a 57% reduction of GAG storage, and histological improvement was also noted in all cell types. Long-term effects of the therapy were not studied so it is not known if there were any improvements in cognitive function. At this time, clinical trials in humans have not been conducted (Kakkis et al., 2004).

In addition, trials are currently being conducted with concomitant use of ERT and HSCT. Patients are infused with ERT before and after transplant with the hope of increasing the chance of engraftment. ERT is tolerated in transplant patients but it is not currently known if it improves outcome (Grewal et al., 2005).

3.2 Mucopolysaccharidosis Type II

Amy Fisher; University of North Carolina

MPS II is an X-linked disorder characterized by multisystemic clinical disease due to deficiency of enzyme iduronate sulfatase (also iduronate-2-sulfatase). The deficiency leads to the progressive accumulation of glycosaminoglycans, resulting in tissue and organ dysfunction. A wide range of clinical presentations, symptom severity, and central nervous system involvement is observed in patients with iduronate sulfatase deficiency. Considering the continuous spectrum of clinical involvement, the multisystemic nature of the condition, and the complex nature of genetic counseling for MPS II individuals and their families, multidisciplinary management of MPS II patients is necessary.

When a MPS condition is suspected, screening testing is often performed according to the practitioner's index of suspicion and to determine the need for additional evaluation and testing. In those with MPS, a skeletal survey will often reveal changes consistent with dysostosis multiplex. Quantitative urinary glycosaminoglycans (GAGs) analysis will allow for the detection of elevated levels of GAGs. Those with MPS II will often have elevated heparan and dermatan sulfate in the urine. However, this finding is not exclusive to MPS II, and additional enzymatic testing is indicated. Deficiency of iduronate sulfatase, performed on serum, plasma, leukocytes, or fibroblasts, is confirmation of MPS II. If iduronate sulfatase deficiency is documented, another sulfatase enzyme should be measured to rule out multiple sulfatase deficiency. Molecular testing for pathogenic mutations in the IDS gene is commercially available. The mutational spectrum is broad, with the majority of individuals with MPS II having private mutations. Once a diagnosis is made, prenatal testing by enzyme or DNA analysis is available for a subsequent pregnancy. Carrier testing for at-risk family members can be reliably performed once a pathogenic DNA mutation is identified in a proband.

Management consists of supportive care and the treatment of complications. The progressive nature of MPS II dictates the need for constant evaluation of clinical status. The goal of the systematic evaluations is to improve the quality of life for the MPS patient and the family. Coordination of care for MPS II individuals involves multiple subspecialty health care services, which often include neurology, cardiology, pulmonology, otolaryngology/audiology, orthopedics, ophthalmology, gastroenterology, dentistry, and genetics. From a neurological perspective, there is risk for developmental delay and mental retardation, communicating hydrocephalus, spinal cord compression, and carpal tunnel syndrome. Individuals with MPS II should have regular brain and spine imaging studies, as well as formal developmental assessment.

A nerve conduction study, even in the absence of clinical symptoms of carpal tunnel is often indicated as early as mid-childhood. Valvular heart disease can be seen in MPS II and should be serially evaluated with regular echocardiograms. Medication may be required to manage heart disease, and valve replacement is a surgical option for advanced disease. Obstructive sleep apnea, restrictive lung disease, and reactive airway disease are significant complications for many with MPS II. A sleep study, bronchoscopy, and/or pulmonary function tests can be used to assess respiratory function. Management for a compromised airway can include oxygen supplementation, asthma medications, positive airway pressure masks, or tracheostomy. Combined sensorineural and conductive hearing loss can be augmented with hearing aids. Joint stiffness and contractures are progressive and ongoing physical therapy, as well as assistive equipment, are recommended to maintain joint range of motion as much as possible.

Historically, bone marrow transplantation has been a widely used treatment for some forms of MPS. However, it was not found to affect the neurological disease in those with MPS II and has not been a recommended treatment. More recently, hematopoetic stem cell transplantation (HSCT) has been used as a therapy for several patients with lysosomal storage disorders and has been performed for a very small number of those with MPS II. Data regarding the success of this intervention for MPS II patients are not yet available. Enzyme replacement therapy is currently in clinical trials. Preliminary data have shown that patients receiving enzyme achieved a statistically significant improvement in the primary efficacy endpoint compared to those who received placebo. Considering these positive data, and enzyme replacement therapy that has been approved for other MPS disorders, FDA approval of ERT for MPS II is anxiously anticipated by practitioners and families. ERT given by peripheral IV infusion is not expected to cross the blood–brain barrier and affect CNS disease.

Although care may be heavily focused on supportive medical care and treatment of complications, care providers must also be accessible and knowledgeable while helping a family through the diagnostic process and general genetic information, life stage transitions, and connecting with other families confronting the same diagnosis. An early challenge to patients and families with MPS II is navigating the diagnostic process. Understanding the difference between screening and diagnostic testing, the time waiting for test completion, and largely unavailable testing to assess clinical prognosis are significant hurdles for any family involved in evaluation for MPS.

Once a diagnosis is made, extensive counseling is required to assist families in coping with the progressive nature of the condition and the need for ongoing medical management. Meeting other individuals with MPS II is crucial for many newly diagnosed families, but can present challenges depending on symptoms and disease progression for each MPS II individual. Parents of an MPS II child need to become strong advocates for their child's care and participate actively in medical decision making. Often families will need to rely heavily on their care providers to supply letters of support for ongoing therapy services, educational modification, adaptive equipment, and insurance approval for procedures and testing. A diagnosis can markedly change family dynamics and special attention is required to recognize the needs of an unaffected sibling(s). Care providers are challenged to counsel a family through the grief process and as they make ongoing decisions about future reproductive planning, medical care, and intervention. For an individual with MPS II that is neurologically intact, counseling through life transitions as he or she gains independence and responsibility for medical care is critical.

Should ERT be approved for commercial use and administered to patients at earlier ages at the time of diagnosis, many current medical concerns and routine evaluations may not be as necessary. However, compliance with therapy and learning more about therapy when applied to a broader MPS II population will continue to alter medical recommendations. Research continues into new therapeutic interventions for MPS II, including HSCT, gene therapy, and alternative ways to deliver ERT. Newborn screening for lysosomal storage disorders is on the horizon, bringing the potential for earlier diagnoses as well as many new challenges. Detailed registry and outcomes data regarding the natural history of MPS II will aid in documenting the historical perspective and impact of the disorder while demonstrating adaptation to new diagnostic methods and therapeutic interventions.

3.3 Mucopolysaccharidosis Type VI

Kim Mooney; Biomarin

Mucopolysaccharidosis type VI (MPS VI or Maroteaux–Lamy syndrome) is an autosomal recessive lysosomal storage disease caused by deficiency of the enzyme arylsulfatase B, (N-acetylgalactosamine 4-sulfatase), one of the enzymes involved in the stepwise catabolism of glycosaminoglycans (GAGs). Specifically, the enzyme hydrolyzes and removes a sulfate from a hexosamine sulfate of dermatan sulfate. Therefore, when this enzyme activity is deficient, dermatan sulfate is stored in the lysosomes of cells and is excreted in the urine.

The prevalence of the disease is estimated at 1 in 250,000. As with the other MPS diseases, the spectrum of clinical phenotypes ranges from mild to severe; however, intelligence is usually normal in MPS VI. Most patients are diagnosed between the ages of one and five years and present with short stature. In the rapidly advancing form of the disease, maximum height achieved ranges from 80–120 cm. Although macrocephaly or a deformed chest may be present at birth, growth can be normal until age six. Other clinical features include coarse facial features, joint contractures (particulary hip dysplasia), pain and stiffness, impaired vision and hearing, corneal clouding, heart valve disease and cardiomyopathy, reduced pulmonary function and frequent infections, upper airway obstruction and sleep apnea, hepatosplenomegaly, increased intracranial pressure, spinal cord compression, and decreased endurance and energy. Patients with the rapidly advancing form of the disease die in the second or third decade of life, whereas mild patients may survive into the fourth or fifth decade.

MPS VI patients reveal predominantly elevated levels of dermatan sulfates in urine MPS screening, but this does not confirm the diagnosis or differentiate between subtypes. The diagnosis is confirmed by lysosomal enzyme assay of leukocytes or fibroblasts demonstrating a deficiency of the enzyme, arylsufatase B. However, carrier testing by enzyme assay is inaccurate. Mutation analysis of the gene at 5q13-q14 is available at select laboratories; however, genotype phenotype correlations are not accurate. Approximately 170 alleles have been identified, including frameshift, nonsense, and missense mutations, which are the most common. Prenatal diagnosis is available by enzyme assay via amniocentesis or CVS for families with a previous affected child.

The management of patients with MPS VI has been symptomatic, with respiratory and cardiovascular systems as the main focus. Medications include treatment for frequent infections, glaucoma, cardiac disease, airway obstruction with CPAP or BiPAP (positive airway pressure), and prophylactic antibiotics before and after dental treatments. The following surgeries may be performed depending on the clinical findings of the individual patient: cervical spinal fusion for skeletal instability, surgical decompression including laminectomy and removal of thickened dura for spinal cord compression, hip replacement for degenerative arthritis, valve replacement for incompetent cardiac valves, tracheostomy and tonsillectomy/adenoidectomy for upper airway obstruction, ventriculoperitoneal shunt placement for increased intracranial pressure, carpal tunnel release, umbilical and inguinal hernia repair, tube (grommet) insertion for chronic otitis media, and corneal transplantation. Caution should be taken when performing these surgeries as there is an increased risk with anesthesia in these patients as with all MPS diseases. Bone marrow transplant has been performed in some patients with varying degrees of success. In addition, enzyme replacement therapy (ERT) has recently become available. Naglazyme™ (galsulfase) received U.S. FDA approval for the treatment of MPS VI on May 31, 2005.

Patients receive weekly IV infusions of 1 mg/kg of Naglazyme. Results from clinical trials revealed improvements in endurance, joint range of motion, pulmonary function, growth, and a reduction in storage of urinary glycosaminoglycans. Infusion-related adverse events commonly included fever, chills/rigors, headache, rash, and mild to moderate urticaria. Severe reactions included angioneurotic edema, hypotension, dyspnea, bronchospasm, respiratory distress, apnea, and urticaria. No patients discontinued Naglazyme infusions for adverse events. Nearly all patients developed antibodies as a result of treatment, but the level of the immune response did not correlate with adverse events or affect the improvements experienced in endurance.

Patients should be monitored in a multidisciplinary clinic with ENT, neurology, ophthalmology, pulmonology, and cardiology on a routine basis. The following evaluations should be performed approximately once a year: electrocardiogram, echocardiogram, vision and hearing tests, sleep studies and pulmonary function tests, skeletal survey or MRI to assess bone involvement, EMG in patients who have evidence of nerve compression to determine the need for carpal tunnel or trigger finger release, and evaluation of joint stiffness to determine the need for physical or occupational therapy. Testing needs to be individualized and more frequent monitoring may be necessary in some patients, specifically cardiac and respiratory status in patients with the rapidly advancing form of the disease.

Families who have children with MPS VI face many challenges. The psychological impact of dysmorphic features, short stature, significant health problems, and the possibility of a shortened life span in a person with normal intellect should be considered when counseling the parents as well as the patient himself. Furthermore, it is difficult to predict the phenotypic outcome of a prenatal diagnosis or even a newborn or young child with a diagnosis of MPS VI and how this outcome will be affected by ERT. In addition, there are new challenges that will arise with ERT, including the time commitment for weekly treatments and insurance coverage concerns. More research is required to gain information on women with MPS VI who become pregnant and whether MPS VI should be added to newborn screening panels.

Resources for families with MPS VI can be found through the MPS Society, NORD, or at www.mpsvi.com, a Web site devoted to MPS VI. In addition, a confidential, Web-based MPS VI registry provides information about the severity, prevalence, and natural progression of the disease (www.mpsvi.org).

4 POMPE DISEASE

Jennifer Sullivan, MS, CGC; Duke University Medical Center

Pompe disease is a metabolic myopathy with a spectrum of clinical symptoms broadly correlated to residual enzyme activity. Pompe disease is typically divided into two groups: infantile onset (presenting at less than one year of age) and late onset (clinical symptoms noted >one year of age). Infantile-onset Pompe disease is uniformly lethal. Affected infants have a classic phenotype with rapid and progressive muscle weakness, a profound cardiomegaly, and death from cardiorespiratory failure usually by one year. Spinal muscular atrophy type 1 (Werdnig–Hoffmann disease) is one of the most frequently cited diseases in the differential diagnosis. Late-onset Pompe disease is characterized by a lack of severe cardiac involvement and a less severe short-term prognosis. Weakness is first noted in large proximal muscles such as the quadriceps. Approximately one third of all adult cases present as respiratory failure. Limb girdle muscular dystrophy and polymyositis are often initial diagnostic considerations (Hagemans et al., 2005; Kishnani and Howell 2004).

As of 2006, rapid technological advances greatly increased diagnostic choices related to Pompe disease. Enzyme-based testing using very limited blood quantity in the form of dried blood spots, is now clinically available for both infantile- and late-onset forms of the disease (Li et al., 2004; Zhang et al., 2006). This advancing technology not only reduces turnaround time for diagnostic testing results, but also provides greater access to testing for Pompe disease in clinical populations where skin or muscle biopsy have previously been the only diagnostic options. Blood-based enzymatic testing has only recently begun widescale clinical use, so clinical correlation, as well as second-tier testing using more established diagnostic methods is currently recommended. Ultimately, especially for late-onset Pompe disease, the "gold standard" for diagnosis is often enzymatic testing on either muscle or cultured fibroblasts. Both of these methodologies have unique benefits and limitations that should be discussed with the family if appropriate. Muscle testing requires invasive sampling, but can yield a rapid result. Alternatively, fibroblast testing is less invasive, but because of the required time for cell growth, results may take up to four to six weeks. In addition, clinical availability of DNA-based testing, via both targeted mutation analysis and full sequencing, offers an alternative additive aspect of diagnostics for both affected and unaffected family members.

Prenatal diagnosis is available in a limited number of laboratories via CVS cells and amniocytes. At present, enzymatic testing is more widely available than DNA-based testing, and can provide information regarding disease status from fetal samples with good reliability, especially in families concerned about the infantile form of Pompe disease. Currently, most laboratories require cultured cells for such testing, which will add to the time from when the prenatal sample is obtained until diagnostic testing results are available for consideration by the family. In general, the turnaround time for prenatal samples is four to six weeks from time of procedure. It is important to note that prenatal enzymatic testing can provide definitive information regarding the status of a fetus even in the absence of known familial mutations. DNA-based prenatal diagnosis is an option in families where familial mutations have been identified.

In April 2006, enzyme replacement therapy (ERT) for Pompe disease was broadly approved for treatment of all forms of the disease in the United States. Much as with other ERT interventions, treatment with Myozyme® involves regular infusions of the product with a dosage based upon body weight (most typically 20 mg/kg every other week). Multiple clinical trials investigating the treatment of Pompe disease with ERT have been reported in recent years (Kishnani et al., 2007; Kishnani et al., 2006). In general, results have been promising, with prolonged survival, a very robust cardiac response, and variable muscular response. However, long-term prognostic information for treated patients is limited, given that initial human clinical trials of ERT began in 1999. At the time of this writing, patients have received ERT for a maximum of seven years. Some patients with classic, infantile-onset Pompe disease have demonstrated dramatic response to treatment, with survivors as old as eight years.

It is important to note that ERT is but one aspect of care. Experience with clinical trials of this product demonstrated that treated individuals are frequently medically fragile, requiring vigilant and aggressive medical management with a multidisciplinary team approach. Furthermore, disease progression may continue, especially during the initial stages of treatment with ERT. In clinical trials a number of infants ultimately died because of disease-related complications. In general, results of the pivotal clinical trials suggest earlier treatment with ERT prior to extensive systems involvement, particularly skeletal muscle, may yield better clinical response with reduced morbidity and mortality.

Because there is multisystem involvement and clinical variability in terms of extent of organ involvement, especially in late-onset Pompe disease, a team approach to medical care is the rule of management. Recently published management guidelines (Kishnani et al., 2006) can provide guidance, especially to newly established clinical care teams. The rarity of this diagnosis means education is of paramount importance for the affected patient, family members, and multidisciplinary medical team. The most frequently involved specialties include cardiology (infantile onset), neurology, pulmonology, intensive care, respiratory therapy, physical therapy, occupational therapy, speech/language pathology, and nutrition. Most frequently, individuals with Pompe disease report the best success when they work closely with a coordinating physician, either primary care or specialist, familiar with the likely disease manifestations.

The best patient response or outcome is often achieved when specialists have an understanding of the unique and complex medical needs specific to individuals with Pompe disease. Therefore, advanced education and consultation among the family, coordinating physician, and specialists planning to work with a particular patient is ideal to ensure that all parties are aware of the role and aim of each specialist's contribution to clinical care (Kishnani and Howell, 2004; Kishnani et al., 2006). Furthermore, a focus on maintaining health and independence related to activities of daily living can have a profound positive impact on quality of life for individuals with Pompe disease (Case and Kishnani, 2006; Hagemans et al., 2004).

For families affected by Pompe disease, the rapidly changing status of treatment options is of particular difficulty because there continue to be many uncertainties. This period of uncertainty in regard to access to treatment is highly frustrating for many families. Many families express frustration about "watching" their family member's disease progression over time, even with the use of supportive and palliative care, as well as ERT. Oftentimes families have unrealistic expectations regarding the ability of ERT to quickly reverse disease manifestations. Furthermore, long-term data regarding outcomes from ERT are limited to information from 1999 or later, and contain minimal data regarding treatment response in late-onset patients.

Families who have experienced the death of a child because of infantile-onset disease must evaluate mixed results in regard to medical status after treatment with ERT. This limited information must then be incorporated into the most common clinical scenario, where the family must decide whether to get pregnant, while considering the use of prenatal diagnosis to determine disease status in a fetus. For the genetic counselor or medical professional working with a family at that critical time period, careful and realistic discussions of the possible outcomes for an infant affected with Pompe disease must be encouraged.

Carrier status for Pompe disease can be accurately determined using DNA-based testing. If mutation analysis has not been performed on the proband, recurrence risk for other family members may be more difficult to determine. Families should be made aware that testing for carrier status of partners of at-risk individuals and/or relatives of affected individuals has only recently become widely available. The sensitivity of testing in individuals with no family history is unknown, and may yield false negative results. Enzymatic analysis is not appropriate for determination of carrier status.

Based upon two population-based studies, the incidence of Pompe disease is estimated to be about 1/40,000 with clinical phenotype spread across the disease spectrum (Ausems et al., 1999; Martiniuk et al., 1998). Pompe disease is panethnic, but with higher incidence of the infantile form noted in African/African American and Southern Chinese/Taiwanese populations and late-onset Pompe disease more frequently observed in the Netherlands.

Studies have revealed that allelic heterogeneity contributes to the wide spectrum of clinical presentations of Pompe disease (Hermans et al., 2004; Hoefsloot et al., 1990). However, there is limited clinical variability noted between affected full siblings who have the infantile-onset form. In families with late-onset disease, the presentation between affected full siblings is often more variable, likely because of the influence of other genetic and environmental factors.

There have been reports of both the infantile-onset and late-onset forms of Pompe disease within the same family (i.e., grandfather and grandchild) caused by segregation of deleterious alleles with the family (Amartino et al., 2006; Hoefsloot et al., 1990). Consequently, when counseling families with new diagnoses, it is important to take a complete family history focused on the symptoms of all clinically affected systems. Encouraging adult patients to share their diagnosis with family members of child-bearing potential allows the use of prenatal diagnosis for interested individuals and may facilitate the early diagnosis of infantile-onset disease. Likewise, astute education regarding the late-onset presentation of Pompe disease may prevent a long diagnostic odyssey for older family members with symptoms of late-onset Pompe disease.

With the recent approval of ERT for use as a treatment of Pompe disease, families will need to adjust to the reality of a lifelong, expensive treatment modality for a chronic disease. Given the trend observed in clinical trials, that earlier intervention with ERT results in better clinical outcome, the medical community as a whole will need to consider the inclusion of Pompe disease as part of newborn screening. Furthermore, families and practitioners alike will likely require additional information regarding prognosis for asymptomatic individuals (Umapathysivam, Hopwood, and Meikle, 2005) as well as methods to monitor response to ERT (An et al., 2005). The natural history of Pompe disease is currently in the process of unfolding. For the future, this translates into an opportunity to learn more about the disease itself, and the implications of diagnosis for families, as well.

REFERENCES

Amartino, H. Painceira, D. Pomponio, R.J. Niizawa, G. Sabio Paz, V. Blanco, M.A. Chamoles, N.A. (2006). Two clinical forms of glycogen-storage disease type II in two generations of the same family. *Clin Genet* 69:187–188.

An, Y. Young, S.P. Kishnani, P.S. Millington, D.S. Amalfitano, A. Corzo, D. Chen, Y.T. (2005). Glucose tetrasaccharide as a biomarker for monitoring the therapeutic response to enzyme replacement therapy for Pompe disease. *Mol Genet Metab* 85:247–254.

Ausems, M.G. Verbiest, J. Hermans, M.P. Kroos, M.A. Beemer, F.A. Wokke, J.H. Sandkuijl, L.A. Reuser, A.J. van der Ploeg, A.T. (1999). Frequency of glycogen storage disease type II in The Netherlands: Implications for diagnosis and genetic counselling. *Eur J Hum Genet* 7:713–716.

Barranger, J.A. Ginns, E.I. (1989) Glucosylceramide lipidosis: Gaucher disease. In: Scriver, C.R. Beaudet, A. Sly, W.S. Valle, D. (Eds). *The Metabolic Basis of Inherited Disease.* 6th ed. New York: McGraw-Hill; pp. 1677–1698.

Barranger, J.A. O'Rourke, E. (2001). Lessons learned from the development of enzyme replacement therapy for Gaucher disease. *J Inherit Metab Dis* 24 Supp2:87–96.

Bennett, R.L, Hart, K. O'Rourke, E, Barranger, J.A. Johnson, J, MacDermot, K.D. Pastores, G.M. Steiner R.D. Tadhani R. (2002). Fabry Disease in Genetic Counseling Practice: Recommendations of the National Society of genetic Counselors. *J Genet Counsel.* 11:121–146.

Brady, R.O. Grabowski, G.A. Thadhani, R. (2000). Fabry disease: α-galactosidase A deficiency. *Gardiner-Caldwell SynderMed* 1–8.

Brady, R.O. Murray, G.J. Moore, D.F. Schiffmann, R.(2001). Enzyme replacement therapy in Fabry disease. *J Inher Metab Dis* 24 Supp2:18–24.

Case, L.E. Kishnani, P.S. (2006). Physical therapy management of Pompe disease. *Genet Med* 8:318–327.

Charrow, J. Andersson, H.C. Kaplan, P. Kolodny, E.H. Pastores, G.M. et al. (2004). Enzyme replacement therapy and monitoring for children with type 1 Gaucher disease: Consensus recommendations. *J Pediatr* 144:112–120.

Cox, T. Lachmann, R. Hollak, C. Aerts, J. van Weely, S. Hrebicek, M. Platt, F.M, Butters, T.D Dwek, R. Moyses, C. Gow, I. Elstein, D. Zimran, A. (2000). Novel oral treatment of Gaucher's disease with N-butyldeoxynojirimycin (OGT 918) to decrease substrate biosynthesis. *Lancet*, 355:1481–1485.

Damiano, A.M.Pastores, G.M. Ware, J.E. (1998) The health related quality of life of adults with Gaucher's disease receiving enzyme replacement therapy: Results from a retrospective study. *Qual Life Res* 7:373–86.

Desnick, R.J. Brady, R.O. Barranger, J.A. Collins, A.J. Germain, D.P. Goldman, M. Grabowski, G.A. Packman, S. Wilcox, W.R. (2003). Fabry disease, an under-recognized multisystemic disorder: Expert recommendations for diagnosis, management, and enzyme replacement therapy. *Ann Intern Med* 138:338–346.

Elstein, D. Hollak, C.E. Aerts, J.M. vanWeely, S. Maas, M. Cox, T.M. Lachmann, R.H. Hrebicek, M. Platt, F.M. Butters, T.D. Dwek, R.A. Zimran, A. (2004). Sustained therapeutic effects of oral miglustat (Zavesca, N-butyldeoxynojirimycin, OGT 918) in type I Gaucher disease. *J Inherit Metab Dis* 27:757–766.

Eng, C.M. Guffon, N. Wilcox, W.R. Germain, D.P, Lee, P. Waldek, S. Caplan, L. Linthorst, G.E. Desnick, R.J. (2001). Safety and efficacy of recombinant human α-galactosidase A replacement therapy in Fabry's disease. *N Eng J Med* 345: 9–16.

Grabowski, G.A. Beutler, E. (2001). Gaucher disease. In: Scriver C, Beaudet A, Sly W, Valle D, Childs B, Kinzler K (Eds). *The Metabolic and Molecular Bases of Inherited Disease*. New York: McGraw-Hill; pp. 3635–3638.

Grewal SS, Wynn R, Abdenur JE, Burton BK, Gharib M, Haase C, Hayashi RJ, Shenoy S, Sillence D, Tiller GE, Dudek ME, van Royen-Kerkhof A, Wraith JE, Woodard P, Young GA, Wulffraat N, Whitley CB, Peters C. (2005). Safety and efficacy of enzyme replacement therapy in combination with hematopoietic stem cell transplantation in Hurler syndrome. *Genet Med* 7: 143–146.

Hagemans, M.L. Janssens, A.C. Winkel, L.P. Sieradzan, K.A. Reuser, A.J. Van Doorn, P.A. Van der Ploeg, A.T. (2004). Late-onset Pompe disease primarily affects quality of life in physical health domains. *Neurology* 63:1688–1692.

Hagemans, M.L. Winkel, L.P. Hop, W.C. Reuser, A.J. Van Doorn, P.A. Van der Ploeg, A.T.(2005). Disease severity in children and adults with Pompe disease related to age and disease duration. *Neurology* 64:2139–2141.

Harmatz, P. Ketteridge, J. Giugliani, R. Guffon, N. Teles, E. Miranda, S, Swiedler, S. Hopwood, J.J. (2005). Direct comparison of measures of endurance, mobility, and joint function during enzyme-replacement therapy of mucopolysaccharidosis VI (Maroteaux-Lamy syndrome): Results after 48 weeks in a phase 2 open-label clinical study of recombinant human N-acetylgalactosamine 4-sulfatase. *Pediatrics* 115, e681–e689.

Harmatz, P. Whitley, C.B. Waber, L. Pais, R. Steiner, R. Plecko, B. Kaplan, P. Simon, J. Butensky, E. Hopwood, J.J. (2004). Enzyme replacement therapy in mucopolysaccharidosis VI (Maroteaux-Lamy syndrome). *J Pediatr* 144(5):574–580.

Hermans, M.M. van Leenen, D. Kroos, M.A. Beesley, C.E. Van Der Ploeg, A.T. Sakuraba, H. Wevers, R. Kleijer, W. Michelakakis, H. Kirk, EP. Fletcher, J. Bosshard, N. Basel-Vanagaite, L. Besley, G. Reuser, A.J. (2004). Twenty-two novel mutations in the lysosomal alpha-glucosidase gene (GAA) underscore the genotype-phenotype correlation in glycogen storage disease type II. *Hum Mutat* 23:47–56.

Hoefsloot, L.H. van der Ploeg, A.T. Kroos, M.A. Hoogeveen-Westerveld, M. Oostra, B.A. Reuser, A.J. (1990). Adult and infantile glycogenosis type II in one family, explained by allelic diversity. *Am J Hum Genet* 46:45–52.

Kakkis E, McEntee M, Vogler C, Le S, Levy B, Belichenko P, Mobley W, Dickson P, Hanson S, Passage M. (2004). Intrathecal enzyme replacement therapy reduces lysosomal storage in the brain and meninges of the canine model of MPS I, *Mol Genet Metab* 83:163–174.

Kishnani, P. Corzo, D. Nicolino, M. Byrne, B. Mandel, H. Hwu, W.L. Leslie, N. Levine, J. Spencer, C. McDonald, M. Li J, DuMontier, J. Michael, H. Chien, Y.H. Hopkin, R. Vijayaraghavan, S. Gruskin, D. Bartholomew, D. van der Ploeg, A. Clancy, J.P. Parini, R. Morin, G. Beck, M. Delagastine, G. S. Jokic, M. Thurberg, B. Richards, S. Bali, D. Davison, M. Worden, M. Chen Y.T. Wraith, J.E. (2007). Recombinant human acid α-glucosidase: Major clinical benefits in infantile-onset Pompe disease. *Neurology* 68:99–109.

Kishnani, P.S. Howell, R.R. (2004). Pompe disease in infants and children. *J Pediatr* 144:S35–43.

Kishnani, P.S. Hwu, W.L. Mandel, H. Nicolino, M. Yong, F. Corzo, D. (2006). A retrospective, multinational, multicenter study on the natural history of infantile-onset Pompe disease. *J Pediatr* 148:671–676.

Kishnani, P.S. Steiner, R.D. Bali, D. Berger, K. Byrne, B.J. Case, L.E. Crowley, J.F. Downs, S. Howell, R.R. Kravitz, R.M. Mackey, J. Marsden, D. Martins, A.M. Millington, D.S. Nicolino, M. O'Grady, G. Patterson, M.C. Rapoport, D.M. Slonim, A. Spencer, C.T. Tifft, C.J. Watson, M.S. (2006). Pompe disease diagnosis and management guideline. *Genet Med* 8:267–288.

Li, Y. Scott, C.R. Chamoles, N.A. Ghavami, A. Pinto, B.M. Turecek, F. Gelb, M.H. (2004). Direct multiplex assay of lysosomal enzymes in dried blood spots for newborn screening. *Clin Chem* 50:1785–1796.

Litjens, T. Hopwood, J.J. (2001) Mucopolysaccharidosis type VI: Structural and clinical implications of mutations in N-acetylgalactosamine-4-sulfatase. *Hum Mutat* 18: 282–295.

MacDermot, K.D, Holmes, A. Miners, A.H. (2001). Anderson-Fabry disease: Clinical manifestations and impact of disease in a cohort of 60 obligate carrier females. *J Med Genet* 38:769–775.

Martiniuk, F. Chen, A. Mack, A. Arvanitopoulos, E. Chen, Y. Rom, W.N. Codd, W.J. Hanna, B. Alcabes, P. Raben, N. Plotz, P. (1998). Carrier frequency for glycogen storage disease type II in New York and estimates of affected individuals born with the disease. *Am J Med Genet* 79(1):69–72.

Meikle, P.J. Hopwood, J.J. Clague, A.E. Carey, W.F. (1999). Prevalence of lysosomal storage disorders. *JAMA* 281:249–254.

Muenzer, J. and Fisher, A. (2004). Advances in the treatment of mucopolysaccharidosis type I. *New Eng J Med* 350:1932–1934.

Neufeld, E.F. Muenzer, J. (2001). The mucopolysaccharidoses. In: Scriver CR, Beaudet AL, Sly WS, Valle D, (Eds). *The Metabolic and Molecular Basis of Inherited Disease*. 8th ed. Vol 3. New York: McGraw-Hill, pp. 3421–3452.

Pastores, G.M. Weinreb, N.J. Aerts, H. (2004). Therapeutic goals in the treatment of Gaucher disease. *Seminars in Hematol* 4.41(4 Suppl5):4–1.

Peters, C. Shapiro, E. Anderson, J. Hensleee-Downey P. Klemperer, M. Cowan, M. Saunders, E. deAlacaron, M. Twist, C. Nachman, J. Hale, G. Harris, R. Rozans M. Kurtzburg, J. Grayson, G. Williams, T. Lenarsky, C. Wagner, J. Krivit, W. and the members of the The Storage Disease Collaborative Study Group, Hurler Syndrome: II. (1998). Outcome of HLA-genotypically identical sibling and HLA-haploidentical related donor bone marrow transplantation in fifty-four children, *Blood* 91:2601–2608.

Swiedler, S.J. Beck, M. Bajbouj, M. Giugliani, R. Schwartz, I. Harmatz, P. Wraith, J.E. Roberts, J. Ketteridge, D. Hopwood, J.J. Guffon, N. Sa Miranda, M.C. Teles, E.L. Berger, K.I. Piscia-Nichols, C. (2005). Threshold effect of urinary glycosaminoglycans and the walk test as indicators of disease progression in a survey of subjects with mucopolysaccharidosis VI (Maroteaux–Lamy syndrome). *Am J Med Genet* 134:144–150.

Terlato, N. Cox, G. (2003). Can mucopolysaccharidosis type I disease severity be predicted based on a patient's genotype? A comprehensive review of the literature, *Genet Med* 5: 286–294.

TKT Reports Positive Top-Line Results of Hunter Syndrome Pivotal Trial. June 20, 2005. www.tktx.com

Umapathysivam, K. Hopwood, J.J. Meikle, P.J. (2005). Correlation of acid alpha-glucosidase and glycogen content in skin fibroblasts with age of onset in Pompe disease. *Clin Chim Acta* 361:191–198.

Wraith, J.E. Clarke, L.A. Beck, M. Kolodny, E.H. Pastores, G.M. Muenzer, J. Rapoport, D.M. Berger, K.I. Swiedler, S.J. Kakkis, E.D. Braakman, T. Chadbourne, E. Walton-Bowen, K. Cox, G.F. (2004). Enzyme replacement therapy for mucopolysaccharidosis I: A randomized, double-blinded, placebo-controlled, multinational study of recombinant human α-L-iduronidase (laronidase), *J Pediatr* 144:581–588.

Zhang, H. Kallwass, H. Young, S.P. Carr, C. Dai, J. Kishnani, P.S. Millington, D.S. Keutzer, J. Chen, Y.T. Bali, D. (2006). Comparison of maltose and acarbose as inhibitors of maltase-glucoamylase activity in assaying acid alpha-glucosidase activity in dried blood spots for the diagnosis of infantile Pompe disease. *Genet Med* 8(5):302–306.

NEURAL STEM CELL THERAPY IN LYSOSOMAL STORAGE DISORDERS

Jean-Pyo Lee[1,2], Dan Clark[1], Mylvaganam Jeyakumar[3], Rodolfo Gonzalez[1], Scott McKercher[1], Franz-Josef Muller[1], Rahul Jandial[4], Rosanne M. Taylor[5], Kook In Park[6], Thomas N. Seyfried[7], Frances M. Platt[3] & Evan Y. Snyder[1,2*]

1 INTRODUCTION

Many lysosomal storage disorders (LSDs) produce neurodegeneration as a prominent feature (Neufeld, 1991). LSDs are autosomal recessive metabolic diseases caused by deficiencies of specific acid hydrolases resulting in accumulation of unmetabolized substrates and macromolecules in lysosomes. There are ~50 diseases that can be classified as LSDs. The precise mechanisms underlying the actual neurodegenerative process remain to be determined, however, it is known that replacement of the absent gene product typically restores normal metabolism to a cell including forestalling neural cell dysfunction, at least in vitro. Nevertheless, there are currently no effective treatments for the neurological manifestations of the infantile-onset forms of the LSDs. The neuropathology of LSDs is characterized not by discrete focal neuropathology, as in Parkinson's disease, but rather by extensive, multifocal, or even "global" neural degeneration or dysfunction. Therapy may require not only therapeutic molecules, such as enzymes, but also widespread neural cell replacement.

Most therapies aim at increasing lysosomal enzyme levels in a widely disseminated manner that makes them accessible to affected CNS cells in order to restore their normal metabolism. These approaches include enzyme replacement therapy (ERT), gene therapy, and cellular therapies (under which bone marrow transplantation [BMT] would fall; Jeyakumar et al., 2002; Schiffmann and Brady, 2002; Takaura et al., 2003). ERT efficacy is limited due to the difficulty of delivering protein molecules across the blood–brain barrier (Sly and Vogler, 2002). In cases where the enzyme is present but at pathologically low levels, certain pharmacological therapies seek to optimize the efficiency of that limited amount via substrate reduction, that is, substrate reduction therapy (SRT).

Neural stem cells (NSCs) are the most primordial and uncommitted cells of the nervous system. Although a precise consensus definition of stem cells remains to be formed, true NSCs can be defined by the following functional properties: (1) a capacity for self-renewal throughout life; (2) multipotency (i.e., the ability to generate differentiated progeny of all three fundamental cell lineages in the nervous system, including

[1] The Burnham Institute, La Jolla, CA, 92037, USA
[2] Department of Pediatrics, UCSD School of Medicine, La Jolla, CA, 92093, USA
[3] Department of Biochemistry, Glycobiology Institute, University of Oxford, Oxford, UK, OX1 3QU
[4] Department of Neurosurgery, UCSD School of Medicine, La Jolla, CA, 92093, USA
[5] University of Sydney, Department of Animal Science, Faculty of Veterinary Science, Sydney NSW 2006, Australia
[6] Department of Pediatrics, Yonsei University College of Medicine, Seoul, 120-749, Korea
[7] Biology Department, Boston College, Chestnut Hill, MA, 02467, USA
* Correspondence to EYS (esnyder@burnham.org)

most, if not all neuronal types, astrocytes, and oligodendrocytes); and (3) the ability to populate the developing and/or repopulate the ablated (e.g., degenerating) neural regions. Cells with a more committed or restricted temporal and/or regional fate are termed "progenitors" or "precursors". The terms often get confused and, unfortunately, in some papers, are used interchangeably and in an imprecise or nonrigorous manner.

Despite their extensive plasticity, NSCs give rise only to cell types appropriate to the nervous system. Furthermore, NSC transplants into both healthy and diseased rodents do not yield neoplasms in immunodeficient mice (Vescovi et al., 1999), suggesting that in vivo tumorigenic potential is low. Upon transplantation, particularly into germinal zones, they can integrate seamlessly within the fabric of the recipient brain, accommodating and assuming neural cell type appropriate to a variety of host regions along the neuroaxis, integrating into normal host circuitry, interacting with host cells, and continuing to express their inherent genetic repertoire. NSCs, because of the aforementioned characteristics, theoretically offer potential as a therapeutic tool not only for prevention, but also for the repair, of various CNS pathology.

This therapeutic potential is particularly appropriate to the treatment of LSD, which might, in fact, constitute, "low hanging fruit" for NSC therapeutic targets. NSCs can mediate the expression of replacement enzyme genes either via their natural expression of lysosomal enzymes (Flax et al., 1998; Snyder, Taylor, and Wolfe, 1995; Yandava, Billinghurst, and Snyder, 1999) or via augmentation by readily achievable ex vivo genetic manipulation. They can also produce and secrete trophic factors and anti-inflammatory agents (again either intrinsically or through the expression of exogenous genes). Although these NSC-mediated "molecular therapies" may be sufficient for LSDs, particularly if instituted early enough, NSCs may also hold the potential of replacing some damaged neural cells. Although NSCs have been reported to be capable of transdifferentiating into hematopoietic cells (Bjornson et al., 1999), endothelial cells (Wurmser et al., 2004), and various other nonneural cell types in vitro without cell fusion, the capacity for transdifferentiation of stem cells in vivo remains quite controversial and likely undependable and inefficient.

Due to their extensive plasticity and migratory capacity, NSC transplantation to the central nervous system (CNS) can be considered in some ways analogous to hematopoietic stem cell-mediated reconstitution of bone marrow (via BMT). However, unlike with BMT, prior irradiation or preconditioning of the recipient is not required before administration. In most inherited metabolic disease, the amount of enzyme required to restore normal metabolism and reduce or prevent CNS disease may be quite small. In fact, NSCs, as normal cells, constitutively express normal amounts of the particular enzymes in question. If necessary, the level of enzyme production can be augmented by ex vivo genetic engineering (by most standard means) to tailor the therapeutic delivery to the needs of the specific disease.

2 NEURAL STEM CELL BIOLOGY

2.1 Location and Function of NSCs in the Central Nervous System

In the early developmental stages of mammalian cerebrogenesis, the appearance of NSCs coincides with the induction of the neural plate following gastrulation (Temple, 2001). NSCs as a pool are likely harbored earliest in a primary germinal region known as the ventricular zone (VZ) which lines the emerging ventricular system (a remnant of the neural tube) and gives rise to most of the neuraxis. In the telencephalon of an embryonic day 10 (E10) mouse, it is estimated that stem cells account for 5 to 20% of the cells in the VZ (He et al., 2001; Kilpatrick and Bartlett, 1993; Qian et al., 2000). Most of the premigratory neural crest, also present at ~E8.5–E10, consists of stem cells (White et al., 2001). During development, the number of uncommitted endogenous NSCs decreases with time, whereas activation of cellular mechanisms leading to NSC differentiation, into neurons or glia, increases.

NSCs divide both symmetrically and asymmetrically. The earliest NSCs divide symmetrically, increasing their number. Later in cerebrogenesis, in response to specific developmental signals, the NSCs divide asymmetrically to generate partially differentiated neural progenitor or precursor cells. These neural progenitor/precursor cells remain proliferative and undergo a finite number of cell divisions before end-differentiation, but, compared to NSCs of the VZ, these cells have a more limited self-renewal capacity and a narrower range of fates. The final fate decision of a given progenitor cell depends on extrinsic morphogenic signals and growth factors that induce specific patterns of transcriptional and post-transcriptional machinery, generally directing neurogenesis prior to gliogenesis (Qian et al., 2000).

Secondary germinal zones develop from the VZ and include the subventricular zone (SVZ) of the forebrain (which also lines the cerebral ventricles), the subgranule layer (SGL) of the hippocampus, and the external germinal layer (EGL) of the cerebellum (which likely emerges from the fourth ventricle). There has been some speculation that the SGL may itself arise from the SVZ, although this remains controversial. The EGL disappears with the completion of cerebellar development by three weeks postnatally in the rodent and by two years of age in humans. There are two well-characterized regions of the adult rodent forebrain in which neurogenesis continues throughout life: the SVZ, which lies directly subjacent to the ciliated ventricular ependyma, and the SGL of the dentate gyrus of the hippocampus. NSCs are believed to populate these areas during early development, possibly in order to assist in the maintenance of homeostasis throughout life.

From the rodent SVZ, newly generated neuronal precursors reach their final destination in the olfactory bulb (OB) after long-distance migration through a well-defined path called the rostral migratory stream (RMS). Under normal conditions these cells primarily replace olfactory neurons, which physiologically likely turn over constantly in the rodent. The functions of NSCs in the hippocampus are less well defined but may play a role in memory acquisition (Gage et al., 1998; Gage, Ray, and Fisher, 1995; Kempermann et al., 2003; Kempermann, Kuhn, and Gage, 1998). Although humans appear also to have proliferative SVZ, it remains unclear whether they have migration along an RMS. The ability of NSCs, from secondary germinal zones, to migrate to areas of pathology, raises the possibility that, in the pediatric and possibly in the adult CNS, they may be recruited for neural preservation and reconstruction.

2.2 Characteristics of NSCs

In 1959, Sidman and again in 1965, Altman and Das provided evidence for neurogenesis in the adult rat brain (Altman and Das, 1965; Sidman, Miale, and Feder, 1959). Neural cells with stemlike qualities in the developing and adult brain and in the peripheral nervous system were subsequently identified and isolated (Anderson, 2001; Gage, 2000; Kuhn and Svendsen, 1999; Lois and Alvarez-Buylla, 1993; Reynolds and Weiss, 1992;

Shen et al., 1998). These NSCs, as they came to be called, could be expanded in vitro either by augmenting their expression of "stemness genes" or by providing tonic exposure to mitogens (e.g., basic fibroblast growth factor [bFGF] and/or epidermal growth factor [EGF]).

It was also observed that clones of these NSCs could be reimplanted into the developing (Snyder et al., 1992) and adult (Snyder et al., 1997; Gage, Ray, and Fisher, 1995) rodent brain (including brains that were abnormal or injured). These studies were characterized by a seamless integration of implanted NSCs, differentiating into integral members of the CNS. Maintaining NSCs in a proliferative state in vitro—whether by epigenetic or genetic manipulation—does not appear to subvert their in vivo ability to migrate and differentiate following transplantation into the developing or diseased brain. The principal differences so far observed between human and mouse NSCs, seem to be the length of the cell cycle (up to four days in the human), the predilection of human NSCs to senesce (after ~100 cell divisions), and their relative responsiveness to various types of mitogens. NSCs can also be derived secondarily in vitro from embryonic stem cells (ESCs) which are derived from the inner cell mass (ICM) of a blastocyst (Reubinoff et al., 2001; Zhang et al., 2001). These ESCs are "pluripotent" in that they can give rise to differentiated progeny of all three embryonic germ layers, including NSCs, but are not typically regarded as "totipotent" because they do not, under normal circumstances give rise to the trophoblast and cannot yield a complete organism de novo. Also, they never senesce, by definition, although they require a great deal of instruction to direct their differentiation fates unequivocally towards a given lineage.

Using ESCs as a source of NSCs would obviate the problems of abstracting a relatively small population of "native" NSCs from a given organ, maximize their proliferation, and avoid senescence. On the other hand, the use of human embryos is currently colored by many political and ethical concerns. Furthermore, being able to "tame" an ESC is still quite challenging.

Immunocytochemical identification of NSCs is not precise; markers for NSCs are selective but not specific. Specificity can be enhanced by using a panel of markers rather than by relying on a single marker. Commonly used NSC markers include: Nestin and Vimentin (intermediate filament proteins); Musashi 1 (an RNA-binding protein); Sox1 (a transcription factor); and AC133 (a surface marker). Ultimately, functional screens (i.e., calling a stem cell such because a clonal population behaves like a stem cell vis-à-vis evincing properties of self-renewal, multipotence, and reconstitution) are still the "gold standard."

2.3 Therapeutic Strategies Involving NSCs

NSCs can be isolated directly from the neuroectoderm or primary germinal zones of fetuses (e.g., the VZ) or from residual secondary germinal zones in adults (e.g., SVZ, SGZ), and expanded in vitro using growth factors. Conventional wisdom has viewed the therapeutic potential of NSCs to be entirely dependent upon modifying them ex vivo to become a particular cell type that might be inserted into the brain (as one might replace a broken part in a car), or to secrete a single gene product that has been identified as absent for a given disease. This view, however, underestimates the rich repertoire of behaviors of the NSC which engages in a complex multifaceted cross-talk with the host in order to help restore a homeostatic milieu to the diseased and injured CNS.

The NSCs express various genes of interest intrinsically; these include neurotrophic factors, lysosomal enzymes, angiogenic factors, anti-inflammatory molecules, and antioxidants. These secreted molecules may help shift the balance between permissive and nonpermissive microenvironments favoring the reacquisition of CNS integrity and function. In fact, their trophic functions, not their cell replacement ability, may account for some of the reported beneficial effects of stem cells from diverse organ systems, including in the CNS. Some of these molecules have been isolated [e.g., glial cell line-derived neurotrophic factor (GDNF), nerve growth factor (NGF), and brain-derived neurotrophic factor (BDNF)] (Ahmed, Reynolds, and Weiss, 1995; Lu et al., 2003), whereas others remain to be identified.

Given the complexities of the CNS, preserving established CNS circuitry is as important and may be more tractable than attempting to reconstruct proper new CNS connectivity. Whether these various molecules are secreted in sufficient amounts and at appropriate times, or whether their production and processing should be altered or augmented, must be assessed empirically and individually for each molecule and each disease, bearing in mind the caveat that changing the regulation of one intracellular molecule might alter others in unanticipated and perhaps undesirable ways (Lu et al., 2003).

Despite the presence of endogenous NSCs in the adult mammalian brain, it is recognized that, upon CNS insult, intrinsic "self-repair" activity, for the most devastating injuries, is inadequate or ineffective. These limits on "regenerative" ability may be due to the restricted location, limited number, or altered responsiveness of adult NSCs and/or to limitations imposed by the surrounding microenvironment (which may not be supportive or instructive for neuronal differentiation or even of NSC survival; Monje et al., 2002). In situations where insufficient quantities of endogenous NSCs in appropriate proximity to the lesion may be a limiting factor to ineffective self-repair, supplemental exogenous NSCs—expanded ex vivo in culture—might be implanted intracerebrally. These additional NSCs may help not only complement inadequacies in the endogenous NSC pool but may offer an opportunity to provide additional supportive molecules that might remodel the "niche," augment host regenerative processes (e.g., angiogenesis, migration), and inhibit obstructive processes (e.g., apoptosis, scarring, inflammation, excitotoxicity, oxidative stress).

NSCs transplanted into the cerebral ventricles in utero or at birth gain access to most of the VZ in the former situation and most of the SVZ in the latter situation including the periventricular regions of the IIIrd and IVth ventricles. These cells are able to circumvent the blood–brain barrier (BBB) and migrate into distant CNS regions. Implanted NSCs participate in the normal development of multiple regions throughout the brain and at multiple stages (from fetus to adult), integrating seamlessly within the parenchyma. Implanted NSCs differentiate appropriately into diverse types of neurons, astrocytes, oligodendrocytes, and even undifferentiated quiescent progenitors, apparently responding to the same spatial and temporal cues as endogenously differentiating neural cell types (Flax et al., 1998; Lacorazza et al., 1996).

Functional integration as neurons of donor NSC-derived cells was confirmed by electrophysiological recordings of transplant-derived neurons in slice cultures of engrafted brains (Auerbach, Eiden, and McKay, 2000) or by assessing the appropriate activation of NSC-derived neurons in vivo within complex physiological networks (e.g., hypothalamic mediation of circadian rhythm (Zlomanczuk et al., 2002). As noted, some transplanted NSCs do remain in an undifferentiated state, yet we believe this fate to be both physiological and important now that we have a better understanding of neural development and the role of the stem cell in that process.

NSCs derived from ESCs (i.e., that have been directed toward a neuroectodermal lineage while retaining stemlike properties; Reubinoff et al., 2001; Tabar et al., 2005; Zhang et al., 2001) may ultimately be valuable in disease models, again with the NSC derived directly from the neuroectoderm as the "gold standard" for safety and efficacy.

In all transplantation paradigms, a pervasive concern is whether donor cells might be immunorejected by the recipient. We are actually beginning to learn that stem cells, in their stemlike state, are more immunotolerated than previously assumed, perhaps because they do not express MHC class II in their immature state (Imitola et al., 2004a). Indeed, in experiments in which we implanted mouse NSCs into mouse hosts, we never use immunosuppression or have evidence of immunorejection. (When one crosses species, immunosuppression is required.) Whether this immunotolerance applies to human stem cells within human hosts is unclear. Hence, there has been a good deal of anticipatory planning for deriving stem cell populations that might be optimally compatible with hosts.

One approach is to attempt to derive stem cells from the potential recipient, for example, from the patient's brain (by biopsing the SVZ) or, more practically, from a more accessible source in the patient (e.g., bone marrow, skin, fat, blood, umbilical cord). Whether these nonneural sources can yield true neural cells remains exceptionally controversial. At the time of publication of this review, the data, when scrutinized most rigorously and critically, speak against such transdifferentiation or the efficient "panning" for ectopic neuroectodermally derived cells residing in these nonneural organs. One theoretical concern that must be raised whenever contemplating an "autograft" for a patient with a genetically based disease (e.g., such hereditary neurodegenerative processes as LSDs), is that the stem cells themselves likely already bear a genetic defect that might make them ineffective or susceptible to early demise and/or poor differentiation.

Another approach being pursued to create immunocompatible graft material for the CNS is the use of somatic cell nuclear transfer (NT), that is, using donor nuclei from a mature somatic cell of the intended recipient (e.g., a fibroblast) to insert into an unfertilized oocyte which, when stimulated, divides to the blastocyst stage from which the ICM can be cultured to yield ESCs; NSCs can, in turn, be theoretically derived from these NSCs, as mentioned above (although better controlled and more efficient strategies for this process will be required prior to therapeutic use). NT for this purpose has recently had its first suggestion of feasibility for humans. Caveats to this approach might be that nuclei from mature somatic cells may have developmentally imposed epigenetic changes, such as specific DNA methylation patterns, that might be difficult to overcome safely. That the oocyte cytoplasm harbors molecules that can mediate a process of dedifferentiation is a fascinating concept that requires greater exploration and may be exploited in contexts even broader than NT, perhaps making "adult cells" act stemlike.

3 NEURAL STEM CELL THERAPY IN LYSOSOMAL STORAGE DISORDERS (LSDs)

3.1 Gene Replacement

In the pediatric CNS, there is the potential for harnessing developmental processes—as represented by stem cell biology—in order to provide adjunctive therapies for heretofore untreatable LSDs. Indeed, because of their monogenetic basis and well-characterized pathobiology, childhood LSDs are actually excellent models for various adult diseases that are phenocopies but are complicated by other factors and other genetic problems. The therapeutic potential of NSCs was, in fact, first demonstrated in a mouse model of one of the LSDs (Snyder, Taylor, and Wolfe, 1995). In that case, the mucopolysaccaridosis type VII (MPS VII) mutant mouse served as a prototype for many diseases that require provision of a therapeutic gene and/or its product.

The MPS VII mouse, like the MPS VII patient, is homozygous for a frame shift mutation for the β-glucuronidase gene and is, hence deficient in the secreted enzyme β̃glucuronidase (GUSB). The enzymatic deficiency results in lysosomal accumulation of undegraded glycosaminoglycan (GAG) in the brain and other tissues, causing a fatal progressive degenerative disorder. Treatments for MPS VII and most other LSDs are designed to provide a source of normal enzyme for uptake by diseased cells, a process termed "cross-correction" (Neufeld and Fratantoni, 1970). The goal of ex vivo gene therapy is to engineer donor cells to express the normal GUSB protein for export to other host cells. The engraftment and integration of GUSB-overexpressing NSCs throughout the newborn MPS VII mutant brain (following transplantation into the lateral ventricles) succeeded in providing a sustained, lifelong, widespread source of cross-correcting enzyme in a manner not previously achievable, reducing CNS pathology (Figure 1(I); Snyder, Taylor, and Wolfe, 1995). These findings have been extended to human neural progenitors transduced with the GUSB gene with similar results (Meng et al., 2003).

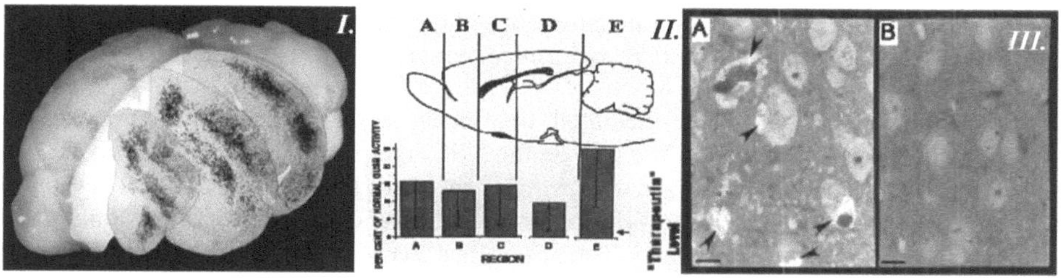

Figure 1. Widespread engraftment of NSCs expressing GUSB throughout the brain of the MPS VII mouse. (I) Brain of a mature MPS VII mouse after receiving a neonatal intraventricular transplant of murine NSCs expressing GUSB. Donor NSC-derived cells, identified by their Xgal histochemical reaction (blue precipitate) for expression of the *lacZ* marker gene, have engrafted throughout the recipient mutant brain. Representative coronal sections, placed at their appropriate level by computer, show these cells to span the rostral–caudal expanse of the brain. (II) Distribution of GUSB enzymatic activity throughout brains of MPS VII NSC transplant recipients. Serial sections were collected from throughout the brains of transplant recipients and assayed for GUSB activity. Sections were pooled to reflect the activity present within the regions demarcated in the schematic. The regions were defined by anatomical landmarks in the anterior-to-posterior plane to permit comparison among animals. The mean levels of GUSB activity for each region (n = 17) are presented as the percentage of average normal levels for each region. Untreated MPS VII mice show no GUSB activity biochemically or histochemically. Enzyme activity of 2% of normal is corrective based on data from liver and spleen. (III) Decreased lysosomal storage in a treated MPS VII mouse brain at eight months of age. [A] Extensive vacuolation representing distended lysosomes (arrowheads) in both neurons and glia in the neocortex of an eight-month-old, untransplanted control MPS VII mouse. [B] Decrease in lysosomal storage in the cortex of an MPS VII mouse treated at birth from a region analogous to the untreated control section in [A]. The other regions of this animal's brain showed a similar decrease in storage compared to untreated, age-matched mutants in regions where GUSB was expressed. Scale bars, 21 μm. (Adopted from Snyder, Taylor, and Wolfe, 1995.)

Although the histograms in Figure 1(II) illustrate the widespread distribution of a lysosomal enzyme, they could similarly reflect both the NSC-mediated distribution of other diffusible (i.e., synthetic enzymes, neurotrophins, and viral vectors) and nondiffusible (i.e., myelin and extracellular matrix) factors, as well as the distribution of "replacement" neural cells (see section below).

Briefly, when NSCs (murine and human) were cocultured with dissociated TSD mouse neural cells for 10 days, hexosaminidase activity increased to a normal level (Figure 3, A–C); all neural cell types from the TSD mouse brain were rescued (Figures 2D–N; Flax et al., 1998). This correction of the metabolic defect could be augmented by engineering the NSCs to overexpress hexosaminidase (e.g., via retroviral transduction of the α-subunit gene). Such NSCs, when transplanted, increased total hexosaminidase levels in vivo (Lacorazza et al., 1996). It is of interest in this case that overexpression of simply the α-subunit did result in an increase in total hexosaminidase. It is also worth noting that excess and/or unregulated expression of a lysosomal enzyme did not lead to dysfunction or additional abnormalities. For some gene products—for example, certain neurotrophic factors—overexpression can be problematic. Lysosomal enzymes by and large do not fall into that category.

The paradigm described above has been emulated by others for other LSDs. For example, Type A Niemann–Pick disease (NPDA) is an LSD caused by a lack of acid sphingomyelinase (ASM). The intracerebral implantation of adult mouse neural progenitor cells (NPCs) transduced with ASM significantly reduced the lysosomal storage of spingomyelin and cholesterol in the ASM null mouse, even when treated as an adult (Shihabuddin et al., 2004).

As suggested above, NSC-mediated treatment of many LSDs may not require genetic manipulation but may rather rely on the NSC's intrinsic production of the enzyme of interest, particularly when (a) that enzyme is secreted, (b) can be readily taken up by abnormal cells via the mannose-6-phosphate receptor, and (c) is required in only small quantities to restore normal metabolism. Such has been our observation in the mouse model of Sandhoff disease (SD), another gangliosidosis, characterized by a mutation of the β-subunit of Hex. SD is a uniformly fatal, rapidly progressive autosomal recessive LSD caused by an absence of both β-hexosaminidase A and B (HexA and HexB) leading to accumulation in the CNS of GM2 and GA2 gangliosides. Preliminary evidence suggests that mouse and human NSCs—whether derived directly from the neuroectoderm or derived secondarily from ESCs—migrate extensively throughout the brain following neonatal transplantation, providing cross-correcting enzyme in a disseminated manner and reducing widespread ganglioside storage (including in cortex). Of note is the fact that the addition of substrate-reducing pharmacological therapy appears to act synergistically with the cellular transplantation. In other words, the combined administration of NSCs and drugs prolonged lifespan longer than each modality individually (greater than simply an additive effect).

The study described above provided an additional insight into the role NSCs may play in LSDs, the provision of anti-inflammatory molecules expressed intrinsically by the NSC. It is becoming recognized that, even in LSDs, inflammation likely plays a significantly inimical role in the disease pathophysiology, including within the CNS. Specifically, CNS inflammation, characterized by microglial activation and macrophage infiltration, has emerged as a hallmark of SD, both in humans and in mice (Jeyakumar et al., 2003;

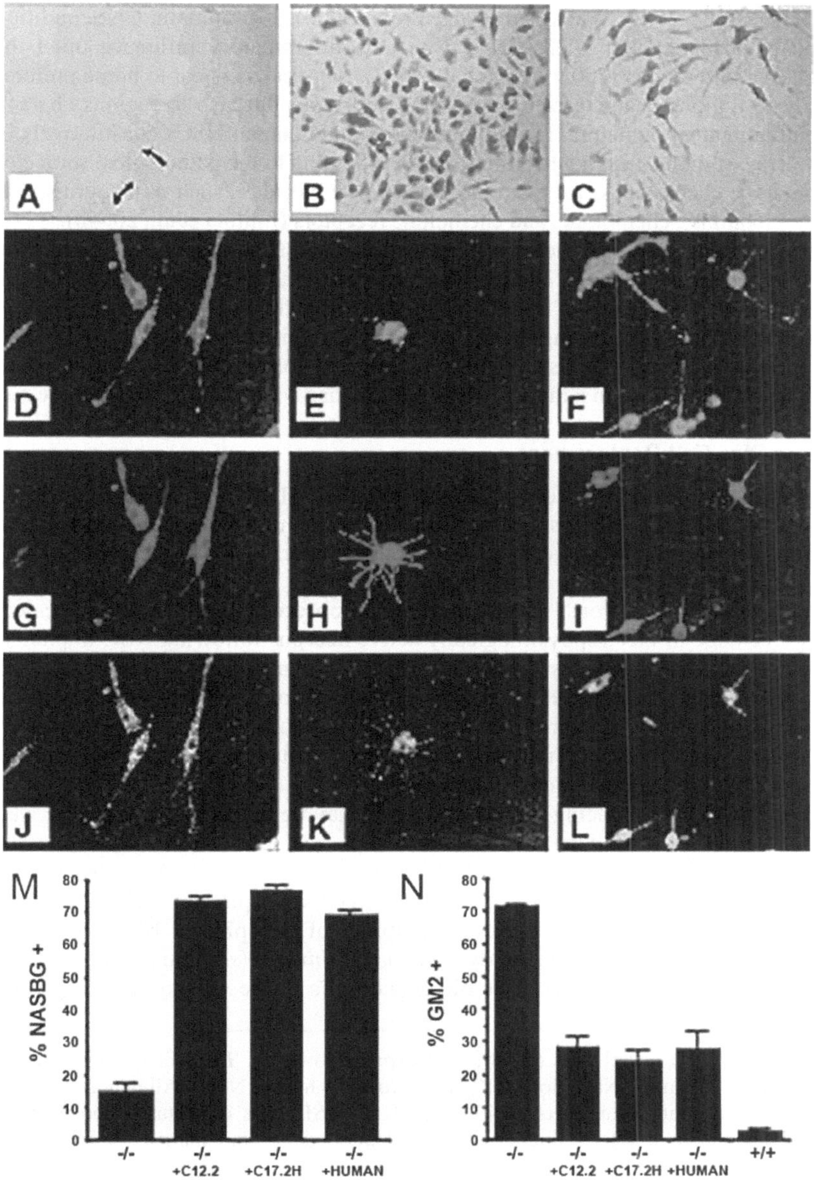

Figure 2. Human NSCs secrete hexosaminidase and cross-correct neural cells from Tay–Sachs disease (TSD). Dissociated brain cells from mice with mutated α-subunit of β-hexosaminidase Tay–Sachs disease cocultured with human NSCs. After 10 days, the mutant neural cells were positive (1) for the presence of hexosaminidase activity determined by NASBG histochemistry [A–C, and M]; (2) for antibody staining to the α-subunit and to CNS cell type markers to determine which TSD neural cells internalized corrective gene product [D–L]; and (3) for reduction in GM2 (Continued)

Myerowitz et al., 2002; Wada, Tifft, and Proia, 2000). In traumatic CNSconditions, we previously observed that NSCs exerted an anti-inflammatory influence upon the host (Park, Teng, and Snyder, 2002; Teng et al., 2002). Stem cells appear to home preferentially not only to regions characterized by missing cell types but also to regions characterized by an inflammatory signature. Indeed, both mouse and human NSCs constitutively express a wide array of molecules that help mediate their homing to the pathological source of these inflammatory chemokines, for example, adhesion molecules (such as integrins, selectins, immunoglobulins), chemokine, and chemokine receptors (Imitola et al., 2004b).

In preliminary studies, we have observed that transplanted NSCs reduced the CNS inflammation in SD brain, as measured using microglia/macrophage activation markers such as F4/80, CD68, and CD11b. When taken together with the expression of total Hex and the reduction of ganglioside storage, these mice had a significantly prolonged lifespan, a delayed onset of disease, and preserved motor function for four months longer than baseline, significant in a mouse model that ordinarily dies within two to three months.

3.2 Gene Plus Cell Replacement

Neural cell loss obviously accompanies many of the LSDs with CNS manifestations. The hope with gene replacement, substrate reduction, and anti-inflammatory interventions, as described above, is to preserve the endogenous cellular population, particularly neurons and oligodendrocytes.

With regard to neurons, although we have observed functional replacement of a small percentage of electrophysiologically active neurons following NSC implantation, it is unlikely that this number is significant to reverse a neurological handicap. It remains unknown whether the number of donor-derived neurons can be safely increased and whether these cells make appropriate connections without making inappropriate connections. Hence, with respect to neurons, protection of extant neurons, rather than replacement, seems to be the most tractable approach.

The prospect of replacing oligodendrocytes may be a bit more achievable, yet, even there, a number of considerations may make this goal less than straightforward. Dys-/demyelination plays an important role in many genetic (i.e., leukodystrophies, inborn metabolic errors) and acquired (i.e., traumatic, infectious, asphyxial, ischemic, and inflammatory) neurodegenerative processes. The ability of transplanted NSCs to remyelinate was first suggested by studies in the dysmyelinated *shiverer* (*shi*) mutant mouse (Yandava, Billinghurst, and Snyder, 1999). The oligodendroglia of the *shi* mouse are dysfunctional

←

storage[N]. [A] TSD neural cells (arrows) not exposed to NSCs. TSD cells exposed to secretory products from [B] murine NSC clone C17.2H or from [C] human NSCs. All neural cell types from the TSD mouse brain were corrected [3D–L]. [D–L] TSD cells co-cultured with human NSCs immunostained with a [D–F] fluorescein-labeled antibody to the human -subunit of –hexosaminidase and [G–I] with antibodies to neural cell type-specific antigens. [G] Neuronal-specific NeuN marker; [H] glial specific GFAP marker; and [I] precursor maker, nestin. [J–L] Dual filter microscopy of the -subunit and cell-type markers. [M] Percentage of -hexosaminidase positive TSD cells; -/-: TSD -subunit–null cells; TSD cells exposed to secretory products from C17.2+ murine NSCs; C17.2H+: murine NSC engineered to overexpress murine hexosaminidase; +human: human NSCs. [N] GM2 accumulation in TSD cells; labels as in [M]; +/+: wild-type mouse brain. The percentage of TSD CNS cells without abnormal GM 2 accumulation was significantly lower in those exposed to secretory products from human NSCs than in untreated TSD cultures ($p < 0.01$), approaching those from wild-type mouse brain. (Adapted from Flax et al., 1998.)

because they bear a deletion mutation of myelin basic protein (MBP), essential for effective myelination. Therapy for this cell-autonomous disorder would, therefore, require widespread replacement with normal MBP-expressing oligodendrocytes.

When undifferentiated murine NSCs were transplanted at birth into the ventricles of newborn *shi* mice, NSCs became integrated throughout the *shi* brain with repletion of significant amounts of MBP (Yandava, Billinghurst, and Snyder, 1999; Figure 3). Accordingly, of the many donor cells that differentiated into oligodendroglia, a subgroup myelinated ~40% of host neuronal processes. In some recipient animals, the symptomatic tremor decreased. In similar experiments, NSCs have been harvested from adult rat brains, expanded with mitogens in culture, differentiated into oligodendrocyte precursors ex vivo, and implanted into the myelin-deficient (md) rat, a rodent in which there is a mutation in phospholipid protein (PLP), another essential component of normal myelin. These NSCs, too, produced robust amounts of new myelin. Recently remyelination experiments in the *shi* mouse have been extended successfully to the use of human oligodendrocyte precursors (Windrem et al., 2004). Taken together, these findings suggest that, in select CNS pathology—in cell-autonomous defects such as these—global cell replacement may be possible when using cells with stemlike properties.

Of significance, there is no human version of "shiverer" disease, the model used most commonly for testing oligodendrocyte replacement strategies. (There is a rare human version of PLP-deficiency called Pelaezius–Merzbacher disease, which is classified as a non-LSD leukodystrophy.) Nevertheless, oligodendrocyte dysfunction (and consequent dysmyelination or demyelination) are prominent features of some important LSDs. For example, patients suffering from Krabbe leukodystrophy, also called globoid cell leukodystrophy (GCL), are devoid of galactosylceramidase activity (also called galactocerebrosidase; GalC; Oehlmann et al., 1993; Wenger et al., 2000; Wenger, Sattler, and Hiatt, 1974). The primary substrate of galactosylceramidase is a constituent of the myelin sheath. GCL is characterized by CNS demyelination, degeneration of oligodendrocytes, and infiltration of macrophages. The twitcher (twi) mutant mouse provides an authentic model of GCL.

There has been a long-standing question regarding the pathophysiology of GCL. Is the observed pathology related to a cell autonomous defect (i.e., a sick cell that dies) or to a toxic environment (a milieu that would theoretically kill not only host cells but any new cells placed into that environment, hence dooming any attempts at cell replacement)? It has been hypothesized that the absence of GalC activity in twi mice (as in human GCL patients) permits build-up of the toxic glycolipid psychosine. Preliminary in vitro experiments suggest that the absence of GalC activity in twi mice predisposes neural cells (especially oligodendrocytes) to be more susceptible to that toxic metabolite.

Wild-type murine NSCs (i.e., expressing a normal amount of GalC), particularly in their undifferentiated state (the state in which we transplant them), tend to be resistant to those toxic effects, certainly more resistant than even end-differentiated wild-type oligodendrocytes. In pilot studies, when transplanted into the ventricles of newborn twi mice, NSCs can survive and differentiate into beautifully myelinating oligodendrocytes throughout the twi cerebral environment. This observation suggests that the environment in the twi brain is not inherently nonpermissive for the implantation, survival, and differentiation (including into functional oligodendrocytes) of NSCs, and, hence, cell replacement strategies seem to be viable approaches.

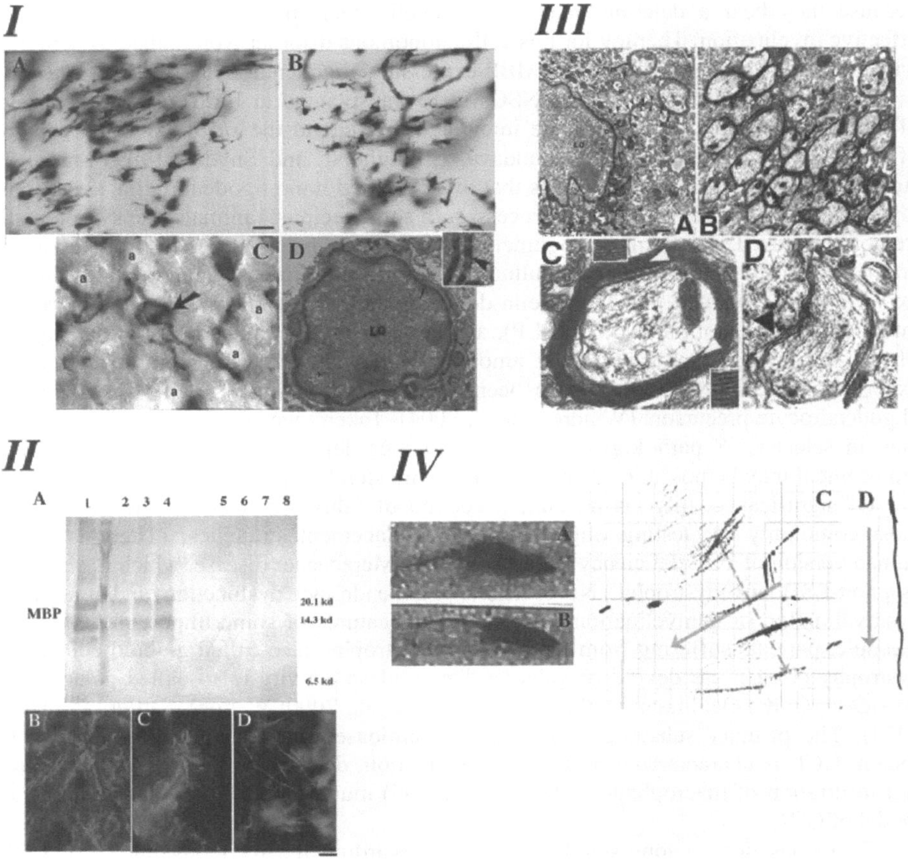

Figure 3. "Global" cell replacement is feasible via NSC transplantation: evidence from the dysmyelinated *shiverer (shi)* mouse brain. (I) NSCs engraft extensively throughout the *shi* dysmyelinated brain, including within white tracts, and differentiate into oligodendrocytes. *LacZ*-expressing, βgalactosidase (βgal)-producing NSCs were transplanted into newborn *shi* mutants and analyzed systematically at intervals between 2–8 wks following engraftment. Coronal sections through the *shi* brain at adulthood demonstrated widely disseminated integration of blue Xgal+ donor-derived cells throughout the neuraxis, similar to the pattern seen in Figure 1(I) in the MPS VII mutant mouse. Donor-derived cells in the *shi* mouse brain are shown at higher magnification and greater detail in [A–D]. [A,B] Donor-derived Xgal+ cells in representative sections through the corpus callosum possessed characteristic oligodendroglial features (small, round, or polygonal cell bodies with multiple fine processes oriented in the direction of the neural fiber tracts). [C] Close-up of a representative donor-derived anti-! gal immunoreactive oligodendrocyte (arrow) extending multiple processes towards and beginning to enwrap large adjacent axonal bundles ("a") viewed on end in a section through the corpus callosum. That cells such as those in [A–C] (and in II(B-D) were oligodendroglia was confirmed by the representative electron micrograph in [D] (and in III) demonstrating their defining ultrastructural features (Yandava, Billinghurst, and Snyder, 1999). A donor-derived Xgal+ oligodendrocyte ("**LO**") can be distinguished by the electron dense Xgal precipitate that is typically localized to the nuclear membrane (arrow), ER (arrowhead), and other cytoplasmic organelles. The area indicated by the arrowhead is magnified in the inset to

demonstrate the unique crystalline nature of individual precipitate particles. (II) MBP expression in mature transplanted and control brains. [A] Western analysis for MBP in whole brain lysates. The brains of three representative transplanted *shi* mutants (lanes 2–4) expressed MBP at levels close to that of an age-matched unaffected mouse (lane 1, positive control), and significantly greater than the amounts seen in untransplanted (lanes 7,8, negative control) or unengrafted (lanes 5,6, negative control) age-matched *shi* mutants. (Identical total protein amounts were loaded in each lane.) [B–D] Immunocytochemical analysis for MBP. [B] The brain of a mature unaffected mouse was immunoreactive to an antibody to MBP (revealed with a Texas Red-conjugated secondary antibody). [C,D] Age-matched engrafted brains from *shi* mice similarly showed immunoreactivity. Because untransplanted *shi* brains lack MBP, MBP immunoreactivity has also classically been a marker for normal donor-derived oligodendrocytes in transplant paradigms. (III) NSC-derived "replacement" oligodendrocytes are capable of myelination of *shi* axons. In regions of MBP-expressing NSC engraftment, *shi* neuronal processes became enwrapped by thick, better compacted myelin. [A] At 2 wks post-transplant, a representative donor-derived, labeled oligodendrocyte ("LO") (recognized by extensive Xgal precipitate ["p"] in the nuclear membrane, cytoplasmic organelles, and processes) was extending processes (a representative one is delineated by arrowheads) to host neurites, and was beginning to ensheathe them with myelin ("m"). [B] If engrafted *shi* regions, such as that in [A], were followed over time (e.g., to 4 wks of age as pictured here), the myelin began to appear healthier, thicker, and better compacted (examples indicated by arrows) than that in age-matched untransplanted control mutants. [C] By 6 wks post-transplant, these matured into even thicker wraps; ~40% of host axons were ensheathed by myelin (a higher-power view of a representative axon is illustrated in [C]) that was dramatically thicker and better compacted than that of *shi* myelin (an example of which is shown in [D] (black arrowhead) from an unengrafted region of an otherwise successfully engrafted *shi* brain). In [C], white arrowheads indicate representative regions of myelin that are magnified in the adjacent insets; MDLs are evident. (IV) Functional and behavioral assessment of transplanted *shi* mutants and controls. The *shi* mutation is characterized by the onset of tremor and a "shivering gait" by the second to third postnatal week. The degree of motor dysfunction in animals was gauged (i) by blindly scoring periods of standardized videotaped cage behavior of experimental and control animals and (ii) by measuring the amplitude of tail displacement from the body's rostra–caudal axis (an objective, quantifiable index of tremor). Video freeze-frames of representative unengrafted and successfully engrafted *shi* mice are seen in [A] and [B], respectively. The whole body tremor and ataxic movement observed in the unengrafted symptomatic animal [A] causes the frame to blur, a contrast with the well-focused frame of the asymptomatic transplanted *shi* mouse [B]. Sixty percent of transplanted mutants evinced nearly normal-appearing behavior as in [B] and attained scores that were not significantly different from normal controls (see Yandava, Billinghurst, and Snyder, 1999 for details). [C,D] depict the manner in which whole body tremor was mirrored by the amplitude of tail displacement (hatched gray arrow in [C]) measured perpendicularly from a line drawn in the direction of the animal's movement (solid gray arrow represents the body's long axis). Measurements were made by dipping the mice tails in India ink, then allowing them to move freely in a straight line on a sheet of graph paper as shown. Large degrees of tremor cause the tail to make widely divergent ink marks away from the midline, representing the body's axis [C]. Absence of tremor allows the tail to make long, straight, uninterrupted ink lines on the paper congruent with the body's axis [D]. The distance between points of maximal tail displacement from the axis was measured and averaged for transplanted and untransplanted *shi* mutants and for unaffected controls (hatched gray arrow). [C] shows data from a poorly engrafted mutant that did not improve with respect to tremor whereas [D] reveals lack of tail displacement in a successfully engrafted asymptomatic mutant. Overall, 64% of transplanted *shi* mice examined displayed at least a 50% decrement in the degree of tremor or "shiver". Several showed 0 displacement (see Yandava, Billinghurst, and Snyder, 1999 for details). (Modified from Yandava, Billinghurst, and Snyder, 1999.)

However, unlike in the *shi* mouse, where implantation of NSCs ameliorated symptoms, cell replacement was unable to forestall disease onset or progression, remediate symptoms, or prolong the life of the twi mouse. Interestingly, however, murine NSCs engineered (via retroviral-mediated transduction) to overexpress GalC (yielding cells that produced an amount high enough to actually cross-correct fibroblasts from Krabbe patients in culture) appeared to be even more resistant to higher levels of psychosine than unengineered immature NSCs. In pilot experiments, in which these GalC-overexpressing NSCs were implanted into the brains of newborn *twi* mice, some of the mice now do seem to have a 2.5–3-fold increase in life span and psychosine levels are diminished to at least 30% of predicted. This phenomenon of nonmutant NSCs being resistant to psychosine toxicity (and more resistant the greater the amount of GalC produced) is also seen preliminarily with NSCs of human origin.

Taken together, these results suggest a number of important points regarding the fundamental pathophysiology of GLC and cellular and molecular approaches for treating it. First, it does validate the psychosine hypothesis of Krabbe pathogenesis. Second, it suggests that GalC not only permits a build-up of psychosine but endows cells with the capacity to resist its toxicity; that is, an absence of GalC makes them more vulnerable whereas an increasing amount of GalC makes them more resistant. Third, a component of treatment will be to cross-correct host cells (by overexpressing GalC) and hence be more resistant. The effectiveness in treatment appears to be less related to the absolute level of GalC achieved but rather to the diminished level of psychosine accumulated. Fourth, and most important, although oligodendrocyte replacement alone is not a sufficient treatment for GLC (even when extensive), the replacement of both cells and of molecules, for example, NSCs that can both become oligodendrocytes and serve as pumps for GalC (to antagonize psychosine), remains a very promising and important basis for multidisciplinary strategies against the disease.

This last point can be used to make an even more important general point regarding LSDs which are frequently noncell autonomous (i.e., the pathology resides not only in the mutant cells but in the mutant environment, as well). In the case of the twi mouse, cell replacement required concomitant gene replacement for an impact. Another more general conclusion may be drawn from studies such as this: NSC-mediated cell replacement is feasible not only in diseases where the disorder is cell intrinsic, but also in disorders that have extrinsic pathology if the NSC is inherently resistant to the toxic milieu or can be rendered resistant via genetic manipulation. This latter condition likely exists in GCL.

4 CELLS OF NONNEURAL ORIGIN FOR A NEUROLOGICAL DISEASE

Are stem cells of neuroectodermal origin the only cells that can achieve some of the above-mentioned goals? Recently, it was reported that, when transplanted within the first two weeks of life into immunosuppressed, marrow-ablated human newborns with infantile Krabbe's disease (presymptomatic), systemically administered umbilical cord cells (UCCs) from unrelated but HLA-matched donors favorably altered the natural history of the disease with a blunting of disease progression. Interestingly, transplantation in babies beyond two weeks of age or after symptoms developed had no significant impact (Escolar et al., 2005).

In preliminary studies, we have sought to determine the mechanism by which the UCCs may have exerted their effect. All leukocytes within the cerebral vasculature of the transplanted patients were donor-derived, suggesting stable chimerism and complete reconstitution of the hematopoietic compartment by donor UCCs, which are composed of hematopoietic stem cells. In addition, numerous cells throughout the brain—around vessels, within the cerebroventricular epithelium, and in white and gray matter—were donor-derived, often juxtaposed with host cells, clearly a desirable finding. Also important, most UCC-derived cells stained with microglial or macrophage markers. No donor cells expressed neuronal, astroglial, oligodendroglial, or neural progenitor markers.

However, globoid bodies, the pathological perivascular signature of Krabbe disease, were essentially eliminated, coincident with the predominant distribution of the UCCs and perhaps attributable to the small spike in GalC activity thus concentrated perivascularly. Globoid cells have traditionally been thought to derive from pathological, storage engorged, microglia. These data suggest the use of UCCs for molecular therapies and/or for conditions where intact microglial/macrophage function is essential (i.e., in many of the LSDs) but not for efficient neural cell replacement. Such observations also forewarn the limited extent to which cells from one germ layer, even embryonic/fetal cells in an injured developing environment, transdifferentiate into cells of another (at least under presently defined conditions).

In short, these preliminary data suggest that nonneural stem cells, such as those obtained from the umbilical cord, may provide therapeutic options even for the neurological manifestations of some LSDs if used appropriately to provide functional macrophages/microglia, to provide secretable enzymes, to provide other molecules that might subserve anti-inflammatory, antiscarring, antiapoptotic, proangiogenic, or protrophic actions. Whether it makes sense to use cells from the actual LSD patient remains to be determined. As noted above, such cells may already harbor a genetic defect that renders them inappropriate. Matched nondisease cells may be more effective. as in the case described above.

5 CONCLUSION

The fundamental biological attributes of NSCs may likely be harnessed to circumvent some of the obstacles that presently exist in treating LSDs with neurological manifestations. Such cells would not be used in isolation but adjunctively with other interventions such as enzyme replacement, substrate reduction, and anti-inflammatory pharmacological interventions, viral vector-mediated gene therapy, and perhaps BMT or UCC infusion. Indeed, many of these interventions work synergistically with each other. For example, substrate reduction works synergistically with BMT to extend lifespan in mouse models of SD. Similarly, NSC transplantation appears to work synergistically with such pharmacology. The reversal of established disease will likely require cell replacement and the reconstruction of a damaged CNS.

Although a good deal of additional research is required before such a strategy falls comfortably within the repertoire of the NSC, this task will be abetted if it is conjoined with other strategies. Intriguingly some of these enabling actions can emanate from the stem cell itself: proangiogenic, anti-inflammatory, antiapoptotic, proregenerative, and neurite-outgrowth-promoting. The rational orchestration of multiple therapeutic modalities for actual patients will require careful planning. The NSC, as central to many fundamental developmental mechanisms, may serve as the "glue" that brings many of these strategies together. The earlier the institution of therapy—ideally pre-symptomatically and/or in utero—the more effective stem cell therapy will be. Early intervention, of course, mandates prenatal and neonatal diagnosis which, in turn, requires either a high level of suspicion and a keen awareness of family history on the part of the clinician and/or a good prenatal/neonatal screening program.

The successful translation of the NSC's therapeutic potential from animal models to human diseases will rely on further characterization of not only stem cell biology but also the pathophysiology of the diseases themselves: what is required to restore homeostasis and preserve neural function? From the point of view of stem cell biology, important questions will include the following. (1) Detailed characterization of the human NSCs to be used including protocols for controlling their propagation and phenotype-specification. (2) Detailed examination of the fate of the human NSCs in vivo following transplantation. For example, do the donor-derived cells become integrated as functional, electrophysiogically active neurons making appropriate synaptic connections without making improper connection? Do they survive as glial cells (including as myelinating cells) or even as undifferentiated progenitors? Do they survive at all in the environment and, if so, for how long? (3) Is the amount of the enzyme expressed intrinsically by the NSC sufficient for the particular disease or must there be additional genetic manipulation to optimize its production, processing, and secretion? (4) Is there behavioral recovery following NSC engraftment and is this long-standing? (5) How might stem cells be administered to address all the neural regions requiring correction? Although NSCs can distribute themselves diffusely throughout the cortex and cerebrum via relatively simple intraventricular infusions, widespread delivery to the cerebellum, to the spinal cord, and to the peripheral nervous system may be necessary as well, particularly for some LSDs of infantile-onset. Methods for obtaining engraftment in those other regions have been devised but require more preclinical animal testing. They usually involve, as for the forebrain, making NSCs accessible to germinal zones.

Notwithstanding the extensive amount of preclinical work still required (some aspects enumerated above), LSDs are, in our estimation, among the "low hanging fruit" of the stem cell field, particularly the neural stem cell field. We have evolved this view for the following reasons: (a) because of the inevitable, unremitting, rapidly progressive, untreatable morbidity and mortality of most LSDs with CNS manifestations; (b) because the etiology and pathophysiology of LSDs, which are typically monogenic, are usually better understood than the more common and famous but polygenically determined or sporadically occurring adult neurodegenerative diseases; (c) because these diseases can usually be diagnosed early, often prenatally and presymptomatically when interventions may blunt progression and preserve cells; (d) because the mutated gene typically encodes a single secretable and diffusible enzyme, repletion of only a small amount of which is capable of restoring normal metabolism to mutant cells via cross-correction; (e) because the ability of the NSC to circumvent the BBB, reside stably, seamlessly, and innocuously throughout the CNS parenchyma while continuing to express its normal lysosomal repertoire (often without the need for additional genetic engineering) precisely addresses some of the present gaps in effective LSD therapy; (f) because the NSC intrinsically exerts additional forces that address other aspects of these diseases (e.g., anti-inflammatory, protrophic actions); (g) because, for the most rudimentary therapeutic needs of these diseases, the particular cell type into which the NSC differentiates is not important, assuming it is not disruptive (e.g., a donor-derived astrocyte acting as a "chaperone" cell is sufficient); (h) because the biology of the NSC meshes well with the biology of the pediatric brain, particularly the fetal or neonatal, where treatment would be optimal and desirable: cells integrate readily into accessible and active germinal zones from which they distribute themselves robustly and efficiently, intermixing with the patient's own progenitors, responding to the same microenvironmental cues, creating chimeric CNS regions in which nonmutant cells intersperse seamlessly with mutant cells, the basis of therapy.

ACKNOWLEDGMENT

The authors gratefully acknowledge the support of the following funding organizations that made much of the work reported here possible: Children's Neurobiological Solutions, A-T Children's Project, National Association for Tay-Sachs and Allied Diseases, Hunter's Hope, NINDS, NIGM, Margot Anderson Brain Restoration Foundation, Viacell, Inc., Late-Onset Tay Sachs Foundation, and The Myelin Project.

REFERENCES

Ahmed, S., Reynolds, B. A., and Weiss, S. (1995). BDNF enhances the differentiation but not the survival of CNS stem cell-derived neuronal precursors. J Neurosci *15*, 5765–5778.

Altman, J., and Das, G. D. (1965). Autoradiographic and histological evidence of postnatal hippocampal neurogenesis in rats. J Comp Neurol *124*, 319–335.

Anderson, D. J. (2001). Stem cells and pattern formation in the nervous system: the possible versus the actual. Neuron *30*, 19–35.

Auerbach, J. M., Eiden, M. V., and McKay, R. D. (2000). Transplanted CNS stem cells form functional synapses in vivo. Eur J Neurosci *12*, 1696–1704.

Bjornson, C. R., Rietze, R. L., Reynolds, B. A., Magli, M. C., and Vescovi, A. L. (1999). Turning brain into blood: a hematopoietic fate adopted by adult neural stem cells in vivo. Science *283*, 534–537.

Escolar, M. L., Poe, M. D., Provenzale, J. M., Richards, K. C., Allison, J., Wood, S., Wenger, D. A., Pietryga, D., Wall, D., Champagne, M., et al. (2005). Transplantation of umbilical-cord blood in babies with infantile Krabbe's disease. N Engl J Med *352*, 2069–2081.

Flax, J. D., Aurora, S., Yang, C., Simonin, C., Wills, A. M., Billinghurst, L. L., Jendoubi, M., Sidman, R. L., Wolfe, J. H., Kim, S. U., and Snyder, E. Y. (1998). Engraftable human neural stem cells respond to developmental cues, replace neurons, and express foreign genes. Nat Biotechnol *16*, 1033–1039.

Gage, F. H. (2000). Mammalian neural stem cells. Science *287*, 1433–1438.

Gage, F. H., Kempermann, G., Palmer, T. D., Peterson, D. A., and Ray, J. (1998). Multipotent progenitor cells in the adult dentate gyrus. J Neurobiol *36*, 249–266.

Gage, F. H., Ray, J., and Fisher, L. J. (1995). Isolation, characterization, and use of stem cells from the CNS. Annu Rev Neurosci *18*, 159–192.

He, W., Ingraham, C., Rising, L., Goderie, S., and Temple, S. (2001). Multipotent stem cells from the mouse basal forebrain contribute GABAergic neurons and oligodendrocytes to the cerebral cortex during embryogenesis. J Neurosci *21*, 8854–8862.

Imitola, J., Comabella, M., Chandraker, A. K., Dangond, F., Sayegh, M. H., Snyder, E. Y., and Khoury, S. J. (2004a). Neural stem/progenitor cells express costimulatory molecules that are differentially regulated by inflammatory and apoptotic stimuli. Am J Pathol *164*, 1615–1625.

Imitola, J., Raddassi, K., Park, K. I., Mueller, F. J., Nieto, M., Teng, Y. D., Frenkel, D., Li, J., Sidman, R. L., Walsh, C. A., et al. (2004b). Directed migration of neural stem cells to sites of CNS injury by the stromal cell-derived factor 1alpha/CXC chemokine receptor 4 pathway. Proc Natl Acad Sci USA *101*, 18117–18122.

Jeyakumar, M., Butters, T. D., Dwek, R. A., and Platt, F. M. (2002). Glycosphingolipid lysosomal storage diseases: therapy and pathogenesis. Neuropathol Appl Neurobiol *28*, 343–357.

Jeyakumar, M., Thomas, R., Elliot-Smith, E., Smith, D. A., van der Spoel, A. C., d'Azzo, A., Perry, V. H., Butters, T. D., Dwek, R. A., and Platt, F. M. (2003). Central nervous system inflammation is a hallmark of pathogenesis in mouse models of GM1 and GM2 gangliosidosis. Brain *126*, 974–987.

Kempermann, G., Gast, D., Kronenberg, G., Yamaguchi, M., and Gage, F. H. (2003). Early determination and long-term persistence of adult-generated new neurons in the hippocampus of mice. Development *130*, 391–399.

Kempermann, G., Kuhn, H. G., and Gage, F. H. (1998). Experience-induced neurogenesis in the senescent dentate gyrus. J Neurosci *18*, 3206–3212.

Kilpatrick, T. J., and Bartlett, P. F. (1993). Cloning and growth of multipotential neural precursors: requirements for proliferation and differentiation. Neuron *10*, 255–265.

Kuhn, H. G., and Svendsen, C. N. (1999). Origins, functions, and potential of adult neural stem cells. Bioessays *21*, 625–630.

Lacorazza, H. D., Flax, J. D., Snyder, E. Y., and Jendoubi, M. (1996). Expression of human beta-hexosaminidase alpha-subunit gene (the gene defect of Tay-Sachs disease) in mouse brains upon engraftment of transduced progenitor cells. Nat Med *2*, 424–429.

Lois, C., and Alvarez-Buylla, A. (1993). Proliferating subventricular zone cells in the adult mammalian forebrain can differentiate into neurons and glia. Proc Natl Acad Sci USA *90*, 2074–2077.

Lu, P., Jones, L. L., Snyder, E. Y., and Tuszynski, M. H. (2003). Neural stem cells constitutively secrete neurotrophic factors and promote extensive host axonal growth after spinal cord injury. Exp Neurol *181*, 115–129.

Meng, X. L., Shen, J. S., Ohashi, T., Maeda, H., Kim, S. U., and Eto, Y. (2003). Brain transplantation of genetically engineered human neural stem cells globally corrects brain lesions in the mucopolysaccharidosis type VII mouse. J Neurosci Res *74*, 266–277.

Monje, M. L., Mizumatsu, S., Fike, J. R., and Palmer, T. D. (2002). Irradiation induces neural precursor-cell dysfunction. Nat Med *8*, 955–962.

Myerowitz, R., Lawson, D., Mizukami, H., Mi, Y., Tifft, C. J., and Proia, R. L. (2002). Molecular pathophysiology in Tay–Sachs and Sandhoff diseases as revealed by gene expression profiling. Hum Mol Genet *11*, 1343–1350.

Neufeld, E. F. (1991). Lysosomal storage diseases. Annu Rev Biochem *60*, 257–280.

Neufeld, E. F., and Fratantoni, J. C. (1970). Inborn errors of mucopolysaccharide metabolism. Science *169*, 141–146.

Oehlmann, R., Zlotogora, J., Wenger, D. A., and Knowlton, R. G. (1993). Localization of the Krabbe disease gene (GALC) on chromosome 14 by multipoint linkage analysis. Am J Hum Genet *53*, 1250–1255.

Park, K. I., Teng, Y. D., and Snyder, E. Y. (2002). The injured brain interacts reciprocally with neural stem cells supported by scaffolds to reconstitute lost tissue. Nat Biotechnol *20*, 1111–1117.

Qian, X., Shen, Q., Goderie, S. K., He, W., Capela, A., Davis, A. A., and Temple, S. (2000). Timing of CNS cell generation: A programmed sequence of neuron and glial cell production from isolated murine cortical stem cells. Neuron *28*, 69–80.

Reubinoff, B. E., Itsykson, P., Turetsky, T., Pera, M. F., Reinhartz, E., Itzik, A., and Ben-Hur, T. (2001). Neural progenitors from human embryonic stem cells. Nat Biotechnol *19*, 1134–1140.

Reynolds, B. A., and Weiss, S. (1992). Generation of neurons and astrocytes from isolated cells of the adult mammalian central nervous system. Science *255*, 1707–1710.

Schiffmann, R., and Brady, R. O. (2002). New prospects for the treatment of lysosomal storage diseases. Drugs *62*, 733–742.

Shen, Q., Qian, X., Capela, A., and Temple, S. (1998). Stem cells in the embryonic cerebral cortex: Their role in histogenesis and patterning. J Neurobiol *36*, 162–174.

Shihabuddin, L. S., Numan, S., Huff, M. R., Dodge, J. C., Clarke, J., Macauley, S. L., Yang, W., Taksir, T. V., Parsons, G., Passini, M. A., et al. (2004). Intracerebral transplantation of adult mouse neural progenitor cells into the Niemann–Pick-A mouse leads to a marked decrease in lysosomal storage pathology. J Neurosci *24*, 10642–10651.

Sidman, R. L., Miale, I. L., and Feder, N. (1959). Cell proliferation and migration in the primitive ependymal zone: An autoradiographic study of histogenesis in the nervous system. Exp Neurol *1*, 322–323.

Sly, W. S., and Vogler, C. (2002). Brain-directed gene therapy for lysosomal storage disease: going well beyond the blood–brain barrier. Proc Natl Acad Sci USA *99*, 5760–5762.

Snyder, E. Y., Deitcher, D. L., Walsh, C., Arnold-Aldea, S., Hartwieg, E. A., and Cepko, C. L. (1992). Multipotent neural cell lines can engraft and participate in development of mouse cerebellum. Cell *68*, 33–51.

Snyder, E. Y., Taylor, R. M., and Wolfe, J. H. (1995). Neural progenitor cell engraftment corrects lysosomal storage throughout the MPS VII mouse brain. Nature *374*, 367–370.

Snyder, E. Y., and Wolfe, J. H. (1996). Central nervous system cell transplantation: A novel therapy for storage diseases? Curr Opin Neurol *9*, 126–136.

Tabar, V., Panagiotakos, G., Greenberg, E. D., Chan, B. K., Sadelain, M., Gutin, P. H., and Studer, L. (2005). Migration and differentiation of neural precursors derived from human embryonic stem cells in the rat brain. Nat Biotechnol *23*, 601–606.

Takaura, N., Yagi, T., Maeda, M., Nanba, E., Oshima, A., Suzuki, Y., Yamano, T., and Tanaka, A. (2003). Attenuation of ganglioside GM1 accumulation in the brain of GM1 gangliosidosis mice by neonatal intravenous gene transfer. Gene Ther *10*, 1487–1493.

Temple, S. (2001). The development of neural stem cells. Nature *414*, 112–117.

Teng, Y. D., Lavik, E. B., Qu, X., Park, K. I., Ourednik, J., Zurakowski, D., Langer, R., and Snyder, E. Y. (2002). Functional recovery following traumatic spinal cord injury mediated by a unique polymer scaffold seeded with neural stem cells. Proc Natl Acad Sci USA *99*, 3024–3029.

Vescovi, A. L., Gritti, A., Galli, R., and Parati, E. A. (1999). Isolation and intracerebral grafting of nontransformed multipotential embryonic human CNS stem cells. J Neurotrauma *16*, 689–693.

Wada, R., Tifft, C. J., and Proia, R. L. (2000). Microglial activation precedes acute neurodegeneration in Sandhoff disease and is suppressed by bone marrow transplantation. Proc Natl Acad Sci USA *97*, 10954–10959.

Wenger, D. A., Rafi, M. A., Luzi, P., Datto, J., and Costantino-Ceccarini, E. (2000). Krabbe disease: Genetic aspects and progress toward therapy. Mol Genet M1–9.

Wenger, D. A., Sattler, M., and Hiatt, W. (1974). Globoid cell leukodystrophy: deficiency of lactosyl ceramide beta-galactosidase. Proc Natl Acad Sci USA *71*, 854–857.

White, P. M., Morrison, S. J., Orimoto, K., Kubu, C. J., Verdi, J. M., and Anderson, D. J. (2001). Neural crest stem cells undergo cell-intrinsic developmental changes in sensitivity to instructive differentiation signals. Neuron *29*, 57–71.

Windrem, M. S., Nunes, M. C., Rashbaum, W. K., Schwartz, T. H., Goodman, R. A., McKhann, G., 2nd, Roy, N. S., and Goldman, S. A. (2004). Fetal and adult human oligodendrocyte progenitor cell isolates myelinate the congenitally dysmyelinated brain. Nat Med *10*, 93–97.

Wurmser, A. E., Nakashima, K., Summers, R. G., Toni, N., D'Amour, K. A., Lie, D. C., and Gage, F. H. (2004). Cell fusion-independent differentiation of neural stem cells to the endothelial lineage. Nature *430*, 350–356.

Yamanaka, S., Johnson, M. D., Grinberg, A., Westphal, H., Crawley, J. N., Taniike, M., Suzuki, K., and Proia, R. L. (1994). Targeted disruption of the Hexa gene results in mice with biochemical and pathologic features of Tay–Sachs disease. Proc Natl Acad Sci USA *91*, 9975–9979.

Yandava, B. D., Billinghurst, L. L., and Snyder, E. Y. (1999). "Global" cell replacement is feasible via neural stem cell transplantation: Evidence from the dysmyelinated shiverer mouse brain. Proc Natl Acad Sci USA *96*, 7029–7034.

Zhang, S. C., Wernig, M., Duncan, I. D., Brustle, O., and Thomson, J. A. (2001). In vitro differentiation of transplantable neural precursors from human embryonic stem cells. Nat Biotechnol *19*, 1129–1133.

Zlomanczuk, P., Mrugala, M., de la Iglesia, H. O., Ourednik, V., Quesenberry, P. J., Snyder, E. Y., and Schwartz, W. J. (2002). Transplanted clonal neural stem-like cells respond to remote photic stimulation following incorporation within the suprachiasmatic nucleus. Exp Neurol *174*, 162–168.

THE GM1 GANGLIOSIDOSES

Gustavo Charria-Ortiz

1 INTRODUCTION

The gangliosidoses are a group of lysosomal storage diseases characterized by the accumulation of these complex glycolipids in multiple organs of the body. They manifest a predominantly neurological phenotype, a fact that is probably related to their high prevalence in nervous tissues. The typical presentation is that of a progressive neuro-degenerative disease with onset in early life followed by loss of acquired developmental milestones, dementia, and worsening neurological deficits. Early death is common, except perhaps in late-onset variants. Systemic manifestations in the form of visceromegaly or skeletal deformities are preferentially seen in the aggressive variants of G_{M1} ganglio-sidosis (i.e., infantile), whereas they are typically absent in G_{M2} gangliosidosis (except for Sandhoff's disease, which may present with mild visceromegaly). Depending upon the severity of the enzymatic defect, they show different rates of clinical progression and involvement, with severe deficiencies leading to the most aggressive forms in an infantile and acute or subacute fashion, whereas milder ones have an onset later in life and slower clinical progression.

Gangliosides were first identified in ganglion cells of the brain by the German scientist Ernst Klenk in 1942 (Klenk, 1942). They are present in almost all known types of animal cells and play a wide range of structural and physiological roles (Colombaioni and Garcia-Gil, 2004; Morales et al., 2004; Zhang and Kiechle, 2004; Bektas and Spiegel, 2004). Gangliosides are one of the most complex groups of glycolipids identified so far, showing an unusually large degree of structural variety; more than 160 of them have been identified to date. They can make up to 5 to 10% of total lipid mass (Mikata and Taniguchi, 1985).

The typical chemical structure of a ganglioside is that of a ceramide attached to an oligosaccharide chain via a glycosidic bond (Figure 1). Ceramide, in turn, consists of a molecule of sphingosine linked to a long fatty acid (usually stearic acid) via an amide bond:

It is the different type of oligosaccharide chain, however, that provides these molecules with such a wide degree of structural and functional variety. These chains, which can be up to eight molecules long, consist of simple hexoses (either glucose or galactose) alternating with more complex ones, such as the amino sugars N-acetylgalacto-samine (GalNac), N-acetylglucosamine (GlcNac), or sialic acid (N-acetyl-neuraminic

Department of Neurology, University of Miami, Leonard Miller School of Medicine; Child Neurologist, Miami Children's Hospital. e-mail: gcharria@med.miami.edu

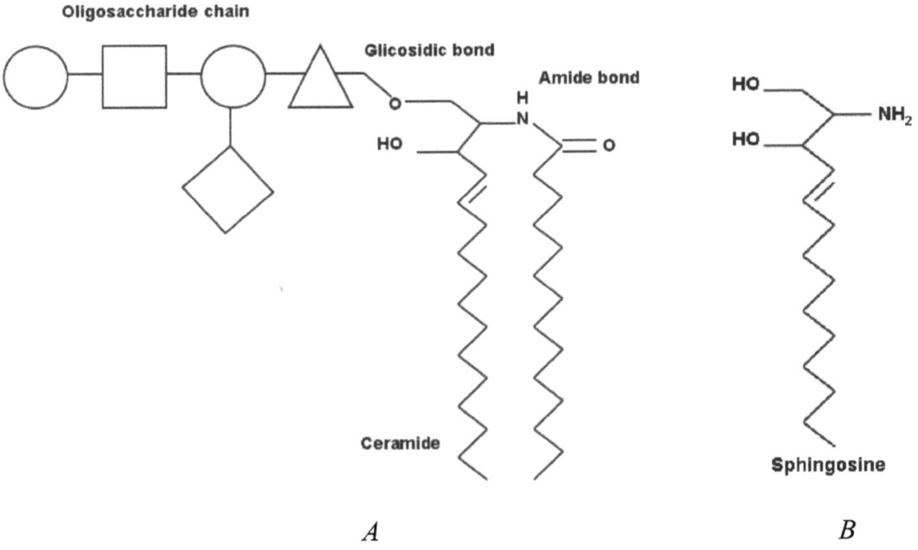

Figure 1. (A) Chemical structure of a ganglioside. Δ: glucose, O: galactose, ◊: N-acetyl-neuraminic acid (sialic acid), □: N-acetylgalactosamine. (B) Sphingosine.

acid or NANA; Figure 2). These oligosaccharide chains are also the main site of their catabolism, which occurs in lysosomes by the sequential removal of the terminal hexoses. Thus, defects in such processes will lead to their intracellular accumulation with resulting pathology.

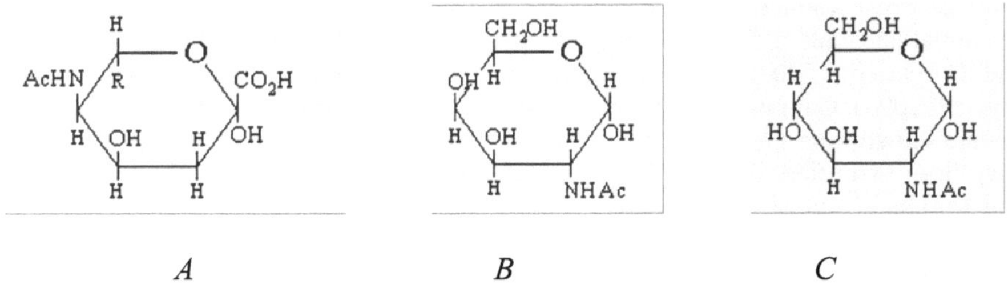

Figure 2. Amino-sugars commonly found in the oligosaccharide chain of gangliosides. (A) N-acetyl-neuraminic acid (NANA or sialic acid); (B) N-acetylgalactosamine (GalNac); (C) N-acetylglucosamine.

Several classifications have been developed in order to deal with the complexity of gangliosides (IUPAC-IUB Joint Commission on Biochemical Nomenclature, JCBN). The one that is more frequently used in the literature was proposed by Lars Svennerholm in 1970 (Svennerholm, 1970). On it, gangliosides are classified according to the number of sialic acid residues present in their terminal oligosaccharide chains. Therefore, they could be either A (a-sialo), M (mono-sialo), D (di-sialo), T (tri-sialo), or Q (tetra-sialo) gangliosides (written as G_A, G_M, G_D, G_T, or G_Q). In addition, numbers (1, 2, or 3) and letters (a, b, c) are used to describe other chemical characteristics, such as their distinct patterns of migration on electrophoresis gels. Following this classification, G_{M1} ganglioside would then consist of a ceramide linked to an oligosaccharide chain containing only one molecule of sialic acid (Figure 1A). G_{M2} and G_{M3} would also have one molecule of sialic acid, however, they will have different electrophoretic migration patterns than G_{M1}. The main gangliosides found in human brain are G_{M1}, G_{D1a}, G_{D1b}, and G_{T1} (Mikata and Taniguchi, 1985).

2 GM1 GANGLIOSIDOSIS

This condition is caused by the defective activity of the enzyme acid β-galactosidase (EC 3.2.1.23; Okada and O'Brien, 1968), a lysosomal enzyme that cleaves the terminal β-galactose moieties of multiple molecules, including G_{M1} ganglioside. Many other compounds also carrying such terminal galactose moieties (such as several oligosaccharides and mucopolysaccharides, in particular keratan sulfate) accumulate in body tissues as well. Nevertheless, G_{M1} ganglioside seems to be by far the predominant one. Defective activity of this same enzyme also causes Morquio disease type B, one of the Mucopolysaccharidoses (Arbisser et al., 1977), which is characterized by multiple skeletal deformities (i.e., "dysostosis multiplex"), cardiomyopathy, and coarse features, but does not usually lead to gross neurological involvement. The reasons for this phenotypic discrepancy are not known.

G_{M1} gangliosidosis is an uncommon disorder, and data about its prevalence in the United States are not available. It is perhaps close to 1 in 100,000 live births. A recent epidemiological survey made in Turkey (Ozkara and Topcu, 2004) found an incidence of 0.45 per 100,000. It is a panethnic disorder. However, and as it occurs with many other metabolic conditions, specific ethnic groups show higher prevalences of some of its variants (see below).

Three variants have been described, which are classified based upon the age of onset and their distinct clinical features. They are the infantile (type 1), late infantile or juvenile (type 2), and adult (type 3) forms. It has been suggested that this variation relates inversely to the residual amount of catalytic activity of the mutant enzyme (William et al., 1998), with patients suffering from the early infantile variant showing activities close to 0.1% (Norden and O'Brien, 1973), whereas patients with the late-onset presentations show values close to 10% (Yoshida et al., 1992).

2.1 Type 1 or Infantile GM1 Gangliosidosis

This is the most severe and aggressive form of the disease, and also the most common one. It presents in the first few months of life with neurodevelopmental arrest and systemic involvement, which includes cardiomyopathy, visceromegaly, skeletal deformities, and coarsening of facial features (Landing et al., 1964). Rare cases of neonatal onset have been reported, presenting in the form of nonimmune hydrops fetalis (Tasso et al., 1996). The degree of CNS involvement is severe, and neuronal degeneration rapidly occurs. Hypotonia and an exaggerated startle response are frequently present. Eventually, patients develop spastic quadriplegia, decerebrate rigidity, seizures, deafness, esotropia, and blindness. A "cherry-red" spot macula occurs in about half of the patients. In regard to its systemic manifestations, there is thickening of the subcutaneous tissues and skeletal abnormalities such as frontal bossing, depressed nasal bridge, stubby hands, broad wrists, anterior beaking of the lumbar vertebrae, thickening of the midshaft of the humerus, and spatulate ribs. The latter findings (also known as "dysostosis multiplex") resemble the clinical features of some of the mucopolysaccharidoses (in fact, this condition used to be called the "pseudo-Hurler" variant of the neurolipidoses, because there is no gross urinary excretion of mucopolysaccharides). Skin lesions in the form of angiokeratomas or dermal melanocytosis can also be seen (Hanson et al., 2003). Vacuolated lymphocytes and foamy histiocytes are frequently found in peripheral smears, and accumulated compounds can also be seen in liver or spleen cells, even renal glomeruli. Isolated cardiomyopathy has also been described as the initial manifestation of the disease (Lin et al., 2000). The disease progresses relentlessly, leading to death within two to three years of its onset.

This particular variant of G_{M1} gangliosidosis has a particularly high incidence in certain areas of Brazil (Severini et al., 1999), the Maltese Islands (Lenicker et al., 1997), and Cyprus (Georgiou et al., 2005).

2.2 Type 2 or Late Infantile/Juvenile GM1 Gangliosidosis

The early development of patients with this variant is usually normal, although an exaggerated startle response or mild hypotonia may be present from early age. The onset of symptoms varies from seven months to three years of age (Derry et al., 1968). Towards the end of the first year, the child may lose his or her ability to crawl, sit, or stand, subsequently showing neurodevelopmental arrest or regression. The plantar responses become extensor and seizures develop. Although mild pallor of the optic discs may be seen, a cherry red spot is usually not present. Corneal clouding and rotatory nystagmus may also occur. In most cases there are no gross systemic findings, the liver is not enlarged, and there are no skeletal abnormalities, although radiographic evidence of this may be found. The condition is progressive, and patients usually survive for only three to ten years after its onset.

2.3 Type 3 or Adult/Chronic Form GM1 Gangliosidosis

The age at onset has been reported as early as 3 years or as late as 30 (mean age of onset: 13.1 years (Muthane et al., 2004). These patients have chronic progressive pyramidal and extrapyramidal disease, which may start with dysarthria or gait disturbances then followed by prominent dystonia (usually facial) or Parkinsonian features (Figure 3). Ataxia has been also reported in up to 12% of patients (Muthane et al., 2004). The degree of CNS involvement is not as extensive as in types 1 or 2, and is mostly localized to basal ganglia mainly caudate nucleus and putamen (Goldman et al., 1981). Skeletal changes are minimal, although X-ray evidence of disease occurs in up to 75% of patients. Visceromegaly does not occur, and eye movements are usually normal.

This variant shows an unusually high incidence in Japan, with almost 75% of the reported cases coming from that geographic region.

3 DIAGNOSIS

The diagnosis is confirmed by measuring deficient activity of β-galactosidase in leukocytes or other cells (such as fibroblasts). Confirmation of diagnosis has also been achieved in a retrospective fashion by testing newborn screen blood samples collected on filter paper (Chamoles et al., 2001).

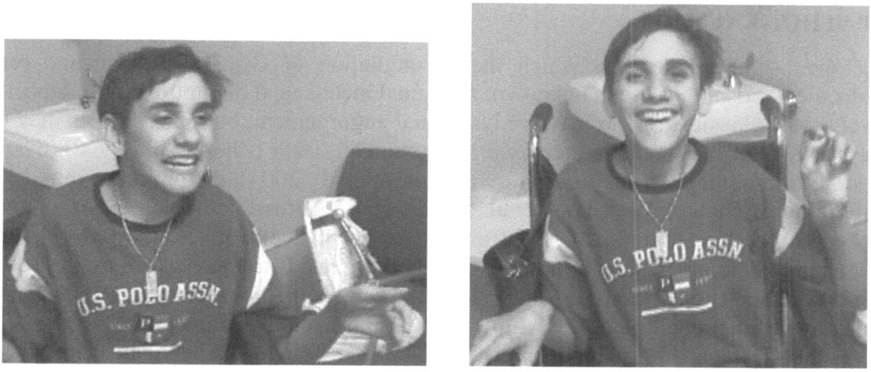

Figure 3. Patient with Type 3 G_{M1} gangliosidosis. Note the prominent dystonia, affecting mainly the face and upper extremities.

Ultrastructural examination of the skin or other organs reveals engorged lysosomes and foamy histiocytes. Galactose-containing oligosaccharides (and occasionally small amounts of keratan sulfate) are excreted in urine. Magnetic resonance imaging studies of the brain reveal delayed myelination and abnormal appearance of the subcortical white matter (Gururaj et al., 2005). Myelination arrest has also been demonstrated in the early-onset variant (Shen, Tsai, and Tsai, 1998). Later, diffuse atrophy develops in the late-onset variants (Figure 4). Routine laboratory tests are usually normal.

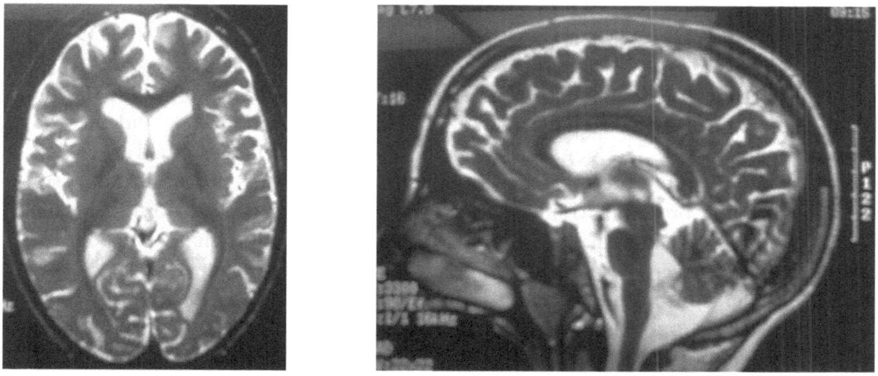

Figure 4. MRI of the brain of a patient with the type 3 variant of G_{M1} gangliosidosis. Note the mild to moderate degree of cerebral and cerebellar atrophy and minimal signal changes in basal ganglia.

Brain FDG-PET scans show a mild decrease in basal ganglia uptake with moderate to severe decrease in thalamic and visual cortex uptake (Al-Essa et al., 1999).

Prenatal diagnosis is available. It may be performed by either measuring enzyme activity in chorionic villi (Minelli et al., 1992) or by analyzing amniotic fluid through chromatography techniques (Booth, Gerbie, and Nadler, 1973; Ramsay et al., 2004).

4 PATHOGENESIS

The precise mechanism by which the accumulation of G_{M1} ganglioside and related molecules leads to disease is unknown. An actual mechanical disruption at the subcellular level seems plausible due to massive lysosomal engorgement, but other factors may be at play. The latter may include increased apoptotic cell death (Zhou et al., 1998), dysregulation of the unfolded protein response (Tessitore et al., 2004), disruption of intracellular trafficking of cholesterol (Glaros et al., 2005), and abnormal activation of inflammatory pathways (Jeyakumar et al., 2003). Clearly, there is some cellular selectivity for the observed neurological involvement, inasmuch as myelin-related structures seem to be predominantly affected (Folkerth et al., 2000). This latter phenomenon seems to be dependent on both severe oligodendrocyte loss (up to 85%) and axonal dysfunction (van der Voorn et al., 2004). Interestingly, no appreciable glycolipid storage has been noted in surviving oligodendrocytes obtained from animal models with this disease.

Acid β-galactosidase, together with two other enzymes (α-neuraminidase and galactosamine 6-sulphate sulphatase) plus the protective protein cathepsin-A form a newly described intralysosomal complex (Hinek et al., 2000). It has been shown that the formation of this complex is essential for the stabilization and adequate posttranslational processing of the enzyme, and also for the proper catabolism of galactose-containing oligosaccharides chains, including G_{M1} (Pshezhetsky and Potier, 1996).

The skeletal and subcutaneous changes are likely related to both oligosaccharide and mucopolysaccharides accumulation. However, an additional factor may be at play because a direct disruption of elastic fiber assembly (due to mutant elastin binding protein or EBP, see below) has also been demonstrated (Pshezhetsky and Ashmarina, 2001).

5 GENETICS

5.1 Clinical and Laboratory Diagnosis

The gene for acid β-galactosidase (GLB1) is located on chromosome 3p21.33 (Takano and Yamanouchi, 1993). It spans about 62.5 kb of genomic DNA and has 16 exons (Oshima et al., 1988). The gene gives rise to two alternatively spliced mRNAs, which encode for two different proteins: acid β-galactosidase per se—the larger transcript—and the elastin binding protein (Privitera et al, 1998). The latter results from the splicing out of three exons (Morreau et al., 1989) and does not possess catalytic activity on its own (it has been called "S-Gal"). EBP functions at the cell surface level, not the lysosome, by binding to the elastin-binding receptor (Hinek and Rabinovitch, 1994). It apparently has a role in delivering tropoelastin to form the extracellular matrix and in the formation of collagen and supportive tissues. It seems plausible to postulate that mutations affecting such mRNA splicing process at different sites may lead to disturbances in the function of either protein,or both, at the same time. This would then correlate with the presence of impaired catalytic activity of β-galactosidase and the integrity of supportive tissues such as bone (Caciotti et al., 2005).

Several types of mutations have been described including missense, nonsense, duplications, insertions, and splice site abnormalities (Pascheke et al., 2001). Although no genotype to phenotype correlation has yet been made, certain types of mutations have been more frequently reported in either the infantile variant (i.e., c.171C→G, c.245+1G→A or c.145C→T; Georgiou et al., 2004) or the adult variant (I51T; Yoshida et al., 1991).

The disease occurs naturally in other species, including dogs (Alroy et al., 1985; Yamato et al., 2003; Kreutzer at al., 2005), sheep (Ryder and Simmons, 2001), and cats (Baker et al., 1971). Animal models are also available (Matsuda et al., 1997; Itoh et al., 2001).

5.2 Differential Diagnosis

Type 1:

I-Cell Disease (Mucolipidosis Type II)
Mucolipidosis Type I (Alpha-Neuraminidase Deficiency-Sialidosis)
Mucopolysaccharidosis Type I (Hurler's disease)
Oligosaccharidoses (e.g., mannosidoses, fucosidoses, sialidoses)

Type 2 and 3:

Niemann–Pick type C,
Wilson's disease,
Glutaric aciduria type 1 and other metabolic diseases with dystonia or prominent extrapyramidal findings.

6 THERAPY

No curative therapy is available. Different methods have been attempted in the past, but they have proven to be unsuccessful. These include transplantation of amniotic tissue (Tylki-Szymanska et al., 1985), enzyme replacement therapy (Reynolds, Baker, and Reynolds, 1978), and bone marrow transplantation (BMT; O'Brien et al., 1990). One of our patients with the type 3 variant underwent this latter procedure, but despite correction of his enzymatic deficiency in leukocytes the clinical picture did not appreciably change. Additional approaches that have also been tried include transplantation of genetically modified bone marrow cells (Sano et al., 2005) and gene therapy (Sena-Esteves et al., 2000; Takaura et al., 2003). The latter has been attempted in several animal and cellular models with various degrees of success.

Symptomatic management should be offered for the extrapyramidal signs and spasticity that is commonly seen in the milder variants of the disease. Agents that are commonly used for this purpose include L-DOPA, lioresal (Baclofen™), trihexyphenidyl (Artane™), and clonazepam. More recently, injections of botulinum toxin in the dystonic areas have also been used. In at least one patient, thalamotomy was attempted for the management of spasticity with good results (Muthane et al., 2004).

Recently, the use of ganglioside synthesis inhibitors such as N-butyldeoxynojirimycin has been attempted in several models of this disease (Kasperzyk et al., 2005, 2004). This "closing-the- faucet" approach (Tifft and Proia, 2000) may be considered as a counterpart to those more traditional "unclogging-the-drain" approaches (Brady and Schiffmann, 2004) used in the past for the treatment of this group of conditions. It also has the advantage of being accessible to the central nervous system because these small compounds can cross the blood–brain barrier. The one which has been more extensively studied is N-butyldeoxygalacto-nojirimycin (NB-DGJ), an imino sugar that inhibits ceramide-specific glucosyltransferase. The latter is the enzyme that catalyzes the first step in GSL biosynthesis. In at least three different studies (Takaura et al., 2003; Kasperzyk et al., 2004, 2005), a decreased amount of total ganglioside accumulation has been reported without affecting other brain lipids. This medication seems to be effective and safe, and has been recently approved by the FDA for the treatment of another lysosomal storage disease (i.e., Gaucher disease or glucocerebrosidase deficiency). Clinical trials are under way for other variants of the gangliosidoses (see G_{M2}).

Finally, an innovative approach that involves the use of chemical "chaperones" has been reported for the treatment of this condition (Matsuda et al., 2003; Tominaga et al., 2001). This concept is based on the fact that certain low-molecular compounds (analogous to galactose) which act as inhibitors of β-galactosidase, may actually increase the residual activity of mutant enzyme when used at low doses. This effect is more prominent when the mutant enzyme has some amount of residual activity, as it occurs in the later onset variants of the disease. It is yet unclear, however, whether such gain of function is enough to reduce ganglioside accumulation to therapeutic levels.

REFERENCES

Achord D, Brot F, Gonzalez-Noriega A, Sly W, Stahl P (1977) Human beta-glucuronidase. II. Fate of infused human placental beta-glucuronidase in the rat. *Pediatr Res.* 11:816.

Al-Essa MA, Bakheet SM, Patay ZJ, Nounou RM, Ozand PT (1999) Cerebral fluorine-18 labeled 2-fluoro-2-deoxyglucose positron emission tomography (FDG PET), MRI, and clinical observations in a patient with infantile G(M1) gangliosidosis. *Brain Dev.* 2:559–562.

Alroy J, Orgad U, Ucci AA, Schelling SH, Schunk KL, Warren CD, Raghavan SS, Kolodny EH (1985) Neurovisceral and skeletal GM1-gangliosidosis in dogs with beta-galactosidase deficiency. *Science* 229: 470.

Arbisser AI, Donnelly KA, Scott CI Jr, DiFerrante N, Singh J, Stevenson RE, Aylesworth AS, Howell RR (1977) Morquio-like syndrome with beta-galactosidase deficiency and normal hexosamine sulfatase activity: mucopolysaccharidosis IV *B. Am. J. Med. Genet.* 1: 195–205.

Baker HJ Jr, Lindsey JR, McKhann GM, Farrell DF (1971) Neuronal GM1 gangliosidosis in a Siamese cat with beta-galactosidase deficiency. *Science* 174: 838.

Bektas M, Spiegel S. (2004) Glycosphingolipids and cell death. *Glycoconj J.*20:39–47.

Booth CW, GerbieAB, Nadler HL. (1973) Intrauterine diagnosis of GM1 gangliosidosis, type 2. *Pediatrics* 52: 521–524

Brady RO, Schiffmann R (2004) Enzyme-replacement therapy for metabolic storage disorders. *Lancet Neurol.* 3: 752–756.

Caciotti A, Donati MA, Boneh A, d'Azzo A, Federico A, Parini R, Antuzzi D, Bardelli T, Nosi D, Kimonis V, Zammarchi E, Morrone A. (2005) Role of beta-galactosidase and elastin binding protein in lysosomal and nonlysosomal complexes of patients with GM1-gangliosidosis. *Hum Mutat.* 25: 285–292

Chamoles NA, Blanco MB, Iorcansky S, Gaggioli D, Specola N, Casentini C (2001) Retrospective diagnosis of GM1 gangliosidosis by use of a newborn-screening card. *Clin Chem.* 47:2068.

Colombaioni L, Garcia-Gil M (2004) Sphingolipid metabolites in neural signalling and function. *Brain Res Brain Res Rev.* 46: 328–355.

Derry DM, Fawcett JS, Andermann F, Wolfe LS (1968) Late infantile systemic lipidosis (major monosialogangliosidosis; delineation of two types). *Neurology* 18: 340–347.
Folkerth RD, Alroy J, Bhan I, Kaye EM (2000) Infantile G(M1) gangliosidosis: Complete morphology and histochemistry of two autopsy cases, with particular reference to delayed central nervous system myelination. *Pediatr Dev Pathol.* 3:73–86.
Georgiou T, Drousiotou A, Campos Y, Caciotti A, Sztriha L, Gururaj A, Ozand P, Zammarchi E, Morrone A, D'Azzo A (2004) Four novel mutations in patients from the Middle East with the infantile form of GM1-gangliosidosis. *Hum Mutat.* 24: 536–537.
Georgiou T, Stylianidou G, Anastasiadou V, Caciotti A, Campos Y, Zammarchi E, Morrone A, D'azzo A, Drousiotou A (2005) The Arg482His mutation in the beta-galactosidase gene is responsible for a high frequency of GM1 gangliosidosis carriers in a Cypriot village. *Genet Test.* 9: 126–132.
Glaros EN, Kim WS, Quinn CM, Wong J, Gelissen I, Jessup W, Garner B (2005) Glycosphingolipid accumulation inhibits cholesterol efflux via the ABCA1/apolipoprotein A-I pathway: 1-phenyl-2-decanoylamino-3-morpholino-1-propanol is a novel cholesterol efflux accelerator. *J Biol Chem.* 280: 4515–4523.
Goldman JE, Katz D, Rapin I, Purpura DP, Suzuki K (1981) Chronic GM1 gangliosidosis presenting as dystonia: I. Clinical and pathological features. *Ann Neurol.* 9: 465–475.
Gururaj A, Sztriha L, Hertecant J, Johansen JG, Georgiou T, Campos Y, Drousiotou A, d'Azzo A (2005) Magnetic resonance imaging findings and novel mutations in GM1 gangliosidosis. *J Child Neurol.* 20: 57–60.
Hanson M, Lupski JR, Hicks J, Metry D (2003) Association of dermal melanocytosis with lysosomal storage disease: Clinical features and hypotheses regarding pathogenesis. *Arch. Derm.* 139: 916–920.
Hinek A, Rabinovitch M (1994) 67-kD elastin-binding protein is a protective "companion" of extracellular insoluble elastin and intracellular tropoelastin. *J Cell Biol.* 126:563–574.
Hinek A, Zhang S, Smith AC, Callahan JW (2000) Impaired elastic-fiber assembly by fibroblasts from patients with either Morquio B disease or infantile GM1-gangliosidosis is linked to deficiency in the 67-kD spliced variant of beta-galactosidase. *Am J Hum Genet.* 67:23–36.
Itoh M, Matsuda J, Suzuki O, Ogura A, Oshima A, Tai T, Suzuki Y, Takashima S (2001) Development of lysosomal storage in mice with targeted disruption of the beta-galactosidase gene: A model of human GM1-gangliosidosis. *Brain Dev.* 23(6, Oct): 379–384.
IUPAC-IUB Joint Commission on Biochemical Nomenclature (JCBN). Nomenclature of glycoproteins, glycopeptides and peptidoglycans (Recommendations 1985). *Eur. J. Biochem.* 159, 1–6 (1986); *Glycoconjugate J.* 3,, 123–134 (1986); *J Biol Chem.* 262, 13–18 (1987); *Pure Appl. Chem.* 60, 1389–1394 (1988); Royal Society of Chemistry Specialist Periodical Report, Amino acids and peptides, vol. 21, p. 329 (1990).
Jeyakumar M, Thomas R, Elliot-Smith E, Smith DA, van der Spoel AC, d'Azzo A, Perry VH, Butters TD, Dwek RA, Platt FM (2003) Central nervous system inflammation is a hallmark of pathogenesis in mouse models of GM1 and GM2 gangliosidosis. *Brain.* 126: 974–987.
Kasperzyk JL, d'Azzo A, Platt FM, Alroy J, Seyfried TN (2005) Substrate reduction reduces gangliosides in postnatal cerebrum-brainstem and cerebellum in GM1 gangliosidosis mice. *J Lipid Res.* 46:744–751.

Kasperzyk JL, El-Abbadi MM, Hauser EC, D'Azzo A, Platt FM, Seyfried TN (2004) N-butyldeoxygalactonojirimycin reduces neonatal brain ganglioside content in a mouse model of GM1 gangliosidosis. *J Neurochem*. 89:645–653.

Klenk EZ (1942) Gangliosides, a new group of sugar-containing brain lipoids. *Physiol Chem* 273: 76.

Kreutzer R, Leeb T, Muller G, Moritz A, Baumgartner W (2005) A duplication in the canine {beta}-galactosidase gene GLB1 causes exon skipping and GM1-gangliosidosis in Alaskan Huskies. *Genetics*. 170: 1857–1861.

Landing BH, Silverman FN, Craig JM, Jacoby MD, Lahey ME, Chadwick DL (1964) Familial neurovisceral lipidosis. An analysis of eight cases of a syndrome previously reported as Hurler-variant, pseudo-Hurler disease and Tay–Sachs disease with visceral involvement. *Am. J. Dis. Child.* 108: 503–522.

Lenicker HM, Vassallo Agius P, Young EP, Attard Monsalto SP (1997) Infantile generalized GM1 gangliosidosis: High incidence in the Maltese Islands. *J Inherit Metab Dis*. 20: 723–724.

Lin HC, Tsai FJ, Shen WC, Tsai CH, Peng CT (2000) Infantile form GM1 gangliosidosis with dilated cardiomyopathy: A case report. *Acta Paediatr*. 89: 880–883.

Matsuda J, Suzuki O, Oshima A, Ogura A, Noguchi Y, Yamamoto Y, Asano T, Takimoto K, Sukegawa K, Suzuki Y, Naiki M (1997) Beta-galactosidase-deficient mouse as an animal model for GM1-gangliosidosis. *Glycoconj J.*14: 729–736.

Matsuda J, Suzuki O, Oshima A, Yamamoto Y, Noguchi A, Takimoto K, Itoh M, Matsuzaki Y, Yasuda Y, Ogawa S, Sakata Y, Nanba E, Higaki K, Ogawa Y, Tominaga L, Ohno K, Iwasaki H, Watanabe H, Brady RO, Suzuki Y (2003) Chemical chaperone therapy for brain pathology in GM1-gangliosidosis. *Proc Natl Acad Sci USA*. 100: 15912–15917.

Mikata, A., Taniguchi, N (1985) Glycosphingolipid. In H. Weigandt (Ed.) *Glycolipids*. New York: Elsevier, pp. 59–82.

Minelli A, Piantanida M, Simoni G, Rossella F, Romitti L, Brambati B, Danesino C (1992) Prenatal diagnosis of metabolic diseases on chorionic villi obtained before the ninth week of pregnancy. *Prenat Diagn*. 12: 959–963.

Morales A, Colell A, Mari M, Garcia-Ruiz C, Fernandez-Checa JC (2004) Glycosphingolipids and mitochondria: Role in apoptosis and disease. *Glycoconj J*. 20: 579–588.

Morreau H, Galjart NJ, Gillemans N, Willemsen R, van der Horst GT, d'Azzo A (1989) Alternative splicing of beta-galactosidase mRNA generates the classic lysosomal enzyme and a beta-galactosidase-related protein. *J Biol Chem*. 264: 20655–20663.

Muthane U, Chickabasaviah Y, Kaneski C, Shankar SK, Narayanappa G, Christopher R, Govindappa SS (2004) Clinical features of adult GM1 gangliosidosis: Report of three Indian patients and review of 40 cases. *Mov Disord*. 19: 1334–1341.

Norden AGW, O'Brien, JS (1973) Ganglioside GM1 β-galactosidase: Studies in human liver and brain. *Arch Biochem Biophys*, 159: 383.

O'Brien JS, Storb R, Raff RF, Harding J, Appelbaum F, Morimoto S, Kishimoto Y, Graham T, Ahern-Rindell A, O'Brien SL (1990) Bone marrow transplantation in canine GM1 gangliosidosis. *Clin. Genet*. 38: 274–280.

Okada S, O'Brien JS (1968) Generalized gangliosidosis: Beta-galactosidase deficiency. *Science* 160: 1002–1004.

Oshima A, Tsuji A, Nagao Y, Saturaba H, Suzuki Y (1988) Cloning, sequencing, and expression of cDNA for human beta-galactosidase. *Biochem. Biophys. Res. Commun.* 157: 238–244.

Ozkara HA, Topcu M (2004) Sphingolipidoses in Turkey. *Brain Dev*. 26(6, Sep): 363–366.

Paschke E, Milos I, Kreimer-Erlacher H, Hoefler G, Beck M, Hoeltzenbein M, Kleijer W, Levade T, Michelakakis H, Radeva B (2001) Mutation analyses in 17 patients with deficiency in acid beta-galactosidase: Three novel point mutations and high correlation of mutation W273L with Morquio disease type B. *Hum Genet.*109: 159–166.

Privitera S, Prody CA, Callahan JW, Hinek A (1998) The 67-kDa enzymatically inactive alternatively spliced variant of beta-galactosidase is identical to the elastin/laminin-binding protein. *J. Biol. Chem.* 273: 6319–6326.

Pshezhetsky AV, Ashmarina M (2001) Lysosomal multienzyme complex: Biochemistry, genetics, and molecular pathophysiology. *Prog Nucleic Acid Res Mol Biol.* 69: 81–114.

Pshezhetsky AV, Potier M (1996) Association of N-acetylgalactosamine-6-sulfate sulfatase with the multienzyme lysosomal complex of beta-galactosidase, cathepsin A, and neuraminidase. Possible implication for intralysosomal catabolism of keratan sulfate. *J Biol Chem.*271: 28359–28365.

Ramsay SL, Maire I, Bindloss C, Fuller M, Whitfield PD, Piraud M, Hopwood JJ, Meikle PJ (2004) Determination of oligosaccharides and glycolipids in amniotic fluid by electrospray ionisation tandem mass spectrometry: In utero indicators of lysosomal storage diseases. *Mol Genet Metab.*83: 231–238.

Reynolds GC, Baker HJ, Reynolds RH (1978) Enzyme replacement using liposome carriers in feline GM1 gangliosidosis fibroblasts. *Nature* 275: 754–755.

Ryder SJ, Simmons MM (2001) A lysosomal storage disease of Romney sheep that resembles human type 3 GM1 gangliosidosis. *Acta Neuropathol (Berl)* 101: 225–228.

Sano R, Tessitore A, Ingrassia A, d'Azzo A (2005) Chemokine-induced recruitment of genetically modified bone marrow cells into the CNS of GM1-gangliosidosis mice corrects neuronal pathology. *Blood.* 106: 2259–2268.

Sena-Esteves M, Saeki Y, Fraefel C, Breakefield XO 2000 Correction of acid beta-galactosidase deficiency in GM1 gangliosidosis human fibroblasts by retrovirus vector-mediated gene transfer: Higher efficiency of release and cross-correction by the murine enzyme. *Hum Gene Ther.* 11: 715–27.

Severini MH, Silva CD, Sopelsa A, Coelho JC, Giugliani R (1999) High frequency of type 1 GM1 gangliosidosis in southern Brazil. Clin Genet. 56: 168–169.

Shen WC, Tsai FJ, Tsai CH (1998) Myelination arrest demonstrated using magnetic resonance imaging in a child with type I GM1 gangliosidosis. *J Formos Med Assoc.* 97: 296–269.

Svennerholm L (1970) *Comprehensive Biochemistry*, vol 18, Amdterdam: Elsevier.

Takano T, Yamanouchi Y (1993) Assignment of human beta-galactosidase-A gene to 3p21.33 by fluorescence in situ hybridization. *Hum. Genet.* 92: 403–404.

Takaura N, Yagi T, Maeda M, Nanba E, Oshima A, Suzuki Y, Yamano T, Tanaka A (2003) Attenuation of ganglioside GM1 accumulation in the brain of GM1 gangliosidosis mice by neonatal intravenous gene transfer. *Gene Ther.* 10: 1487–1493.

Tasso MJ, Martinez-Gutierrez A, Carrascosa C, Vazquez S, Tebar R (1996) GM1-gangliosidosis presenting as nonimmune hydrops fetalis: A case report. *J Perinat Med.* 24: 445–449.

Tessitore A, del P Martin M, Sano R, Ma Y, Mann L, Ingrassia A, Laywell ED, Steindler DA, Hendershot LM, d'Azzo A (2004) GM1-ganglioside-mediated activation of the unfolded protein response causes neuronal death in a neurodegenerative gangliosidosis. *Mol Cell.*15: 753–766.

Tifft CJ, Proia RL (2000) Stemming the tide: Glycosphingolipid synthesis inhibitors as therapy for storage diseases. Glycobiology. 10: 1249–1258.

Tominaga L, Ogawa Y, Taniguchi M, Ohno K, Matsuda J, Oshima A, Suzuki Y, Nanba E (2001) Galactonojirimycin derivatives restore mutant human beta-galactosidase activities expressed in fibroblasts from enzyme-deficient knockout mouse. *Brain Dev.* 23:284–287.

Tylki-Szymanska A, Maciejko D, Kidawa M, Jablonska-Budaj U, Czartoryska B (1985) Amniotic tissue transplantation as a trial of treatment in some lysosomal storage diseases. *J Inherit Metab Dis.* 8: 101–104.

van der Voorn JP, Kamphorst W, et al. (2004) The leukoencephalopathy of infantile GM1 gangliosidosis: Oligodendrocytic loss and axonal dysfunction. *Acta Neuropathol (Berl)* 107(6, Jun): 539–545.

William L., Md. aNyhan, Pinar T., Md. Ozand (1998) *Atlas of Metabolic Diseases*. 1st ed., Arnold,, p. 534.

Yamato O, Masuoka Y, Yonemura M, Hatakeyama A, Satoh H, Kobayashi A, Nakayama M, Asano T, Shoda T, Yamasaki M, Ochiai K, Umemura T, Maede Y (2003) Clinical and clinico-pathologic characteristics of Shiba dogs with a deficiency of lysosomal acid beta-galactosidase: A canine model of human GM1 gangliosidosis. *J Vet Med Sci.* 65: 213–217.

Yoshida K, Oshima A, Sakuraba H, Nakano T, Yanagisawa N, Inui K, Okada S, Uyama E, Namba R, Kondo K, et al. (1992) GM1 gangliosidosis in adults: Clinical and molecular analysis of 16 Japanese patients. *Ann Neurol.* 31: 328–332.

Yoshida K, Oshima A, Shimmoto M, Fukuhara Y, Sakuraba H, Yanagisawa N, Suzuki Y (1991) Human beta-galactosidase gene mutations in GM1-gangliosidosis: A common mutation among Japanese adult/chronic cases. *Am J Hum Genet.* 49: 435–442.

Zhang X, Kiechle FL (2004) Review: Glycosphingolipids in health and disease. *Ann Clin Lab Sci.* 34: 3–13.

Zhou J, Shao H, Cox NR, Baker HJ, Ewald SJ (1998) Gangliosides enhance apoptosis of thymocytes. *Cell Immunol.* 183: 90–98.

THE GM2 GANGLIOSIDOSES

Gustavo A. Charria-Ortiz, MD

1 INTRODUCTION

In this group of conditions, G_{M2} ganglioside and related compounds accumulate in lysosomes due to enzymatic deficiencies in their degradation pathways. Their typical presentation is that of a pure neurodegenerative disorder characterized by developmental arrest, worsening neurological deficits and a shortened life span (Kolodny, 1966), although one of its variants, infantile Sandhoff's disease, may manifest systemic involvement (see below). The extent and type of these manifestations, however, vary widely, ranging from a relatively acute presentation to a more slowly progressive syndrome characterized by either motor deficits—in the form of worsening extrapyramidal or motor neuron disease—or behavioral changes—such as psychiatric symptoms or dementia in young adulthood. This variation in the clinical picture seems to be associated with different degrees of residual enzyme activity, which lead to corresponding severities of their clinical manifestations (i.e., the most severe deficits showing the most aggressive and early phenotypes. This situation is very similar to that of G_{M1} gangliosidosis. The prototype of this group of conditions is Tay–Sachs disease.

The biochemical degradation of G_{M2} ganglioside is slightly more complex than that of G_{M1} (Hechtman, 1977), and therefore, the same applies to the molecular and metabolic defects leading to this group of disorders. Such catalytic process depends upon the interaction of two molecules: the actual enzyme (β-hexosaminidase A; Brady, 1967) and a small polypeptide called the "G_{M2} activator" (Hechtman, 1977). This latter molecule presents the substrate to the enzyme, making it more accessible and soluble for catalysis (Meier et al., 1991). In addition, β-hexosaminidase A (also named HEXA, EC 3.2.1.52) is a trimer composed by two different subunits (1α and 2β) (Beutler et al., 1975), which are coded for by different genes (Korneluk et al., 1986), whereas the G_{M2} activator protein is a single gene product (Schroder et al., 1989). We have therefore three different gene products interacting during this process, a situation that leads to several possible combinations of defects (see below).

The β-hexosaminidases degrade amino-hexose moieties (i.e., N-acetyl-glucosamine and N-acetyl-galactosamine) that are linked in a β-glycosidic fashion to the terminal part of glycoproteins, proteoglycans and glycolipids (Mark et al., 2003). The latter compounds include G_{M2} ganglioside, which has a terminal N-acetylgalactosamine. This molecule becomes the terminal amino-hexose in the oligosaccharide chain of G_{M2} upon removal of the terminal galactose of G_{M1} ganglioside by β-galactosidase (see Figure 1).

Department of Neurology, University of Miami, Leonard Miller School of Medicine; Child Neurologist, Miami Children's Hospital. e-mail: gcharria@med.miami.edu

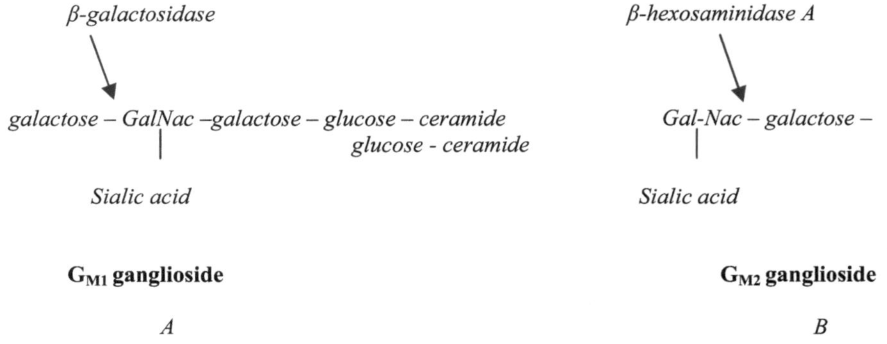

Figure 1. A: G_{M1} (A) and G_{M2} (B) ganglioside.

Two additional β-hexosaminidases have been described (HEXB and HEXS) (Beutler et al., 1976), which are also formed by a combination of these subunits (β-β and α-α, respectively). Each subunit has specific enzymatic activity against certain types of substrates. As such, the combinations of such subunits will determine the respective activity of the whole compound. The α-subunit can cleave negatively charged substrates (including sulfated ones), whereas the β-subunit predominantly hydrolyzes neutral substrates (Besley, Broadhead, and Young, 1987). However, it is only HEXA (i.e., the 1α-2β trimer) that can cleave G_{M2} ganglioside, and can do so only in the presence of the G_{M2} activator. This action has been shown to depend upon the proper dimerization of the α- and β-subunits, a fact that was discovered by creating chimeric hexosaminidases and exchanging their analogous regions (Pennybacker et al., 1996). This study also determined that there are two noncontiguous sequences in the α-subunit (from amino acids 1 to 191 and from amino acids 403 to 529) that, when substituted into analogous positions in the β-subunit, conferred activity against the sulfated substrate. It was also found that the sequence between amino acids 225 to 556 of the β-subunit is required for the action of the G_{M2} activator protein. These differences may help explain the specific biochemical findings noted in the diverse clinical conditions produced by specific genetic defects of these subunits.

HEXB (β-β) cleaves terminal amino-hexoses of a variety of other compounds, including several mucopolysaccharides, oligosaccharides, and globoside (a major red blood cell glycolipid). These compounds will therefore accumulate in the body when deficient (i.e., Sandhoff's disease; see below). HEXS (α-α) is a normal constituent of plasma and degrades a wide range of glycoconjugates containing β-linked N-acetylhexosaminyl residues. Again, none of them has significant activity against G_{M2} ganglioside, and therefore, their presence does not counteract this particular catalytic deficiency of HEXA. These differences between the activities of HEXA and HEXB against certain artificial substrates become important during the laboratory evaluation of these enzymes (see below).

In the most typical variant of the G_{M2} gangliosidoses, (i.e., infantile Tay–Sachs disease, TSD), mutations in the α-subunit lead to an almost total absence of HEXA activity (Okada and O'Brien, 1969). Because the structure of the β chain is normal, HEXB activity will remain normal. TSD manifests in early life with massive and severe neuronal degeneration but no significant extraneural involvement. If those mutations affecting the α-subunit leave HEXA with some degree of residual activity, the phenotype will be that of a milder disease with later onset, which can be either juvenile or adult. These latter variants are characterized by enzymatic activities of 5 to 7% and 10 to 15%, respectively, although there is no a clear-cut relationship between such values and the clinical presentation in a given individual patient.

An additional manifestation of the genetic defects in this subunit is the so-called B-1 variant (Kytzia et al., 1983). On it, a specific mutation (a G→A substitution in nucleotide 533 leading to the amino acid change R178H; Ohno and Suzuki, 1988) leads to the production of a mutant HEXA that despite being practically identical to wild-type HEXA from the structural viewpoint (i.e., it will form a dimer and will have the same isoelectric point), will be devoid of catalytic activity against natural G_{M2} ganglioside or sulfated substrates. It has been shown (Tutor, 2004) that despite the occurrence of its proper dimerization, the actual function of this HEXA variant is devoid of activity against G_{M2} ganglioside. The similarities between this mutant HEXA and the wild-type are so impressive that they even have the same activity against some other nonsulfated artificial substrates and, therefore, the diagnosis of this variant may be missed during routine laboratory evaluations if this fact is not taken into consideration.

A final variant of the genetic changes affecting this subunit is a state called "pseudodeficiency," a condition whereby some individuals show low activity levels of HEXA in vitro, but have no evidence of disease or G_{M2} ganglioside accumulation (Triggs-Raine et al., 1992). This is frequently associated with the R247W or R249W polymorphisms (Cao et al., 1997), and is an important fact to consider during genetic counseling.

If the genetic defect affects the β-subunit, then the activities of both HEXA and HEXB will be affected. This leads to widespread accumulation of G_{M2}, but also of several mucopolysaccharides, oligosaccharides, and globoside, which lead to variable degrees of systemic involvement. This latter condition is known as Sandhoff's disease (Sandhoff, Andreae, and Jatzkewitz, 1968), and it commonly manifests in its infantile variant. Interestingly, later onset forms of this disease show little or no systemic involvement at all.

Defects can also affect the G_{M2} activator protein (Schepers et al., 1996). This rare condition presents with a phenotype that is identical to infantile Tay-Sachs disease, including the absence of systemic involvement. The laboratory measured activity of HEXA and HEXB is normal. The author is not aware of any reports of late-onset variants of this disease.

The G_{M2} gangliosidoses are a relatively uncommon group of disorders, with a global incidence of about 1:310,000 live births (Kaback, Rimoin, and O'Brien, 1977). However, these values vary greatly between specific ethnic groups, with some of them (such as the Ashkenazi Jews) showing numbers as high as 1:3000 (Yokoyama, 1979) before the advent of widespread carrier detection programs. These conditions have been reported in almost every existing ethnic group, except perhaps the Eskimo, Gypsy, and Mongolian populations. Data about its incidence and carrier prevalence in different populations are widely available (see below, Genetics).

2 CLINICAL DESCRIPTION

2.1 Mutations in the α-Subunit of HEXA

2.1.1 Infantile Tay–Sachs Disease (ITSD)

This condition is caused by severe deficiencies in the catalytic activity of HEXA, which in the vast majority of patients is close to zero. ITSD is the prototype of the G_{M2} gangliosidoses, and also their most common variant. Onset occurs in the first few months of life, with neurodevelopmental delays followed by arrest, regression, and worsening neurological deficits leading to early death (Sachs, 1887; Gravel et al., 1995). Affected infants are typically born normal, showing no evidence of disease until around three to four months of age. The initial manifestation is usually an increased startle response, which is later followed by generalized muscle weakness and hypotonia. Between the ages of three to six months, patients become less responsive to their environment and show decreased facial expression. Their visual fixation is poor. By the age of one year, seizures and blindness develop. The disease progresses relentlessly, and patients are never able to reach the ability to crawl or stand. After two years of age, episodes of autonomic dysfunction with poor regulation of basic bodily functions appear, which tends to be an ominous sign of central nervous system involvement. Patients eventually develop spastic quadriplegia and a state of decerebrate rigidity that leads to death within the first few years of life. These numbers, however, have recently changed due to improvements in nursing care. There are no systemic manifestations.

Neurological examination reveals an encephalopathic-looking child with decreased responsiveness to environmental stimuli, marked hypotonia, long tract signs in the form of hypereflexia, and a positive Babinski sign. Megalencephaly is also common, usually developing after the first year of life. A "cherry-red" spot with a white perimacular halo can be observed in both optic fundi by around the age of four months. This classical finding is due to the accumulation of lipids around the macula, which leads to peripheral pallor and macular pseudo-prominence, not to increased macular pigmentation. A depressed gag reflex is also common.

On laboratory evaluation, most routine studies are normal. Standard neuroimaging evaluations contribute little to the diagnosis, and can be even normal in the early stages of the disease. The most commonly reported abnormalities include signal changes in the basal ganglia and cerebral white matter on T2-weighted sequences on magnetic resonance imaging (MRI) studies (Fukumizu et al., 1992; Yoshikawa, Yamada, and Sakuragawa, 1992; Brismar et al., 1990). Later on, diffuse cerebral atrophy becomes evident, although this is not a specific finding.

Nuclear magnetic resonance spectroscopy (NMRS) studies (Aydin et al., 2005) have revealed an increase in the myoinositol/creatine and choline/creatine ratios together with a decrease in the N-acetyl aspartate (NAA)/creatine ratio. These latter findings are compatible with a process of demyelination, gliosis, and neuronal loss.

Prior to the widespread availability of biochemical and genetics testing, skin (Takahashi, Naito, and Suzuki, 1987), rectal—for analysis of myenteric plexus neurons—(Yamano et al., 1982), or even brain (Ul-Haque, 1995) biopsies were used for confirmation of the diagnosis. Such studies revealed the characteristic membranous cytoplasmic body inclusions (MCBs) seen on electron microscopy in neuronal cells and fibroblasts (Shirabe, Hirokawa, and Asaki, 1980). The latter are composed of glycolipids, cholesterol and phospholipids.

Autopsy studies reveal severe cerebral and cerebellar atrophy, neuronal degeneration, and reactive gliosis. The latter seems to be responsible for the observed megalencephaly (Johnson et al., 1980). Pathological examination of postmortem tissue reveals the presence of ballooned neurons and widespread neuronal loss (Moriwaki et al., 1977).

2.1.2 Juvenile Tay–Sachs Disease

Onset usually occurs between the ages of three to five years, mostly in the form of speech disturbances, ataxia, or deterioration of daily activities (Brett et al., 1973; Specola et al., 1990). Worsening motor functions and cognitive abilities occur, leading to progressive neurological dysfunction, spasticity, and a state of decerebrate rigidity a few years later. One of the author's patients, a six-year-old girl from mixed ethnic background, presented with behavioral changes consisting of excessive anxiety, language disturbances, and ataxia at the age of five years. About a year later, she became highly disabled in her gross motor functions, with mild generalized weakness and an increased startle response. She did not exhibit retinal abnormalities. Her condition slowly progressed over the years leading to neurological regression, including loss of sphincter control.

Patients usually succumb to the disease by age 15 to 20, after a prolonged period in a persistent vegetative state. In some cases, the disease follows a particularly aggressive course culminating in death within two to four years of its onset. Dystonia has also been reported as the main manifestation (Nardocci et al., 1992). No systemic manifestations are noted.

Loss of vision occurs later than in the acute infantile form, and a cherry-red spot is not consistently observed. Instead, optic atrophy and retinitis pigmentosa may be seen later in the course of the disease.

Most routine laboratory studies are normal. Neuroimaging studies (MRI) reveal minimally increased T2 signal in the periventricular white matter and a mild degree of cerebral and cerebellar atrophy, but these findings are nonspecific.

2.1.3 Late-Onset or Adult Tay–Sachs Disease (LOTSD)

LOTSD typically presents in young adults in the form of a slowly progressive neurodegenerative disorder, mean age of onset: 18 years (Neudorfer et al., 2005) with or without cognitive involvement (Neudorfer et al., 2005; Frey, Ringel, and Filley, 2005). Dysarthria, ataxia, and gait disturbances, also related to a neurogenic type of muscle involvement, are common manifestations. The most prominent findings are related to lower motor neuron disease, and include weakness, proximal muscle wasting, cramps, and fasciculations (Harding, Young, and Schon, 1987; Federico, 1987). Psychiatric disease, which has also been described as a presenting symptom (Turpin and Baumann, 2003), may be prominent, mostly in the form of psychosis (McQueen, Rosebush, and Mazurek, 1998), catatonia (Rosebush et al., 1995), or recurrent depression (Hamner, 1998). Atypical presentations include a clinical picture that is highly similar to late-onset spinal muscular atrophy (Navon et al., 1997; Rubin et al., 1988), spinocerebellar diseases (Forster, Heuss, and Claus, 1999), amyotrophic lateral sclerosis (Drory et al., 2003), or even Friedreich's ataxia (Perlman et al., 2002). Up to 44% of patients manifest cognitive involvement (Frey, Ringel, and Filley, 2005), which may present in the form of either a static or progressive dementia. These latter features are associated to worsening neurological defects. Seizures are uncommon.

On examination, patients look alert and (unless gross dementia occurs) properly oriented and responsive. Their posture, gait, and body attitude are consistent with a slowly progressive neuromuscular disease with early distal atrophy of the extremities. A neurogenic pattern of muscle weakness (i.e., predominantly distal) is seen, but proximal deficits are also common. Ataxia, dysmetria, and dysarthria are frequently observed. The plantar responses become extensor and a positive Babinski sign is frequently encountered. Dystonia and Parkinsonism can also occur. There are no systemic manifestations.

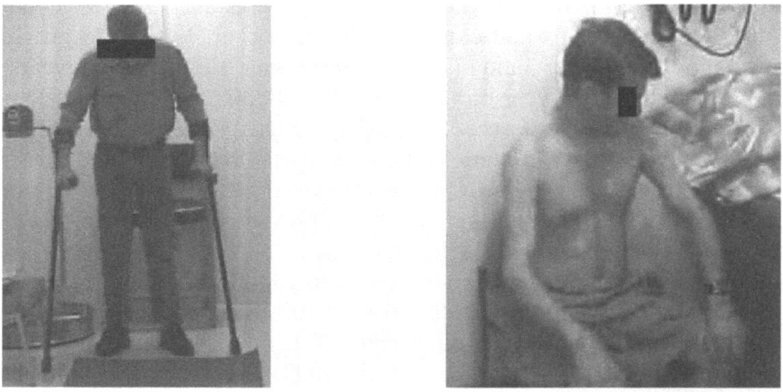

A B

Figure 2. Patients with LOTSD. Note the neurogenic type of muscle disease, including diffuse proximal and distal weakness, distal atrophy, and decreased muscle bulk of the shoulder girdle. Patient B was markedly dysarthric.

Although these patients do not present with the typical "cherry red" spot that characterizes this group of conditions, neuro-ophthalmological manifestations are common, in particular, abnormal saccadic movements (Rucker et al., 2004). However, in general, visual functions and other cranial nerves are preserved.

Electrophysiological studies reveal neurogenic myopathy with denervation and reinervation changes. Typical neuroimaging findings include prominent cerebellar atrophy. Reduced levels of NAA can be identified by NMRS in gray and white cerebral matter, even before the appearance of atrophy (Inglese et al., 2005). Severe changes in brain phosphorus metabolism have been described (Felderhoff-Mueser et al., 2001) by means of ^{31}Phosphorus MRS, which shows a decreased amount of phosphodiesters and membrane-bound phosphates. The clinical implications of this latter finding are not yet clear. Unusual manifestations include sensory (Barnes et al., 1991; Chow, Clarke, and Banwell, 2001) and autonomic system abnormalities (Salman et al., 2001).

The vast majority of patients described with this variant are of Ashkenazi Jewish ancestry and carry the specific mutation G269S (Neudorfer et al., 2005; Paw, Kaback, and Neufeld, 1989), although exceptions do occur (De Gasperi et al., 1996). This particular allelic variation seems to confer some residual enzymatic activity to HEXA, and when present in the homozygous state is associated with a milder clinical picture of later onset and slower progression.

This condition should be considered among young patients presenting with slowly progressive problems in speech, gait, or balance, and in those with otherwise unexplained new onset of psychiatric symptoms

2.1.4 B1 Variant of Tay–Sachs Disease

This highly unusual presentation of HEXA deficiency has been reported to occur in two forms: juvenile (Ribeiro et al., 1996; Maia et al., 1990) and early infantile (Grosso et al., 2003). The phenotype is very similar to TSD (in both forms), including the presence of megalencephaly and a macular "cherry-red spot" in the early-onset presentations. The degree of deterioration, however, seems to be faster once it appears, and death usually occurs within a few years. Neuroimaging studies are more remarkable in the infantile variant, revealing diffuse and symmetrical hyperintensities of the cerebral and cerebellar white matter, posterior thalami, and corpus striatum (Grosso et al., 2003). The corpus callosum looks thinner than normal, with residual and diffuse hypomyelination leading to cortical atrophy at a later stage.

This is one of the least common types of G_{M2} gangliosidoses, but has an exceedingly high incidence in Portugal (dos Santos et al., 1991), where specific mutations have been reported.

Because all the previously described disorders affecting the α-subunit do not interfere with HEXB activity, they have been named as the "B variants" of these diseases.

2.2 Mutations in the β-Subunit of HEXA and HEXB

2.2.1 Infantile Sandhoff Disease

In its typical form, this condition is almost indistinguishable from infantile Tay–Sachs disease, except for the presence of variable systemic involvement (Sandhoff, Andreae, and Jatzkewitz, 1968). A previously healthy infant begins manifesting diffuse muscle weakness in the first six months of life, associated with increased startle response and progressive mental and motor deterioration. Early blindness, hypomimic facies, and macular "cherry-red" spots are also common. Macrocephaly is also seen at around the same time as in ITSD. Systemic manifestations include coarsening of facies, visceromegaly, or even cardiomegaly (Venugopalan and Joshi, 2002). Death usually occurs by the age of three years.

Typical neuroimaging findings include signal intensity changes in the thalamus and high signal intensity of the cerebral white matter on MRI T2-weighted sequences (Yuksel et al., 1999; Stalker and Han, 1989). NMRS studies (Alkan et al., 2003) have revealed findings suggestive of widespread demyelination and neuroaxonal loss with secondary increases in anaerobic metabolism.

Sandhoff's disease can also be distinguished from other forms of G_{M2} gangliosidosis by the presence of foam cells in the bone marrow (Kolodny and Charria, 2003). The most common neuropathological findings are related to delayed myelination or demyelination and the accumulation of ganglioside in multiple types of cells (Krivit et al., 1972). The degree of G_{M2} accumulation is more severe than in TSD.

This condition is even rarer than TSD, but has an unusually large incidence in specific ethnic groups such as a Creole group residing in Argentina (Dodelson et al., 1985), the Maronite community of Cyprus (Drousiotou et al., 2000), and the Metis Indians of Saskatchewan (Lowden et al., 1978).

2.2.2 Juvenile Sandhoff Disease

This is an extremely rare variant of this disease, with only around 20 cases reported in the medical literature (Unnikrishnan, Danda, and Seshadri. 2001). One of the first reported cases was a ten-year-old boy with ataxia, spasticity, and psychomotor retardation (McLeod et al., 1977). Speech abnormalities seem to be a common manifestation. Constipation and urinary incontinence are frequent, and may be related to autonomic neuropathy (Hendriksz et al., 2004). A "cherry-red" spot is not noted in the vast majority of cases. Physical findings include increased lower limb reflexes and other long tract signs. Even rarer cases have presented with an atypical picture of ataxia, extrapyramidal signs, and seizures, even as late as 12 years of age (Beck, Sieber, and Goebel, 1998).

2.2.3 Adult Sandhoff's Disease

Similarly to a subset of patients with LOTSD, the adult-onset variant of this condition presents with a slowly progressive lower motor neuron or a spinocerebellar syndrome (Yoshikawa et al., 2002; Oonk, Van der Helm, and Martin, 1979). One of the patients reported by Kohno et al. (2001) was a 35-year old Japanese male who presented with progressive weakness in the thighs since the age of 15. The disease eventually led to upper and lower motor neuron disturbances by the age of 28, triggering the presumptive diagnosis of atypical amyotrophic lateral sclerosis. However, because he presented with sensory disturbances in the lower extremities, additional studies were made and confirmatory for the diagnosis. There were no signs of mental deterioration or cerebellar ataxia. The activities of HEXA and HEXB were 7–15% of controls.

Imaging studies have been nonspecific.

In one patient who underwent a rectal biopsy at age 24, membranous cytoplasmic bodies were noted in submucosal ganglion cells (Cashman et al., 1986). No systemic manifestations have been reported.

In as much as both HEXA and HEXB are affected in this subgroup of disorders, they have been called the "0 variant" of the disease.

2.3 Mutations in the GM2 Ganglioside Activator Protein

This extremely rare condition presents in a very similar fashion as to classic infantile TSD (de Baecque et al., 1975; Li, Hirabayashi, and Li, 1981), although some differences in the neuropathological findings have been reported, including heterogeneous inclusions in astrocytes and oligodendrocytes (Kotagal et al., 1986). The condition was initially described by Sandhoff and colleagues in 1971 (Sandhoff et al., 1971). One of the author's patients was a 22-month-old girl born to an indigenous couple of Central America who presented in the first months of life with neurodevelopmental arrest, evidence of progressive neurodegeneration (including intractable seizures), and a "cherry-red" spot (Figure 2). HEXA activity was normal on laboratory assays, but chromatography studies of her CSF showed massive amounts of GM2 ganglioside. No further studies were available, but the clinical picture and laboratory findings seemed highly suggestive of this condition. The patient is still alive at two years of life but severely impaired on her neurological development, similarly to three of her previously deceased siblings.

Figure 3. Optic fundus of a ten-month-old girl with a presumed diagnosis of G_{M2} activator deficiency.

MRI studies showed increased signal density in the periventricular white matter and altered signal density in the basal ganglia (Figure 4). Because both β-hexosaminidases are active in this unusual form of G_{M2} gangliosidoses, this has been called the "AB variant" of this disease.

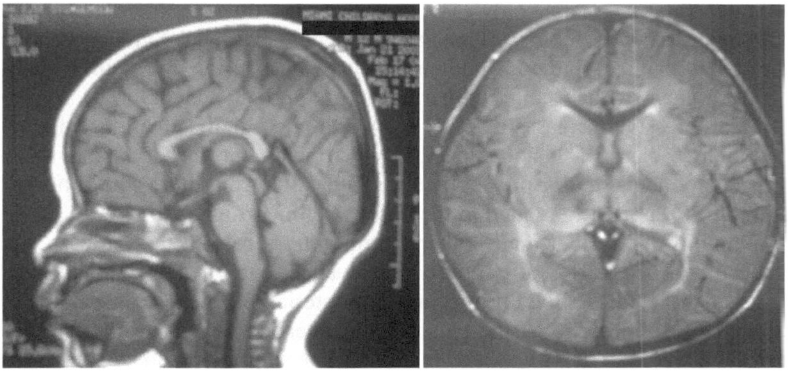

Figure 4. MRI of the same patient as in the previous figure. Note the mild thinning of the corpus callosum and the increased signal on subcortical white matter and basal ganglia.

Almost all variants of the G_{M2} gangliosidoses have been reported to occur naturally in other species such as deer (Fox et al., 1999), dogs (Yamato et al., 2002), cats (Martin et al., 2004; Yamato et al., 2004), and even flamingos (Kolodny, 2006). Animal models are also available for all of them (Proia, 2001; Tifft and proia 1997; i.e., TSD (Taniike et al, 1995; Cohen-Tannoudji et al., 1995; Yamanaka et al., 1994; Sango et al., 1995), LOTSD (Jeyakumar et al., 2002; Miklyaeva et al., 2004), Sandhoff's disease (Sango et al., 1995, 1996; Suzuki et al., 1997), and the AB variant (Liu et al., 1997), although, interestingly, the mouse model for TSD is behaviorally indistinguishable from the wild-type until at least one year of age (Phaneuf et al., 1996). Thereafter, it resembles the LOTSD variant.

2.3.1 Pathogenesis

Although the most likely explanation for the subcellular derangements caused by the lack of ganglioside degradation is the mechanical disruption produced by engorged lysosomes (which may occupy the majority of cell volume in pathology specimens), it is highly likely that other factors may play additional roles as well. This may be due to the wide variety of activities performed by gangliosides in nervous cell function (Walkley, Zervas, and Wiseman, 2000). Some of those postulated factors include improper activation of inflammatory (Jeyakumar et al., 2003; Wada, Tifft, and Proia, 2000) or apoptotic (Huang et al., 1997) pathways, decreased phospholipid synthesis (Buccoliero et al., 2004), abnormalities in intracellular calcium homeostasis (Pelled et al., 2003), abnormal accumulation of aggregation-prone proteins (Suzuki et al., 2003), and even the induction of autoantibodies against gangliosides (Yamaguchi et al., 2004). The relative importance of each of these factors is not yet known, although some of them, in particular inflammation, have been postulated as potential targets for therapeutic intervention (Wada, Tifft, and Proia, 2000; Jeyakumar et al., 2004; Wu and Proia, 2004). Changes at the structural level, in particular growth rates of neurite-related areas have been also found (Pelled et al., 2003; Sango et al., 2005), although the evidence is still controversial at this time (Sango et al., 2002).

The actual molecular pathways disrupted by ganglioside accumulation are not fully known, although in at least one study (Myerowitz et al., 2002) an attempt was made by using mRNA microarray technology. Through this technique, global patterns of expression of multiple genes can be studied simultaneously by measuring the degree of mRNA production from each of them and comparing it to controls (Schena et al., 1995). This study showed progressive increases in genes related to microglial activation, inflammatory markers (including class II histocompatability antigens), proinflammatory cytokines, complement components, and a prostaglandin synthase. Furthermore, the time profiling of such expression was clearly related to the onset of clinical signs.

It seems highly likely that many other factors play a role in this process, and that some of them may have different degrees of relevance, even at the regional level. For instance, one of the studies showing increased apoptotic cell death in an animal model (Huang et al., 1997) revealed that the patterns and intensity of cell death were different across several brain regions. This was a relatively unexpected finding due to the widespread accumulation of nonmetabolized compounds that is normally seen in these disorders.

2.3.2 Genetics

All three components of the enzymatic complex responsible for the degradation of G_{M2} ganglioside (i.e., the HEXA α- and β-subunits and the G_{M2} activator protein) may be affected by genetic defects (Cordeiro, Hechtman, and Kaplan, 2000). The types and extent of such defects, however, vary among different ethnic groups, a fact that is responsible for the specific incidences seen in some of them. The best example is Tay–Sachs disease, which prior to the establishment of current carrier detection programs (see below) was about 100 times more frequent among people of Ashkenazi Jewish (AJ) extraction than in the general population (Kaback, Rimoin, and O'Brien, 1977).

The genes coding for all of these three components are located on autosomal loci. They all follow the pattern of heterozygous asymptomatic carriers, and are inherited in an autosomal recessive fashion.

2.3.2.1 α-Subunit of HEXA

The gene coding for this subunit is located on chromosome 15q23-q24 (Nakai et al., 1991). It is about 35 kb long and contains 14 exons (Proia and Soravia, 1987), with an open reading frame corresponding to 529 amino acids (Myerowitz et al., 1985). More than 75 mutations have been identified (Myerowitz, 1997), and newer ones are being constantly reported (Montalvo et al., 2005). They occur more frequently among Ashkenazi Jews (AJ) (Kaback, Rimoin, and O'Brien, 1977; Petersen et al., 1983), people of French-Canadian ancestry (Palomaki et al., 1995; Prence et al., 1997; Andermann et al., 1977), including the Louisiana Cajuns (McDowell et al., 1992) and Acadians (Thurmon, 1993), the Pennsylvania Dutch (Mules et al., 1992), and Iraqi Jews (Karpati et al., 2004). In these ethnic groups the carrier frequency can be as high as between 1 in 25. These values are clearly greater than those observed in the general population (1 in 283). Other ethnic groups seem to carry specific mutations (Rozenberg et al., 2004; Tanaka et al., 1993), but in this latter case, the majority of them are considered as being private pedigrees.

In AJ, the most common mutation responsible for infantile TSD (80% of cases) is a 4 base-pair insertion in exon 11 (1278insTATC) (Myerowitz and Costigan, 1988). This leads to a frameshift reading and a premature stop signal. Additional mutations having higher frequencies in this ethnic group include the so-called "null alleles" (+1IVC12, +1IVS9; Strasberg et al., 1997), which are intervening sequences also leading to premature stop codons. Mutations in this subunit responsible for LOTS also occur more frequently in the AJ population, the most common one being Gly269-->Ser (Neudorfer et al., 2005; Paw, Kaback, and Neufeld, 1989).

The most common mutation in people of French-Canadian ancestry is a deletion of about 5 to 8 kb at the 5' end (Myerowitz and Hogikyan, 1986; De Braekeleer et al., 1992). In Japanese, the major mutation is a G-to-T transversion at the 3'-splice site of intron 5 (Tamasu et al., 1999). All other mutations are relatively rare.

2.3.2.2 β-Subunit of HEXA

The gene coding for this subunit is located on 5q13.2 (Fox et al., 1984). It has 14 exons, and spans about 40 kb of DNA (O'Dowd et al., 1985), with a reading open frame corresponding to a polypeptide of 556 amino acids. There is a high degree of similitude between both the α- and the β-subunits, which has led some to postulate that they may derive from a common ancestor (Myerowitz et al., 1985; Proia, 1988).

The prevalence of mutations in this subunit has been estimated between 1 in 1000 in Jews and 1 in 600 in non-Jews (Cantor and Kaback, 1985). As mentioned earlier, mutations in this subunit occur more frequently in certain ethnic groups such as the Creole/Spanish community of Cordoba, Argentina (Dodelson et al., 1985), the Maronite community of Cyprus (Drousiotou et al., 2000), and the Metis Indians of Saskatchewan (Lowden et al., 1978). In these groups, the carrier frequency may be as high as 1 in 7.

Overall, Sandhoff disease occurs in 1 of every 309,000 newborns, and such values are even lower in the Jewish population (1 per 1 million).

2.3.2.3 The GM2 Activator Protein

The gene coding for this protein is located on chromosome 5q31.3-33.1 (Heng et al., 1993). It is about 16 Kb long and has 4 exons (Chen et al., 1999). Only a handful of mutations have been described (Schroder et al., 1993), and such analysis is not easily available.

2.4 Carrier Testing and Prenatal Diagnosis

The advances obtained in elucidating the genetic and metabolic defects responsible for the G_{M2} gangliosidoses (in particular TSD) were paramount for the development of medical genetics, in particular for the study and understanding of lysosomal storage diseases. Perhaps the most impressive of such developments was the identification of the common genetic defects responsible for their high incidence in the Ashkenazi Jewish population, which resulted in quick and widespread applications for prenatal diagnosis (Callahan et al., 1990; Sakuraba et al., 1993) and asymptomatic carrier detection (Ben-Yoseph et al., 1985; Clarke, Skomorowski, and Zuker, 1989; Triggs-Raine et al., 1990; Scriver and Clow, 1990; Blitzer and McDowell, 1992; Kaback et al., 1993; Bach et al., 2001).

These programs started in the 1970s, and the results speak for themselves: more than 1.4 million individuals worldwide have been screened (voluntarily), with more than 1400 couples identified as being at risk and more than 3200 pregnancies being examined (Kaback, 2000). It has been estimated that through these interventions, the births of over 600 infants with this uniformly fatal neurodegenerative disease were prevented. Only in North America, the incidence of this condition has been reduced by more than 90%. This, of course, has also raised several important issues, such as the ethical (Ekstein, 2004) and financial (Shabat, 1999) implications of these types of programs. It should also be mentioned that most of such programs have been the result of admirable joined efforts from both health care providers and parents of affected children (Ekstein and Katzenstein, 2001) that have paved the road for the creation of regional and national organizations devoted to the development of such programs and their funding. The degree of success has been of such magnitude that many other genetic conditions also having increased incidence in specific ethnic groups can now be approached in the same manner with excellent results (Blitzer and McDowell, 1992; Kaplan, 1998).

Recently, a comprehensive review of some of these programs has been published (ACOG Committee on Genetics, 2004), and specific guidelines have been suggested. Several other reviews are also available for the interested reader (Leib et al., 2005; Gason et al., 2003). The historical implications of these genetic findings have been discussed elsewhere (Slatkin, 2004; Frisch et al., 2004). The interested reader is forwarded to the available references.

Preimplantation diagnosis (i.e., screening of embryos prior to in vitro fertilization) has also been reported with excellent success rates (Verlinsky and Kuliev, 1996; Gibbons et al., 1995; Hansis and Grifo, 2001).

Prenatal diagnosis is also available for some of the other G_{M2} gangliosidoses (Lemos et al., 1995; Sheth, Bhattacharya, and Sheth, 2002).

2.5 Differential Diagnosis

The differential diagnosis of these conditions is limited due to their stereotypical clinical presentation and the absence of systemic involvement, except perhaps in the early-onset variants of Sandhoff disease. Traditionally, the clinical finding most commonly pursued is the macular cherry-red spot, but this is not a pathognomonic one, and can be also found in other neurolipidoses such as G_{M1} gangliosidosis, sialidosis type I, metachromatic leukodystrophy, Niemann–Pick disease, Farber's disease, some of the mucopolysaccharidoses (i.e., Hurler's disease or Sly disease), galactosialidosis, or Gaucher's disease. Again, in the majority of cases, the absence of extraneural or peripheral nervous system involvement usually alerts about the diagnosis. There are other neurometabolic disorders (not lipidoses) in which this finding has been reported, including pantothenate kinase deficiency or PKAN (previously known as Hallervorden–Spatz disease) and even some variants of the neuronal ceroid lipofuscinoses. The cherry-red spot has been also reported in patients without neurometabolic diseases such as incontinentia pigmenti (Goldberg, 1994) and the Norman–Roberts syndrome (Caksen et al., 2004; lissencephaly and intracranial calcifications).

Alexander and Canavan diseases can also present with gross macrocephaly and neurodevelopmental arrest in early life, but they are considered more as being leukodystrophies (white matter diseases), thus, a cherry red spot is not normally seen.

The syndromes that can be mimicked by the late-onset variants of these disorders have been already discussed, and include amyotrophic lateral sclerosis, spinal muscular atrophy, spinocerebellar degenerations syndromes, and Friedreich's ataxia.

3 MANAGEMENT

There is no curative therapy available for any of these disorders, so their management is limited to the symptomatic care of their complications. For the infantile variants, proper nutritional and respiratory support is indicated once a proper discussion of their implications for prognosis and prolongation of life is made with caretakers. Patients with the late-onset variants may benefit from medications aimed at decreasing spasticity or extrapyramidal findings (such as Lioresal (Baclofen™) or some of the benzodiazepines. Seizure management is usually difficult, but it should be attempted with agents such as Valproate, benzodiazepines, or Zonisamide because they carry the greatest efficacy against myoclonic seizures. There are only a few reports of either controlled or anecdotal trials for the management of such symptoms, and only one of them is available (Gazulla et al., 2002) in which the GABAergic drugs Gabapentin and Tiagabine were successfully used for the management of myokymia and ataxia seen in a patient with the late-onset form of the B1 variant. Proper bed care and rehabilitation therapies are also indicated.

An important point to keep in mind during the behavioral pharmacotherapy of these conditions is that the addition of classical psychiatric medications such as phenothiazines (Mellaril™) or butyrophenones (such as Haloperidol) or even some of the more recently marketed atypical neuroleptics may be contraindicated for the management of their psychiatric complications because they may actually worsen such symptoms (Ovsiew, 1993). This is also a commonly encountered fact in clinical practice (Kolodny, 2004). The reason for this is not clear, although the possibility of some degree of "biochemical competition" between these compounds and the remaining enzyme at the lysosomal level has been mentioned. The best alternatives are traditional antiepileptic drugs such as Carbamazepine, Valproate, or Lamotrigine.

Nevertheless, alternative therapeutic strategies are under development aiming at getting an actual cure for these disorders. Some of them were attempted in the past, including bone marrow (Hoogerbrugge et al., 1995) or amniotic tissue (Sakuragawa,

Yoshikawa, and Sasaki, 1992) transplantation, and even direct intravenous injection of HEXA (Johnson et al., 1973) into affected patients. Unfortunately, all of them were marked by poor success, although bone marrow transplantation (BMT) is still being used in animal models of these diseases (Norflus et al., 1998). More innovative experimental approaches that have been recently attempted include "substrate reduction therapies" (SRT) and gene therapy (see below).

In regard to the first one (SRT), it involves the use of compounds known to inhibit the synthesis of gangliosides by antagonizing ceramide glucosyltransferase (Platt et al., 2000, 2003). This approach is based on the fact that assuming some residual amounts of enzyme activity (such as the one seen in the late-onset variants of these diseases) they may be enough to achieve a decrease in the total amount of accumulated substrate (a "closing-the-faucet" approach instead of the "unplugging-the-drain" approach used with enzyme replacement therapy). Such an approach also has the potential of utilizing compounds that can cross the blood–brain barrier (BBB) making them suitable for the management of the neurological manifestations associated with these conditions (current approaches using enzyme replacement therapy have not shown good results for this particular type of complications in lysosomal storage diseases). In addition, some of these compounds may also serve as "chemical chaperones" (Tropak et al., 2004), a finding that seems to be based in the fact that the majority of disease-associated mutations do not affect the active site of the enzyme but, rather, alter its ability to retain its native fold in the endoplasmic reticulum, which leads to its improper retention and accelerated degradation. This was noted when fibroblasts of TSD patients showed a paradoxical increase in the amount of residual protein and activity levels (Tropak et al., 2004) when they were grown in culture medium containing known inhibitors of HEXA. Those levels rose well above the critical 10% of baseline levels, suggesting a potential therapeutic target for this type of conditions.

The compound (and its derivatives) that has been most used is the imino sugar deoxynojirimycin (Platt et al., 1994). In one study (Platt et al., 1997), it was shown that its administration led to prevention of G_{M2} accumulation in brains of mice with TSD. Improvements have also been noted at the clinical level in an animal model of Sandhoff's disease (Andersson et al., 2004; Jeyakumar et al., 1999). These compounds have also been used in combination with other modalities such as BMT with relatively good results in animal models (Jeyakumar et al., 2001). Unfortunately, this combination of therapies did not induce any changes in the natural course of the disease when attempted in a child with the early juvenile variant of TSD (Jacobs et al., 2005). In this particular report, a three-year-old child was treated by BMT followed a year and a half later by SRT. This poor outcome was noticed despite the fact that she was fully asymptomatic at the time of transplantation, including nearly normal neuroimaging studies, electroencephalogram, and neuropsychological evaluation. Interestingly, final enzyme levels in leukocytes were about three times higher than at baseline, although they were still clearly below the reference ranges for normal.

Side effects of these compounds include weight loss and gastrointestinal disturbances. Clinical trials are currently being designed in order to evaluate the efficacy of this approach in patients with LOTS (Kolodny, 2004).

In regard to gene therapy, it has been attempted in several experimental models, although their possible therapeutic applications are still at very early stages. Initial studies were performed in vitro, mostly by transducing fibroblasts of patients with TSD or Sandhoff's disease with different viral vectors and examining the degree of enzymatic correction (Akli et al., 1996; Guidotti et al., 1998). They used either adenoviral or retroviral plasmids, which carried the cDNA for the alpha-subunit of human HEXA. According to those reports, the obtained level of correction ranged between 40 to 84% of normal, with the "corrected" cells secreting up to 25 times more HEXA α-subunit than controls. The protein product was also shown to normalize the impaired degradation of G_{M2} ganglioside in TSD fibroblasts, and to be correctly transported into lysosomes. More recently, lentiviral vectors have been used in fibroblasts of an animal model of Sandhoff disease (Arfi et al., 2005). Through this technique, researchers achieved a significant restoration of HEXB activity against synthetic substrates together with a significant decrease (20%) of the accumulated natural substrate (G_{M2} ganglioside). In this latter case, lentiviral vectors contained both HEXA and HEXB cDNAs. Cross-correction of human fibroblasts was also obtained.

A somehow related approach has been to use cells that have the potential for further differentiation and/or migration into the CNS and transduce them with viral vectors containing the required HEXA subunits. In one of these studies (Martino et al., 2002), bone marrow-derived stromal cells from an animal model of adult TSD were transduced with the α-subunit of HEXA, and then transplanted into animals. Reportedly, those cells reached enzymatic levels that were comparable with the wild-type mice, and were also able to degrade G_{M2} ganglioside in a normal fashion. This study, however, measured only corrected enzymatic levels in vitro. Other types of cells that have been used for this latter approach include multipotent neural cell-lines (Lacorazza et al., 1996). In this latter study, such transduced cells progenitors expressed the protein in a stable fashion, and secreted high levels of biologically active HEXA in vitro while cross-correcting the metabolic defect in a human Tay–Sachs fibroblast cell-line. These cells were later transplanted into brains of normal fetal and newborn mice, and the engrafted brains analyzed at various ages after transplant. Reportedly, they were still able to produce substantial amounts of the α-subunit for human HEXA, which was enzymatically active at therapeutic levels.

As mentioned earlier, one of the difficulties faced by researchers while developing innovative therapies for these (and other) neurological lysosomal storage diseases is the presence of the blood–brain barrier, which clearly limits the amount of either gene, protein, or vector transfer into the CNS. One approach to overcome this has been to attempt the systemic administration of vectors (either intraperitoneally or intravenously) with lentiviral or adenoviral vectors, which have a theoretical advantage for migration into the CNS. In one such study (Kyrkanides et al., 2005), neonatal pups of an animal model for Sandhoff disease were injected intraperitoneally with one of these vectors carrying HEXB. Reportedly, there was an excellent transfer of the lentiviral vector into CNS, with HEXB being expressed in periventricular areas of the cerebrum as well as in the cerebellar cortex. In addition, such treatment resulted in reduction of G_{M2} storage along with attenuation of the brain inflammation and amelioration of the observed neuromuscular deterioration.

In another study (Guidotti et al., 1999), it was found that the maximal expression of target genes (HEXA) was obtained when the cDNAs for both the α- and β-subunits were used. The highest levels of expression were obtained in liver. However, because this led to clear increases on HEXA serum levels, it was assumed that they will eventually help with enzymatic restoration in other organs. The authors emphasized the need for overexpression of both subunits in order to obtain higher levels of protein secretion. Intramuscular injections did not lead to the same effect, and no correction of HEXA levels was noted in the brain. There was no specific mention of clinical outcomes observed in the animals studied.

Additional approaches have included the concomitant injection of mannitol in order to enhance the migration of viral vectors inside the CNS by disrupting the blood–brain barrier, or by performing intracerebral injections (Bourgoin et al., 2003). This latter technique has been also attempted through the direct injection of transduced vectors into the CNS (Martino et al., 2005). In this latter study, a specific vector (a nonreplicating herpes simplex virus) transduced with the gene coding for the alpha-subunit of HEXA was used. Reportedly, injections were directly applied to the cerebral internal capsule and led to the re-establishment of normal HEXA activity and total removal of G_{M2} storage. Such changes were also evident in the contralateral cerebral hemispheres, cerebellum, and spinal cord within a month time. No adverse effects were observed.

It seems clear that the pathway for the clinical applications of gene therapy in the G_{M2} gangliosidoses is still a long and winding one. Several technical considerations need to be overcome, including valid ethic concerns. The author is not aware of any specific gene therapy trials involving human subjects at this time.

REFERENCES

ACOG Committee on Genetics. ACOG committee opinion (2004). Prenatal and preconceptional carrier screening for genetic diseases in individuals of Eastern European Jewish descent. Obstet Gynecol. 104: 425–428.

Akli S, Guidotti JE, Vigne E, Perricaudet M, Sandhoff K, Kahn A, Poenaru L (1996). Restoration of hexosaminidase A activity in human Tay–Sachs fibroblasts via adenoviral vector-mediated gene transfer. Gene Ther. 3:769–774.

Alkan A, Kutlu R, Yakinci C, Sigirci A, Aslan M, Sarac K (2003). Infantile Sandhoff's disease: multivoxel magnetic resonance spectroscopy findings. J Child Neurol. 8:425–428.

Andermann E, Scriver CR, Wolfe LS, Dansky L, Andermann F (1977). Genetic variants of Tay–Sachs disease and Sandhoff's disease in French-Canadians, juvenile Tay–Sachs disease in Lebanese Canadians, and a Tay–Sachs screening program in the French-Canadian population. In: Kaback MM, Rimoin DL, O'Brien JS (Eds.), *Tay–Sachs Disease: Screening and Prevention*. Alan R. Liss, New York, pp. 161–168.

Andersson U, Smith D, Jeyakumar M, Butters TD, Borja MC, Dwek RA, Platt FM (2004). Improved outcome of N-butyldeoxygalactonojirimycin-mediated substrate reduction therapy in a mouse model of Sandhoff disease. Neurobiol Dis. 16:506–15.

Arfi A, Bourgoin C, Basso L, Emiliani C, Tancini B, Chigorno V, Li YT, Orlacchio A, Poenaru L, Sonnino S, Caillaud C (2005). Bicistronic lentiviral vector corrects beta-hexosaminidase deficiency in transduced and cross-corrected human Sandhoff fibroblasts. Neurobiol Dis. 20:583–593.

Aydin K, Bakir B, Tatli B, Terzibasioglu E, Ozmen M (2005). Proton MR spectroscopy in three children with Tay–Sachs disease. Pediatr Radiol. 35:1081–1085.

Bach G, Tomczak J, Risch N, Ekstein J (2001). Tay–Sachs screening in the Jewish Ashkenazi population: DNA testing is the preferred procedure. Am J Med Genet. 99:70–75.

Barnes D, Misra VP, Young EP, Thomas PK, Harding AE (1991). An adult onset hexosaminidase A deficiency syndrome with sensory neuropathy and internuclear ophthalmoplegia. J Neurol Neurosurg Psychiatry. 54:1112–1113.

Beck M, Sieber N, Goebel HH (1998). Progressive cerebellar ataxia in juvenile GM2-gangliosidosis type Sandhoff. Eur J Pediatr. 157:866–7.

Ben-Yoseph Y, Reid JE, Shapiro B, Nadler HL (1985). Diagnosis and carrier detection of Tay–Sachs disease: direct determination of hexosaminidase A using 4-methylumbelliferyl derivatives of beta-N-acetylglucosamine-6-sulfate and beta-N-acetylgalactosamine-6-sulfate. Am J Hum Genet. 37:733–740.

Besley GT, Broadhead DM, Young JA (1987). GM2-gangliosidosis variant with altered substrate specificity: Evidence for alpha-locus genetic compound. J Inherit Metab Dis. 10:403–404.

Beutler E, Kuhl, W, Comings D (1975).Hexosaminidase isozyme in type O GM2 gangliosidosis (Sandhoff–Jatzkewitz disease). Am. J. Hum. Genet. 27: 628–638.

Beutler E, Yoshida A, Kuhl W, Lee JE (1976). The subunits of human hexosaminidase A. Biochem J. 159(3):541–543.

Blitzer MG, McDowell GA (1992). Tay–Sachs disease as a model for screening inborn errors. Clin Lab Med. 12:463–80.

Bourgoin C, Emiliani C, Kremer EJ, Gelot A, Tancini B, Gravel RA, Drugan C, Orlacchio A, Poenaru L, Caillaud C (2003). Widespread distribution of beta-hexosaminidase activity in the brain of a Sandhoff mouse model after coinjection of adenoviral vector and mannitol. Gene Ther.10:1841–1849.

Brady RO (1967). Enzymatic abnormalities in diseases of sphingolipid metabolism. Clin Chem. 13:565–577.

Brett EM, Ellis RB, Haas L, Ikonne JU, Lake BD, Patrick AD, Stephens R (1973). Late onset GM2-gangliosidosis. Clinical, pathological, and biochemical studies on 8 patients. Arch Dis Child. 48:775–785.

Brismar J, Brismar G, Coates R, Gascon G, Ozand P (1990). Increased density of the thalamus on CT scans in patients with GM2 gangliosidoses. AJNR Am J Neuroradiol. 11:125–130.

Buccoliero R, Bodennec J, Van Echten-Deckert G, Sandhoff K, Futerman AH (2004). Phospholipid synthesis is decreased in neuronal tissue in a mouse model of Sandhoff disease. J Neurochem. 90:80–88.

Caksen H, Tuncer O, Kirimi E, Fryns JP, Uner A, Unal O, Cinal A, Odabas D (2004). Report of two Turkish infants with Norman–Roberts syndrome. Genet Couns. 15: 9–17.

Callahan JW, Archibald A, Skomorowski MA, Shuman C, Clarke JT (1990). First trimester prenatal diagnosis of Tay–Sachs disease using the sulfated synthetic substrate for hexosaminidase A. Clin Biochem. 23: 533–536.

Cantor RM, Kaback M (1985). Sandhoff disease (SHD) heterozygote frequencies (HF) in North American (NA) Jewish (J) and non-Jewish (NJ) populations: Implications for carrier (C) screening (Abstract) Am. J. Hum. Genet. 37: A48.

Cao Z, Petroulakis E, Salo T, Triggs-Raine B (1997). Benign HEXA mutations, C739T(R247W) and C745T(R249W), cause beta-hexosaminidase A pseudodeficiency by reducing the alpha-subunit protein levels. J Biol Chem. 272:14975–14982.

Cashman NR, Antel, JP, Hancock LW, Dawson G, Horwitz AL, Johnson WG, Huttenlocher PR, Wollmann RL (1986). N-acetyl-beta-hexosaminidase beta locus defect and juvenile motor neuron disease: A case study. Ann. Neurol. 19: 568–572.

Chen B, Rigat B, Curry C, Mahuran DJ (1999). Structure of the GM2A gene: identification of an exon 2 nonsense mutation and a naturally occurring transcript with an in-frame deletion of exon 2. Am J Hum Genet. 65:77–87.

Chow GC, Clarke JT, Banwell BL (2001). Late-onset GM2 gangliosidosis presenting as burning dysesthesias. Pediatr Neurol. 25:59–61.

Clarke JT, Skomorowski MA, Zuker S (1989). Tay–Sachs disease carrier screening: follow-up of a case-finding approach. Am J Med Genet. 34:601–605.

Cohen-Tannoudji M, Marchand P, Akli S, Sheardown SA, Puech JP, Kress C, Gressens P, Nassogne MC, Beccari T, Muggleton-Harris AL, et al. (1995). Disruption of murine Hexa gene leads to enzymatic deficiency and to neuronal lysosomal storage, similar to that observed in Tay–Sachs disease. Mamm Genome. 6:844–849.

Cordeiro P, Hechtman P, Kaplan F (2000). The GM2 gangliosidoses databases: allelic variation at the HEXA, HEXB, and GM2A gene loci. Genet Med. 2:319–2327.

de Baecque CM, Suzuki K, Rapin I, Johnson AB, Whethers DL (1975). GM2-gangliosidosis, AB variant: Clinico-pathological study of a case. Acta Neuropathol (Berl). 33:207–26.

De Braekeleer M, Hechtman P, Andermann E, Kaplan F (1992). The French Canadian Tay–Sachs disease deletion mutation: Identification of probable founders. Hum Genet. 89:83–87.

De Gasperi R, Gama Sosa MA, Battistini S, Yeretsian J, Raghavan S, Zelnik N, Leshinsky E, Kolodny EH (1996). Late-onset GM2 gangliosidosis: Ashkenazi Jewish family with an exon 5 mutation (Tyr180–>His) in the Hex A alpha-chain gene. Neurology. 47:547–552.

Dodelson de Kremer R, Boldini CD, Capra AP, Levstein IM, Bainttein N, Hidalgo PK, Hliba H (1985). Sandhoff disease: 36 cases from Cordoba, Argentina. J. Inherit. Metab. Dis. 8: 46.

dos Santos MR, Tanaka A, sa Miranda MC, Ribeiro MG, Maia M, Suzuki K (1991). GM2-gangliosidosis B1 variant: Analysis of beta-hexosaminidase alpha gene mutations in 11 patients from a defined region in Portugal. Am J Hum Genet. 49:886–890.

Drory VE, Birnbaum M, Peleg L, Goldman B, Korczyn AD (2003). Hexosaminidase A deficiency is an uncommon cause of a syndrome mimicking amyotrophic lateral sclerosis. Muscle Nerve. 28(1, July):109–112.

Drousiotou A, Stylianidou G, Anastasiadou V, Christopoulos G, Mavrikiou E, Georgiou T, Kalakoutis G, Oladimeja A, Hara Y, Suzuki K, Furihata K, Ueno I, Ioannou PA, Fensom AH (2000). Sandhoff disease in Cyprus: Population screening by biochemical and DNA analysis indicates a high frequency of carriers in the Maronite community. Hum Genet. 107:12–17.

Ekstein J (2004). The rabbi's dilemma. New Sci. 181:44–47.

Ekstein J, Katzenstein H (2001). The Dor Yeshorim story: Community-based carrier screening for Tay–Sachs disease. Adv Genet.44:297–310.

Federico A (1987). GM2 gangliosidosis with a motor neuron disease phenotype: Clinical heterogeneity of hexosaminidase deficiency disease. Adv Exp Med Biol. 209:19–23.

Felderhoff-Mueser U, Sperner J, Konstanzcak P, Navon R, Weschke B. (2001). 31Phosphorus magnetic resonance spectroscopy in late-onset Tay–Sachs disease. J Child Neurol. 16:377–380.
Forster A, Heuss D, Claus D (1999). Hexosaminidase deficiency as differential spinocerebellar diseases. Nervenarzt. 70:162–166.
Fox J, Li YT, Dawson G, Alleman A, Johnsrude J, Schumacher J, Homer B (1999). Naturally occurring GM2 gangliosidosis in two Muntjak deer with pathological and biochemical features of human classical Tay–Sachs disease (type B GM2 gangliosidosis). Acta Neuropathol (Berl). 97:57–62.
Fox MF, DuToit DL, Warnich L, Retief AE (1984). Regional localization of alpha-galactosidase (GLA) to Xpter-q22, hexosaminidase B (HEXB) to 5q13-qter, and arylsulfatase B (ARSB) to 5pter-q13. Cytogenet. Cell Genet. 38: 45–49.
Frey LC, Ringel SP, Filley CM (2005). The natural history of cognitive dysfunction in late-onset GM2 gangliosidosis. Arch Neurol. 62:989–994.
Frisch A, Colombo R, Michaelovsky E, Karpati M, Goldman B, Peleg L (2004). Origin and spread of the 1278insTATC mutation causing Tay–Sachs disease in Ashkenazi Jews: Genetic drift as a robust and parsimonious hypothesis. Hum Genet. 114:366–76.
Fukumizu M, Yoshikawa H, Takashima S, Sakuragawa N, Kurokawa T (1992). Tay–Sachs disease: Progression of changes on neuroimaging in four cases. Neuroradiology. 34:483–486.
Gason AA, Sheffield E, Bankier A, Aitken MA, Metcalfe S, Barlow Stewart K, Delatycki MB (2003). Evaluation of a Tay–Sachs disease screening program. Clin Genet. 63:386–392.
Gazulla Abio J, Benavente Aguilar I (2002). Adult GM2 gangliosidosis: Improvement of ataxia with GABAergic drugs. Neurologia.17:157–161.
Gibbons WE, Gitlin SA, Lanzendorf SE, Kaufmann RA, Slotnick RN, Hodgen GD (1995). Preimplantation genetic diagnosis for Tay–Sachs disease: Successful pregnancy after pre-embryo biopsy and gene amplification by polymerase chain reaction. Fertil Steril. 63:723–728.
Goldberg MF (1994). The blinding mechanisms of incontinentia pigmenti. Trans Am Ophthalmol Soc 92:167–176.
Gravel RA, Clarke JTR, Kaback MM, Mahuran D, Sandhoff K, Suzuki K (1995). The GM2 gangliosidosis. In: Scriver CR, Beaudet AL, Sly WS, Valle D (Eds.) *The Metabolic and Molecular Bases of Inherited Disease*. McGraw-Hill, New York, pp. 2847–2869.
Grosso S, Farnetani MA, Berardi R, Margollicci M, Galluzzi P, Vivarelli R, Morgese G, Ballestri P (2003). GM2 gangliosidosis variant B1, neuroradiological findings. J Neurol. 250:17–21.
Guidotti J, Akli S, Castelnau-Ptakhine L, Kahn A, Poenaru L (1998). Retrovirus-mediated enzymatic correction of Tay–Sachs defect in transduced and non-transduced cells. Hum Mol Genet. 7:831–838.
Guidotti JE, Mignon A, Haase G, Caillaud C, McDonell N, Kahn A, Poenaru L (1999). Adenoviral gene therapy of the Tay–Sachs disease in hexosaminidase A-deficient knock-out mice. Hum Mol Genet. 8:831–838.
Hamner MB (1998). Recurrent psychotic depression associated with GM2 gangliosidosis. Psychosomatics. 39:446–448.
Hansis C, Grifo J (2001). Tay–Sachs disease and preimplantation genetic diagnosis. Adv Genet. 44:311–315.

Harding AE, Young EP, Schon F (1987). Adult onset supranuclear ophthalmoplegia, cerebellar ataxia, and neurogenic proximal muscle weakness in a brother and sister: another hexosaminidase A deficiency syndrome. J Neurol Neurosurg Psychiatry. 50:687–690.

Hechtman P (1977). Characterization of an activating factor required for hydrolysis of GM2 ganglioside catalyzed by hexosaminidase A. Can J Biochem.55:315–324.

Hendriksz CJ, Corry PC, Wraith JE, Besley GT, Cooper A, Ferrie CD (2004). Juvenile Sandhoff disease–nine new cases and a review of the literature. J Inherit Metab Dis. 27:241–249.

Heng HHQ, Xie B, Shi XM, Tsui, LC, Mahuran DJ (1993). Refined mapping of the GM2 activator protein (GM2A) locus to 5q31.3-q33.1, distal to the spinal muscular atrophy locus. Genomics 18: 429–431.

Hoogerbrugge PM, Brouwer OF, Bordigoni P, Ringden O, Kapaun P, Ortega JJ, O'Meara A, Cornu G, Souillet G, Frappaz D, et al. (1995). Allogeneic bone marrow transplantation for lysosomal storage diseases. The European Group for Bone Marrow Transplantation. Lancet. 345:1398–402.

Huang JQ, Trasler JM, Igdoura S, Michaud J, Hanal N, Gravel RA (1997). Apoptotic cell death in mouse models of GM2 gangliosidosis and observations on human Tay–Sachs and Sandhoff diseases. Hum Mol Genet. 6:1879–1885.

Inglese M, Nusbaum AO, Pastores GM, Gianutsos J, Kolodny EH, Gonen O (2005). MR imaging and proton spectroscopy of neuronal injury in late-onset GM2 gangliosidosis. AJNR. Am J Neuroradiol. 26:2037–2042.

Jacobs JF, Willemsen MA, Groot-Loonen JJ, Wevers RA, Hoogerbrugge PM (2005). Allogeneic BMT followed by substrate reduction therapy in a child with subacute Tay–Sachs disease. Bone Marrow Transplant. 36:925–926.

Jeyakumar M, Butters TD, Cortina-Borja M, Hunnam V, Proia RL, Perry VH, Dwek RA, Platt FM (1999). Delayed symptom onset and increased life expectancy in Sandhoff disease mice treated with N-butyldeoxynojirimycin. Proc Natl Acad Sci U S A. 96:6388–6393.

Jeyakumar M, Norflus F, Tifft CJ, Cortina-Borja M, Butters TD, Proia RL, Perry VH, Dwek RA, Platt FM (2001). Enhanced survival in Sandhoff disease mice receiving a combination of substrate deprivation therapy and bone marrow transplantation. Blood. 97:327–329.

Jeyakumar M, Smith D, Eliott-Smith E, Cortina-Borja M, Reinkensmeier G, Butters TD, Lemm T, Sandhoff K, Perry VH, Dwek RA, Platt FM (2002). An inducible mouse model of late onset Tay–Sachs disease. Neurobiol Dis. 10:201–210.

Jeyakumar M, Smith DA, Williams IM, Borja MC, Neville DC, Butters TD, Dwek RA, Platt FM (2004). NSAIDs increase survival in the Sandhoff disease mouse: synergy with N-butyldeoxynojirimycin. Ann Neurol. 56:642–649.

Jeyakumar M, Thomas R, Elliot-Smith E, Smith DA, van der Spoel AC, d'Azzo A, Perry VH, Butters TD, Dwek RA, Platt FM (2003). Central nervous system inflammation is a hallmark of pathogenesis in mouse models of GM1 and GM2 gangliosidosis. Brain. 126:974–987.

Johnson WG, Cohen CS, Miranda AF, Waran SP, Chutorian AM (1980). Alpha-locus hexosaminidase genetic compound with juvenile gangliosidosis phenotype: Clinical, genetic, and biochemical studies. Am J Hum Genet. 32:508–518.

Johnson WG, Desnick RJ, Long DM, Sharp HL, Krivit W, Brady B, Brady RO (1973). Intravenous injection of purified hexosaminidase A into a patient with Tay–Sachs disease. Birth Defects Orig Artic Ser. 9:120–124.

Kaback M, Lim-Steele J, Dabholkar D, Brown D, Levy N, Zeiger K (1993). Tay-Sachs disease carrier screening, prenatal diagnosis, and the molecular era. An international perspective, 1970 to 1993. The International TSD Data Collection Network. JAMA. 270:2307–2315.

Kaback MM (2000). Population-based genetic screening for reproductive counseling: The Tay–Sachs disease model. Eur J Pediatr. 159 Suppl 3:S192–195.

Kaback MM, Rimoin DL, O'Brien JS (1977). *Tay–Sachs Disease: Screening and Prevention.* New York: Alan R. Liss.

Kaplan F (1998). Tay–Sachs disease carrier screening: A model for prevention of genetic disease. Genet Test. 2:271–292.

Karpati M, Gazit E, Goldman B, Frisch A, Colombo R, Peleg L (2004). Specific mutations in the HEXA gene among Iraqi Jewish Tay–Sachs disease carriers: Dating of founder ancestor. Neurogenetics. 5:35–40.

Kohno Y, Yoshizawa T, Ohkoshi N, Tamaoka A, Shoji S (2001). Adult Sandhoff disease presented as a motor neuron disease phenotype with slow progression. Rinsho Shinkeigaku. 41:36–39.

Kolodny EH (2004, 2006). Personal communication.

Kolodny EH (1966). The GM2 gangliosidoses. In Rosenberg RN, Prusiner SB, et al. (Eds.), *The Molecular and Genetic Basis of Neurological Disease*. 2nd ed. Boston: Butterworth Heinemann.

Kolodny EH, Charria-Ortiz, GA (2003). Lysosomal storage diseases. In Nathan and Oski (Eds.), *Hematology of Infancy and Childhood*, 6th edition, Elsevier, USA.

Korneluk RG, Mahuran DJ, Neote K, Klavins MH, O'Dowd BF, Tropak M, Willard HF, Anderson MJ, Lowden JA, Gravel RA (1986). Isolation of cDNA clones coding for the alpha-subunit of human beta-hexosaminidase. Extensive homology between the alpha- and beta-subunits and studies on Tay–Sachs disease. *J Biol Chem*. 261:8407–8413.

Kotagal S, Wenger DA, Alcala H, Gomez C, Horenstein S (1986). AB variant GM2 gangliosidosis: Cerebrospinal fluid and neuropathologic characteristics. Neurology. 36:438–440.

Krivit W, Desnick RJ, Lee J, Moller J, Wright F, Sweeley CC, Snyder PD, Sharp HL (1972). Generalized accumulation of neutral glycosphingolipids with G(m2) ganglioside accumulation in the brain. Sandhoff's disease (variant of Tay–Sachs disease). Am. J. Med. 52: 763–770.

Kyrkanides S, Miller JH, Brouxhon SM, Olschowka JA, Federoff HJ (2005). Beta-hexosaminidase lentiviral vectors: Transfer into the CNS via systemic administration. Brain Res Mol Brain Res. 133:286–298.

Kytzia HJ, Hinrichs U, Maire I, Suzuki K, Sandhoff K (1983). Variant of GM2-gangliosidosis with hexosaminidase A having a severely changed substrate specificity. EMBO J 2:1201–1205.

Lacorazza HD, Flax JD, Snyder EY, Jendoubi M (1996). Expression of human beta-hexosaminidase alpha-subunit gene (the gene defect of Tay–Sachs disease) in mouse brains upon engraftment of transduced progenitor cells. Nat Med.2:424–429.

Leib JR, Gollust SE, Hull SC, Wilfond BS (2005). Carrier screening panels for Ashkenazi Jews: Is more better? Genet Med. 7:185–190.

Lemos M, Pinto R, Ribeiro G, Ribeiro H, Lopes L, Sa Miranda MC (1995). Prenatal diagnosis of GM2-gangliosidosis B1 variant. Prenat Diagn.15:585–588.

Li SC, Hirabayashi Y, Li YT (1981). A new variant of type-AB GM2-gangliosidosis. Biochem. Biophys. Res. Commun. 101: 479–485.

Liu Y, Hoffmann A, Grinberg A, Westphal H, McDonald MP, Miller KM, Crawley JN, Sandhoff K., Suzuki K, Proia RL (1997). Mouse model of GM2 activator deficiency manifests cerebellar pathology and motor impairment. Proc. Nat. Acad. Sci. 94: 8138–8143.

Lowden JA, Ives EJ, Keene DL, Burton AL, Skomorowski MA, Howard F (1978). Carrier detection in Sandhoff disease. Am. J. Hum. Genet. 30: 38–45.

MacQueen GM, Rosebush PI, Mazurek MF (1998). Neuropsychiatric aspects of the adult variant of Tay–Sachs disease. J Neuropsychiatry Clin Neurosci. 10:10–19.

Maia M, Alves D, Ribeiro G, Pinto R, Sa Miranda MC (1990). Juvenile GM2 gangliosidosis variant B1: Clinical and biochemical study in seven patients. Neuropediatrics. 21:18–23.

Mark BL, Mahuran DJ, Cherney MM, Zhao D, Knapp S, James MN (2003). Crystal structure of human beta-hexosaminidase B: Understanding the molecular basis of Sandhoff and Tay–Sachs disease. J Mol Biol. 327:1093–1109.

Martin DR, Krum BK, Varadarajan GS, Hathcock TL, Smith BF, Baker HJ (2004). An inversion of 25 base pairs causes feline GM2 gangliosidosis variant. Exp Neurol. 187:30–37.

Martino S, Cavalieri C, Emiliani C, Dolcetta D, Cusella De Angelis MG, Chigorno V, Severini GM, Sandhoff K, Bordignon C, Sonnino S, Orlacchio A (2002). Restoration of the GM2 ganglioside metabolism in bone marrow-derived stromal cells from Tay–Sachs disease animal model. Neurochem Res. 27:793–800.

Martino S, Marconi P, Tancini B, Dolcetta D, De Angelis MG, Montanucci P, Bregola G, Sandhoff K, Bordignon C, Emiliani C, Manservigi R, Orlacchio A (2005). A direct gene transfer strategy via brain internal capsule reverses the biochemical defect in Tay–Sachs disease. Hum Mol Genet. 14:2113–2223.

McDowell GA, Mules EH, Fabacher P, Shapira E, Blitzer MG (1992). The presence of two different infantile Tay–Sachs disease mutations in a Cajun population. Am. J. Hum. Genet. 51:1071–1077.

McLeod PM, Wood S, Jan JE, Applegarth DA, Dolman CL (1977). Progressive cerebellar ataxia, spasticity, psychomotor retardation and Hexosaminidase deficiency in a 10-year-old child: Juvenile Sandhoff disease. Neurology 27:571–573.

Meier EM, Schwarzmann G, Furst W, Sandhoff K (1991). The human GM2 activator protein. A substrate specific cofactor of beta-hexosaminidase A. J Biol Chem. 266:1879–1887.

Miklyaeva EI, Dong W, Bureau A, Fattahie R, Xu Y, Su M, Fick GH, Huang JQ, Igdoura S, Hanai N, Gravel RA (2004). Late onset Tay–Sachs disease in mice with targeted disruption of the Hexa gene: Behavioral changes and pathology of the central nervous system Brain Res. 1001:37–50.

Montalvo AL, Filocamo M, Vlahovicek K, Dardis A, Lualdi S, Corsolini F, Bembi B, Pittis MG (2005). Molecular analysis of the HEXA gene in Italian patients with infantile and late onset Tay–Sachs disease: Detection of fourteen novel alleles. Hum Mutat. 26:282.

Moriwaki S, Takashima S, Yoshida H, Kawano N, Goto M (1977). Histological observation of the brain of Tay–Sachs disease with seizure and chronic DPH intoxication—Report of an autopsy case. Acta Pathol Jpn. 27:387–407

Mules EH, Hayflick S, Dowling CE, Kelly TE, Akerman BR, Gravel RA, Thomas GH (1992). Molecular basis of hexosaminidase A deficiency and pseudodeficiency in the Berks County Pennsylvania Dutch. Hum. Mutat. 1: 298–302.

Myerowitz R, Costigan FC (1988). The major defect in Ashkenazi Jews with Tay–Sachs disease is an insertion in the gene for the alpha-chain of beta-hexosaminidase. J. Biol. Chem. 263: 18587–18589.

Myerowitz R, Hogikyan ND (1986). Different mutations in Ashkenazi Jewish and non-Jewish French Canadians with Tay–Sachs disease. Science. 232:1646–8.

Myerowitz R, Lawson D, Mizukami H, Mi Y, Tifft CJ, Proia RL (2002). Molecular pathophysiology in Tay–Sachs and Sandhoff diseases as revealed by gene expression profiling. Hum Mol Genet.11:1343–50.

Myerowitz R, Piekarz R, Neufeld, EF, Shows, TB, Suzuki K (1985). Human beta-hexosaminidase alpha chain: Coding sequence and homology with the beta chain. Proc Nat Acad Sci. 82: 7830–7834.

Myerowitz RT (1997). Tay–Sachs disease-causing mutations and neutral polymorphisms in the Hex A gene. Hum Mutat. 9:195–208.

Nakai H, Byers MG, Nowak N J, Shows T (1991). Assignment of beta-hexosaminidase A alpha-subunit to human chromosomal region 15q23-q24. Cytogenet. Cell Genet. 56: 164.

Nardocci N, Bertagnolio B, Rumi V, Angelini L (1992). Progressive dystonia symptommatic of juvenile GM2 gangliosidosis. Mov Disord. 7:64–67.

Navon R, Khosravi R, Melki J, Drucker L, Fontaine B, Turpin JC, N'Guyen B, Fardeau M, Rondot P, Baumann N (1997). Juvenile-onset spinal muscular atrophy caused by compound heterozygosity for mutations in the HEXA gene. Ann Neurol. 41:631–638.

Neudorfer O, Pastores GM, Zeng BJ, Gianutsos J, Zaroff CM, Kolodny EH (2005). Late-onset Tay–Sachs disease: Phenotypic characterization and genotypic correlations in 21 affected patients. Genet Med. 7:119–123.

Norflus F, Tifft CJ, McDonald MP, Goldstein G, Crawley JN, Hoffmann A, Sandhoff K, Suzuki K, Proia RL (1998). Bone marrow transplantation prolongs life span and ameliorates neurologic manifestations in Sandhoff disease mice. J Clin Invest. 101:1881–1888.

O'Dowd BF, Quan F, Willard HF, Lamhonwah AM, Korneluk RG, Lowden JA, Gravel RA, Mahuran DJ (1985). Isolation of cDNA clones coding for the beta subunit of human beta-hexosaminidase. Proc Nat Acad Sci. 82: 1184–1188

Ohno K, Suzuki K (1988). Mutation in GM2-gangliosidosis B1 variant. J Neurochem. 50:316–318.

Okada S, OBrien JS (1969). Tay–Sachs disease: Generalized absence of a beta-D-N-acetylhexosaminidase component. Science 165: 698–700, 1969.

Oonk JG, Van der Helm HJ, Martin JJ (1979). Spinocerebellar degeneration: hexosaminidase A and B deficiency in two adult sisters. Neurology 29: 380–384.

Ovsiew F (1993). Contraindications to the use of antipsychotic drugs. Am J Psychiatry 150:1753–1754.

Palomaki GE, Williams J, Haddow JE, Natowicz MR (1995). Tay–Sachs disease in persons of French-Canadian heritage in northern New England. Am J Med Genet. 56:409–412.

Paw BH, Kaback MM, Neufeld EF (1989). Molecular basis of adult-onset and chronic GM2 gangliosidoses in patients of Ashkenazi Jewish origin: Substitution of serine for glycine at position 269 of the alpha-subunit of beta-hexosaminidase. Proc Natl Acad Sci U S A. 86:2413–2417.

Pelled D, Lloyd-Evans E, Riebeling C, Jeyakumar M, Platt FM, Futerman AH (2003). Inhibition of calcium uptake via the sarco/endoplasmic reticulum Ca2+-ATPase in a mouse model of Sandhoff disease and prevention by treatment with N-butyldeoxy-nojirimycin. J Biol Chem. 278:29496–29501.

Pelled D, Riebeling C, van Echten-Deckert G, Sandhoff K, Futerman AH (2003). Reduced rates of axonal and dendritic growth in embryonic hippocampal neurones cultured from a mouse model of Sandhoff disease. Neuropathol Appl Neurobiol. 29:341–349.

Pennybacker M, Liessem B, Moczall H, Tifft CJ, Sandhoff K, Proia RL (1996). Identification of domains in human beta-hexosaminidase that determine substrate specificity. J Biol Chem. 271:17377–17382.

Perlman SL (2002). Late-onset Tay-Sachs disease as a Friedreich ataxia phenocopy. Arch Neurol. 59:1832.

Petersen GM, Rotter JI, Cantor RM, Field LL, Greenwald S, Lim JS, Roy C, Schoenfeld V, Lowden JA, Kaback MM. (1983). The Tay–Sachs disease gene in North American Jewish populations: Geographic variations and origin. Am J Hum Genet. 35: 1258–1269.

Phaneuf D, Wakamatsu N, Huang JQ, Borowski A, Peterson AC, Fortunato SR, Ritter G, Igdoura SA, Morales CR, Benoit G, Akerman BR, Leclerc D, Hanai N, Marth, JD, Trasler JM, Gravel RA (1996). Dramatically different phenotypes in mouse models of human Tay–Sachs and Sandhoff disease. Hum Molec Genet. 5: 1–14.

Platt FM, Butters TD (2000). Substrate deprivation: A new therapeutic approach for the glycosphingolipid lysosomal storage diseases. Expert Rev Mol Med. 2000:1–17.

Platt FM, Jeyakumar M, Andersson U, Heare T, Dwek RA, Butters TD (2003). Substrate reduction therapy in mouse models of the glycosphingolipidoses. Philos Trans R Soc Lond B Biol Sci. 358:947–54.

Platt FM, Neises GR, Dwek RA, Butters TD (1994). N-butyldeoxynojirimycin is a novel inhibitor of glycolipid biosynthesis. J Biol Chem. 18,269(Mar 11):8362–8365.

Platt FM, Neises GR, Reinkensmeier G, Townsend MJ, Perry VH, Proia RL, Winchester B, Dwek RA, Butters TD (1997). Prevention of lysosomal storage in Tay–Sachs mice treated with N-butyldeoxynojirimycin. Science. 276(5311, Apr 18):428–431.

Prence EM, Jerome CA, Triggs-Raine BL, Natowicz MR (1997). Heterozygosity for Tay–Sachs and Sandhoff diseases among Massachusetts residents with French Canadian background. J Med Screen. 4:133–136.

Proia R, Soravia E (1987). Organization of the gene encoding the human beta-hexosaminidase alpha-chain. J Biol Chem. 262:5677–5681.

Proia RL (1988). Gene encoding the human beta-hexosaminidase beta chain: Extensive homology of intron placement in the alpha- and beta-chain genes. Proc Nat Acad Sci. 85:1883–1887.

Proia RL (2001). Targeting the hexosaminidase genes: Mouse models of the GM2 gangliosidoses. Adv Genet. 44:225–231.

Ribeiro MG, Sonin T, Pinto RA, Fontes A, Ribeiro H, Pinto E, Palmeira MM, Sa Miranda MC (1996). Clinical, enzymatic, and molecular characterisation of a Portuguese family with a chronic form of GM2-gangliosidosis B1 variant. J Med Genet. 33:341–343.

Rosebush PI, MacQueen GM, Clarke JT, Callahan JW, Strasberg PM, Mazurek MF (1995). Late-onset Tay–Sachs disease presenting as catatonic schizophrenia: diagnostic and treatment issues. J Clin Psychiatry. 56:347–353.

Rozenberg R, Martins AM, Micheletti C, Mustacchi Z, Pereira LV (2004). Tay–Sachs disease in Brazilian patients: Prevalence of the IVS7+1g>c mutation. J Inherit Metab Dis. 27:109–110.

Rubin M, Karpati G, Wolfe LS, Carpenter S, Klavins MH, Mahuran DJ (1988). Adult onset motor neuronopathy in the juvenile type of hexosaminidase A and B deficiency. J Neurol Sci. 87:103–119.

Rucker JC, Shapiro BE, Han YH, Kumar AN, Garbutt S, Keller EL, Leigh RJ (2004). Neuro-ophthalmology of late-onset Tay–Sachs disease (LOTS). Neurology. 63:1918–1926.

Sachs B (1887). On arrested cerebral development, with special reference to its cortical pathology. J Nerv Ment Dis. 14:541–553.

Sakuraba H, Itoh K, Kotani M, Tai T, Yamada H, Kurosawa K, Kuroki Y, Suzuki H, Utsunomiya T, Inoue H, et al. (1993). Prenatal diagnosis of GM2-gangliosidosis. Immunofluorescence analysis of ganglioside GM2 in cultured amniocytes by confocal laser scanning microscopy. Brain Dev. 15:278–82.

Sakuragawa N, Yoshikawa H, Sasaki M (1992). Amniotic tissue transplantation: Clinical and biochemical evaluations for some lysosomal storage diseases. Brain Dev. 14:7–11.

Salman MS, Clarke JT, Midroni G, Waxman MB (2001). Peripheral and autonomic nervous system involvement in chronic GM2-gangliosidosis. J Inherit Metab Dis. 24:65–71.

Sandhoff K, Andreae U, Jatzkewitz H (1968). Deficient hexosaminidase activity in an exceptional case of Tay–Sachs disease with additional storage of kidney globoside in visceral organs. Life Sci. 7:283–288.

Sandhoff K, Harzer K, Wassle W, Jatzkewitz, H (1971). Enzyme alterations and lipid storage in three variants of Tay–Sachs disease. J Neurochem. 18:2469–2489.

Sango K, McDonald MP, Crawley JN, Mack ML, Tifft CJ, Skop E, Starr CM, Hoffmann A, Sandhoff K, Suzuki K, Proia RL (1996). Mice lacking both subunits of lysosomal beta-hexosaminidase display gangliosidosis and mucopolysaccharidosis. Nature Genet. 14:348–352.

Sango K, Takano M, Ajiki K, Tokashiki A, Arai N, Kawano H, Horie H, Yamanaka S (2005). Impaired neurite outgrowth in the retina of a murine model of Sandhoff disease. Invest Ophthalmol Vis Sci. 46:3420–3425.

Sango K, Yamanaka S, Ajiki K, Tokashiki A, Watabe K (2002). Lysosomal storage results in impaired survival but normal neurite outgrowth in dorsal root ganglion neurones from a mouse model of Sandhoff disease. Neuropathol Appl Neurobiol. 28:23–34.

Sango K, Yamanaka S, Hoffmann A, Okuda Y, Grinberg A, Westphal H, McDonald MP, Crawley JN, Sandhoff K, Suzuki K, Proia RL (1995). Mouse models of Tay–Sachs and Sandhoff diseases differ in neurologic phenotype and ganglioside metabolism. Nature Genet. 11:170–176, 1995.

Schena M, Shalon D, Davis RW, Brown PO (1995). Quantitative monitoring of gene expression patterns with a complementary DNA microarray. Science. 270:457–470.

Schepers U, Glombitza G, Lemm T, Hoffmann A, Chabas A, Ozand P, Sandhoff K. (1996). Molecular analysis of a GM2-activator deficiency in two patients with GM2-gangliosidosis AB variant. Am J Hum Genet. 59: 1048–1056.

Schroder M, Klima H, Nakano T, Kwon H, Quintern LE, Gartner S, Suzuki K, Sandhoff K (1989). Isolation of a cDNA encoding the human GM2 activator protein. FEBS Lett. 251:197–200.

Schroder M, Schnabel D, Hurwitz R, Young E, Suzuki K, Sandhoff K (1993). Molecular genetics of GM2-gangliosidosis AB variant: A novel mutation and expression in BHK cells. Hum Genet. 92:437–440.

Scriver CR, Clow CL (1990). Carrier screening for Tay–Sachs disease. Lancet. 335:856.

Shabat S (1999). Economics of screening programs for Tay–Sachs disease. Health Policy. 47:183–4.

Sheth J, Bhattacharya R, Sheth F (2002). Prenatal diagnosis of Tay–Sachs B1 variant in a Maharashtrian family. Indian Pediatr. 39:704–706.

Shirabe T, Hirokawa M, Asaki H (1980). An autopsy case of Tay–Sachs disease—with special reference to axonal swellings of the central nervous system and freeze-fracture replication studies of the membranous cytoplasmic bodies. Folia Psychiatr Neurol Jpn. 34:515–523.

Slatkin M (2004). A population-genetic test of founder effects and implications for Ashkenazi Jewish diseases. Am J Hum Genet. 75:282–293.

Specola N, Vanier MT, Goutieres F, Mikol J, Aicardi J (1990). The juvenile and chronic forms of GM2 gangliosidosis: Clinical and enzymatic heterogeneity. Neurology. 40:145–150.

Stalker HP, Han BK (1989). Thalamic hyperdensity: A previously unreported sign of Sandhoff disease. AJNR Am J Neuroradiol. 10:S82.

Strasberg P, Warren I, Skomorowski MA, Feigenbaum A (1997). Homozygosity for the common Ashkenazi Jewish Tay–Sachs +1 IVS-12 splice-junction mutation: first report. Hum Mutat. 10(1):82–83.

Suzuki K, Iseki E, Katsuse O, Yamaguchi A, Katsuyama K, Aoki I, Yamanaka S, Kosaka K (2003). Neuronal accumulation of alpha- and beta-synucleins in the brain of a GM2 gangliosidosis mouse model. Neuroreport. 14:551–554.

Suzuki K, Sango K, Proia RL, Langaman C (1997). Mice deficient in all forms of lysosomal beta-hexosaminidase show mucopolysaccharidosis-like pathology. J Neuropathol Exp Neurol. 56:693–703.

Takahashi K, Naito M, Suzuki Y (1987). Lipid storage disease: Part III. Ultrastructural evaluation of cultured fibroblasts in sphingolipidoses. Acta Pathol Jpn. 37:261–272.

Tamasu S, Nishio H, Ayaki H, Lee MJ, Mizutori M, Takeshima Y, Nakamura H, Matsuo M, Maruo T, Sumino K (1999). Prenatal diagnosis of a Japanese family at risk for Tay–Sachs disease. Application of a fluorescent competitive allele-specific polymerase chain reaction (PCR) method. Kobe J Med Sci. 45:259–270.

Tanaka A, Sakuraba H, Isshiki G, Suzuki K (1993). The major mutation among Japanese patients with infantile Tay–Sachs disease: A G-to-T transversion at the acceptor site of intron 5 of the beta-hexosaminidase alpha gene. Biochem Biophys Res Commun. 192:539–546.

Taniike M, Yamanaka S, Proia RL, Langaman C, Bone-Turrentine T, Suzuki K (1995). Neuropathology of mice with targeted disruption of Hexa gene, a model of Tay–Sachs disease. Acta Neuropath. 89:296–304.

Thurmon TF (1993). Tay–Sachs genes in Acadians. Am J Hum Genet. 53:781–783.

Tifft CJ, Proia RL (1997). The beta-hexosaminidase deficiency disorders: Development of a clinical paradigm in the mouse. Ann Med. 29:557–561.

Triggs-Raine BL, Feigenbaum AS, Natowicz M, Skomorowski MA, Schuster SM, Clarke JT, Mahuran DJ, Kolodny EH, Gravel RA (1990). Screening for carriers of Tay-Sachs disease among Ashkenazi Jews. A comparison of DNA-based and enzyme-based tests. N Engl J Med. 323:6–12.
Triggs-Raine BL, Mules EH, Kaback MM, Lim-Steele JST, Dowling CE, Akerman BR, Natowicz MR, Grebner EE, Navon R, Welch JP, Greenberg CR, Thomas GH, Gravel RA (1992). A pseudodeficiency allele common in non-Jewish Tay–Sachs carriers: Implications for carrier screening. Am J Hum Genet. 51:793–801.
Tropak MB, Reid SP, Guiral M, Withers SG, Mahuran D (2004). Pharmacological enhancement of beta-hexosaminidase activity in fibroblasts from adult Tay–Sachs and Sandhoff Patients. J Biol Chem. 279:13478–13487.
Turpin JC, Baumann N (2003). Presenting psychiatric and cognitive disorders in adult neurolipidoses. Rev Neurol (Paris). 159:637–647.
Tutor JC (2004). Biochemical characterization of the GM2 gangliosidosis B1 variant. Braz J Med Biol Res. 37(6):777–783.
Ul-Haque A (1995). Fine needle aspiration cytology of Tay–Sachs disease. A case report. Acta Cytol. 39:762–765.
Unnikrishnan A, Danda S, Seshadri MS (2001). Juvenile Sandhoff disease. Indian Pediatrics 38:89–92 .
Venugopalan P, Joshi SN (2002). Cardiac involvement in infantile Sandhoff disease. J Paediatr Child Health. 38:98–100.
Verlinsky Y, Kuliev A (1996). Preimplantation polar body diagnosis. Biochem Mol Med. 58:13–7.
Wada R, Tifft CJ, Proia RL (2000). Microglial activation precedes acute neurodegeneration in Sandhoff disease and is suppressed by bone marrow transplantation. Proc Natl Acad Sci U S A. 97:10954–10959.
Walkley SU, Zervas M, Wiseman S (2000). Gangliosides as modulators of dendritogenesis in normal and storage disease-affected pyramidal neurons. Cereb Cortex. 10:1028–1037.
Wu YP, Proia RL (2004). Deletion of macrophage-inflammatory protein 1 alpha retards neurodegeneration in Sandhoff disease mice. Proc Natl Acad Sci U S A. 101: 8425–8430.
Yamaguchi A, Katsuyama K, Nagahama K, Takai T, Aoki I, Yamanaka S (2004). Possible role of autoantibodies in the pathophysiology of GM2 gangliosidoses. J Clin Invest. 113:200–208.
Yamanaka S, Johnson MD, Grinberg A, Westphal H, Crawley JN, Taniike M, Suzuki K, Proia RL (1994). Targeted disruption of the Hexa gene results in mice with biochemical and pathologic features of Tay–Sachs disease. Proc Natl Acad Sci USA. 91:9975–9979.
Yamano T, Shimada M, Okada S, Yutaka T, Kato T, Yabuuchi H (1982). Ultrastructural study of biopsy specimens of rectal mucosa. Its use in neuronal storage diseases. Arch Pathol Lab Med. 106:673–677.
Yamato O, Matsuki N, Satoh H, Inaba M, Ono K, Yamasaki M, Maede Y (2002). Sandhoff disease in a golden retriever dog. J Inherit Metab Dis. 25:319–320.
Yamato O, Matsunaga S, Takata K, Uetsuka K, Satoh H, Shoda T, Baba Y, Yasoshima A, Kato K, Takahashi K, Yamasaki M, Nakayama H, Doi K, Maede Y, Ogawa H (2004). GM2-gangliosidosis variant 0 (Sandhoff-like disease) in a family of Japanese domestic cats. Vet Rec. 155:739–44.

Yokoyama S (1979). Role of genetic drift in the high frequency of Tay–Sachs disease among Ashkenazic Jews. Ann Hum Genet. 43:133–136.
Yoshikawa H, Yamada K, Sakuragawa N (1992). MRI in the early stage of Tay–Sachs disease. Neuroradiology 34:394–395.
Yoshizawa T, Kohno Y, Nissato S, Shoji S (2002). Compound heterozygosity with two novel mutations in the HEXB gene produces adult Sandhoff disease presenting as a motor neuron disease phenotype. J Neurol Sci. 195:129–138.
Yuksel A, Yalcinkaya C, Islak C, Gunduz E, Seven M (1999). Neuroimaging findings of four patients with Sandhoff disease. Pediatr Neurol. 21:562–565.

ACID SPHINGOMYELINASE-DEFICIENT NIEMANN–PICK DISEASE

Edward H. Schuchmann, Margaret Mc Govern, Calogera M. Simonaro, Melissa P. Wasserstein and Robert J. Desnick

1 INTRODUCTION AND OVERVIEW

The first case of Niemann–Pick disease (NPD) was described in 1914 by the German pediatrician, Albert Niemann (Niemann, 1914). Over the ensuing years, numerous reports of infants with similar clinical manifestations appeared, and in 1927 Ludwig Pick distinguished this disorder from infantile Gaucher disease based on the differential appearance of the bone marrow foam cells (Pick, 1927). The first adults with this disease were described in 1946 (Pflander, 1946; Dusendschon, 1946). By 1925, NPD was known as a storage disease, and the storage material was thought to consist primarily of phospholipid and cholesterol (Bloom, 1925; Sobotka, Epstein, and Lichtenstein, 1930). Later Klenk (1934) identified the phospholipid as sphingomyelin. A deficiency of acid sphingomyelinase (ASM; EC 3.1.4.12) activity was first demonstrated in human tissue samples obtained from NPD patients in 1966 (Brady et al., 1966; Schneider and Kennedy, 1967; Figure 1).

In 1961, Crocker classified NPD into four clinical entities, types A to D (Crocker, 1961). The infantile, neurodegenerative phenotype originally described by Niemann was termed type A. Type B NPD was distinguished from this severe neuronopathic phenotype by the absence of primary neurological involvement, later onset of hepatosplenomegaly, and survival into adulthood. Types C and D, initially thought to be allelic forms of types A and B (based on similar morphological and clinical findings), are now known to be distinct disorders (Pentchev et al., 1984; Carstea et al., 1997). This chapter discusses only types A and B NPD, which are also referred to as ASM-deficient NPD.

2 EPIDEMIOLOGY AND DEMOGRAPHICS

Although ASM-deficient NPD is a panethnic disorder, the majority of reported type A cases are found in individuals of Ashkenazi Jewish ancestry. Three common ASM mutations (R496L, L302P, fsP330) account for over 90% of all Ashkenazi Jewish individuals with type A NPD (see below), and the combined carrier frequency for these mutations within this group is between 1:80 and 1:100 (Li et al., 1997). Precise information on the frequency of type A NPD in other ethnic groups is not available.

Departments of Human Genetics, Mount Sinai School of Medicine, New York, NY. USA.
Address correspondence to: Dr. Edward Schuchman; Department of Human Genetics, Box 1498.Mount Sinai School of Medicine.1425 Madison Avenue, Room 14-20A
New York, NY 10029. Tel: (212) 659-6711 Fax: (212) 849-2447 Email: Edward.Schuchman@mssm.edu

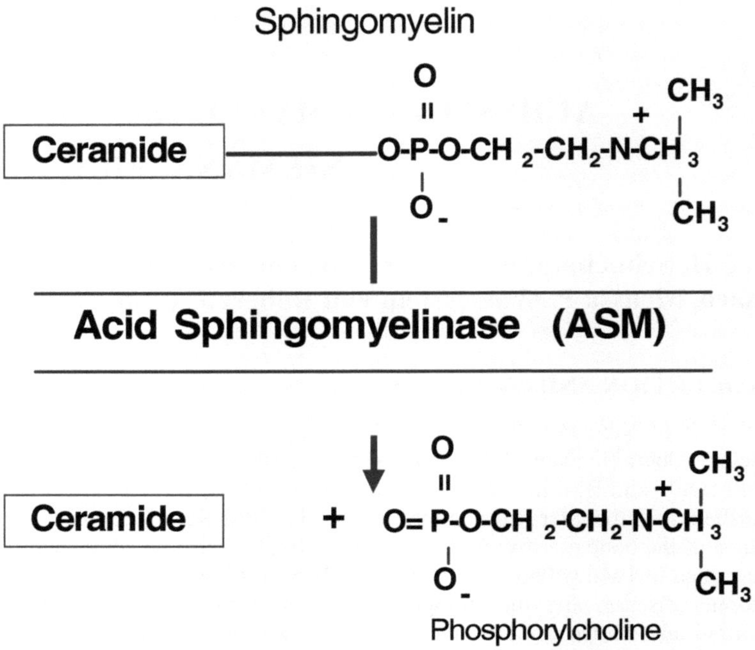

Figure 1. Metabolic disease in acid sphingomielinase (ASM) deficient Niemann–Pick disease (NPD). Types A and B NPD are caused by the marked deficiency of ASM activity, leading to the accumulation of sphingomyelin and other lipids in the lysosome.

By contrast, type B NPD does not appear to have an Ashkenazi Jewish predilection. Areas where this form of the disease is relatively common include the Maghreb region (countries such as Tunisia, Algeria, and Morocco; Vanier et al., 1993), Turkey, and several countries within the Arabian Peninsula (e.g., Saudi Arabia). Although the exact frequency of type B NPD in these groups is not known, several common type B NPD mutations have been identified. One prominent mutation, ! R608, for example, may account for nearly 90% of the mutant alleles in the Maghreb region14 and for about 30% in the United States (Schuchman et al., unpublished results). It must be noted, however, that because the presentation of type B NPD is variable and symptoms may not occur until adulthood, it is likely that this disorder is underdiagnosed by clinicians and therefore the true frequency remains unknown.

3 GENETICS

3.1 The Human ASM cDNA

The full-length cDNA encoding human ASM has an open reading frame of 1890 base pairs, encoding a 629-amino acid precursor protein (Schuchman et al., 1991b). Two in-frame initiation codons are present, and both may be functional. Six *N*-glycosylation sites were predicted in the mature ASM polypeptide, five of which are known to be occupied

(Ferlinz et al., 1997). Several other ASM cDNAs were identified from alternatively spliced ASM mRNAs (Quintern et al., 1989), but none expressed functional protein

3.2 The Human ASM Gene

3.2.1 Genomic Structure

The human ASM gene locus (designated *SMPD1*) has been assigned to the chromosomal region 11p15.1–p15.4 (Da Veiga et al., 1991). The gene is about 5 kb long and the coding sequence contains six exons (Figure 2; Schuchman et al., 1992) Exon 2 is unusually large, encoding about 44% of the mature ASM polypeptide (258 amino acids). Within intron 2 is a single Alu 1 repeat element inserted in reverse orientation. The regulatory region upstream of the ASM coding sequence is GC rich and contains putative promoter elements, including SP1, TATA, CAAT, NF-1, and AP-1 binding sites. Recently, it was found that the *SMPD1* gene is paternally imprinted (Simonaro et al., 2006). Preferential expression of the maternal allele could have important implications in ASM-deficient NPD patients, particularly for those carrying maternal mutations that express residual activity (see below).

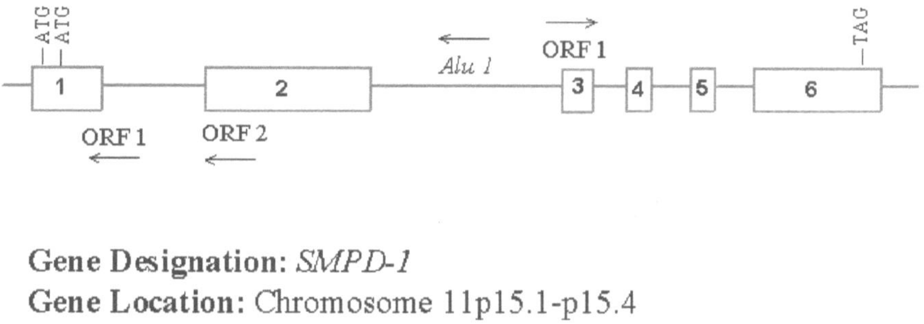

Figure 2. Structure of the human acid sphingomyelinase (ASM) gene. The human ASM gene is divided into six exons and five introns. The locations of the putative translation initiation (ATG) and stop (TAG) codons are indicated, as is the location of the Alu 1 repeat. The locations and transcriptional orientations of three other open reading frames (ORF) within the ASM genomic region are also shown.

3.2.2 Polymorphisms in the Human ASM Gene

Two common polymorphisms have been identified within the ASM gene, leading to amino acid substitutions at codons 322 and 506. The common allele for each codon is Thr 322 (ACA) and Gly 506 (GGG; allele frequencies of 0.6 and 0.8, respectively; Schuchman et al., 1991a, 1992). The less common alleles are Ile 322 (ATA) and Arg 506 (AGG). In addition, the number of alanine/leucine repeats within the putative ASM signal peptide region varies within the normal population (Wan and Schuchman, 1995).

3.2.3 Mutations in the ASM Gene: The Molecular Genetics of Types A and B NPD

Over 90 mutations causing ASM deficiency have been identified in the *SMPD1* gene (reviewed in Simonaro et al., 2002). Among these, 21 are frame-shift or termination (nonsense) mutations that likely result in the production of a nonfunctional, mutant protein that is rapidly degraded within cells. Most of the remaining mutations are point mutations, resulting in single amino acid changes in the mutant ASM polypeptide. Several of these mutations have been found at high frequency in specific ethnic groups (Table 1), and some genotype/phenotype correlations may therefore be made based on the clinical evaluation of multiple affected patients. In particular, several mutations appear to be neuroprotective (Table 2). Of particular note, Q292K is associated with an intermediate phenotype characterized by a protracted neurological course (Sikora et al., 2003), emphasizing the wide spectrum of disease caused by ASM deficiency.

Table 1. Typical features of ASM-deficient Niemann–Pick disease

Feature	Type A	Type B
Age at onset/diagnosis	Early infancy	Infancy to adulthood
Neurodegenerative course	Always	Unusual
Retinal stigmata	>50%	~1/3 (Not associated with neurodegeneration)
Hepatosplenomegaly	+	+
Marrow NPD cells	+	+
Pulmonary involvement	±	+
Age at death	2–3 years	Childhood/adulthood
Autosomal recessive	+	+
Ashkenazi Jewish predilection	+	–
Acid sphingomyelinase activity	<5%	<10%

+, typically present; –, not typically present; ±, may be present.

Table 2. Common NPD mutations in specific ethnic group (adapted from Simonaro et al., 2002)

Ethnic Group (Disease Type)	No. of NPD Alleles Studied	Mutation Designation	Allele Frequency (%)
Ashkenazi Jewish (Type A)	>200	R496L L302P fsP330	36 24 32
Israeli Arabs (Type A)	>20	fsC226	100
Turkish (Intermediate)	20	L137P L549P fsP189	37 20 16.6
Saudi Arabian (Intermediate)	28	H421Y K576N	71.4 13.2
Scottish/British (Type B)	30	A196P	42.3
Southern Chile (Type B)	18	A357D	100

Brazilian/Portuguese (Type B)	24	ΔR608	20.8
		S379P	16.6
		R441X	12.5
		R474W	12.5
		F480L	12.5
Other (Type B)	>250	ΔR608	24

4 BIOCHEMISTRY

Sphingomyelin is the major lipid that accumulates in the cells and tissues of patients with ASM-deficient NPD (Vanier, 1983). Cells of the monocyte–macrophage system, particularly in the spleen and lymph nodes, accumulate the most sphingomyelin. Tissue cholesterol levels also are almost always increased in this disorder (Ludatscher et al., 1981), and the distribution of cholesterol storage mirrors that of sphingomyelin, with cells of the monocyte–macrophage system accumulating the most. The other major lipid to accumulate in the tissues of ASM-deficient NPD patients is bis (monoacylglycero) phosphate (Rouser et al., 1968). There have also been various reports of glycosphingolipid accumulation, including glucocerebroside and the gangliosides, G_{M2} and G_{M3}, (Kamoshita et al., 1969) present; the role of glycolipid accumulation in the pathogenesis of types A and B NPD remains unclear.

Human ASM has been purified from numerous sources, many of the physical and kinetic properties are known, and several assay procedures have been developed to monitor its activity (Gal et al., 1975; Vanier et al., 1985; He et al., 2003) However, the reported levels of residual ASM activity in NPD vary widely due to differences in assay procedures and the source of the enzyme. Therefore, clinical predictions cannot be made based on residual enzyme activity levels. Monospecific antibodies raised against human ASM also have been used to analyze the level of residual ASM protein in ASM-deficient NPD patients (Weitz et al., 1985). In most cases, the amount of ASM-cross-reacting material was within the normal range, consistent with a predominance of point mutations rather than frame-shift or truncation mutations.

In addition to the storage of various lipids, patients with ASM deficiency are defective in one pathway of ceramide production. The role of ceramide in cell signaling is well documented (Kolesnik and Kronke, 1998; Hannun, 1997), although the exact source of ceramide used in this process is incompletely understood. The recent findings that ASM is actively secreted by many cell types, can hydrolyze sphingomyelin at physiological pH, and can be rapidly reinternalized and sequestered in endosomal compartments, support a role for this enzyme in signal transduction. Moreover, mouse models of ASM deficiency (see below) are defective in several of these pathways (Santana et al., 1996) suggesting that the role of abnormal ceramide signaling in NPD patients should be investigated further.

5 ANIMAL MODELS

Two mouse models of ASM-deficient NPD were constructed in 1995 by gene targeting (Horinouchi et al., 1995; Otterbach and Stoffel, 1995). Although the precise targeting events differed in the two models, the phenotypes were essentially identical. The NPD knockout mice (referred to as ASMKO mice) appeared healthy at birth and developed normally until around three months of age, when mild ataxia became apparent. This progressed rapidly, leading to a severe gait abnormality by five months of age. Affected animals died between six and eight months of age.

Histological analysis of the ASMKO mice revealed that the infiltration of NPD cells throughout the reticuloendothelial system was evident by three months of age and progressed rapidly until death. In the central nervous system, there was an almost complete absence of Purkinje cells, as well as evidence of lipid storage vacuoles in neurons. Biochemical analyses of various tissues revealed elevated sphingomyelin levels of between 5- and 40-fold above those found in normal, age-matched littermates. As ASMKO mice develop features of both type A and type B NPD, they offer an excellent model in which to evaluate various therapeutic strategies.

Recently, a new murine model specific for type B NPD was constructed (Marathe et al., 2000) by introducing a modified ASM transgene onto the ASMKO background. Although progressive lipid storage was evident in visceral organs, this was slower than in ASMKO animals. The levels of residual ASM activity in the tissues from these animals varied from about 2 to 12% of normal levels. Interestingly, despite a residual ASM activity of only about 10% in the brain, the animals were completely intact neurologically and lived a normal lifespan, providing proof of principle that small amounts of residual ASM activity can have a major impact on the progression of central nervous system disease in NPD.

6 CLINICAL COURSE

ASM deficiency causes a wide spectrum of disease. Historically, ASM-deficient patients were divided into two subtypes, type A and type B, based on the presence (type A) or absence (type B) of central nervous system involvement (Table 3). However, patients with intermediate neurological phenotypes have now been described, and even among patients without primary neurological disease, the visceral organ manifestations vary widely. To maintain historical context, the type A and B designations have been retained throughout this chapter. However, the reader should be aware that the distinction between these two types is in many cases arbitrary, and that we are in fact discussing a single disorder (ASM deficiency) with a wide spectrum of clinical manifestations.

Table 3. Common, neuroprotective[a] NPD mutations

Genotype	No. of Patients[b]	Average Age
L137P	6	17
A196P	10	35
A357D	9	?
P474Ω	6	18
ΔR608	>75	17

[a]Defined as "neuroprotective" if found in patients without neurological involvement as one mutation in combination with another mutation known to cause type A NPD.
[b]Total number of patients hetero- or homozygous for mutation.

6.1 The Type A Phenotype

The clinical presentation and course of type A NPD is quite uniform. Typically, in the first few months of life, the abdomen of affected infants will become protuberant and hepatosplenomegaly will be noted on physical examination. Bone marrow examination reveals the histochemically characteristic Niemann–Pick foam cells. Pancytopenia occurs later in the disease, and thrombocytopenia may be particularly prominent. Early neurological manifestations include hypotonia and muscle weakness. Failure to thrive, feeding difficulties, recurrent vomiting, and chronic constipation are frequent complications. Cardiac function is typically normal. Most infants with type A NPD have minimal respiratory difficulties in the first year of life, with the exception of repeated bronchitis and intercurrent or aspiration pneumonias. However, infiltration of the alveoli is apparent on X-ray films as a uniform, diffuse reticular or finely nodular pattern (Grunebaum, 1976). Death usually occurs by about three years of age.

6.2 The Type B Phenotype

The clinical presentation and disease course of patients with type B NPD are very variable. Most patients are diagnosed in childhood, when liver and/or spleen enlargement is detected during a routine physical examination. At diagnosis, there is often evidence of mild pulmonary involvement. Typically, hepatosplenomegaly is particularly prominent in childhood, but with increasing linear growth the abdominal protuberance becomes less conspicuous. In less severely affected patients, it may not be noted until adulthood (Dawson and Dawson, 1982; Lever and Ryder, 1983). Although leukopenia and thrombocytopenia secondary to hypersplenism are common, splenectomy is not recommended as splenectomized patients have been noted to undergo subsequent rapid deterioration of their pulmonary status.

Few type B NPD patients have been identified beyond the fifth decade of life, suggesting that the disorder is not compatible with a normal lifespan. However, the cause(s) of mortality have been reported for only a few patients, and most of these individuals succumbed to liver failure. Dyslipidemia is also very common in type B NPD, characterized by elevated total cholesterol, LDL cholesterol and triglycerides, and low HDL cholesterol. However, it is not known if this dyslipidemic phenotype results in an increased risk of coronary artery disease.

Although type B NPD was originally described as a nonneurological disorder, there are exceptions and individuals with protracted neurological disease have been described (Sogawa et al., 1978; Elleder and Cihula, 1983). It is likely that these patients represent the expected clinical spectrum between the typical type A and B NPD phenotypes. Presumably, their ASM activities are sufficient to preclude the development of severe type A symptoms, yet enough neuronal substrate accumulates to cause mild to moderate neurological complications.

7 DIAGNOSIS

Most ASM-deficient patients are diagnosed during childhood with hepatosplenomegaly. The appearance of NPD foam cells in the bone marrow may assist in the diagnosis, but the only definitive test is to measure markedly deficient (<5% of normal) ASM activity in WBCs and/or cultured cells. As noted above, the age of onset and severity of clinical findings in type B NPD patients is highly variable. Early neurological symptoms (during the first year of life) are indicative of type A NPD. Although ASM activity levels can be used to diagnose affected patients, they cannot reliable predict the clinical course nor can they reliably detect heterozygous individuals. DNA analysis may be useful in this regard, particularly in populations where common NPD mutations are known. ASM activity also can be measured reliably in cultured amniocytes or chorionic villi, facilitating prenatal diagnosis (Vanier et al., 1985; Patrick et al., 1977)

8 TREATMENT

8.1 Current Status

At present, no specific treatment is available for ASM-deficient NPD. Orthotopic liver transplantation and amniotic cell transplantation in several NPD patients have been attempted with little or no success (Daloze et al., 1977; Scaggiante et al., 1987). Bone marrow transplantation (BMT) may reduce spleen and liver volumes and lung infiltration, but the morbidity and mortality associated with the procedure limits its usefulness. Also, the effect of BMT on the progression of neurological disease remains unknown (see animal studies below).

8.2 Studies in the ASMKO Mouse; Lessons Learned

BMT, hematopoietic stem cell-mediated gene therapy, and enzyme replacement have been evaluated in the ASMKO mouse model (Miranda et al., 1995b, 1998, 2000; He et al., 1999, Jin et al., 2002; Jin and Schuchman, 2003). BMT and haematopoietic stem cell gene therapy led to an almost complete correction of the histological and biochemical phenotype in the reticuloendothelial system organs, delayed the onset of ataxia by several months, and led to a near doubling of the life expectancy. Despite these positive findings, however, the treated mice still died prematurely of severe neurological disease, casting doubt on the usefulness of these approaches in the treatment of type A NPD. Recent experiments in which genetically engineered bone marrow cells were injected directly into the CNS improved this outcome (Jin et al., 2002, Jin and Schuchman, 2004), but also did not prevent neurodegeneration.

As noted above, although BMT may have positive effects on the clinical course of type B NPD, it is associated with a significant degree of mortality and morbidity. Patients considering this difficult procedure should consult carefully with a specialist physician and counselor. Furthermore, the limited availability of HLA-compatible donor marrow severely hampers the use of this technique.

Type B NPD would appear to be an excellent candidate for enzyme replacement therapy, because, like type 1 Gaucher disease, where this approach is already used, the disease primarily affects macrophages in visceral organ systems. The isolation of the human ASM cDNA has allowed the stable expression of high levels of recombinant human protein in mammalian expression systems (He et al., 1999). Evaluation of this recombinant human ASM in the ASMKO mouse model demonstrated its effectiveness in both the prevention and reversal of the visceral organ pathology. Clinical trials of enzyme therapy in type B NPD patients are planned for the near future. Regarding type A NPD, enzyme replacement therapy is not likely to prove successful for the neurological component with existing technologies, inasmuch as the injected enzyme is not expected to cross the blood–brain barrier. Thus, new technologies must continue to be investigated.

ACKNOWLEDGMENTS

The authors acknowledge the contributions of the many individuals in our laboratories who participated in research from our department that is mentioned in this chapter, and the many patients and families who have contributed valuable time and materials. These studies were supported by research grants from the National Institutes of Health (1 RO1 HD28607), a grant (1 P30 HD28822) for the Mount Sinai Child Health Research Center from the National Institutes of Health, and a grant (5 MO1 RR00071) for the Mount Sinai General Clinical Research Center from the National Center for Research Resources, National Institutes of Health. MW is the recipient of a Mentored Patient-Oriented Research Career Development Award (K23 RR16052-01) from the National Institutes of Health (NIH). The authors also acknowledge support from the Genzyme Corporation for clinical and basic research studies.

REFERENCES

Bloom W. (1925). Splenomegaly (type Gaucher) and lipoid-histiocytosis (type Niemann), Am J Pathol 1, 595.

Brady RO, Kanfer JN, Mock MB, Fredrickson DS. (1966) The metabolism of sphingomyelin. II. Evidence of an enzymatic deficiency in Niemann–Pick disease. Proc Natl Acad Sci U S A 55, 366–369.

Carstea ED, Morris JA, Coleman KG, Loftus SK, Zhang D, Cummings C, Gu J, Rosenfeld MA, Pavan WJ, Krizman DB, Nagle J, Polymeropoulos MH, Sturley SL, Ioannou YA, Higgins ME, Comly M, Cooney A, Brown A, Kaneski CR, Blanchette-Mackie EJ, Dwyer NK, Neufeld EB, Chang TY, Liscum L, Strauss JF 3rd, Ohno K, Zeigler M, Carmi R, Sokol J, Markie D, O'Neill RR, van Diggelen OP, Elleder M, Patterson MC, Brady RO, Vanier MT, Pentchev PG, Tagle DA. (1997) Niemann–Pick C1 disease gene: Homology to mediators of cholesterol homeostasis. Science 277, 228–231.

Crocker AC. (1961) The cerebral defect in Tay–Sachs disease and Niemann–Pick disease. J Neurochem 17, 69.

Daloze P, Delvin EE, Glorieux FH, Corman JL, Bettez P, Toussi T. (1977) Replacement therapy for inherited enzyme deficiency: Liver orthotopic transplantation in Niemann–Pick disease type A. Am J Med Genet 1, 229–239.

da Veiga Pereira L, Desnick RJ, Adler DA, Disteche CM, Schuchman EH. (1991) Regional assignment of the human acid sphingomyelinase gene (SMPD1) by PCR analysis of somatic cell hybrids and in situ hybridization to 11p15.1-p15.4. Genomics 9, 229–234.

Dawson PJ, Dawson G. (1982) Adult Niemann–Pick disease with sea-blue histiocytes in the spleen. Hum Pathol 13, 1115–1120.

Dusendschon A. (1946) Deux cas familiaux de maladie de Niemann–Pick chez adulte [Thesis], Geneva, Faculte de Medecine.

Elleder M, Cihula J. (1983) Niemann–Pick disease (variation in the sphingomyelinase-deficient group). Neurovisceral phenotype (A) with an abnormally protracted clinical course and variable expression of neurological symptomatology in three siblings. Eur J Pediatr 140, 323–328.

Ferlinz K, Hurwitz R, Moczall H, Lansmann S, Schuchman EH, Sandhoff K. (1997). Functional characterization of the N-glycosylation sites of human acid sphingomyelinase by site-directed mutagenesis. Eur J Biochem 243, 511–517.
Gal AE, Brady RO, Hibbert SR, Pentchev PG. (1975) A practical chromogenic procedure for the detection of homozygotes and heterozygous carriers of Niemann–Pick disease. N Engl J Med 293, 632–636.
Grunebaum M. (1976) The roentgenographic findings in the acute neuronopathic form of Niemann–Pick disease. Br J Radiol 49, 1018–1022.
Hannun YA. (1997) Sphingolipid second messengers: tumor suppressor lipids. Adv Exp Med Biol 400A, 305–312.
He X, Chen F, Dagan A, Gatt S, Schuchman EH. (2003) A fluorescence-based, high-performance liquid chromatographic assay to determine acid sphingomyelinase activity and diagnose types A and B Niemann–Pick disease. Anal Biochem 314, 116–120.
He X, Miranda SR, Xiong X, Dagan A, Gatt S, Schuchman EH. (1999) Characterization of human acid sphingomyelinase purified from the media of overexpressing Chinese hamster ovary cells. Biochim Biophys Acta 1432, 251–264.
Horinouchi K, Erlich S, Perl DP, Ferlinz K, Bisgaier CL, Sandhoff K, Desnick RJ, Stewart CL, Schuchman EH. (1995) Acid sphingomyelinase deficient mice: A model of types A and B Niemann–Pick disease. Nat Genet 10, 288–293.
Jin HK, Carter JE, Huntley GW, Schuchman EH. (2002) Intracerebral transplantation of bone marrow-derived mesenchymal stem cells into acid sphingomyelinase deficient mice delays the onset of neurological abnormalities and extends their lifespan. J Clin Invest 109, 1183–1191.
Jin HK, Schuchman EH. (2003) Combined bone marrow and intracerebral mesenchymal stem cell transplantation leads to synergistic visceral and neurological improvements in Niemann–Pick disease mice. Mol Ther 26, 775–785.
Kamoshita S, Aron AM, Suzuki K, Suzuki K. (1969) Infantile Niemann–Pick disease. A chemical study with isolation and characterization of membranous cytoplasmic bodies and myelin. Am J Dis Child 117, 379–394.
Klenk E. (1934) Uber die Natur der Phosphatide der Milz bei Niemann–Pickscher Krankheit, Z Physiol Chem 229, 151 (1934).
Kolesnick RN, Kronke M. (1998) Regulation of ceramide and apoptosis. Ann Rev Physiol 60, 643–645.
Lever AM, Ryder JB. (1983) Cor pulmonale in an adult secondary to Niemann–Pick disease. Thorax 38, 873.874.
Li L, Caggana M, Robinowitz, Shabeer J, Desnick RJ, Eng CM. (1997) Prenatal genetics screening in the Ashkenazi Jewish population: A pilot program of multiple option testing for five disorders. Am J Hum Genet 61(Suppl), A24 (1997).
Ludatscher RM, Naveh Y, Auslaender L, Gellei B. (1981) Electron microscopic studies in lipid storage disease. Isr J Med Sci 17, 323–330.
Marathe S, Miranda SR, Devlin C, Johns A, Kuriakose G, Williams KJ, Schuchman EH, Tabas I. (2000) Creation of a mouse model for non-neurological (type B) Niemann-Pick disease by stable, low level expression of lysosomal sphingomyelinase in the absence of secretory sphingomyelinase: Relationship between brain intra-lysosomal enzyme activity and central nervous system function. Hum Mol Genet 9, 1967–1976.
Miranda SR, Erlich S, Friedrich VL Jr, Gatt S, Schuchman EH. (2000a) Hematopoietic stem cell gene therapy leads to marked visceral organ improvements and a delayed

onset of neurological abnormalities in the acid sphingomyelinase deficient mouse model of Niemann–Pick disease. Gene Ther 7, 1768–1776.

Miranda SR, Erlich S, Friedrich VL Jr, Haskins ME, Gatt S, Schuchman EH. (1998) Biochemical, pathological, and clinical response to transplantation of normal bone marrow cells into acid sphingomyelinase deficient mice. Transplantation 65, 884–892.

Miranda SR, He X, Simonaro CM, Gatt S, Dagan A, Desnick RJ, Schuchman EH. (2000b) Infusion of recombinant human acid sphingomyelinase into Niemann–Pick disease mice leads to visceral, but not neurological, correction of the pathophysiology. FASEB J 14, 1988–1995.

Niemann A. (1914) Ein unbekanntes Krankheitsbild, Jahrb Kinderheilkd 79, 1.

Otterbach B, Stoffel W. (1995) Acid sphingomyelinase-deficient mice mimic the neuro-visceral form of human lysosomal storage disease (Niemann–Pick disease). Cell 81, 1053–1061.

Patrick AD, Young E, Kleijer WJ, Niermeijer MF. (1977) Prenatal diagnosis of Niemann–Pick disease type A using chromogenic substrate. Lancet 2, 144.

Pentchev PG, Boothe AD, Kruth HS, Weintroub H, Stivers J, Brady RO. (1984) A genetic storage disorder in BALB/C mice with a metabolic block in esterification of exogenous cholesterol. J Biol Chem 259, 5784–5791.

Pflander U. (1946) La maladie de Niemann–Pick dans le cadre des lipoidoses, Schweiz Med Wochenschr 76.

Pick L. (1927) Uber die lipoidzellige Splenohepatomegalie Typus Niemann–Pick als Stoffwechselerkrankung, Med Klin 23, 1483.

Quintern LE, Schuchman EH, Levran O, Suchi M, Ferlinz K, Reinke H, Sandhoff K, Desnick RJ. (1989) Isolation of cDNA clones encoding human acid sphingomyelinase: Occurrence of alternatively processed transcripts. EMBO J 8, 2469–2473.

Rouser, G., Kritschevsky, G., Yamamoto, A., Knudson, A. G., Jr., Simon, G. (1968) Accumulation of a glycerolphospholipid in classical Niemann–Pick disease. Lipids 3, 287–290.

Santana P, Pena LA, Haimovitz-Friedman A, Martin S, Green D, McLoughlin M, Cordon-Cardo C, Schuchman EH, Fuks Z, Kolesnick R. (1996) Acid sphingomyelinase deficient mice and human lymphoblasts are defective in radiation-induced apoptosis. Cell 86, 189–200.

Scaggiante B, Pineschi A, Sustersich M, Andolina M, Agosti E, Romeo D. (1987) Successful therapy of Niemann–Pick disease by implantation of human amniotic membrane. Transplantation 44, 59–61.

Schneider PB, Kennedy EP. (1967) Sphingomyelinase in normal human spleens and in spleens from subjects with Niemann–Pick disease. J Lipid Res 8, 202–209.

Schuchman EH, Levran O, Pereira LV, Desnick RJ. (1992) Structural organization and complete nucleotide sequence of the gene encoding human acid sphingomyelinase (SMPD1). Genomics 112, 197–205.

Schuchman EH, Levran O, Suchi M, Desnick RJ. (1991a) An MspI polymorphism in the human acid sphingomyelinase gene (SMPD1). Nucleic Acid Res 19, 3160.

Schuchman EH, Suchi M, Takahashi T, Sandhoff K, Desnick RJ. (1991b) Human acid sphingomyelinase. Isolation, nucleotide sequence and expression of the full-length and alternatively spliced cDNAs. J Biol Chem 266, 8531–8539.

Sikora J, Pavlu-Pereira H, Elleder M, Roelofs H, Wevers RA. (2003) Seven novel acid sphingomyelinase genemutations in Niemann-Pick type A and B patients. Ann Hum Genet 67,63–70.

Simonaro CM, Desnick RJ, McGovern MM, Wasserstein MP, Schuchman EH. (2002) The demographics and distribution of type B Niemann-Pick disease: Novel mutations lead to new genotype/phenotype correlations. Am J Hum Genet 71: 1413–1419.

Simonaro CM, Park JH, Eliyahu E, Shtraizent N, McGovern MM, Schuchman EH. (2006) Imprinting at the SMPD-1 gene: Implications for acid sphingomyelinase-deficient Niemann–Pick disease. Am J Hu. Ge. 78, 79–84.

Sobotka H, Epstein E, Lichtenstein L. (1930) The distribution of lipoid in a case of Niemann-Pick's disease associated with amaurotic family idiocy, Arch Pathol 10, 677.

Sogawa H, Horino K, Nakamura F, Kudoh T, Oyanagi K, Yamanouchi T, Minami R, Nakao T, Watanabe A, Matsuura Y. (1978) Chronic Niemann–Pick disease with sphingomyelinase deficiency in two brothers with mental retardation. Eur J Pediatr 128, 235–240.

Vanier MT. (1983). Biochemical studies in Niemann–Pick disease. I. Major sphingolipids of liver and spleen. Biochim Biophys Acta 750, 178–184.

Vanier MT, Boue J, Dumez Y. (1985) Niemann–Pick disease type B: First-trimester prenatal diagnosis on chorionic villi and biochemical study of a foetus at 12 weeks of development. Clin Genet 28, 348–354.

Vanier MT, Ferlinz K, Rousson R, Duthel S, Louisot P, Sandhoff K, Suzuki K. (1993) Deletion of arginine (608) in acid sphingomyelinase is the prevalent mutation among Niemann–Pick disease type B patients from northern Africa. Hum Genet 92, 325–330.

Vanier MT, Rousson R, Garcia I, Bailloud G, Juge MC, Revol A, Louisot P. (1985) Biochemical studies in Niemann–Pick disease. III. In vitro and in vivo assays of sphingomyelin degradation in cultured skin fibroblasts and amniotic fluid cells for the diagnosis of the various forms of the disease. Clin Genet 27, 20–32.

Victor S, Coulter JB, Besley GT, Ellis I, Desnick RJ, Schuchman EH, Vellodi A. (2003) Niemann–Pick disease type B: 16 year follow-up after allogenic bone marrow transplantation. J Inherit Met Dis 26, 775–785.

Wan Q and Schuchman EH. (1995) A novel polymorphism in the human acid sphingomyelinase gene due to size variation of the signal peptide region. Biochim Biophys Acta 1270, 207–210.

Weitz G, Driessen M, Brouwer-Kelder EM, Sandhoff K, Barranger JA, Tager JM, Schram AW. (1985) Soluble sphingomyelinase from human urine as antigen for obtaining anti-sphingomyelinase antibodies. Biochim Biophys Acta 838, 92–97.

KRABBE DISEASE (GLOBOID CELL LEUKODYSTROPHY)

Junko Matsuda and Kunihiko Suzuki

Krabbe disease (globoid cell leukodystrophy, GLD) can be caused by genetic defects either in a lysosomal enzyme, galactosylceramidase (*galc*), or in its natural activator/protective protein, saposin A. The latter was first established in a mouse model, but the first human GLD patient due to saposin A deficiency has very recently been discovered. The primary natural substrate of the enzyme, galactosylceramide, is nearly exclusively localized in the myelin sheath. Consequently, the disease is one of the two classical genetic leukodystrophies, the other being metachromatic leukodystrophy due to arylsulfatase A (sulfatide sulfatase) deficiency. Mode of inheritance is autosomal recessive. Typically, the disease occurs among infants and takes a rapidly fatal course, but rarer late-onset forms also exist. Clinical manifestations are exclusively neurological with prominent and progressive white matter signs.

A normally insignificant but highly cytotoxic metabolite, galactosylsphingosine (psychosine), is also a substrate of galactosylceramidase. It appears to play a critical role in the pathogenesis of GLD. A set of neuropathology is unique and pathognomonic; a rapid and nearly complete disappearance of myelin and myelin-forming cells, the oligodendrocytes in the CNS and the Schwann cells in the PNS, reactive astrocytic gliosis, and infiltration of often multinucleated macrophages ("globoid cells") that contain strongly PAS-positive materials. A large number of disease-causing mutations have been identified. The only human patient known so far with saposin A deficiency had an in-frame three-base deletion in the saposin A domain of the prosaposin gene. Either assays for galactosylceramidase activity or nucleic acid-based methodologies can be used for definitive clinical and prenatal diagnosis when the genetic defect is in the *galc* gene. The standard galactosylceramidase assay in the presence of detergents cannot diagnose the saposin A deficiency state.

1 HISTORICAL OVERVIEW

The history of Krabbe disease dates back to 1916, when a Danish physician, Knud Krabbe, described clinical and pathological findings in five infants from two families who died of an "acute infantile familial diffuse brain sclerosis" (Krabbe, 1916). These infants developed episodes of violent crying and irritability beginning at age 4 to 6 months, followed by progressive muscular rigidity, tonic spasms evoked by such stimuli

Institute of glycotechnology; Future Science and Technology Research Center. Tokai Universtity.
Address correspondence to: Kunihiko Suzuki, M.D. Future Science and Technology Joint Research Center. Institute of Glycotechnology, Tokai University. 1117 Kita-kaname, Hiratsuka. Kanagawa-ken 259-1292 JAPAN
Tel +81-463-58-1211, ext 4656; Fax +81-463-50-2432
e-mail: kuni.suzuki@attglobal.net

as noise, light, and touching. Death occurred between 11 months to 1.5 years. He provided a detailed description of the globoid cells, the histologic hallmark of the disease. Thus, the disease carries his name. The disease remained a clinical and pathological entity for a half century.

In 1970, deficiency of galactosylceramidase activity was identified as the underlying genetic cause (Suzuki and Suzuki, 1970) that made noninvasive antemortem and prenatal diagnosis possible (Suzuki and Suzuki, 1971; Suzuki et al., 1971). In 1972, the toxic effect of a related metabolite, galactosylsphingosine (psychosine), was proposed as the critical factor in the biochemical pathogenesis (Miyatake and Suzuki, 1972). The psychosine hypothesis has since been generally substantiated both in human disease and in animal models (Suzuki, 1998). Human galactosylceramidase cDNA was cloned in 1993–1994 (Chen et al., 1993; Sakai et al., 1994), and more than 60 disease-causing mutations have been identified (Wenger et al., 2001).

Effective treatment to "cure" the disease is not available, although bone marrow transplantation can alleviate clinical conditions and retard progression of the disease, particularly in patients with late-onset forms (Krivit et al., 1998). Specific deficiency of saposin A, the galactosylceramidase activator/protective protein, was first demonstrated in a mouse model to cause a disease with a clinical, pathological, and biochemical phenotype that was qualitatively identical with human Krabbe disease (Matsuda et al., 2001a). The first human patient with saposin A deficiency was reported in 2005 (Spiegel et al., 2005). Relatively recent comprehensive reviews are available (Wenger et al., 2001; Suzuki, 2004a).

2 INCIDENCE

The incidence of the typical infantile form is estimated as 1 in 100,000 births in the United States, 2 in 100,000 births in Sweden, and 0.5 to 1 in 100,000 births in Japan. There are pockets of populations where the incidence is unusually high; for example, the Druze community in Israel has an incidence of 6 in 1000 births. In contrast, no Jewish patients have ever been reported. Late-onset and adult forms of the disease are even rarer but are being reported in increasing frequency. Because definitions of "late-onset" and "adult" forms are not necessarily standardized, precise incidence of the later-onset forms is difficult to assess. In contrast to the infantile form, which appears to have a distinctly higher incidence in Nordic countries, patients with the late-onset form appear to be more common in Southern Europe (Fiumara et al., 1990; Barone et al., 1996). The frequency of late-onset cases was estimated to be approximately 10% of the nearly 350 GLD patients diagnosed by Wenger (Wenger et al., 2001). The only known patient due to saposin A deficiency was of Arab–Muslim extraction (Spiegel et al., 2005).

3 CLINICAL MANIFESTATIONS

3.1 Infantile GLD

Clinical phenotype of the classical infantile Krabbe disease is stereotypic. Contrary to the prediction based on the mouse model, the first human patient due to saposin A deficiency had an infantile phenotype (Spiegel et al., 2005). The general clinical picture is that of a steady and rapidly progressive white matter disorder. The classical account of Hagberg, Sourander, and Svennerholm (1963) divided the clinical course into three stages. Stage I is characterized by generalized hyperirritability, hyperesthesia, episodic fever of unknown origin, and some stiffness of the limbs. The child, apparently normal for the first few months after birth, becomes hypersensitive to auditory, tactile, or visual stimuli and begins to cry frequently without apparent cause. Slight retardation or regression of psychomotor development, vomiting with feeding difficulty, and convulsive seizures may occur as initial clinical symptoms. The cerebrospinal fluid protein level is already highly increased.

In stage II, rapid and severe motor and mental deterioration develops. There is marked hypertonicity, with extended and crossed legs, flexed arms, and the backward-bent head. Tendon reflexes are hyperactive. Minor tonic or clonic seizures occur. Optic atrophy and sluggish pupillary reactions to light are common. Stage III is the "burnt-out" stage, sometimes reached within a few months. The infant is decerebrate and blind and has no contact with the surroundings. Deafness may appear. Patients rarely survive for more than a few years.

Clinical examination may not always reveal neuropathy, especially in the early stages, because symptoms and signs of central nervous system involvement are overwhelming. Krabbe (1916), however, noted in his original five patients that knee jerks could not be elicited and that stiffness passed into a flaccid state toward the end of the disease. The typical pathology is always present in the peripheral nerves, and careful clinical examination combined with appropriate electrophysiological studies should reveal presence of PNS involvement. Spinal fluid protein is invariably highly elevated in patients with infantile GLD. The symptoms and signs are, for all practical purposes, confined to the nervous system. No visceromegaly is present.

3.2 Late Onset GLD

Earlier, patients with late-onset GLD were considered exceedingly rare and often misdiagnosed during life. The definite diagnosis could be established only by histological examination with the possible exception of peripheral nerve biopsy. Since the advent of the enzymatic diagnosis, late-onset GLD has been reported in increasing frequency. Most patients develop initial clinical signs and symptoms by 10 years of age, but some may develop neurological signs after 40 years of age.

Late-onset GLD is often divided into two types; late infantile (or early childhood) and juvenile (late childhood). In the late infantile group (onset six months to three years), irritability, psychomotor regression, stiffness, ataxia, and loss of vision are frequent initial symptoms. The course is progressive, resulting in death in approximately two to three years after the onset. In the juvenile group (onset three to eight years), patients commonly develop loss of vision, together with hemiparesis, ataxia, and psychomotor regression. Most patients with the juvenile form show rapid deterioration initially, followed by a more gradual progression possibly lasting for years. The number of reports on adult cases is also increasing steadily. Adult patients may develop slowly progressive spastic paraparesis or slow, unsteady, stiff, and wide-based gait during life. Progressive and generalized neurological deterioration may not be apparent until 40 years of age. Some adult patients may have a normal lifespan. The cerebrospinal fluid protein is normal or only mildly elevated in the juvenile or adult type patients. Peripheral nerve conduction velocity is generally reduced in late infantile patients.

4 PATHOLOGY

Pathology is, for all practical purposes, limited to the nervous system (Suzuki and Suzuki, 2002). In the most common infantile type, the brain is grossly atrophic with firm rubbery gliotic white matter. At the terminal stage, loss of myelin is nearly complete with the possible exception of the subcortical intergyral arcuate U-fibers. Microscopically, extensive fibrillary gliosis, infiltration of numerous macrophages, often multinucleated ("globoid cells"), and nearly complete loss of myelin are the histological hallmark of infantile Krabbe disease (Figure 1). The globoid cells are abundant in the region of active demyelination and often clustered around blood vessels.

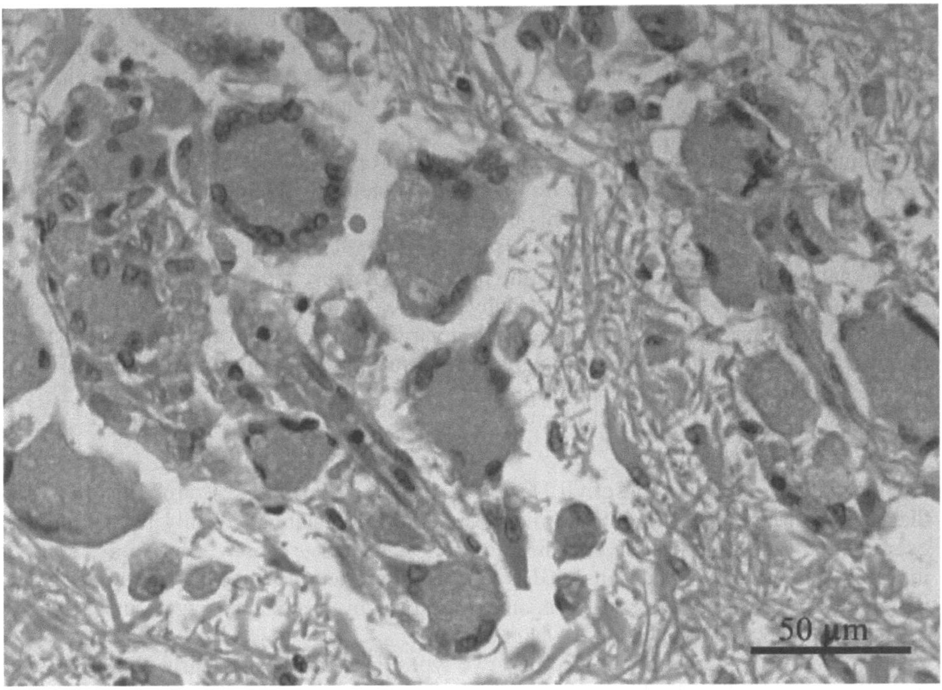

Figure 1. Typical histology of the white matter of a patient with infantile Krabbe disease. Hematoxylin-eosin stain. The cellular architecture is highly disrupted with nearly complete loss of myelin and the oligodendrocytes, which are replaced by severe astrocytic gliosis and many characteristic multinucleated macrophages ("globoid cells"). (Courtesy, Dr. Kinuko Suzuki.)

Oligodendrocytes are markedly reduced. Correlative MRI and neuropathological studies showed that the areas of marked hyperintensity on the T2-weighted MRI images corresponded to the areas of demyelination with globoid cell infiltration (Percy et al., 1994). Globoid cells contain PAS-positive storage materials. On the ultrastructural level, the globoid cells contain tubular and filamentous structures with polygonal cross-sections that are structurally identical with chemically pure galactosylceramide (Yunis and Lee, 1970). In long surviving cases, the white matter may be totally gliotic and devoid of macrophages ("globoid cell leukodystrophy ohne globoid cells"). The optic nerves are usually atrophic but, in some cases, they are markedly enlarged with extensive gliosis. The peripheral nerves are often grossly enlarged and firm with marked endoneurial fibrosis, segmental demyelination, and evidence of remyelination process with onion bulb formation. Quantitative analyses demonstrated a severe loss of large myelinated fibers with relative preservation of unmyelinated fibers. Endoneurial macrophages and also Schwann cells contain tubular inclusions similar to those in the globoid cells in the cerebral white matter.

Neuropathological reports of late-onset cases are limited. In a meeting abstract, Choi and Enzmann (1993) described the neuropathology of the adult-onset GLD in 18-year-old twins. Their clinical symptoms developed 12 and 7 months prior to their death, respectively. Both died of severe graft-versus-host disease two months after allogeneic bone marrow transplantation. The brains showed degeneration of the optic radiation and frontoparietal white matter with corticospinal tract degeneration. Multiple necrotic foci with calcium deposits were found within the lesion. Globoid cell infiltration was present in actively degenerating white matter. In the peripheral nerves, loss of myelinated fibers, disproportionately thin myelin sheaths, and inclusions in Schwann cells were described.

5 BIOCHEMISTRY

Galactosylceramidase, the genetically defective enzyme in all so-far known human GLD patients except for the single patient with saposin A deficiency, is a degradative enzyme with an acid pH optimum localized in the lysosome. Thus, the disease conceptually belongs to the category of the inherited lysosomal disease as originally defined 50 years ago by Hers (1966). The enzyme is fairly specific for glycolipids with a terminal galactose moiety in β anomeric configuration. Quantitatively by far the major natural substrate is galactosylceramide, which is highly localized in the myelin sheath. Other known natural substrates are psychosine (galactosylsphingosine), monogalactosyldiglyceride, and the precursor of seminolipid (1-alkyl, 2-acyl-, 3-galactosyl glycerol). In vivo degradation of these substrates requires, in addition to the enzyme, galactosylceramidase, an activator/protective protein, saposin A. The mouse model of saposin A deficiency suggested that saposin A is not only an activator protein but also is a protecting (stabilizing) protein for galactosylceramidase (Matsuda et al., 2001a). Apparently consistent with the protecting ability of saposin A is the discrepant galactosylceramidase activities in leukocytes (deficient) and in cultured fibroblasts (normal) in the recently discovered first human patient with saposin A deficiency (Spiegel et al., 2005).

Essentially all lysosomal diseases are "storage diseases," in which subtrates of the genetically defective enzymes accumulate to abnormally high levels. However, the unique biochemical feature of Krabbe disease is lack of abnormal accumulation of galactosylceramide in the brain, contrary to what is expected from the enzymatic defect (Vanier and Svennerholm, 1975; Svennerholm, Vanier, and Månsson, 1980). This paradoxical phenolmenon results from the unique localization of galactosylceramide in the myelin sheath and very rapid and early disappearance of the myelinating cells in the process of the disease. Because the myelinating cells disappear at a very early stage of myelination and because no further synthesis of galactosylceramide occurs, it does not accumulate beyond the level attained at the early stage of myelination. Instead, however, a related toxic metabolite, psychosine (galactosylsphingosine) does accumulate abnormally and is considered the key compound in the pathogenesis of the disease (see below; Miyatake and Suzuki, 1972; Vanier and Svennerholm, 1976; Suzuki, 2003).

Although there is no abnormal accumulation of galactosylceramide in the brain tissue as a whole, localized accumulation of galactosylceramide does occur within the characteristic globoid cells. Biochemical analysis of a fraction enriched with globoid cells contained relatively large amounts of galactosyceramide (Austin, 1963). Galactosylceramide has the unique capacity to elicit infiltration of globoid cells when it is implanted into the brain (Austin and Lehfeldt, 1965) and such experimentally induced globoid cells appear morphologically identical to those seen in patients with Krabbe disease (Andrews and Menkes, 1970; Suzuki, 1970). Biochemical abnormalities are essentially limited to the nervous system, at least in human patients. In the mouse models, a conspicuous accumulation of galactosylceramide occurs in the kidney (Igisu et al., 1983; Ida et al., 1982) but not in human patients (Suzuki, 1971).

6 PATHOPHYSIOLOGY

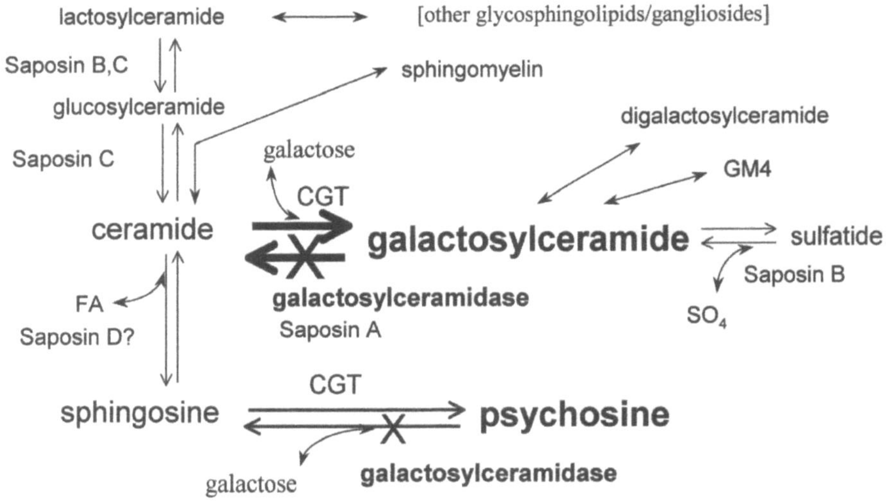

Figure 2. Metabolic pathways pertinent to galactosylceramide and related compounds. In the synthetic pathway, sphingosine is first acylated to ceramide, which in turn is galactosylated by UDP-galactose:ceramide galactosyltransferase (CGT) to form galactosylceramide. The same enzyme can galactosylate sphingosine directly to generate psychosine. Both galactosylceramide and psychosine are degraded by galactosylceramidase, which is genetically deficient in Krabbe disease. In vivo degradation of galactosylceramide requires, in addition to the enzyme, a sphingolipid activator protein, saposin A. Galactosylceramide is further sulfated to form sulfatide. Both galactosylceramide and sulfatide are characteristic myelin glycolipids.

6.1 Distribution and Metabolism of Myelin Galactolipids

Galactosylceramide has a uniquely restricted tissue distribution. It is mostly, but not exclusively, localized in the myelin sheath and thus almost exclusively synthesized within the oligodendrocytes and the Schwann cells. Its sulfate ester, sulfatide, is synthesized only by sulfation of galactosylceramide (Figure 2). Thus, both galactosylceramide and sulfatide are characteristically the glycolipids of the myelin sheath and are virtually absent in the brain before myelination and are present at abnormally low concentrations in any pathological conditions where severe loss of myelin occurs. It is practically absent in systemic organs except in the kidney, which normally contains appreciable amounts of galactosylceramide, although much less than in the nervous system.

It is understandable in view of the unusually high concentrations of galactosylceramide and sulfatide in the myelin sheath that the metabolic diseases involving these lipids (Krabbe disease and metachromatic leukodystrophy) are primarily disorders of white matter and peripheral nerves. The most significant metabolic features of CNS myelin are its high rate of formation and turnover during a relatively short period of brain development and its slow turnover in the adult brain. The period of most active myelination in humans probably extends from the perinatal period to about age 18 months. Myelination does not stop after this period, and in the human brain, it may not be complete until age 20 years.

The amount of galactosylceramide in immature brain is very low. Activity of galactosylceramide synthase, UDP-galactose:ceramide galactosyltransferase (CGT) peaks sharply at 20–25 days after birth in rodent brains well correlating with the most active period of myelination (Costantino-Ceccarini and Morell, 1972). The recent cloning of the rat and mouse CGT confirmed that the peak levels of the corresponding mRNA occur during the period of most active myelination and that relatively high mRNA levels are found in the brain and kidney (Stahl et al., 1994). Synthesis and turnover of galactosylceramide occurs at much lower rates in the adult brain.

6.2 Pathogenesis

Some aspects of the chemistry and metabolism of galactosylceramide should be kept in mind when the pathogenetic mechanism of GLD is considered. (1) Galactosylceramide consists of sphingosine, fatty acid, and galactose. (2) Galactosylceramide is the direct and only known precursor of sulfatide. (3) Both galactosylceramide and sulfatide are highly concentrated in the myelin sheath. (4) Galactosylceramidase degrades galactosylceramide to ceramide and galactose with the help of saposin A as the essential activator/protective protein. (5) A few related galactolipids also serve as substrates for galactosylceramidase, including psychosine, monogalactosyldiglyceride, seminolipid precursor, and lactosylceramide. (6) Biosynthesis of galactosylceramide reaches a peak during the most active period of myelination (during the first year and a half in humans and 15 days to 25 days in rodents), when myelin also turns over relatively rapidly. (7) Once formed, adult myelin is relatively stable metabolically. (8) Galactosylceramide is uniquely capable of inducing infiltration of globoid cells when implanted into the brain but does not appear toxic, whereas another normally insignificant substrate, psychosine, is highly cytotoxic.

6.2.1 Two Separate but Related Pathogenetic Mechanisms

Three of the most characteristic pathological features of Krabbe disease are (1) the infiltration of macrophages that are often multinucleated and contain strongly PAS-positive materials ("globoid cells"), (2) the rapid and almost complete disappearance of the oligodendrocytes, and (3) lack of abnormal tissue accumulation of the primary substrate of the defective enzyme, galactosylceramide, contrary to what is expected in a "storage disease" due to genetic defect in degradative enzymes. These phenotypic characteristics must be explained as consequences of the underlying genetic defect. Defective degradation of two substrates, galactosylceramide and psychosine (galactosylsphingosine) appears to play critical roles in the pathogenesis. Although these mechanisms are fundamentally distinct from each other, they are closely intertwined to result in the unique phenotype of the disease.

6.2.1.1 Globoid Cells

The genetic defect in degradation of galactosylceramide clearly is a major factor for the unique pathological feature of the disease, the globoid cells. It has long been known that free galactosylceramide has a specific capacity to elicit infiltration of macrophages into the brain (Austin and Lehfeldt, 1965; Suzuki et al., 1976). Once in the brain, they phagocytize galactosylceramide and are transformed to multinucleated globoid cells. The characteristic inclusions in the globoid cells have morphological appearance identical to galactosylceramide itself (Yunis and Lee, 1970). No other agent is known to have a similar capacity in vivo. The globoid cell reaction can be reconstructed in the following way. Once the active period of myelination begins, turnover of already formed myelin also begins. In patients' brain, however, galactosylceramide cannot be degraded due to the underlying galactosylceramidase or saposin A deficiency. Undegraded galactosylceramide thus elicits infiltration of macrophophages, which become the characteristic PAS-positive, often multinucleated globoid cells.

6.2.1.2 Psychosine Hypothesis

On the other hand, the devastating early destruction of the myelin-forming cells is difficult to explain on the basis of undegradable galactosylceramide because galactosylceramide implanted in the brain does not exhibit any functionally detrimental capacity other than eliciting the globoid cell reaction. There is no experimental evidence that galactosylceramide is a metabolic toxin. A closely related metabolite, psychosine (galactosylsphingosine), is highly cytotoxic (Taketomi and Nishimura, 1964) and causes fatal hemorrhagic infarct when implanted into the brain (Miyatake and Suzuki, 1972).

At least in mammalian tissues, psychosine can be generated only by galactosylation of sphingosine by galactosylceramide synthase, UDP-galactose:ceramide galactosyltransferase (CGT), but not by de-acylation of galactosylceramide. Because CGT is nearly exclusively localized in the myelin-forming cells, synthesis of psychosine should also occur only in the oligodendrocytes and Schwann cells. Psychosine is detectable in normal brain with highly sensitive analytical methods but its concentration is minuscule (less than 10 picomoles/mg protein). It appears to be a dead-end product, which is normally degraded immediately. However, psychosine is degraded also by galactosylceramidase. Therefore, patients with Krabbe disease cannot degrade psychosine.

A hypothesis, known as the psychosine hypothesis, was first proposed on the basis of this enzymological consideration (Miyatake and Suzuki, 1972) and then its abnormal accumulation in the brain of patients was analytically demonstrated (Vanier and Svennerholm, 1976; Svennerholm, Vanier, and Månsson, 1980) and in canine and murine models (Igisu and Suzuki, 1984). The psychosine hypothesis postulates that, in globoid cell leukodystrophy, not only the primary substrate of the defective enzyme, galactosylceramide, but also the toxic metabolite, galactosylsphingosine (psychosine), cannot be degraded and the consequent abnormal accumulation of psychosine causes the uniquely rapid destruction of the myelin-forming cells. The hypothesis initially met considerable skepticism but has survived the intervening 30 years (Suzuki, 1998). In fact, the basic premise of the hypothesis has been extended to other sphingolipidoses (Hannun and Bell, 1987). For varieties of reasons, however, its plausibility for other disorders is not as firm as it is for GLD, with possible exceptions of neuronopathic form of Gaucher disease and Niemann–Pick type A disease.

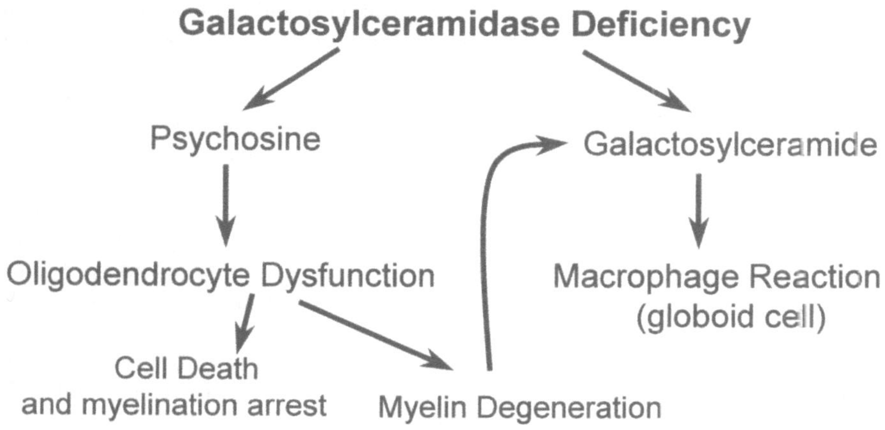

Figure 3. Pathogenetic mechanisms operating in Krabbe disease as understood at this time. See text for explanation.

6.2.1.3 Overall pathogenesis

Close interactions of the above two pathogenetic mechanisms can explain the most important aspects of the characteristic phenotype of Krabbe disease (Figure 3). The fundamental cause of the disease is the defective galactosyceramide degradation due to abnormalities in the enzyme or in saposin A. Because galactosylceramide synthesis is limited to actively myelinating cells, the disease process does not begin until the active myelination period. Once myelination begins, its metabolic turnover also starts. This generates free galactosylceramide in the brain of patients because of the inability to degrade galactosylceramide, which in turn elicits the characteristic globoid cell reaction.

Galactosylceramide synthase also synthesizes psychosine within the actively myelinating cells. Normally, it is immediately degraded and never reaches beyond barely detectable levels. In Krabbe disease, however, an abnormal accumulation of psychosine occurs to the level toxic to cellular metabolism. This causes the other characteristic feature of the disease, a rapid and almost complete disappearance of the oligodendrocytes. Psychosine is as potent an apoptosis inducer as C6 ceramide (Tohyama, Matsuda, and Suzuki, 2001). The cellular death results in further destruction of already formed myelin, which contributes more free galactosylceramide that in turn further elicits the globoid cell infiltration. On the other hand, myelination ceases at a very early stage due to the near-complete loss of the oligodendrocytes. This explains the paradoxical characteristics of the disease that the primary substrate of the defective enzyme, galactosylceramide, does not accumulate abnormally.

7 MOLECULAR GENETICS

7.1 Gene Structure

7.1.1 Galactosylceramidase Gene

Human galactosylceramidase gene, *galc* (GenBank database Accession No. 119970), was localized to the region of 14q24.3-q32.1 by linkage analysis (Oehlmann et al., 1993) and later further narrowed to 14q31 by in situ hybridization (Cannizzaro et al., 1994). Using N-terminal amino acid sequence information, the human cDNA was cloned in 1993–1994 (Chen et al., 1993; Sakai et al., 1994). The full-length cDNA consists of 3795 bp, including 2007 bp of the coding region, 47 base pairs of 5' untranslated sequence, and 1741 bp of 3' untranslated sequence. The base and amino acid sequences have no similarities and thus no suggestion of evolutionary relationship with the other β-galactosidases or any other known genes.

The encoded protein consists of 669 amino acids with six potential glycosylation sites. The first 26 amino acids have the characteristic of a leader sequence. The precursor protein is approximately 80–85 kD, which is processed to 50–52 kD and 30 kD subunits. The coding sequence for the 50–52 kD subunit is at the 5' end and the 30 kD subunit is at the 3' end of the coding region. The organization of the human gene was characterized in 1995 (Luzi, Rafi, and Wenger, 1995). It consists of 17 exons spread over about 56 kb. Other than exons 1 and 17, the other exons range in size from 39 to 181 nucleotides. The 200 nucleotides preceding the initiation codon and the 5' end of intron 1 are GC-rich, including 13 GGC trinucleotides. The 5' flanking region includes a YYI element and one potential SP1 binding site.

No consensus TATA box or CAAT box is present among the 800 nucleotides preceding the initiation codon. A construct containing nucleotides –176 to –24 had the strongest promoter activity. However, evidence for inhibitory sequences was found just upstream of the promoter region and also at the 5' end of intron 1. There is another potential initiation codon located 48 nucleotides upstream, however, there is no evidence that it is utilized or if it might play a role in tissue-specific expression.

7.1.2 Saposin A

Sphingolipid activator proteins (saposins A, B, C, D) are small heat-stable glycoproteins required for in vivo degradation of some sphingolipids with short carbohydrate chains (Sandhoff, Kolter, and Harzer, 2001). They are derived from a common precursor protein (prosaposin), which is encoded by a single gene located on human chromosome 10 (10q21–q22). Prosaposin includes saposin A, B, C, and D domains in tandem. It is transported to the lysosome and proteolytically processed to the four saposins. These four saposins are all homologous to each other, having six conserved cysteines and one common glycosylation site. In spite of these structural similarities, their activator functions are specific, with some overlaps, for individual sphingolipid hydrolases.

7.2 Disease-Causing Mutations

Over 60 disease-causing missense, nonsense mutations, deletions, and insertions have been identified in the human galactosylceramidase gene (Wenger et al., 2001). A major deletion of 30 kb from the middle of intron 10 to beyond the end of the gene that always occurs on a 502T polymorphic background (502T/del) is common among patients from Northern Europe and the United States, including those with Mexican ancestry. The 30 kb deletion eliminates all of the coding region for the 30 kD subunit and about 15% of the coding region for the 50–52 kD subunit. A survey conducted among patients within the Dutch population and from other parts of Europe confirmed that 502T/del mutation makes up about 50% of the total mutant alleles. In infantile Swedish patients, this mutation makes up 75% of the mutant alleles. This mutation probably initially occurred in Sweden and was transmitted from there throughout Europe, Near Asia, and the United States. It has not been found among Japanese patients.

Two other mutations (C1538T and A1652C) make up an additional 10–15% of the mutant alleles in infantile patients with European ancestry. Three mutations (635del+ins, A198G, and T1853C) have been found in multiple unrelated Japanese patients. In Israel there are two populations with an extremely high carrier rate for Krabbe disease, and they have different mutations. All infantile patients in the Druze population in Northern Israel are homozygous for the T→G transversion at nucleotide 1748 (I583S), and patients from a Moslem village near Jerusalem are homozygous for the G→A transition at nucleotide 1582 (D528N).

The only human patient so far known with specific saposin A deficiency had an in-frame 3 bp deletion in the saposin A domain resulting in removal of a conserved valine (Spiegel et al., 2005).

7.3 Polymorphisms

There is a relatively broad range of galactosylceramidase activities in the "normal" population and among the obligate heterozygotes. This makes enzyme-based carrier testing in the general population nearly impossible. Also, there are normal individuals, including obligate heterozygotes, who have galactosylceramidase activity sufficiently low for diagnosis for Krabbe disease but who are clinically normal. These phenomena can be explained at least partially by the presence of polymorphisms in the galactosylceramidase gene that result in amino acid substitutions. The C502T polymorphism is widespread but the A865G polymorphism has been reported only among Japanese. These polymorphisms generate galactosylceramidase proteins that are less active than the most common type. It has been observed by several groups that these polymorphisms occur on the same alleles as disease-causing mutations at a higher than expected frequency. Some "disease-causing" mutations may in fact be deleterious only when the polymorphism is present on the same allele. Polymorphisms may also play a role in the development of clinical disease when inherited either in multiple copies, on the same allele with another mutation, or together with a known disease-causing mutation on the other chromosome.

8 TREATMENT

Only supportive care is available for patients with the classical infantile form of the disease, who are diagnosed too late for hematopoietic stem cell transplantation. For patients with either late-onset, slowly progressive disease or infantile disease prior to the onset of neurological manifestations, clinical improvements can occur by bone marrow transplantation (Krivit et al., 1998). Experimental studies using hematopoietic stem cells and viral vectors to transduce transplantable cells are currently under way with animal models (Krivit, 2004; De Gasperi et al., 2004; Yagi et al., 2005). In high-risk populations of Krabbe disease, prevention by prenatal diagnosis and genetic counseling is realistic. If there is no family history, then the infants are not diagnosed until months after the onset of clinical symptoms. Then, it is too late for hematopoietic stem cell transplantation if "cure" is the goal. Development of specific neonatal screening program for GLD would be important to prevent misdiagnosis and to provide optimal care and treatment (Li et al., 2004).

9 ANIMAL MODELS

Genetic galactosylceramidase deficiency (Krabbe disease) occurs naturally in the mouse (twitcher), sheep, dogs (West Highland white terriers and Cairn terriers, blue-tick hound, and beagles), and Rhesus monkeys. Clinical and pathological features of these models are similar to those of the human disease. Galactosylceramidase cDNA was cloned and disease-causing mutations have been identified in the mouse (Sakai et al., 1996), West Highland and Cairn terriers (Victoria, Rafi, and Wenger, 1996), and the Rhesus monkey (Luzi et al., 1997). For more details about animal models due to galactosylceramidase defects, readers are referred to elsewhere (Suzuki, 2004b). As already referred to above, a mouse model of saposin A deficiency was generated experimentally by the gene targeting technology (Matsuda et al., 2001a,b;Yagi et al., 2004).

REFERENCES

Andrews, J. M. and Menkes, J. H. 1970, Ultrastructure of experimentally produced globoid cells in the rat. *Exp Neurol* **29,** 483–493.

Austin, J. H. 1963, Studies in globoid (Krabbe) leukodystrophy II. Controlled thin-layer chromatographic stuides of globoid body fractions in seven patients. *J Neurochem* **10,** 921–930.

Austin, J. H. and Lehfeldt, D. 1965, Studies in globoid (Krabbe) leukodystrophy. III. Significance of experimentally-produced globoid-like elements in rat white matter and spleen. *J Neuropathol Exp Neurol* **24,** 265–289.

Barone, R., Brühl, K., Stoeter, P., Fiumara, A., Pavone, L., and Beck, M. 1996, Clinical and neuroradiological findings in classic infantile and late-onset globoid-cell leukodystrophy (Krabbe disease). *Am J Med Genet* **63,** 209–217.

Cannizzaro, L. A., Chen, Y. Q., Rafi, M. A., and Wenger, D. A. 1994, Regional mapping of the human galactocerebrosidase gene (GALC) to 14q31 by in situ hybridization. *Cytogenet Cell Genet* **66,** 244–245.

Chen, Y. Q., Rafi, M. A., De Gala, G., and Wenger, D. A. 1993, Cloning and expression of cDNA encoding human galactocerebrosidase, the enzyme deficient in globoid cell leukodystrophy. *Hum Mol Genet* **2,** 1841–1845.

Costantino-Ceccarini, E. and Morell, P. 1972, Biosynthesis of brain sphingolipids and myelin accumulation in the mouse. *Lipids* **7,** 656–659.

De Gasperi, R., Friedrich, V. L., Perez, G. M., Senturk, E., Wen, P. H., Kelley, K., Elder, G. A., and Gama Sosa, M. A. 2004, Transgenic rescue of Krabbe disease in the twitcher mouse. *Gene Ther* **11,** 1188–1194.

Fiumara, A., Pavone, L., Siciliano, L., Tine, A., Parano, E., and Innico, G. 1990, Late-onset globoid cell leukodystrophy. Report on seven new patients. *Childs Nerv Syst* **6,** 194–197.

Hagberg, B., Sourander, P., and Svennerholm, L. 1963, Diagnosis of Krabbe's infantile leukodystrophy. *J Neurosurgery Psychiatry* **26,** 195–198.

Hannun, Y. A. and Bell, R. M. 1987, Lysosphingolipids inhibit protein kinase C: Implications for the sphingolipidoses. *Science* **235,** 670–674.

Hers, H. G. 1966, Inborn lysosomal disease. *Gastroenterology* **48,** 625–633.

Ida, H., Umezawa, F., Kasai, E., Eto, Y., and Maekawa, K. 1982, An accumulation of galactocerebroside in kidney from mouse globoid cell leukodystrophy (twitcher). *Biochem Biophys Res Commun* **109,** 634–638.

Igisu, H. and Suzuki, K. 1984, Progressive accumulation of toxic metabolite in a genetic leukodystrophy. *Science* **224,** 753–755.

Igisu, H., Takahashi, H., Suzuki, K., and Suzuki, K. 1983, Abnormal accumulation of galactosylceramide in the kidney of twitcher mouse. *Biochem Biophys Res Commun* **110,** 940–944.

Krabbe, K. 1916, A new familial, infantile form of diffuse brain sclerosis. *Brain* **39,** 74–114.

Krivit, W. 2004, Allogeneic stem cell transplantation for the treatment of lysosomal and peroxisomal metabolic diseases. *Springer Semin Immunopathol* **26,** 119–132.

Krivit, W., Shapiro, E. G., Peters, C., Wagner, J. E., Cornu, G., Kurtzberg, J., Wenger, D. A., Kolodny, E. H., Vanier, M. T., Loes, D. J., Dusenbery, K., and Lockman, L. A. 1998, Hematopoietic stem-cell transplantation in globoid-cell leukodystrophy. *N Engl J Med* **338,** 1119–1126.

Li, Y., Brockmann, K., Turecek, F., Scott, C. R., and Gelb, M. H. 2004, Tandem mass spectrometry for the direct assay of enzymes in dried blood spots: Application to newborn screening for Krabbe disease. *Clin Chem* **50,** 638–640.

Luzi, P., Rafi, M. A., and Wenger, D. A. 1995, Structure and organization of the human galactocerebrosidase (GALC) gene. *Genomics* **26,** 407–409.

Luzi, P., Rafi, M. A., Victoria, T., Baskin, G. B., and Wenger, D. A. 1997, Characterization of the rhesus monkey galactocerebrosidase (GALC) cDNA and gene and identification of the mutation causing globoid cell leukodystrophy (Krabbe disease) in this primate. *Genomics* **42,** 319–324.

Matsuda, J., Vanier, M. T., Saito, Y., Suzuki, K., and Suzuki, K. 2001b, Dramatic phenoltypic improvement during pregnancy in a genetic leukodystrophy: Estrogen appears to be a critical factor. *Hum Mol Genet* **10,** 2709–2715.

Matsuda, J., Vanier, M. T., Saito, Y., Tohyama, J., Suzuki, K., and Suzuki, K. 2001a, A mutation in the saposin A domain of the sphingolipid activator protein (prosaposin) gene causes a late-onset, slowly progressive form of globoid cell leukodystrophy in the mouse. *Hum Mol Genet* **10,** 1191–1199.

Miyatake, T. and Suzuki, K. 1972, Globoid cell leukodystrophy: Additional deficiency of psychosine galactosidase. *Biochem Biophys Res Commun* **48,** 538–543.

Oehlmann, R., Zlotogora, J., Wenger, D. A., and Knowlton, R. G. 1993, Localization of the Krabbe disease gene (GALC) on chromosome 14 by multipoint linkage analysis. *Am J Hum Genet* **53,** 1250–1255.

Percy, A. K., Odrezin, G. T., Knowles, P. D., Rouah, E., and Armstrong, D. D. 1994, Globoid cell leukodystrophy: Comparison of neuropathology with magnetic resonance imaging. *Acta Neuropathol (Berl)* **88,** 26–32.

Sakai, N., Inui, K., Fujii, N., Fukushima, H., Nishimoto, J., Yanagihara, I., Isegawa, Y., Iwamatsu, A., and Okada, S. 1994, Krabbe disease: Isolation and characterization of a full-length cDNA for human galactocerebrosidase. *Biochem Biophys Res Commun* **198,** 485–491.

Sakai, N., Inui, K., Tatsumi, N., Fukushima, H., Nishigaki, T., Taniike, M., Nishimoto, J., Tsukamoto, H., Yanagihara, I., Ozone, K., and Okada, S. 1996, Molecular cloning and expression of cDNA for murine galactocerebrosidase and mutation analysis of the twitcher mouse, a model of Krabbe's disease. *J Neurochem* **66,** 1118–1124.

Sandhoff, K., Kolter, T., and Harzer, K. 2001, Sphingolipid activator proteins, in *The Metabolic and Molecular Basis of Inherited Disease* C. R. Scriver, A. L. Beaudet,

W. S. Sly, D. Valle, B. Childs, and B. Vogelstein, Eds. McGraw-Hill, New York.pp. 3371–3388.

Spiegel, R., Bach, G., Sury, V., Mengistu, G., Meidan, B., Shalev, S., Shneor, Y., Mandel, H., and Zeigler, M. 2005, A mutation in the saposin A coding region of the prosaposin gene in an infant presenting as Krabbe disease: First report of saposin A deficiency in humans. *Mol Genet Metab* **84,** 160–166.

Stahl, N., Jurevics, H., Morell, P., Suzuki, K., and Popko, B. 1994, Isolation, characterization, and expression of cDNA clones that encode rat UDP-galactose:ceramide galactosyltransferase. *J Neurosci Res* **38,** 234–242.

Suzuki, K. 1970, Ultrastructural study of experimental globoid cells. *Lab Invest* **23,** 612–619.

Suzuki, K. 1971, Renal cerebrosides in globoid cell leukodystrophy (Krabbe's disease). *Lipids* **6,** 433–436.

Suzuki, K. 1998, Twenty five years of the psychosine hypothesis: A personal perspective of its history and present status. *Neurochem Res* **23,** 251–259.

Suzuki, K. 2003, Evolving perspective of the pathogenesis of globoid cell leukodystrophy (Krabbe disease). *Proceedings of the Japan Academy, series B* **79,** 1–8.

Suzuki, K. 2004a, Krabbe disease, in *Myelin Biology and Disorders* R. A. Lazzarini, J. Griffin, H. Lassman, K.-A. Nave, R. Miller, and B. Trapp, Eds. Elsevier/Academic Press, San Diego, CA, pp. 841–850.

Suzuki, K. 2004b, Mouse models of globoid cell leukodystrophy, in *Myelin Biology and Disorder* R. A. Lazzarini, J. Griffin, H. Lassman, K.-A. Nave, R. Miller, and B. Trapp, Eds. Elsevier/Academic Press, San Diego, CA, pp. 1101–1113.

Suzuki, K. and Suzuki, K. 2002, Lysosomal disease, in *Greenfield's Neuropathology* D. I. Graham and P. L. Lantos, Eds. Edward Arnold, London, pp. 653–735.

Suzuki, K. and Suzuki, Y. 1970, Globoid cell leucodystrophy (Krabbe's disease): Deficiency of galactocerebroside β-galactosidase. *Proc Natl Acad Sci , USA* **66,** 302–309.

Suzuki, K., Schneider, E. L., and Epstein, C. J. 1971, *In utero* diagnosis of globoid cell leukodystrophy. *Biochem Biophys Res Commun* **45,** 1363–1366.

Suzuki, K., Tanaka, H., and Suzuki, K. 1976, Studies on the pathogenesis of Krabbe's leukodystrophy: Cellular reaction of the brain to exogenous galactosylsphingosine, monogalactosyl diglyceride and lactosylceramide, in *Current Trends in Sphingolipidoses and Allied Disorders* B. W. Volk and L. Schneck, Eds. Plenum Press, New York, pp. 99–113.

Suzuki, Y. and Suzuki, K. 1971, Krabbe's globoid cell leukodystrophy: Deficiency of galactocerebrosidase in serum, leukocytes, and fibroblasts. *Science* **171,** 73–75.

Svennerholm, L., Vanier, M.-T., and Månsson, J.-E. 1980, Krabbe disease: A galactosylsphingosine (psychosine) lipidosis. *J Lipid Res* **21,** 53–64.

Taketomi, T. and Nishimura, K. 1964, Physiological activity of psychosine. *Jap J Exp Med* **34,** 255–265.

Tohyama, J., Matsuda, J., and Suzuki, K. 2001, Psychosine is as potent an inducer of cell death as C6-ceramide in cultured fibroblasts and in MOCH-1 cells. *Neurochem Res* **26,** 667–671.

Vanier, M.-T. and Svennerholm, L. 1975, Chemical pathology of Krabbe's disease. III. Ceramide hexosides and gangliosides of brain. *Acta Paediat Scand* **64,** 641–648.

Vanier, M.-T. and Svennerholm, L. 1976, Chemical pathology of Krabbe disease: The occurrence of psychosine and other neutral sphingoglycolipids, in *Current Trends in*

Sphingolipidoses and Allied Disorders, B. W. Volk and L. Schneck, Eds. Plenum Press, New York, pp. 115–126.

Victoria, T., Rafi, M. A., and Wenger, D. A. 1996, Cloning of the canine GALC cDNA and identification of the mutation causing globoid cell leukodystrophy in west highland white and cairn terriers. *Genomics* **33,** 457–462.

Wenger, D. A., Suzuki, K., Suzuki, Y., and Suzuki, K. 2001, Galactosylceramide lipidosis: Globoid cell leukodystrophy (Krabbe disease), in *The Metabolic and Molecular Basis of Inherited Disease* C. R. Scriver, A. L. Beaudet, W. S. Sly, D. Valle, B. Childs, and B. Vogelstein, Eds. McGraw-Hill, New York, pp. 3669–3694.

Yagi, T., Matsuda, J., Takikita, S., Mohri, I., Suzuki, K., and Suzuki, K. 2004, Comparative clinico-pathological study of saposin-A-deficient (SAP-A$^{-/-}$) and Twitcher mice. *J Neuropathol Exp Neurol* **63,** 721–734.

Yagi, T., Matsuda, J., Tominaga, K. Suzuki, K., and Suzuki, K. 2005, Hematopoietic cell transplantation ameliorates clinical phenotype and progression of the CNS pathology in the mouse model of late onset Krabbe disease. *J Neuropathol Exp Neurol*, **64,** 565–575.

Yunis, E. J. and Lee, R. E. 1970, Tubules of globoid cell leukodystrophy: A right-handed helix. *Science* **169,** 64–66.

METACHROMATIC LEUKODYSTROPHY

Volkmar Gieselmann

Metachromatic leukodystropy is a sphingolipid storage disease caused by mutations in the arylsulfatase A gene. Arylsulfatase A is a lysosomal enzyme involved in the degradation of various sulfated glycolipids. Its major substrate is 3-O-sulfo-galactosylceramide (sulfatide). Arylsulfatase A deficiency leads to sulfatide storage most importantly in oligodendrocytes and Schwann cells of the nervous system. The pathological hallmark of the disease is a progressive demyelination, which results in multiple, finally lethal neurologic symptoms. This chapter summarizes clinical, genetic, diagnostic, and biochemical aspects of this disease.

1 CLINICAL ASPECTS OF METACHROMATIC LEUKODYSTROPHY

1.1 Symptoms

Symptoms of metachromatic leukodystrophy may start at almost any age. In the typical, most frequent form of disease, the symptoms become apparent in between the age of 6 months to 4 years, with most patients becoming symptomatic between 18 to 24 months of age. This type is designated as late infantile type of metachromatic leukodystrophy. Children lose previously aquired capabilities. They develop gait disturbances, which initially are due to muscular hypotonia, which later progresses into a spastic tetraparesis. In the course of the disease ataxia, optic atrophy, nystagmus, epileptic seizures, and dementia develop. These numerous neurological symptoms are caused by progressive demyelination and patients frequently die in a decerebrated state within a few years after the onset (for details see von Figura, Gieselmann, and Jaeken, 2001).

Juvenile metachromatic leukodystrophy starts in between the age of 4–16 years and adults may develop first symptoms even beyond the age of 60 (Bosch and Hart, 1978; Duyff and Weinstein, 1996). These late-onset forms – juvenile and adult – frequently do not present with neurological symptoms but initially show deterioration in school performance or develop psychiatric symptoms. Frequently, late-onset patients are diagnosed as having attention deficit disorders or being schizophrenic (Hyde, Zeigler, and Weinberger, 1992; Fukutani et al., 1999; Halsall et al., 1999). In former times metachromatic leukodystrophy was only considered when the neurologic symptoms became slowly apparent in the course of the disease. MRI has been a turning point in the diagnosis of leukodystrophies and today almost always leads to a rapid proper diagnosis.

Institute for Physiological Chemistry, Universitat Bonn, Bonn, Germany. E-mail: gieselmann@institut.physiochem.uni-bonn.de

Adults usually present with a gradual decline of intellectual capabilities, emotional instablity, memory deficits, and behavioural abnormalities. Later symptoms resemble those of the late-infantile patients, but progression is usually much slower. Metachromatic leukodystrophy affects the central as well as the peripheral nervous system. Therefore a reduction in nerve conduction velocity is an important diagnostic criterion. A number of late-onset cases were described with isolated affection of the peripheral nervous system (Felice et al., 2000; Comabella et al., 2001; Coulter Mackie et al., 2002) at the time when the patients were admitted. Therefore metachromatic leukodystrophy must also be considered in individuals presenting with a polyneuropathy only, even if MRI appears normal (Felice et al., 2000). In contrast, in other patients the peripheral nervous system may be unaffected, but central nervous system involvement dominates the clinical picture (Cengiz et al., 2002; Marcao et al., 2004).

Whereas the clinical course of late infantile metachromatic leukodystrophy seems to be rather uniform, variations are frequent in the late-onset forms. Substantial variations can also be found among siblings, demonstrating that other genetic or epigenetic factors have a strong impact on the course of the disease (Clarke, Skomorowski, and Chang, 1989; Arbour et al., 2000). None of these modifying factors has so far been identified. The clinical variability makes correlations of particular phenotypes with mutations difficult. Such correlations are only significant if a number of individuals with identical genotype and similar clinical course have been identified. Intrafamilial variability also renders the interpretation of results of bone marrow transplantations difficult, when treated and untreated siblings are compared (see Section 5).

Sulfatide storage is not restricted to the nervous system but is also found in bile ducts, gall bladder, and kidney. Some patients also display cholecystitis (Clarke, Skomorowski, and Chang, 1989), which is related to the inability of the gallbladder to contract because of involvement of the autonomic nervous system. Although storage is also severe in the kidney, renal problems do not occur in metachromatic leukodystrophy patients.

1.2 CT, MR, and MR Spectroscopy

The development of computer tomography (CT) and in particular magnetic resonance imaging (MRI) had great impact on the diagnostics of metachromatic leukodystrophy and of leukodystrophies in general. MRI especially greatly facilitates the diagnosis and differential diagnosis of these diseases.

CT of late infantile metachromatic leukodystrophy patients shows low-density areas in the cerebral white matter, related to the areas affected by demyelination. The periventricular regions are affected initially. The low density areas expand symmetrically, in particular into the parietal and occipital parts of the hemispheres in the further course of the disease. In adult patients the frontal lobes tend to be affected most prominently. Progredient demyelination is accompanied by cerebral atrophy with widening of the lateral ventricles and to a lesser degree the subarachnoid spaces. The atrophy is particularly prominent in adult patients (Schipper and Seidel, 1984).

MRI of late infantile patients shows symmetric, confluent high signal intensity regions in the periventricular region, and the centrum ovale on T2 weighted images (Faerber, Melvin, and Smergel, 1999; Kim et al., 1997). Within the abnormal white matter radiating low signal intensity stripes are seen, which are characteristic for metachromatic leukodystrophy and Krabbe disease. At microscopic examination, this radiating pattern is explained by the accumulation of products of myelin breakdown in perivascular macrophages and by sparing of some myelin sheaths (van der Voorn et al., 2004). Posterior occipital regions are predominantly affected and the demyelination progresses in a dorsofrontal direction. In contrast, in later-onset patients the frontal white matter is usually most prominently affected. The corpus callosum is affected already in an early stage of the disease. Depending on the stage of disease in most patients demyelination can also be seen in the posterior but not the anterior limbs of the internal capsule, descending pyramidal tracts, and cerebellar white matter. The subcortical white matter is spared until a late stage of the disease. As has also been reported in pathological descryptions, the subcortical U fibers are spared at least in early stages of disease (Kim et al., 1997) There is no contrast enhancement of the abnormal white matter.

Proton MR spectroscopy of metachromatic leukodystrophy patients shows an elevation of choline within the abnormal white matter, related to the enhanced membrane turnover associated with demyelination. A reduction of N-acetylaspartate is seen in white and grey matter, but tends to be most severe in the abnormal white matter. The reduction is related to axonal and neuronal damage and loss. Myo-inositol is increased, related to gliosis (Kruse et al., 1993).

1.3 Neuropsychological Examinations

The peripheral neuropathy results in reduced nerve conduction velocity, prolonged peak latency, and reduced amplitude of compound muscle action potentials. Although reduced nerve conduction velocity is considered as a characteristic feature of the disease, it should be kept in mind that some patients lack signs of peripheral polyneuropathy, although within the central nervous system the disease is already advanced (Gallo et al., 2004; Marcao et al., 2004).

A delay in the interpeak latencies of brain stem auditory evoked potentials reflects delayed conduction in the brainstem (Brown et al., 1981). Alterations are usually seen in late infantile patients, but depending on the stage of disease may be absent in juvenile and adult patients (Clark , Miller, and Vidgoff, 1979; Wulff and Trojaborg, 1985).

1.4 Pathology

Hallmarks of metachromatic leukodystrophy are demyelination and metachromatic sulfatide inclusions. In the nervous system inclusions can be found in Schwann cells, oligodendrocytes, and also neurons. There are numerous macrophages, displaying larger 15–20 μm sulfatide deposits. Demyelination is widespread, also involving brain stem, spinal cord, and peripheral nerves. Surprisingly, the cortical U fibers are frequently spared, which can also be seen in MRI (see Section 1.2; for further details and references see von Figura, Gieselmann, and Jaeken, 2001). An astrogliosis accompanies the disease.

2 BIOCHEMISTRY AND FUNCTION OF SULFATIDE

Sulfatide is found in the plasma membrane of various cell types such as bile duct, respiratory, gastric, and distal renal tubule epithelia. The highest amount of this lipid, however, is found in myelin of the nervous system, where it constitutes 4–5% of the total myelin lipids. It is synthesized in a two-step process: an endoplasmic reticulum located galactosyltransferase transfers a galactose residue from UDP-galactose to ceramide and subsequently a Golgi located 3'-phosphoadenosine 5'-phosphosulfate dependent sulfotransferase sulfonylates this galactose residue in position 3 (Honke et al., 2002). Mice lacking the sulfotransferase are not able to synthesize sulfatide. These sulfatide-deficient mice are born apparently healthy and surprisingly the overall appearance of their myelin sheaths is normal. Ultrastructural analysis, however, revealed a disruption of the paranodal axoglial junctions and a disorganized termination of the lateral loop of the node of Ranvier (Honke et al., 2002). Immunohistochemical analysis showed an altered distribution of voltage gated Na^+ and K^+ channels, which are usually clustered in the nodal or juxta-paranodal region, respectively.

In sulfatide-deficient mice the number of clusters of these channels was reduced, because most Na^+ channels were dispersed along the axolemma and K^+ channels were abnormally localized in the paranodal region (Ishibashi et al., 2002). Thus, myelin sulfatide seems to play an important role in the organization of paranodal axoglial junctions and the proper localization of Na^+ and K^+ channels at the axonal surface. The altered distribution may explain the neurologic symptoms such as hindlimb weakness, which the mice develop at the age of six weeks.

Sulfatide also plays an important role in oligodendrocyte differentiation. Sulfatide is expressed in oligodendrocyte precursors at a time point when these cells start to enter the differentiation phase. Exposure of such cells to antisulfatide antibodies leads to a reversible arrest of oligodendrocyte differentiation. When oligodendrocytes of the sulfatide-deficient mouse model were investigated, the number of terminally differentiated oligodendrocytes in vivo and in vitro was two- to threefold higher than in wild-type mice. Thus, sulfatide is an important negative regulator of oligodendrocyte differentiation (Hirahara et al., 2004).

Another sulfated glycolipid that is degraded by arylsulfatase A is 3-O-monogalactosylalkylacylglycerol or briefly called seminolipid. This lipid contains the same carbohydrate structure as sulfatide and is sulfonylated by the same sulfotransferase. Therefore it can also not be synthesized in sulfotransferase-deficient mice. As the name of this lipid already suggests, it plays an important role in spermatogenesis. Mice lacking this lipid show a block of spermatogenesis and are sterile (Honke et al., 2002).

Although sulfatide is the predominant substrate of arylsulfatase A it must be emphasized that the enzyme is involved in the degradation of various sulfated glycolipids. In the kidneys of arylsulfatase A-deficient mice an elevation of sulfatide, lactosylceramide, and gangliotetraosylceramide bis-sulfate was found (Sandhoff et al., 2002). In addition, seminolipid accumulates in testis. This proves that arylsulfatase A is involved in the degradation of all these sulfolipids. Whether lipids other than sulfatide contribute to pathogenesis is currently unclear.

3 BIOCHEMISTRY OF ARYLSULPHATASE A

Arylsulfatase A is synthesized at the rough endoplasmic reticulum as a precursor of 507 amino acids (Stein et al., 1989). Upon translocation into the lumen of the endoplasmic reticulum the signal peptide is cleaved. The amino acid sequence of arylsulfatase A harbours three potential N-glycosylation sites all of which are used (Sommerlade et al., 1994). The glycosylated enzyme has a molecular weight of ~62 kDa and it forms noncovalently linked dimers in the endoplasmic reticulum. As do all other sulfatases, arylsulfatase A undergoes a unique posttranslational modification within the endoplasmic reticulum. This modification converts a cysteine residue located in the active center into a formylglycine residue, which is essential for the enzymatic catalysis (Waldow et al., 1999). If the sulfatase-modifying enzyme is defective all sulfatases are nonfunctional, which results in multiple sulfatase deficiency (Dierks et al., 2003). This rare disease combines aspects of various disorders due to deficiencies of single sulfatases, among those, metachromatic leukodystrophy.

In the Golgi apparatus the first and the third N-linked oligosaccharide side chain aquire mannose-6-phosphate residues, whereas the second site is only weakly phosphorylated (Sommerlade et al., 1994). From the Golgi mannose-6-phosphate receptors mediate the vesicular transport of the enzyme to the lysosomes. Upon arrival in the lysosome the acidic pH favors the formation of arylsulfatase A octamers (von Bülow et al., 2002). In contrast to many other lysosomal enzymes arylsulfatase A does usually not undergo further proteolytic processing. In some cells, however, processing into a 54 kDa form has been described (Fujii et al., 1992). Arylsulfatase A alone is unable to degrade sulfatide. It needs a small activator protein, called saposin B, which is generated proteolytically from a larger precursor (Sandhoff, Kolter, and Harzer, 2001).

Saposin B binds sulfatide and presents it in a one-to-one complex to the enzyme. Then arylsulfatase A catalyzes the desulfation of the galactose moiety, which is the first step in the degradation pathway of sulfatide. Only after desulfation can the galactose be removed and subsequently the ceramide is cleaved. Thus, if arylsulfatase A is deficient sulfatide is stored as such, because no other breakdown products can be generated. Saposin B is neither specific for sulfatide nor arylsulfatase A. It can also bind, for example, globotriaosylceramide for the enzymatic degradation by α-galactosidase. Deficiencies of saposin B have also been described and cause disorders which resemble juvenile metachromatic leukodystrophy (Sandhoff, Kolter, and Harzer, 2001).

Arylsulfatase A has been crystallized and its three-dimensional structure determined. It consists of a central ß-pleated sheet, which is decorated on both sites by various helices (Lukatela et al., 1998). Although no sequence homologies were apparent, the three-dimensional structure of arylsulfatase A is very similar to that of bacterial alkaline phosphatase. This structural similarity reflects the similarity of the hydrolytic cleavage of sulfate and phosphoester bonds.

4 GENETICS OF ARYLSULPHATASE A ALLELES

4.1 Inheritance, Incidence, Important Polymorphisms and Mutations

Metachromatic leukodystrophy is inherited in an autosomal recessive mode. Early determinations of incidence suggested a frequency of 1 in 40,000 newborns (Gustavson and Hagberg, 1971). This study, however, was based on the examination of Swedish cases only and may thus have not been representative. More recent determinations focusing on German patients allowed the calculation of an incidence of 1 in 160,000 in Germany (Heim et al., 1997). The arylsulfatase A gene is located on chromosome 22q13. It encompasses only about 3 kb and consists of 8 exons. The small size of the gene has facilitated the detection of mutations. So far about 100 different mutations have been described. The majority are missense mutations; a few represent splice site mutations, nonsense mutations, and deletions (von Figura, Gieselmann, and Jaeken, 2001). In addition, some arylsulfatase A alleles have been described, which bear two deleterious mutations on the same allele (Kappler et al., 1994).

Only three of the defective alleles are frequent (Polten et al., 1991, Lugowska et al., 2002). Most mutations have only been described in single or few families. Among the frequent alleles is a splice donor site mutation at the border of exon/intron 2 (459+1G>A) and two missense mutations causing a P426L or I179S substitution, respectively. The frequencies of these alleles among Caucasian patients of different nationalities were investigated in various studies. For the 459+1G>A splice site allele frequencies among patients vary from 14 to 43%. The P426L allelle represents ~25% and the I179S allele 11% of all metachromatic leukodystrophy alleles examined (Berger et al., 1997; Polten et al., 1991, Regis et al., 1996; Barth, Fensom, and Harris, 1993) The frequency may vary substantially depending on the nationality of the Caucasian patients. These alleles were not found in a sample of Japanese patients. Among Japanese a missense mutation causing a G99D substitution is frequent and accounts for about 50% of all metachromatic leukodystrophy alleles (Kurosawa, Ida, and Eto, 1998)

Using various intronic polymorphism it has been shown that the 459+1G>A splice site allele and the P426L allele have a fixed haplotype being in complete linkage disequilibrium (Coulter-Mackie and Gagnier, 1997; Zlotogora et al., 1994a; Gort, Coll, and Chabas, 2000).

When genotypes of patients bearing these frequent alleles in homo- or heterozygosity were correlated with their clinical phenoytpe a simple genotype–phenotype correlation became apparent (Polten et al., 1991). Patients homozygous for the 459+1G>A splice site allele always suffer from the most severe late infantile form of metachromatic leukodystrophy. Patients who are heterozygous for the 459+1G>A splice site allele and the P426L mutation almost always suffer from the juvenile form of disease. The majority of patients homozygous for the P426L allele display the adult type of metachromatic leukodystrophy. The clinical variations among patients with identical phenotype, however, are substantial, so that the genotype does not allow predictions about the individual course of disease.

The genotype–phenotype correlation can be explained by different amounts of residual enzyme activity associated with the 459+1G>A and P426L alleles. Whereas no detectable arylsulfatase A mRNA and thus no arylsulfatase A protein can be generated from the 459+1G>A allele it represents a null mutation with complete lack of enzyme activity. Therefore homozygosity leads to the severe late infantile form of metachromatic leukodystrophy. The P426L allele is still associated with low amounts (<5%) of residual enzyme activity (for explanation see Section 4.3). Therefore one P426L allele mitigates the course to juvenile metachromatic leukodystrophy whereas two copies frequently allow for the mildest adult phenotype of disease (Polten et al., 1991; Berger et al., 1997).

The third of the frequent alleles, I179S, is also associated with residual enzyme activity (Fluharty et al., 1991) and has interestingly so far only been found in heterozygosity with null alleles. In spite of its frequency so far no patient homozygous for this allele has been described. Many of the patients carrying a single copy of this allele have late juvenile or adult metachromatic leukodystrophy (Tylki-Szymanska et al., 1996), suggesting that the I179S allele is associated with rather high amounts of enzyme activity and one may speculate that homozygous individuals may be healthy and remain undetected.

Adult metachromatic leukodystrophy may start with neurological symptoms but frequently psychiatric symptoms prevail in the adults. The correlation of clinical data with genetic data has recently revealed another interesting genotype–phenotype correlation in metachromatic leukodystrophy. Whereas homozygotes for the P426L allele mostly develop neurological symptoms initially, the I179S heterozygotes always start with psychiatric symptoms (Tylki-Szymanska et al., 1996; Baumann et al., 2002; Gomez Lira et al., 1998).

4.2 Arylsulphatase A Pseudo-Deficiency

About 1–2% of the normal population have a substantial arylsulfatase A deficiency. These individuals have only about 10% of normal enzyme activity (Shen et al., 1993), but these individuals do not develop disease. Obviously this low amount of enzyme is sufficient to sustain a normal metabolism. The term pseudo-deficiency is misleading, inasmuch as it suggests an only apparent deficiency, however, the deficiency is real. Arylsulfatase A pseudo-deficieny is due to homozygosity for an arylsulfatase A allele, which is characterized by two polymorphisms (Gieselmann et al., 1989): one polymerphism (N350S) causes the loss of one of the three N-glycosylation sites of the enzyme. Therefore pseudo-deficient arylsulfatase A has only two oligosaccharide side chains, whereas normal arylsulfatase A has three (see above, Section 3.). The other polymorphism leads to the loss of a polyadenylation signal (2723A>G), which is required for proper termination of synthesis of the smallest 2.1 kb arylsulfatase A mRNA.

There have been conflicting results about the relative contribution of these polymerphisms to arylsulfatase A deficiency. An initial study demonstrated that the loss of the polyadenylation signal causes the loss of 90% of arylsulfatase A poly(A)+ mRNA. This correlated well with the reduction of arylsulfatase A synthesis and enzyme activity each by 90% (Gieselmann et al., 1989). Other studies have shown that the loss of the N-glycosylation site may cause reduced stability and missorting of pseudo-deficiency arylsulfatase A, and attributed the reduction of enzyme activity mainly to the loss of the N-glycosylation site (Ameen et al., 1990; Harvey, Carey, and Morris, 1998). The matter was resolved when leukocyte arylsulfatase A activity of various individuals with different arylsulfatase A genotypes was compared. A number of individuals were identified who were heterozygous for an arylsulfatase A allele, which only displayed the N-glycosylation site mutation, but which had retained the polyadenylation signal. In some studies normal arylsulfatase A activities were found (Leistner, Young, and Winchester, 1995; Barth et al., 1994); in others activity was reduced by about 50% (Shen et al., 1993). This proves that in vivo predominantly the polyadenylation signal contributes to arylsulfatase A activity reduction and that the loss of the N-glycosylation site also has an effect which is, however, minor.

The frequency of the pseudo-deficiency allele has been determined in various populations and ranges from 6% in Denmark (Salamon, Christensen, and Schwartz, 1994) to 23% in Yemenite Jews (Zlotogora et al., 1994b). The average value is 11% when all published studies are included. In contrast, the frequency of the pseudo-deficiency allele in Africans, Chinese, and Asians is very low (Ott, Waye, and Chang 1997; Hwu et al., 1996; Ricketts et al., 1996).

The pseudo-deficiency allele is characterized by two polymorphisms, but alleles carrying these polymorphisms independently have also been described. Thus, the frequency of an allele, which has lost the N-glycosylation site only is similar to that of the pseudo-deficiency allele in Caucasians, but in Africans and Asians this allele shows frequencies of up to 40% (Ott, Waye, and Chang, 1997). Haplotype analysis shows that this N-glycosylation site allele can occur on at least three different haplotypes, one of which is identical to the unique pseudo-deficiency allele haplotype. This indicates that the N-glycosylation site polymorphism has occurred independently on different haplotypes and that this allele is older than the pseudo-deficiency allele (Coulter-Mackie and Gagnier, 1997; Zlotogora et al., 1994b).

Interestingly, alleles that only bear the polyadenylation site polymorphism also exist (Ott, Waye, and Chang, 1997; Ricketts et al., 1996; Gort, Coll, and Chabas, 1999). Surprisingly this allele occurs on a haplotype different from the pseudo-deficiency allele. Because the independent formation of alleles carrying the same polymorphisms is unlikely to be a random process, it has been speculated that heterozygotes may have an evolutionary advantage. So far, however, no reasonable hypothesis explaining a potential advantage has been brought forward.

In the literature the term pseudo-deficiency allele has always been used for the allele described above. However, any allele that expresses low but metabolically sufficient arylsulfatase A activity can be considered a pseudo-deficiency allele. Recently, a family was identified in which three siblings and the father had enzyme activities in the range of MLD patients, but only one of the siblings had developed late infantile metachromatic leukodystrophy. The classical N-glycosylation/polyadenylation site pseudo-deficiency allele was not detected in any of the family members. Further investigations revealed an allele with a polymorphism coding for a A464V substitution. The mode in inheritance, different phenotypes, and residual arylsulfatase A activities of the family members allowed the conclusion that this allele also fulfills the pseudo-deficiency criteria (Berger et al., 1999).

4.3 Biochemical Consequences of Mutations

Although about 100 defective arylsulfatase A alleles were identified so far, only about a dozen of these have been investigated with respect to the biochemical consequences of the encoded amino acid substitutions.

As for many other genetic diseases the most frequent cause of enzyme deficiency in metachromatic leukodystrophy is the misfolding of the enzyme, its retention in the endoplasmic reticulum, and subsequent degradation (e.g., Kafert et al., 1995; Hess et al., 1996). In all of the investigated cases the misfolded enzymes have also lost their enzymatic activity. Homozygosity for such mutations is therefore frequent among severely affected late infantile patients.

The most thoroughly examined mutant arylsulfatase A is encoded by the frequent P426L allele. The P426L substituted arylsulfatase A is active, synthesized in normal amounts, leaves the endoplasmic reticulum, and is correctly sorted to the lysosomes (von Figura et al., 1983). As mentioned above, arylsulfatase A forms dimers in the endoplasmic reticulum and these dimers form octamers at the intralysosomal low pH. Cristallographic and biochemical examination of the P426L arylsulfatase A revealed that the substitution affects the ability of arylsulfatase A to octamerize with the consequence that most of the enzyme remains dimeric in the lysosome (von Bülow et al., 2002). The dimers are rapidly degraded due to enhanced susceptibility to proteolytic cleavage by cathepsin L and have thus a severely reduced half life. This instability is the cause for the enzyme deficiency in patients carrying this allele. Because the enzyme is active, however, a low level of functional intralysosomal arylsulfatase A remains, so that this allele is associated with low amounts of residual enzyme activity. This explains the comparatively mild phenotype of patients homozygous for this allele (see also Section 4.1).

Defective oligomerization as a cause of enzyme deficiency is not unique for the P426L allele, but has also been demonstrated for other missense mutations (Marcao et al., 2003).

Furthermore, the severity of a mutation can be influenced by polymorphisms present on the same allele (Regis et al., 2002). Thus, the P426L allele, which is usually associated with late-onset forms of metachromatic leukodystrophy is occasionally found in late infantile patients (Berger et al., 1997; Regis et al., 2002). Sequencing of the arylsulfatase A gene in one of these patients revealed that in contrast to adult patients the respective mutation in this individual occurred on the background of the pseudo-deficiency allele. Polymorphisms of this allele cause a reduction of enzyme expression (see above, Section 4.2.) by a factor of about 10 compared to the normal allele. This converts the P426L substitution, which is mild on the normal allele, to a severe mutation on the background of the pseudo-deficiency allele. Similar results were reported for an E253K mutation, which is aggravated by the presence of a T391S polymorphism on the same allele (Regis et al., 2002).

5 BIOCHEMICAL AND GENETIC DIAGNOSIS OF METACHROMATIC LEUKODYSTROPHY

5.1 Biochemical Diagnosis

The diagnosis of metachromatic leukodystrophy must be based on the demonstration of an arylsulfatase A deficiency. The enzyme can be determined preferably in leukocytes or cultured fibroblasts. Determination in urine is less reliable due to highly variable enzyme levels. In the routine assay either p-nitrocatechol sulfate or 4-methylumbelliferyl sulfate is used as artifical substrates. Although a number of modifications of this assay were described with the aim to make this procedure specific for arylsulfatase A, it must be kept in mind that these artificial substrates are not specific for arylsulfatase A, but will also be hydrolysed by other sulfatases to a varying extent. Thus, in the routine assays there is always a background of non-arylsulfatase A sulfatase activity. This is the main reason why in the range of low enzyme activities artificial substrates cannot distinguish between an arylsulfatase A deficiency causing metachromatic leukodystrophy and benign arylsulfatase A pseudo-deficiency. Based on enzymatic assays this is only possible when radioactively labeled sulfatide is used as a natural substrate. This can either be used to measure activity in vitro in cell homogenates or can also be applied to cultured fibroblasts to determine their in vivo capability to degrade sulfatide. This assay, however, is laborious and not suited as a routine diagnostic procedure. Its use should be restricted to particular diagnostic problems. Prenatal diagnosis is also possible.

Differentiation of arylsulfatase A deficiencies can in most cases be accomplished by the exclusion of pseudo-deficiency via genetic testing for the respective polymorphisms and determination of sulfatide in urine. Because sulfatide is also massively stored in kidney, lipid-containing epithelial cells appear in urine. The demonstration of sulfatide in urine is therefore usually considered as a proof of metachromatic leukodystrophy. It should be noted, however, that increased sulfatide excretions were also found in individuals who were heterozygous for a pseudo-deficiency and a metachromatic leukodystrophy allele (Lugowska et al., 1997), but the amount of urine sulfatide in these individuals was lower than in patients.

5.2 Diagnostic Problems Caused by Arylsulphatase A Pseudo-Deficiency

Arylsulfatase A pseudo-deficiency is not just of academic interest but its existence causes problems in the diagnosis and in genetic counselling (Rafi et al., 2003; Shen et al., 1993). It is important to realize that an arylsulfatase A deficiency does not prove metachromatic leukodystrophy, because it may be due to homozygosity for the pseudo-deficiency allele.

Inasmuch as arylsulfatase A pseudo-deficiency is frequent in the general population, it is also frequent among patients with neurological symptoms of unknown origin. Thus, although it is very tempting to ascribe neurological symptoms to metachromatic leukodystrophy when a patient is arylsulfatase A deficient, it is important to realize that he may be arylsulfatase A pseudo-deficient and his symptoms unrelated to metachromatic leukodystrophy. Arylsulfatase A pseudo-deficiency cannot be reliably distinguished from a disease causing deficiency using the routine enzymatic assays based on artificial substrate (Rafi et al., 2003). Enzymatically a distinction is only possible when the natural radioactively labelled substrate is used.

Because this substrate is not available commercially its use is restricted to a few specialised laboratories. For this reason arylsulfatase A pseudo-deficiency is usually excluded by a genetic test, detecting the two polymorphisms characteristic for the arylsulfatase A pseudo-deficiency allele (Gieselmann, 1991, Nelson, Carey, and Morris, 1991). Although this genetic test is important in the exclusion of arylsulfatase A pseudo-deficiency it must be emphasized that a number of disease-causing mutations have been found, which occur on the background of the arylsulfatase A pseudo-deficiency allele (e.g., Gieselmann et al., 1991; Regis et al., 1998; Regis et al., 2004). This complicates the diagnosis further. The presence of these polymorphisms does not exclude the existence of further mutations on the same allele and thus cannot exclude metachromatic leukodystrophy definitely. The situation can be very complicated in genetic counselling when metachromatic leukodystrophy alleles occur in heterozygosity with pseudo-deficiency alleles. Space limitations do not allow us to discuss the problem here. The reader is referred to Rafi et al., 2003; Shen et al., 1993; Li, Waye, and Chang, 1992; Francis et al., 1993).

There have been several reports about increased frequencies of the pseudo-deficiency allele among patients with various neurological diseases other than metachromatic leukodystrophy. Whereas some reports claimed associations to certain neurological syndromes (Alessandri, De Vito, and Fornai, 2002), others found none (Hagemann et al., 1995, Francis et al., 1993). Another study examined 16 individuals who had one metachromatic leukodystrophy allele and one pseudo-deficiency allele. These individuals therefore had only about 50% of the activity of pseudo-deficient individuals. None of these displayed neurological abnormalities even at advanced age (Penzien et al., 1993). Indications of micro-organic brain damage in metachromatic leukodystrophy/pseudo-deficiency heterozygotes were found by Tylki-Szymanska et al. (2002) but these did not lead to any progressive overt clinical symptoms. Taken together these studies allow us to conclude that even arylsulfatase activities well below the pseudo-deficiency level imply no major health risk. The matter is still controversial, but so far none of the data provide convincing evidence which could support a role of arylsulfatase A pseudo-deficiency in the pathogenesis of a specific neurological disease.

5.3 Genetic Diagnosis

Because there are so many different mutant alleles, genetic testing for mutations causing metachromatic leukodystrophy has little importance in the verification of the diagnosis. The classical biochemical assays – enzyme activity and urine sulfatide determination – are reliable and superior to genetic testing. Genetic assays may only be helpful in genetic counselling, where it is frequently difficult to determine whether the carrier status of the probands is simply enzyme activity determinations. If a carrier must be identified, the most frequent alleles (see above) can be excluded via PCR/restriction enzyme assays

(Berger et al., 1997). If these frequent alleles are not found, the entire ASA gene of the index case in the respective family should be sequenced if DNA is available, because it is too laborious to specifically search for the numerous known mutations. After mutations are identified in the index case the carrier status of the relatives can be clarified.

6 THERAPY

Treatment of metachromatic leukodystrophy is largely confined to symptomatic therapies of spasticity, pain sensations, and seizures. Over the last 20 years several dozens of patients were treated by bone marrow transplantation. Until some years ago the efficacy of this treatment was debated controversially, but experience, which has accumulated over time, now allows for generally accepted conclusions. Thus, the treatment of patients with the late infantile form of disease is not recommended, because this type of metachromatic leukodystrophy progresses too rapidly (e.g., Malm et al., 1996). The same applies to late-onset patients, who have already developed substantial neurologies symptoms. In late-onset patients, however, who are in the initial possibly presymptomatic phases of disease with bone marrow transplantation progression appears to be decelerated perhaps even arrested (e.g., Kidd et al., 1998; Kapaun et al., 1999). For this group of patients bone marrow transplantation is a therapeutic option.

7 ANIMAL MODELS

7.1 Phenotype of Arylsulphatase A Deficient Mice

Whereas naturally occurring animal models exist for most lysosomal storage diseases, metachromatic leukodystrophy has only been described in humans. This has made histopathological and biochemical studies on the course and development of the disease impossible, because such investigations in humans would rely on repeated brain biopsies. For this reason an arylsulfatase A deficient mouse was generated (Hess et al., 1996). In these mice sulfatide is stored in distal tubules of kidney, bile duct epithelia, glial, and neuronal cells of the nervous system (Wittke et al., 2004). The tissue pattern of storage resembles that in humans.

The mice, however, do not show the widespread demyelination characteristic for metachromatic leukodystrophy patients. Only in old animals mild focal demyelination and degeneration of nerve fibers and cerebellar white matter develop. In addition, mild degenerative alterations may also be seen in the cerebellar white matter in animals older than 18 months. Others have described demyelination (Biffi et al., 2004) and degeneration of hippocampal neurons (Consiglio et al., 2001) already in animals below the age of one year. Both results, however, could not be confirmed by various groups working independently with the same animal model (Hess et al., 1996; Wittke et al., 2004; Mansson, unpublished). Although the histopathological phenotype of the mice is mild, animals develop progressive neurological symptoms in particular in the second year of life. However, symptoms do not limit the lifespan of the animals. Although this animal model does not resemble the severity of human disease it can be considered as a mild early stage of metachromatic leukodystrophy and has been used for pathogenetic and therapeutic studies.

7.2 Pathogenic Studies in Arylsulphatase A Deficient Mice

The molecular pathogenesis of metachromatic leukodystrophy is still unknown. It is unclear why and how sulfatide storage affects Schwann cells and oligodendrocytes.

Glycosphingolipids in general have been shown to play an essential role in the formation of lipid rafts. These rafts have important roles in signal transduction processes and intracellular sorting (Degroote, Wolthoorn, and van Meer, 2004). Many myelin proteins are at least partially associated with lipid rafts. Thus, if the accumulation of the glycolsphingolipid sulfatide may interfere with the amount and/or raft/nonraft distribution of membrane proteins, which are functionally important for myelination, this may have implications for the pathogenesis of metachromatic leukodystrophy. For this reason the sulfatide and galactosylceramide levels were investigated in myelin of aged arylsulfatase A deficient mice. In these animals sulfatide was increased to 140% of normal in whole brain extracts, whereas galactosylceramide was decreased to 85% of controls (Saravanan et al., 2004).

The same alterations were found when isolated myelin was examined. Sulfatide accumulation and decrease of galactosylceramide is therefore not restricted to the lysosomes but is also found in the myelin/plasma membrane of cells. Investigation of most myelin proteins, however, did not show abnormalities with respect to amount and raft localization, with the exception of the myelin protein MAL, which was reduced in arylsulfatase A deficient mice. In sulfatide-storing cells this protein is mislocalized to the lysosomal compartment, which may explain its reduced amount (Saravanan et al., 2004). Interestingly, this protein is known to bind strongly to sulfatide, but its exact function in myelin is still unknown (Frank, 2000). Thus, pathogenetic studies so far yield no explanation for the demyelination occuring in metachromatic leukodystrophy.

As described above (see Section 2), sulfatide plays a role in the organization of paranodal axoglial junctions. Ultrastructural examination, however, showed no abnormalities in the arylsulfatase A deficient mice (Schaeren-Wiemers and Gieselmann, unpublished).

7.3 Therapeutic Studies in Arylsulphatase A Deficient Mice

Arylsulfatase A deficient mice were used to investigate the potential feasibility of various therapeutic approaches. Two studies involved a transplantation of genetically modified hematopoetic stem cells (Matzner et al., 2002; Biffi et al., 2004). The cells were modified with retroviral vectors to overexpress arylsulfatase A. The arylsulfatase A secreted by these cells can be endocytosed via mannose-6-phosphate receptors by deficient cells and cross-correct the metabolic defect. It can, however, not be expected that arylsulfatase A secreted by hematopoetic cells into the blood can pass the blood–brain barrier and reach the glial and neuronal cells, which are the main target of any therapy. It has been shown in other studies that part of the brain microglia is derived from the monocytic lineage of hematopoetic cells. The rationale of the approach to transplant genetically modified hematopoetic cells is that arylsulfatase A overexpressing cells derived from the bone marrow become resident in the brain, differentiate into microglia, and secrete arylsulfatase A, which may be accessible to oligodendrocytes and neurons.

Two studies used this approach, but the results were inconsistent. Matzner et al. (2002) saw improvement of sulfatide storage in liver and kidney in some but not all animals, although high amounts of arylsulfatase A were expressed from the integrated retroviral vector over a period of more than a year after transplantation. Pathologically peripheral nerves improved, but sulfatide storage in brain was unchanged. In contrast, Biffi et al. (2004) claimed correction of the disease using a similar approach, which only differed in the use of lentiviral vectors. This study claimed improvement of demyelination in the cerebellar white matter and peripheral nerves of treated animals. Unfortunately, this report did not provide any biochemical data showing the reduction of sulfatide storage in any organ nor about arylsulfatase A activities in the organs of the treated animals. Thus, a biochemical proof of therapeutic efficacy was not performed. Thus, the feasibility of bone marrow stem cell-based gene therapy for treatment of metachromatic leukodystrophy remains highly questionable.

In vivo gene therapy is an alternative approach. A viral vector conferring the expression of high amounts of arylsulfatase A to the infected cells is directly injected into the brain. Lentiviral vectors encoding human arylsulfatase A cDNA were injected into the hippocampal region of the brain of arylsulfatase A deficient mice (Consiglio et al., 2001). This group claimed a rescue of hippocampal degeneration on the treated mice and a concomitant improvement of hippocampus-related learning abilities. However, hippocampal degeneration was not detected in this mouse model by various other groups. Most surprisingly, from the data provided it can be calculated that the enzyme levels in the brain of the treated mice were less than 1% of wild-type. Such a low level leads certainly to metachromatic leukodystrophy in humans and it remains obscure why such low levels of arylsulfatase A are curative in mice.

8 FUTURE PERSPECTIVES

The ultimate goal of research in the field of metachromatic leukodystrophy is the curative treatment of this disease. The development of rationale therapies requires an eventually complete revelation of disease pathogenesis. Understanding the pathogenic mechanisms may yield new therapeutic targets that have not been considered so far. Such studies can be done with cell cultures of sulfatide-storing glial or neuronal cells, which can be derived from the mouse model. Depending on the results their in vivo significance may also be tested in the animal model. Even without a complete understanding of pathogenesis enzyme replacement and gene therapy based trials in the mouse model should proceed, because it cannot be predicted which of the approaches will be the first to yield benefits for the patients.

ACKNOWLEDGEMENTS

I thank Dr. Marjo van der Knaap, Amsterdam, and Dr. Alfried Kohlschütter, Hamburg, for helpful comments on this chapter

REFERENCES

Alessandri MG, De Vito G, Fornai F, 2002, Increased prevalence of pervasive developmental disorders in children with slight arylsulfatase A deficiency, *Brain Dev.* **24**:688.

Ameen M, Lazzarino DA, Kelly BM, Gabel CA, Chang PL, 1990, Deficient glycosylation of arylsulfatase A in pseudo arylsulfatase-A deficiency, *Mol Cell Biochem.* **92**:117.

Arbour LT, Silver K, Hechtman P, Treacy EP, Coulter-Mackie MB, 2000, Variable onset of metachromatic leukodystrophy in a Vietnamese family, *Pediatr Neurol.* **23**:173.

Aula N, Jalanko A, Aula P, Peltonen L, 2002, Unraveling the molecular pathogenesis of free sialic acid storage disorders: Altered targeting of mutant sialin, *Mol Genet Metab.* **77**:99.

Aula N, Kopra O, Jalanko A, Peltonen L, 2004, Sialin expression in the CNS implicates extralysosomal function in neurons, *Neurobiol Dis.* **15**:251.

Aula N, Salomaki P, Timonen R, Verheijen F, Mancini G, Mansson JE, Aula P, Peltonen, L, 2000, The spectrum of SLC17A5-gene mutations resulting in free sialic acid-storage diseases indicates some genotype-phenotype correlation, *Am J Hum Genet.* **67**: 832.

Aula P, Autio S, Raivio K, Rapola J, Thoden C, Koskela S, Yamashina I, 1979, "Salla disease." A new lysosomal storage disorder, *Arch Neurol.* **36**:88.

Aula P, Gahl W, 2001, Disorders of free sialic acid storage. In The Metabolic and Molecular Bases of Inherited Disease, C Scriver, A Beaudet, W Sly, and D Valle, Eds. (New York, McGraw-Hill), pp. 5109–5120.

Barth ML, Fensom A, Harris A, 1993, Prevalence of common mutations in the arylsulphatase A gene in metachromatic leukodystrophy patients diagnosed in Britain, *Hum Genet.* **91**:73.

Barth ML, Ward C, Harris A, Saad A, Fensom A, 1994, Frequency of arylsulphatase A pseudodeficiency associated mutations in a healthy population, *J Med Genet.* **31**:667.

Baumann N, Turpin JC, Lefevre M, Colsch B, 2002, Motor and psycho-cognitive clinical types in adult metachromatic leukodystrophy: Genotype/phenotype relationships? *J Physiol Paris.* **96**:301.

Berger J, Gmach M, Mayr U, Molzer B, Bernheimer H, 1999, Coincidence of two novel arylsulfatase A alleles and mutation 459+1G>A within a family with metachromatic leukodystrophy: Molecular basis of phenotypic heterogeneity, *Hum Mutat.* **13**:61.

Berger J, Loschl B, Bernheimer H, Lugowska A, Tylki-Szymanska A, Gieselmann V, Molzer B, 1997, Occurrence, distribution, and phenotype of arylsulfatase A mutations in patients with metachromatic leukodystrophy, *Am J Med Genet.* **69**:335.

Biancheri R, Verbeek E, Rossi A, Gaggero R, Roccatagliata L, Gatti R, van Diggelen O, Verheijen FW, Mancini GM, 2002, An Italian severe Salla disease variant associated with a SLC17A5 mutation earlier described in infantile sialic acid storage disease, *Clin Genet.* **61**:443.

Biffi A, De Palma M, Quattrini A, Del Carro U, Amadio S, Visigalli I, Sessa M, Fasano S, Brambilla R, Marchesini S, Bordignon C, Naldini L, 2004, Correction of metachromatic leukodystrophy in the mouse model by transplantation of genetically modified hematopoietic stem cells, *J Clin Invest.* **113**:1118.

Bonifacino J, Dell'Angelica E, 1999, Molecular bases for the recognition of tyrosine-based sorting signals, *J Cell Biol.* **145**:923.

Bosch EP, Hart MN, 1978, Late adult-onset metachromatic leukodystrophy. Dementia and polyneuropathy in a 63-year-old man, *Arch Neurol.* **35**:475.

Brown FR 3rd, Shimizu H, McDonald JM, Moser AB, Marquis P, Chen WW, Moser HW, 1981, Auditory evoked brainstem response and high-performance liquid chromatography sulfatide assay as early indices of metachromatic leukodystrophy, *Neurology.* **31**:980.

Cengiz N, Ozbenli T, Onar M, Yildiz L, Ertas B, 2002, Adult metachromatic leukodystrophy: Three cases with normal nerve conduction velocities in a family, *Acta Neurol Scand.* **105**:454.

Cherqui S, Kalatzis V, Trugnan G, Antignac C, 2001, The targeting of cystinosin to the lysosomal membrane requires a tyrosine-based signal and a novel sorting motif, *J Biol Chem.* **276**:13314.

Clark JR, Miller RG, Vidgoff JM, 1979, Juvenile-onset metachromatic leukodystrophy: biochemical and electrophysiologic studies, *Neurology.* **29**:346.

Clarke JT, Skomorowski MA, Chang PL, 1989, Marked clinical difference between two sibs affected with juvenile metachromatic leukodystrophy, *Am J Med Genet.* **33**:10.

Clements P, Taylor J, Hopwood J, 1988, Biochemical characterization of patients and prenatal diagnosis of sialic acid storage disease for three families., *J Inherit Metab Dis.* **11**:30.

Comabella M, Waye JS, Raguer N, Eng B, Dominguez C, Navarro C, Borras C, Krivit W, Montalban X, 2001, Late-onset metachromatic leukodystrophy clinically presenting as isolated peripheral neuropathy: Compound heterozygosity for the IVS2+1G-->A mutation and a newly identified missense mutation (Thr408Ile) in a Spanish family, *Ann Neurol.* **50**:108.

Consiglio A, Quattrini A, Martino S, Bensadoun JC, Dolcetta D, Trojani A, Benaglia G, Marchesini S, Cestari V, Oliverio A, Bordignon C, Naldini L, 2001, In vivo gene therapy of metachromatic leukodystrophy by lentiviral vectors: Correction of neuropathology and protection against learning impairments in affected mice, *Nat Med.* **7**:310.

Coulter-Mackie M, Gagnier L, 1997, Two new polymorphisms in the arylsulfatase A gene and their haplotype associations with normal, metachromatic leukodystrophy and pseudodeficiency alleles, *Am J Med Genet.* **73**:32.

Coulter-Mackie MB, Applegarth DA, Toone JR, Gagnier L, Anzarut AR, Hendson G, 2002, Isolated peripheral neuropathy in atypical metachromatic leukodystrophy: A recurrent mutation, *Can J Neurol Sci.* **29**:159.

Daneman A, Stringer D, Reillly B, 1983, Neonatal ascites due to lysosomal storage disease, *Radiology.* **149**:463.

Degroote S, Wolthoorn J, van Meer G, 2004, The cell biology of glycosphingolipids, *Semin Cell Dev Biol.* **15**:375.

Dierks T, Schmidt B, Borissenko LV, Peng J, Preusser A, Mariappan M, von Figura K, 2003, Multiple sulfatase deficiency is caused by mutations in the gene encoding the human C(alpha)-formylglycine generating enzyme, *Cell.* **113**:435.

Duyff RF, Weinstein HC, 1996, Late-presenting metachromatic leukodystrophy, *Lancet.* **348**:1382.

Faerber EN, Melvin JJ, Smergel EM, 1999, MRI appearances of metachromatic leukodystrophies, *Pediatr Radiol* **29**:669.

Felice KJ, Gomez Lira M, Natowicz M, Grunnet ML, Tsongalis GJ, Sima AA, Kaplan RF, 2000, Adult-onset MLD: A gene mutation with isolated polyneuropathy, *Neurology.* **55**:1036.

Fluharty AL, Fluharty CB, Bohne W, von Figura K, Gieselmann V, 1991, Two new arylsulfatase A (ARSA) mutations in a juvenile metachromatic leukodystrophy (MLD) patient, *Am J Hum Genet* **49**:1340.

Francis GS, Bonni A, Shen N, Hechtman P, Yamut B, Carpenter S, Karpati G, Chang PL, 1993, Metachromatic leukodystrophy: Multiple nonfunctional and pseudodeficiency alleles in a pedigree: Problems with diagnosis and counseling, *Ann Neurol* **34**:212.

Frank M, 2000, MAL, a proteolipid in glycosphingolipid enriched domains: Functional implications in myelin and beyond, *Prog Neurobiol* **60**:531.

Fujii T, Kobayashi T, Honke K, Gasa S, Ishikawa M, Shimizu T, Makita A, 1992, Proteolytic processing of human lysosomal arylsulfatase A, *Biochim Biophys Acta* **1122**:93.

Fukutani Y, Noriki Y, Sasaki K, Isaki K, Kuriyama M, Kurosawa K, Ida H, 1999, Adult-type metachromatic leukodystrophy with a compound heterozygote mutation showing character change and dementia, *Psychiatry Clin Neurosci.* **53**:425.

Gallo S, Randi D, Bertelli M, Salviati A, Pandolfo M, 2004, Late onset MLD with normal nerve conduction associated with two novel missense mutations in the ASA gene, *J Neurol Neurosurg Psychiatry.* **75**:655.

Gieselmann V, 1991, An assay for the rapid detection of the arylsulfatase A pseudo-deficiency allele facilitates diagnosis and genetic counseling for metachromatic leukodystrophy, *Hum Genet.* **86**:251.

Gieselmann V, Fluharty AL, Tonnesen T, von Figura K, 1991, Mutations in the aryl-sulfatase A pseudodeficiency allele causing metachromatic leukodystrophy, *Am J Hum Genet.* **49**:407.

Gieselmann V, Polten A, Kreysing J, von Figura K, 1989, Arylsulfatase A pseudo-deficiency: loss of a polyadenylylation signal and N-glycosylation site, *Proc Natl Acad Sci U S A.* **86**:9436.

Gillan JE, Lowden JA, Gaskin K, Cutz E, 1984, Congenital ascites as a presenting sign of lysosomal storage disease, *J Pediatrics.* **104**:225.

Gomez-Lira M, Perusi C, Mottes M, Pignatti PF, Manfredi M, Rizzuto N, Salviati A, 1998, Molecular genetic characterization of two metachromatic leukodystrophy patients who carry the T799G mutation and show different phenotypes; description of a novel null-type mutation, *Hum Genet.* **102**:459.

Gort L, Coll MJ, Chabas A, 1999, Identification of 12 novel mutations and two new polymorphisms in the arylsulfatase A gene: Haplotype and genotype-phenotype correlation studies in Spanish metachromatic leukodystrophy patients, *Hum Mutat.* **14**:240.

Gort L, Coll MJ, Chabas A, 2000, Metachromatic leukodystrophy: a novel mutation (c237delC) and extension of the haplotype associated with the P426L mutation, *Hum Mutat.* **16**:375.

Gustavson KH, Hagberg B, 1971, The incidence and genetics of metachromatic leuco-dystrophy in northern Sweden, *Acta Paediatr Scand.* **60**:585.

Haataja L, Schleutker J, Laine AP, Renlund M, Savontaus ML, Dib C, Weissenbach J, Peltonen L, Aula P, 1994, The genetic locus for free sialic acid storage disease maps to the long arm of chromosome 6, *Am J Hum Genet.* **54**:1042.

Hageman AT, Gabreels FJ, de Jong JG, Gabreels-Festen AA, van den Berg CJ, van Oost BA, Wevers RA, 1995, Clinical symptoms of adult metachromatic leukodystrophy and arylsulfatase A pseudodeficiency, *Arch Neurol.* **52**:408.

Halsall DJ, Halligan EP, Elsey TS, Cox TM, 1999, Metachromatic leucodystrophy: A newly identified mutation in arylsulphatase A, D281Y, found as a compound heterozygote with I179L in an adult onset case, *Hum Mutat.* **14**:447.

Hancock L, Horwitz A, Dawson G, 1983, N-acetylneuraminic acid and sialoglyco-conjugate metabolism in fibroblasts from a patient with generalized N-acetylneur-aminic acid storage disease, *Biochem Biophys Acta.* **760**:42.

Harvey JS, Carey WF, Morris CP, 1998, Importance of the glycosylation and polyadeny-lation variants in metachromatic leukodystrophy pseudodeficiency phenotype, *Hum Mol Genet.* **7**:1215.

Havelaar AC, Mancini GM, Beerens CE, Souren RM, Verheijen FW, 1998, Purification of the lysosomal sialic acid transporter. Functional characteristics of a monocarboxy-late transporter, *J Biol Chem.* **273**:34568.

Heim P, Claussen M, Hoffmann B, Conzelmann E, Gartner J, Harzer K, Hunneman DH, Kohler W, Kurlemann G, Kohlschutter A, 1997, Leukodystrophy incidence in Germany, *Am J Med Genet.* **71**:475.

Hess B, Kafert S, Heinisch U, Wenger DA, Zlotogora J, Gieselmann V, 1996, Characterization of two arylsulfatase A missense mutations D335V and T274M causing late infantile metachromatic leukodystrophy, *Hum Mutat.* **7**:311.

Hess B, Saftig P, Hartmann D, Coenen R, Lullmann-Rauch R, Goebel HH, Evers M, von Figura K, D'Hooge R, Nagels G, De Deyn P, Peters C, Gieselmann V, 1996, Phenotype of arylsulfatase A-deficient mice: Relationship to human metachromatic leukodystrophy, *Proc Natl Acad Sci USA.* **93**:14821.

Hirahara Y, Bansal R, Honke K, Ikenaka K, Wada Y, 2004, Sulfatide is a negative regulator of oligodendrocyte differentiation: Development in sulfatide-null mice, *Glia.* **45**:269.

Honke K, Hirahara Y, Dupree J, Suzuki K, Popko B, Fukushima K, Fukushima J, Nagasawa T, Yoshida N, Wada Y, Taniguchi N, 2002, Paranodal junction formation and spermatogenesis require sulfoglycolipids, *Proc Natl Acad Sci. USA.* **99**:4227.

Hwu WL, Tsai LP, Wang WC, Chuang SC, Wang PJ, Wang TR, 1996, Arylsulfatase A pseudodeficiency in Chinese, *Hum Genet.* **97**:148.

Hyde TM, Zeigler JC, Weinberger DR, 1992, Psychiatric disturbances in metachromatic leukodystrophy, insights into the neurobiology of psychosis, *Arch Neurol.* **49**:401

Irie A, Koyama S, Kozutsumi Y, Kawasaki T, Suzuki A, 1998, The molecular basis for the absence of N-glycolylneuraminic acid in humans, *J Biol Chem.* **273**:15866.

Ishibashi T, Dupree JL, Ikenaka K, Hirahara Y, Honke K, Peles E, Popko B, Suzuki K, Nishino H, Baba H, 2002, A myelin galactolipid, sulfatide, is essential for maintenance of ion channels on myelinated axon but not essential for initial cluster formation, *J Neurosci.* **22**:6507.

Kafert S, Heinisch U, Zlotogora J, Gieselmann V, 1995, A missense mutation P136L in the arylsulfatase A gene causes instability and loss of activity of the mutant enzyme, *Hum Genet.* **95**:201.

Kapaun P, Dittmann RW, Granitzny B, Eickhoff W, Wulbrand H, Neumaier-Probst E, Zander A, Kohlschuetter A, 1999, Slow progression of juvenile metachromatic leukodystrophy 6 years after bone marrow transplantation, *J Child Neurol.* **14**:222.

Kappler J, Sommerlade HJ, von Figura K, Gieselmann V, 1994, Complex arylsulfatase A alleles causing metachromatic leukodystrophy, *Hum Mutat.* **4**:119.

Kidd D, Nelson J, Jones F, Dusoir H, Wallace I, McKinstry S, Patterson V, 1998, Long-term stabilization after bone marrow transplantation in juvenile metachromatic leukodystrophy, *Arch Neurol.* **55**:98.

Kim TS, Kim IO, Kim WS, Choi YS, Lee JY, Kim OW, Yeon KM, Kim KJ, Hwang YS, 1997, MR of childhood metachromatic leukodystrophy, *AJNR Am J Neuroradiol.* **18**:733.

Kleta R, Aughton DJ, Rivkin MJ, Huizing M, Strovel E, Anikster Y, Orvisky E, Natowicz M, Krasnewich D, Gahl WA, 2003, Biochemical and molecular analyses of infantile free sialic acid storage disease in North American children, *Am J Med Genet.* **120A**:28.

Kleta R, Morse R, Orvisky E, Krasnewich D, Alroy J, Ucci A, Bernardini I, Wenger D, Gahl W, 2004, Clinical, biochemical, and molecular diagnosis of a free sialic acid storage disease patient of moderate severity, *Mol Genet Metab.* **82**:137.

Kornfeld S, Kornfeld R, Neufeld E, O'Brien P, 1964, The feedback control of sugar nucleotide biosynthesis in liver, *Proc Natl Acad Sci USA.* **52**:371.

Kruse B, Hanefeld F, Christen HJ, Bruhn H, Michaelis T, Hanicke W, Frahm J, 1993, Alterations of brain metabolites in metachromatic leukodystrophy as detected by localized proton magnetic resonance spectroscopy in vivo, *J Neurol.* **241**:68.

Kurosawa K, Ida H, Eto Y, 1998, Prevalence of arylsulphatase A mutations in 11 Japanese patients with metachromatic leukodystrophy: Identification of two novel mutations, *J Inherit Metab Dis.* **21**:781.

Lake B, Young E, Nicolaides K, 1989, Prenatal diagnosis of infantile sialic acid storage disease in a twin pregnancy, *J Inherit Metab Dis* **12**:152.

Leistner S, Young E, Winchester B, 1995, Pseudodeficiency of arylsulfatase A: Strategy for clarification of genotype in families of subjects with low ASA activity and neurologic symptoms, *J Inherit Metab Dis.* **18**:710.

Lemyre E, Russo P, Melancon SB, Gagne R, Potier M, Lambert M, 1999, Clinical spectrum of infantile free sialic acid storage disease, *Am J Med Genet.* **82**:385.

Li ZG, Waye JS, Chang PL, 1992, Diagnosis of arylsulfatase A deficiency, *Am J Med Genet.* **43**:976.

Lugowska A, Berger J, Tylki-Szymanska A, Czartoryska B, Loschl B, Molzer B, 2002, High prevalence of I179S mutation in patients with late-onset metachromatic leukodystrophy, *Clin Genet.* **61**:389.

Lugowska A, Tylki-Szymanska A, Berger J, Molzer B, 1997, Elevated sulfatide excretion in compound heterozygotes of metachromatic leukodystrophy and ASA pseudodeficiency allele, *Clin Biochem.* **30**:325.

Lukatela G, Krauss N, Theis K, Selmer T, Gieselmann V, von Figura K, Saenger W, 1998, Crystal structure of human arylsulfatase A: The aldehyde function and the metal ion at the active site suggest a novel mechanism for sulfate ester hydrolysis, *Biochemistry.* **37**:3654.

Malm G, Ringden O, Winiarski J, Grondahl E, Uyebrant P, Eriksson U, Hakansson H, Skeldal O, Mansson JE, 1996, Clinical outcome in four children with metachromatic leukodystrophy treated by bome marrow transplantation, *Bone Marrow Transpl.* **17**:1003.

Mancini G, Beerens C, Galjaard H, Verheijen F, 1992, Functional reconstitution of the lysosomal sialic acid carrier into proteoliposomes., *Proc Natl Acad Sci. USA.* **89**:6609.

Mancini G, de Jonge H, Galjaard H, Verheijen F, 1989, Characterization of a proton-driven carrier for sialic acid in the lysosomal membrane. Evidence for a group-specific transport system for acidic monosaccharides, *J Biol Chem.* **264**:15247.

Mancini G, Verheijen F, Galjaard H, 1986, Free N-acetylneuraminic acid storage disorders: evidence for defective NANA transport across the lysosomal membrane, *Hum Genet.* **73**:214.

Marcao A, Azevedo JE, Gieselmann V, Sa Miranda MC, 2003, Oligomerization capacity of two arylsulfatase A mutants: C300F and P425T, *Biochem Biophys Res Commun.* **306**:293.

Marcao A, Wiest R, Schindler K, Wiesmann U, Schroth G, Sa Miranda C, Polten A, Berna L, Sturzenegger M, Gieselmann V, 2004, Adult onset metachromatic leucodystrophy without peripheral nervous system involvement: A case with a new mutation in the arylsulfatase A gene, *Arch. Neurol.* **62**:309.

Matzner U, Hartmann D, Lullmann-Rauch R, Coenen R, Rothert F, Mansson JE, Fredman P, D'Hooge R, De Deyn PP, Gieselmann V, 2002, Bone marrow stem cell-based gene transfer in a mouse model for metachromatic leukodystrophy: Effects on visceral and nervous system disease manifestations, *Gene Ther.* **9**:53.

Mendla K, Baumkotter J, Rosenau C, Ulrich-Bott B, and Cantz M, 1988, Defective lysosomal release of glycoprotein-derived sialic acid in fibroblasts from patients with sialic acid storage disease, *Biochem J.* **250**:261.

Nelson PV, Carey WF, Morris CP, 1991, Population frequency of the arylsulphatase A pseudo-deficiency allele, *Hum Genet.* **87**:87.

Ott R, Waye JS, Chang PL, 1997, Evolutionary origins of two tightly linked mutations in arylsulfatase-A pseudodeficiency, *Hum Genet.* **101**:135.

Parazzini C, Arena S, Marchetti L, Menni F, Filocamo M, Verheijen F, Mancini G, Triulzi F, Parini R, 2003, Infantile sialic acid storage disease: Serial ultrasound and magnetic resonance imaging features, *AJNR Am J Neuroradiol.* **24**:398.

Penzien JM, Kappler J, Herschkowitz N, Schuknecht B, Leinekugel P, Propping P, Tonnesen T, Lou H, Moser H, Zierz S, 1993, Compound heterozygosity for metachromatic leukodystrophy and arylsulfatase A pseudodeficiency alleles is not associated with progressive neurological disease, *Am J Hum Genet.* **52**:557.

Polten A, Fluharty AL, Fluharty CB, Kappler J, von Figura K, Gieselmann V, 1991, Molecular basis of different forms of metachromatic leukodystrophy, *N Engl J Med.* **324**:18.

Pueschel S, O'Shea P, Alroy J, Ambler M, Dangond F, Daniel P, Kolodny E, 1988, Infantile sialic acid storage disease associated with renal disease, *Pediatr Neurol.* **4**:207.

Rafi MA, Coppola S, Liu SL, Rao HZ, Wenger DA, 2003, Disease-causing mutations in cis with the common arylsulfatase A pseudodeficiency allele compound the difficulties in accurately identifying patients and carriers of metachromatic leukodystrophy, *Mol Genet Metab.* **79**:83.

Rapoport I, Chen Y, Cupers P, Shoelson S, Kirchhausen T, 1998, Dileucine-based sorting signals bind to the beta chain of AP-1 at a site distinct and regulated differently from the tyrosine-based motif-binding site., *EMBO J* **17**:2148.

Regis S, Corsolini F, Ricci V, Di Duca M, Filocamo M, 2004, An unusual arylsulfatase A pseudodeficiency allele carrying a splice site mutation in a metachromatic leukodystrophy patient, *Eur J Hum Genet.* **12**:150.

Regis S, Corsolini F, Stroppiano M, Cusano R, Filocamo M, 2002, Contribution of arylsulfatase A mutations located on the same allele to enzyme activity reduction and metachromatic leukodystrophy severity, *Hum Genet.* **110**:351.

Regis S, Filocamo M, Stroppiano M, Corsolini F, Caroli F, Gatti R, 1998, A 9-bp deletion (2320del9) on the background of the arylsulfatase A pseudodeficiency allele in a metachromatic leukodystrophy patient and in a patient with nonprogressive neurological symptoms, *Hum Genet.* **102**:50.

Regis S, Filocamo M, Stroppiano M, Corsolini F, Gatti R, 1996, Molecular analysis of the arylsulphatase A gene in late infantile metachromatic leucodystrophy patients and healthy subjects from Italy, *J Med Genet.* **33**:251.

Renlund M, Aula P, 1987, Prenatal detection of Salla disease based upon increased free sialic acid in amniocytes, *Am J Med Genet.* **28**:377.

Renlund M, Aula P, Raivio KO, Autio S, Sainio K, Rapola J, Koskela SL, 1983a, Salla disease: A new lysosomal storage disorder with disturbed sialic acid metabolism, *Neurology* **33**:57.

Renlund M, Chester MA, Lundblad A, Aula P, Raivio KO, Autio S, Koskela SL, 1979, Increased urinary excretion of free N-acetylneuraminic acid in thirteen patients with Salla disease, *Eur J Biochem.* **101**:245.

Renlund M, Chester MA, Lundblad A, Parkkinen J, Krusius T, 1983b, Free N-acetylneuraminic acid in tissues in Salla disease and the enzymes involved in its metabolism, *Eur J Biochem.* **130**:39.

Renlund M, Kovanen PT, Raivio KO, Aula P, Gahmberg CG, Ehnholm C, 1986a, Studies on the defect underlying the lysosomal storage of sialic acid in Salla disease. Lysosomal accumulation of sialic acid formed from N-acetyl-mannosamine or derived from low density lipoprotein in cultured mutant fibroblasts, *J Clin Invest.* **77**:568.

Renlund M, Tietze F, Gahl WA, 1986b, Defective sialic acid egress from isolated fibroblast lysosomes of patients with Salla disease, *Science* **232**:759.

Ricketts MH, Goldman D, Long JC, Manowitz P, 1996, Arylsulfatase A pseudodeficiency-associated mutations: Population studies and identification of a novel haplotype, *Am J Med Genet.* **67**:387.

Salamon MB, Christensen E, Schwartz M, 1994, Searching for mutations in the arylsulphatase A gene, *Metab Dis.* **17**: 311

Salomaki P, Aula N, Juvonen V, Renlund M, Aula P, 2001, Prenatal detection of free sialic acid storage disease: Genetic and biochemical studies in nine families, *Prenat Diagn.* **21**:354.

Sandhoff K, Kolter T, Harzer K, 2001, Sphingolipid activator protein, in *The Metabolic & Molecular Bases of Inherited Disease*, Scriver CR, Beaudet AL, Sly, WS, Valle D, Eds, McGraw Hill, New York, 3371–3388.

Sandhoff R, Hepbildikler ST, Jennemann R, Geyer R, Gieselmann V, Proia RL, Wiegandt H, Grone HJ, 2002, Kidney sulfatides in mouse models of inherited glycosphingolipid disorders: Determination by nano-electrospray ionization tandem mass spectrometry, *J Biol Chem.* **277**:20386.

Saravanan K, Schaeren-Wiemers N, Klein D, Sandhoff R, Schwarz A, Yaghootfam A, Gieselmann V, Franken S, 2004, Specific downregulation and mistargeting of the lipid raft-associated protein MAL in a glycolipid storage disorder, *Neurobiol Dis.* **16**:396.

Schauer R, Sommer U, Kruger D, van Unen H, Traving C, 1999, The terminal enzymes of sialic acid metabolism: Acylneuraminate-pyruvate-lyases, *Biosci Rep.* **19**:373.

Schipper HI, Seidel D, 1984, Computed tomography in late-onset metachromatic leucodystrophy, *Neuroradiology.* **26**:39.

Schleutker J, Laine AP, Haataja L, Renlund M, Weissenbach J, Aula P, Peltonen L, 1995, Linkage disequilibrium utilized to establish a refined genetic position of the Salla disease locus on 6q14-q15, *Genomics* **27**:286.

Seyrantepe V, Poupetova H, Froissart R, Zabot M, Maire I, Pshezhetsky A, 2003, Molecular pathology of NEU1 gene in sialidosis, *Hum Mutat.* **22**:343.

Shen N, Li ZG, Waye JS, Francis G, Chang PL, 1993, Complications in the genotypic molecular diagnosis of pseudo arylsulfatase A deficiency, *Am J Med Genet.* **45**:631.

Sommerlade HJ, Selmer T, Ingendoh A, Gieselmann V, von Figura K, Neifer K, Schmidt B, 1994, Glycosylation and phosphorylation of arylsulfatase A, *J Biol Chem.* **269**:20977.

Sperl W, Gruber W, Quatacker J, Monnens L, Thoenes W, Fink FM, Paschke E, 1990, Nephrosis in two siblings with infantile sialic acid storage disease, *Eur J Pediatr.* **149**:477.

Stein C, Gieselmann V, Kreysing J, Schmidt B, Pohlmann R, Waheed A, Meyer HE, O'Brien JS, von Figura K, 1989, Cloning and expression of human arylsulfatase A, *J Biol Chem.* **264**:1252.

Stevenson R, Lubinsky M, Taylor H, Wenger D, Schroer R, Olmstead P, 1983, Sialic acid storage disease with sialuria: Clinical and biochemical features in the severe infantile type, *Pediatrics.* **72**:441.

Stone DL, Sidransky E, 1999, Hydrops fetalis: Lysosomal storage disorders in extremis, *Adv Pediatr.* **46**:409.

Thomas G, Scocca J, Libert ., Vamos E, Miller C, Reynolds L, 1983, Alterations in cultured fibroblasts of sibs with an infantile form of free (unbound) sialic acid storage disorder, *Pediatr Res.* **17**: 307.

Tietze F, Seppala R, Renlund M, Hopwoo, JJ, Harper GS, Thomas GH, Gahl WA, 1989, Defective lysosomal egress of free sialic acid (N-acetylneuraminic acid) in fibroblasts of patients with infantile free sialic acid storage disease, *J Biol Chem.* **264**: 15316.

Tondeur M, Libert J, Vamos E, Van Hoof F, Thomas GH, Strecker G, 1982, Infantile form of sialic acid storage disorder: clinical, ultrastructural, and biochemical studies in two siblings, *Eur J Pediatr.* **139**:142.

Tylki-Szymanska A, Berger J, Loschl B, Lugowska A, Molzer B., 1996, Late juvenile metachromatic leukodystrophy (MLD) in three patients with a similar clinical course and identical mutation on one allele, *Clin Genet.* **50**:287.

Tylki-Szymanska A, Lugowska A, Chmielik J, Kotowicz J, Jakubowska-Winecka A, Zobel M, Berger J, Molzer B., 2002, Investigations of micro-organic brain damage (MOBD) in heterozygotes of metachromatic leukodystrophy, *Am J Med Genet.* **110**:315.

Vamos E, Libert J, Elkhazen N, Jauniaux E, Hustin J, Wilkin P, Baumkotter J, Mendla K, Cantz M, Strecker G, 1986, Prenatal diagnosis and confirmation of infantile sialic acid storage disease, *Prenat Diagn.* **6**: 437.

van der Voorn J, Pouwels P, Kamphorst W, Powers J, Lammens M, Barkhof F, van der Knaap M, 2004, Histopathologic correlates of radial stripes on mr imaging in lysosomal storage disorders, *Am. J. Neurorad. in press*

Varho T, Jaaskelainen S, Tolonen U, Sonninen P, Vainionpaa L, Aula P, Sillanpaa M, 2000, Central and peripheral nervous system dysfunction in the clinical variation of Salla disease, *Neurology* **55**:99.

Varho T, Komu M, Sonninen P, Holopainen I, Nyman S, Manner T, Sillanpaa M, Aula P, Lundbom N, 1999, A new metabolite contributing to N-acetyl signal in 1H MRS of the brain in Salla disease. *Neurology* **52**:1668.

Varh, TT, Alajoki LE, Posti KM, Korhonen TT, Renlund MG, Nyman SR, Sillanpaa ML, Aula PP, 2002, Phenotypic spectrum of Salla disease, a free sialic acid storage disorder, *Pediatr Neurol.* **26**:267.

Verheijen FW, Verbeek E, Aula N, Beerens CE, Havelaar AC, Joosse M, Peltonen L, Aula P, Galjaard H, van der Spek PJ, Mancini GM, 1999, A new gene, encoding an anion transporter, is mutated in sialic acid storage diseases, *Nat Genet.* **23**:462.

.von Bulow R, Schmidt B, Dierks T, Schwabauer N, Schilling K, Weber E, Uson I, von Figura K, 2002, Defective oligomerization of arylsulfatase a as a cause of its instability in lysosomes and metachromatic leukodystrophy, *J Biol Chem.* **277**:9455.

von Figura K, Gieselmann V, Jaeken J, 2001, Metachromatic Leukodystrophy, in *The Metabolic & Molecular Bases of Inherited Disease*, Scriver CR, Beaudet AL, Sly, WS, Valle D, Eds., McGraw-Hill, New York, pp. 3695–3724.

von Figura K, Steckel F, Hasilik A, 1983, Juvenile and adult metachromatic leukodystrophy: partial restoration of arylsulfatase A (cerebroside sulfatase) activity by inhibitors of thiol proteinases, *Proc Natl Acad Sci. USA.* **80**:6066.

Waldow A, Schmidt B, Dierks T, von Bulow R, von Figura K, 1999, Amino acid residues forming the active site of arylsulfatase A. Role in catalytic activity and substrate binding, *J Biol Chem.* **274**:12284.

Weiss P, Tietze F, Gahl W, Seppala R, Ashwell G, 1989, Identification of the metabolic defect in sialuria, *J Biol Chem.* **264**:17635

Wittke D, Hartmann D, Gieselmann V, Lullmann-Rauch R, 2004, Lysosomal sulfatide storage in the brain of arylsulfatase A-deficient mice: Cellular alterations and topographic distribution, *Acta Neuropathol (Berl).*

Wulff CH, Trojaborg W, 1985, Adult metachromatic leukodystrophy: Neurophysiologic findings, *Neurology.* **35**:1776.

Zlotogora J, Furman-Shaharabani Y, Goldenfum S, Winchester B, von Figura K, Gieselmann V, 1994a, Arylsulfatase A pseudodeficiency: A common polymorphism which is associated with a unique haplotype, *Am J Med Genet.* **52**:146.

Zlotogora J, Furman-Shaharabani Y, Harris A, Barth ML, von Figura K, Gieselmann V, 1994b, A single origin for the most frequent mutation causing late infantile metachromatic leucodystrophy, *J Med Genet.* **31**:672.

FABRY DISEASE

Roscoe O. Brady

1 CLINICAL MANIFESTATIONS

Fabry disease is the second most prevalent metabolic storage disorder of humans. Patients with this condition were originally described in 1898 by the dermatologists William Anderson in England and Johannes Fabry in Germany because of the occurrence of pigmented angiokeratomas on their skin, a common but not universal finding in this condition. It is an X-linked genetic condition frequently termed recessive, but heterozygous females often become symptomatic. With time, the kidneys, heart, brain, peripheral nerves, gastrointestinal tract, eyes, ears, lungs, and even the skeleton may become involved. A surprising aspect of this disorder is the nonuniformity of presentation of the signs and symptoms in various patients, even in siblings. Many patients experience pain in their hands and feet because of the peripheral neuropathy associated with this condition. It is frequently the initial sign of the disease in young affected males and frequently occurs somewhat later in heterozygous females (Brady and Schiffmann, 2005). The diagnosis of Fabry disease is not infrequently made by ophthalmologists because of the presence of corneal whorls (verticillata) and lens opacities along with tortuosity of the blood vessels of the retina. Vision is not appreciably impaired by these manifestations.

More seriously incapacitating aspects of the disorder occur as patients age. Some hemizygous males exhibit proteinuria in their late teens that eventuates in frank renal failure generally by their early 40s. The episodic pain in the hands and feet increases in intensity, and a number of patients become unable to work even with antipain medication. The pain usually becomes worse with exercise and heat. Signs of vascular system damage become apparent, and transient ischemia attacks and strokes occur in the late 20s or even earlier (Ries et al., 2005). An unusual type of vasculopathy occurs in Fabry patients that is manifested by hyperperfusion of the posterior portion of the brain (Moore et al., 2001a). Some patients have premature myocardial infarctions in their early 30s.

Hypertrophy of the wall of the left ventricle and conduction abnormalities are not uncommon. Most of the patients experience increasingly severe gastrointestinal difficulties, primarily manifested as frequent bouts of diarrhea. Some patients exhibit pulmonary difficulties, and many suffer hearing loss. With time, renal involvement proceeds to frank kidney failure in many patients (Branton et al., 2002). Principal clinical manifestations are summarized in Table 1. Heterozygous females may present with many of these signs and symptoms, but usually at a later stage in life. It should be re-emphasized that not all of these manifestations occur in all patients with Fabry disease. Interfamilial variation in the type and extent of organs involved and time of onset of clinical signs is common. This statement is particularly applicable to female carriers. The quality of life for most hemizygous males with Fabry disease is quite poor, and the same is true for a number of the

Developmental and Metabolic Neurology Branch, National Institute of Neurological Disorders and Stroke National Institutes of Health, Bethesda, MD 20892-1260. Tel (301) 496-3285; Fax: (301) 496-9480; e-mail rb57v@nih.gov

Table 1. Frequency of Signs and Symptoms of Fabry disease in Hemizygous Males.

	Almost always	Frequent	Occasional	Rare
Acroparesthesias	+			
Corneal verticillata	+			
Gastrointestinal problems	+			
Hypohydrosis	+			
Lenticular opacities	+			
Premature mortality	+			
Proteinuria	+			
Angiokeratomas		+		
Chronic fatigue		+		
Electrocardiographic changes		+		
End stage renal failure		+		
Hearing loss		+		
Increased RBC sedimentation rate		+		
Large and small vessel strokes		+		
Left ventricle wall hypertrophy		+		
Morphometric changes in the face		+		
Myocardial infarction		+		
Stroke		+		
Tortuoisty of retinal vessels		+		
Transient ischemic attacks		+		
White matter lesions in the brain		+		
Bone changes in the hands and feet			+	
Mitral valve prolapse			+	
Pulmonary involvement			+	
Anemia			+	
Hypothyroidism				+
Low serum vitamin C				+

female heterozygotes. However, some carriers documented by genotyping may remain asymptomatic. These divergent findings have been well documented in excellent overviews of Fabry disease (MacDermot et al., 2001a,b; Desnick et al., 2005).

2 PATHOLOGY

One of the major early signs is the appearance of verticillata in the cornea (Figure 1a). Evidence of lipid deposition is variable but widespread throughout the body of patients with Fabry disease. It is particularly apparent in the kidney where concentric birefringent lamellated accumulations are present in podocytes of the glomerulus, capsular epithelium, interstitial cells, and epithelial cells of distal tubules (Figure 1b). The heart is also involved with lipid accumulation in myocardial cells and in extensive occlusion of coronary blood vessels (Figure 1c). The peripheral neuropathy is likely due to the accumulation of lipid in endothelial and perithelial cells of endoneurial capillaries (Figure 1d). Kupffer cells in the liver may contain some lipid, but they are not nearly so involved as in patients with Gaucher disease. Hepatocytes are minimally involved. The blood vessels of the gastro-intestinal tract are involved in lipid storage to a significant extent. Neuronal cells of the Meissner's plexus and smooth muscle cells of the lamina propria as well as cells in Brunner's glands accumulate lipid to a varying degree.

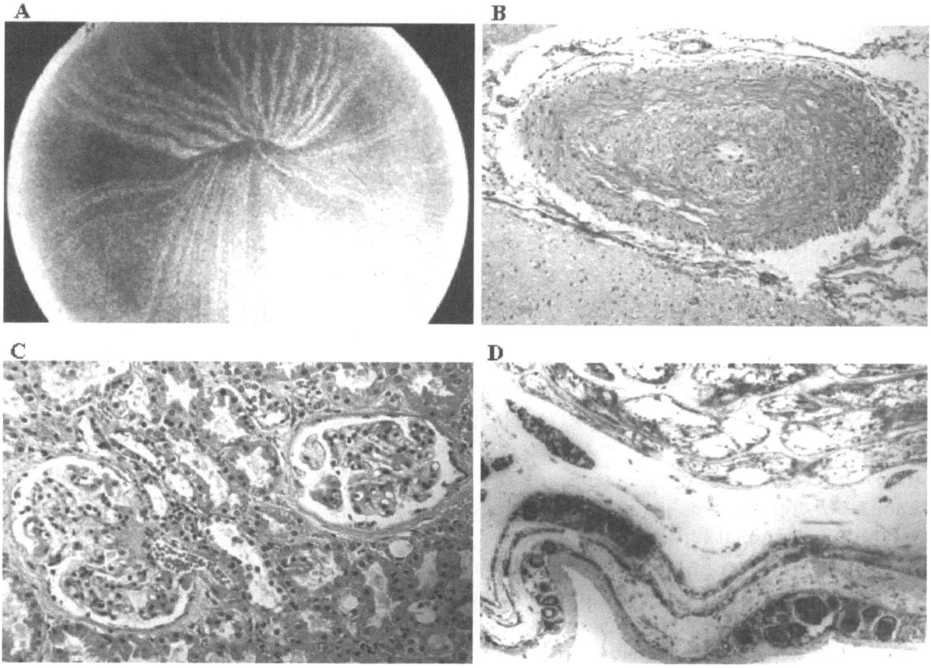

Figure 1. Representative organ and tissue alterations in Fabry disease. A. Cornea verticillata. B. Occluded coronary vessel. C. Lipid deposits in the kidney. D. Lipid deposits (black inclusions) in peripheral nerve.

3 ENZYMATIC DEFECT

The accumulation of globotriaosylceramide Gb3 (Figure 2) in patients with Fabry disease was demonstrated by Sweeley and Klionsky in 1963. About that time, my colleagues and I had demonstrated the nature of the enzymatic defects in Gaucher disease and in Niemann–Pick disease. I postulated that a deficiency of α–galactosidase that catalyzed the hydrolytic cleavage of the terminal molecule of galactose from Gb3 was the metabolic defect in Fabry disease (Brady, 1966). Investigations with unlabeled Gb3 did not reveal the nature of the enzymatic defect because of the insensitivity of these determinations. Because the chemical synthesis of Gb3 had not been accomplished at that time, my colleagues and I labeled Gb3 with 3H throughout the molecule by exposing natural Gb3 to 3H in a sealed tube for a week (the Wilzbach technique). Andrew Gal working with me was able to purify Gb3 labeled in this fashion sufficiently so it was a useful substrate. With it, we demonstrated that the enzymatic defect in Fabry disease was a deficiency of the terminal galactosidase required to initiate the catabolism of Gb3 (Brady et al., 1967a).

Several years later, Kint (1970) showed that the affected enzyme catalyzed the hydrolysis of the α-anomeric linkage between the two molecules of galactose, and the enzyme is therefore designated as an α-galactosidase. Two α-galactosidases exist in human tissues, and the enzyme involved in Fabry disease is designated α-galactosidase A. The catalytic activity of the other isozyme, α-galactosidase B is inhibited by N-acetylgalactosamine (Kusiak, Quirk, and Brady, 1978) whereas α-galactosidase A is unaffected by the presence of N-acetylgalactosamine. This differential effect of N-acetylgalactosamine on the two isozymes provided the basis for the development of a specific assay for α-galactosidase A using an artificial fluorogenic substrate in unfractionated tissues and cells that is widely used for the diagnosis of patients with Fabry disease (Mayes et al., 1981).

4 PATHOPHYSIOLOGY

In simplistic terms, organ and vascular abnormalities in patients with Fabry disease are viewed as the result of excessive accumulation of Gb3 (Figure 2a) along with galabiosylceramide (GbA; Figure 2b) in certain organs such as the kidney as well as lipids of the blood group B series. Gb3 is synthesized in most of the organs and tissues of the body. It seems likely that a major source of accumulating Gb3 is globoside (Figure 2c) that arises from the biodegradation of lipids in the stroma of senescent erythrocytes. This deduction is supported by unanticipated findings in the murine model of Fabry disease where the extent of Gb3 accumulation is considerably less than that in patients with Fabry disease (Ohshima et al., 1997). The principal reason for this discrepancy appears to be the absence of globoside in mouse erythrocytes (Ohshima et al., 1999). These deductions are consistent with a substantially altered phenotype in the α-galactosidase A knock-out mouse that has a normal life span and does not exhibit strokes, myocardial infarctions, or renal failure.

It also follows in humans that Gb3 that arises from globoside catabolism in macrophages in the spleen and bone marrow and Kupffer cells in the liver apparently escapes from these cells to a large extent and enters the circulation from which it is widely distributed to various organs and tissues and blood vessels such as the coronary, cerebral, and renal arteries and probably the vaso nervorum. Small intestinal tissue was found to have the highest specific α-galactosidase A activity among all of the systemic organs

A. Structure of Globotriosylceramide (Gb3)

```
                 ↓
Sphingosine—Glucose—Galactose—Galactose
|
Fatty Acid
```

B. Structure of Galabiosylceramide (GbA)

```
                 ↓
Sphingosine—Galactose—Galactose
|
Fatty Acid
```

↓ = Site of enzymatic defect in Fabry disease

C. Structure of Globoside

```
Sphingosine—Glucose—Galactose—Galactose—N-Acetylgalactosamine
|
Fatty Acid
```

Figure 2. Enzymatic defects in Fabry disease.

examined (Brady et al., 1967b), and its reduction or absence is presumably involved in the frequently severe gastrointestinal difficulties experienced by patients with Fabry disease. The teleology of the high α-galactosidase A activity in this organ and its natural substrate in this tissue (presumably Gb3) are still unknown.

A particularly important consideration is the nature of pathophysiological changes in blood vessels in patients with Fabry disease. Despite the demonstration of narrowing of arteries, the posterior circulation of the brain is characteristically hyperdynamic (Moore, Scott, and Galdwin, 2001). This alteration appears to be due to an abnormality of a reactive oxygen species (Moore, Ye, and Brennan, 2004). These changes have potential physiological importance, and they have proven to be useful for monitoring therapy trials in patients with Fabry disease (v.i.).

4.1 The α-Galactosidase A Gene

The gene is located in the Xq22.1 region of the X-chromosome. The nucleotide sequence of the gene was determined in 1989 (Kornreich, Desnick, and Bishop). At this time, 356 mutations of the α-galactosidase A gene are known (Desnick, Ioannou, and Eng, 2005; Schäfer, Baron, and Widmer, 2005; Shabbeer, Robinsin, and Desnick, 2005; http://archive.uwcm.ac.uk/uwcm/mg/hgmd0.html). Mutations in many patients are called "private" and appear to have arisen spontaneously in the families in which they appear. A high percentage of the mutations preclude formation of the enzyme. The time of onset of the manifestations of the disease is earlier and the rate of progression is more rapid in such patients than in those with detectable α-galactosidase A activity (Branton et al., 2002). Crystallization of human α-galactosidase A permitted the determination of its molecular structure (Garman and Garboczi, 2004; Figure 3). The enzyme exists as a dimer.

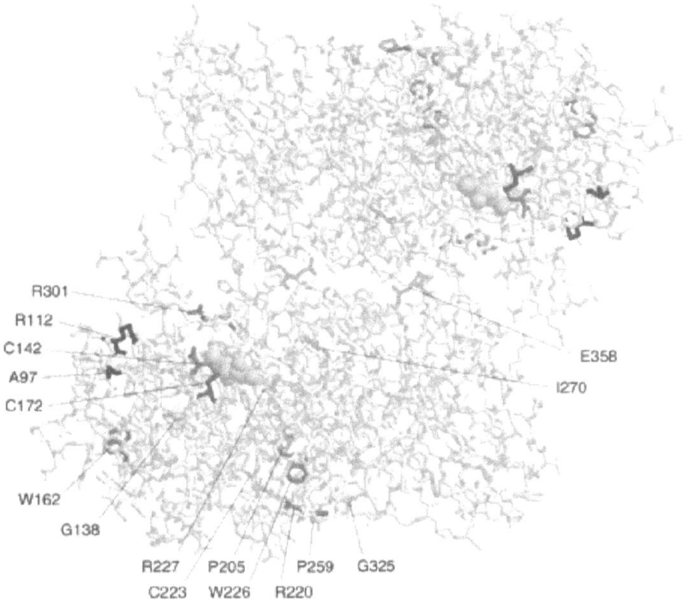

Figure 3. Positions of selected mutations superimposed on the crystallographic structure of the dimer of α-galactosidase A. (Reproduced with permission from Ries, M., et al. *Pediatrics* 2005; 115: e352.)

Mutations occur in all of the exons of the gene. Missense mutations that are partially exposed on the surface of the dimer and away from the active site of the enzyme were associated with residual catalytic activity in a cohort of pediatric patients with Fabry disease (Ries et al., 2005). Patients with residual α-galactosidase A activity generally have less rapid progression of the disease than those undetectable α-galactosidase A. Some patients whose α-galactosidase A activity is $\geq$12% of normal may primarily exhibit abnormalities of the cardiovascular system. Such patients have been classified as having the "cardiac variant" form of Fabry disease (Desnick, Ioannou, and Eng, 2005).

5 TREATMENT

5.1 Enzyme Replacement Therapy

Soon after the defects in Gaucher disease and Fabry disease were discovered, I postulated that supplementation or replacing the deficient enzyme might help patients with these and other metabolic storage disorders (Brady, 1966). I wished to use a human tissue as source of the missing enzyme, and it occurred to me that the placenta might be a useful tissue. My colleague William Johnson found that placenta did have ceramidetrihexosidase (α-galactosidase A) activity, and he isolated small quantities of the enzyme from this source (Johnson and Brady, 1972). We felt that the preparation was sufficiently pure that we might infuse it intravenously into patients with Fabry disease. We found that caused a rapid reduction of Gb3 in the blood, but it returned to the preinfusion level in approximately three days (Brady et al., 1973). We were not permitted to obtain tissue samples in the course of this investigation. Two additional preparations of placental α-galactosidase A were obtained, but technical difficulties were encountered that precluded an assessment of the effect of the enzyme on Gb3 in tissues of the recipients.

Many years passed before two biotechnical corporations simultaneously began large-scale preparations of α-galactosidase A. The enzyme was produced in cultured human skin fibroblasts by Transkaryotic Therapies, Inc., Cambridge, MA. Injection of the enzyme into the knock-out mouse model of Fabry disease caused a complete restoration of elevated Gb3 to normal levels in the liver and spleen, a 90% reduction in the heart, and a significant reduction in the kidneys (Brady et al., 2001). A related experiment with recombinant human α-galactosidase A produced in Chinese hamster ovary cells by the Genzyme Corporation, Cambridge, MA, (agalsidase beta) revealed clearance of Gb3 from the liver, heart, spleen, and kidneys of α-galactosidase A knock-out mice (Ioannou et al., 2001).

A phase 1 safety and dose-response trial was carried out in ten patients with Fabry disease with α-galactosidase A produced in skin fibroblasts (Schiffmann et al., 2000). The enzyme caused a reduction of Gb3 in the liver of nine patients, and nine patients showed a decrease in urine sediment at 28 days after infusion. The tenth patient had a reduction of Gb3 in the urinary sediment at 21 days. A phase 1/2 open label trial of agalsidase beta revealed that infusion of this enzyme preparation caused a reduction of Gb3 in the plasma, liver, skin, heart, and kidney in patients with Fabry disease (Eng et al., 2001a).

These investigations were followed by a double-blind, placebo-controlled trial of enzyme replacement in 26 hemizygous males with Fabry disease employing α-galactosidase A produced in a continuous human cell-line by Transkaryotic Therapies' gene activation technique (agalsidase alpha; Schiffmann et al., 2001). Patients experienced a reduction in neuropathic pain, improved pathology of the kidney, restoration of the abnormal regional blood flow in the brain (Moore, Hescovitch, and Schiffmann, 2001; Moore et al., 2002), beneficial effects on sweating and warm and cold sensation (Schiffmann et al., 2003), and stabilization of glomerular filtration rate in subgroups of patients with stage 1 (GFR > 90 mL/min) and stage 2 renal disease (GFR 60–90 mL/min; Schiffmann et al., 2006). Studies with agalsidase beta showed that administration of this preparation caused clearance of Gb3 from renal capillary endothelial cells in 69% of the treated group compared with no reduction in the placebo group along with reduction of Gb3 in microvascular endothelial cells in the heart and skin and reduction of plasma Gb3 to an undetectable level (Eng et al., 2001b). These findings formed the basis of approval of both agalsidase alpha and agalsidase beta for the treatment of patients with Fabry disease in countries of the European Union in August 2001, and agalsidase beta in the United States in April, 2003.

A number of reports concerning the clinical efficacy of these enzyme preparations have appeared with regard to overall beneficial effects in hemizygous males and heterozygous females (Beck et al., 2004; Baehner et al., 2003), improvement of cardiac function (Weidemann, et al., 2003), gastrointestinal difficulties (Dehout et al., 2004; Hoffmann, Reinhardt, and Koeetzko, 2004), nerve function (Hilz et al., 2004), and quality of life (Hoffmann et al., 2005). Administration of the enzyme is sometimes accompanied by infusion-related reactions such as facial flushing, shortness of breath, tachycardia, and rigors. These are generally well controlled by increasing the time of infusion of the enzyme, H_1 and H_2 receptor antagonists, and/or premedication with steroids (Schiffmann et al., 2006). Many of the patients developed IgG antibodies which is not surprising inasmuch as approximately 45% of the patients under our clinical care are cross-reacting immunologic material negative (CRIM-). In time, most of the titers decline, and few if any patients develop IgM antibodies to α-galactosidase A. However, a potentially serious immunological aspect is the development of neutralizing antibodies that reduce the catalytic activity of the administered enzyme (Linthorst et al., 2004). Fortunately procedures are available to overcome this potential impediment (Brady et al., 1997), and clinicians are optimistic about the long-term prospects of enzyme replacement therapy for Fabry disease.

5.2 Substrate Reduction Therapy

An alternative strategy currently in consideration for the treatment of patients with metabolic storage disorders is the use of small molecular weight compounds to block for biosynthesis of accumulating glycosphingolipids. This approach has been examined quite extensively in Gaucher disease (Cox et al., 2000). My colleagues and I have initiated an ongoing investigation using N-butyldeoxynojirimycin (NBDNJ) in patients with Type 3 (chronic neuronopathic) Gaucher disease along with intravenously administered enzyme replacement therapy. This strategy was developed with the hope that a sufficient quantity of orally administered NBDNJ would cross the blood–brain barrier and reduce the pathological quantities of glucocerebroside in neurons.

Investigations of substrate reduction have been carried out in the α-galactosidase A knock-out mouse model of Fabry disease (Abe et al., 2000). A more potent inhibitor of glucocerebroside formation D-*threo*-1-elthylenedioxyphenyl-2-palmitoylamino-3-pyrrolidino-propanol (D-t-EtDO-P4)) was injected intraperitoneally in this experiment. All of the recipient animals had an approximately 50% reduction of Gb3 in the kidney, liver, and heart. Electron microscopy revealed a significant reduction of lamellated inclusion bodies in the proximal tubular epithelial cells of the treated mice. At this time, however, no data are available on the effect of substrate reduction in patients with Fabry disease.

5.3 Molecular Chaperone Therapy

It has been shown that certain inhibitors of α-galactosidase A at subinhibitory levels can associate with nascent α-galactosidase A within cells and stabilize the transport of these molecules from the endoplasmic reticulum to lysosomes. Under these circumstances, the residual catalytic activity of certain mutated forms of α-galactosidase A can be increased (Fan et al., 1999). This observation has stimulated a trial of oral 1-deoxy-galactonojirimycin (DGJ) in patients with Fabry disease when the residual activity of α-galactosidase A can be increased in lymphoblasts and fibroblasts in vitro in the presence of this active site-directed molecular chaperone. Low catalytic activity of α-galactosidase A having amino acid substitutions A97V and R301Q and other missense mutations was significantly enhanced in the presence of subinhibitory concentrations of DGJ (~20 μM). Patients with enhanceable α-galactosidase A activity will be enrolled in clinical trials to determine the safety, pharmacokinetics, and pharmacodynamics of DGJ. It is presumed that data collected in these investigations will provide the basis for clinical efficacy trials of this molecular chaperone in patients with Fabry disease.

5.4 Gene Editing

Considerable initiative has been devoted to the correction of genetic mutations by the preparation of self-complementary oligonucleotides that fold into a double hairpin configuration (Liu, Parekh-Olmedo, and Kmiec, 2003). When prepared with a chimeric DNA and 2'-O-methyl RNA backbone, such a construct called a chimeraplast can serve as a template for the repair of point mutations in genes (Gamper et al., 2000). This technique gained a degree of credence when it was reported that gene editing could correct defects in the α-glucosidase gene in vitro and in vivo that cause the glycogen-storage disorder Pompe disease (Lu, Lin, and Lin, 2003). However, it has been reported that chimeraplasty was not useful for correcting the frequently encountered c.1448C->T (L444P) mutation of glucocerebrosidase in patients with Gaucher disease (Diaz-Font et al., 2003). Therefore, it seems unlikely that a conclusion can be reached at this point concerning the general utility of this form of gene editing for the correction of genetic defects in lysosomal storage disorders.

5.5 Gene Therapy

Based on the premise that a continuous supply of α-Gal A is produced in the body of patients, it might be inferred that a constant supply of enzyme produced by gene therapy may be more beneficial than intermittent infusions of this enzyme. Much of the accumulating Gb_3 in Fabry disease appears to arise from the catabolism of sphingolipids that arise from the stroma of senescent erythrocytes. This biodegradation occurs primarily in reticuloendothelial cells that arise from the bone marrow. Introduction of the functioning gene into bone marrow stem or progenitor cells may arrest, and possibly reverse, the accumulation of Gb_3. We carried out pivotal investigations of gene therapy in the mouse α-galactosidase A knockout model of Fabry disease with highly encouraging results (Takenaka et al., 1999a,b, 2000). Improved results were obtained by preselecting cells that contained the transgene (Qin et al., 2001). Later investigations using adeno-associated viral vectors brought about long-term enzymatic and functional correction in multiple organs of the Fabry mice (Jung et al., 2001; Park et al., 2003). No immune response was elicited in the knockout mice which are CRIM- for α-galactosidase A similar to the major proportion of patients with Fabry disease. A still more prolonged beneficial response in Fabry mice was reported by Ziegler et al. (2004) who also employed an adeno-associated virus(34). Moreover, induction of immune tolerance to α-galactosidase A was observed in this study.

Despite such findings that appear to augur well for gene therapy for Fabry disease, caution must be exercised in the use of viral vectors for gene transfer. Mice that lack β-glucuronidase injected with adeno-associated viral vectors containing the human glucuronidase gene were found to have hepatocellular carcinomas and angiosarcomas (Donsante et al., 2001). Several types of neoplasms have been reported in one line of mice that overexpressed β–glucuronidase, but the cause of the tumors is uncertain. Recombinant AAV vectors with the same promoter employed in these studies have been used in studies with other genes without evidence of tumor formation (Kay, 2003). Tumor production appeared to depend on the overexpression of glucuronidase as well as the strain of mice employed. Hot spots for the integration of AAV have been identified including chromosomal DNA breaks (Miller, Petek, and Russell, 2004) and in the region of gene regulatory sequences (Nakia et al., 2005). Possible cancer-related genes were involved at a frequency of 3.5%. Considerably more information is obviously required concerning the safety of such vectors. It is hoped that procedures will soon be developed to target vectors for the genes of lysosomal enzymes to noninjurious sites in the human genome.

REFERENCES

Abe, A., Gregory, S., Lee, L., Brady, R.O., Kulkarni, A., and Shayman, J.A., 2000, Reduction of globotriaosylceramide in Fabry disease mice by substrate deprivation. *J Clin Invest.* 105: 1563.

Baehner, F., Kampmann, C., Whybra, C., et al., 2003, Enzyme replacement therapy in heterozygous females with Fabry disease: Results of a phase IIIB study. *J. Inherit. Metab. Dis.* 26: 617.

Beck, M., Ricci, R., Widmer, U., et al., 2004, Fabry disease: Overall effects of agalsidase alpha treatment. *Eur. J. Clin. Invest.* 34: 838.

Brady, R.O., 1966, The sphingolipidoses. *N. Engl J. Med.* 275: 312.

Brady, R.O., and Schiffmann, R., 2005, Fabry's disease, in: Peripheral Neuropathy, Fourth Edition, P.J. Dyck and P.K. Thomas, Eds., Elsevier, Philadelphia, pp. 1893–1904.

Brady, R.O., Gal, A.E., Bradley, R.M., et al., 1967a, Enzymatic defect in Fabry's disease. Ceramidetrihexosidase deficency. *N. Engl. J. Med.* 276: 1163.

Brady, R.O., Gal, A.E., Bradley, R.M., Martensson E., 1967b, The metabolism of ceramidetrihexosides. I. Purification and properties of an enzyme which cleaves the terminal galactose molecule of galactosylgalactosylglucosylceramide. *J. Biol. Chem.*, 242: 1021.

Brady, R.O., Murray, G.J., Moore, D.F., and Schiffmann, R., 2001, Enzyme replacement therapy in Fabry disease: *J Inher Metab Dis.* 24: S2 18.

Brady, R.O., Murray, G.J., Oliver, K.L., et al., 1997, Management of neutralizing antibody to Ceredase in a patient with type 3 Gaucher disease. *Pediatr*, 100: e11-e14.

Brady, R.O., Tallman, J.F., Johnson, W.G., et al., 1973, Replacement therapy for inherited enzyme deficiency: Use of purified ceramidetrihexosidase in Fabry's disease. *N. Engl. J. Med.* 289: 9.

Branton, M.H., Schiffmann, R., Sabnis, S.G., et al., 2002, Natural history of Fabry disease: Influence of α-galactosidase activity and genetic mutations on clinical course. *Medicine* 81: 122.

Cox, T., Lachmann. R., Hollak, C., et al., 2000, Novel oral treatment of Gaucher's disease with N-butyldeoxynojirimycin (OGT 918) to decrease substrate biosynthesis. *Lancet* 355: 1481.

Dehout, F., Roland, D., Treille de Granseigne, S., et al., 2004, Relief of gastrointestinal symptoms under enzyme replacement therapy in patients with Fabry disease. *J. Inherit. Metab. Dis.* 27: 499.

Desnick, R.J., Ioannou, Y.A., and Eng, C.M., 2005, α–Galactosidase A deficiency: Fabry disease. In The Metabolic & Molecular Bases of Inherited Disease, ninth edition. C. R. Scriver, A.L. Beaudet, W.S. Sly, and D. Valle, Eds. McGraw-Hill, New York.

Diaz-Font, A., Cormand, B., Chabas, A., Vilageliu, L., and Grinberg, D., 2003, Unsuccessful chimeraplast strategy for the correction of a mutation causing Gaucher disease. *Blood Cell Mol Dis.* 31: 183.

Donsante, A., Vogler, C., Muzyczka, N., et al., 2001, Observed incidence of tumorigenesis in long-term rodent studies of rAAV vectors. *Gene Ther.* 8: 1343.

Eng, C.M., Banikazemi, M., Gordon, R.E., et al., 2001a, A phase 1/2 clinical trial of enzyme replacement in Fabry disease: Pharmacokinetic, substrate clearance, and safety studies. *Am. J. Hum. Genet.* 68: 711.

Eng, C.M., Guffon, N., Wilcox, W.R., et al., 2001b, Safety and efficacy of recombinant human α-galactosidase A replacement therapy in Fabry's disease. *N Engl J Med.* 345: 9.

Fan, J-Q, Ishgii, S., Asano, N., and Suzuki, Y., 1999, Accelerated transport and maturation of lysosomal α-galactosidase A in Fabry lymphoblasts by an enzyme inibitor, *Nature Med.* 5: 112.

Gamper, Jr., H.B., Cole-Trasuss, A., Metz, R., et al., 2000, A plausible mechanism for gene correction by chimeric oligonucleotides. *Biochemistry* 39: 5808.

Garman, S.C., and Garboczi, D.N., 2004, The molecular defect leading to Fabry disease: Structure of human α-galactosidase. *J Mol Biol.* 337: 319.

Hilz, M.J., Brys, M., Marthol, H., Stemper, B., and Dutsch, M., 2004, Enzyme replacement therapy improves function of C-, Adelta-, and Abeta-nerve fibers in Fabry neuropathy. *Neurology*, 62: 1066.

Hoffmann, B., de Lorenzo, A.G., Mehta, A, Beck, M., Widmer, U., and Ricci, R., 2005, Effects of enzyme replacement therapy on pain and health related quality of life in patients with Fabry disease: Data from FOS (Fabry Outcome Survey). *J Med Genet.* 42: 247.

Hoffmann, B., Reinhardt, D., and Koeetzko, B., 2004, Effect of enzyme-replacement therapy on gastrointestinal symptoms in Fabry disease. *Eur. J. Gastroenter. Hepatol.* 16: 1067.

Ioannou, Y.A., Zeidner, K.M., Gordon, R.E., and Desnick, R.J., 2001, Fabry disease: Preclinical studies demonstrate the effectiveness of α-galactosidase A replacement in enzyme-deficient mice. *Am. J. Hum. Genet.*, 68: 14.

Johnson, W.G., and Brady, R.O., 1972, Ceramidetrihexosidase from human placenta. *Methods Enzymol.* XXVIII: 849.

Jung, S-C., Han, I.P., Limaye, A., et al., 2001, Adeno-associated viral vector-mediated gene transfer results in long-term enzymatic and functional correction in multiple organs of Fabry mice. *Proc Nat Acad Sci USA.* 98: 2676.

Kay, M.A., 2003, Looking into the safety of AAV vectors. *Nature* 424: 251.

Kint, J.A., 1970, Fabry's disease: α-Galactosidase deficiency, *Science* 167: 1268.
Kornreich, R., Desnick, R.J., and Bishop, D.F., 1989, Nucleotide sequence of the human alpha-galactosidase A gene. *Nucleic Acids Res*., 17: 3301.
Kusiak, J.W., Quirk, J.M., and Brady, R.O., 1978, Purification and properties of the two major isozymes of α-galactosidase from human placenta. *J Biol Chem*. 253: 184.
Linthorst, G.E., Holalck, C.E.M., Donker-Koopoman, W.E., et al., 2004, Enzyme therapy for Fabry disease: Neutralizing antibodies toward agalsidase alpha and beta. *Kidney Internat*., 66: 1589.
Liu, L., Parekh-Olmedo, H., and Kmiec, E.B., 2003, The development and regulation of gene repair. *Nature Rev Genet*., 4: 679.
Lu, I-L., Lin, C-Y., Lin, S-B., et al., 2003, Correction/mutation of acid α-D-glucosidase gene by modified single-stranded oligonucleotides: In vitro and in vivo studies. *Gene Ther*. 10: 1910.
MacDermot, K.D., Holmes, A., and Miners, A.H., 2001a, Anderson-Fabry disease: Clinical manifestations and impact of disease in a cohort of 98 hemizygous males. *J. Med. Genet*. 38: 750.
MacDermot, K.D., Holmes, A., and Miners, A.H., 2001b, Anderson-Fabry disease: Clinical manifestations and impact of disease in a cohort of 60 obligate carrier females. *J. Med. Genet*., 38: 769.
Mayes, J.S., Scheerer, J.B., Sifers, R.N., and Donaldson, M.L., 1981, Differential assay for lysosomal alpha-galactosidases in human tissues and its application to Fabry's disease. *Clin Chim Acta*. 112: 247.
Miller, D.G., Petek, L.M., and Russell, D.W., 2004, Adeno-associated virus vectors integrate at chromosome breakage sites. *Nat. Genet*. 36: 767l.
Moore, D.F., Altarescu, G., Ling, G.S., et al., 2002, Elevated cerebral blood flow velocities in Fabry disease with reversal after enzyme replacement therapy. *Stroke*, 33: 525.
Moore, D.F., Hescovitch, P., and Schiffmann, R. 2001, Selective arterial distribution of cerebral hyperperfusion in Fabry disease. *J. Neuroimaging* 11: 303.
Moore, D.F., Scott, L.T., Galdwin, M.T., et al., 2001, Regional cerebrl hyperperfusion and nitric oxide pathway dysregulation in Fabry disease: Reversal by enzyme replacement therapy. Circulation 104: 1506.
Moore, D.F., Ye, F.Q., Brennan, M., et al., 2004, Ascorbate decreases Fabry cerebral hyperperfusion suggesting a reactive oxygen species abnormality: An arterial spin labeling study. *JMRI*, 20: 674.
Nakia, H., Wu, X., Fuess, S., et al., 2005, Large-scale molecular characterization of adeno associated virus vector integration in mouse liver. *J Virol*. 79: 3605.
Ohshima, T., Murray, G.J., Swain, W.D., et al., 1997, α-Galactosidase A deficient mice: a model of Fabry disease. *Proc. Natl. Acad. Sci. USA* 94: 2540.
Ohshima, T., Schiffmann, R., Murray, G.J., et al., 1999, Aging accentuates and bone marrow transplantation ameliorates metabolic defects in Fabry disease mice. *Proc. Natl. Acad. Sci. USA* 96: 6423.
Park, J., Murray, G.J., Limaye, A., et al., 2003, Long-term correction of globotriaosylceramide storage in Fabry mice by recombinant adeno-associated virus-mediated gene transfer. *Proc Nat Acad Sci. USA* 100: 3450.
Qin, G., Takenaka, T., Telsch, K., et al., 2001, Preselective gene therapy for Fabry disease. *Proc Natl Acad. Sci. USA* 98: 3428.
Ries, M., Gupta, S., Moore D.F., et al., 2005, Pediatric Fabry disease, *Pediatr*. 115: e344.

Schäfer, E., Baron, K., Widmer, U., et al., 2005, Thirty-four novel mutations of the GLA gene in 121 patients with Fabry disease. *Hum. Mutation, Mutation in Brief* #798 (2005) Online.

Schiffmann, R., Floeter, M.K, Dambrosia, J.M., et al., 2003, Enzyme replacement therapy improves peripheral nerve and sweat function in Fabry disease. *Muscle and Nerve* 28: 703.

Schiffmann, R., Kopp, J.B., Austin, H.A., et al., 2001, Enzyme replacement therapy in Fabry disease. A randomized controlled trial. *JAMA*, 285: 2743.

Schiffmann, R., Murray, G.J., Treco, D., et al., 2000, Infusion of α-galactosidase A reduces tissue globotriaosylceramide storage in patients with Fabry disease. *Proc. Natl. Acad. Sci. USA* 97: 365.

Schiffmann, R., Ries, M., Flaherty, J.T., and Brady, R.O., 2006, Long-term therapy with agalsidase alpha for Fabry disease: Safety and effects on renal function. In preparation.

Shabbeer, J, Robinsin, M., and Desnick, R.J., 2005, Detection of α-galactosidase A mutations causing Fabry disease by denaturing high performance liquid chromatography. *Hum. Mutat.*, 25: 299.

Sweeley, C.C., and Klionsky, B., 1963, Fabry's disease: Classification as a sphingolipidoses and partial characterization of a novel glycolipid. *J. Biol. Chem.* 238: 3148.

Takenaka, T., Hendrickson, D.S., Tworek, D.M., et al., 1999a, Enzymatic and functional correction along with long-term enzyme secretion from transduced bone marrow hematopoietic stem/progenitor and stromal cells derived from patients with Fabry disease. *Exp Hematol.* 27: 1149.

Takenaka, T., Murray G.J., Quin, G., et al., 2000, Long-term enzyme correction and lipid reduction in multiple organs of primary and secondary transplanted Fabry mice receiving transduced bone marrow cells. *Proc. Natl. Acad. Sci. USA* 97: 7515.

Takenaka, T., Quin, G., Brady, R.O., and Medin, J.A., 1999b, Circulating alpha-galactosidase A derived from transduced bone marrow cells: relevance for corrective gene transfer for Fabry disease, *Hum Gene Ther.* 10: 1931.

Weidemann, F., Breunig, F., Deer, M., et al., 2003, Improvement of cardiac function during enzyme replacement therapy in patients with Fabry disease: A prospective strain rate imaging study. *Circulation* 108: 1299.

Wilcox, W.R., Banikazemi, M., Gufon, N., et al., 2004, Long term safety and efficacy, of enzyme replacement theray for Farby disease. *Am. J. Hum. Genet.* 75: 65.

Ziegler, R.J., Lonning, S.M., Armentano, D., et al., 2004, AAV2 vector harboring a liver- restricted promoter facilitates sustained exression of the therapeutic levels of α-galactosidase A and the induction of immune tolerance in Fabry mice. *Mol. Ther.* 9: 231.

GAUCHER DISEASE: REVIEW AND PERSPECTIVES ON TREATMENT

Mario A. Cabrera-Salazar[1] and John A. Barranger[2*].

1 INTRODUCTION

Gaucher disease is an autosomal recessive disease and the most prevalent lysosomal storage disorder with an incidence of about 1 in 20,000 live births. Despite the fact that GD consists of a phenotypic spectrum with varying degrees of severity, it has been subdivided in three subtypes according to the presence or absence of neurological involvement. It is also the most common genetic disease among Ashkenazi Jews, with a carrier frequency of 1 in 10 (Barranger and Ginns, 1989). This panethnic disease involves many organ systems (summarized in Table 1).

The disease is highly variable as a consequence of modifier genes whose identities remain unknown. However, Gaucher disease is progressive in all of its forms. Genotype/phenotype correlations are not reliable with the exception that the N370S allele, even if present in a single dose, protects from a neurodegenerative course (Tsuji et al., 1988).

1.1 Historical Background

Even though Gaucher did not recognize the multisystemic implications of the disease that would be named after him, his studies increased the curiosity and the will of those who would find answers to this rare disease. The biochemical understanding of Gaucher disease progressed with the identification of the stored substance as a cerebroside by Lieb in 1924, and its further characterization as glucocerebroside by Aghion in 1934. The historical perspectives on the lysosome have been outlined in chapters 1 and 2 of this book and those details are beyond the scope of this chapter. Briefly, after the identification of lysosomes, lysosomal storage, and the formulation of the enzyme replacement concept by Dr. De Duve, Dr. Brady and colleagues identified the specific enzymatic deficiency, explaining glucocerebroside accumulation as secondary to an enzyme deficiency in its degradative pathway (Brady et al., 1966). A long series of efforts led to the purification, modification, and efficient targeting of placental glucocerebrosidase, leading to regulatory approval in 1991 and to the development of a recombinant enzyme, Cerezyme, in 1994. The development of these two products has become a template for the development of treatments for other lysosomal storage disorders, seven of which are either approved or are in clinical trials.

---

1. Applied Discovery Research, Genzyme Corporation. 31 New York Avenue. Framingham MA. 01701. USA.
2. Department of Human Genetics, University of Pittsburgh. Pittsburgh PA 15261. USA.
e-mail: *please address correspondence to: jbarrang@hgen.pitt.edu

Table 1. Clinical features of Gaucher disease

	Type 1 (1 in 20.000 – 1 in 40.000)	**Type 2 (1 in >100.000)**	**Type 3 (1 in 100.000)**
Hematological 94% of non-splenectomized patients have anemia and thrombocytopenia	**Thrombocytopenia:** Platelet counts between 10–100,000, bruising,, bleeding. Increased surgical risk, epistaxis Abnormal clotting times **Anemia** Hgb <12 mg/dl in 94% before ERT **Leukopenia**	Anemia and thrombocytopenia, severe hypersplenism	Similar to type 1 Gaucher disease
Visceral Visceral involvement is variable but always progressive	**Hepatomegaly** 1.5 to 2.5 times normal size, clotting factor deficiencies, **Splenomegaly** 5 to 50 times normal size Prior to enzyme replacement therapy surgical removal was common because of hyperesplenism, functional impairment or necrosis (~30% of untreated patients) Hepatopulmonary syndrome	Massive visceromegaly begins in the first few months of life coincident with neurological symptoms	Similar to type 1. Visceral involvement observed earlier in life, similar to severe type 1 cases
Skeletal 96% of patients have at least one radiologic abnormality	**Pain** Acute or chronic pain in nearly all cases **Skeletal abnormalities** Remodeling defects (Erlenmeyer flask deformity) Osteopenia BM infiltration Avascular necrosis Bone infarcts Fractures Lytic lesions	Early death precludes skeletal involvement	Onset in childhood with severe deformity; kyphoscoliosis is common, as well as all the lesions in Type 1 disease
Neurologic	No primary neurodegenerative sings or symptoms N370S allele protects against neurological involvement	Symptoms at 3–6 months of age. Cranial nerve and extrapyramidal tract involvement Trismus, strabismus, and retroflexion of the head Spasticity , hyperreflexia, dysphagia Seizures Apnea	Type 3a: aggressive and extensive neurologic disease. Type 3b: supranuclear oculomotor abnormalities Type 3c: neurodegenerative course and valvular heart disease
Life span	Shortened life span in untreated, poor quality of life due to skeletal and hematological involvement	Death before 2 years of age	Survival between 2nd to 4th decade of life

1.2 Incidence and Inheritance

All of the forms of Gaucher disease are inherited as an autosomal recessive trait (Barranger and Ginns, 1989). The collective subtypes of Gaucher disease constitute the most prevalent form of the sphingolipidoses and the most common lysosomal storage disorder. Approximately 90% of the cases are type 1, or the chronic form. Although panethnic, this subtype is much more common among Ashkenazi Jews (Weinreb et al., 2002). In this group, it occurs with an incidence of 1 in 450, making it the most common Jewish genetic disorder. The incidence of type 2, also known as acute neuronopathic Gaucher disease, is much less frequent. The other subtypes are also infrequent and have no ethnic predilection. These subtypes are so rare that no collective incidence figure is available, but it is probably less than 1 in 50,000.

2 PATHOPHYSIOLOGY

After glucocerebrosidase deficiency was demonstrated as the cause of Gaucher disease (Brady et al., 1966), the biology of the missing enzyme and the effects of substrate storage became relevant as described in the following sections.

2.1 Stored Substances

2.1.1 Glucocerebroside

Glucocerebrosidase deficiency results in the accumulation of glucocerebroside (GL-1), a compound consisting of a glucose moiety esterified to the C-1 of ceramide in a beta-glucosidic linkage. Cerebrosides are composed of ceramide esterified to a variety of different substituents at C-1. This carbon may participate in reactions with phosphorylcholine to produce the sphingomyelins, or an unsubstituted monosaccharide or oligosaccharide needed to produce the neutral glycosphingolipids. The common unit among these compounds is ceramide. Ceramide is derived from a long-chain base named sphingosine (D(+)-erythro-1,3-dihydroxy-2-amino-4-transoctadecene, or C18 sphingosine). This lipid is joined by an amide bond at C-2 to a long-chain fatty acid to form ceramide. The fatty acid chain length varies. In general, the neutral glycosphingolipids and sphingomyelins contain C22 to C24 fatty acids, whereas the gangliosides contain C18 fatty acids. It is from sphingosine that the group of disorders of lipid catabolism obtains its name (i.e., sphingolipidoses) because the accumulating lipid compounds are derived from it.

Glucocerebroside (GL-1) is at the end of the glycosphingolipid catabolic pathway (Figure 1). The higher glycosphingolipids and gangliosides are degraded in a stepwise fashion by specific acid hydrolases, resulting in GL1, which is normally degraded to ceramide and glucose by glucocerebrosidase. The compounds that contribute to the pool of glucocerebroside in peripheral organs are globoside, globotriose, and lactosylceramide. These are derived from the degradation of membranes, the major source of which is white blood cells (Kattlove et al., 1969).

It is important to note that the glucocerebroside (GL-1) found in spleen, liver, kidney, plasma, and red cells contains fatty acids with chain length of approximately C22 to C24 (Fredrickson and Sloan, 1978). The glucocerebroside in normal brain is primarily C18 (steric acid), but in type 2 and 3 brain, it is C22-24 glucocerebrosidase (Svennerholm, 1967; Nilsson et al., 1985). These data have been interpreted to mean that the glucocerebroside accumulating in brain derives from sources outside the brain itself.

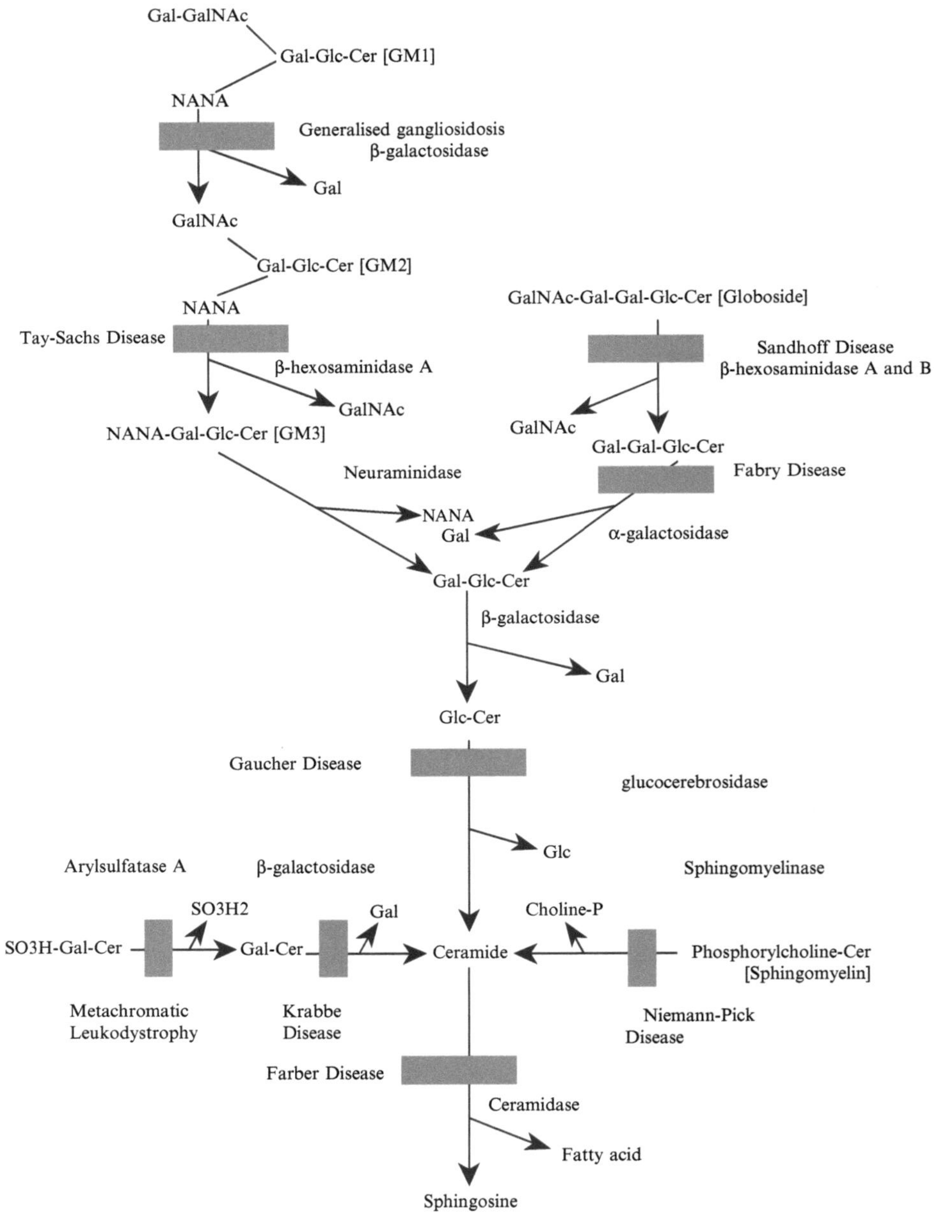

Figure 1. Glycosphingolipid catalysis is a complex metabolic process carried out by a multi-enzyme complex. Individual enzymatic deficiencies are responsible for each one of the sphingolipidoses.

If the data are confirmed, they have important consequences, especially because it has been suggested that levels of plasma and tissue glucocerebroside increase following splenectomy in Norrbottnian cases (Nilsson et al., 1985). They also provide important biological guidance to the utility of novel therapies including enzyme therapy, gene transfer, chaperones, and substrate synthesis inhibitors.

2.1.2 Glucosylsphingosine (Psychosine)

Glucosylsphingosine has been implicated in toxic damage to neural cells. This metabolic product has been shown to accumulate in the cerebral and cerebellar cortex of patients with type 2 and 3 Gaucher disease (Nilsson 1982; Nilsson and Svennerholm 1982; Nilsson et al., 1985) and Gaucher mice. Hannun and Bell (1987) have shown that lysosphingolipids, including glucosylsphingosine, inhibit protein kinase C, mitochondrial cytochrome c oxidase, and CTP phosphocoline-cytidyltransferase, are compounds that interfere with signal transduction, cellular differentiation, and apoptosis.

Glucosylsphingosine is not increased in patients with type 1 Gaucher disease, however, its levels are markedly increased in patients with type 3 (36-fold on average) and type 2 disease (225-fold). The highest levels of this metabolite were found in brain tissue from two fetuses presenting with hydrops fetalis associated with profound glucocerebrosidase deficiency as a result of a null allele (Orvisky et al., 2002).

2.2 Deficient Enzyme

2.2.1 The Glucocerebrosidase Gene

In order to more completely understand the nature of the defects responsible for the heterogeneity in Gaucher disease, isolation and characterization of the gene for glucocerebrosidase was achieved by Barranger and his colleagues (Ginns et al., 1984; Tsuji et al., 1986). A second clone carrying a polymorphism was also published (Sorge et al., 1985).

The glucocerebrosidase (*GC*) gene spans a 7.2 kb fragment on locus 1q21, consisting of 11 exons and 10 introns. Located 16 kb downstream is a highly homologous (approximately 95%) pseudo-gene sequence (Horowitz et al., 1989). Recombination between the gene and the pseudo-gene leads to complex mutations ("rec" mutations). These fusion genes are either not transcribed or so poorly transcribed and translated as to be of no significance in enzyme or protein analysis. Consequently, these would play no role in the proteosome degradation pathway recently implicated in lysosomal enzyme folding mutants (Sawkar et al., 2005).

The GC cDNA contains 1548 base pairs encoding human glucocerebrosidase, a molecule of 58 Kd with four glycosylation sites (Asp-X-Ser/Thr; Takasaki et al., 1984). In addition to the amino acid sequence of the structural protein, an additional 19 amino acid signal polypeptide has been described (Erickson, Ginns, and Barranger, 1985). This characteristic motif is responsible for cotranslational translocation of the precursor into the lumen of the ER from the cytosolic ribosome.

To date, more than 160 mutations in the GC gene have been registered in the human mutation database at Cardiff, however, most of these mutations are private and do not have value in genotype–phenotype correlation analysis (http://archive.uwcm.ac.uk/uwcm/mg/hgmd0.html).

Study of the glucocerebrosidase gene led to the discovery of prevalent mutations such as N370S and L444P. The N370S allele is the most frequent allele in Gaucher disease and is associated with only type 1 Gaucher disease. However, its presence does not allow the clinician to determine the severity of the disease due to the wide phenotypic diversity within all genotypes. Presence of an N370S allele in a patient precludes the development of neurological manifestations of the disease.

X-ray structure studies of native glucocerebrosidase has permitted a theoretical three-dimensional structure of mutant GC allowing the analysis of the relations among mutations, residual activity, and severity of the disease. Most of the 200 reported GC mutations either reduce the catalytic activity or destabilize the molecule. The N370S mutation is responsible for only minor structural changes, but results in reduced catalytic activity (Erikson, Ginns, and Barranger, 1985). In contrast to N370 data, L444P, an amino acid frequently mutated to arginine or proline is associated with neuronopathic disease, especially in its homozygous state. L444 is located near the hydrophobic core of the molecule and its mutation might cause a conformational change that disrupts the core by altered folding of the protein. Although the L444P mutant protein is unstable, its catalytic activity is completely normal (Morimoto et al., 1990; Hong et al., 1990). Other mutations such as H311R, A341T, and C342G, located near the active site, have effects on catalytic activity.

2.3 Affected Organs

2.3.1 Reticuloendothelial System

It is postulated that GL1 storage in the cells of the reticuloendothelial tissue may induce an inflammatory response mediated by the secretion of proinflammatory cytokines such as macrophage colony stimulating factor (M-CSF), soluble CD14 (sCD14), interleukin 8 (Hollak et al., 1997), and other cells of the immune system. Extensive tissue damage would in turn ensue, particularly in the liver, spleen, and bones. A recent study of the gene expression profile in Gaucher disease patients showed enhanced expression of genes associated with inflammatory reactions in the affected spleen (Moran et al., 2000). In particular, the transcript abundance of the cDNAs representing cysteine proteinases (cathepsins B, K, and S) was greatly increased. These proteins are known to participate in tissue modeling, antigen presentation, and in the case of cathepsin K, bone matrix destruction.

As part of the reticuloendothelial system, the spleen is one of the key organs involved in the pathophysiology of Gaucher disease. Painless splenomegaly is usually the earliest sign in all types of Gaucher disease. The rate of enlargement/reduction of the spleen is helpful in judging the rate of progression of the disease and the response to therapies. The preferred modalities for measuring organ volume are volumetric MRI or computed tomography (Charrow et al., 1998).

Because the main pathophysiological implication of the spleen in GD is hypersplenism, partial or total splenectomy was used prior to the availability of enzyme replacement therapy to ease the hematological abnormalities related to hypersplenism, however, worsening of bone pathology, and the risk of severe overwhelming sepsis have made splenectomy an obsolete procedure for GD for whom ERT is available.

Histologically, fibrosis with distortion of the splenic architecture is commonly observed. The spleen is enlarged and firm and contains pale areas caused by infarcts. The red pulp of the spleen is replaced by white collections of Gaucher cells. Dark-purple nodules can be observed and are consistent with foci of extramedullary hematopoiesis (Barranger and Ginns, 1989).

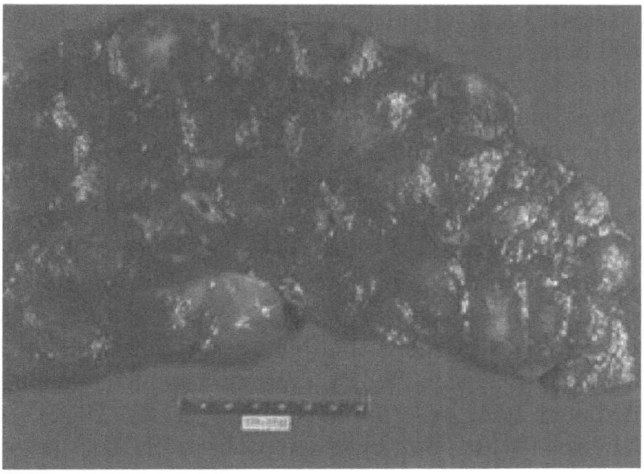

Figure 2. Nodular enlargement of the spleen in untreated Gaucher disease.

Although hepatomegaly is often noted at the time splenomegaly is observed, the liver may not become enlarged until later in the course of the disease. Moderate hepatic dysfunction is discerned by elevation of liver enzymes in serum, reduced sulfobromophthalein clearance, or reduced and nonhomogeneous uptake of radionuclide tracers. Histologically, all patients have some degree of hepatic fibrosis (James et al., 1981). Despite the frequent occurrence of hepatomegaly in Gaucher disease, hepatic failure occurs infrequently. Gaucher cells are seen within the sinusoids. In the more severely affected cases, fibrosis distorts the architecture, forming small regenerating nodules that are infiltrated by Gaucher cells. These cases have been described as cirrhotic. In contrast to other lipidoses (such as Niemann–Pick disease), hepatocytes are not involved in storage but are affected by the toxic action of molecules released from storage cells. The enlarged liver may contain sites of extramedullary hematopoiesis. The majority of patients have abnormal liver function tests.

Marked portal hypertension and consequent complications such as ascites and esophageal varices do occur, and are common in severe disease. Radionuclide scans show a shift in tracer from liver to spleen in the majority of cases, indicating a degree of portal hypertension. Jaundice is a serious sign in this disease and represents either intercurrent infectious, chronic active hepatitis, or hepatic failure. In the authors' experience, only four cases out of several hundred have died of hepatic failure. In cases with liver involvement, early intervention with enzyme replacement therapy is indicated. Hepatic transplantation is required in patients with end-stage liver disease. Hepatic failure can occur in untreated patients. Portal hypertension leading to esophageal varices is a known complication, occurring in patients with severe type 1 and type 3 disease. Fortunately, appropriate treatment with ERT has made this series of complications rare.

2.3.1.2 Skeletal System

The presence of lipid-laden macrophages in the bone marrow and the consequences of glucocerebroside storage have been proposed as mechanisms of bone involvement in GD. This dysfunction leads to abnormal remodeling of the bones, osteopenia, ostrosclerosis, osteonecrosis, and bone crises.

Abnormal bone remodeling can be observed as Erlenmeyer flask deformity of the distal femur and proximal tibia (Molino Trinidad et al., 1979).

An increase in bone resorption is a contributing cause of osteopenia-osteoporosis in patients with Gaucher disease. These pathological features are due to the increased osteoclastic activity. Osteoclastic activation in Gaucher disease is a consequence of the release of stimulating cytokines (M-CSF, IL-8, and soluble CD 14; Hollak et al., 1997). Such cytokines have a direct impact in the activation of cells from the monocyte–macrophage lineage from which osteoclasts are also derived. Osteoclasts respond to increases in such cytokines by increasing bone resorption. The levels of tartrate-resistant (nonprostatic) acid phosphatase (TRAP), especially the isoenzyme 5b, a specific marker of osteoclast-mediated bone resorption (Halleen et al., 2002), is increased in some patients with Gaucher disease (Barranger and Cabrera-Salazar, personal communication). Severity Score Index, a clinical semiquantitative assessment of the burden of Gaucher disease is significantly correlated with the elevation of surrogate markers of the disease, especially with chitotriosidase, a specific marker of macrophage activation (Cabrera-Salazar et al., 2004). These mechanisms may explain why bisphosphonates are useful adjuncts to ERT in treating the bone complications of GD.

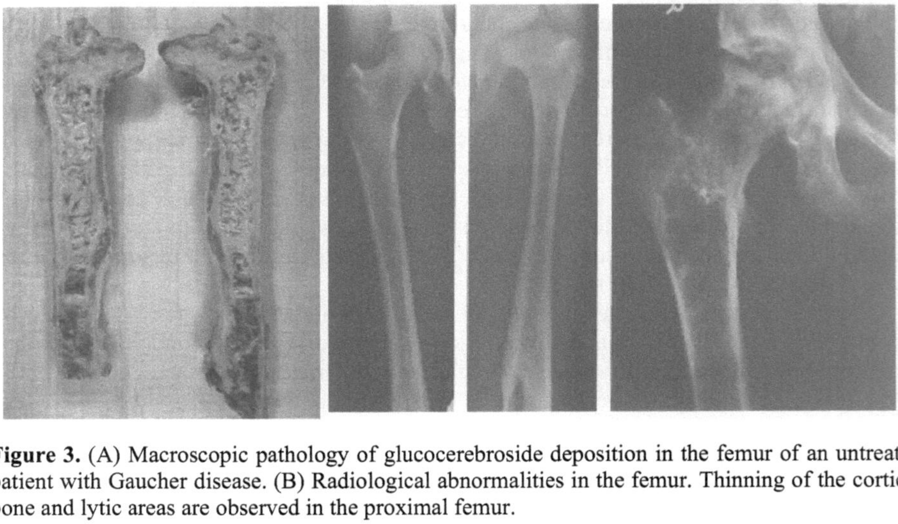

Figure 3. (A) Macroscopic pathology of glucocerebroside deposition in the femur of an untreated patient with Gaucher disease. (B) Radiological abnormalities in the femur. Thinning of the cortical bone and lytic areas are observed in the proximal femur.

Metabolic and endocrinologic studies suggest an imbalance in calcium homeostasis; however, the entire skeleton is not affected uniformly. On the contrary, the lesions consist of collections of Gaucher cells scattered throughout the skeleton. Bone infarcts probably result from a toxic process around these foci, which then leads secondarily to edema, vascular compromise, and infarction. There is a correlation between the severity of type 1 disease and serum levels of macrophage-derived cytokines. (Barranger and Ginns, 1989; Hollak et al., 1997).

The natural history of the bone involvement in Gaucher disease has been redefined with the introduction and with the continued use of enzyme replacement therapy in the last 14 years. Early diagnosis and treatment have allowed patients to have less bone-related symptoms and complications increasing their quality of life (see Table 1).

Degenerative changes in the skeleton are the leading cause of disability in patients with type 1 disease. Some degree of osteopenia and osteolysis occurs in virtually all patients. However, the extent of bone disease is variable (Matoth and Fried, 1965; Goldblatt, Sacks, and Beighton, 1978; Wenstrup et al., 2002). Very few patients have neither radiographic nor scintigraphic evidence of bone involvement. Most affected individuals have a substantial burden of bone disease that is progressive and leads to clinical presentation. Others have such severe involvement that they are confined to a wheelchair early in life because of pain, pathologic fracture, or skeletal instability. Many patients experience episodic pain lasting for days to months in the hips, legs, back, and shoulders. These episodes have been referred to as "bone crises."

2.3.1.3 Central Nervous System

Understanding of the involvement of the brain in Gaucher disease has improved in the last several years but is far from complete because of the limited number of postmortem cases available for study. Type 1 Gaucher disease patients do not have clinical symptoms or signs referable to the nervous system. In this group, anatomic and biochemical examinations of the brain have been infrequent. Consequently, little is certain about the pathology of the brain in type 1 Gaucher disease. The earlier reports of hypophyseal and leptomeningeal Gaucher cells in type 1 GD require confirmation (Teilum 1944; Chang-Lo, Yam, and Rubenstone, 1967). Glucocerebroside is not elevated in the brains of patients with this type of Gaucher disease, except in the periadventitial spaces (Virchow–Robin spaces) around blood vessels (Nilsson et al., 1985). Recent suggestions that the incidence of pulmonary disease is increased in both carriers and affected individuals may provide some new insight into the disease (Sidransky. 2004; Aharon-Peretz, Rosenbaum, and Gershoni-Baruch, 2004).

In type 2 patients' brains, free Gaucher cells have been demonstrated within the parenchyma accompanied by gliosis and microglial nodules. These changes are present but much less frequent in type 3 patients' brains. (Kaye et al., 1986). However, increased levels of glucocerebroside have been reported in brains of both type 2 and type 3 patients. Significant neuropathologic abnormalities have been observed in the brains of type 2 patients. They are more subtle in type 3 brain. Neuronal storage of lipid has been suggested in several reports, but this has not been confirmed ultrastructurally in any case of Gaucher disease. In type 2 disease, neuronophagia and neuronal cell death in the deeper layers of the cortex, thalamus, basal ganglia, brainstem nuclei, cerebellum, and spinal cord have been reported. Variable degrees of demyelination have been described in brains of type 2 patients. From the available information, one would have to conclude that the accumulation of glucocerebroside in the brain produces dysfunction in surrounding cells long before discrete pathologic changes are seen (Kaye et al., 1986). Even then, the lesions described are small, focal, and not of the profound nature seen in other neurovisceral storage disorders. These facts have prompted the hypothesis that there must be some toxic effect on the brain which may be glucosyl-sphingosine.

Neonatal GD is the most severe end of the spectrum. This rare and unique form of Gaucher disease presents with hydrops fetalis and collodion skin (Tayebi, Stone, and Sidransky, 1999). In an extensive molecular analysis of 31 patients with this type of conatal Gaucher disease, homozygosity for a recombinant allele that included the mutation L444P was associated with early death (Stone et al., 2000). An extensive review of the clinical and hematological features of neonatal Gaucher disease has been published (Mignot et al., 2003). Dysmorphic features are observed in 22% of cases of neonatal GD characterized by low-set ears, small nose with a flat bridge, and anteverted nares.

It must be emphasized that children with type 1 Gaucher disease are not infrequently diagnosed before two years of age. They may have rapid progression in bone, liver, and spleen manifestations. Some of these cases have been erroneously diagnosed as type 2. It is essential that central nervous system involvement be documented and established as an associated finding prior to making a diagnosis of type 2 disease. In these cases, genotyping studies will help the clinician determine if the patient will be affected by neuronopathic GD as well as the potential for treatment.

2.3.1.4 Biochemical Abnormalities

The need of useful markers that allow the clinician to predict the course of the disease, and also to assess the therapeutical response in patients with this disease, led to the discovery of surrogate markers that, among others, are routinely used to monitor Gaucher disease patients such as angiotensin converting enzyme (ACE), tartrate-resistant acid phosphatase (TRAP), chitotriosidase (CHITO), and recently the chemokine CCL18.

Serum tartrate-resistant acid phosphatase (TRAP) was the first surrogate marker used for the diagnosis of type 1 Gaucher disease (Tuchman et al., 1956). This marker is also included in the standards of care for these patients and it is actually used to monitor disease progression/regression in patients receiving enzyme replacement therapy (ERT) with recombinant human glucocerebrosidase.

Acid phosphatase, angiotensin-converting enzyme, lysosomal hydrolases, lysozyme, and immunoglobulins are elevated in the plasma of Gaucher patients (Tuchman, Suna, and Carr, 1956; Ockerman and Kohlin, 1969; Lieberman and Beutler, 1976; Silverstein and Friedland, 1977; Moffitt et al., 1978; Hultberg et al., 1980; Robinson and Glew, 1980). Chitotriosidase is a particularly sensitive indicator of macrophage storage in patients with Gaucher disease (Hollak et al., 1994). Because of liver involvement, patients may have prolonged partial thromboplastin, prothrombin, and bleeding times. In addition, it has been suggested that increased amounts of plasma glucocerebroside may interfere with the clotting cascade (Boklan and Sawitsky, 1976). Glucocerebroside is elevated in the plasma and may be as high as ten times normal. Although the level of plasma glucocerebroside has been shown to increase following splenectomy in some Norrbottnian cases, this has not been reported in cases of other subtypes of the disease.

The Gaucher cell is also a prominent source of the chemokine CCL18/PARC. As such, this chemokine has been determined to be an indicator of Gaucher disease burden and the response to treatment, especially in patients presenting a deficiency of chitotriosidase, a marker to which CCL18 presents a close correlation (Boot et al., 2004).

3 CLINICAL DIAGNOSIS OF GAUCHER DISEASE

The diagnosis of Gaucher disease should be considered in any case of unexplained splenomegaly with or without a bleeding diathesis or other manifestations of the disease in the skeleton or liver. In Jews, the diagnosis should be the first rule-out in patients with splenomegaly. Type 2 Gaucher disease should be considered likely in any infant with early-onset hepatosplenomegaly and a neurodegenerative course. Elevation of tartrate-resistant acid phosphatase in serum is very suggestive of the disease and an elevation of chitotriosidase is nearly pathognomonic (Hollak et al., 1994,1997). Although the observation of characteristic Gaucher cells in bone marrow biopsies narrows the diagnostic possibilities, bone marrow biopsy is not the recommended procedure for the diagnosis of Gaucher disease. In Jews, the disease is so common that bone marrow biopsy should not be done. A definitive diagnosis is made by assay of glucocerebrosidase in leukocytes, fibroblasts, chorionic villi, or urine (Barranger and Ginns, 1989).

3.1 Hematologic Workup

Hemoglobin, platelet count, and biochemical surrogate markers of the disease (tartrate-resistant acid phosphatase, angiotensin converting enzyme, and chitotriosidase) should be obtained when the diagnosis is established. These measurements are recommended every 3 months for patients on enzyme replacement therapy and at least every 12 months for those patients who choose not to receive ERT or are unable to receive it. Any of these three parameters is useful to evaluate the response to treatment and to monitor the activity of the disease (Cabrera-Salazar et al., 2004).

3.2 Visceral Studies

Spleen and liver volume should also be asessed at the time of diagnosis. The recommended technique is MRI or volumetric CT scan. These tests are recommended every 12 months in patients receiving ERT and every 24 months in those patients who achieve a stable state.

3.3 Skeletal Assessment

Computerized tomography, radionuclide scan, and magnetic resonance imaging have been useful in assessing the extent of bone abnormalities. The recommended method for routine assessment of bone disease is T1- and T2-weighted MRI of the entire femora (Charrow et al., 1998).

Bone density should also be assessed in patients with Gaucher disease because of decreased bone mineral density. DEXA studies are useful to determine and assess the risk of pathological fractures. Spine, femur, forearm, and whole body densitometries can be performed in these patients. These studies should be performed at baseline and every 12 months in patients receiving ERT.

3.4 Neurological Evaluation

Neurological complications develop by three to six months of age; constant features of type 2 disease are trismus, strabismus, and retroflection of the head with progressive spasticity, hyperreflexia, positive Babinski signs, and other pathological reflexes. Dysphagia and difficulty in handling secretions develop and are often followed by aspiration pneumonia.

Seizures may occur. As neurological deterioration proceeds, the child usually becomes apathetic and motionless. Death occurs either from apnea or aspiration pneumonia at an average age of nine months, with a range of one month to two years. Confirmation of the diagnosis should be made by determining the in vitro activity of glucocerebrosidase in either fibroblast or peripheral blood leukocytes.

The clinical features of type 3 Gaucher disease, apart from those referable to the nervous system, are common to the other types of Gaucher disease. Hepatosplenomegaly usually precedes neurological abnormalities. The variability in systemic organ involvement is similar to that seen in type 1.

In the well-documented and biochemically proven cases, of which the Swedish collection is the largest, there is marked variation in age of onset and severity of organ involvement.

3.5 Undertreated Gaucher Disease

Severe complications of type 1 Gaucher disease were seen in the epoch prior to the advent of enzyme replacement therapy. In areas of the world where heightened awareness of Gaucher disease has not been achieved or the value of ERT not completely understood, these complications can occur.

4 TREATMENT OF GAUCHER DISEASE

4.1 Development of Therapies

The development of therapies for lysosomal storage disorders started with the discovery of the lysosome and identification of the accumulated substrates and the determination of their position in human metabolic pathways (De Duve, 1969; Brady et al., 1974). The discovery of the unique biology of the lysosome provided key findings that proved critical to the design of therapy (Table 2). These discoveries allowed the development of therapeutic approaches including bone marrow transplantation (BMT), enzyme replacement therapy (ERT), substrate reduction therapy (SRT), gene transfer, and recently chaperone therapy. These approaches are described briefly.

Table 2. Historical milestones for the development of enzyme replacement therapy for Gaucher disease

Discovery	Researcher/Year
Description of the lysosome	De Duve, 1955
Deficiency of glucocerebrosidase	Brady, 1966
Secretion-reuptake hypothesis	Fratantoni, 1968
Discovery of receptors for glycoproteins (lectins)	Ashwell and Morell, 1974
Purification of glucocerebrosidase.	Furbish et al., 1977
Identification of mannose-specific lectin on Kupffer cells	Ashwell and Kawasaki, 1978. Archord et al., 1977
Determination of glucocerebrosidase structure, glycosylation and terminal sugars	Barranger et al., 1978
Demostration of glucocerebrosidase uptake by Kupffer cells via mannose-specific lectin	Furbish et al., 1978
Demostration of the first clinical response	Barranger et al., 1980

4.1.1 Enzyme Replacement Therapy

ERT has had excellent therapeutic success, especially in GD. The relative inconvenience of biweekly infusions and the cost of therapy have encouraged the development of other therapeutic approaches. Gene transfer is a potentially curative treatment that has not yet achieved its potential. Recent developments using AAV vectors in animal models (Marshall et al., 2002; McEachern et al., 2006) and using lentiviral vectors are discussed by Drs. Biffi and Naldini in this book.

The description of the degradative function of the lysosome and the demonstration that lysosomal storage is secondary to enzymatic deficiencies, spawned the concept of enzyme replacement as a strategy for treatment (illustrated in Figure 4). Its practical success followed a series of discoveries as outlined in Table 2.

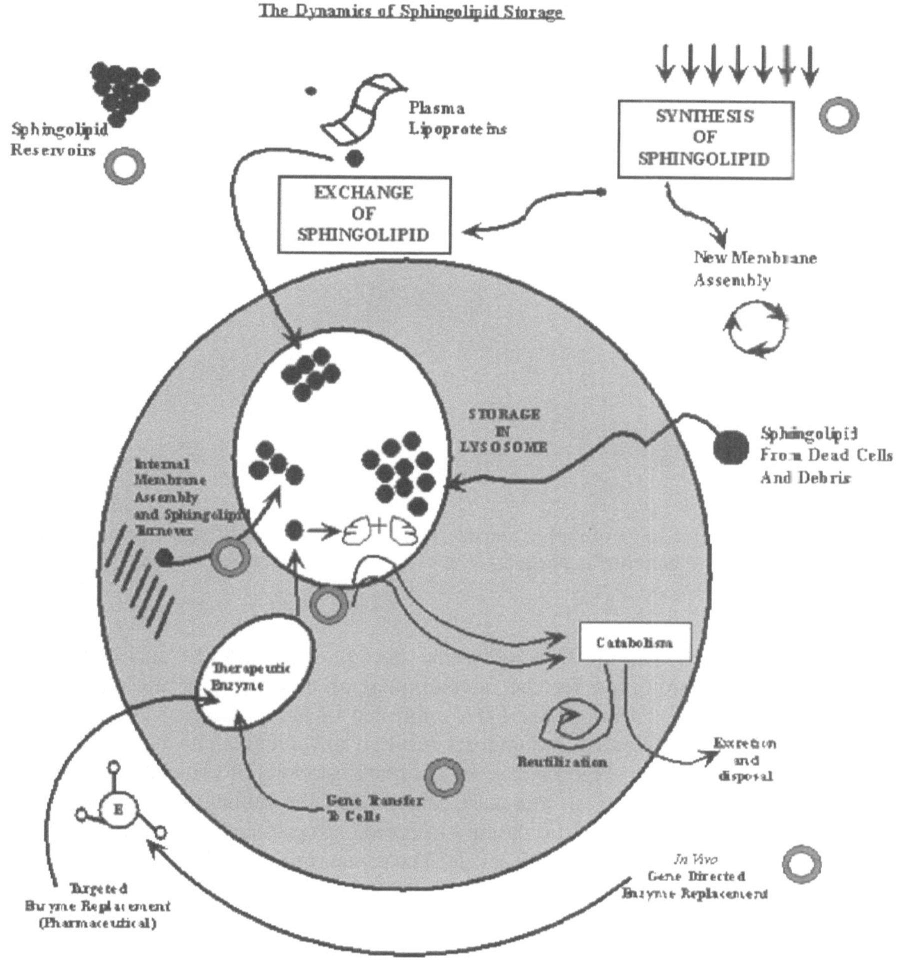

Figure 4. Pathogenesis of sphyngolipid storage approaches to treatment.

Advances in the isolation of glucocerebrosidase from placenta led to clinical studies that were unsuccessful at first because the amount of enzyme taken up in vivo was not sufficient to produce a measurable clinical response (Brady et al., 1974). The identification of mannose moieties in the carbohydrate chains of glucocerebrosidase (see Figure 5) and the discovery of a mannose-specific uptake system on the membranes of macrophages provided the clues that lead to successful enzyme replacement therapy. This receptor allows efficient binding and facilitated uptake of the enzyme specifically by Kupffer cells. This drug delivery system was exploited to achieve dramatic clinical results (Ashwell and Morel, 1974; Furbish et al., 1977; Barton et al., 1990).

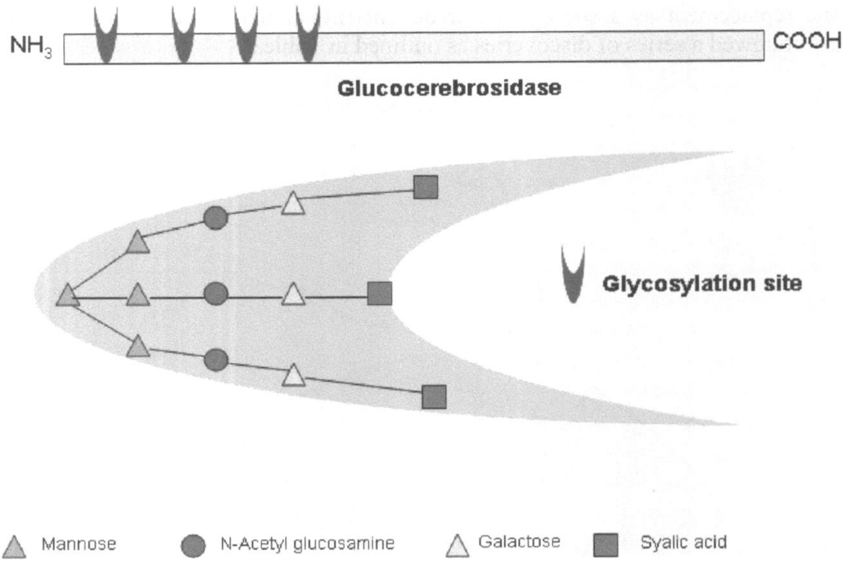

Figure 5. Carbohydrate units of native glucocerebrosidase. Removal of syalic acid, galactose, and N-acetyl glucosamine moieties is needed for appropriate mannose-mediated uptake of glucocerebrosidase by macrophages.

Gaucher disease was the first lysosomal storage disorder to be successfully treated by ERT. It is the prototype for the development of treatment for the entire group of diseases, seven of which are either FDA approved or are in clinical trials. Application of the principles of facilitated endocytosis resulted in the excellent therapeutic effect of mannose-terminated glucocerebrosidase on the manifestations of Gaucher disease, initially demonstrated in one patient. In this case, anemia and thrombocytopenia were corrected within months. Interruption of the therapy triggered a regression of the improved clinical parameters to the previous abnormal levels. The reinstitution of ERT led to further and an improved clinical benefit as observed in Figure 6. The demonstration of this consistent and sustained therapeutic effect led to more extensive clinical trials (Barton et al., 1991) that resulted in FDA approval of purified GC (Alglucerase, Ceredase). This commercial form was replaced by GC produced by recombinant DNA technology (Imiglucerase, Cerezyme) in 1994. The recombinant and placental enzyme are equal in their clinical effects (Grabowski et al., 1995).

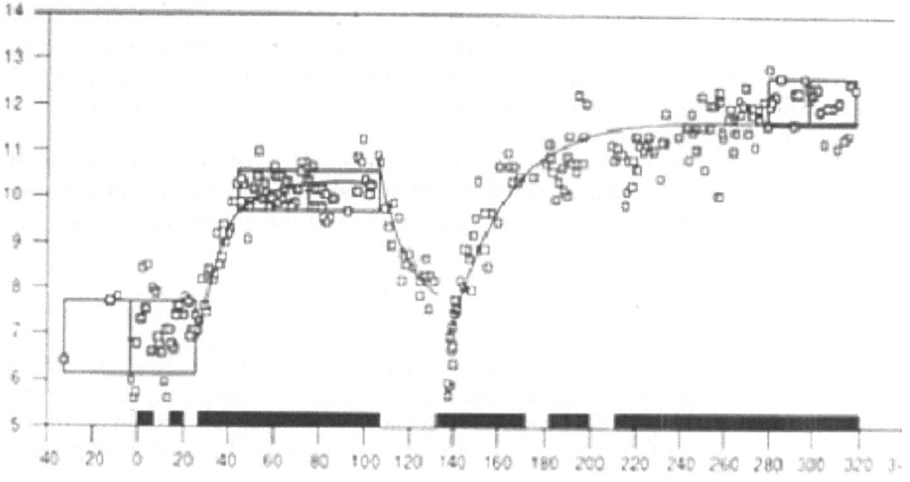

Figure 6. Hemoglobin response to glucocerebrosidase treatment and withdrawal: *y*-axis indicates hemoglobin (mg/dL). Solid lines in the *x*-axis indicate the time (weeks) at which the enzyme was administered/withdrawn.

4.1.1.2 Outcomes

Treatment with glucocerebrosidase provides significant improvements in hematological, biochemical, visceral, skeletal and in the quality of life as outlined in Table 3.

Table 3. Clinical response to enzyme replacement therapy in Gaucher disease

Organ/Manifestation	Response to ERT After 1 Year
Visceral	
• Liver	• Volume decreases by approximately 20%
• Spleen	• Volume decreases to near normal in children and by 50–70% in adults
Hematological	
• Anemia	• Increase of 1 to 7 gm% in hemoglobin
• Thrombocytopenia	• Normalization to over 100.000
Biochemical/Surrogate markers	
• Chitotriosidase, angiotensin converting enzyme and tartrate resistant acid phosphatase	• Decrease and stabilization • Reduction in bone resorption markers (Ciana et al., 2003)
Bone	• Elimination of bone infarcts • Dramatic reduction in bone pain
Quality of life	• Greatly improved

The clinical symptoms related to skeletal deterioration respond well to ERT. Skeletal complications can be avoided if patients are treated before clinical bone disease is evident (Bembi et al., 2002). Bone pain is relieved and the incidence of bone infarct is reduced dramatically to near zero in patients treated with 60 U/Kg of recombinant human glucocerebrosidase every two weeks (Barranger et al., 1978; Wenstrup et al., 2002). As is the case generally in the treatment of GD, avoidance of bone complications requires careful individual dose adjustments guided by clinical monitoring (NIH Tech conference, No author, 1996; Weinreb et al., 2004; Charrow et al., 1998). Thorough assessment at baseline, repeated at three-month intervals is essential to dose adjustment in the initial period. Less frequent observations are needed after satisfactory therapeutic goals are achieved. Recommended baseline and monitoring guidelines are available at http://www.cerezyme.com/healthcare/docs/ICGGMonitoring. When these guidelines are strictly applied in young patients, a disease-free state can be achieved.

4.1.1.3 Adverse Effects of ERT

The adverse effects of enzyme replacement therapy for Gaucher disease are few and rarely a reason for discontinuing the therapy. Antibodies to the enzyme have been reported in approximately 15% of patients with Gaucher disease, however, most of these antibodies do not neutralize the enzyme and the titers tend to decrease with continued therapy. The incidence of serious allergic complications such as anaphylactic reactions to glucocerebrosidase or the production of neutralizing antibodies to the enzyme is very low (Rosenberg et al., 1999).

4.2 Substrate Reduction Therapy

Other approaches, focused primarily on the treatment of neurodegenerative lysosomal diseases include the development of compounds capable of reducing the rate of synthesis of the glycosphingolipid (GSL) substrates that accumulate. Inhibition of GSL biosynthesis by pharmacological means would allow the residual enzyme that is present in the lysosomes to either reduce substrate storage (Platt et al., 2001) or stabilize the residual enzyme by working as chemical chaperones, preventing their degradation in the endoplasmic reticulum and increasing the residual enzymatic activity of the cells. Chemical protection of enzyme activity has been proven in cells homozygous for the N370S mutant allele, in which there is residual enzymatic activity. In diseased cells, a twofold increase of enzyme activity was observed (Sawkar et al., 2002). These results suggest a potential treatment in those patients bearing mutant alleles which retain some catalytic activity and at least some residual protein. It would have no value for diseases caused by null mutations.

4.2.1 Inhibition of Glycosphingolipid Biosynthesis

Glycosphingolipid biosynthesis is nonspecifically inhibited by iminosugars. The most studied compound is N-Butyldeoxynojirimycin (DNJ), is also known as Zevesca, Vevesca, or Miglustat.

The compound is an inhibitor of the ceramide-specific glucosyltransferase that initiates the glycosphingolipid biosynthetic pathway and catalyses the formation of glucocerebroside (Figure 1). The ability of DNJ to reduce GSL biosynthesis was first studied in mouse models of Tay–Sachs and Sandhoff disease. The inhibition of the glucocerebroside synthase (GCS) by DNJ was proven to be feasible in an in vitro Gaucher disease model by clearing of the lysosomal accumulation of glucocerebroside. This fact is a consequence of the ability of the residual enzyme to catabolize the accumulated substrate and partial prevention of its synthesis. A more detailed review of this therapeutic approach was written by Dr. Platt as part of this book.

4.2.2 Clinical Trials of SRT

In Gaucher disease, several clinical trials have been carried out using DNJ. In a group of adult patients naïve to ERT, a marginal decrease in splenomegaly, and no improvement in hematologic indexes was observed after 12 months of treatment (Cox et al., 2000). Extension of the study for 36 months revealed a further decrease in spleen size, but insignificant improvement in hematologic indexes. These responses to SRT are slow to occur and do not have a real impact on the disease of type-1 patients as a single therapy. In addition, the drug has side effects that are not insignificant including acroparesthesias and excessive diarrhea which is not dose-dependent (Moyses, 2003). Although DNJ may not be a suitable drug for most Gaucher disease patients, other uses of this drug are important to note. It may have more impact in the neuronopathic forms of Gaucher disease, particularly if it is used in combination with ERT. Enzyme therapy will reduce the total burden of the GSL and may reduce their concentration in the central nervous system, a place in which ERT is not effective because of its inability to cross the blood–brain barrier. The combination of reduced substrate synthesis and reduced total body lipid burden could have benefit for the most devastating forms of Gaucher disease providing some potential for the treatment of sphingolipidoses with central nervous system involvement.

Substrate reduction therapy for Gaucher disease using DNJ received regulatory approval by the European Agency for the Evaluation of Medicinal Products (EMEA) in 2002 and later by the United States Food and Drug Administration (U.S. FDA). However, these approvals are limited to symptomatic patients with mild to moderate type 1 GD for whom ERT is unsuitable (EMEA) or ERT is not an option (FDA). A review of this therapy was done by a group of experts working under the auspices of the European Working Group on Gaucher Disease (EWGDD; Weinreb et al., 2005).

4.3 Gene Transfer

The effectiveness of BMT/ERT supports the rationale for permanent somatic cell gene transfer strategies aimed at transducing autologous BM stem cells. In Gaucher disease, these cells could provide enzyme-competent macrophages to repopulate organs. They may also secrete the enzyme locally and into the circulation for subsequent macrophage uptake. Success in long-term expression of the gene encoding GC in mouse haematopoietic cells has been demonstrated, as well as high-efficiency retroviral transduction of human CD34 + progenitor/stem cells obtained from patients with Gaucher disease (Ohashi et al., 1992; Bahnson et al., 1994).

Three clinical trials of hematopoietic stem cell gene transfer and autologous transplantation without myeloablation for type 1 Gaucher disease were performed using retroviral vectors. Two employed a protocol consisting of transduction of autologous bone marrow stroma in the presence of IL-3, IL-6, and stem cell factor (with or without IL-1), protamine-sulfate and exposure to glucocerebrosidase-expressing viral supernatant for three to five days. Engraftment of the transduced cells in the patients was not observed in one clinical trial. In another trial, a low level of corrected cells (0.02%) was detected in the peripheral blood, which did not result in incremental glucocerebrosidase expression or clinical benefit.

In a third clinical study, CD34+ cells derived from four patients underwent a one-day prestimulation in long-term bone marrow culture medium containing IL-3, IL-6, and stem cell factor and transduction with glucocerebrosidase-expressing viral supernatants by centrifugal enhancement (Barranger and Novelli, 2001). The transduction efficiency averaged 20%, and enzymatic activity of glucocerebrosidase in transduced CD34+ cells increased more than tenfold over baseline. Total peripheral blood leukocytes (PBL) and CD34+ cells carried the glucocerebrosidase transgene as demonstrated by polymerase chain reaction of FACS sorted cells, and an increase in enzymatic activity of PBL was measurable following transplantation. In one patient, the glucocerebrosidase activity of total peripheral blood lymphocytes rose to a level as high as 80% of control, and the transgene was detected in all fluorescence-activated cell sorted lineages including lymphocytes, polymorphonuclear leukocytes, and monocytes. These results permitted a gradual withdrawal of enzyme replacement therapy over 12 months. During this time and for additional 5 months, the enzymatic activity in peripheral blood lymphocytes remained substantially above deficient levels. Despite this initial improvement the transduced cells gradually disappeared. Clinical and laboratory parameters returned to pretreatment levels. Enzyme replacement therapy had to be resumed in the patient.

From these studies, it can be concluded that CD34+ cells from Gaucher disease patients can be safely transduced with a retroviral vector and transplanted in nonmyeloablated recipients. In one patient, transduced cells and the correction of the enzymatic deficit persisted for approximately two years without a decline in clinical status. Improvements in the gene transfer protocol may lead to consistent, long-term efficacy, but at the present time it is not successful because of the inefficiency of HSC transduction. At present this strategy, although viable, needs to overcome several hurdles that impair a consistent response and could determine the long-term engraftability of the transduced cells such as quiescence, expression of cellular markers for homing into the bone marrow stroma, and cell cycle status. In addition, in contrast to retrovirally transduced cells from patients with severe combined immunodeficiency (SCID), glucocerebrosidase-transduced CD 34 cells do not have a competitive advantage over normal cells in the medullar stroma and therefore marrow reconstitution and engraftment would either be observed at a slower rate or be unsuccessful. Some of these variables must be assessed and controlled to achieve the engraftment and enzyme production required for therapeutic benefits

Because it is believed that totipotential stem cells are nondividing cells, there is increasing interest in the development of vectors with a significant capacity to transduce quiescent cells, especially lentiviral vectors. Equine infectious anemia virus (EIAV) and human immunodeficiency virus (HIV) have been shown to transduce cell-lines from various species with efficiency (Ikeda et al., 2002). Gene transfer with these lentiviral vectors has been attempted successfully showing efficient transduction and persistent transgene expression in MPS fibroblasts. Interesting results have been achieved in studies aimed at treating metachromatic leukodystrophy with these vectors. In vivo improvement of an MLD mouse model has been achieved (Consiglio et al., 2001). Intracranial administration of these vectors, near the hippocampal cortex generated sustained enzyme expression avoiding neuronal degeneration, protecting the MLD mouse against learning impairments and behavioral deficits. Additionally, this platform based on lentiviral gene transfer presents widespread transduction of adult, mammalian neural stem cells, bringing hope for improvements in the neuropathic phenotypes of lysosomal storage disease (Consiglio et al., 2004).

An interesting study of gene transfer of the glucocerebrosidase gene, delivered by a lentiviral vector has demonstrated the introduction of the GC gene into human fibroblasts, integration in transduced cells and release of glucocerebrosidase, which was taken up by GC-deficient cells. In this study, vascular and intraportal administration of lentivirus vector produced therapeutic levels of GC in the serum of mice, providing an increase in GC activity in liver, spleen, heart, lung, and kidney (Kim et al., 2004). Further testing in murine models of Gaucher disease and evaluation of the reversal of the pathological manifestations of the disease is needed before considering clinical development.

New approaches using adeno-associated vectors encoding glucocerebrosidase have shown that it is possible to increase enzymatic levels and induce immune tolerance by using liver-specific promoters (McEachern et al., 2006) This study in the murine model of Gaucher disease proved that transgene expression and secretion at therapeutic levels is achievable, together with correction of the pathology in the animal model of Gaucher disease that harbors the mutation D409V. Moreover, the secreted product presents similar targeting properties and therapeutic effects to those exhibited by commercially available high mannose recombinant human enzymes, providing proof of concept for liver directed gene therapy in Gaucher disease.

4.4 Mouse Models of Gaucher Disease

The transgenic glucocerebrosidase-deficient mouse model obtained by targeted knockout of the glucocerebrosidase gene exhibits a severe phenotype with prenatal or perinatal death. The inconveniences of chemical induction and the failure of earlier knockout models have been circumvented with the generation of newer targeted mutant models. These mutants are obtained by the introduction of point mutations in the glucocerebrosidase locus by cre/lox recombinants (Xu et al., 2003). Interestingly N370S, the most common mutation in type I Gaucher disease proved to be lethal in mice that are homozygous for this mutation. When D409H or D409V mutant alleles are combined with null alleles, storage cells could be seen as early as three months after birth. These animals provide a useful model for future approaches. The availability of these mice is an important fact because it will enhance the knowledge about the pathophysiology of the disease, facilitate discovery of surrogate markers of the disease, improve preclinical testing of forthcoming therapies, and facilitate studies of treatments that are currently available. An improved animal model of type 1 Gaucher disease has recently been developed. These mice closely resemble the human disease presenting hematological and visceral abnormalities similar to the disease observed in humans (Enquist et al., 2006).

5 CONCLUSIONS

The development of ERT for Gaucher disease is an impressive accomplishment. The clinical results warrant the development of a definitive cure. Several barriers have been overcome and make the possibility of providing the cure by gene transfer. The excellent results and the high safety profile that have been achieved with ERT ensure that patients with Gaucher disease will continue receiving the benefits of this therapy. Application and use of SRT as an adjuvant to ERT, especially in cases where the blood–brain barrier cannot be overcome, provides a role in the near future for small molecules that can safely inhibit or reduce the rate of substrate synthesis, however, these compounds, although commercially available for the treatment of patients with GD, do not present the safety and efficacy profiles that enzyme replacement therapy has provided for more than a decade, making it the treatment of choice. Further research in this area will provide more efficient compounds with fewer side effects and better therapeutic performance.

REFERENCES

Achord, D., Brot, F., Gonzalez-Noriegam, A., Sly, W., Stahl, P. (1977) Human beta-glucuronidase. II. Fate of infused human placental beta-glucuronidase in the rat. *Pediatr Res.* 11:816.

Aharon-Peretz, J., Rosenbaum, H., Gershoni-Baruch, R. (2004) Mutations in the glucocerebrosidase gene and Parkinson's disease in Ashkenazi Jews. *N Engl J Med.* 351:1972

Ashwell, G., Kawasaki, T. (1978) A protein from mammalian liver that specifically binds galactose-terminated glycoproteins. *Methods Enzymol.* 50:287.

Ashwell, G., Morell A.G. (1974) The role of surface carbohydrates in the hepatic recognition and transport of circulating glycoproteins. *Adv Enzymol Relat Areas Mol Biol.* 99.

Bahnson, A.B., Dunigan, J.T., Baysal, B.E., Mohney, T., Atchison, R.W., Nimgaonkar, M.T., Ball, E.D., Barranger, J.A. (1995) Centrifugal enhancement of retroviral mediated gene transfer. *J Virol Methods.* 54:131

Barranger, J.A., Ginns, E. (1989) Glucosylceramide lipidoses: Gaucher disease. In: Scriver,C.R., Sly, W.S., Valle, D. (Eds.), *The Metabolic Basis of Inherited Disease.* New York: McGraw-Hill, pp. 1677–1698.

Barranger, J.A., Pentchev, P.G., Furbish, F.S., Steer, C.J., Jones, E.A., Brady, R.O. (1978) Studies of lysosomal function: I. Metabolism of some complex lipids by isolated hepatocytes and Kupffer cells. *Biochem Biophys Res Commun.* 83:1055.

Barton, N.W., Brady, R.O., Dambrosia, J.M., Di Bisceglie, A.M., Doppelt, S.H., Hill, S.C., Mankin, H.J., Murray, G.J., Parker, R.I., Argoff, C.E., et al. (1991) Replacement therapy for inherited enzyme deficiency: Macrophage-targeted glucocerebrosidase for Gaucher's disease. *N Engl J Med.* 324:1464–1470.

Bembi, B., Ciana, G., Mengel, E., Terk, M.R., Martini, C., Wenstrup, R.J. (2002) Bone complications in children with Gaucher disease. *Br J Radiol.* 75 Suppl 1:A37–44.

Boklan, B.F., Sawitsky, A. (1976). Factor IX deficiency in Gaucher's disease. An in vitro phenomenon. *Arch Intern Med.* 136:489–92.

Boot, R.G.,Verhoek, M., de Fost, M., Hollak, C.E., Maas, M., Bleijlevens, B., van Breemen, M.J., van Meurs, M., Boven, L.A., Laman, J.D., Moran, M.T., Cox, T.M., Aerts, J.M. (2004) Marked elevation of the chemokine CCL18/PARC in Gaucher disease: A novel surrogate marker for assessing therapeutic intervention. Blood. 103:33

Brady, R.O., Barranger, J.A., Gal, A.E., Pentchev, P.G.,Furbish, F.S. (1980). Status of enzyme replacement therapy for Gaucher disease. *Birth Defects Orig Artic Ser.* 16:361

Brady, R.O., Kanfer, J.N., Bradley, R.M., Shapiro, D. (1966) Demonstration of a deficiency of glucocerebroside-cleaving enzyme in Gaucher's disease. *J Clin Invest.* 45:1112

Brady R.O., Pentev, P.G., Gal, A.E., Hibbert, S.R., Dekeban, A.S. (1974) Replacement therapy for inherited enzyme deficiency: Use of purified glucocerebrosidase in Gaucher's disease. *N Engl J Med.* 291:989–993.

Cabrera-Salazar, M.A., O'Rourke, E., Henderson, N., Wessel, H., Barranger, J.A. (2004) Correlation of surrogate markers of Gaucher disease. Implications for long-term follow up of enzyme replacement therapy. *Clin Chim Acta.* 344:101–107.

Chang-Lo, M., Yam, L.T., Rubenstone, A.I. (1967) Gaucher's disease. *Am J Med Sci* 254:303–315.

Charrow, J., Esplin, J.A., Gribble, T.J., Kaplan, P., Kolodny, E.H., Pastores, G.M., Scott, C.R., Wappner, R.S., Weinreb, N.J., Wisch, J.S. (1998) Gaucher disease: Recommendations on diagnosis, evaluation, and monitoring. *Arch Intern Med.* 158:1754–1760.

Ciana, G., Martini, C., Leopaldi, A., Tamaro, G., Katouzian, F., Ronfani, L., Bembi, B. (2003) Bone marker alterations in patients with type 1 Gaucher disease. *Calcif Tissue Int.* 72:185–189

Consiglio, A., Gritti, A.,Dolcetta, D., Follenzi, A., Bordignon, C.,Gage, F.H., Vescovi, A.L., Naldini, L. (2004) Robust in vivo gene transfer into adult mammalian neural stem cells by lentiviral vectors. *Proc Natl Acad Sci USA.* 101:14835–14840.

Consiglio, A., Quattrini, A., Martino, S., Bensadoun, J.C., Dolcetta, D., Trojani, A., Benaglia, G., Marchesini, S., Cestari, V., Oliverio, A., Bordignon, C., Naldini, L. (2001) In vivo gene therapy of metachromatic leukodystrophy by lentiviral vectors: Correction of neuropathology and protection against learning impairments in affected mice. *Nat Med.* 7:310–316.

Cox, T., Lachmann, R., Hollak, C., Aerts, J., van Weely, S., Hrebicek, M., Platt F.M., Butters, T.D., Dwek, R., Moyses, C., Gow, I., Elstein, D., Zimran, A. (2000) Novel oral treatment of Gaucher's disease with N-butyldeoxynojirimycin (OGT 918) to decrease substrate biosynthesis. *Lancet.* 355:1481–1485.

De Duve, C. (1969). The lysosome in retrospect. In: J.T. Dingle, H.B. Fell (Eds.) *Lysosomes in Biology and Pathology*, Amsterdam: North-Holland, Vol. 1, pp. 3–40.

De Duve, C., Pressman, B.C., Gianetto, R., Wattiaux, R., Appelmans, F. (1955) Tissue fractionation studies. 6. Intracellular distribution patterns of enzymes in rat-liver tissue. *Biochem J.* 60:604–617.

Enquist, I.B., Nilsson, E., Ooka, A., Mansson, J.E., Olsson, K., Ehinger, M., Brady, R.O., Richter, J., Karlsson, S. (2006) Effective cell and gene therapy in a murine model of Gaucher disease. *Proc Natl Acad Sci USA.* 103:13819–13824.

Erickson, A.H., Ginns, E.I., Barranger, J.A. (1985) Biosynthesis of the lysosomal enzyme glucocerebrosidase. *J Biol Chem.* 260:14319–14324.

European Public Assessment Report (EPAR). Zavesca [miglustat]. (2002) Committee for Proprietary Medicinal Products EPAR. CPMP/3795/02. London, England: The European Agency for the Evaluation of Medicinal Products. http://www.emea.eu.int/humandocs/Humans/EPAR/zavesca/zavesca.htm.

Fratantoni, J.C., Hall, C.W., Neufeld, E.F. (1968) Hurler and Hunter syndromes: Mutual correction of the defect in cultured fibroblasts. *Science.* 162:570–572.

Fredrickson, D.S., Sloan, H.R. (1978) In: Stanburg, J.B. Wyngaarden, J.B. Fredrickson, D.S. (Eds.) *The Metabolic Basis of Inherited Disease.* 4th ed. New York: McGraw-Hill.

Furbish, F.S., Blair, H.E., Shiloach, J., Pentchev, P.G., Brady, R.O. (1977) Enzyme replacenment therapy in Gaucher disease. Large Scale purification of glucocerebrosidase suitable for human administration. *Proc Natl Acad Sci USA.* 74:3560

Furbish, F.S., Steer, C.J., Barranger, J.A., Jones, E.A., Brady, R.O. (1978) The uptake of native and desialylated glucocerebrosidase by rat hepatocytes and Kupffer cells. *Biochem Biophys Res Commun.* 81:1047–1053.

Ginns, E.I., Choudary, P.V., Martin, B.M., Winfield, S., Stubblefield, B., Mayor, J., Merkle-Lehman, D., Murray, G.J., Bowers, L.A., Barranger, J.A. (1984) Isolation of cDNA clones for human beta-glucocerebrosidase using the lt11 expression system. *Biochem Biophys Res Commun.* 123:574–580.

Goldblatt, J., Sacks, S., Beighton, P. (1978) The orthopedic aspects of Gaucher disease. *Clin Orthop.* 137:208–214.

Grabowski, G.A., Barton, N.W., Pastores, G., Dambrosia, J.M., Banerjee, T.K., McKee, M.A., Parker, C. , Schiffmann, R., Hill, S.C., Brady, R.O. (1995) Enzyme therapy in Type 1 Gaucher disease: Comparative efficacy of mannose-terminated glucocerebrosidase from natural and recombinant sources. *Ann Intern Med.* 122:33–39.

Halleen, J.M., Ylipahkala, H., Alatalo, S.L., Janckila, A.J., Hekkinen, J.E., Suominen, H., Cheng, S., Vaananen, H.K. (2002) Serum Tartrate-Resistant Acid Phosphatase 5b, but not 5a, Correlates with other markers of bone turnover and bone mineral density. *Calcif Tissue Int.* 71:20–25.

Hannun, Y.A., Bell, R.M. (1987). Lysosphingolipids inhibit protein kinase C: Implications for the sphingolipidoses. *Science.* 235:670–674.

Hollak, C.E., Evers, L., Aerts, J.M.,van Oers, M.H. (1997) Elevated levels of M-CSF, sCD14 and IL8 in type 1 Gaucher disease. *Blood Cells Mol Dis.* 23:201–212.

Hollak, C.E., van Weely, S., van Oers, M.H., Aerts, J.M. (1994) Marked elevation of plasma chitotriosidase activity. A novel hallmark of Gaucher disease. *J Clin Invest.* 93:1288.

Hong, C.M., Ohashi, T., Yu, X.J., Weiler, S., Barranger, J.A. (1990) Sequence of two alleles responsible for Gaucher disease. *DNA Cell Biol.* 9:233–41.

Horowitz, M., Wilder, S., Horowitz, Z., Reiner, O., Gelbart, T., Beutler, E. (1989) The human glucocerebrosidase gene and pseudogene: Structure and evolution. *Genomics.* 4:87–96.

Hultberg, B., Isaksson, A., Sjoblod, S., Ockerman, P.A. (1980) Acid hydrolases in serum from patients with lysosomal disorders. *Clin Chim Acta.* 100:33–38.

Ikeda, Y., Collins, M.K., Radcliffe, P.A., Mitrophanous, K.A.,Takeuchi, Y. (2002) Gene transduction efficiency in cells of different species by HIV and EIAV vectors. *Gene Ther.* 9:932–938.

James, S.P., Stromeyer, F.W., Chang, C.S.C., Barranger, J.A. (1981) Liver abnormalities in Gaucher's Disease. *Gastroenterology* 80:126–133.

Kattlove, H.E., Williams, J.C., Gaynor, E., Spivack, M., Bradley, R.M., Brady, R.O. (1969) Gaucher cells in chronic myelocytic leukemia: An acquired abnormality. *Blood.* 33:379–390.

Kaye, E.M., Ullman, M.D., Wilson, E.R., Barranger, J.A. (1986) Type 2 and type 3 Gaucher disease: a morphological and biochemical study. *Ann Neurol.* 20:223–230

Kim, E.Y., Hong, Y.B., Lai, Z., Kim, H.J., Cho, Y.H, Brady, R.O., Jung, S.C. (2004) Expression and secretion of human glucocerebrosidase mediated by recombinant lentivirus vectors in vitro and in vivo: Implications for gene therapy of Gaucher disease. *Biochem Biophys Res Commun.* 318:381–390.

Lieb, H. (1924) Der zucker im cerebrosid der milz bei der Gaucher krankheit. Hoppe-Seyler's Z *Physiol Chem.* 271:211.

Lieberman, J., Beutler, E. (1976) Elevation of serum angiotensin-converting enzyme in Gaucher's disease. *N Engl J Med.* 294:1442–1444.
Marshall, J., McEachern, K.A., Kyros, J.A., Nietupski, J.B., Budzinski, T., Ziegler, R.J., Yew, N.S., Sullivan, J. Scaria, A., van Rooijen, N., Barranger, J.A., Cheng, S.H. (2002) Demonstration of feasibility of in vivo gene therapy for Gaucher disease using a chemically induced mouse model. *Mol Ther.* 6:179–89.
Matoth, Y., Fried, K. (1965) Chronic Gaucher's disease: clinical observations on 34 patients. *Isr J Med Sci.* 1:521.
McEachern K.A., Nietupski, J.B., Chuang, W.L., Armentano, D., Johnson, J., Hutto, E., Grabowski, G.A., Cheng, S.H., Marshall, J. (2006) AAV8-mediated expression of glucocerebrosidase ameliorates the storage pathology in the visceral organs of a mouse model of Gaucher disease. J Gene Med. 8:719–729.
Mignot, C., Gelot, A., Bessieres, B.,Daffos, F., Voyer, M., Menez, F., Fallet-Bianco, C., Odent, S., Le Duff, D., Loget, P., Fargier, P., Costil, J., Josset, P., Roume, J., Vanier, M.T., Maire, I., Billette, de Villemeur, T. (2003) Perinatal-lethal Gaucher disease. *Am J Med Genet.* 120:338–344.
Moffitt, K.D., Chambers, J.P., Diven, W.F., Glew, R.H., Wenger, D.A., Farrell, D.F. (1978) Characterization of lysosomal hydrolases that are elevated in Gaucher's disease. *Arch Biochem Biophys.* 190:247–260.
Molino-Trinidad, C., Alvarez-Moro, F.J., Salvador-Aznar, A., Junyent Pares, P. (1979) Roentgenologic characteristics of Gaucher's disease. *Med Clin (Barc).* 72:299–302.
Moran, M.T., Schofield, J.P., Hayman, A.R., Shi, G.P., Young, E., Cox, T.M. (2000) Pathologic gene expression in Gaucher disease: Up-regulation of cysteine proteinases including osteoclastic cathepsin K. *Blood* 96:1969–1978.
Morimoto, S., Kishimoto, Y., Tomich, J., Weiler, S., Ohashi, T., Barranger, J.A., Kretz, K.A., O'Brien, J.S. (1990) Interaction of saposins, acidic lipids, and glucosylceramidase. *J Biol Chem.* 265:1933–1937.
Moyses, C. (2003) Substrate reduction therapy: clinical evaluation in type 1 Gaucher disease. *Philos Trans R Soc Lond B Biol Sci.* 358:955–960.
Nilsson, O. (1982) Glycolipid changes in Gaucher disease [thesis]. Goteborg,
Nilsson, O., Grabowski, G.A., Ludman, M.D., Desnick, R.J., Svennerholm, L. (1985) Glycosphingolipid studies of visceral tissues and brain from type 1 Gaucher disease variants. *Clin Genet.* 27:443–450.
Nilsson, O., Svennerholm, L. (1982) Accumulation of glucosylceramide and glucosylsphingosine (psychosine) in cerebrum and cerebellum in infantile and juvenile Gaucher disease. *J Neurochem.* 39:709–718.
No author. (1996) Gaucher disease. Current issues in diagnosis and treatment. NIH Technology Assessment Panel on Gaucher Disease. *JAMA.* 275:548–553.
Ockerman, P.A, Kohlin, P. (1969) Acid hydrolases in plasma in Gaucher's disease. *Clin Chem.* 15:61–64.
Ohashi, T., Boggs, S., Robbins, P., Bahnson, A., Patrene, K., Wei, F.S., Wei, J.F., Li, J., Lucht, L., Fei, Y. (1992) Efficient transfer and sustained high expression of the human glucocerebrosidase gene in mice and their functional macrophages following transplantation of bone marrow transduced by a retroviral vector. *Proc Natl Acad Sci USA.*89:11332–11336.
Orvisky, E., Park, J.K., LaMarca, M.E., Ginns, E.I., Martin, B.M., Tayebi, N., Sidransky, E. (2002) Glucosylsphingosine accumulation in tissues from patients with Gaucher disease: correlation with phenotype and genotype. *Mol Genet Metab.* 76:262–270.

Platt, F.M., Jeyakumar, M., Andersson, U., Priestman, D.A., Dwek, R.A., Butters, T.D., Cox, T.M, Lachmann, R.H., Hollak, C., Aerts, J.M., Van Weely, S., Hrebicek, M., Moyses, C., Gow, I., Elstein, D., Zimran, A. (2001) Inhibition of substrate synthesis as a strategy for glycolipid lysosomal storage disease therapy. *J Inherit Metab Dis.* 24:275–290.

Robinson, D.B., Glew, R.H. (1980) Acid phosphatase in Gaucher's disease. *Clin Chem.* 26:371–382.

Rosenberg, M., Kingma, W., Fitzpatrick, M.A., Richards, S.M. (1999) Immunosurveillance of alglucerase enzyme therapy for Gaucher patients: Induction of humoral tolerance in seroconverted patients after repeat administration. *Blood.* 93:2081–2088.

Sawkar, A.R., Adamski-Werner, S.L., Cheng, W.C., Wong, C.H., Beutler, E., Zimmer, K.P., Kelly, J.W. (2005) Gaucher disease-associated glucocerebrosidases show mutation-dependent chemical chaperoning profiles. *Chem Biol.* 12:1235–1244.

Sawkar, A.R., Cheng, W.C., Beutler, E., Wong, C.H., Balch, W.E., Kelly, J.W. (2002) Chemical chaperones increase the cellular activity of N370S beta -glucosidase: A therapeutic strategy for Gaucher disease. *Proc Natl Acad Sci USA.* 99:15428–15433.

Sidransky, E. (2004) Gaucher disease: complexity in a "simple" disorder. *Mol Genet Metab.* 83:6–15

Silverstein, E., Friedland, J. (1977) Elevated serum and spleen angiotensin converting enzyme and serum lysozyme in Gaucher's disease. *Clin Chim Acta.* 74:21–25.

Sorge, J., West, C., Westwood, B., Beutler, E. (1985) Molecular cloning and nucleotide sequence of human glucocerebrosidase cDNA. *Proc Natl Acad Sci USA.* 82:7289–7293.

Stone, D.L., Tayebi, N., Orvisky, E., Stubblefield, B., Madike, V., Sidransky, E. (2000) Glucocerebrosidase gene mutations in patients with type 2 Gaucher disease. *Hum Mutat.* 15:181–188.

Svennerholm, L. (1967) The metabolism of gangliosides in cerebral lipidoses. In: Aronson, S.M., Volk, B.W. (Eds.) Inborn disorders of sphingolipid metabolism. New York: Pergamon Press, pp. 169.

Takasaki, S., Murray, G.J., Furbish, F.S.,Brady, R.O., Barranger, J.A., Kobata A. (1984) Structure of the N-asparagine-linked oligosaccharide units of human placental glucocerebrosidase. *J Biol Chem.* 259:10112–10117.

Tayebi, N., Stone, D.L., Sidransky, E. (1999) Type 2 Gaucher disease: An expanding phenotype. *Mol Genet Metab.* 68:209–19.

Teilum, G. (1944) Die Gauchersche krankheit mit der beschreibung eines falles, der veranderungen in der hypophyse und im hypothalamus zeigte. *Acta Med Scand* 116:170.

Tsuji, S., Choudary, P.V., Martin, D.M., Winfield, S., Barranger, J.A., Ginns, E.I. (1986) Nucleotide sequence of cDNA containing the complete coding sequence for human lysosomal glucocerebrosidase. *J Biol Chem.* 261:50–53.

Tsuji, S., Martin, B.M., Barranger, J.A., Stubblefield, B.K., LaMarca, M.E., Ginns, E.I. (1988) Genetic heterogeneity in type 1 Gaucher disease: Multiple genotypes in Ashkenazic and non-Ashkenazic individuals. *Proc Natl Acad Sci USA.* 85:2349–2352.

Tuchman, L.R., Suna, H., Carr, J.J. (1956) Elevation of serum acid phosphatase in Gaucher's disease. *J Mt Sinai Hosp.* 23:277.

U.S. FDA Medical Review. http://www.fda.gov/cder/foi/nda/(2003)/21–348_Zavesca.htm.

Weinreb, N.J., Aggio, M.C., Andersson, H.C., Andria, G., Charrow, J., Clarke, J.T., Erikson, A., Giraldo, P., Goldblatt, J., Hollak, C.E., Ida, H., Kaplan, P., Kolodny, E.H., Mistry, P., Pastores, G.M., Pires, R., Prakash-Cheng, A., Rosenbloom, B.E., Scott, C.R., Sobreira, E., Tylki-Szymanska, A., Vellodi, A., vom Dahl, S., Wappner, R.S., Zimran, A. (2004) International Collaborative Gaucher Group (ICGG). Gaucher disease type 1: Revised recommendations on evaluations and monitoring for adult patients. *Semin Hematol.* 4:15–22.

Weinreb, N.J., Barranger, J.A., Charrow, J., Grabowski, G.A., Mankin, H.J., Mistry, P. (2005) Guidance on the use of miglustat for treating patients with type 1 Gaucher disease. *Am J Hematol.* 80:223–229.

Weinreb, N.J., Charrow, J., Andersson, H.C., Kaplan, P., Kolodny, E.H., Mistry, P., Pastores, G., Rosenbloom, B.E., Scott, C.R., Wappner, R.S., Zimran, A. (2002) Effectiveness of enzyme replacement therapy in 1028 patients with type 1 Gaucher disease after 2 to 5 years of treatment: A report from the Gaucher Registry. *Am J Med.* 113:112–119.

Wenstrup, R.J., Roca-Espiau, M., Weinreb, N.J., Bembi, B. (2002) Skeletal aspects of Gaucher disease: a review. *Cell Mol Life Sci.* 59:694–707.

Xu, Y.H., Quinn, B., Witte, D., Grabowski, G.A. (2003) Viable mouse models of acid beta-glucosidase deficiency: the defect in Gaucher disease. *Am J Pathol.* 163:2093–2101.

THERAPEUTIC GOALS IN THE TREATMENT OF GAUCHER DISEASE

Neal Weinreb

1 INTRODUCTION

Although first identified in 1882 by French physician Philippe C. E. Gaucher, for many years little was known about Gaucher disease beyond its debilitating and often fatal manifestations in patients who were recognized as having the disorder. By the mid-twentieth century, Gaucher disease came to be understood as an inborn biochemical disorder characterized by an accumulation of glucocerebroside predominantly within organs such as the bone marrow, spleen, liver, and lungs (Aghion, 1934), and its nature as a hereditary autosomal recessive disorder had been clarified. Two major clinical subtypes of Gaucher disease were recognized: the relatively rare variants with neurological deterioration (Oberling et al., 1927; Hillborg, 1959), and the more common nonneuronopathic illness now known as Type 1 disease that was erroneously believed to be largely confined to the Ashkenazic Jewish population. The identification, in 1965, of deficient activity of the lysosomal enzyme glucocerebrosidase as the cause of the disease (Brady, 1965), paved the way for definitive biochemical diagnosis using easily obtained blood or other tissue samples (Beutler and Kuhl, 1970; Ho, 1972), for purification and crystallization of the enzyme and elucidation of its structural configuration and active site (Dvir et al., 2003; Premkumar et al., 2005), for genetic diagnosis after characterization of the GBA gene and adjacent pseudo-gene (Ginns et al., 1984), and for the discovery of currently more than 200 mutated alleles associated with variably defective glucocerebrosidase activity (Beutler and Grabowski 2001; Grabowski, 2005). This chapter in the history of Gaucher disease might be entitled "Description and Discovery: From Phenotype to Genotype," although the ending is yet to be written as an understanding of phenotype–genotype correlations and of clinical heterogeneity within genotype continues to pose a substantial investigative challenge.

From a patient perspective, the development of intravenous enzyme replacement therapy with mannose-terminated, macrophage-targeted human glucocerebrosidase was undoubtedly the most important outcome of the basic research efforts (Furbish et al., 1984). Human placenta-derived glucocerebrosidase (Ceredase® [alglucerase injection]) was introduced in 1991. In 1994, recombinant enzyme (Cerezyme® [imiglucerase injection]) became available and has since become the standard of care for patients with both type 1 and type 3 Gaucher disease. Because alglucerase was approved for clinical use in the United States based on demonstrated improvement in anemia, thrombocytopenia, and hepatosplenomegaly in a pivotal trial involving only 12 patients (Barton et al., 1991), most of this chapter in Gaucher disease history (possibly entitled: "From Bench to Bedside:

University Research Foundation for Lysosomal Storage Diseases and Northwest Oncology Hematology Associates PA, Coral Springs, Florida. E-mail: boneal@winning.com

Enzyme Replacement Therapy") has been devoted to determining indications for treatment, dosage schedules, parameters of response, definition of efficacy, safety profiles, categorization of adverse events, and immunological reactivity. The results of this multinational but disparate clinical research effort demonstrating the efficacy and safety of enzyme replacement therapy are extensively documented in the medical literature in the form of reports primarily involving relatively small patient cohorts emanating from either individual treatment centers or from single country registries (see multiple citations in Weinreb et al., 2002). The worldwide aggregate number of patients with type 1 Gaucher disease who are being treated with enzyme replacement therapy currently exceeds 4000.

The introduction of enzyme replacement therapy also provided the impetus for the formation of the International Collaborative Gaucher Group (sponsored by Genzyme Corporation, Cambridge, MA) and for the development of the International Gaucher Registry, the purpose of which is to gather clinical data on all patients with Gaucher disease, irrespective of treatment status (Charrow et al., 2000). With its current roster of more than 4000 patients (of whom approximately 75% are receiving imiglucerase), the Gaucher Registry is an important resource for understanding the natural history of Gaucher disease as well as the response to enzyme treatment.

Once an apparently successful treatment becomes established as a standard of care, there is risk of intellectual stagnation that can impede continuing critical appraisal as to whether the intervention in question is truly applied for optimal clinical effect. A relevant example is the long reign of the now obsolete Halstead radical mastectomy as standard treatment for breast carcinoma. Alternatively, the focus of investigation may shift to the development of new, it is hoped, improved, treatment strategies even in the absence of evidence that standard treatment has reached its maximal potential. In the case of Gaucher disease, this attitude is expressed in the certainly important current interest in new treatment approaches such as substrate inhibition, molecular chaperones, and gene therapy somewhat to the neglect of what should still be the central question relevant to present clinical practice: how, using available enzyme replacement therapy, can each individual with Gaucher disease achieve the best possible health outcome? The ability to answer this question is best promoted through development of a disease management system incorporating the following components.

1. Measures of overall and organ-specific disease burden applicable to initial assessment and subsequent monitoring of clinical outcomes
2. Short-term treatment-dependent endpoints hereafter referred to as therapeutic goals that are quantitative in an established timeframe
3. Longer-term outcome measures, presumably related to attainment of therapeutic goals, such as quality of life and longevity
4. Documentation of the process of care including initial dose selection, frequency of follow-up assessment, and dose adjustment and modification.

These components can be effectively integrated into a disease management algorithm as shown in Figure 1 (Pastores et al., 2004).

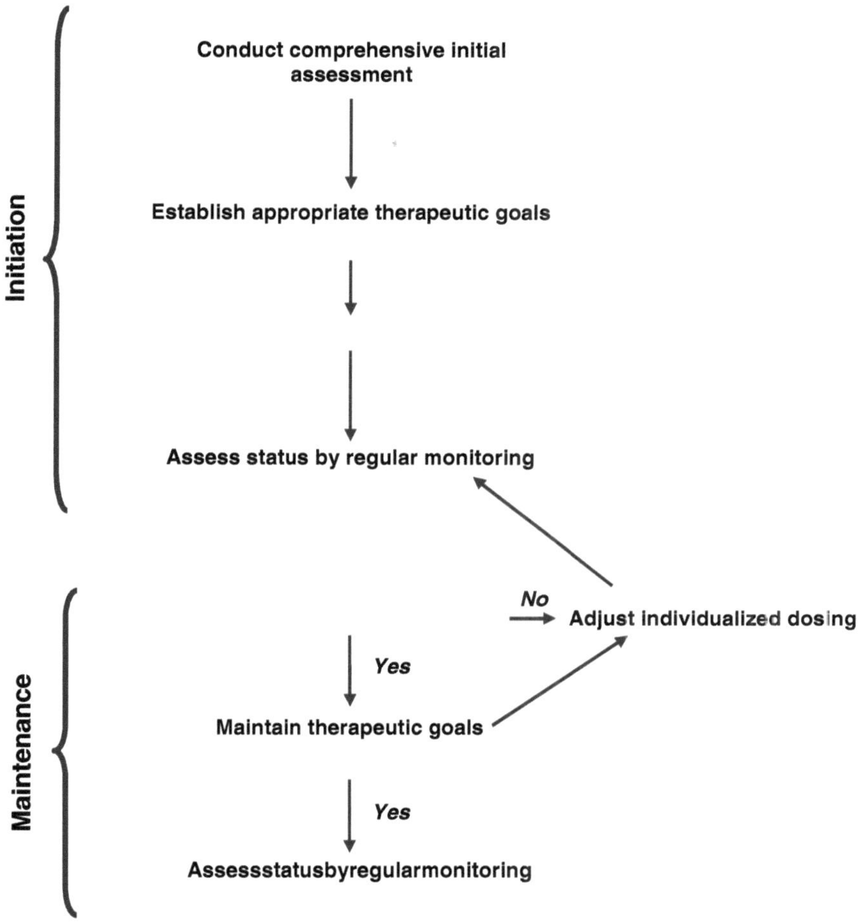

Figure 1. A treatment model algorithm for type 1 Gaucher disease incorporating assessment and monitoring guidelines, therapeutic goals, individualized dosing, and a patient registry.

2 COMPREHENSIVE ASSESSMENT AND MONITORING

As has been demonstrated in previous chapters, Gaucher disease is heterogeneous with respect to organ system involvement, presentation, and rate of progression. The clinical course and life expectancy is most variable in type 1 (nonneuronopathic) disease, although increasing familiarity with the worldwide spectrum of type 3 disease indicates that chronic neuronopathic disease also demonstrates considerable phenotypic heterogeneity (Vellodi et al., 2001). There is no question that genotype is the most significant predictor of clinical expression of disease, and the pattern of distribution of genotypes best explains the distinctive ethnic and geographic phenomena that are increasingly evident as awareness of Gaucher disease extends beyond that of Ashkenazic Jewry.

However, Gaucher disease is also characterized by considerable intragenotype variability, the causes of which are as yet undefined. Thus, patients who are homozygous for the N370S genotype encompass a spectrum of expression ranging from disease that is overtly manifested in childhood to an indolent, virtually asymptomatic disorder that may not be discovered until the patient is elderly (Maaswinkel-Mooij et al., 2000). Furthermore, this variability is not family-specific, but may be evident among siblings or even between identical twins (Lachmann et al., 2004). Phenotypic variation has been described for other genotypes as well (Goker-Alpan et al., 2005).

Aside from interpatient variability, for each individual, disease progression may be inexorable, or may be slow and erratic, punctuated by periods of rapid exacerbation and clinical crises interspersed with sometimes lengthy periods of quiescence lasting for months or even many years. In addition, for each patient, disease severity may be unevenly distributed according to organ compartments. Some patients have little or no enlargement of the liver, minimal splenomegaly, and few if any hematological abnormalities, but suffer from severe and crippling skeletal disease. Others may have few overt skeletal manifestations but can have marked splenomegaly, hypersplenism, anemia, and thrombocytopenia.

Because of clinical heterogeneity, comprehensive and reproducible evaluation and monitoring of all clinically relevant aspects of Gaucher disease are vital to assess patterns of Gaucher disease and effective management of patients (Weinreb et al., 2004). In addition to confirmation of β-glucosidase deficiency and genotyping, the recommended primary assessments are outlined in Table 1. These include:

- Physical examination
- Hematologic (hemoglobin and platelet count)
- Biochemical (chitotriosidase, CCL-18 if available, angiotensin-converting enzyme, and/or tartrate-resistant acid phosphatase)
- Visceral (liver and spleen volumes)
- Skeletal (bone marrow infiltration, osteonecrosis, and osteopenia)
- Patient-reported Quality of Life Survey (SF-36).

The physical examination is a highly important and readily accessible clinical assessment that should not be overlooked. Significant information about the severity, rate of progression, and response to therapy may be derived through observation of the patient's general physical appearance and demeanor, mood, affect, weight and height, gait, range of motion, muscle strength, bone tenderness, cardiopulmonary and abdominal findings, skin, and sexual development. Thorough initial and serial neurological and eye movement examinations are important for detecting evidence of neuronopathic disease in patients with suspect genotypes, and for identifying phenomena such as tremor and other extrapyramidal movement abnormalities, peripheral neuropathy, and age-inappropriate cognitive dysfunction that may possibly be part of the natural history even in patients believed to be free of neurological risk (Tayebi et al., 2003). Serial quantification of pain using a recognized "pain assessment tool" is also highly recommended. As for all patients with a sustained chronic illness, periodic examination for clinical depression is also indicated.

Table 1. Comprehensive assessment of patients with type 1 Gaucher disease: Recommended laboratory and imaging procedures

Comprehensive Assessment
• Skeletal – X-rays of femora, spine, symptomatic sites – MRI of the femora – Dual energy X-ray absorptiometry (DEXA)
• Hematological – Hemoglobin, platelet count, biomarkers
• Visceral – Volumetric CT or MRI
• Pulmonary – Doppler ECHO of heart, chest X-ray, ECG

Recommended assessment schedules for other pertinent laboratory and imaging evaluations for untreated patients (either asymptomatic or adults with minimal signs and symptoms of disease) and those on enzyme replacement therapy (or other treatment) are presented in Table 2. The detailed evidence and rationale for these guidelines have been published (Weinreb et al., 2004).

Table 2. Schedule of recommended assessment for patients with type 1 Gaucher disease

No ERT	**On ERT, not at goal**	**On ERT, at goal**
• PE, blood tests yearly • Viscera and bone assessments every 12–24 mo	• Blood tests every 3 mo • PE, QoL, viscera and bone assessments every 12 mo	• Yearly PE and QoL assessment • Blood tests, viscera and bone assessments every 12–24 mo
All Patients • Assess compartments at the time of a dose change or emergence of a significant clinical complication • Repeat pulmonary testing every 12–24 mo in patients with borderline or increased pulmonary pressures, patients w/o spleen, and females • Assess pulmonary pressure every four to five years otherwise		

3 DECISION TO TREAT AND INITIAL TREATMENT (Andersson et al., 2005)

Because Gaucher disease is a multisystemic, chronic heterogeneous disorder, an individualized treatment plan should be established based on each patient's unique clinical status. Many variables need to be considered when making treatment decisions including the severity and rate of disease progression, the likelihood for continued or new-onset complications, and the impact of disease manifestations on quality of life and life expectancy. As mentioned above, some patients, particularly but not exclusively those who are homozygous for the N370S allele, may have subclinical disease for their entire lives, or may have manifestations of apparently minimal clinical significance such as mild thrombocytopenia or modest splenomegaly. In evaluating such patients, it is important to be sure that bone involvement is also minimal, being confined to asymptomatic bone marrow infiltration and/or age-appropriate osteopenia. For such patients, a careful "watchful waiting" approach may be justified with respect to ERT although other medical intervenetions such as bisphosphonates or other antibone resorptive treatment may be indicated (Wenstrup et al., 2004). For all other adult patients, initiation of ERT is indicated. Although some investigators continue to favor the use of "low dose protocols" for imiglucerase (10–15 units/kg body weight every two weeks), the consensus of opinion favors a higher initial dose (30–60 units/kg every two weeks) depending on assessment of clinical risk (see Table 3).

Table 3. Individualized dosing: assessment of clinical risk in adults

Risk	**Manifestations**	**Initial Dose**
Lower	Meets all: normal visceral function, minimal decrease in QoL, no recent or rapid deterioration, mild skeletal disease, moderate anemia, mild thrombocytopenia, moderate hepatomegaly, moderate splenomegaly	30 U/kg/2 wk
Higher	Meets one: symptomatic skeletal disease, impaired QoL, cardiopulmonary disease, severe thrombocytopenia, transfusion dependency, significant hepatic, splenic, or renal disease, concomitant condition complicating Gaucher disease	60 U/kg/2 wk

Secure identification of the "benign" phenotype in children is problematic. Of 113 patients enrolled in the Gaucher Registry who are younger than 18 years of age and N370S homozygous, 54% are currently receiving enzyme replacement therapy as dictated by their clinical presentation. Furthermore, we cannot conclude that ERT is not medically necessary in all the 52 children who are currently not being treated in view of the observation that 77 children with less favorable genotypes and active symptoms have not yet started ERT, presumably due to socioeconomic rather than medical reasons. Therefore, the general consensus is that all children with any relevant physical signs or manifestations of Gaucher disease should be treated with ERT. The consensus initial imiglucerase dose in children with a low risk assessment is 30–60 units/kg every two weeks (Table 4) whereas children at higher risk should receive 60 units/kg. It should be noted that both in children and adults, the decision to treat with ERT and the selection of initial dose are based on the results of comprehensive clinical assessment. An arbitrary decision based on genotype alone is never indicated.

For the physician prescribing ERT, the clinical challenge is to find the dose that will achieve and sustain the optimal clinical benefit of therapy for each patient. This challenge is best met in the context of a treatment plan that defines specific, evidence-based quantitative and qualitative therapeutic goals. The treatment plan is considered successfully

Table 4. Individualized dosing: Assessment of clinical risk in children

Risk	Manifestation	Initial Dose
Lower	Any relevant physical sign	30 to 60 U/kg/2 wk
Higher	Physical signs plus >1 of: symptomatic disease, growth failure, skeletal involvement, severe thrombocytopenia, severe anemia, impaired QoL	60 U/kg/2 wk

implemented only when all goals are achieved and maintained in all affected organ systems. Using the cumulative experience with ERT in more than 4000 patients worldwide, particularly as documented in the international Gaucher Registry, an expert panel has proposed outcome-based therapeutic goals as a guide for clinicians treating patients with type 1 Gaucher disease (Pastores et al., 2004).

4 THERAPEUTIC GOALS

Because type 1 Gaucher disease is a clinically heterogeneous chronic disorder, definition of therapeutic goals is a complex endeavor that must take account of each manifestation of disease. In this regard, it is useful to look at Gaucher disease in the context of other prototypical medical conditions (Figure 2). In diseases with acute onset such as otitis media or postoperative pain, the goal of treatment is to achieve a rapid response and full recovery. Treatment is short term and maintenance is not required. In the case of otitis media, antibiotic therapy is generally standardized with regards to dose and duration whereas for patients suffering postoperative pain, patient heterogeneity with respect to pain tolerance and responsiveness to analgesics dictates that treatment be tailored to individual need and modified as necessary according to the observed response.

This model could well be applied to the management of an acute Gaucher bone crisis, but it is too simplistic to be relevant to the treatment of complex chronic disorders. Similarly, for chronic, relatively homogeneous disorders such as hypothyroidism, achievement of a single treatment goal (i.e., sustained normalization of TSH) usually directly translates into resolution of all clinical symptoms. Gaucher disease management, on the other hand, is more analogous to the model of type 2 diabetes mellitus in which treatment goals include not only achievement of 24-hour blood glucose control and normalization of hemoglobin A_1C, but also correction of proteinuria and prevention of complications such as nephropathy, retinopathy, neuropathy, and cardiovascular disease. Accomplishment of these various goals requires an individualized treatment plan that incorporates combinations of multimodality interventions including hypoglycemic agents, ACE inhibitors, diet, exercise, and weight loss (American Diabetes Association, 2005).

In much the same fashion, successful treatment of type 1 Gaucher disease is predicated on the identification and achievement of multiple therapeutic goals using an individuallized, multimodality treatment regimen that incorporates not only ERT, but also controlled exercise, and, as necessary, physical therapy, orthopedic intervention, and adjunctive treatments such as antiresorptive therapy, analgesics for acute and chronic pain, or rarely, splenectomy for life-threatening bleeding, shunting for intractable portal hypertension, and vasodilator treatment for pulmonary hypertension. The proposed treatment goals for Gaucher disease incorporate the following principles.

Disease phenotype

		Homogeneous	*Heterogeneous*
Disease course	*Acute*	**Acute Otitis Media** Goal: Eliminate infection Treatment:: Standard dosing • Amoxicillin 250 mg po tid x 10 days	**Post-operative Pain** Goal: Eliminate pain Treatment: Individualized dosing • Morphine 2-15 mg IV q2-6 hours PRN
	Chronic	**Hypothyroidism** Goal: • Normal TSH Treatment: Standard dosing • Levothyroxine 1.5 mcg/kg/d	**Type 2 Diabetes** Goals: • HbA1c<7, BG 80-120 • Prevent or treat: o Nephropathy o Retinopathy o Neuropathy Treatment: Individualized regimen • Diet, exercise, weight loss • Oral antidiabetic drugs +/- insulin • ACE inhibitors

Figure 2. Gaucher disease: a chronic heterogeneous disorder. Lessons from the rest of medicine.

Goals address all current and known potential manifestations of the disease process.

Goals are adaptable to individual disease manifestations and course of disease.

Goals are realistic.

They reflect the expectations of the physician and the patient.

For any given parameter, response may not solely be dependent on ERT dosage, but may also be a function of and limited by pretreatment disease severity.

Goals include a quantitative or qualitative objective and an expected timeframe for response that is consistent with accepted standards of care.

Goals for the following manifestations are discussed with an accompanying presentation of supporting evidence: Anemia, thrombocytopenia, splenomegaly, hepatomegaly, bone disease, growth retardation, pulmonary disease, biomarkers, and quality of life. While working towards these goals, it is important to remember that patients may not be measuring their progress in the same way their physicians do. For example, some patients are likely to be less interested in clinical measurements than in the quality of life issues related to general health, vitality, physical function, bodily pain, physical and emotional role, social function, and mental health.

4.1 Anemia

Anemia is a common manifestation of Gaucher disease. The most frequent presentation among anemic patients is fatigue. Older patients, particularly those with coexistent cardiopulmonary disease, may also present with effort-related dyspnea or angina. Data from the International Gaucher Registry show that anemia usually responds rapidly (i.e., within 6–12 months) to ERT (Figure 3). Therefore, the therapeutic goal for treating anemia in patients with Gaucher disease is to:

- Increase hemoglobin levels to age-appropriate normal levels (>11.0 g/dL for women and children, >12.0 g/dL for adult men) within 12 to 24 months of initiation of ERT.
- In patients with severe anemia who have required blood transfusions, eliminate the need for future transfusions.
- Eliminate symptoms of anemia.
- Maintain the improvement in hemoglobin concentration.

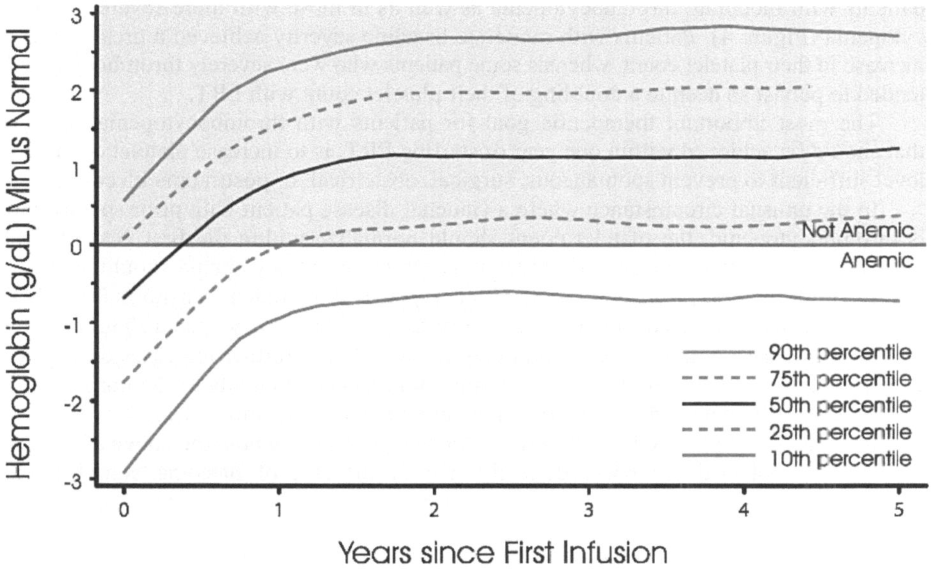

Figure 3. Hemoglobin response to ERT in type 1 patients abnormal at first infusion*,†,‡ (667 patients, 9534 observations). Gaucher Registry results: *2004 Annual Report*. *"At first infusion" is defined as the date closest to first infusion -8 to +2 weeks. † "Abnormal" is defined as anemic according to age and gender norms for hemoglobin concentrations, as follows: <12 g/dL for males older than 12 years; <11 g/dL for females older than 12 years; <10.5 g/dL for children ages >2 to 12 years; <9.5 for children ages 6 months to 2 years; <10.1 g/dL for children younger than 6 months of age. As a result of curve smoothing and the omission of data points in the window of first infusion but prior to the actual first infusion date, the 75th and 90th percentile lines appear above 0 at first infusion. ‡ All lines are smoothed curves.

In more severely affected patients (Hb < 8.0 g/dL), although significant increases in hemoglobin concentration are expected during the first six months of ERT, mild anemia may persist, especially among those with pretreatment massive splenomegaly and hypersplenism. However, in all patients with persistent anemia, evaluation should be undertaken to exclude coexistent etiologies, including iron deficiency, vitamin B12 deficiency, anemia of chronic disease, and, especially in older individuals, myelodysplasia or immunoproliferative disorders. Similar considerations, including studies to rule out either acute or chronic blood loss, pertain to patients in whom anemia recurs after correction with ERT. It should be remembered that serum ferritin levels are frequently elevated in patients with Gaucher disease and that iron deficiency is not ruled out in the presence of normal or elevated ferritin concentrations (Morgan et al., 1983). As of this writing, there is yet no published information about serum ferroportin or urinary hepcidin levels in patients with Gaucher disease.

4.2 Thrombocytopenia

Thrombocytopenia justifying initiation of ERT is defined as repeated platelet counts of less than 100–120,000/μL. The magnitude and rapidity of response to ERT is influenced by pretreatment severity of thrombocytopenia, pretreatment spleen volume, and splenectomy status. In the Gaucher Registry database, platelet responses to ERT were seen in patients with moderate thrombocytopenia as well as in those with more severe thrombocytopenia (Figure 4). Patients with moderate baseline severity achieved a greater overall increase in their platelet count, whereas some patients who were severely thrombocytopenic tended to persist so despite a doubling of their platelet count with ERT.

The most important therapeutic goal for patients with thrombocytopenia, and one that should be achieved within one year of starting ERT, is to increase platelet counts to a level sufficient to prevent spontaneous, surgical, obstetrical, or posttrauma bleeding.

In the unusual circumstance where a Gaucher disease patient with prior splenectomy is thrombocytopenic, the platelet count should normalize within the first year of ERT. Patients with an intact spleen and moderate baseline thrombocytopenia should achieve a 1.5–2.0-fold increase in platelet count within 1 year and approach near-normal levels by year 2. Nonsplenectomized patients with severe baseline thrombocytopenia should achieve a 1.5-fold increase within 1 year, a doubling by year 2, and further increases during year 2–5. However, in these circumstances, restoration of platelet counts to the normal range is not necessarily expected and is most likely not clinically harmful.

It should be emphasized that thrombocytopenic patients in whom the above therapeutic goal is accomplished are not necessarily free of the risk of bleeding complications, particularly during pregnancy, parturition, and postdelivery when thrombocytopenia may recur due to the pregnancy state itself. In addition, some patients with Gaucher disease may have qualitative platelet defects or other coagulation abnormalities that may contribute to bleeding either in the obstetrical setting or in association with surgery or invasive diagnostic procedures (Gillis et al., 1999). When these clinical situations arise, it is prudent for the treating physician to request hematological consultation. Should patients fail to have improvement in thrombocytopenia with ERT, or should previous improvement not be sustained in the absence of ERT interruption or dose reduction, coexistent conditions such as immune thrombocytopenia with antiplatelet antibodies, phospholipid antibody syndromes, HIV-associated thrombocytopenia, and myelodysplasia should be ruled out. In this circumstance, appropriate bone marrow diagnostic studies may be necessary.

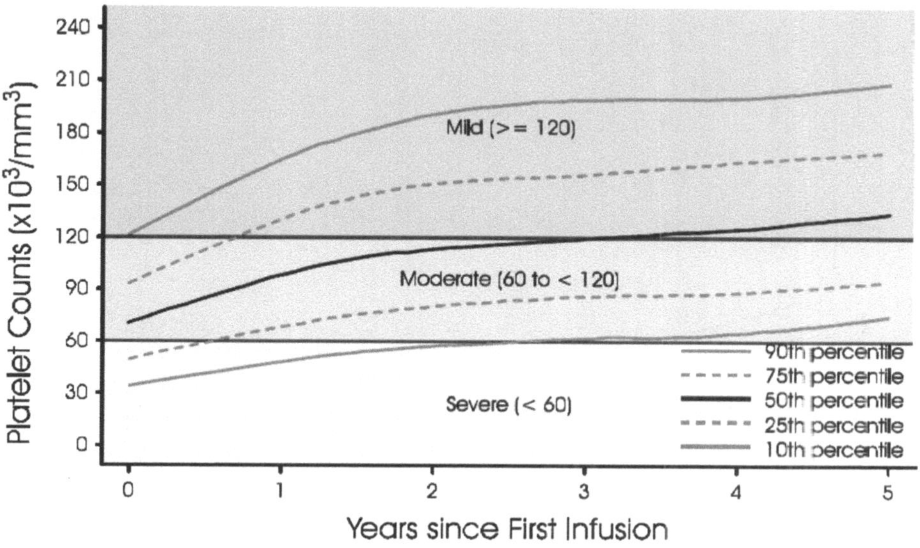

Figure 4. Platelet count response to ERT in type 1 patients abnormal at first infusion*,†,‡,§ (848 patients, 11292 observations). Gaucher Registry results: *2004 Annual Report* * "At first infusion" is defined as the date closest to first infusion -8 to +2 weeks. † "Abnormal" is defined as a platelet count of <120,000 × $10^3/mm^3$. As a result of curve smoothing and the omission of data points in the window of first infusion but prior to the actual first infusion date, the 90th percentile line appears above 120,000 at first infusion. ‡ All lines are smoothed curves. § 92 patients with partial or total splenectomy were excluded.

4.3 Splenomegaly

Splenomegaly is defined as a splenic volume greater than the normal 0.2% of total body weight in kilograms. Spleen enlargement is arbitrarily considered moderate when >5 times normal but ≤15 times the normal volume; the enlargement is considered severe if the volume exceeds 15 times normal. The pretreatment spleen volume exceeds 5 multiples of normal in 90% of all symptomatic nonsplenectomized patients with type 1 disease (Charrow et al., 2000).

International Gaucher Registry data reveal substantial decreases in spleen volume with ERT in patients with either moderate or severe pretreatment splenomegaly (Figure 5). Overall, splenomegaly decreased by 30–50% within 12 months and by 50–60% over two to five years. Responses were influenced by pretreatment volume. Fewer patients with severe splenomegaly than with moderate enlargement achieved <5 times normal volume.

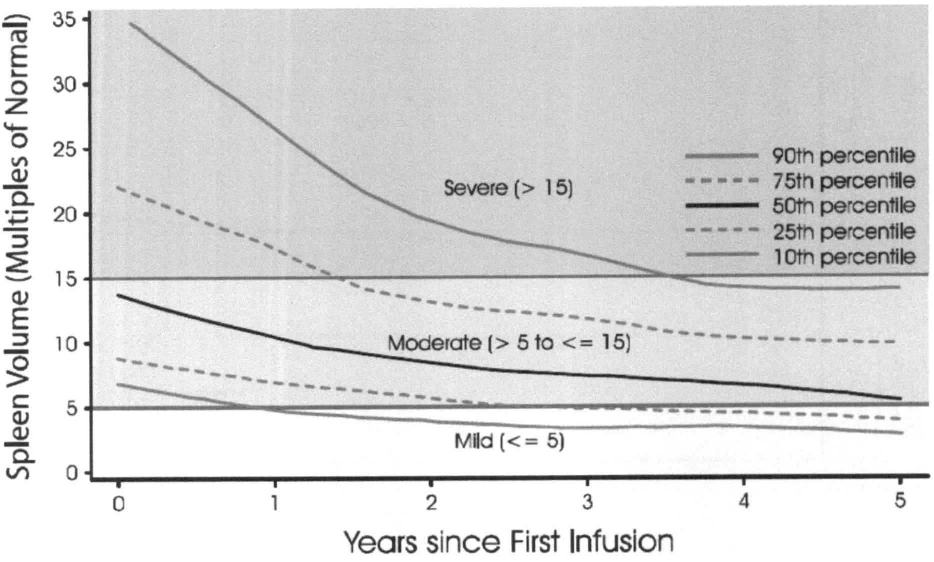

Figure 5. Spleen multiples of normal response to ERT in type 1 patients abnormal at first infusion*,†,‡,§ (565 patients, 2089 observations). Gaucher Registry results: *2004 Annual Report.* * "At first infusion" is defined as the date closest to first infusion -6 months to +6 weeks. † "Abnormal" is defined as a spleen volume greater than 5 MN. ‡ All lines are smoothed curves. § 22 patients with partial or total splenectomy were excluded.

As shown in Figure 6, therapeutic goals for patients with splenomegaly are to reduce spleen volume to ≤2 to 8 times normal and alleviate discomfort due to splenic enlargement (abdominal distension, early satiety). Additional goals focus on ameliorating hypersplenism and alleviating abdominal pain due to recurrent episodes of splenic infarction. Volume normalization is not expected in patients with severe baseline splenomegaly. Indeed, many patients will have some residual enlargement with long-term treatment, probably as a result of pre-existing postinfarction fibrotic scars or nodule formation. There is still insufficient information regarding the effects of ERT on the risk for splenic rupture, and whether enzyme therapy addresses persistent intrasplenic sanctuary sites that may be associated with chronic immune and/or cytokine stimulation that may create risk for development of hematologic malignancy (e.g., lymphoma, myeloma, amyloidosis).

4.4 Hepatomegaly

Hepatomegaly is defined as a liver volume >1.25 times the normal 2.5% of total body weight in kilograms. Liver enlargement is properly assessed by quantitative imaging, preferably using MRI or CT. ERT-induced reduction in hepatomegaly is affected by

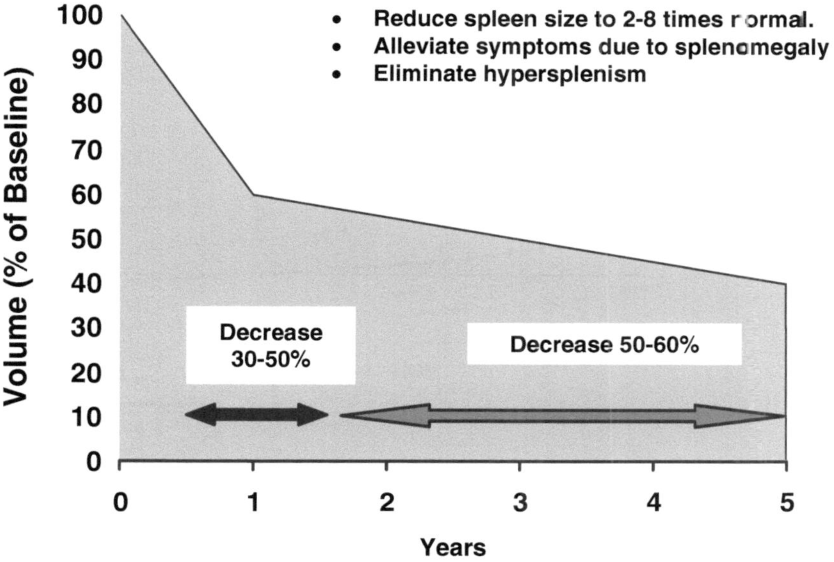

Figure 6. Therapeutic goals for patients with splenomegaly.

pretreatment liver volume. Patients with moderate hepatomegaly (liver volume >1.25 and ≤2.5 times normal) are more likely to achieve liver volume normalization than those with severe hepatomegaly (>2.5 times normal). Reduction in hepatomegaly is greatest during the first 2 years of ERT, but liver volumes in those with enlarged livers continue to decrease at a slower rate during years 3 to 5 (Figure 7).

Therapeutic goals for hepatomegaly are to reduce and then maintain the liver volume to 1.0–1.5 times normal. Liver volume should decrease by 20–30% within years 1 to 2 and by 30–40% by years 3 to 5. Volume normalization is generally not possible when hepatomegaly is severe, presumably because of existence of pretreatment fibrosis. Data on the impact of ERT for prevention of hepatic infarction, fibrotic scarring, progression to cirrhosis, portal hypertension, and hepatopulmonary syndrome are currently limited.

If therapeutic goals for hepatomegaly are not met, screening for intercurrent conditions (e.g., viral hepatitis, chronic hepatitis, autoimmune liver disease, iron overload) should be performed, especially in patients with highly abnormal liver function test results and/or advanced liver disease. Some patients with established cirrhosis and hepatic failure will require allogeneic hepatic transplantation.

4.5 Skeletal Pathology

The goals for liver, spleen, and peripheral blood pathology are primarily directed at reversal of existent pretreatment abnormalities and maintenance of achieved therapeutic

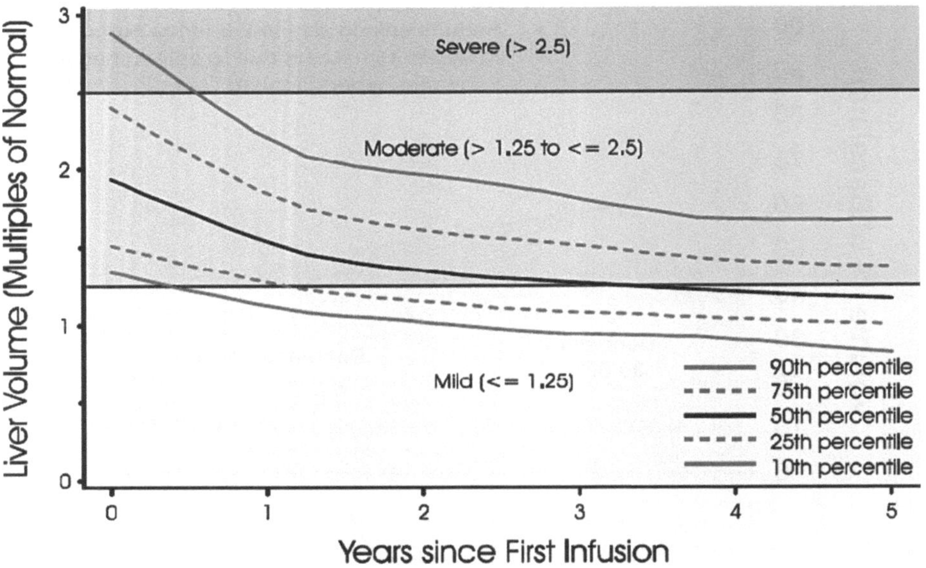

Figure 7. Liver multiples of normal response to ERT in type 1 patients abnormal at first infusion*,†,‡ (602 patients, 2438 observations). Gaucher Registry results: *2004 Annual Report.* * "At first infusion" is defined as the date closest to first infusion -6 months to +6 weeks. † "Abnormal" is defined as a liver volume greater than 1.25 MN. ‡ All lines are smoothed curves.

gains. In contrast, the goals for skeletal manifestations of Gaucher disease are largely focused on prevention of both bone complications that have not yet occurred as well as recurrence and progression of existent skeletal pathology. Although it is also hoped that treatment will alleviate pain and discomfort caused by Gaucher bone disease, achievement of this goal is often limited by the irreversible damage associated with skeletal events such as infarction, avascular necrosis, fractures, and joint destruction and deformity.

The skeletal pathologies of Gaucher disease, which affect most patients with non-neuronopathic disease, can be the most debilitating aspect of the disease. Skeletal pathology and severe clinical symptoms can occur in patients with relatively minor organomegaly and normal hematological parameters. From the International Gaucher Registry data, 66% of patients receiving ERT had a pretreatment complaint of bone pain, 29% had suffered prior bone crises, 50% had osteopenia, 35% infarction, 34% avascular necrosis, 28% new fractures before ERT, and 14% had undergone joint replacement (Charrow et al., 2000). Because skeletal pathology is often progressive, yet unpredictable, treatment should be started as early as possible to prevent the development of irreversible pathology (Wenstrup et al., 2002).

After inception of ERT, bone pain resolved in 50% of symptomatic patients within 1 to 2 years, and new onset of bone pain in asymptomatic patients occurred in only 4%. During the first 2 years of ERT 80–90% with prior bone crises had no recurrences, and the incidence of de novo crises during the first 2 years of ERT was less than 1% (Weinreb et al., 2002). After 2 to 4 years of ERT, physician assessment of overall bone involvement (clinical and radiological) indicated improvement in 30% to 40% of pediatric patients (age <18 years), progression in <5% of pediatric patients, improvement in 20% to 30% of adult patients, and progression in 5% to 10% of adult patients. Bone mineral density (BMD) has been shown to improve with ERT, although the response may take at least 2 years to manifest in children and longer in adults (Rosenthal et al., 1995; Ciana et al., 2005). For some adult patients, adjuvant therapy with bisphosphonates may be required for those who fail to demonstrate clinically significant improvement in BMD after 2 to 4 years of ERT (Poll et al., 2002; Bembi et al., 2002; Grabowski and Hopkin, 2003). An additional response of BMD is achieved by adding bisphosphonates in significantly osteopenic/osteoporotic patients on ERT (Wenstrup et al., 2004). Decreased bone marrow infiltration, as assessed experimentally by MRI and quantitative chemical shift imaging (QCSI), has been observed after as little as 1 year of ERT but more commonly takes at least 3 years (Rosenthal et al., 1995; Terk et al., 2000; Poll et al., 2001, 2002; Hollak et al., 2001a), and is more apparent in more active bone, that is, vertebral marrow (Maas et al., 2003). However, these techniques are not yet standardized or correlated with clinical outcomes in large numbers of patients Therefore, they have not yet been included among the routine recommended assessments and therapeutic goals.

The all-encompassing goal for children and adults with skeletal pathology is to retain skeletal function by preventing the onset of new skeletal complications and to relieve and prevent recurrence of acute and chronic bone pain and bone crises. Pediatric and adult patients have differences in skeletal physiology that must be taken into consideration when therapeutic goals are set. For children the aim should be to achieve ideal peak skeletal mass and prevent skeletal pathology whereas in adults, the aim is to maintain or improve the skeleton and retain, preserve, or improve function. Patients with advanced bone disease at the time of ERT initiation also may benefit from orthopedic intervention, physiotherapy, and adjunctive treatment (e.g., bisphosphonates for severe bone mineral density loss (Wenstrup et al., 2004).

The specific therapeutic goals for skeletal disease are to lessen or eliminate bone pain within 1 to 2 years of treatment, to eliminate bone crises within 1 to 2 years of treatment, to prevent osteonecrosis and subsequent subchondral joint collapse, and with respect to bone mineral density, to attain peak or ideal skeletal mass for pediatric patients and to achieve quantitative improvement in BMD for adults (Table 5).

4.6 GROWTH AND DEVELOPMENT

Approximately half of the children with Gaucher disease exhibit growth retardation (Zevin et al., 1993; Kaplan et al., 1996; Kauli et al., 2000), and approximately 25% are shorter than expected compared with mid-parental height (Charrow, 1998, 2000). Children with Gaucher disease who exhibit markedly stunted growth also tend to have severe visceral involvement and may experience delayed puberty (Kauli et al., 2000). In otherwise mildly affected children, growth retardation should prompt appropriate assessments by an endocrinologist. Studies indicate that with ERT, children with Gaucher disease exhibiting growth retardation can achieve normal height, according to population averages (Kaplan et al., 1996; Ida et al., 2001). The therapeutic goal for children with Gaucher disease is to normalize growth such that a normal height is achieved according to population standards within three years of starting ERT and the onset of puberty occurs at a normally expected age.

Table 5. Therapeutic goals for skeletal disease

Patients	Goals	Timeframe
All Patients	• Lessen or eliminate bone pain • Prevent bone crises • Prevent osteonecrosis and subchondral joint collapse	1 to 2 years
Pediatric Patients	• Attain normal or ideal peak skeletal mass • Increase cortical and trabecular bone mineral density	Year 2
Adults	• Increase cortical and trabecular bone mineral density	3 to 5 years

4.7 Pulmonary Involvement

One to two percent of patients with type 1 Gaucher disease exhibit overt pulmonary manifestations in the form of interstitial lung disease, severe pulmonary hypertension, and/or hepatopulmonary syndrome (Dawson et al., 1996; Mistry et al., 2002). Development of pulmonary hypertension is a grave sign because it is recognized to be an important cause of premature death in patients with type 1 Gaucher disease (Lee et al., 1982; Harats et al., 1997; Elstein et al., 1998). Pulmonary hypertension responds to ERT in combination with vasodilators (e.g., prostacyclin and bosentan). Interstitial lung disease also responds to ERT in some patients with Gaucher disease (Pastores et al., 1993). In addition, there are reports of dramatic reversal of hepatopulmonary syndrome with ERT (Dawson et al., 1996). The vascular lesions underlying hepatopulmonary syndrome may coexist with plexogenic vasculopathy that cause severe pulmonary hypertension (Sirrs et al., 2002), and therefore reversal of heaptopulmonary syndrome may unmask underlying pulmonary hypertension (Dawson et al., 1996). The latter observations underscore the importance of maintaining optimal ERT with adjuvant therapies as required after the resolution of hepatopulmonary syndrome. If the clinical signs of hepatopulmonary syndrome fail to resolve with ERT, underlying cirrhosis or presence of large anatomic intrapulmonary shunts should be excluded.

Patients with overt, symptomatic pulmonary involvement may suffer from sudden life-threatening deterioration. The goal of ERT and adjuvant therapies is to prevent sudden death or rapid and inexorable clinical deterioration, and to improve the patient's functional capacity and quality of life. For hepatopulmonary syndrome and interstitial disease, the goal of ERT is to reverse this syndrome and obviate dependency on oxygen. The goal of treatment of severe pulmonary hypertension with ERT and adjuvant therapies is to improve hemodynamic and functional status as well as to prolong life. Almost all cases of severe pulmonary involvement in type 1 Gaucher disease relate to splenectomized patients. An important goal of ERT is prevention of these complications through timely institution of ERT and avoidance of splenectomy.

4.8 Functional Health and Well-Being

The World Health Organization defines health as "a state of complete physical, mental, and social well-being and not merely the absence of disease or infirmity." In a retrospective study of 212 patients with Gaucher disease (mean age 45.1 ± 17 years) receiving ERT, health-related quality of life was assessed using the SF-36 Health Survey (Damiano, et al., 1998). Scores for Physical Functioning, Role Physical, Bodily Pain, General Health, and Vitality were significantly lower for not-yet treated patients with Gaucher disease compared with those for age- and sex-matched control subjects. The benefits of ERT can be observed from the results of a survey involving 55 patients (mean age 46.8 years) with Gaucher disease (Figure 8). Scores for most domains approximated those of the general population and were better than those reported by patients who were not receiving ERT. Patients receiving ERT had better scores for domains related to physical health (Physical Function, Role Physical, Bodily Pain, General Health, and Vitality). Domains related to psychological health (Social Function, Role Emotional, and Mental Health) were similar for the untreated patients and the general population. Comparable findings have been reported in Spanish patients (Giraldo et al., 2005).

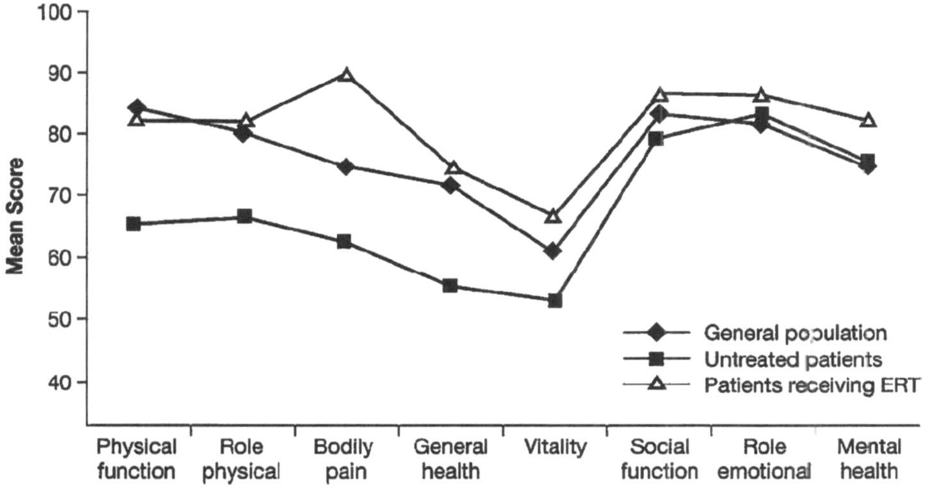

Figure 8. Domains of the SF-36 Health Survey for the general population, untreated patients with Gaucher disease ($N = 212$), and patients receiving ERT ($N = 55$; mean duration: 6.3 y, range: 0.1 – 11.8 y) for Gaucher disease. (Adapted with kind permission of Kluwer Academic Publishers.)

The therapeutic goals for health-related quality of life are to improve and restore physical function so that patients are able to carry out their normal daily activities and fulfill their functional roles within their families and society. Patients should show improvement in scores from baseline of a validated quality-of-life instrument such as the SF-36 (Ware, 1993).

4.9. Biomarkers

Many endogenous serum proteins such as transcobalamin-2, lysozyme, and ferritin are variably elevated in patients with Gaucher disease but their levels do not reliably relate to the clinical presentation (Gilbert and Weinreb, 1976; Silverstein and Friedland, 1977; Morgan et al., 1983). Chitotriosidase, angiotensin-converting enzyme (ACE), and tartrate-resistant acid hosphatase (TRAP) do appear to more closely correlate with disease involvement and these commercially available tests may be useful as adjuncts to clinical observations for monitoring patient responses to ERT (Hollak et al., 2001b; Whitfield et al., 2002, Cabrera-Salazar et al., 2004). Serum levels of the chemokine CCL18 are reported to correlate with the magnitude of hepatomegaly and splenomegaly (Boot et al., 2003) and applicability of CCL18 measurement and that of another potential biomarker, lysosomal-associated membrane protein (LAMP), are being investigated (Whitfield et al., 2002; Deegan et al., 2005).

Of the serum proteins known to be elevated in Gaucher disease, chitotriosidase has been studied most extensively. Chitotriosidase is secreted by activated macrophages (Korolenko et al., 2000), and activity in plasma is markedly increased in patients with Gaucher disease (Hollak et al., 1994). Chitotriosidase activity has been seen to decrease in response to ERT (Young et al., 1997), and the decrease was related to dose and correlated with other indicators of positive clinical responses (e.g., improvements in hematologic and visceral parameters; Giraldo et al., 2001; Hollak et al., 2001b). Chitotriosidase activity is absent in a small proportion of patients (6–8%) because of a mutation in the chitotriosidase gene (Giraldo et al., 2001). For such patients, assessment of ACE, TRAP, or CCL18 (where available) can be substituted. Absolute levels of chitotriosidase, ACE, or TRAP are not indicative of disease severity in individual patients, are not useful for comparisons among patients, and have no prognostic significance. Single measurements should not be included among criteria for starting ERT or selecting the initial dose.

For some patients, serial increases in chitotriosidase activity may be an early indicator of clinical relapse following dose reduction or after a treatment interruption (Czartoryska et al., 2000), but poor correlation between changes in plasma chitotriosidase activity and clinical course has also been reported (Zhao et al., 2003). Increases in chitotriosidase activity should prompt investigation of the patient's disease status and compliance with therapy (if the patient is receiving ERT). However, alterations to therapy must be supported by a complete assessment of disease status and are not to be made solely on the basis of change in the activity of chitotriosidase or any other biomarker. Quantitative goals for biochemical markers are not presented because of insufficient data on clinical correlations. Serum and urine biomarkers that are associated with perturbations in either bone formation or bone resorption have been examined in patients with Gaucher disease but demonstrate no consistent pattern of abnormality or obvious clinical utility (Ciana et al., 2005).

5 APPLICATION OF THERAPEUTIC GOALS TO ERT DOSE MODIFICATION

If therapeutic goals are not achieved within the expected timeframe, or should patients relapse after an initial favorable response, dose adjustment may need to be considered provided that the possibility of neutralizing antibodies (very rarely seen), irreversible pathology, or intercurrent illness have been excluded. The amount of imiglucerase administered may be augmented by increasing either the unit dose or the frequency of infusion. Although some patients, particularly those with type 3 disease, have routinely been treated with doses of imiglucerase as high as 120 units/kg every two weeks, progressive increase of the unit dose may be inefficient if the binding capacity of the macrophage mannose receptors is exceeded resulting in nonspecific absorption of imiglucerase by hepatocytes or other nontargeted cells (Sato et al.,1993; Bijsterbosch et al.,

1996). Therefore, when, due to a failure to achieve or maintain therapeutic goals, a higher dose is indicated for patients who are being infused biweekly with 60 units/kg or higher, it may be prudent to increase the frequency of administration (e.g., weekly) rather than the unit dose. Although dexamethasone-mediated up-regulation of the mannose receptor was shown to improve the delivery of recombinant glucocerebrosidase to cultured macrophages (Zhu et al., 2004), clinical trials of corticosteroids in Gaucher disease patients with suboptimal responses to standard doses of ERT have not as yet been reported. In some circumstances, rather than a dose increase, alternative strategies should also be considered such as adjuvant therapy with bisphosphonates for adults who fail to achieve significant improvement in osteopenia in the expected timeframe (Wenstrup et al., 2004).

When all therapeutic goals are achieved, a clinician's decision to decrease the dose of imiglucererase should be implemented based on the history and objective evidence of disease status and course. For example, in patients with severe skeletal disease (moderate to severe osteopenia, chronic bone pain, bone crises, avascular necrosis, pathological fractures, and joint replacements) who remain stable or show only modest improvements, dose reductions may not be appropriate until significant improvements are achieved and maintained for at least one year. There is as yet no evidence from controlled clinical trials for a schema of dose reductions for patients who have achieved all therapeutic goals. The following recommendations represent a consensus opinion of some experts based on clinical experience and review of the literature (Andersson et al., 2005).

Adult patients with increased risk at baseline or at any subsequent time (Table 3) and who have achieved all therapeutic goals can have the dose decreased in small decrements (~15–25%) until their next scheduled evaluation (in three to six months). For patients who maintain all therapeutic goals, similar decrements in dose may then be considered. However, in increased-risk adults with severe disease and all children, the minimum recommended long-term maintenance dose is 30 U/kg every two weeks. This dose has been reported as a threshold for improving versus worsening radiological manifestations of skeletal disease (Mota, 2004). Adult patients with less severe disease at baseline (Table 3) may tolerate larger dose reductions (e.g., 25–50% per dose), when on an every other week regimen. However, the minimum recommended long-term maintenance dose for adult patients with less severe disease is no less than 20 U/kg every two weeks (Altarescu et al., 2000). The safety and efficacy of an every four week imiglucerase maintenance regimen for lower-risk patients is the subject of an on-going, randomized clinical trial, compared with the usual every two weeks infusion regimen.

At the time of this writing, insufficient information is available concerning dose reduction in the pediatric population. Caution should be exercised when considering dose reductions in children because it is unknown what the appropriate dose is to prevent long-term Gaucher disease complications.

Following a dose reduction, it is essential that the patient be closely monitored to be sure that previously achieved goals are maintained. Criteria that are useful for defining therapeutic failure following a dose reduction are indicated in Table 6. Any of these signs necessitates return to the treatment regimen previously known to be effective.

6 DISEASE MANAGEMENT AND THE ROLE OF THE GAUCHER REGISTRY

At a National Institutes of Health (United States) Technology Assessment Conference, the consensus noted that it would be advantageous to enter all patients with Gaucher

Table 6. Criteria for failure to maintain therapeutic goals following imiglucerase dose reductions

Any of these signs necessitates return to dose regimen used to achieve/maintain goals	
• ↓ Hemoglobin (2 readings 2 wk apart) – 1.25 g/dL for women & children – 1.5 g/dL for men • Fatigue recurrence due to anemia • Platelet count ↓ 25% of prior level or <80,000/mm3 (2 readings 2 wk apart) • Liver or spleen volume increase by 20%, recurrence of abdominal pain, or appetite decrease	• Skeletal disease progression • Bone pain: ↑ severity or frequency • Bone crisis—recurrence or ↑ frequency • Quality-of-life decrease • Development of Gaucher disease pulmonary symptoms • Delay in growth/development or loss of previously achieved milestones

disease into a registry. Such a registry would provide a valuable resource for increasing our knowledge of the natural history of the disease, help to identify predictors of response, and facilitate clinical trials to answer specific questions about therapy (NIH, 1996). Partly as a condition for the initial approval of ERT with alglucerase, the International Gaucher Registry database of the International Collaborative Gaucher Group was launched by the Genzyme Corporation in 1991. For clinical investigators, the Registry data are a foundation for research on disease management and outcomes and on new innovative treatments. For the participating physician, the Registry offers an opportunity to exchange data with other physicians to facilitate clinical decision making, and provides access to a rich repository of data, reports, and information on current guidelines and practice patterns. For the individual patient, whose informed consent is an absolute condition for enrollment, the serial patient case reports generated by the Registry can be reviewed with the treating physician as a tool for tracking progress toward achievement and subsequent maintenance of therapeutic goals. A sample protocol demonstrating how the Registry patient case report form can be adapted to clinical practice is illustrated in Table 7.

7 SUMMARY

Gaucher disease is a highly heterogeneous disorder, requiring an individualized treatment plan for each patient designed to achieve a maximal response in all affected disease compartments. The construction of a comprehensive treatment plan by defining specific treatment goals is an approach that is useful not only for guiding the treating physician, consulting specialists, and allied health personnel, but also for educating patients and families, establishing reasonable expectations, and creating a partnership in caring. Most patients will have multiple therapeutic goals that must be achieved within the expected timeframes and maintained throughout their lives. Successful application of this method requires a comprehensive initial assessment of all potentially affected organ systems and

Table 7. Integrating the Gaucher Registry patient report form into routine clinical practice

Day before patient visit	Coordinator reviews appointment schedule Logs on to Gaucher Registry Prints out updated PCR for each patient Attaches PCR to front of medical chart
Day of patient visit	Physician reviews PCR with patient Provides PCR to patient Records data from visit in registry Orders lab and radiology assessments
Following patient visit	Coordinator enters new assessment data into Gaucher Registry

regular ongoing monitoring to assess the global response to therapy, to adjust the treatment plan when goals are not met, and to ensure the maintenance of attained goals. This treatment approach is similar to that for other chronic diseases with heterogeneous manifestations and responses to therapy, thus requiring individualized dosing and ongoing assessment and monitoring. A patient registry, such as the Gaucher Registry, provides a platform to track the progress of patients individually and collectively. Such an approach has been presented in Figure 1.

Treatment of Gaucher disease is lifelong, and the establishment of therapeutic goals may encourage compliance with treatment and minimization of treatment interruptions. Compliance with ERT during long-term therapy has been approximately 90% and even higher for patients who receive home care (Genzyme Corporation, Cambridge, MA. Data on file). Some patients may desire a temporary break from treatment because of personal issues or changes in lifestyle. Although some patients may remain stable during treatment interruptions, many patients experience disease progression (e.g., varying combinations of decreased platelet counts and/or hemoglobin levels, increasing organomegaly, recurrent or heightened pain, exacerbation of skeletal pathology, reversal of quality of life gains, and progressively increased biomarkers when treatment is stopped (Elstein et al., 2000; Schwartz et al., 2001; vom Dahl et al., 2001; Grinzaid et al., 2002; Toth et al., 2003). Few Few patients, if any, are expected to maintain therapeutic goals in the prolonged absence of treatment, and there are no proven methods for identifying these patients (Weinreb, 2001). Therefore, treatment interruptions are not recommended, and, when unavoidable, necessitate close monitoring of the patients. Therapeutic goals, once achieved, require continued compliance with treatment in order to maintain the goals over time.

The achievement of therapeutic goals also can serve as a benchmark for assessing the consequences of changes or adjustments in the treatment regimen. Modifications such as dose reductions or decreases in dosing frequency should be considered only after all relevant therapeutic goals have been achieved. All therapeutic goals should be maintained with any modification in the treatment regimen. Maintenance of the goals should be monitored according to a regular comprehensive monitoring schedule such as outlined in the Gaucher Registry guidelines (Gaucher Registry, 2004; Weinreb et al., 2004). Reversal of the attainment of any therapeutic goal indicates a failure of the treatment modification, although the possibility of an intercurrent condition that may confound the course of Gaucher disease should be considered as well. Inadequate long-term maintenance therapy may be first noticed by changes in biomarkers, decreased hematologic parameters, fatigue, and increases in organ volumes. The effects of suboptimal treatment in the skeleton may not become apparent until the development of an irreversible complication such as an osteonecrotic joint collapse. Therefore, any changes in the treatment regimen of an otherwise stable patient should be made cautiously.

Enzyme replacement therapy is expensive, but it has dramatically improved the function and well-being of several thousand patients worldwide with type 1 Gaucher disease. The goal-oriented, individualized approach to ERT with imiglucerase described above, when combined with the appropriate use of adjuvant pharmacological and nonpharmacological interventions under the guidance of an experienced multidisciplinary team at an established center for the treatment of this disorder, is the most efficient, and cost-effective method for improving the outcome and health of patients with type 1 Gaucher disease.

REFERENCES

Aghion E (1934). La maladie de Gaucher dans l'enfance. These, Faculte de Medicine, Paris.

Altarescu G, Schiffmann R, Parker CC, Moore DF, Kreps C, Brady RO, et al. (2000). Comparative efficacy of dose regimens in enzyme replacement therapy of type I Gaucher disease. Blood Cells Mol Dis 26: 285–290.

American Diabetes Association (2005). Standards of medical care in diabetes. Diabetes Care; 28 S1: 540–536.

Andersson HC, Charrow J, Kaplan P, Mistry P, Pastores GM, Prakesh-Cheng A, Rosenbloom BE, Scott CR, Wappner RS, Weinreb NJ, for the International Collaborative Gaucher Group U.S. Regional Coordinators (2005). Individualization of long-term enzyme replacement therapy for Gaucher disease. Genet Med 7: 105–110.

Barton NW, Brady RO, Dambrosia JM et al. (1991). Replacement therapy for inherited enzyme deficiency—macrophage-targeted glucocerebrosidase for Gaucher's disease. N Eng J Med 324: 1464–1470.

Bembi B, Ciana G, Mengel E, Terk MR, Martini C, Wenstrup RJ (2002). Bone complications in children with Gaucher disease. Br J Radiol 75 suppl 1: A37–A43.

Beutler E, Grabowski GA (2001). Gaucher disease. In Scriver CR, Beaudet AC, Sly WS, Valle D, Eds. *The Metabolic and Molecular Bases of Inherited Disease*. 8th ed. New York: McGraw-Hill, pp. 3635–3668.

Beutler E, Kuhl W (1970). The diagnosis of the adult type of Gaucher's disease and its carrier state by demonstration of a deficiency of beta-glucosidase avtivity in peripheral blood leukocytes. J Lab Clin Med 76: 747–755.

Bijsterbosch MK, Donker W, van de Bilt H, van Weely S, van Berkel TJ, Aerts JM (1996). Quantitative analysis of the targeting of mannose-terminal glucoderebrosidase. Predominant uptake by liver endothelial cells. Eur J Biochem 237: 344–349.

Boot RG, Verhoek M, De Fost M, Hollak CE, Maas M, Bleijlevens B, et al. (2003). Marked elevation of the chemokine CCL18/PARC in Gaucher disease: A novel surrogate marker for assessing therapeutic intervention. Blood 103: 33–39.

Brady RO, Kanfer J, Shapiro D (1965). Metabolism of glucocerebrosides II. Evidence of an enzymatic deficiency in Gaucher's disease. Biochem Biophys Res Comm 18: 221–225.

Cabrera-Salazar MA, O'Rourke E, Henderson N, Wessel H, Barranger JA (2004). Correlation of surrogate markers of Gaucher disease. Implications for long-term follow up of enzyme replacement therapy. Clin Chim Acta 344: 101–107.

Charrow J, Andersson HC, Kaplan P, Kolodny EH, Mistry P, Pastores G, et al. (2000). The Gaucher registry: Demographics and disease characteristics of 1698 patients with Gaucher disease. Arch Intern Med 160: 2835–2843.

Charrow J, Esplin JA, Gribble TJ, Kaplan P, Kolodny EH, Pastores GM, et al. (1998). Gaucher disease: Recommendations on diagnosis, evaluation, and monitoring. Arch Intern Med 158: 1754–1760.

Ciana G, Addobbati R, Tamaro G, Leopaldi A, Nevyjel M, Ronfani L, Vidoni L, Pittis MG, Bembi B (2005). Gaucher disease and bone: Laboratory and skeletal mineral density variations during a long period of enzyme replacement therapy. J Inherit Metab Dis 28: 723–732.

Czartoryska B, Tylki-Szymanska A, Lugowska A (2000). Changes in serum chitotriosidase activity with cessation of replacement enzyme (cerebrosidase) administration in Gaucher disease. Clin Biochem 33: 147–149.

Damiano AM, Pastores GM, Ware JE (1998). The health-related quality of life of adults with Gaucher's disease receiving enzyme replacement therapy: Results from a retrospective study. Qual Life Res 7: 373–386.

Dawson A, Elias DJ, Rubenson D, Bartz SH, Garver PR, Kay AC, et al. (1996). Pulmonary hypertension developing after alglucerase therapy in two patients with type 1 Gaucher disease complicated by the hepatopulmonary syndrome. Ann Intern Med 125: 901–904.

Deegan PB, Moran MT, McFarlane I, Schofield JP, Boot RG, Aerts JM, Cox TM (2005). Clinical evaluation of chemokine and enzymatic biomarkers of Gaucher disease. Blood Cels Mol Dis 35: 259–267.

Dvir H, Harel M, McCarthy AA, Toker L, Silman I, Futerman AH, Sussman JL (2003). X-ray structure of human acid-betga-glucosidase, the defective enzyme in Gaucher disease. EMBO Rep 4: 704–709.

Elstein D, Abrahamov A, Hadas-Halpern I, Zimran A (2000). Withdrawal of enzyme replacement therapy in Gaucher's disease. Br J Haematol 110: 488–492.

Elstein D, Klutstein MW, Lahad A, Abrahamov A, Hadas-Halpern I, Zimran A (1998). Echocardiographic assessment of pulmonary hypertension in Gaucher's disease. Lancet 351: 1544–1546.

Furbish FS, Blair HE, Shiloach J et al. (1984). Interaction of human placental glucocerebrosidaes with hepatic lectins. In Barranger JA & Brady RO, Eds. *Molecular Basis of Lysosomal Storage Disorders*. Orlando, FL: Academic Press, pp. 219–232.

Gaucher Registry (2004). Available at: http://www.lsdregistry.net/gaucherregistry.

Gilbert HS, Weinreb NJ (1976). Increased circulatory levels of transcobalamin II in Gaucher's disease. New Eng J Med 295: 1096–1101.

Gillis S, Hyam E, Abrahamov A, Elstein D, Zimran A (1999). Platelet function Abnormalities in Gaucher disease patients. Am J Hematol 61: 103–106.

Ginns EI, Choudary PV, Tsiji S, et al. (1984). Isolation of cDNA clones for human β-glucocerebrosidase using the λgt11 expression system. Biochem Biophys Res Comm 123: 574–580.

Giraldo P, Cenarro A, Alfonso P, Perez-Calvo JI, Rubio-Felix D, Giralt M, et al. (2001). Chitotriosidase genotype and plasma activity in patients type 1 Gaucher's disease and their relatives (carriers and non carriers). Haematologica 86: 977–984.

Giraldo P, Solano V, Perez-Calvo JI, Giralt M, Rubio-Felix D, Spanish Group of Gaucher Disease (2005). Quality of life related to type 1 Gaucher disease: Spanish experience. Qual Life Res 14: 453–462.

Goker-Alpan O, Hruska KS, Orvisky E, Kishnani PS, Stubblefield BK, Schiffmann R, Sidransky E (2005). Divergent phenoypes in Gaucher diseae implicate the roe of modifiers. J Med Genet 42: e37.

Grabowski GA (2005). Recent clinical progress in Gaucher disease. Curr Opin Pediatr 17: 519–524.

Grabowski GA, Hopkin RJ (2003). Enzyme therapy for lysosomal storage disease: Principles, practice, and prospects. Annu Rev Genomics Hum Genet 4: 403–436.

Grinzaid KA, Geller E, Hanna SL, Elsas LJ 2nd (2002). Cessation of enzyme replacement therapy in Gaucher disease. Genet Med 4: 427–433.

Harats D, Pauzner R, Elstein D, Many A, Klutstein MW, Kramer MR, et al. (1997). Pulmonary hypertension in two patients with type I Gaucher disease while on alglucerase therapy. Acta Haematol 98: 47–50.

Hillborg PO (1959). Morbus Gaucher: Norbotten. Nordisk Medicin. 61: 303–306.

Ho MW, Seck J, Schmidt D, Veath ML, Johnson W, Brady RO, O'Brien JS (1972). Adult Gaucher's disease: Kindred studies and demonstration of a deficiency of acid-glucosidase in cultured fibroblasts. Am J Hum Genet 24: 37–45.

Hollak C, Maas M, Akkerman EM, den Heeten GJ, Aerts H (2001a). Dixon quantitative chemical shift imaging is a sensitive tool for the evaluation of bone marrow responses to individualised doses of enzyme supplementation therapy in type 1 Gaucher disease. Blood Cells Mol Dis 27: 1005–1012.

Hollak CE, Maas M, Aerts JM (2001b) Clinically relevant therapeutic endpoints in type I Gaucher disease. J Inherit Metab Dis 24 suppl 2: 9097–9105.

Hollak CE, van Weely S, van Oers MH, Aerts JM (1994). Marked elevation of plasma chitotriosidase activity. A novel hallmark of Gaucher disease. J Clin Invest 93: 1288–1292.

Ida H, Rennert OM, Kobayashi M, Eto Y (2001). Effects of enzyme replacement therapy in thirteen Japanese paediatric patients with Gaucher disease. Eur J Pediatr 160: 21–25.

Kaplan P, Mazur A, Manor O, Charrow J, Esplin J, Gribble TJ, et al. (1996). Acceleration of retarded growth in children with Gaucher disease after treatment with alglucerase. J Pediatr 129: 149–153.

Kauli R, Zaizov R, Lazar L, Pertzelan A, Laron Z, Galatzer A, et al. (2000). Delayed growth and puberty in patients with Gaucher disease type 1: Natural history and effect of splenectomy and/or enzyme replacement therapy. *Isr Med Assoc J* 2: 158–163.

Korolenko TA, Zhanaeva SY, Falameeva OV, Kaledin VI, Filyushina EE, Buzueva II, et al. (2000). Chitotriosidase as a marker of macrophage stimulation. Bull Exp Biol Med 130: 948–950.

Lachmann RH, Grant IR, Halsall D, Cox TM (2004): Twin pairs showing discordance of phenotype in adult Gaucher's disease. Q J Med 97: 199–204.

Lee RE (1982). The pathology of Gaucher disease, in Desnick RJ, Gatt S, Grabowski GA, Eds. *Gaucher Disease: A Century of Delineation and Research*. New York: Liss, pp. 177–217.

Maas M, van Kuijk C, Stoker J, Hollak CE, Akkerman EM, Aerts JF, et al. (2003). Quantification of bone involvement in Gaucher disease: MR imaging bone marrow burden score as an alternative to Dixon quantitative chemical shift MR imaging—Initial experience. Radiology 229: 554–561.

Maaswinkel-Mooij P, Hollak C, van Eysden-Plaisier M, Prins M, Aerts H, Poll R (2000). The natural course of Gaucher disease in The Netherlands: Implications for monitoring of disease manifestations. J Inherit Metab Dis 23: 77–82.

Mistry PK, Sirrs S, Chan A, Pritzker MR, Duffy TP, Grace ME, et al. (2002). Pulmonary hypertension in type 1 Gaucher's disease: Genetic and epigenetic determinants of phenotype and response to therapy. Mol Genet Metab 77: 91–98.

Morgan MA, Hoffbrand AV, Laulicht M, Luck W, Knowles S (1983). Serum feritin concentration in Gaucher's disease. Br Med J 286: 1864.

Mota R (2004). Bone disease as a parameter of evaluation of therapy, progression, and effectiveness in Gaucher disease. Paper presented at the Latin American Symposium for Lysosomal Storage Diseases. Santiago, Chile, August 7–8, 2004.

NIH Technology Assessment Panel on Gaucher Disease (1996). Gaucher disease: Current issues in diagnosis and treatment. JAMA. 275: 548–553.

Oberling C, Woringer P (1927). La maladie de Gaucher chez la nourisson. Revue Francais de Pediatrie. 3: 475–532.

Pastores GM, Sibille AR, Grabowski GA (1993). Enzyme therapy in Gaucher disease type 1: Dosage efficacy and adverse effects in 33 patients treated for 6 to 24 months. Blood 82: 408–416.

Pastores GM, Weinreb NJ, Aerts H, Andria G, Cox TM, Giralt M, Grabowski GA, Mistry PK, Tylki-Szymanska A (2004). Therapeutic goals in the treatment of Gaucher disease. Semin Hematol 41(4 Suppl 5): 4–14.

Poll LW, Koch JA, vom Dahl S, Willers R, Scherer A, Boerner D, et al. (2001). Magnetic resonance imaging of bone marrow changes in Gaucher disease during enzyme replacement therapy: First German long-term results. Skeletal Radiol 30: 496–503.

Poll LW, Maas M, Terk MR, Roca-Espiau M, Bembi B, Ciana G, et al. (2002). Response of Gaucher bone disease to enzyme replacement therapy. Br J Radiol 75 suppl 1: A25–A36.

Premkumar L, Sawkar AR, Boldin-Adamsky S, Toker L, Silman I, Kelly, JW, Futerman AH, Sussman JL (2005). X-ray structure of human acid-beta-glucosidase covalently bound to conduritol-B-epoxide. Implications for Gaucher disease. J Biol Chem 280: 23815–23819.

Rosenthal DI, Doppelt SH, Mankin HJ, Dambrosia JM, Xavier RJ, McKusick KA, et al. (1995). Enzyme replacement therapy for Gaucher disease: Skeletal responses to macrophage-targeted glucocerebrosidase. Pediatrics 96: 629–637.

Sato Y, Beutler E (1993). Binding, internalization, and degradation of mannose-terminated glucocerebrosidase by macrophages. J Clin Invest 91: 1909–1917.

Schwartz IV, Karam S, Ashton-Prolla P, Michelin K, Coelho J, Pires RF, et al. (2001). Effects of imiglucerase withdrawal on an adult with Gaucher disease. Br J Haematol 113: 1089.

Silverstein E, Friedland J (1977). Elevated serum and spleen angiotensin converting enzyme and serum lysozyme in Gaucher's disease. Clin Chim Acta 74: 21–25.

Sirrs S, Irving J, McCauley G, Gin K, Munt B, Pastores G, et al. (2002). Failure of resting echocardiography and cardiac catheterization to identify pulmonary hypertension in two patients with type I Gaucher disease. J Inherit Metab Dis 25: 131–132.

Tayebi N, Walker J, Stubblefield B, Orvisky E, LaMarca ME, Wong K, et al. (2003). Gaucher disease with Parkinsonian manifestations: Does glucocerebrosidase deficiency contribute to a vulnerability to Parkinsonism? Mol Genet Metab 79: 104–109.

Terk MR, Dardashti S, Liebman HA (2000). Bone marrow response in treated patients with Gaucher disease: Evaluation by T1-weighted magnetic resonance images and correlation with reduction in liver and spleen volume. Skeletal Radiol 29: 563–571.

Toth J, Erdos M, Marodi L (2003). Rebound hepatosplenomegaly in type 1 Gaucher disease. Eur J Haematol 70: 125–128.

Vellodi A, Bembi B, de Villemeur TB, Collin-Histed T, Erikson A, Mengel E, et al. (2001). Management of neuronopathic Gaucher disease: A European consensus. J Inherit Metab Dis 24: 319–327.

vom Dahl S, Poll LW, Haussinger D (2001). Clinical monitoring after cessation of enzyme replacement therapy in M. Gaucher. Br J Haematol 113: 1084–1087.

Ware JE (1993) *SF-36 Health Survey Manual and Interpretation Guide*. Boston, MA: The Health Institute, New England Medical Center.

Weinreb NJ (2001). Interruption in enzyme replacement therapy for Gaucher disease. Br J Haematol 113: 1087–1089.

Weinreb NJ, Aggio MC, Andersson HC, Andria G, Charrow J, Clarke JTR, Erikson A, Giraldo, Goldblatt J, Hollak C MD, Ida H, Kaplan P, Kolodny EH, Mistry P, Pastores GM, Pires R, Prakesh-Cheng A, Rosenbloom BE, Scott CR, Sobreira E, Tylki-Szymanska A, Vellodi A, vom Dahl S, Wappner RS, Zimran A (2004). Gaucher disease type 1: Revised recommendations on evaluations and monitoring for adult patients. Seminars in Hematology; 41 S5: 15–22.

Weinreb NJ, Charrow J, Andersson HC, et al. (2002). Effectiveness of enzyme replacement therapy in 1,028 patients with type 1 Gaucher disease after 2-5 years of treatment: A report from the Gaucher Registry. Am J Med. 113: 112–119.

Wenstrup RJ, Bailey L, Grabowski GA, Moskovitz J, Oestreich AE, Wu W, Sun S (2004). Gaucher disease: Alendronate disodium improves bone mineral density in adults receiving enzyme therapy. Blood; 104: 1253–1257.

Wenstrup RJ, Roca-Espiau M, Weinreb NJ, Bembi B (2002). Skeletal aspects of Gaucher disease: A review. Br J Radiol 75 Suppl 1: A2–A12.

Whitfield PD, Nelson P, Sharp PC, Bindloss CA, Dean C, Ravenscroft EM, et al. (2002). Correlation among genotype, phenotype, and biochemical markers in Gaucher disease: Implications for the prediction of disease severity. Mol Genet Metab 75: 46–55.

Young E, Chatterton C, Vellodi A, Winchester B (1997). Plasma chitotriosidase activity in Gaucher disease patients who have been treated either by bone marrow transplantation or by enzyme replacement therapy with alglucerase. J Inherit Metab Dis 20: 595–602.

Zevin S, Abrahamov A, Hadas-Halpern I, Kannai R, Levy-Lahad E, Horowitz M, et al. (1993). Adult-type Gaucher disease in children: Genetics, clinical features and enzyme replacement therapy. Q J Med 86: 565–573.

Zhao H, Bailey LA, Grabowski GA (2003). Enzyme therapy of Gaucher disease: Clinical and biochemical changes during production of and tolerization for neutralizing antibodies. Blood Cells Mol Dis 30: 90–96.

Zhu Y, Li X, Schuchman EH, Desnick RJ, Cheng SH (2004). Dexamethasone-mediated up-regulation of the mannose receptor improves the delivery of recombinant glucocerebrosidase to Gaucher macrophages. J Pharmacol Exp Ther 308: 705–711.

THE NEURONAL CEROID LIPOFUSCINOSES: CLINICAL FEATURES AND MOLECULAR BASIS OF DISEASE

Beverly L Davidson[1*] Mario A. Cabrera-Salazar[2] and David A Pearce[3]

The neuronal ceroid lipofuscinoses (NCLs) are among the most common groups of fatal neurodegenerative diseases affecting children, being estimated at 1 in 12,000 live births. To date, all NCLs, collectively known as Batten disease, have recessive modes of inheritance. The original classification of the NCLs relied on a combination of histological qualities of tissues harvested from patients, and symptoms of disease. More recently, classical genetics and biochemical approaches have revealed the molecular basis of six of the seven loci causative of the NCLs. In this chapter we describe the genetic and biochemical distinctions among the NCLs and representative animal models, and review what is known currently about the proteins encoded.

1 INTRODUCTION

The neuronal ceroid lipofuscinoses (NCLs), collectively known as Batten's disease, were described first by Dr. Batten in 1903, and by Dr. Vogt in 1905. However the first known entry of a family afflicted with the NCLs was in 1826 by Dr. Stengel. In the early 1900s, the NCLs were grouped with Tay–Sachs, an error that was ultimately corrected by ultrastructural and biochemical studies.

A characteristic of the NCLs is the accumulation of autofluorescent ceroid and lipofuscin (storage) within cells, a recessive mode of inheritance, devastating neurological deterioration, and premature death. Neurological symptoms include seizures and progressive visual, cognitive, and physical deficits. The variable time of disease onset, coupled with differences in the ultrastructural features of the storage deposits visualized by electron microscopy, underlay the standard classification of the NCLs, and ranges from the infantile form to the rarer, adult-onset disease (Table 1).

In recent years scientists have capitalized on biochemical and genetic approaches to define the molecular basis for six of the eight NCL subtypes originally classified as distinct diseases based on clinical data and histopathology, notably *CLN*1-3, *CLN*5-6, and CLN8. We review the clinical, biochemical, and genetic hallmarks of the various NCLs and representative animal models. Treatments being developed to halt or cure the NCLs are also discussed.

[1]Departments of Internal Medicine, Neurology, Physiology & Biophysics, University of Iowa, Iowa City, IA, 52242 Tel: (319) 353-5511 / Fax: (319) 353-5572
Email: beverly-davidson@uiowa.edu
[2] Applied Discovery Research, Genzyme Corporation, Framingham MA 01701
[3] Center for Aging and Developmental Biology and Department of Biochemistry and Biophysics, University of Rochester School of Medicine, Rochester NY 14642
* to whom correspondence should be addressed.

Table 1. The neuronal ceroid lipofuscinoses

NCL Classification	Gene	Protein	Age of Onset (yrs)	Storage	Animal Models	References
Infantile	*CLN 1*	Palmitoyl-protein thioesterase (PPT1)	0– 2	Granular osmophilic deposits – Saposins A and D	PPT1 KO[b] PPT$^{\square Ex4}$	(Vesa et al. 1995; Gupta et al. 2001; Jalanko et al. 2005)
Late Infantile	*CLN 2*	Tripeptidyl protease 1 (TPP1)	2– 4	Curvilinear profiles – ATP synthase subunit c	TPP1 null	(Sleat et al. 1997; Sleat et al. 2004)
Finnish variant	*CLN 5*	CLN5	2– 6	Fingerprint/Rectilinear profiles – ATP synthase subunit c	CLN5 KO	(Savuloski et al. 1998; Kopra et al. 2004)
Variant	*CLN 6*	CLN6	2 – 7	Fingerprint/Rectilinear profiles – ATP synthase subunit c	*nclf*[c]	Gao et al. 2002; Wheeler et al. 2002)
Juvenile	*CLN 3*	CLN3	5 – 10	Fingerprint profile – ATP synthase subunit c	CLN3 KO CLN3$^{\square Ex7/8}$	(Lerner et al. 1995; Mitchison et al. 1999; Cotman et al. 2002)
Adult	*CLN 4*	Unknown	Variable	Variable	–	
Northern Epilepsy[d]	*CLN 8*	CLN8	5 – 10	Rectilinear profile – ATP synthase subunit c	*mnd*[c]	(Bolivar et al. 2002; Ranta et al. 1999)
Turkish variant	*CLN 8*[e]	Unknown	1 – 6	Fingerprint/Rectilinear profiles – constituents unknown	–	(Ranta et al. 2004)

[a]For additional review, see (Ezaki et al., 2004; Mole, 2004).
[b]KO, knockout
[c]The *nclf* and *mnd* mice are naturally occurring mouse models
[d] Also known as epilepsy with progressive mental retardation
[e] A subset of Turkish variants originally classified as CLN7 have been shown to harbor mutations in CLN8

2 CLINICAL FEATURES OF THE NCLs

2.1 Infantile NCL (INCL)

Santavuori–Haltia disease or INCL affects infants or children before two years of age. Initial presentation is marked by loss or failure of the infant to reach developmental milestones followed by visual failure and seizures. Both magnetic resonance spectroscopy and magnetic resonance imaging show abnormal findings prior to symptom onset consisting of cerebral atrophy and hypointensities in the thalamus and basal ganglia observed in T2 sequences (Sinha et al., 2004). Cerebral atrophy, although observed in all clinical subtypes, is substantially more evident in INCL. Clinical studies on affected individuals show hypoperfusion of the cortex and loss of benzodiazepine receptors coincident with clinical signs (Vanhanen et al., 2004). Patients rapidly lose their vision and acquired psychomotor skills, and become comatose by three years of age. Death generally occurs in the first decade of life. Examination of patient tissue reveals characteristic ultrastructural changes in cells when examined by electron microscopy. These intracellular deposits in INCL are granular in nature, and biochemical studies show that they contain saposins A and D (Tyynela et al., 1993).

2.2. Late Infantile NCL (LINCL)

Onset of late infantile NCL (LINCL) or Jansky–Bielschowsky disease usually occurs by years 2–4. Children present most commonly with tonic-clonic seizures, but absence, partial, or secondary generalized seizures may also be present. By this time regression of developmental milestones is evident (Wisniewski, Zhong, and Philippart, 2001). The disease progresses to include vision loss and motor deficits, ataxia, and myoclonus. Patients have progressive motor dysfunction and lose their ability to walk and become wheelchair bound between 3–6 years of age. Progression of the disease leads to death between 7 and 15 years of age.

Visual impairment, a cardinal feature of the disease is observed at four to six years of age. Electroretinograms, electroencephalograms, and visual evoked potentials are progressively altered and ERG activity declines. Reactive gliosis, cerebellar, and cerebral neurodegeneration are progressive and widespread. MRI images are similar to those observed in INCL patients and are characterized by severe cortical atrophy and marked dilatation of the ventricular system, however, in LINCL there is no compromise of the basal ganglia (Figure 1).

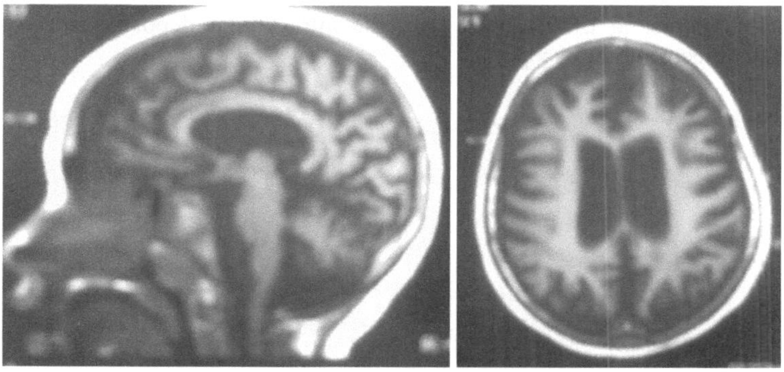

Figure 1. MRI of a 5-year-old patient with LINCL; note the thinning of cortical structures, massive enlargement of the ventricles, and atrophy of the corpus callosum. (Image kindly provided by Dr. Gustavo Charria-Ortiz.)

Examination of tissue reveals autofluorescent storage deposits in lysosomes in LINCL patient cells. Under EM, the deposits appear curvilinear, in contrast to infantile and juvenile NCL. One major component of the storage material has been identified as subunit c of mitochondrial F_1F_0 ATP synthase (Ezaki, Wolfe, and Kominami, 1997).

2.2.1 Variants of LINCL

Variants of LINCL have also been described and are now attributed to three different genetic disorders. Clinically they are referred to as Finnish variant of LINCL (fvLINCL), variant LINCL (vLINCL), and Turkish variant of LINCL. FvLINCL, also called variant-Jansky–Bielshowsky disease, is found almost exclusively in Finland or in patients of Swedish and Finnish decent. Initial disease symptoms of motor incoordination and lack of concentration appear from 2–6 years of age, followed by progressive visual loss and seizures. Disease progression is slower than classical LINCL, with death occurring from 13 to 30 years of age (Santavuori et al., 1991). Magnetic resonance imaging studies show signal intensity changes, and cerebral and cerebellar atrophy (Holmberg et al., 2000).

2.3 Juvenile NCL

The first patients described with Batten disease by Dr. Stengel in 1826 and Batten in 1903 had juvenile onset disease. Thus, Batten disease often refers to Juvenile NCL alone. JNCL, additionally called Spielmeyer–Vogt disease or ceroid lipofuscinoses type III, is the most common of the NCLs in North America. JNCL typically presents with progressive loss of vision from ages 5–8, and decline in cognitive function with seizure onset typically between 10–20 years of age. Progressive decline in psychomotor skills along with ataxia and speech disturbances are evident. Magnetic resonance imaging studies show progressive cerebral atrophy (Figure 2). Disease progression is more protracted relative to INCL and LINCL with death often occurring in the second or third decades, although patients can live longer. Similar to LINCL, JNCL patients show subunit c of mitrochondrial F_1F_0 ATP synthase as a major constituent of intracellular deposits. In contrast to LINCL, however, the intracellular deposits have a fingerprint profile under electron microscopy.

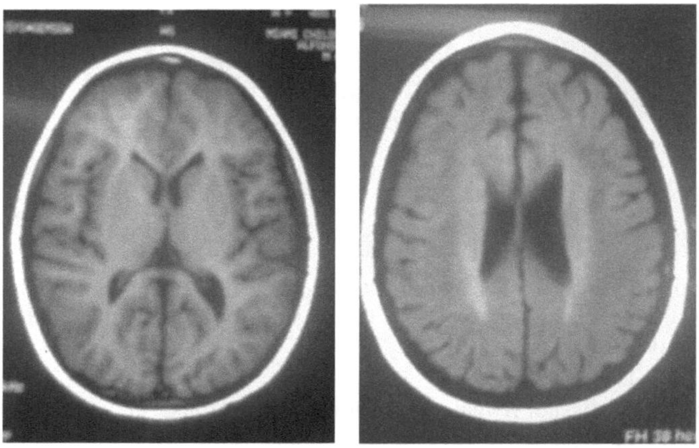

Figure 2. T2 MRI of a patient with juvenile neuronal ceroid lipofuscinosis, In this patient, cortical atrophy and hyperintensity of the subventricular white matter are the most prominent radiological observed. (Image kindly provided by Dr. Gustavo Charria-Ortiz.)

2.4 Northern Epilepsy or EPMR

Northern epilepsy, also called progressive epilepsy with mental retardation (EPMR), is a NCL with age of onset similar to JNCL. Generalized tonic-clonic seizures present between ages five and ten, which subsequently decline during adulthood. Patients have a normal life span but display slowly progressive mental retardation that initiates upon symptom onset.

2.5 Adult NCL

Adult-onset NCL (ANCL), also known as Kufs' disease, has also been described. The spectrum of clinical symptoms, and the severity and progression of symptoms, is very broad. Although two cases of ANCL have been attributed to mild mutations in the gene causative of INCL, most patients have not been shown to harbor mutations in genes underlying the known NCLs. Symptoms of Kufs' disease patients may overlap substantially with other neurological and psychiatric disorders, impairing disease diagnosis.

The observation that lysosomal enzymes secreted into the systemic circulation are capable of supporting metabolic cross-correction of distal cells suggests that the genetic modification of a small number of depot cells engineered to produce the deficient enzyme should allow for broad therapeutic correction of affected cells. Although this concept has obvious implications for treating the visceral manifestations associated with the LSD, gratifyingly, this has also been shown to apply, at least in part, to the treatment of the CNS disease as well. Several of the lysosomal enzymes produced in the brain parenchyma following stereotaxic injections of various viral vectors exhibited surprisingly good diffusion capacity to surrounding cells. Additionally, some AAV serotypes are capable of undergoing retrograde axonal transport providing yet another route to broader distribution of therapy in the CNS.

This mechanism of viral uptake by terminal axons, and retrograde transport was first evidenced by Kaspar et al. (2002) following injection of axon terminal fields in the hippocampus and striatum with AAV vectors. They observed transgene expression not only at the site of injection but also at distal sites such as the entorhinal cortex and substantia nigra. Examples of the effectiveness of this approach are given when discussing treatment of LSD that affect the CNS. The possibility of using a common gene-delivery platform such as AAV to treat both the neurological and nonneurological aspects of LSD is an attractive consideration.

3 DIAGNOSIS OF THE NEURONAL CEROID LIPOFUSCINOSES

Diagnosis of the NCLs is most reliably done from sequencing patient DNA at suspected loci (*cln2, cln3*, etc.). Samples can be obtained noninvasively from cheek swabs, blood, or tissue specimens. A list of sites that offer this diagnostic procedure with emphasis on NCL is listed in Appendix 1.

In addition, enzymatic assays can be done to confirm the diagnosis for patients with PPT1 or TPP1 mutations. Artificial fluorescence-based assays have been developed for both enzymes (Tian et al., 2006; Sohar et al., 1999).

Although the NCLs are rare there has been progress to document the natural history of the disease. With recent advances in imaging technologies, further refinement of our understanding of how the brain regionally responds to the mutations underlying the NCLs will become critical. Of particular importance is a newly developed clinical rating scale for the NCLs, developed by Marshall and colleagues (Marshall et al., 2005).

3.1 Differential Diagnosis

The diagnosis of NCL presents a challenge because it is a rare disease, and many clinicians have not seen the disorder previously, inhibiting their ability to make a rapid diagnosis. Other lysosomal storage disorders that are also confined to the CNS must be ruled out, for example, metachromatic leukodystrophy, Krabbe disease, GM2 gangliosidoses, and neuraminidase deficiency (sialidoses). Also, progressive myoclonic epilepsy, such as Unverricht–Lundborg disease, Lafora disease, mitochondrial disorders such as myoclonic epilepsy with ragged red fibers (MERRF), and peroxisomal disorders such as X-Linked adrenoleukodystrophy must be ruled out. Again, rapid diagnosis based on sequencing loci known to be mutant in these disorders is critical. If sequencing reveals no known mutation, tissue biopsies and analysis of the storage material will be helpful, although not definitive.

3.2 Management of Patients with NCLs

Current treatment consists of symptomatic relief of seizures, feeding problems such as gastric reflux, behavior disorders, rigidity, and visual loss. In a study of 60 patients with JNCL, lamotrigine and valproate were effective in decreasing seizure frequency in 80% of treated patients (Aberg et al., 2000). Similarly lamotrigine in monotherapy or in combination with clonazepam or clorazepate may provide good seizures control in JNCL patients.

Bone marrow transplantation as a means of providing normal repopulating microglia to the brain has been proposed. However, thus far, this approach has been unsuccessful in LINCL and JNCL patients (Lake et al., 1997). Experimental approaches to therapy attempting to either block N-methyl-D-aspartate-induced neuronal apoptosis (Dhar et al., 2002) or break down the storage material by supplying the missing transgene by intracerebral injections of recombinant adeno-associated vectors are currently being evaluated.

4 MOLECULAR BASIS OF THE NCLs

The preceding section was organized in respect to age of onset of the NCLs. The following is organized based on genotype name and number, which for the NCLs are *CLN*1–8 (see Table 1). Mutations in the genes *CLN1-3* cause infantile, late-infantile, and juvenile forms of the disease, respectively. Adult NCL has been named *CLN*4, although a gene specific to this patient population has not been identified. Finnish LINCL and vLINCL have mutations in *CLN5* and *CLN6*, respectively. Patients with northern epilepsy have mutations in *CLN8*. Mutations in *CLN8* also underlie Turkish variant LINCL, originally ascribed to a loci coined *CLN7* (adult NCL and TvNCL are not discussed further because the basis for these disorders remains unclear). Diagnosis of the NCLs is done by genetic (*CLN*1–3, *CLN*5–6, *CLN*8) and biochemical (PPT1 and TPP1 enzyme assays for *CLN*1 and *CLN*2) methods. Sites for genetic and biochemical testing are listed by the Batten Disease Support and Research Association at www.bdsra.org and the appendix.

4.1 CLN1

Patients with INCL, have mutations in *CLN1*, the gene encoding palmitoyl protein thioesterase 1 (PPT1) on chromosome 1p32 (Vesa et al., 1995). PPT1, a nonmembrane-bound lysosomal protein (Hellsten et al., 1996; Verkruyse et al., 1996), functions as a monomer, and removes palmitate from S-acetylated proteins in vitro. Mutations in PPT1 are distributed throughout the nine coding exons, and include splice site mutations, deletions, missense mutations, and nonsense mutations. The most common PPT1 mutations are R122W and R151X (Das et al., 1998; Das, Lu, and Hoffman, 2001). The former is most common in Finland, whereas the latter is found in populations of diverse ethnicity, suggestive of a founder effect. Other mutations can be enriched in some populations.

The age of onset and rapidity of clinical course is variable, and correlates directly with percent loss of enzyme activity (Das et al., 1998; Bellizzi et al., 2000; Das, Lu, and Hoffman, 2001). The crystallization of PPT1 defined residues critical for its function (Figure 1; Bellizzi et al., 2000), including a catalytic histidine (H289) in the COOH-terminal region. PPT2 is a second lysosomal hydrolase with partial amino acid identity and high structural homology to PPT1 (Calero et al., 2003). No human NCL cases have been genetically linked to PPT2, and biochemically, the enzymes are nonredundant (Calero et al., 2003).

Mutations in PPT1 cause loss of enzyme activity by inducing structural perturbations, decreasing enzyme stability, inhibiting protein maturation, or inducing nonsense-mediated mRNA decay (Vesa et al., 1995; Das, Lu, and Hoffman, 2001; Salonen et al., 2001). Some mutations cause clinical symptoms more closely resembling the late infantile or juvenile NCLs. Data on residual activity of mutant PPT1, and the associated clinical course, will help guide researchers developing methods of enzyme replacement on the levels required for normalization of phenotypes.

PPT1 is localized to cerebral cortical neurons, cerebellar Purkinje cells, and Bergman glia (Suopanki et al., 1999). PPT1 is expressed in proliferating cell populations within the ventricular zone at the onset of corticogenesis in humans and mice (Heinonen et al., 2000). The nonhomogeneous nature of PPT1 expression during CNS development (Isosomppi et al., 1999) could underlie the region-specific neurodegeneration seen in early disease in INCL patients. PPT1 deficient mouse models (Gupta et al., 2001; Jalanko et al., 2005) will be useful for assessing the impact of PPT1 loss on normal patterning in the brain, and the reversibility of the disease process. PPT1 KO mice and PPT1 mice with exon 4 removed show progressive neuropathology, seizures, shortened lifespan, and autofluorescent storage material, characteristics common to human INCL. [For additional review of mouse NCL models see Mitchison, Lim, and Cooper (2004).]

Therapy for PPT1 deficiency remains supportive, but gene replacement strategies are being tested in the knockout mouse model (Griffey et al., 2004). In studies using adeno-associated viral vectors expressing PPT1, injected into newborn PPT1 KO mice, brain mass was increased and autofluorescent storage levels were reduced. Although encouraging, the data suggest that the infantile onset of disease may require prenatal enzyme replacement for full reversal of symptoms.

4.2 *CLN2*

LINCL is caused by mutations in *CLN2* on chromosome 11p15, which encodes the lysosomal enzyme tripeptidyl protease 1 (TPP1; Sleat et al., 1997). TPP1 is a lysosomal, serine carboxyl peptidase (Lin et al., 2001b) and activity is decreased or absent in LINCL patients. Substrates for TPP1 may include subunit c of mitochondrial ATP-synthase (Ezaki et al., 1999) and neuropeptides (Kopan, Sivasubramaniam, and Warburton, 2004). At least 52 different disease-associated mutations have been identified, with two common mutations found in nearly 60% of *CLN2* mutant chromosomes: Arg208X and a splice site mutation in intron 5 (Sleat et al., 1999). Compound heterozygosity with one of these mutations and an Arg447His point mutation confers a protracted, juvenile-onset phenotype.

TPP1 has not been crystallized, and has no structural homology to the proteosome-associated cytoplasmic tripeptidyl protease II (Tomkinson, 1999). Ser475 is the primary catalytic residue in the TPP1 active site (Lin et al., 2001b; Walus et al., 2005). TPP1 is synthesized as a preproenzyme with a 19 amino acid signal peptide that is posttranslationally removed, and a 176 amino acid prodomain that is lost during maturation by autoactivation in lysosomes (Sleat et al., 1997; Liu et al., 1998; Lin et al., 2001b; Golabek et al., 2003, 2005). Glycosylation is required for proper targeting and folding (Wujek et al., 2004). The glycosaminoglycans heparin sulfate and dermatan sulfate have been shown to stabilize the mature enzyme against alkaline pH induced inactivation (Golabek et al., 2005). The possibility that extracellular matrix constituents may protect secreted TPP1 from degradation may have beneficial implications on enzyme replacement therapies. As with most lysosomal enzymes, a proportion of TPP1 is capable of trafficking through a secretory pathway. TPP1 can re-enter cells via mannose-6-phosphate receptors (Lin and Lobel, 2001a; Haskell et al., 2003). This can be taken advantage of from a therapeutic perspective; genetically corrected cells can secrete recombinant protein, providing enzyme replacement to a larger cell population by cross-correction.

In the brain, TPP1 expression appears in neurons during late gestation and increases gradually during postnatal development, paralleling neuronal differentiation and maturation (Kida et al., 2001). TPP1 is also expressed in endothelial cells, the choroid plexus, microglial cells, and the ependyma. The adult pattern of TPP1 expression is generally established by the age of two to four, coinciding with the typical age of onset of clinical symptoms for LINCL.

A mouse model of LINCL was generated by a knockin strategy to reproduce a human point mutation (Arg447His) in *CLN2* on mouse chromosome 7 (Sleat et al., 2004). The model was originally designed to be a hypomorph (after removal of the neomycin selection cassette), however, the mouse described here retains *NEO*, effectively knocking out the function of TPP1. In this mouse TPP1 is undetectable by western blot, immunohistochemistry, or in vitro activity assay. The LINCL mouse develops tremors by 7 weeks of age and a visible ataxic gait by 15 weeks of age. The mice have an average lifespan of 20–25 weeks, and show extensive pathology in the CNS including gliosis and neuronal cell loss. There is also progressive accumulation of autofluorescent storage material in the cerebral cortex, hippocampus, cerebellum, and brain stem.

As yet, no therapy exists for TPP1 deficiency. However, preclinical and clinical studies have been or are being done to test small molecules (Dhar et al., 2002) and enzyme replacement via gene therapy (Passini et al., 2006). In the latter case, recombinant adeno-associated viral vectors expressing TPP1 have been injected intracranially into LINCL patients (Crystal et al., 2004). In this study design, four patients with severe disease will be treated. If well tolerated, therapy will be administered to children that are earlier in the disease course.

4.3 *CLN3*

Mutations in *CLN3* cause JNCL (Lerner et al., 1995). *CLN3* encodes a hydrophobic lysosomal protein with multiple membrane-spanning regions. Biochemical studies of *CLN3* in cell-free systems and mammalian cells concur that *CLN3* is glycosylated, and possesses COOH-terminal and likely NH2-terminal cytosolic domains (Ezaki et al., 2003; Mao et al., 2003a,b).

The function of *CLN3* is currently unknown. *CLN3* is highly conserved among humans, dogs, rats, mice, *Drosophila*, *C. elegans*, and yeast; however, *CLN3* shares no significant sequence homology with any other protein. Proposed functions for *CLN3* are various with most evidence pointing to a role in lysosomal pH balance (Pearce et al., 1999; Puranam et al., 1999; Golabek et al., 2000; Holopainen et al., 2001; Persaud-Sawin et al., 2002; Kim et al., 2003; Padilla-Lopez and Pearce, 2006).

Like all known NCLs, JNCL is a recessive genetic disease. The most common allele in JNCL patients is a 1 kb deletion that introduces a frameshift. Translational read-through of the novel 28 amino acids postframeshift produces a truncated *CLN3* of 181 amino acids. In cells overexpressing the 1 kb deletion product, *CLN3* is retained in the ER/Golgi (Järvelä et al., 1999). In contrast, other point mutations associated with Batten disease do not affect trafficking but do impair function as assessed by complementation of *CLN3*-deficient yeast (Haskell et al., 2000).

Studies evaluating the distribution of *CLN3* among mammalian tissues are not consistent. Within human or mouse brain, *CLN3* has been described in endothelia, astrocytes, and neurons of the cerebral cortex, hypothalamus, and cerebellum (Margraf et al., 1999; Luiro et al., 2001). In human peripheral tissues, Margraf and colleagues also reported immunoreactivity in the peripheral nervous system, pancreatic islet cells, and seminiferous tubules in the testis, with no detectable protein in the liver, lung, bowels, kidneys, as examples (Margraf et al., 1999). In mouse, recent work demonstrates that *CLN3* mRNA and protein expression is low in the central nervous system in general, but with focal areas of high expression (Chattopadhyay and pearce, 2000; Stein and Davidson, unpublished observation). Murine *CLN3* expression has also been shown to be very high in areas of the gastrointestinal tract, kidney, and glandular/secretory tissues (Chattopadhyay and pearce, 2000; Stein and Davidson, unpublished observation).

Three mouse models for JNCL have been published. In the Mitchison model (Mitchison et al., 1999) exons 2–6 and most of exon 1 (including the ATG) was deleted, presumably knocking out *CLN3* expression. Homozygous-deficient mice show cellular pathology including accumulation of autofluorescent pigments and loss of phenotypic markers in some neurons. Gliosis and cortical atrophy occurs, but seizures, common in the human disorder, are not apparent (Mitchison et al., 1999; Pontikis et al., 2004). In a separate study, Katz and colleagues also introduced a neomycin expression cassette into exons 7 and 8 (Katz et al., 1999). As are the Mitchison mice, the Katz mice are viable and display characteristic neuropathology reminiscent of JNCL. The MacDonald laboratory (Cotman et al., 2002) generated mice that phenocopied the major human mutation, the 1 kb deletion. *CLN3*$^{ex\Delta 7/8}$ mice show more extensive disease, with degenerative changes in the retina, cerebral cortex, and cerebellum. The mice have measurable gait deficits and reduced survival. Interestingly, *CLN3*$^{ex\Delta 7/8}$ mice revealed the presence of alternative *CLN3* transcripts, most notably a mutant cDNA comprising exons 1–4, a novel intervening sequence encoding a frameshifted exon 6, and exons 9–15. A human homologue of this transcript from a JNCL patient was identified, suggesting that alternative splicing of *CLN3* may occur as a consequence of the 1 kb deletion.

To date, treatment for *CLN3* patients is supportive. Curative therapy for JNCL will be a challenge and will likely not involve gene replacement unless distinct populations of cells key to disease pathogenesis are identified and targetable. This is because, unlike PPT1 and TPP1, *CLN3* is not soluble and protein correction cannot occur via cross-correction. The mouse models described above, as well as worm, fly, and fish models, will be critical to developing small molecule therapies for the JNCLs.

4.4 *CLN5*

Finnish variant LINCL is due to mutations in *CLN5*, which encodes a 407 amino acid lysosomal protein with two predicted membrane-spanning regions (Savukoski et al., 1998; Isosomppi et al., 2002). The most common Finnish variant is a 2 bp deletion in exon 4, causing production of a truncated, 391 amino acid protein (Savukoski et al., 1998; Holmberg et al., 2000).

In situ hybridization and immunohistochemical studies in mice (Holmberg et al., 2004) demonstrate *CLN5* expression by embryonic day 15, with intense staining seen in the ganglionic eminence, the developing hippocampus, and ventricular zone. *CLN5* expression in the cerebellum is notable at postnatal day 5, consistent with maturation of this brain region after birth. In the adult brain, *CLN5* is restricted to the hippocampus, cerebral cortex, Purkinje cells, and cerebellar granule cells.

The mouse model of FvLINCL was generated through targeted disruption of exon 3 (Kopra et al., 2004). *CLN5*-deficient mice show characteristic storage and progressive pathology including loss of some GABAergic neurons. The *CLN5* deficient model will be important to test putative therapies and help delineate the function and importance of CLN5 in brain.

4.5. *CLN6*

Variant LINCL is caused by mutations in *CLN6* (Gao et al., 2002; Wheeler et al., 2002), which encodes an ER resident protein with 6–7 transmembrane domains (Mole et al., 2004). The Costa Rican haplotype has a nonsense mutation in exon 3, and the Venezuelan haplotype contains a tyrosine deletion in exon 5 (Gao et al., 2002). Other mutations have also been identified (Wheeler et al., 2002). Although *CLN6* function is unknown, recent work shows that *CLN6* mutations impair lysosome function (Heine et al., 2004).

The *nclf* mouse is a naturally occurring animal model of vLINCL (Bronson et al., 1998; Gao et al., 2002; Wheeler et al., 2002). *Nclf* mice show progressive retinal atrophy, gliosis, and degeneration of motor neurons leading eventually to paralysis and death by 9 months (Bronson et al., 1998). The mutation in the *nclf* mouse is a cytosine insertion within a string of cytosines in exon 4, resulting in a frameshift, a mutation also found in three Pakistani families (Wheeler et al., 2002). NCL in the Southern Hampshire and Merino sheep likely arise due to mutations in the orthologue to human *CLN6*, although mutations have not yet been reported (Broom and Zhou, 2001; Tammen et al., 2001).

4.6 *CLN8*

Northern Epilepsy, or epilepsy with progressive mental retardation (EPMR) is due to mutations in *CLN8* (Ranta et al., 1999). The *CLN8* protein ~ 33 kDa, and like *CLN6*, resides in the ER. In contrast to *CLN6*, however, *CLN8* also partially localizes to the ER-Golgi intermediate compartment (ERGIC; Lonka et al., 2000). The human EPMR mutation, R24G, does not affect localization. As yet, the function of *CLN8* is not known.

Mutations in *CLN8* have also been linked to a subset of Turkish variant LINCL, extending the number of known mutations to five (EPMR and these four; Ranta et al., 2004). These mutations are more severe than the R24G mutation found in EPMR, presumably underlying the more severe clinical disease in this subpopulation. The effects of the mutations on protein trafficking have not been reported.

Mnd mice have a spontaneous 1 bp insertion in *CLN8* inducing a frameshift and a truncated protein product (Ranta et al., 1999). The motor incoordination, learning and memory impairments, retinal degeneration, and hindlimb weakness mirror human *CLN8* deficiency. However, *mnd* mice do not display seizures, and most symptoms arise in adulthood (Bolivar, Scott Ganus, and Messer, 2002). Nonetheless, the *mnd* mice will be invaluable for determining *CLN8* function in the ER and ERGIC, and deciphering how mutations in *CLN8* cause lysosomal storage and neuronal degeneration.

In summary, mutations in various proteins underlie the NCLs. At first pass, the NCLs appear biochemically distinct. However, they may be metabolically interconnected in manners not yet understood. Work in cell and small and large animal models will help determine how mutations in the *CLNs* induce devastating and most often fatal neurologic disease.

ACKNOWLEDGMENTS

The authors would like to thank the BDSRA, members of their laboratory for critical review, the Roy J Carver Charitable Trust, and Dr. Gustavo Charria-Ortiz for his revision and clinical images.

REFERENCES

Aberg LE, Backman M, Kirveskari E, Santavuori P, (2000). Epilepsy and antiepileptic drug therapy in juvenile neuronal ceroid lipofuscinosis. Epilepsia 41(10): 1296–1302.

Bellizzi JJ 3rd, Widom J, Kemp C, Lu JY, Das AK, Hofmann SL, Clardy J, (2000). The crystal structure of palmitoyl protein thioesterase 1 and the molecular basis of infantile neuronal ceroid lipofuscinosis. Proc Natl Acad Sci USA 97: 4573–4578.

Bolivar VJ, Scott Ganus J, Messer A, (2002). The development of behavioral abnormalities in the motor neuron degeneration (mnd) mouse. Brain Res 937: 74–82.

Boustany RM, Alroy J, Kolodny EH, (1988). Clinical classification of neuronal ceroid-lipofuscinosis subtypes. Am J Med Genet Suppl. 5:47–58.

Bronson RT, Donahue LR, Johnson KR, Tanner A, Lane PW, Faust JR, (1998). Neuronal ceroid lipofuscinosis (nclf), a new disorder of the mouse linked to chromosome 9. Am J Med Genet 77: 289–297.

Broom MF, Zhou C, (2001). Fine mapping of ovine ceroid lipofuscinosis confirms orthology with CLN6. Eur J Paediatr Neurol 5 Suppl A: 33–35.

Calero G, Gupta P, Nonato MC, Tandel S, Biehl ER, Hofmann SL, Clardy J, (2003). The crystal structure of palmitoyl protein thioesterase-2 (PPT2) reveals the basis for divergent substrate specificities of the two lysosomal thioesterases, PPT1 and PPT2. J Biol Chem 278: 37957–37964.

Chattopadhyay S and Pearce DA, (2000). Neural and extraneural expression of the neuronal ceroid lipofuscinoses genes CLN1, CLN2, and CLN3: functional implications for CLN3. Mol Genet Metab 71: 207–211.

Cotman SL, Vrbanac V, Lebel LA, Lee RL, Johnson KA, Donahue LR, Teed AM, Antonellis K, Bronson RT, Lerner TJ, MacDonald ME, (2002). Cln3(Deltaex7/8) knock-in mice with the common JNCL mutation exhibit progressive neurologic disease that begins before birth. Hum Mol Genet 11: 2709–2721.

Crystal RG, Sondhi D, Hackett NR, Kaminsky SM, Worgall S, Stieg P, Souweidane M, Hosain S, Heier L, Ballon D, Dinner M, Wisniewski K, Kaplitt M, Greenwald BM, Howell JD, Strybing K, Dyke J, Voss H, (2004). Clinical protocol. Administration of a replication-deficient adeno-associated virus gene transfer vector expressing the human CLN2 cDNA to the brain of children with late infantile neuronal ceroid lipofuscinosis. Hum Gene Ther 15: 1131–1154.

Das AK, Becerra CH, Yi W, Lu JY, Siakotos AN, Wisniewski KE, Hofmann SL, (1998). Molecular genetics of palmitoyl-protein thioesterase deficiency in the U.S. J Clin Invest 102(2): 361–370.

Das AK, Lu JY, Hoffman SL, (2001). Biochemical analysis of mutations in palmitoyl-protein thioesterase causing infantile and late-onset forms of neuronal ceroid lipofuscinosis. Hum Mol Genet 10:1431–1439.

Dhar S, Bitting RL, Rylova SN, Jansen PJ, Lockhart E, Koeberl DD, Amalfitano A, Boustany RM, (2002). Flupirtine blocks apoptosis in batten patient lymphoblasts and in human postmitotic CLN3- and CLN2-deficient neurons. Ann Neurol 51: 448–466.

Ezaki J, and Kominami E, 2004. The intracellular location and function of proteins of neuronal ceroid lipofuscinoses. Brain Pathol 14: 77–85.

Ezaki J, Takeda-Ezaki M, Koike M, Ohsawa Y, Taka H, Mineki R, Murayama K, Uchiyama Y, Ueno T, Kominami E, (2003). Characterization of Cln3p, the gene product responsible for juvenile neuronal ceroid lipofuscinosis, as a lysosomal integral membrane glycoprotein. J Neurochem 87: 1296–1308.

Ezaki J, Tanida I, Kanehagi N, Kominami E, (1999). A lysosomal proteinase, the late infantile neuronal ceroid lipofuscinosis gene (CLN2) product, is essential for degradation of a hydrophobic protein, the subunit c of ATP synthase. J Neurochem 72: 2573–2582.

Ezaki J, Wolfe LS, Kominami E, (1997). Decreased lysosomal subunit c-degrading activity in fibroblasts from patients with late infantile neuronal ceroid lipofuscinosis. Neuropediatrics 28: 53–55.

Gao H, Boustany RM, Espinola JA, Cotman SL, Srinidhi L, Antonellis KA, Gillis T, Qin X, Liu S, Donahue LR, Bronson RT, Faust JR, Stout D, Haines JL, Lerner TJ, MacDonald ME, (2002). Mutations in a novel CLN6-encoded transmembrane protein cause variant neuronal ceroid lipofuscinosis in man and mouse. Am J Hum Genet 70: 324–335.

Golabek AA, Kida E, Walus M, Wujek P, Mehta P, Wisniewski KE, (2000). CLN3 protein regulates lysosomal pH and alters intracellular processing of Alzheimer's amyloid-beta protein precursor and cathepsin D in human cells. Mol Genet Metab 70: 203–213.

Golabek AA, Kida E, Walus M, Wujek P, Mehta P, Wisniewski KE, (2003) Biosynthesis, glycosylation, and enzymatic processing in vivo of human tripeptidyl-peptidase I. J Biol Chem 278: 7135–7145.

Golabek AA, Walus M, Wisniewski KE, Kida E, (2005). Glycosaminoglycans modulate activation, activity, and stability of tripeptidyl-peptidase I in vitro and in vivo. J Biol Chem 280: 7550–7561.

Griffey M, Bible E, Vogler C, Levy B, Gupta P, Cooper J, Sands MS, (2004). Adeno-associated virus 2-mediated gene therapy decreases autofluorescent storage material and increases brain mass in a murine model of Infantile Neuronal Ceroid Lipofuscinosis (INCL). Neurobiol Dis 16: 360–369.

Griffey MA, Wozniak D, Wong M, Bible E, Johnson K, Rothman SM, Wentz AE, Cooper JD, Sands MS, (2006).CNS-directed AAV2-mediated gene therapy ameliorates functional deficits in a murine model of infantile neuronal ceroid lipofuscinosis. Mol Ther. 13: 538–547.

Gupta P, Soyombo AA, Atashband A, Wisniewski KE, Shelton JM, Richardson JA, Hammer RE, Hofmann SL, (2001). Disruption of PPT1 or PPT2 causes neuronal ceroid lipofuscinosis in knockout mice. PNAS, USA 98: 13566–13571.

Haskell RE, Carr CJ, Pearce DA, Bennett MJ, Davidson BL (2000). Batten Disease: Evaluation of CLN3 mutations on protein trafficking and function. Hum Mol Genet 9: 735–744.
Haskell RE, Hughes SM, Chiorini JA, Alisky JM, Davidson BL, (2003). Viral-mediated delivery of the late-infantile neuronal ceroid lipofuscinosis gene, TPP I to the mouse central nervous system. Gene Ther 10: 34–42.
Heine C, Koch B, Storch S, Kohlschutter A, Palmer DN, Braulke T, (2004). Defective endoplasmic reticulum-resident membrane protein CLN6 affects lysosomal degradation of endocytosed arylsulfatase A. J Biol Chem 279: 22347–22352.
Heinonen O, Salonen T, Jalanko A, Peltonen L, Copp A, (2000). CLN-1 and CLN-5, genes for infantile and variant late infantile neuronal ceroid lipofuscinoses, are expressed in the embryonic human brain. J Comp Neurol 426: 406–412.
Hellsten E, Vesa J, Olkkonen VM, Jalanko A, Peltonen L, (1996). Human palmitoyl protein thioesterase: Evidence for lysosomal targeting of the enzyme and disturbed cellular routing in infantile neuronal ceroid lipofuscinosis. EMBO J 15: 5240–5245.
Holmberg V, Jalanko A, Isosomppi J, Fabritius AL, Peltonen L, Kopra O, (2004). The mouse ortholog of the neuronal ceroid lipofuscinosis CLN5 gene encodes a soluble lysosomal glycoprotein expressed in the developing brain. Neurobiol Dis 16: 29–40.
Holmberg V, Lauronen L, Autti T, Santavuori P, Savukoski M, Uvebrant P, Hofman I, Peltonen L, Jarvela I, (2000). Phenotype-genotype correlation in eight patients with Finnish variant late infantile NCL (CLN5). Neurology 55: 579–581.
Holopainen JM, Saarikoski J, Kinnunen PK, Jarvela I, (2001). Elevated lysosomal pH in neuronal ceroid lipofuscinoses (NCLs). Eur J Biochem 268: 5851–5856.
Isosomppi J, Heinonen O, Hiltunen JO, Greene ND, Vesa J, Uusitalo A, Mitchison HM, Saarma M, Jalanko A, Peltonen L, (1999). Developmental expression of palmitoyl protein thioesterase in normal mice. Brain Res Dev Brain Res 118: 1–11.
Isosomppi J, Vesa J, Jalanko A, Peltonen L, (2002). Lysosomal localization of the neuronal ceroid lipofuscinosis CLN5 protein. Hum Mol Genet 11: 885–891.
Jalanko A, Vesa J, Manninen T, von Schantz C, Minye H, Fabritius AL, Salonen T, Rapola J, Gentile M, Kopra O, Peltonen L, (2005). Mice with Ppt1Deltaex4 mutation replicate the INCL phenotype and show an inflammation-associated loss of interneurons. Neurobiol Dis 18: 226–241.
Jarvela I, Lehtovirta M, Tikkanen R, Kyttala A, Jalanko A, (1999). Defective intracellular transport of CLN3 is the molecular basis of Batten disease (JNCL). Hum Mol Genet 8(6): 1091–1098.
Kaspar BK, Erickson D, Schaffer D, Hinh L, Gage Fh, Peterson DA (2002). Targeted retrograde gene delivery for neuronal protection. Mol Ther 5: 50–56.
Katz ML, Shibuya H, Liu PC, Kaur S, Gao CL, Johnson GS, (1999). A mouse gene knockout model for juvenile ceroid-lipofuscinosis (Batten disease). J Neurosci Res 57: 551–556.
Kida E, Golabek AA, Walus M, Wujek P, Kaczmarski W, Wisniewski KE, (2001). Distribution of tripeptidyl peptidase I in human tissues under normal and pathological conditions. J Neuropathol Exp Neurol 60: 280–292.
Kim M, Mao Q, Davidson BL, Wiemer DF, (2003). Tripeptide probes for tripeptidyl protease I production via gene transfer. J Med Chem 46: 1603–1608.
Kopan S, Sivasubramaniam U, Warburton MJ, (2004). The lysosomal degradation of neuromedin B is dependent on tripeptidyl peptidase-I: evidence for the impairment of neuropeptide degradation in late-infantile neuronal ceroid lipofuscinosis. Biochem Biophys Res Commun 319: 58–65.

Kopra O, Vesa J, von Schantz C, Manninen T, Minye H, Fabritius AL, Rapola J, van Diggelen OP, Saarela J, Jalanko A, Peltonen L, (2004). A mouse model for Finnish variant late infantile neuronal ceroid lipofuscinosis, CLN5, reveals neuropathology associated with early aging. Hum Mol Genet 13: 2893–2906.

Lake BD, Steward CG, Oakhill A, Wilson J, Perham TG, (1997). Bone marrow transplantation in late infantile Batten disease and juvenile Batten disease.Neuropediatrics. 28: 80–81.

Lerner TJ, Boustany RM, Anderson JW, D'Arigo KL, Schlumpf K, Bucker AJ, Gusella JF, Haines JL, Kremmidiotis G, Lensink IL, Sutherland GR, Callen DF, Taschner PE, De Vos N, Van Ommen GJ, Breuning MH, Doggett NA, Meincke LJ, Liu Z, Goodwin LA, Tesmer JG, Mitchson HM, O'Rawe AM, Munroe PB (1995). Isolation of a novel gene underlying Batten disease, CLN3. Cell 82: 949–957.

Lin NL, Lobel P (2001)a. Production and characterization of recombinant human CLN2 protein for enzyme-replacement therapy in late infantile neuronal ceroid lipofus. Biochem J 357:49–55.

Lin NL, Sohart I, Lackland H, Lobel P (2001)b. The human CLN2 protein/tripeptidyl-peptidase I is a serine protease that autoactivates at acidic pH. J Biol Chem 276: 2249–2255.

Liu CG, Sleat DE, Donnelly RJ, Lobel P (1998). Structural organization and sequence of CLN2, the defective gene in classical late infantile neuronal ceroid lipofuscinosis. Genomics 50: 206–212.

Lonka L, Kyttala A, Ranta S, Jalanko A, Lehesjoki AE (2000). The neuronal ceroid lipofuscinosis CLN8 membrane protein is a resident of the endoplasmic reticulum. Hum Mol Genet 9: 1691–1697.

Luiro K, Kopra O, Lehtovirta M, Jalanko A (2001). CLN3 protein is targeted to neuronal synapses but excluded from synaptic vesicles: New clues to Batten disease. Hum Mol Genet 10: 2123–2131.

Mao Q, Foster BJ, Xia H, Davidson BL, (2003a). Membrane topology of CLN3, the protein underlying Batten disease. FEBS Lett 541: 40–46.

Mao Q, Xia H, Davidson BL, (2003b). Intracellular trafficking of CLN2, the protein underlying the childhood neurodegenerative disease, Batten disease. FEBS Lett 555: 351–357.

Margraf LR, Boriack RL, Routheut AA, Cuppen I, Alhilali L, Bennett CJ, Bennett MJ, (1999). Tissue expression and subcellular localion of CLN3, the Batten disease protein. Mol Gen Metab 66: 283–289.

Marshall FJ, De Blieck EA, Mink JW, Dure L, Adams H, Messin, S, Rothberg PG, Levy E, McDonough T, DeYoung J, Wang M, Ramirez-Montealegre D, Kwon JM, Pearce DA, (2005) A clinical rating scale for Batten disease: Reliable and relevant for clinical trials. Neurology 65(2): 275–279.

Mitchison HM, Bernard DJ, Greene ND, Cooper JD, Junaid MA, Pullarkat RK, de Vos N, Breuning MH, Owens JW, Mobley WC, Gardiner RM, Lake BD, Taschner PE, Nussbaum RL (1999). Targeted disruption of the Cln3 gene provides a mouse model for Batten disease. The Batten Mouse Model Consortium (corrected). Neurobiol Dis 6: 321–334.

Mitchison HM, Lim MJ, Cooper JD, (2004). Selectivity and types of cell death in the neuronal ceroid lipofuscinoses. Brain Pathol 14: 86–96.

Mole SA, Michaux G, Godlin S, Wheeler RB, Sharp JD, Cutler DF, (2004). CLN6, which is associated with a lysosomal storage disease, is an endoplasmic reticulum protein. Exp Cell Res 298: 399–406.

Mole SE (2004). The genetic spectrum of human neuronal ceroid-lipofuscinoses. Brain Pathol 14: 70–76.

Padilla-Lopez S, Pearce DA, (2006). Saccharomyces cerevisiae lacking Btn1p modulate vacuolar ATPase activity in order to regulate pH imbalance in the vacuole. J Biol Chem. 281: 10273–10280

Passini MA, Dodge JC, Bu J, Yang W, Zhao Q, Sondhi D, Hackett NR, Kaminsky SM, Mao Q, Shihabuddin LS, Cheng SH, Sleat DE, Stewart GR, Davidson BL, Lobel P, Crystal RG, (2006). Intracranial delivery of CLN2 reduces brain pathology in a mouse model of classical late infantile neuronal ceroid lipofuscinosis. J Neurosci. 26: 1334–1342.

Pearce DA, Ferea T, Nosel SA, Das B, Sherman F (1999). Action of Btn1p, the yeast ortholog of the human Batten disease gene. Nat Genet 22: 55–58.

Persaud-Sawin DA, Vandongen A, Boustany RM, (2002). Motifs within the CLN3 protein: Modulation of cell growth rates and apoptosis. Hum Mol Genet 11: 2129–2142.

Pontikis CC, Cella CV, Parihar N, Lim MJ, Chakrabarti S, Mitchison HM, Mobley WC, Rezaie P, Pearce DA, Cooper JD, (2004). Late onset neurodegeneration in the Cln3-/- mouse model of juvenile neuronal ceroid lipofuscinosis is preceded by low level glial activation. Brain Res 1023: 231–242.

Puranam KL, Guo WX, Qian WH, Nikbakht K, Boustany RM (1999). CLN3 defines a novel antiapoptotic pathway operative in neurodegeneration and mediated by ceramide. Mol Gen Metab 66: 294–308.

Ranta S, Topcu M, Tegelberg S, Tan H, Ustubutun A, Saatci I, Dufke A, Enders H, Pohl K, Alembik Y, Mitchell WA, Mole SE, Lehesjoki AE, (2004). Variant late infantile neuronal ceroid lipofuscinosis in a subset of Turkish patients is allelic to Northern epilepsy. Hum Mutat 23: 300–305.

Ranta S, Zhang Y, Ross B, Lonka L, Takkunen E, Messer A, Sharp J, Wheeler R, Kusumi K, Mole S, Liu W, Soares MB, Bonaldo MF, Hirvasniemi A, de la Chapelle A, Gilliam TC, Lehesjoki AE, (1999). The neuronal ceroid lipofuscinoses in human EPMR and mnd mutant mice are associated with mutations in CLN8. Nat Genet 23: 233–236.

Salonen T, Heinonen-Kopra O, Vesa J, Jalanko A, (2001). Neuronal trafficking of palmitoyl protein thioesterase provides an excellent model to study the effects of different mutations which cause infantile neuronal ceroid lipofuscinocis. Mol Cell Neurosci 18: 131–140.

Santavuori P, Rapola J, Nuutila A, Raininko R, Lappi M, Launes J, Herva R, Sainio K, (1991). The spectrum of Jansky-Bielschowsky disease. Neuropediatrics 22: 92–96.

Savukoski M, Klockars T, Holmberg V, Santavuori P, Lander ES, Peltonen L, (1998). CLN5, a novel gene encoding a putative transmembrane protein mutated in Finnish variant late infantile neuronal ceroid lipofuscinosis. Nat Genet 19: 286–288.

Shahwan A, Farrell M, Delanty N, (2005). Progressive myoclonic epilepsies: A review of genetic and therapeutic aspects. Lancet Neurol. 4: 239–248.

Sinha S, Satishchandra P, Santosh V, Gayatri N, Shankar SK (2004). Neuronal ceroid lipofuscinosis: a clinicopathological study. Seizure 13: 235–240.

Sleat DE, Donnelly RJ, Lackland H, Liu CG, Sohar I, Pullarkat RK, Lobel P, (1997). Association of mutations in a lysosomal protein with classical late-infantile neuronal ceroid lipofuscinosis. Science 277: 1802–1805.

Sleat DE, Gin RM, Sohar I, Wisniewski K, Sklower-Brooks S, Pullarkat RK, Palmer DN, Lerner TJ, Boustany RM, Uldall P, Siakotos AN, Donnelly RJ, Lobel P, (1999). Mutational analysis of the defective protease in classic late-infantile neuronal ceroid lipofuscinosis, a neurodegenerative lysosomal storage disorder. Am J Hum Genet 64: 1511–1523.

Sleat DE, Wiseman JA, El-Banna M, Kim KH, Mao Q, Price S, Macauley SL, Sidman RL, Shen MM, Zhao Q, Passini MA, Davidson BL, Stewart GR, Lobel P, (2004). A mouse model of classical late-infantile neuronal ceroid lipofuscinosis based on targeted disruption of the CLN2 gene results in a loss of tripeptidyl-peptidase I activity and progressive neurodegeneration. J Neurosci 24: 9117–9126.

Sohar I, Sleat DE, Jadot M, Lobel P, (1999). Biochemical characterization of a lysosomal protease deficient in classical late infantile neuronal ceroid lipofuscinosis (LINCL) and development of an enzyme-based assay for diagnosis and exclusion of LINCL in human specimens and animal models. J Neurochem. 73: 700–711.

Suopanki J, Tynela J, Baumann M, Haltia M, (1999). Palmitoyl-protein thioesterase, an enzyme implicated in neurodegeneration, is localized in neurons and is developmentally regulated in rat brain. Neurosci Lett 265: 53–56.

Tammen I, Cook RW, Nicholas FW, Raadsma HW, (2001). Neuronal ceroid lipofuscinosis in Australian Merino sheep: a new animal model. Eur J Paediatr Neurol 5 Suppl A: 37–41.

Tian Y, Sohar I, Taylor JW, Lobel P, (2006). Determination of the substrate specificity of tripeptidyl-peptidase I using combinatorial peptide libraries and development of improved fluorogenic substrates. J Biol Chem. 281: 6559-6572.

Tomkinson B, (1999). Tripeptidyl peptidases: enzymes that count. Trends Biochem Sci 24: 355–359.

Tyynela J, Palmer DN, Baumann M, Haltia M Storage of saposins A and D in infantile neuronal ceroid-lipofuscinosis. FEBS Lett. (1993) 330: 8–12.

Vanhanen SL, Puranen J, Autti T, Raininko R, Liewendahl K, Nikkinen P, Santavuori P, Suominen P, Vuori K, Hakkinen AM, (2004). Neuroradiological findings (MRS, MRI, SPECT) in infantile neuronal ceroid-lipofuscinosis (infantile CLN1) at different stages of the disease. Neuropediatrics 35: 27–35.

Verkruyse LA and Hoffman SL, (1996). Lysosomal targeting of palmitoyl-protein thioesterase. J Biol Chem 271: 15831–15836.

Vesa J, Hellsten E, Verkruyse LA, Camp LA, Rapola J, Santavuori P, Hofmann SL, Peltonen L, (1995). Mutations in the palmitoyl protein thioesterase gene causing infantile neuronal ceroid lipofuscinosis. Nature 376: 584–587.

Walus M, Kida E, Wisniewski KE, Golabek AA, (2005). Ser475, Glu272, Asp276, Asp327, and Asp360 are involved in catalytic activity of human tripeptidyl-peptidase I. FEBS Lett 579: 1383–1388.

Wheeler RB, Sharp JD, Schultz RA, Joslin JM, Williams RE, Mole SE, (2002). The gene mutated in variant late-infantile neuronal ceroid lipofuscinosis (CLN6) and in nclf mutant mice encodes a novel predicted transmembrane protein. Am J Hum Genet 70: 537–542.

Wisniewski KE, Zhong N, Philippart M, (2001).Pheno/genotypic correlations of neuronal ceroid lipofuscinoses. Neurology. 57: 576–581.

Wujek P, Kida E, Walus M, Wisniewski KE, Golabek AA, (2004). N-glycosylation is crucial for folding, trafficking, and stability of human tripeptidyl-peptidase I. J Biol Chem 279: 12827–12839.

APPENDIX

Diagnostic Centers for the NCLs

DNA Diagnosis, Carrier and Prenatal

Batten Disease Diagnostic and Clinical Research Center
Division of Pediatric Neurology
University of Rochester Medical Center
601 Elmwood Avenue, Box 631
Rochester NY, 14642, USA
Tel: (585) 275-4762

http://dbb.urmc.rochester.edu/labs/pearce/bddcrc/index.htm

Batten@urmc.rochester.edu

Neurogenetics DNA Diagnostic Laboratory
Massachusetts General Hospital
East 149 - 13th St.
Charlestown, MA 02129, USA
Katherine B. Sims, M.D. - Director
Winie Xin, Ph.D., Laboratory Supervisor
Ph: (617) 726-5721

Molecular genetics Laboratory -Rm 3421 CLN1,2,3,5,6,8
The Hospital for Sick Children
555 University Ave
Toronto, CANADA, M5G1X8
Peter N. Ray, PhD, Director
Contact: Leslie Steele, MSc, Coordinator
Ph: (416) 813-6590
Fax: (416) 813-7732
email: leslie.steele@sickkids.ca
www.sickkids.ca/molecular

Enzyme Laboratory (for PPT1 and TPP1 enzyme testing)
Camelia Botnar Laboratories, Institute of Child Health (& Great Ormond Street Hospital for Children)
30 Guilford Street
London WC1N 1EH, UNITED KINGDOM
Ph: + 020 7242 9789 x 2509/2440
Fax: + 020 7404 6191
Contact: Dr. Ying Foo, Consultant Biochemist
Ph: + 020 7813 8321 / +020 7794 0500 bleep 397
Email: fooy@gosh.nhs.uk

Department of Histopathology (for EM histopathology)
Great Ormond Street Hospital for Children
Great Ormond Street
London WC1N 3JH, UNITED KINGDOM
Contact: Glenn Anderson
Ph: + 020 7813 1170
Fax: + 020 7829 8663
Email: anderg@gosh.nhs.uk

Laboratory of Molecular Genetics
Helsinki University Hospital, Finland
PL 140, Haartmaninkatu 2
Helsinki, FINLAND 00290
Irma Järvelä, Laboratory Supervisor
Genetic testing for CLN1, CLN3, CLN5 and CLN8 and analysis for the major mutation of each. Part of an on-line European Directory of DNA diagnostic Laboratories (EDDNAL) website:
www.EDDNAL.com
Email: eddnal@skypro.be

Electronmicroscopy Diagnosis, Carrier and Prenatal

Ultrastructural Diagnostic Laboratory
IBR Specialty Clinic Laboratories
1050 Forest Hill Rd.
Staten Island, NY 10314, USA
Krystyna E. Wisniewski, MD, PhD, Director
Ph: (718) 494-5202
Email: battenkw@aol.com

CENTRO DE ESTUDIO DE LAS METABOLOPATIAS CONGENITAS (CEMECO)
Director: Prof. Dra. Raquel Dodelson de Kremer,
Pediatrist, Medical Genetician
Address: Hospital de Niños de la Provincia de Córdoba
Bajada Pucará y Ferroviarios
5000 Córdoba/ARGENTINA
Ph: + 54 351 4586 473
Fax: + 54 351 4586 439
This facility can presently do electronmicroscopy only. The BDSRA grant will enable them to eventually do DNA and enzyme testing.

DNA and Enzyme Assay for Diagnosis, Carrier and Prenatal:
Molecular Neurogenetics Diagnostic Laboratory
IBR-Specialty Clinic Laboratories
1050 Forest Hill Rd.
Staten Island, NY 10314, USA
Nanbert A. Zhong, M.D. Director
Ph: (718) 494-5242/4810
Fax: (718) 494-4882
Email: ibrsclmndl@aol.com

Women's and Children's Hospital
72 King William Rd.
North Adelaide, 5006, AUSTRALIA
Michael Fietz, Ph.D. ˇ Head
Ph: 00-61-8-8204-8062
Fax: 00-61-8-8204-7100
Email: michael.fietz@adelaide.edu.au

The Netherlands
Department of Human and Clinical Genetics
DNA Diagnostic Section
Wassenaarseweg 72
NL-2333 AL Leiden, THE NETHERLANDS
Professor E Bakker, Head
Ph: +31 71 527 60 82
Fax: +31 71 527 16 01
Email: e.bakker@lumc.nl

Enzyme Activity Analysis for Infantile (CLN1) and Classic Late Infantile (CLN2)
Department Clinical Genetics
Erasmus University
Dr. Molewaterplein 50
3015 GE Rotterdam, THE NETHERLANDS
Dr. O. P.van Diggelen, Director
Tel: +31-10-408 7224
Fax: +31-10-408 7200

Enzyme Assay - Late Infantile - CLN2
Child & Parent Resource Inst
Biochemical Genetics Lab
600 Sanatorium Rd
London N6H 3W7 ON, CANADA
Tony Rupar, PhD
Ph: (519) 858-2774 x 2204
Fax: (519) 858-3913
Email: trupar@uwo.ca

Infantile & Late Infantile Enzyme Activity Assay
Children's Hospital and Medical Center
Laboratory CH-37
4800 Sand Point Way NE
Seattle, WA 98105, USA
Rhona Jack, Ph.D. - Director
Ph: (206) 526-2216
Email: rjack@chmc.org

MUCOPOLYSACCHARIDOSIS I

Lorne A. Clarke

1 OVERVIEW

Mucopolysaccharidosis type I (MPS I) has historically been considered to represent the prototypical generalized storage disease. As such, insights provided by clinical observations, therapeutic attempts, and the understanding of the molecular basis of disease pathophysiology of this disease, are likely to be applicable to most of the generalized lysosomal disorders. The detailed delineation of the natural history of this disorder has been particularly instructive. Deficiency of the lysosomal enzyme α-L-iduronidase (IDUA; EC 3.2.1.76) is the primary defect in MPS I. Interestingly, the initial clinical classification of the MPSs included a separate group of patients classified as MPS V (McKusick et al., 1965). It was not until the lysosomal enzyme iduronidase was discovered that it soon became known that MPS I (Hurler syndrome) and MPS V (Scheie syndrome) patients had the same primary metabolic defect (Wiesmann and Neufeld, 1970). The clinical distinctiveness of the two groups related to the fact that they represented opposite ends of a wide clinical disease spectrum.

This concept of "disease spectrum" is now considered universal in the lysosomal storage disorders (LSDs). For most of the LSDs this disease spectrum is caused primarily by different mutations within the gene coding for the deficient hydrolase or transporter, that is, allelic heterogeneity. On the other hand, observations in Gaucher disease (Beutler et al., 2004; Zhao et al., 2003; Beutler 2001) and Fabry disease (Germain et al., 2002; Ashton-Prolla et al., 2000; Knol et al., 1999) indicate that significant clinical heterogeneity can be seen for patients that have identical mutations. These disorders indicate the likelihood of modifier genes that modulate the disease phenotype. Detailed understanding of the molecular basis of the LSDs will reveal important insights into the various mechanisms that underlie clinical heterogeneity. The genotype–phenotype observations for MPS I have been particularly instructive in this regard as will further insights into the identification of factors that modify the disease severity of other lysosomal disorders.

2 CLINICAL DISEASE PHENOTYPE

Extensive and comprehensive clinical descriptions of MPS I have been previously publicshed and represent excellent resources for detailed clinical disease delineation (Neufeld and Muenzer, 2001). The discussion that follows centers on disease complications and disease progression, as well as aspects of disease heterogeneity in an attempt to highlight novel approaches and newer insights into the pathophysiology of MPS I.

Centre for Biomedical Research, Department of Biology, University of Victoria, Canada. E-mail: lclarke@cw.bc.ca

MPS I is a panethnic autosomal recessive disorder with an incidence of approximately 1/90,000 in most populations (Lowry and Renwick, 1971; Poorthuis et al., 1999; Nelson, 1997; Meikle et al., 1999). The traditional clinical entities used to characterize the wide clinical spectrum associated with MPS I disease, namely Hurler-, Hurler–Scheie- and Scheie-syndromes (from the most severe to less severe forms), do not adequately reflect the tremendous variation in clinical symptoms manifested by MPS I patients (Roubicek, Gehler, and Spranger, 1985). In addition, there are many patients who do not fit precisely into any of these three clinical entities; moreover, the clinical phenotypes are not distinguishable biochemically by routine diagnostic procedures. Because there is no clear delineation between the syndromes, patients are best described by the term MPS-I with severe, intermediate, or attenuated disease.

Although the term "attenuated" may more ably be used to describe any patient that is not in the severe disease category, a "binary" disease nosology leads to the "attenuated" form representing a large, very heterogeneous grouping of patients. Therefore the addition of the term "intermediate" allows for separation of this group. The designation of intermediate should be reserved for patients that have childhood onset of significant and clinically relevant disease but clearly a modulated phenotype from that of severe patients. This then leaves the attenuated group to represent patients that do not manifest clinically significant disease until after the childhood years. With this designation, the greatest heterogeneity of symptoms are manifested by individuals exhibiting MPS I disease of the intermediate form. Particularly instructive in this classification of individual patients is the fact that all severe MPS I patients present or have readily detectable symptoms in the first year of life, have rapidly progressive multisystem disease in early childhood, and have obvious CNS manifestations by the age of two years. Attenuated patients may not present until adolescence and although they have significant disease manifestations resulting in considerable morbidity, they can have a normal life span and intelligence. The intermediate patients fit between these extremes and manifest with progressive multisystem disease with associated early mortality and morbidity. Conservative estimates of disease prevalence suggest that at least 80% of patients fall within the severe end of the disease spectrum.

2.1 Severe MPS I

Individuals with severe MPS I exhibit a chronic and progressive disease course with involvement of multiple organs and tissues with death occurring prior to the tenth birthday. The specific clinical features seen are highly dependent on the age of the patient (Table 1). Infants with severe MPS I appear normal at birth but if examined carefully, already have evidence of mild to moderate hepatomegaly, with many having inguinal or umbilical hernias. The characteristic coarse facial features are usually obvious by the latter part of the first year of life. The age of diagnosis for severe MPS I is highly variable but in one report the mean age of a cohort of MPS I children was reported as approximately 9 months (Cleary and Wraith, 1995), with the majority of patients diagnosed before 18 months. The most common presenting features tend to be skeletal in origin, for example, joint restriction or gibbous deformity but on history, most severely affected children have early onset, chronic otitis media, and upper respiratory symptomatology. The most debilitating aspects of severe MPS I are the progressive CNS, skeletal, and cardiorespiratory manifesttations.

Table 1. Clinical manifestations and complications of severe MPS I

Early in Disease Course: (Birth to 24 Months)

General:

- Coarse facial features – enlarged scaphocephalic skull, hypertrichosis, straight coarse textured hair, thicken lips and tongue.
- Umbilical and inguinal hernia often recurrent.
- Hepatosplenomegaly.

CNS/PNS:

- Hydrocephalus.
- Developmental delay.

Skeletal

- Gibbous deformity varus and valgus deformity of feet.
- Moderate joint restriction – most noticeable in wrist and hip.
- Generalised dysostosis multiplex.

Cardiorespiratory

- Cardiac murmur – mild valveular disease.
- Fatal endocardiofibroelastosis.
- Recurrent and persistent upper respiratory tract infections – chronic rhinorrhea and otitis media.

Ophthalmological

- Corneal clouding.

ENT

- Otitis media, nasal discharge.

Later in Disease Course: (24 Months to 10 Years)

General:

- Progressive facial coarseness with large and prominent tongue, thick lips, facial hypertrichosis, broad and thickened gingival.
- Wide-spaced peg shaped teeth, stiffness to aural and nasal cartilage.
- Thatchlike hair.
- Massive hepatosplenomegaly.

CNS/PNS:

- Hydrocephalus.
- Spinal root entrapment.
- Pachymeningitiscervicalis.
- Arrested development with limited development of language skills followed by progressive developmental decline to profound mental retardation – usually associated with placid rather than aggressive behaviour.
- Hearing loss.
- Peripheral nerve entrapment.

Skeletal:

- Progressive dysostosis multiplex.
- Short stature.
- Kyphosis and scoliosis.
- Arthropathy – progressive joint restriction leading to progressive fixed deformities of knees, hips, chest, jaw and characteristic claw hand deformity.

Cardiorespiratory

- Cardiac valveular damage.
- Ischemic myocardial damage.
- Obstructive airway disease – noisy breathing, sleep apnea.

Ophthalmological

- Corneal clouding.
- Glaucoma.
- Retinal degeneration.
- Optic atrophy.

2.2 Intermediate and Attenuated MPS I

Currently no generally accepted simple criteria exist to subclassify patients into intermediate or attenuated MPS I. If development is normal at 24 months of age and if somatic involvement is evident in childhood and is of a moderate degree, patients should be classified as intermediate. This is not to say that intermediate patients will all be intellectually normal, but if intelligence is affected, the onset is later and the rate of progression is considerably slower than that seen in severe patients. Most intermediate patients have near normal intelligence, however, comprehensive neuropsychological assessment of intermediate patients has not been reported. Onset of significant symptoms in intermediate MPS I is usually between three and eight years of age and survival to adulthood is common. Although clinical disease onset is in mid-childhood, careful evaluations performed earlier in the disease course would reveal moderate hepatomegaly, mild corneal clouding, and mild dysostosis multiplex.

Unfortunately many patients go through considerable complex and lengthy diagnostic evaluations and testing before the final diagnosis of MPS I is made. This delay in diagnosis relates to the fact that most intermediate patients do not have the classic coarse facial features, obvious hepatosplenomegaly, and severe joint disease as seen in patients with severe disease. Individuals with attenuated MPS I are usually not diagnosed until after the age of 15 years and generally have normal intellect, normal stature, and a normal lifespan; it is conceivable that there may be undiagnosed adults with this disorder. Although usually diagnosed in adulthood retrospectively, these individuals are surely to have evidence of disease in late childhood.

The physical appearance of individuals with intermediate MPS I is variable. Coarseness of facial features is less obvious in childhood and in some patients may be very mild even in late childhood and adolescence. The most important aspect that separates this series of patients from the severe form is the relative preservation of intellectual skills. The most significant morbidity for intermediate patients relates to progressive and debilitating arthropathy secondary to intrinsic joint involvement as well as dysostosis multiplex progressive cardiorespiratory involvement, leading to apnea and cardiac failure; spinal cord compression at any level; and peripheral nerve entrapment. Attenuated patients tend to not have obvious coarseness to the facial features and have significant disease onset usually in the teenage or later years. The most significant aspects of disease morbidity relate to progressive valveular heart disease, corneal clouding, progressive multijoint arthropathy, and progressive skeletal deformation Intelligence is not affected in this group of patients.

2.3 Disease Complications

The importance of anticipatory guidance in the care of MPS I cannot be overstated. Due to fact that intermediate and attenuated patients live longer, they require specific and formal evaluations in order to prevent or abrogate possible disease complications. Close follow-up directed towards early diagnosis and prevention of disease complications (Table 2) will likely result in better outcomes for these patients particularly now that enzyme replacement therapy is available. This requires a multidisciplinary approach to patient care with specific testing centered on the possible disease complications.

Table 2. Medical concerns and approaches to disease complications in intermediate and attenuated MPS I patients

Disease Complication	Approach
General	
• Physical limitations related to many aspects of this disease can lead to low self esteem and depression :	• Direct counselling and discussion related to this is often helpful.
CNS/PNS:	
• Hydrocephalus	• Early detection and VP shunting. This can occur at any stage of the disease
• Carpel tunnel syndrome	• Frequent neurological exams including ERGs as patients do not present with classic early pain. Early neurosurgical decompression.
• Spinal compression	• Spinal cord compression and nerve entrapment can occur at any level: Frequent neurological exams and early intervention is key.
Skeletal	
• Dysostosis multiplex	• Progressive bone deformation leads to various orthopaedic complications incluing: valgus deformity of the lower limbs, scoliosis, and kyphosis and various hip deformities. Extensive and frequent orthopaedic evaluations with early surgical intervention are critical. • Progressive arthropathy involving all joints leads to significant morbidity for patients. Frequent evaluations and assessments with particular emphasis on the role of physiotherapy, activity and pain relief are important.
Cardiorespiratory	
• Cardiac valveular damage • Conduction defects and cardiomyopathy	• Progressive valveular heart disease is seen in all patients, although all valves can be affected the aortic valve is

	commonly affected first followed by the mitral value. Valve replacement should be considered where appropriate. SBE prophylaxis is indicated in all patients.
• Obstructive airway disease	• Sleep apnea commonly develops and can be an important aspect of progression of CNS disease as well as cardiac involvement. Formal sleep studies should be performed frequently in patients with intervention when indicated: BiPAP, CPAP, and tracheotomy.
Ophthalmological	
• Corneal clouding. • aucoma. • Retinal degeneration. • Optic atrophy.	• Progressive corneal clouding requiring corneal transplantation. • Glaucoma is common and should be monitored frequently. • Patients can show evidence of retinal degeneration leading to severe visual impairment prior to corneal transplantation this should be addressed.
ENT	
• Frequent otitis media and nasal discharge	• Early evaluation and consideration of early permanent grommet placement.
• Hearing loss	• Early detection and hearing aid placement.

2.4 Prediction of Disease Severity

The advent of enzyme replacement for MPS I has brought to center stage the importance of early diagnosis and commencement of therapy. The wide spectrum of disease severity in MPS I remains an important challenge in the development and assessment of therapies as well as accurate disease course prognostication. It is clear that current approaches to ERT are unlikely to significantly affect the CNS manifestations of disease and therefore the ability to predict the disease severity after early diagnosis is important.

Although the exact pathophysiology of disease is not accurately known (see below), at any point in the disease state there likely exist aspects of disease that are reversible by therapy, further progressive disease complications that are preventable by commencement of therapy, and aspects of disease that not reversible nor preventable by commencement of therapy. In addition, as early screening programs and even neonatal screening programs for the LSDs are implemented, some means to accurately predict disease severity will be important for the appropriate counseling of newly diagnosed families. Although no single accurate method exists to predict the disease severity for patients, the consideration of age, genotype, degree of somatic involvement, and careful neuropsychological evaluation will often provide an accurate disease severity prediction. Despite these evaluations, disease prediction may remain elusive in some patients. In these cases emerging biochemical studies may be helpful.

2.5 Genotype–Phenotype correlations

With the isolation and characterization of the IDUA gene by Scott et al., in 1992 (Scott et al., 1991, 1993) it was hoped that specific genotype–phenotype correlations would be found for MPS I. Although genotype–phenotype correlations do indeed exist for MPS I, the large number of mutations that have been uncovered limits the practical application of genotype–phenotype prediction in individual patients (Scott et al., 1993b; Clarke et al., 1994; Bunge et al., 1995; Scott et al., 1995; Tsui et al., 1995; Beesley et al., 2001; Li, Wood, and Thompson, 2002; Matte et al., 2003; Terlato et al., 2003). Although recurrent mutations have been described, many patients have at least one "private mutation" which limits the usefulness of the genotype in predicting the phenotype. Conceptually the following pattern of genotype–phenotype correlation is emerging for MPS I.

1. There does not appear to be strong evidence for the involvement of "other" genes in the modulation of the phenotype. Therefore it appears that it is the different IDUA sequence in individual patients which predicts the disease severity. It is important to note that although the mutations predict the overall severity of the disease, prediction of the individual's exact disease complications is impossible.
2. Most patients are compound heterozygote for two separate mutations at the IDUA locus. Therefore both mutations need to be recurrent mutations before phenotype can be predicted. Large deletions or insertions in the IDUA gene are not common.
3. Individual mutation frequencies vary considerably in individual populations.
4. Severe patients are most often compound heterozygous or homozygous for one of the following recurrent mutations: W402X, Q70X, 474-2a-g, P533R, A327P, A75T, and L218P.
5. Intermediate severity patients most often have one allele containing a mutation that has been associated with severe disease and the other allele represents a private mutation that is invariably a missense mutation.
   - This plethora of private missense mutations in the intermediate patients is what underlies the vast clinical heterogeneity of this group. Interestingly these observations point out the fine modulation of disease phenotype that is conferred by very small modulations of residual activity. Using conventional IDUA activity assays the differences between intermediate and severe patient's IDUA activity cannot be detected.
6. Patients with attenuated disease can be homozygous for a mutation that confers attenuated disease but may also be compound heterozygous for a common severe allele and an allele which modulates the phenotype. Although two common mutations have been found to underlie the attenuated phenotype, 678-7a-g and R89Q, these only represent 30% of attenuated alleles (Tsui et al., 1995).

The large number of mutations underlying MPS I and the large number of protein polymorphisms (reference the mutation Web site or Scott paper; Li, Wood, and Thompson,

2002; Scott et al., 1993) make genotype–phenotype correlations difficult to interpret. Nevertheless only one study has reported on the practical use of these data for prognostication. In a study involving 85 families with MPS I, Beesley et al. (2001), using a combination of direct mutation analysis and mutation scanning identified both IDUA mutations in 81 (95%) families, one IDUA mutation in three (3.5%) families, and none in one (1.1%) family, with an overall mutation detection rate for mutant alleles of 97%. In this patient series; 75% of the mutant alleles represented previously characterized mutations, and thus could be used for phenotype prediction. However, previously identified mutations accounted for both alleles in only 55 of the 85 families, thus phenotype could not be predicted by genotype in 35% of these cases. A similar study conducted in Italy (Venturi et al., 2002) identified 93% of the alleles in 30 individuals. Of the 23 different mutations found, 13 were novel and 40% of patients had at least one novel allele.

The analysis of genotype–phenotype correlations for MPS I and the realization that missense mutations at the IDUA locus confer modulation of the phenotype despite lack of significant detectable residual IDUA activity, would indicate that minute amounts of IDUA activity are needed to modulate the disease. The most significant clinical difference between the severe and intermediate patients relates to the early progressive CNS disease that is seen in the former. Therefore it appears that CNS disease manifestations are exquisitely sensitive to small amounts of IDUA activity. On the other hand, the progressive cardiac and joint involvement seen in attenuated patients indicate that considerably more residual IDUA activity is required to protect these organs from damage.

2.6 Mutant Protein Considerations

The fact that misssense mutations are the most common modulators of disease would indicate the likelihood that either altered catalytic capacity of the resultant mutant protein and/or altered mutant protein stability underlies the enzyme deficiency. A number of mutations have been described which alter the active site residues of the protein and thus clearly interfere with catalytic capacity (Brooks et al., 2001); these predictably lead to a severe phenotype. Some correlations have been recognized between the amount of residual activity and enzyme mass with disease severity; this approach utilizes techniques to measure IDUA activity and mass which are easily performed by most laboratories (Ashton et al., 1992; Bunge et al., 1998).

Recent advancements in the ability to measure dermatan sulfate and heparan sulfate derived oligosaccharides in cultured fibroblasts, utilizing electrospray ionization-tandem mass spectrometry, indicate that when combined with an accurate residual IDUA activity measurement, clear distinction could be seen between patients whom had the presence or absence of CNS involvement (Fuller et al., 2005). How practical these measures will be in the clinical setting remains to be determined. These studies are promising and clearly indicate that the disease modulation of the disease phenotype is directly related to "residual" flux through this degradation pathway.

Although IDUA has yet to be crystallized, the modulation of disease severity by missense mutations is unlikely to be directly related to the effect of these mutations on the catalytic capacity of the enzyme (Rempel et al., 2005). It is now clear that posttranslational processing of lysosomal enzymes and ultimate targeting of the proteins to the lysosomal compartment is highly dependent on normal protein folding during translation in the rough endoplasmic reticulum. There exists a complex cellular "surveillance" system of chaperones and chaperonins which promote the normal folding of proteins and also recognize altered protein folding, reviewed in Slavotinek and Biesecker (2001). The latter process directs proteins to a degradative pathway rather than to be further targeted to the lysosome.

It is therefore likely that the main effect of misssense mutations in IDUA and perhaps other lysosomal disorders, is related to mutant proteins not folding in a "normal" manner and thus being directed to a degradative pathway. Once in the lysosome, however, the mutant protein may be stable and catalytically active, thus phenotype modulation by missense mutations is likely related to the very small amount of mutant protein which happens to be transported to the lysosomal compartment. This hypothesis, if true, opens the door to possible therapies directed to altering the amount of mutant protein which is degraded or so-called protein stabilization therapy.

Substrate analogues, which paradoxically can also be potent enzyme inhibitors, are potential small molecules which may serve this purpose (Fan, 2003). These molecules, also termed active site-specific chaperones, have been developed for Fabry disease and have been found to be effective both in vitro and in vivo (Fan et al., 1999). As these represent small molecules there is the potential for access into many more tissues than recombinant proteins. Active site chaperones for MPS I have yet to be developed; clearly the large number of missense mutations underlying MPS I and the likelihood that small increases in enzyme activity could drastically alter the disease phenotype would make this an attractive therapeutic approach.

3 CURRENT THERAPIES

At this time no single therapeutic approach alleviates all aspects of MPS I. Nevertheless, the experiences gathered by hematopoietic stem cell transplantation in severe MPS I and the recent attempts at direct enzyme replacement indicate that phenotype modulation is possible in this disorder. Hematopoietic stem cell transplantation is the only therapy which has been shown to alter the CNS aspects of severe MPS I disease. A salient observation related to HSCT for severely affected patients, is the fact that stabilization of the CNS disease is observed in patients only when transplantation is performed within the first 18 to 24 months of life. This indicates that relatively small amounts of enzyme are required to be delivered to the CNS compartment to modulate disease and that there appears to be a temporal threshold for this effect. This observation ties in with the observation that missense mutations which confer virtually undetectable residual IDUA activity tend to drastically modulate the CNS phenotype.

3.1 Bone Marrow Transplantation

There has now been considerable experience with BMT for this disorder (Krivit et al., 1992; Whitley et al., 1993; Peters et al., 1996; Vellodi et al., 1997; Guffon et al., 1998; Peters, Shapiro, and Krivit, 1998; Peters et al., 1998; Souillet et al., 2003; Staba et al., 2004) and the following conclusions can be made based on the published reports of patient outcomes. The importance of detailed clinical evaluation of the patient and the identification of the underlying IDUA mutations cannot be overstated here.

1. HSCT is the only known therapeutic approach shown to alter the course of severe MPS I.
2. Dramatic somatic responses are seen in post-HSCT survivors with normalization of spleen size, liver size, and reduction of facial coarseness.

3. Developmental outcome is dependent upon the degree of developmental disability at the initiation of HSCT. Stabilization of developmental effects of MPS I by HSCT appear to be seen only in patients that are transplanted prior to the age of 18–24 months and who have near normal to low normal intelligence prior to HSCT. Therefore detailed neuropsychological and neurological evaluation of patients should be performed prior to consideration of HSCT.
4. HSCT does not appear to drastically alter the progressive skeletal nor ophthalmological manifestations of disease.
5. Post-HSCT follow-up greater than 15 years is limited and thus the true long-term effects of HSCT on MPS I are unknown.

It is important to provide families with accurate and specific counseling in relation to possible complications and outcomes post-HSCT. This discussion should include issues directly related to transplantation risks (i.e., graft versus host disease, GVHD), direct morbidity and mortality as well as the ultimate responsiveness of disease.

Whether differences in transplant complications may occur with the use of cord blood or HSC-mediated procedures has yet to be elucidated. It is unlikely that the different approaches to bone marrow transplantation will result in different outcomes for the primary disease process. Although noted in the literature, it is uncertain whether GVHD, pulmonary hemorrhage, or other HSCT complications are more common in MPS I as compared to other indications for transplant (Gassas et al., 2003).

Although the phenotype of disease is modified by HSCT, long-term follow-up of severe patients post-HSCT suggests that orthopaedic complications of disease remain a significant issue as does cardiac valveular involvement and corneal involvement (Weisstein et al., 2004; Kachur and Del Maestro, 2002; Masterson et al., 1996; Braunlin et al., 2003, 2001; Gullingsrud, Krivit, and Summers, 1998). Therefore it is critical that MPS I patients who have received HSCT be followed carefully with aggressive and early management of joint, skeletal, and cardiac manifestations of disease.

3.2 Enzyme Replacement Therapy

Data pertaining to the efficacy of direct enzyme replacement using CHO cell produced human IDUA has been published after an open label phase I/II (Kakkis et al., 2001) study and a phase III double-blinded study (Wraith et al., 2004). In the latter study, 45 individuals were part of a multinational double-blind placebo-controlled study of weekly intravenous therapy with 0.58 mgs/kg of human IDUA. During this short 26 week trial, significant improvement in pulmonary function, hepatosplenomegaly, severe sleep apnea, joint range of motion, and urinary GAG excretion was seen. This clinical effect was seen in a relatively short period of time despite the significant disease burden. One year follow-up of patients in the phase I/II study showed a similar robust clinical responsiveness. Both studies have shown direct enzyme replacement to be safe and although IgG antibodies to IDUA can be detected, most patients showed evidence of developing tolerance by 52 weeks on therapy (Kakavanos et al., 2003).

Long-term follow-up of treated patients as well as experience with the use of enzyme replacement therapy begun early in the course of disease is needed to determine the ultimate responsiveness of MPS I to direct enzyme replacement. In addition, the role that enzyme replacement therapy has in the treatment of severe MPS I and whether there may be added benefit from combined enzyme replacement and hematopoietic stem cell transplantation remain to be explored. Initial studies have shown combined ERT and transplantation to be safe (Grewal et al., 2005).

4 INSIGHTS INTO DISEASE PATHOPHYSIOLOGY

It is important to note that there is very little known about the true pathophysiology of MPS I or for that matter most of the mucopolysaccharidoses. Although the primary enzyme deficiency underlying each MPS disorder is known and for the most part the glycosaminoglycan substrates are also known, the exact mechanisms by which the altered substrate metabolism leads to the disease state is unknown. There are likely many factors that are at play here and detailed understanding of them will be required to appropriately assess the effects of therapeutic approaches and the development of newer therapeutic approaches. Factors that are likely to underlie the pathophysiology of disease are considered below. There are likely to be primary factors (i.e., events that are directly related to GAG storage) and secondary factors (i.e., events that take place secondary to GAG storage). This distinction is likely to be important as secondary events may not be reversible or may be more slowly reversed by therapies that directly replace the enzyme.

4.1 Primary Events

Direct accumulation of substrate: It is likely that direct accumulation of glycosaminoglycans within the lysosome of cells is responsible for the following clinical symptoms. It is therefore not surprising that some signs such as hepatosplenomegaly and GAG storage in the oropharinx are rapidly reversible with direct enzyme replacement.

4.2 Secondary Events

Many possible secondary effects of GAG accumulation can occur. These can involve interference with other pathways that reside within the lysosome or may relate to more "down stream" effects on heparan and dermatan sulfate containing proteoglycans. The latter events have yet to be analyzed in great detail but may represent important components of disease pathophysiology.

4.2.1 Interference with Other Pathways Within the Lysosome

Analysis of brain tissue in both humans with MPS I as well as animal models of the MPSs, have revealed increased levels of gangliosides (GM2 and GM3) as well as cholesterol esters. Subcellular localization studies have revealed that the gangliosides are sequestered within a different compartment than the glycosaminoglycans, thus suggesting that altered synthesis and/or trafficking of gangliosides may underlie the increases. This has led to the attractive hypothesis that the increases in gangliosides and cholesterol may be reflective of defects in the composition, trafficking, and recycling of membrane rafts within neurons and thus may be an important underlying mechanism of neuronal dysfunction in the MPSs.

4.2.2 Alteration of Extracellular Proteins

Proteoglycans: Glycosaminoglycans make up an important part of the extracellular matrix. This matrix is a dynamic environment that is key to the establishment of cell-to-cell

Table 3. Proteoglycans which may play a role in the secondary effects of MPS I:choice examples, not exclusive, of proteoglycans which may be involved in the pathogenesis of MPS I

Proteoglycan	Function
Syndicans 1–4	Cell adhesion, ligand recognition at the cell surface, ligand receptor interactions, microbial pathogenesis (Bernfield et al., 1999)
Glypicans 1–6	Cell adhesion, ligand recognition at the cell surface, ligand receptor interactions, microbial pathogenesis, membrane stability
Fibromodulin	Tendon and connective tissue morphology and function (Benevides et al., 2004)
Decorin	Tendon and connective tissue morphology and function
Versican	Cellular proliferation and migration (Kinsella et al., 2004)
Biglycan	Cellular proliferation, migration, tensile strength (Grande-Allen et al., 2004

communication, cell differentiation, and organ structure and function. Glycosaminoglycans are present in the extracellular matrix as both free GAGs but most commonly are associated with proteins. The most studied proteins are the heparan sulfate proteoglycans (Bernfield et al., 1999). These proteins represent a vast heterogeneous group of proteins that have diverse functions including the sequestration of proteins in the secretory vesicles, linking of proteins within the extracellular matrix and the binding of proteins to cell surface receptors. As such, heparan sulfate proteoglycans represent important components of the interaction of cells and the way that cells respond to receptor ligands. The latter represent important aspects of cell differentiation.

Although yet to be comprehensively explored, there may be important alterations of cell surface and extracellular proteoglycans in the MPS disorders. These changes would be secondary to the primary defect of glycosaminoglycan degradation. Table 3 lists possible proteoglycans which may be involved in the pathogenesis of MPS I. Changes in these key proteins may underlie the pathophysiology of dysostosis multiplex, some aspects of neuronal dysfunction, as well as cardiovascular and connective tissue effects of the MPSs.

Other proteins: Hinek and Wilson (2000) have shown impairment of elastic fiber assembly in fibroblasts from MPS I patients with normal fiber assembly in MPS III fibroblast. This difference has been shown likely to be caused by dermatan sulfate containing moieties which cause functional inactivation of a key protein (elastin binding protein) involved in processing of tropoelastin (Urban and Boyd, 2000). This results in impaired elastogenesis and likely accounts for the poorly developed system of elastic fibers in various tissues. Interestingly these changes can be reversed in tissue culture by overexpression of versican, a chondroitin sulfate proteoglycan (Hinek et al., 2004).

In addition to alterations of elastic fiber networks there have been observations of abnormalities in collagen organization in various tissues of MPS patients. Most notable is the observation of altered collagen matrix in the cornea of animal models of MPS (Alroy, Haskins, and Birk, 1999) as well as a single observation of accumulation of dermatan sulfate proteoglycans and altered collagen fibrils in the cornea of a 14-year-old MPS I patient who received a bone marrow transplantation at the age of 2 (Huang et al., 1996). This secondary effect on collagen is an interesting observation and may indicate that other matrix proteins may also be altered in this condition. Accumulation of intracellular amyloid-beta peptide (A beta 1–40) has also been noted in the brain of both MPS I as well as MPS III humans. Despite significant accumulation of this peptide there did not appear to be formation of plaques or tangles.

5 FUTURE PROSPECTS

The advancements in the development of therapeutics and phenotype delineation in MPS I illustrate the great potential impact of understanding a complex disease at the molecular and biochemical level. There is no doubt that further advancements in therapies, particularly those directed to the central nervous system, will be realized once a more thorough understanding of the pathophysiology of the CNS components of disease is realized. MPS I and other lysosomal disorders will continue to serve as excellent model systems for the study of the impact of gene-based therapies on complex genetic disease.

REFERENCES

Alroy, J., Haskins, M., and Birk, D. E., 1999, Altered corneal stromal matrix organization is associated with mucopolysaccharidosis I, III and VI. *Exp. Eye. Res*. **68** 523.

Ashton, L. J., Brooks, D. A., McCourt, P. A., Muller, V. J., Clements, P. R., and Hopwood, J. J., 1992, Immunoquantification and enzyme kinetics of alpha-L-iduronidase in cultured fibroblasts from normal controls and mucopolysaccharidosis type I patients. *Am. J. Hum. Gene*. **50**:787.

Ashton-Prolla, P., Tong, B., Shabbeer, J., Astrin, K. H., Eng, C. M., and Desnick R. J., 2000, Fabry disease: Twenty-two novel mutations in the alpha-galactosidase A gene and genotype/phenotype correlations in severely and mildly affected hemizygotes and heterozygotes. *J. Investig. Med*. **48**:227.

Beesley, C. E., Meaney, C. A., Greenland, G., Adams, V., Vellodi, A., Young, E.P., Winchester, B.G., 2001, Mutational analysis of 85 mucopolysaccharidosis type I families: Frequency of known mutations, identification of 17 novel mutations and in vitro expression of missense mutations. *Hum. Genet*. **109**:503.

Benevides, G., Pimentel, E., Toyama, M., Novello, J. C., Marangoni, S., and Gomes, L., 2004, Biochemical and biomechanical analysis of tendons of caged and penned chickens. *Connect Tissue Res*. **45**:206.

Bernfield, M., Götte, M., Park, P. W., Reizes, O., Marilyn L. Fitzgerald, M. L., Lincecum, J., and Zako, M., 1999, Functions of cell surface heparan sulfate proteoglycans. *Annu. Rev. Biochem*. **68**:729.

Beutler E., 2001, Discrepancies between genotype and phenotype in hematology: an important frontier. *Blood*. **98**:2597.

Beutler, E., Beutler, L., and West, C., 2004, Mutations in the gene encoding cytosolic beta-glucosidase in Gaucher disease. *J. Lab. Clin. Med*. **144**:65.

Braunlin, E. A., Rose, A. G., Hopwood, J. J., Candel, R. D., and Krivit, W., 2001, Coronary artery patency following long-term successful engraftment 14 years after bone marrow transplantation in the Hurler syndrome. *Am. J. Cardiol*. **88**:1075.

Braunlin, E. A., Stauffer, N. R., Peters, C. H., Bass, J. L., Berry, J. M., Hopwood, J. J., and Krivit, W., 2003, Usefulness of bone marrow transplantation in the Hurler syndrome. *Am. J. Cardiol*., **92**:882.

Brooks, D. A., Fabrega, S., Hein, L. K., Parkinson, E. J., Durand, P., Yogalingam, G., Matte, U., Giugliani, R., Dasvarma, A., Eslahpazire, J., Henrissat, B., Mornon, J. P., Hopwood, J. J., and Lehn, P., 2001, Glycosidase active site mutations in human alpha-L-iduronidase. *Glycobiology*. **11**:741.

Bunge, S., Clements, P. R., Byers, S., Kleijer, W. J., Brooks, D.A., and Hopwood, J. J., 1998, Genotype-phenotype correlations in MPS I using enzyme kinetics, immunoquantification and in vitro turnover studies. *Bioch. Biophy. Acta.* **1407**:249.

Bunge, S., Kleijer, W. J., Steglich, C., Beck, M., Schwinger, E., and Gal, A., 1995, Mucopolysaccharidosis type I: Identification of 13 novel mutations of the α-L-iduronidase gene. *Hum. Mutat.* **6**:91.

Clarke, L. A., Nelson, P. V., Warrington, C. L., Morris, C. P., Hopwood, J. J., and Scott, H. S., 1994, Mutation analysis of 19 North American mucopolysaccharidosis type I patients: Identification of two additional frequent mutations. *Hum. Mutat.* **3**:275.

Cleary, M. A., and Wraith, J. E., 1995, The presenting features of mucopolysaccharidosis type IH (Hurler syndrome). *Acta. Paediatr.* **84**:337.

Fan, J.-Q., 2003, A contradictory treatment for lysosomal storage disorders: Inhibitors enhance mutant enzyme activity. Trends *Pharm. Sci.* **24**:355.

Fan, J. Q., Ishii, S., Asano, N., and Suzuki, 1999, Accelerated transport and maturation of lysosomal a-galactosidase A in Fabry lymphoblasts by an enzyme inhibitor. *Nat. Med.* **5**:112.

Fuller, M., Brooks, D. A., Evangelista, M., Hein, L. K., Hopwood, J. J., and Meikle, P. J., 2005, Prediction of neuropathology in mucopolysaccharidosis I patients. *Mol. Genet. Metab.* **84**:18.

Gassas, A,, Sung, L,, Doyle, J. J., Clarke, J. T., and Saunders, E. F., 2003, Life-threatening pulmonary hemorrhages post bone marrow transplantation in Hurler syndrome. Report of three cases and review of the literature. *Bone Marrow Transplant.* **32**:213.

Germain, D. P., Shabbeer, J., Cotigny, S., Desnick, R. J., 2002, Fabry disease: twenty novel alpha-galactosidase A mutations and genotype-phenotype correlations in classical and variant phenotypes. *Mol. Med.* **8**:306.

Ginsberg, S. D., Galvin, J. E., Lee, V. M., Rorke, L. B., Dickson, D. W., Wolfe, J. H., Jones, M. Z., and Trojanowski, J. Q., 1999, Accumulation of intracellular amyloid-beta peptide (A beta 1-40) in mucopolysaccharidosis brains. *J. Neuropathol. Exp. Neurol.* **58**:815.

Grande-Allen, K. J., Calabro, A., Gupta, V., Wight, T. N., Hascall, V. C., and Vesely, I., 2004, Glycosaminoglycans and proteoglycans in normal mitral valve leaflets and chordae: association with regions of tensile and compressive loading. *Glycobiology.* **14**:621.

Grewal, S. S., Wynn, R., Abdenur, J. E., Burton, B. K., Gharib, M., Haase, C., Hayashi, R. J., Shenoy, S., Sillence, D., Tiller, G. E., Dudek, M. E., van Royen-Kerkhof, A., Wraith, J. E., Woodard, P., Young, G. A., Wulffraat, N., Whitley, C. B., and Peters, 2005, Safety and efficacy of enzyme replacement therapy in combination with hematopoietic stem cell transplantation in Hurler syndrome. *Genet Med.* **7**:143.

Guffon, N., Souillet, G., Maire, I., Straczek, J., and Guibaud, P., 1998, Follow-up of nine patients with Hurler syndrome after bone marrow transplantation. *J. Pediatr.* **133**:119.

Gullingsrud, E.O, Krivit, W, and Summers, C. G., 1998, Ocular abnormalities in the mucopolysaccharidoses after bone marrow transplantation longer follow up. Ophthal. **105**:1099.

Hinek, A., and Wilson, S. E., 2000, Impaired elastogenesis in Hurler disease: Dermatan sulfate accumulation linked to deficiency in elastin-binding protein and elastic fiber assembly. *Am. J. Path.* **156**:925.

Hinek, A., Braun, K. R., Liu, K., Wang, Y., and Wight, T. N., 2004, Retrovirally mediated overexpression of versican v3 reverses impaired elastogenesis and heightened prolixferation exhibited by fibroblasts from Costello syndrome and Hurler disease patients. *Am. J. Pathol.* **164**:119.

Huang, Y., Bron, A . J., Meek, K. M., Vellodi, A., and McDonald, B., 1996, Ultrastructural study of the cornea in a bone marrow-transplanted Hurler syndrome patient. *Exp. Eye. Res.* **62**:377.

Kakavanos, R. Turner, C. T., Hopwood, J. J., Kakkis, E. D., and Brooks, D. A., 2003, Immune tolerance after long-term enzyme-replacement therapy among patients who have mucopolysaccharidosis I. *Lancet.* **361**:1608.

Kakkis, E. D., Muenzer, J., Tiller, G, E,, Waber, L., Belmont, J., Passage, M., Izykowski, B., Phillips, J., Doroshow, R., Walot, I., Hoft, R., and Neufeld, E. F., 2001, Enzyme-replacement therapy in mucopolysaccharidosis I. *N. Engl. J. Med.* **344**:182.

Kinsella, M. G., Bressler, S. L., and Wight, T. N., 2004, The regulated synthesis of versican, decorin, and biglycan: Extracellular matrix proteoglycans that influence cellular phenotype. *Crit. Rev. Eukaryot. Gene Expr.* **14**:203.

Knol, I. E,, Ausems, M. G., Lindhout, D., van Diggelen, O. P., Verwey, H., Davies, J., Ploos, van Amstel J. K., Poll-The, B. T., 1999, Different phenotypic expression in relatives with Fabry disease caused by a W226X mutation. *Am. J. Med. Genet.* **82**:436.

Krivit, W., Shapiro, E., Hoogerbrugge, P. M., and Moser, H. W., 1992, State of the art review. Bone marrow transplantation treatment for storage diseases. *Bone Marrow Transplant* **10** Suppl. 1:87.

Li, P, Wood, T, and Thompson, J. N., 2002, Diversity of mutations and distribution of single nucleotide polymorphic alleles in the human alpha-L-iduronidase (IDUA) gene. *Genet. Med.* **4**:420.

Lowry, R. B., and Renwick, D. H., 1971, Relative frequency of the Hurler and Hunter syndromes. *N Engl J Med.* **284**:221.

Matte, U., Yogalingam, G., Brooks, D., Leistner, S., Schwartz, I., Lima, L., Norato, D. Y., Brum, J.M., Beesley, C., Winchester, B., Giugliani, R., and Hopwood, J. J., 2003, Identification and characterization of 13 new mutations in mucopolysaccharidosis type I patients. *Mol. Genet. Metab.* **78**:37.

McKusick, V. A., Kaplan, D., Wise, D., Hanley, W. B., Suddarth, S. B., Sevick, M. E., and Maumanee, A. W., 1965, The genetic mucopolysaccharidoses. *Medicine.* **44** 445.

Meikle, P. J., Hopwood, J. J., Clague, A. E., and Carey, W. F., 1999, Prevalence of lysosomal storage disorders. *JAMA.*, **281**:249.

Nelson, J., 1997, Incidence of the mucopolysaccharidoses in Northern Ireland. *Hum Genet.* **101**:355.

Neufeld, E.F., and Muenzer, J., 2001, The mucopolysaccharidoses, in: *The Metabolic and Molecular Bases of Inherited Diseases*, Scriver, C. R., Beaudet, A.L., Sly, W. S., Valle, D., Childs, R., Kinzler, K. W., 8th ed., McGraw-Hill, New York, pp. 3421–3452.

Peters, C., Balthazor, M., Shapiro, E. G., King, R. J., Kollman, C., Hegland, J. D., Henslee-Downey, J., Trigg, M. E., Cowan, M. J., Sanders, J., Bunin, N., Weinstein, H., Lenarsky, C., Falk, P., Harris, R., Bowen, T., Williams, T. E., Grayson, G. H,, Warkentin, P., Sender, L., Cool, V. A., Crittenden, M., Packman, S., Kaplan, P., Lockman, L. A., et al., 1996, Outcome of unrelated donor bone marrow transplantation in 40 children with Hurler syndrome. *Blood* **87**:4894.

Peters, C., Shapiro, E. G., and Krivit, W., 1998, Hurler syndrome: Past, present, and future. *J. Pediatr.* **133**:7.

Peters, C., Shapiro, E. G., Anderson, J., Henslee-Downey, P. J., Klemperer, M. R., Cowan, M. J., Saunders, E. F., deAlarcon, P. A., Twist, C., Nachman, J. B., Hale, G. A., Harris, R. E., Rozans, M. K., Kurtzberg, J., Grayson, G. H., Williams, T. E., Lenarsky, C., Wagner, J. E., and Krivit, W., 1998, Hurler syndrome: II. Outcome of HLA-genotypically identical sibling and HLA-haploidentical related donor bone marrow transplantation in fifty-four children. The Storage Disease Collaborative Study Group. *Blood* **91**:2601.

Poorthuis, B. J., Wevers, R. A., Kleijer, W. J., Groener, J.E., de Jong, J. G., van Weely, S., Niezen-Koning, K. E., and van Diggelen, O. P., 1999, The frequency of lysosomal storage diseases in The Netherlands. *Hum Genet.* **105**:151.

Rempel, B. P., Clarke, L. A., and Withers, S. G., 2005, A homology model for human α-L-iduronidase: insights into human disease, *Hum. Mol. Genet.* **85**:28–37.

Roubicek, M., Gehler, J., and Spranger, J., 1985, The clinical spectrum of a-L-iduronidase deficiency. *Am. J. Med. Genet.* **20**:471.

Scott, H. S., Anson, D. S., Orsborn, A. M., Nelson, P. V., Clements, P. R., Morris, C. P., and Hopwood, J. J., 1991, Human alpha-L-iduronidase: cDNA isolation and expression. *Proc. Natl. Acad. Sci. U S A* **88**:9695.

Scott, H. S., Bunge, S., Gal, A., Clarke, L. A., Morris, C. P., and Hopwood, J. J., 1995, Molecular genetics of mucopolysaccharidosis type I: Diagnostic, clinical, and biological implications. *Hum. Mutat.* **6**:288.

Scott, H. S., Guo, X. H., Hopwood, J. J., and Morris, C. P., 1992, Structure and sequence of the human alpha-L-iduronidase gene. *Genomics*, **13**:1311.

Scott, H. S., Litjens, T., Nelson, P. V., Thompson, P. R., Brooks, D. A., Hopwood, J. J., and Morris, C. P. 1993b, Identification of mutations in the α-L-iduronidase gene (IDUA) that cause Hurler and Scheie syndromes. *Am. J. Hum. Genet.* **53**:973.

Scott, H. S., Nelson, P. V., Litjens, T., Hopwood, J. J., and Morris, C. P., 1993, Multiple polymorphisms within the a-L-iduronidase gene (IDUA): Implications for a role in modification of MPS-I disease phenotype. *Hum. Mol.Genet.* **2**:1471.

Slavotinek, A. M., and Biesecker, L. G., 2001, Unfolding the role of chaperones and chaperonins in human disease. *Trends Genet.* **17**:528.

Souillet, G., Guffon, N., Maire, I., Pujol, M., Taylor, P., Sevin, F., Bleyzac, N., Mulier, C., Durin, A., Kebaili, K., Galambrun, C., Bertrand, Y., Froissart, R., Dorche, C., Gebuhrer, L., Garin, C., Berard, J., and Guibaud, P., 2003, Outcome of 27 patients with Hurler's syndrome transplanted from either related or unrelated haematopoietic stem cell sources. *Bone Marrow Transplant.* **31**:1105.

Staba, S. L., Escolar, M. L., Poe, M., Kim, Y., Martin, P. L., Szabolcs, P., Allison-Thacker, J., Wood, S., Wenger, D. A., Rubinstein, P., Hopwood, J. J., Krivit, W., and Kurtzberg J., 2004, Cord-blood transplants from unrelated donors in patients with Hurler syndrome. *N. Eng. J. Med.* **350**:1960.

Terlato, N. J., and Cox, G. F., 2003, Can mucopolysaccharidosis type I disease severity be predicted based on a patient's genotype? A comprehensive review of the literature. *Genet Med.* **5**:286.

Tsui, P. T., Bach, G., Matynia, A., Hwang, M., and Neufeld, E. F., 1995, Four novel mutations underlying mild or intermediate forms of alpha-L-iduronidase deficiency (MPS I and MPS I H/S). *Hum. Mutat.* **6**:55.

Urban, Z., and Boyd, C. D., 2000, Elastic fiber pathologies: Primary defects in assembly and secondary disorders in transport and delivery. *Am. J. Hum. Genet.* **67**:4.

Vellodi, A., Young, E. P., Cooper, A., Wraith, J. E., Winchester, B., Meaney, C., Ramaswami, U., and Will, A., 1997, Bone marrow transplantation for mucopolysaccharidosis type I: Experience of two British centres. *Arch. Dis. Child.* **76**:92.

Venturi, N., Rovelli, A., Parini, R., Menni, F., Brambillasca, F., Bertagnolio, F., Uziel, G., Gatti, R., Filocamo, M., Donati, M. A., Biondi. A., and Goldwurm, S., 2002, *Hum. Mutat.* **20**:231.

Weisstein, J. S., Delgado, E., Steinbach, L. S., Hart, K., and Packman, S., 2004, Musculoskeletal manifestations of Hurler syndrome. *J. Pediatr. Orthop.* **24**:97.

Whitley, C. B., Belani K. G., Chang, P. N., Summers, C. G., Blazar, B. R. Tsai, M. Y., Latchaw, R. E., Ramsay, N. K., and Kersey, J. H., 1993, Long-term outcome of Hurler syndrome following bone marrow transplantation. *Am. J. Med. Genet.* **46**:209.

Wiesmann, U. N. and Neufeld, E. F., 1970, Scheie and Hurler syndromes: apparent identity of the biochemical defect. *Science.* **169**:72.

Wraith, J. E., Clarke, L. A., Beck, M., Kolodny, E. H., Pastores, G. M., Muenzer, J., Rapoport, D. M., Berger, K. I., Swiedler, S. J., Kakkis, E. D., Braakman, T., Chadbourne, E., Walton-Bowen, K., and Cox G. F., 2004, Enzyme replacement therapy for Mucopolysaccharidosis I: A randomized, double-blind, placebo-controlled, multinational study of recombinant human a-L-iduronidase (Laronidase). *J. Pediatr.* **144**:581.

Zhao, H., Keddache, M., Bailey, L., Arnold, G., and Grabowski, G., 2003, Gaucher's disease: Identification of novel mutant alleles and genotype-phenotype relationships. *Clin. Genet.* **64**:57.

MUCOPOLYSACCHARIDOSIS II

Lorne A. Clarke

1. OVERVIEW

Mucopolysaccharidosis type II or Hunter syndrome is the only mucopolysaccharide storage disease to be inherited as an X-linked recessive disorder. Deficiency of the lysosomal enzyme iduronate-2-sulfatase (IDS; EC 3.1.6.13) is the primary defect in MPS I. As is seen with most X-linked potentially lethal conditions, allelic heterogeneity is significant and thus most families have individual or "private" mutations at the iduronate-2-sulfatase locus. In addition, unlike other LSDs, there is considerable heterogeneity in the molecular mechanisms underlying iduronate-2-sulfatase gene alteration in this condition. As such, this disorder, as with most of the MPSs, shows considerable clinical heterogeneity with the clinical phenotype spanning a spectrum of disease severity from early-onset childhood lethal disease to attenuated disease leading to a near normal life expectancy (Neufeld and Muenzer, 2001). In addition, patients with severe MPS II can have atypical symptomatology which likely relates to more complex genomic rearrangements underlying the disease pathogenesis.

The incidence of MPS II appears to be variable in populations which have been studied, and range from 1 in 165,000 (Western Australia) to 1 in 34,000 (Israel) male live births; issues related to incomplete ascertainment in populations is noted in most studies (Lowry and Renwick, 1971; Schaap and Bach, 1980; Lowry et al., 1990; Poorthuis et al., 1999; Nelson, 1997; Meikle et al., 1999; Nelson et al., 2003). Therefore it is likely that MPS II is a panethnic disorder with an incidence in the order of 1 in 75,000 male live births. Increased incidence is expected to be seen in small populations where a founder effect would be anticipated.

2 CLINICAL DISEASE PHENOTYPE

The literature abounds with clinical descriptions and comprehensive discussion of the various features of Hunter syndrome, many which attempt to classify this disease into two rigid severity levels (Neufeld and Muenzer, 2001; Young and Harper, 1979, 1982, 1983; Young et al., 1982a,b; Wraith et al., 1991) . The first published clear clinical description of Hunter syndrome came from Charles Hunter in 1917 (Hunter, 1917). Dr. Hunter was a Scottish-Canadian medical internist who reported the clinical features of two Canadian brothers with what appears to have been an X-linked form of mucopolysaccharidosis. This was followed in 1919 by the description of Hurler syndrome by

Centre for Biomedical Research, Department of Biology, University of Victoria, Canada. E-mail: lclarke@cw.bc.ca

Gertrude Hurler (Hurler, 1919). These reports were similar but clear differences in intelligence as well as the absence of corneal clouding in Hunter's cases are noted.

As MPS II is an X-linked recessive disease, the majority of affected individuals represent hemizygous males but many reports of manifesting females have been noted. In the female cases that have been studied, most had either structural X-chromosome abnormalities or evidence of severely skewed X-inactivation (Broadhead et al., 1986; Clarke et al., 1991; Mossman et al., 1983; Sukegawa et al., 1998; Winchester et al., 1992). Unlike female carriers of OTC deficiency or Fabry disease, there does not appear to be evidence that heterozygous carrier females are at significant risk of disease manifestations. The clinical discussions that follow in this chapter pertain to those seen in affected males. It is clear that patients with this disease fit into a linear clinical spectrum that directly relates the differential effects that mutations have on the iduonate-2-sulfatase gene and surrounding genes. Tables 1 and 2 of Chapter 24 of this book highlight the clinical features and disease complications seen in MPS I.

The clinical features, disease complications, and progression of disease in MPS II are very similar to those noted for MPS I with the exception that the cornea is not involved in MPS II, and skin findings are common in MPSII with a pebbly papular appearance to the skin on the back often being pathognomonic (Sapadin and Friedman, 1998). In addition, the progression of disease, particularly of severely affected patients appears to be somewhat slower in MPS II than in MPS I. This latter statement needs to be taken in context with the more heterogeneous mutations which underlie MPS II. A key difference between MPS I and MPS II relates to the differing molecular mechanisms underlying the respective gene alterations. MPS II patients for the most part have private mutations at the iduronate-2-sulfatase locus and thus the exact disease course and rate of progression are highly variable. The lack of "common" mutations clearly makes the exact prediction of phenotype difficult. This being said, the modulation of the phenotype seen in MPS II individuals by different mutations and mutational mechanisms attests to the importance of residual enzyme activity in the alteration of the disease phenotype.

2.1 Clinical Features of Severe MPS II

By way of definition, severe MPS II represents the subgroup of MPS II patients whom have early-onset of disease and evidence of involvement of the central nervous system. The clinical features of severe MPS II are similar to those described in MPS I (Tables 1 and 2, Chapter 24, this book). A comprehensive reporting of the "natural history" of severe MPS II by Young and Harper (1983) illustrates the important differences in the phenotype of severe MPS II in comparison to that of MPS I. Severely affected MPS II patients appear to have more slowly progressive but more pervasive CNS involvement than that seen in MPS I. Severely affected MPS II patients often survive past age ten years and in the latter stages of disease, have profound mental retardation with significant behavioral manifestations.

In addition, end-stage neurological manifestations consisting of extensive pyramidal tract disease and muscle wasting appear to be common in MPS II. Additional neurological features that separate MPS II from MPS I are the higher frequency of seizures in MPS II as well as the evolution of severe intellectual compromise leading to loss of speech and progression to a bedridden state. The latter is commonly seen in MPS III but is less commonly seen in the end stages of severe MPS I. Gastrointestinal manifestations with the occurrence of persistent unexplained diarrhea is also common in MPS II. The latter disease complication can be particularly distressing and is similarly seen in MPS III but is more uncommon in MPS I.

2.2 Clinical Features of Attenuated MPS II

As stated above, MPS II should be considered as a disease spectrum; although patients may be classified as severe based on the early involvement of the central nervous system, the remainder of the patients represent a continuous spectrum of disease involvement. Interestingly, as is seen in MPS I, patients with attenuated disease tend to have preservation of the CNS with varying degrees and rates of progression of somatic manifestations of disease. The extent and rate of progression of somatic disease manifestations are highly variable and there are often significant differences between the extent of disease involvement in different organ systems. Therefore complete and thorough evaluation of each patient is needed to ensure appropriate evaluation of disease impact.

2.3 Disease Complications

Anticipatory guidance and multidisciplinary care represent the key strategic approach to the care of MPS patients. Table 2 of Chapter 24 in this book lists the important clinical disease complications and approaches to intervention. MPS II patients require constant and consistent follow-up by a knowledgeable and experienced health care team who are willing to provide early intervention in order to aid in the prevention of disease complications. Although corneal clouding does not occur in MPS II patients, retinal dysfunction can be observed and thus ophthalmological follow-up should not be ignored.

2.4 Predictions of Disease Severity

With the advent of ERT strategies for lysosomal storage diseases which will soon be available for MPS II, prediction of disease severity and the development of reliable procedures for early diagnosis of MPS II will be a significant aspect of improving the outcome of MPS II patients. The underlying allelic heterogeneity in MPS II limits somewhat the ability to accurately predict phenotype from genotype. Interestingly, as has been noted for other lysosomal storage diseases, the age of onset of significant disease appears to be a very strong indicator of disease severity. These clinically based prognostic factors though will not be helpful in predicting disease severity with the advent of neonatal or other early screening methods.

3 BIOCHEMICAL CONSIDERATIONS

It is clear that the main mechanism which modulates the severity of disease is the amount of residual iduronate-2-sulfatase (IDS) activity. Iduronate-2-sulfatase, unlike other lysosomal enzymes, is detectable in high concentration in the circulation. Plasma IDS is highly sialylated thus ensuring a circulating pool of enzyme by preventing rapid immune clearance of the protein (Archer, Harper, and Wusterman, 1982). The purpose of circulating IDS is unknown but this may be an important factor in the differential responsiveness of MPS II to therapies such as BMT. Unfortunately the currently available methods for the measurement of residual activity, mutant protein quantity, and protein processing are too complex for routine use (Millat et al., 1997; Parkinson et al., 2004). Complex studies directed towards the measurement of the amount of immunodectable mutant protein in the plasma of patients have shown no apparent correlation with disease severity (Parkinson et al., 2005). Augmentation of these studies by selective epitope mapping of mutant protein in plasma to determine the degree of structural change of the mutant IDS protein has shown some ability to classify patients based on these measures. These studies require complex analysis and are currently not readily available for clinical use (Parkinson-Lawrence et al., 2005). Nevertheless these studies do reaffirm the view that the modulation of the disease phenotype in individual patients is likely directly related to characteristics of the mutant protein.

4 GENOTYPE–PHENOTYPE CONSIDERATIONS

The IDS gene has been localized to the very gene-rich Xq28 region (Mossman et al., 1983; Wilson et al., 1990, 1991; Flomen et al., 1993). This genomic region is complex with many neighboring genes and pseudo-genes (Timms et al., 1995; Rathmann et al., 1995). An IDS pseudo-gene is located 20 kb telomeric to the active IDS gene; IDS gene–pseudo-gene exchange has been shown to be responsible for a number of major genomic alterations which underlie MPS II in individual patients. To date over 300 IDS mutations have been logged at the Human Gene Mutation Database (http://www.hgmd.cf.ac.uk). An estimated 15–20% of patients have major structural alterations, sometimes interfering with neighboring genes, with the remainder of patients having point mutations, small deletions, or insertions (Froissart et al., 2002; Rathmann et al., 1996; Timms et al, 1997; Wilson et al., 1991; Vafiadaki et al., 1998; Lualdi et al., 2005; Lagerstedt et al., 2000). The following conclusions can be drawn from the various published mutation reports.

- Large-scale genomic alterations; total deletions (gene and pseudo-gene), partial deletions, gene–pseudo-gene rearrangements, and small deletions, insertions, or nonsense mutations that predictably alter the reading frame and occur early in the coding sequence, are most commonly associated with a severe phenotype.
- Patients with atypical symptoms which have included ptosis, more rapid Hurler-like disease progression, and seizures commonly have large-scale deletions which extend into the FMR2 locus which is proximal to IDS on the X chromosome.
- Missense and splice site mutations have variable effects on the catalytic activity, stability, and/or processing of IDS and therefore lead to unpredictable clinical effects.
- There are very few recurrent mutations. In one study of 98 patients with small gene alterations, 65 mutations were found in only one family with only 10 mutations being found in more than one family with a single mutation never being found in more than seven unrelated families.

Females hemizygous for IDS mutations do not commonly show clinical effects. Hemizygous females do not appear to be at increased risk of disease manifestations unless there is evidence of skewed X inactivation or major structural abnormalities of the X chromosome which would interfere with random X inactivation.

5 CURRENT THERAPIES

Currently there are no specific therapies which alleviate all of the symptomatology of MPS II. Recent advances in the development of recombinant enzyme replacement strategies for

MPS II are encouraging with human clinical trial data soon to be available (Muenzer et al., 2002). It is anticipated that ERT for MPS II should have similar effects to those noted for MPS I. Clearly the current approaches to ERT do not allow for delivery of recombinant enzymes to the CNS and thus future studies directed at CNS delivery will be critical to significantly alter the course of severe MPS II. There is also no doubt that ERT is likely to be more effective at the prevention of disease complications rather than reversal of existing symptoms, thus early identification of cases and prompt instigation of therapy will be critical to ensure optimal outcome.

Bone marrow transplantation has been attempted in MPS I and has resulted in similar but not identical outcomes to those seen for MPS I. The following conclusions can be made from the limited published studies to date (Bergstrom et al., 1994; McKinnis et al., 1996; Li et al., 1996; Vellodi et al., 1999; Guffon et al., 2001).

- BMT does not appear to alter the ultimate CNS disease in severe MPS II although there is evidence in selected cases that the rate of progression of CNS disease was slower after BMT. Once CNS disease is well advanced there is no benefit from BMT.
- The ultimate effect of early BMT on the rate of progression of CNS disease in severe MPS II patients is not known with any certainty. The association of atypical symptoms with deletions of neighboring genes would question whether a signify-cant BMT effect would be expected.

BMT definitely alters the somatic features of MPS II and appears to have positive but noncurative effects on the skeletal manifestations of disease.

6 FUTURE PROSPECTS

The advent of enzyme replacement therapy for MPS II will mark an important time in the history of MPS disorders. The understanding of the molecular basis of MPS II and the unraveling of the complex genomic alterations which underlie the disease have served well for the future advances that will be made in the tailoring of enzyme replacement strategies to the CNS as well as the eventual development and testing of gene-based therapy for this disease. In addition, advances in the understanding of the pathophysiology of the other MPSs and the identification of biomarkers of MPS disease, are certain to shed light on the potential of alternative therapeutic strategies for MPS II.

REFERENCES

Archer, I. M., Harper, P. S., and Wusterman, F.S., 1982, Multiple forms of iduronate-2-sulfatase in human tissues and body fluids. Biochem. Biophys. Acta. **708**:134.

Bergstrom, S. K, Quinn, J. J., Greenstein, R., and Ascensao, J., 1994, Long-term follow-up of a patient transplanted for Hunter's disease type IIB: A case report and literature review. Bone Marrow Transplant. **14**:653.

Broadhead, D. M., Kirk, J. M., Burt, A. J., Gupta,V., Ellis, P. M., and Besley, G. T. N., 1986, Full expression of Hunter's disease in a female with an X-chromosome deletion leading to non-random inactivation. Clin. Genet. **30**:392.

Clarke, J. T. R., Greer, W. L., Strasberg, P. M., Pearce, R. D., Skomorowski, M. A., and Ray, P. N., 1991, Hunter disease (mucopolysaccharidosis type II) associated with unbalanced inactivation of the X chromosomes in a karyotypically normal girl. Am. J. Hum. Genet. **49**:289.

Flomen, R. H., Green, E. P., Green, P. M., Bentley, D. R., and Giannelli, F., 1993, Determination of the organisation of coding sequences within the iduronate sulphate sulphatase (IDS) gene. Hum. Mol. Genet. **2**:5.

Froissart, R., Moreira da Silva, I., Guffon, N., Bozon, D., and Maire, I., 2002, Mucopolysaccharidosis type II–genotype/phenotype aspects. Acta. Paediatr. Suppl. **91**:82.

Guffon, N., Froissart, R., Philippe, N., and Maire, I., 2001, Outcome of bone marrow transplantation in eight patients with Hunter disease. J. Inherit. Metab. Dis. **24**(Suppl 2):172

Hunter, C., 1917, A rare disease in two brothers. Proc. R. Soc. Med. **10**:104.

Hurler, G., 1919, Über einen Typ multipler Abartungen, vorwiegend am Skelettsystem. Zeitschrift für Kinderheilkunde. **24**:220.

Kakkis, E. D., Muenzer, J., Tiller, G. E., Waber, L., Belmont, J., Passage, M., Izykowski, B., Phillips, J., Doroshow, R., Walot, I., Hoft, R., and Neufeld, E. F., 2001, Enzyme-replacement therapy in mucopolysaccharidosis I. N. Engl. J. Med. **344**:182.

Lagerstedt, K., Carlberg, B. M., Karimi-Nejad, R., Kleijer, W. J., and Bondeson, M. L., 2000, Analysis of a 43.6 kb deletion in a patient with Hunter syndrome (MPSII): Identification of a fusion transcript including sequences from the gene W and the IDS gene. Hum. Mutat. **15**:324.

Li, P., Thompson, J. N., Hug, G., Huffman, P., and Chuck, G., 1996, Biochemical and molecular analysis in a patient with the severe form of Hunter syndrome after bone marrow transplantation. Am. J. Med. Genet. **64**:531.

Lowry, R. B., and Renwick, D. H., 1971, Relative frequency of the Hurler and Hunter syndromes. N. Engl. J. Med. **284**:221.

Lowry, R. B., Applegarth, D. A., Toone, J. R., MacDonald, E., and Thunem, N. Y., 1990, An update on the frequency of mucopolysaccharide syndromes in British Columbia. Hum. Genet. **85**:389.

Lualdi, S., Regis, S., Di Rocco, M., Corsolini, F., Stroppiano, M., Antuzzi, D., and Filocamo, M., 2005, Characterization of iduronate-2-sulfatase gene-pseudogene recombinations in eight patients with Mucopolysaccharidosis type II revealed by a rapid PCR-based method. Hum. Mutat. **25**:491.

McKinnis, E. J., Sulzbacher, S., Rutledge, J. C., Sanders, J., and Scott C. R.,1996, Bone marrow transplantation in Hunter syndrome. J. Pediatr. **129**:145.

Meikle, P. J., Hopwood, J. J., Clague, A. E., and Carey, W. F., 1999, Prevalence of lysosomal storage disorders. JAMA **281**:249.

Millat, G., Froissant, R., Maire, I., and Bozon, D., 1997, Characterization of iduronate sulfatase mutants affecting N-glycosylation sites and the cysteine-84 residue. Biochem. J. **326**:243.

Mossman, J., Blunt, S., Stephens, R., Jones, E. E., and Pembrey, M., 1983, Hunter's disease in a girl: Association with X:5 chromosomal translocation disrupting the Hunter gene. Arch. Dis. Child. **58**:911.

Muenzer, J., Lamsa, J. C., Garcia, A., Dacosta, J., Garcia, J., Treco, D. A., 2002, Enzyme replacement therapy in mucopolysaccharidosis type II (Hunter syndrome): A preliminary report. Acta. Paediatr. Suppl. **91**:98.

Nelson, J., 1997, Incidence of the mucopolysaccharidoses in Northern Ireland. Hum. Gen. **101**:355.

Nelson, J., Crowhurst, J., Carey, B., and Greed, L., 2003, Incidence of the mucopolysaccharidoses in Western Australia. Am. J. Med. Gen. **123**:310.

Neufeld, E. F., and Muenzer, J., 2001, The mucopolysaccharidoses, in: *The Metabolic and Molecular Bases of Inherited Diseases*, Scriver, C. R., Beaudet, A. L., Sly, W. S., Valle, D., Childs, R., Kinzler, K. W., 8th ed., McGraw-Hill, New York, pp. 3421–3452.

Parkinson-Lawrence, E., Turner, C., Hopwood, J., and Brooks, D., 2005, Analysis of normal and mutant iduronate-2-sulphatase conformation. Biochem. J. **386**:395.

Poorthuis, B. J., Wevers, R. A., Kleijer, W. J., Groener, J. E., de Jong, J. G., van Weely, S., Niezen-Koning, K. E., and van Diggelen, O. P., 1999, The frequency of lysosomal storage diseases in The Netherlands. Hum Genet. **105**:151.

Rathmann, M., Bunge, S., Beck, M., Kresse, H., Tylki-Szymanska, A., and Gal, A., 1996, Mucopolysaccharidosis type II (Hunter syndrome): Mutation "hot spots" in the iduronate-2-sulfatase gene. Am. J. Hum. Genet. **59**:1202.

Rathmann, M., Bunge, S., Steglich, C., Schwinger, E., and Gal, A., 1995, Evidence for an iduronate-sulfatase pseudogene near the functional Hunter syndrome gene in Xq27.3-q28. Hum Genet. **95**:34.

Sapadin, A. N., and Friedman, I. S., 1998, Extensive Mongolian spots associated with Hunter syndrome. J. Am. Acad. Derm. **39**:1013.

Schaap, T., and Bach, G., 1980, Incidence of mucopolysaccharidoses in Israel: Is Hunter disease a 'Jewish disease'? Hum. Genet. **56**:221.

Sukegawa, K., Matsuzaki, T., Fukuda, S., Masuno, M., Fukao, T., Kokuryu, M., Iwata, S., Tomatsu, S., Orii, T., and Kondo, N., 1998, Brother/sister siblings affected with Hunter disease: Evidence for skewed X chromosome inactivation. Clin. Genet. **52**:96.

Timms, K. M., Bondeson, M. L., Ansari-LarI, M. A., Lagerstedt, K., Muzny, D. M., Dugan-Rocha, S. P., Nelson, D. L., Pettersson, U., and Gibbs, R. A., 1997, Molecular and phenotypic variation in patients with severe Hunter syndrome. Hum. Mol. Genet. **6**:479.

Timms, K. M., Lu, F., Shen, Y., Pierson, C. A., Muzny, D. M., Gu, Y., Nelson, D. L., Gibbs, R. A., 1995, 130 kb of DNA sequence reveals two new genes and a regional duplication distal to the human iduronate-2-sulfate sulfatase locus. Genome Res. **5**:71.

Vafiadaki, E., Cooper, A., Heptinstall, L. E., Hatton, C. E., Thornley, M., and Wraith, J. E., 1998, Mutation analysis in 57 unrelated patients with MPS II (Hunter's disease). Arch. Dis. Child. **79**:237.

Vellodi, A., Young, E., Cooper, A., Lidchi, V., Winchester, B., and Wraith, J. E., 1999, Long-term follow-up following bone marrow transplantation for Hunter disease. J. Inherit. Metab. Dis. **22**:638.

Wilson, P. J., Morris, C. P., Anson, D. S., Occhiodoro, T., Bielicki, J., Clements, P. R., and Hopwood, J. J., 1990, Hunter syndrome: Isolation of an iduronate-2-sulfatase cDNA clone and analysis of patient DNA. Proc. Nat. Acad. Sci. **87**:8531.

Wilson, P. J., Suthers, G. K., Callen, D. F., Baker, E., Nelson, P. V., Cooper, A., Wraith, J. E., Sutherland, G. R., Morris, C. P., and Hopwood, J. J., 1991, Frequent deletions at Xq28 indicate genetic heterogeneity in Hunter syndrome. Hum. Genet. **86**:505.

Winchester, B., Young, E., Geddes, S., Genet, S., Hurst, J., Middleton-Price, H., Williams, N., Webb, M., Habel, A., and Malcolm, S., 1992, Female twin with Hunter disease due to nonrandom inactivation of the X-chromosome: A consequence of twinning. Am. J. Med. Genet. **44**:834.

Wraith, J. E., Clarke, L. A., Beck, M., Kolodny, E. H., Pastores, G. M., Muenzer,J., Rapoport, D. M., Berger, K. I., Swiedler, S. J., Kakkis, E. D., Braakman, T., Chadbourne, E., Walton-Bowen, K., and Cox G. F., 2004, Enzyme replacement therapy for Mucopolysaccharidosis I: A randomized, double-blind, placebo-controlled, multinational study of recombinant human a-L-iduronidase (Laronidase). J. Pediatr. **144**:581.

Wraith, J. E., Cooper, A., Thornley, M., Wilson, P. J., Nelson, P. V., Morris, C. P., and Hopwood, J. J., 1991, The clinical phenotype of two patients with a complete deletion of the iduronate-2-sulphatase gene (mucopolysaccharidosis II–Hunter syndrome). Hum. Genet. **87**:205.

Young, I. D., and Harper, P. S., 1979, Long-term complications in Hunter's syndrome. Clin. Genet. **16**:125.

Young, I. D., and Harper, P. S., 1982, Mild form of Hunter's syndrome: Clinical delineation based on 31 cases. Arch. Dis. Child. **57**:828.

Young, I. D., and Harper, P. S., 1983, The natural history of the severe form of Hunter's syndrome: a study based on 52 cases. Dev. Med. Child Neurol. **25**:481.

Young, I. D., Harper, P. S., Archer, I. M., and Newcombe, R. G.,1982b, A clinical and genetic study of Hunter's syndrome. 1. Heterogeneity. J. Med. Gen. **19**:401.

Young, I. D., Harper, P. S., Newcombe, R. G., and Archer, I. M., 1982a, A clinical and genetic study of Hunter's syndrome. 2. Differences between the mild and severe forms. J. Med. Gen. **19**:408.

SANFILIPPO SYNDROME: CLINICAL GENETIC DIAGNOSIS AND THERAPIES

John J. Hopwood

1 FOREWORD

This chapter aims to briefly review the natural history, pathophysiology, clinical features, and molecular genetics of Sanfilippo syndrome, together with current and emerging advances and future perspectives, particularly to achieve early diagnosis and effective therapy.

Sanfilippo syndrome has a systematic name of mucopolysaccharidosis type III (MPS III). It is an autosomal-recessive lysosomal storage disorder (LSD) that affects about 1 in 70,000 live births (Meikle et al., 1999; Poorthuis et al., 1999). Sanfilippo syndrome is the most common MPS, and represents four biochemically distinct disorders, each of which results from a deficiency of one of four enzymes required for the lysosomal degradation of the glycosaminoglycan (GAG) heparan sulfate (HS): sulfamidase (MPS IIIA); α-N-acetylglucosaminidase (MPS IIIB); acetyl Co:A glucosamine N-acetyl transferase (MPS IIIC); and glucosamine-6-sulfatase (MPS IIID). Fragments of partially degraded HS accumulate in lysosomes and are elevated in the urine of all Sanfilippo patients. Genes for all but one (type C) of the four MPS III types have been characterized and used to identify mutations that lead to a deficiency of the enzyme product, storage of HS fragments, and the Sanfilippo phenotype.

No currently available therapies are yet able alter the natural history of Sanfilippo syndrome. The grouping and in-depth study of the genes and enzyme systems involved in this syndrome have provided extensive insight and appreciation of cellular processes involved in the systematic turnover of HS. This insight is slowly but surely identifying options for the development of potential therapies.

2 HISTORY

Sanfilippo syndrome is a rare and catastrophic genetic disorder. It takes its name from Dr. Sylvester Sanfilippo, who, with colleagues in 1963 (McKusick, 1966), first described a patient with severe neurological dysfunction manifested by loss of learned abilities and deterioration in behavior. Kaplan (1969) reported the classification of Sanfilippo syndrome on the basis of excessive urinary excretion of HS. Skin fibroblasts cultured from Sanfilippo syndrome patients were shown to have excessive accumulation of sulfated mucopolysaccharide that resulted from a deficit of a "corrective" protein factor (Kresse et al.,

---

Lysosomal Diseases Research Unit, Department of Genetic Medicine, Children, Youth and Women's Health Service, North Adelaide, Australia. E-mail: john.hopwood@adelaide.edu.au

1971). As a result of different corrective factors, Kresse and Neufeld (1972) were able to show that Sanfilippo patients fall into more than one genetic group. Subsequently, the corrective factors were shown to be lysosomal enzymes required for the degradation of HS.

3 CLINICAL DIAGNOSIS

The clinical phenotype alone is often enough to distinguish Sanfilippo syndrome from the other MPS disorders, which mostly show skeletal pathology in combination with CNS deterioration. However, diagnosis based on clinical presentation alone can lead to errors: for example, the absence of corneal clouding is common in both Sanfilippo syndrome and MPS II (Hunter syndrome). Delayed or regressing language, often associated with behavioral problems, were changes that alerted clinicians to the possibility of a Sanfilippo diagnosis.

Sanfilippo syndrome usually presents in early childhood. In common with all LSD, Sanfilippo syndrome patients may present within a broad spectrum of severity with clinical detection as early as the first year of life through to the third or fourth decade. The diagnosis of Sanfilippo syndrome usually results from the concern of parents at the increase in behavioral disturbance, loss of language skills, or delayed development they observe in their children. Generally, at the time of clinical presentation, minimal somatic pathology is observed in Sanfilippo patients. There are, however, exceptions where patients have presented with somatic features and these potentially provide opportunity to investigate the natural history of Sanfilippo syndrome. Typically, these cases present with somatic changes such as facial coarsening, hepatosplenomegaly, and skeletal changes without loss of neurological development (Barone et al., 2001). The classical Sanfilippo phenotype usually develops later.

Each of the four Sanfilippo syndrome subtypes (A, B, C, and D) result in similar clinical presentation, typically characterized by progressive, severe central nervous system (CNS) degeneration and mild somatic pathology. Presenting symptoms often include behavioral problems, speech delay, poor articulation, and hearing loss; coarse hair and hirsutism can also be part of the presenting symptoms. Noticeable, and therefore diagnostic, neurological degeneration occurs in most Sanfilippo patients by six years of age followed by a relatively rapid deterioration of behavior and the loss of learned skills. Behavioral problems include hyperactivity, aggression, temper tantrums, and poor attention spans. Affected children are physically strong with good mobility under the age of ten years, making the second phase of the illness the most difficult for caregivers to manage. Behavior in this middle stage is characterized by frequent severe temper tantrums, hyperactivity and aggression, rapid loss in attention span, and severe sleep disturbance. Early onset of puberty is common. There is minimal skeletal involvement; with coarse facial features not a prominent component of the syndrome. The majority of Sanfilippo children suffer from sleep problems. Recurrent diarrhea is common in the early stages of the disorder, but usually improves in older patients.

In the final and quieter stage of the disorder, general physical health and strength deteriorate. Progressive dementia, in the form of withdrawal and lost contact with the environment, is the usual course for most patients. A lack of balance makes falls more common; seizures become common in older patients. Impaired chewing and swallowing mechanisms lead to feeding difficulties. Degenerative joint disease and increasing spasticity severely impair mobility. Death occurs in severely affected children in their mid- to late-teenage years, usually as a result of respiratory infection. Each stage of the disorder can last for several years.

3.1 Heparan Sulfate-Uria

A deficiency in activity of any one of the four MPS III enzymes that lead to lysosomal accumulation of HS fragments and the Sanfilippo phenotype is best reflected in the resulting HS-uria. Care is needed in selecting a method to determine HS-uria. HS and chondroitin sulfates are common GAGs found in unaffected control urine. Although the total amount of GAGs decreases with age, particularly in the first year of life, the relative amount of HS in urine is seen to increase with age, such that in adults the amount of HS may represent approximately 50% of the total load compared to less than 10% HS found in the urine of infants. There are numerous reports of missed diagnoses of Sanfilippo syndrome resulting from the use of insensitive methods to detect total and/or specific GAG levels. Rather than the measurement of total GAG, a classical screening method for MPS III is the presence of elevated HS in urine, and a number of methods are recommended by which to measure relative HS-uria. We developed and continue to use a urine screening method that evaluates the MPS pattern (Figure 1; Hopwood and Harrison, 1982) MPS III patients with elevated HS are clearly distinguished from normal controls with relatively low levels of HS to chontroitin sulfate, and other MPS types with relatively elevated levels of heparan, dermatan sulfates, or keratan sulfate (Figure 1).

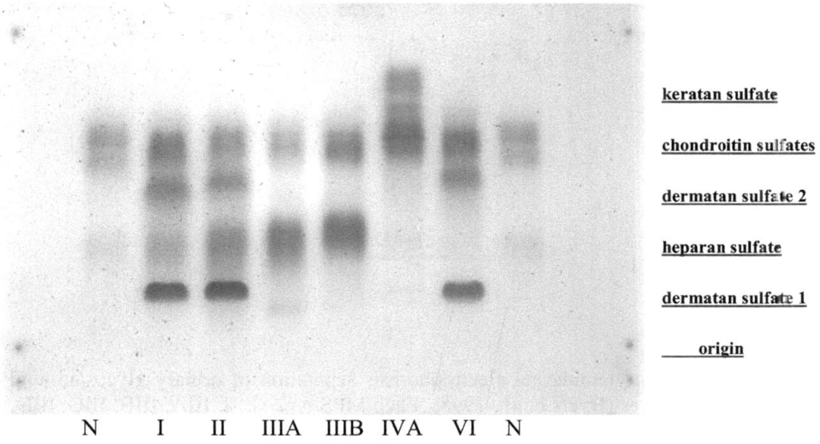

Figure 1. MPS electrophoresis on cellulose acetate (Hopwood and Harrison, 1982). Key: N, normal control; MPS types I, II, IIIA, IIIB, IVA, VI.

In over 10,000 assays this method has not missed a patient who was subsequently shown to have MPS III by confirmatory enzyme analysis. In those individuals shown to have HS-uria by this screening method, enzyme activities are then measured in peripheral blood white cells and/or cultured skin fibroblasts. The demonstration of HS-uria by this urine screening method has always led to a finding of a deficiency of one or more of the *exo*-enzymes listed below. Importantly, we have found some late-presenting patients with total GAG levels in the normal range for age but with an obvious elevation in HS relative to chondroitin sulfate. In the presence of classic Sanfilippo symptoms but no MPS-uria we always offer to determine the activity of the four MPS III enzymes.

Following clinical presentation and confirmation by the presence of HS-uria, a diagnosis of Sanfilippo syndrome is achieved by determining a deficiency or absence of one of the four enzymes involved in this MPS, in peripheral blood leukocytes and/or cultured skin fibroblasts. In the case of a deficiency of sulfamidase or glucosamine-6-sulfatase, at least one other sulfatase activity needs to be measured to eliminate the possibility of multiple sulfatase deficiency, which is known to result in reduced activity of these two enzymes (Hopwood and Ballibio, 2001). A further complication arises from the presence of HS-uria with normal activity of each of the four enzymes involved in Sanfilippo syndrome, and the possibility that multiple sulfatase has been excluded.

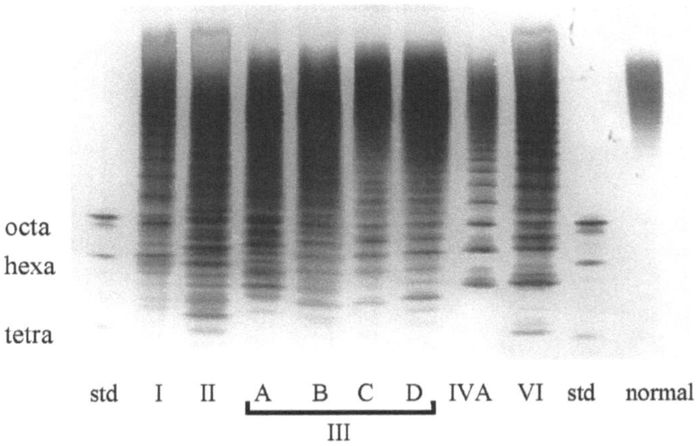

Figure 2. Gradient polyacrylamide gel electrophoresis separation of urinary glycosaminoglycans from a variety of MPS types (Byers et al., 1998). Each MPS type (I, II, IIIA, IIIB, IIIC, IIID, IVA, VI) has unique glycosaminoglycan oligosaccharide patterns resulting from the action of specific endohydrolases followed by further digestion with *exo*-enzyme activities (Figures 3C and 4) to oligosaccharides having specific nonreducing terminals consisting of the substrate specific for the enzyme deficient in that particular MPS disorder.

Alternatively, the observed HS-uria may result from a deficiency of glucuronate-2-sulfatase or glucosamine 3-sulfatase: both enzymes are required for HS degradation and have been characterized (Freeman and Hopwood, 1991). A patient syndrome that results from a deficiency of either activity has yet to be described.

We have recently developed a new urinary screening method for the detection of Sanfilippo patients based on findings showing patterns of oligosaccharides that are characteristic of each MPS type (Figure 2; Byers et al., 1998). The HS that accumulates in MPS III is a sequential series of oligosaccharides from disaccharides through to higher oligosaccharides of more than eight repeat disaccharides to full length glycosaminoglycans (Figure 3; Fuller, Meikle, and Hopwood, 2004a; Fuller et al., 2004b; King et al., 2006; Mason et al., 2006). This method uses mass spectrometry to detect sulfated oligosaccharides shown to accumulate in this disorder; it is able to discriminate between the four Sanfilippo subtypes based on the presence of different oligosaccharide structures coming from the *endo*-hydrolysis of HS; these cannot be further degraded because of a deficiency in one of the four *exo*-enzymes required to reduce these oligosaccharides to sulfate and monosaccharides (Figure 3C). Once a Sanfilippo outcome is established the next step is often to identify the molecular genetics that cause the enzyme deficiency (Yogalingam and Hopwood, 2001).

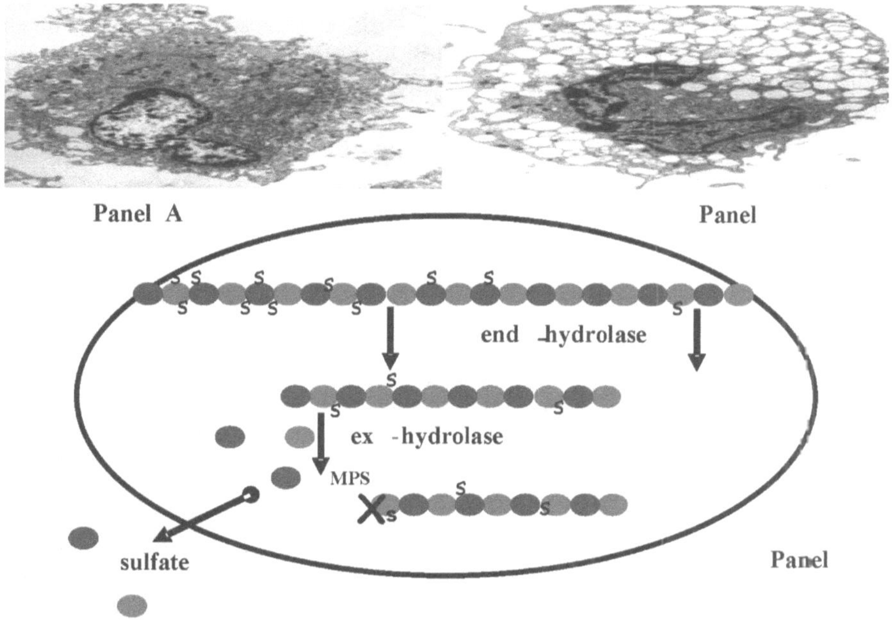

Figure 3. Cultured cells from normal control (Panel A); cultured cell from a MPS patient showing storage vacuoles (Panel B); a scheme (Panel C) illustrating heparan sulfate consisting of sulfated repeating uronic acid/glucosamine disaccharides undergoing initial degradation with *endo*-hydrolases to oligosaccharides in endosomes that are further degraded in the lysosome to monosaccharides and inorganic sulfate by the action of the *exo*-enzyme activities (*exo*-sulfatases, *exo*-glycosidases, and N-acetyltransferase). The monosaccharides and inorganic sulfate produced are actively transported out of the lysosome (the oval shape) into the cytoplasm for reuse.

3.2 Enzymology/Molecular Genetic Diagnosis

A deficiency or absence of in any one of the following four *exo*-enzymes required for the lysosomal degradation of HS leads to the clinical presentation of Sanfilippo syndrome: sulfamidase (Sanfilippo type A; MPS IIIA); α-N-acetyglucosaminidase (Sanfilippo type B; MPS IIIB); acetyl Co:A glucosamine N-acetyl transferase (Sanfilippo type C; MPS IIIC); and glucosamine-6-sulfatase (Sanfilippo type D; MPS IIID). At least five other lysosomal *exo*-enzymes are involved in HS turnover: α-L-iduronidase (Hurler–Scheie syndrome; MPS I); iduronate-2-sulfatase (Hunter syndrome; MPS II); β-glucuronidase (Sly syndrome; MPS VII); glucuronate-2-sulfatase (unknown MPS); and glucosamine-3-sulfatase (unknown MPS). The first three are also required for the degradation of dermatan sulfate and therefore their deficiency also leads to the accumulation of dermatan sulfate and HS. The latter two may only be involved in the degradation of HS but a clinical presentation resulting from a deficiency of either activity is yet to be described.

HS is synthesized as linear sulfated GAG attached to a protein core to yield HS proteoglycans. HS proteoglycans are ubiquitously present on the surface and in the matrix of all cells and dramatically influence the function and regulation of many cells. The GAG chains are composed of up to several hundred alternating uronic acid and N-acetylglucosamine residues that are posttranslationally modified via a series of enzymically driven reactions, to yield a complex pattern of O-sulfation (to produce C-3 and/or C-6 sulfate esters), N-sulfation, N-acetylated and free glucosamine residues, together with epimerization of some glucuronic acid residues to iduronic acid. These reactions produce HS chains with specific sequence and ability to recognize specific growth factors and present them to cell surface receptors.

3.3 Carrier Detection

Mutation analysis has enabled the detection of carriers in affected families. This is of particular concern when pregnancy is being considered (Yogalingam and Hopwood, 2001).

3.4 Prenatal Diagnosis

The accurate diagnosis of all four Sanfilippo subtypes is possible using chorionic villi samples (CVS), cultured CV cells, or amniocentesis (Hopwood, 2005). Measurement of direct enzyme activity or mutation analysis has been reported for all four subtypes. The choice of the diagnostic method depends on the genetic details available on the index case and the gestational age.

Preimplantation genetic diagnosis is possible for all Sanfilippo subtypes except type C. This diagnostic option would enable embryos that develop from in vitro fertilization to be tested for the disorder before they enter the uterus (Thornhill and Snow, 2002). In the near future it may be possible to use less invasive sampling procedures for recovering fetal DNA or cells from the maternal circulation and testing these for the presence of pathology-causing mutations (Dhallan et al., 2004)

3.5 *Endo*-Hydrolase Activity Toward HS

HS proteoglycans are internalized from the cell surface through an endocytic pathway; here the protein core is proteolically cleaved to release single HS chains in early endosomes that are then clipped to shorter oligosaccharide fragments by the action of a number of *endo*-hydrolyases (or heparanase; Figure 3C). These *endo*-hydrolyases have been described as having selective substrate specificity toward HS sequence to clip either β-1,4 linked glucuronic acid or α-1,4 linked N-acetylglucosamine residues. It has been proposed that *endo*-activities act in early endosome compartments to clip HS polysaccharides to oligosaccharides that are then transported to lysosomal compartments where degradation continues with the action of up to nine *exo*-enzymes. A deficiency inactivity of one of four *exo*-enzymes may lead to the primary accumulation of HS-oligosaccharide fragments and the expression of the Sanfilippo phenotype. These are discussed below.

3.6 *Exo*-Hydrolase Activity Toward Glucosamine Residues in HS

3.6.1 Heparan N-Sulfatase or Sulfamidase

3.6.1.1 Diagnostic Enzyme Activity Measurement

A deficiency of sulfamidase may lead to MPS IIIA. A number of diagnostic substrates been developed to assess sulfamidase activity in biological samples. Hopwood and Elliott (1982a) developed a sensitive radiolabelled tetrasaccharide substrate to measure the activity of this sulfatase to hydrolyze sulfamate esters from the nonreducing end glucosamine residues present in HS. Freeman and Hopwood (1986) reported other more sensitive oligosaccharide substrates. Fluorometric substrates (4-methylumbelliferyl-α-D-N-sulfoglucosaminide and 4-methylcoumarin-α-D- N-sulfoglucosaminide) are also available to specifically measure this sulfatase activity (Karpova et al., 1996; Dasgupta and Masada, 2002). These fluorometric substrates require the action of a yeast α-glucosaminidase on the de-N-sulfated product to release fluorescence for measurement. Substrates that are able to separate from product by electrospray ionization mass spectroscopy have been prepared and validated (Gerber et al., 2001).

3.6.1.2 Mutations

Sulfamidase has been purified to homogeneity from a number of human tissues (Freeman and Hopwood, 1986); amino acid sequence used to isolate cDNA encoding this sulfatase predicts a 502 amino acid polypeptide with five potential N-glycosylation sites (Scott et al., 1995). The gene is located on 17q25.3 (Scott et al., 1995), spans about 11 kb, and includes 8 exons (Karageorgos et al., 1996). Over 60 mutations in the sulfamidase gene have been reported to lead to a Sanfilippo phenotype (Yogalingam and Hopwood, 2001). Some common mutations have been reported that vary between different populations. For example, the R74C mutation is common in Poland (Bunge et al., 1997); R245H is common in Australia, Germany, and The Netherlands (Weber et al., 1998); S66W in Italy (DiNatale et al., 1998); and 1091delC in Spain (Montfort et al., 1998). These concentrations of mutations in particular geographic locations represent founder effects.

Attenuated MPS IIIA patients have been described (Wisniewski et al., 1985; Lindor et al., 1994; Date et al., 1998; Miyazaki et al., 2002; van Hove et al., 2003; Di Natale et al., 2005; Gabrielli et al., 2005). These patients generally show mild neurological signs that develop at a rate that is much slower than the classical Sanfilippo phenotype. Van Hove et al. (2003) reported a patient who presented at 45 years of age with cardiomyopathy but normal neurological function. Mutations, such as R206P, have been reported to reduce sulfamidase activity to about 8% of normal, and cause a "nonevolving" HS-uria and mental retardation in a 20-year-old female patient (Gabrielli et al., 2005). This patient presented at 6 years of age with short stature and mental retardation that did not appear to change or progress further.

Perkins et al. (1999, 2001) have shown that sulfamidase protein as well as activity is reduced in cultured skin fibroblasts from most MPS IIIA patients. This reduction in activity reflects the clinical spectrum from no detectable enzyme in R245H homozygotes, to approximately 10% activity in a patient who presented at 45 years of age. A fivefold variation in specific activity of the mutant sulfamidase was determined for 35 MPS IIIA patients (Perkins et al., 2001).

3.6.2 a-N-Acetylglucosaminidase

3.6.2.1 Diagnostic Activity Measurement

A deficiency of NAGLU may lead to MPS IIIB. A number of diagnostic substrates have been developed to assess NAGLU activity in biological samples. Hopwood and Elliott (1982a) developed a sensitive radiolabelled disaccharide substrate derived from heparin to measure the activity of this glycosidase to hydrolyze α 1,4 linked GlcNAc residues from the nonreducing end of HS. A fluorometric substrate (4-methylumbelliferyl-α-D-N-acetyglucosaminide) is a convenient and sensitive substrate that has become the choice for diagnosis (Marsh and Fensom, 1985). Substrates that are able to separate from product by electrospray ionization mass spectroscopy have been prepared and validated (Gerber et al., 2001).

3.6.2.2 Mutations

NAGLU has been purified from a number of human tissues. The cDNA encoding this hydrolase has been cloned to identify a polypeptide of 743 amino acids (Zhao et al., 1996; Weber et al., 1996). Over 80 mutations have been described in MPS IIIB patients leading to lysosomal storage of HS and endolytic oligosaccharide products (Yogalingam and Hopwood, 2001). There are no common mutations leading to MPS IIIB, with most communities showing a variety of different mutations.

3.6.3 N-Acetyl Transferase

3.6.3.1 Diagnostic Enzyme Activity Measurement

A deficiency of this enzyme may lead to MPS IIIC. N-Acetyl transferase is the only lysosomal enzyme known to be involved in making a bond rather than in its hydrolysis. This enzyme catalyses N-acetylation of the amino group of the nonreducing end glucosamine product of the previous enzyme activity in the sulfamidase pathway. N-Acetyl transferase has not been purified to homogeneity but its partial purification has revealed some of the kinetic mechanisms used by this unusual enzyme (Meikle, Whittle, and Hopwood, 1995). Radiolabelled glucosamine has been used with acetyl-CoA as a substrate for this enzyme (Hopwood and Elliott, 1981), but other diagnostic substrates have been developed (He et al., 1994). The cDNA encoding this protein is yet to be isolated. Recently, the gene has been located to the pericentromeric region of chromosome 8 (Ausseil et al., 2004).

3.6.4 Glucosamine-6-Sulfatase

3.6.4.1 Diagnostic Enzyme Activity Measurement

A deficiency of this enzyme may lead to MPS IIID. The natural substrate for this sulfatase is nonreducing end C6 sulfated N-acetylglucosamine or N-sulfateglucosamine residues in HS or keratan sulfate. However, only HS is stored in lysosomes and secreted in excess in the urine of MPS IIID patients. An excess of keratan sulfate is not observed in these patients as β-N-acetylhexosaminidase is able to remove β-linked N-acetylglucosamine-6-sulfate from the nonreducing end of keratan sulfate to yield the monosaccharide N-acetyl-glucosamine-6-sulfate, which is found as a stored and secreted component in MPS IIID.

A number of radiolabelled oligosaccharide substrates have been prepared for this sulfatase (Freeman, Clements, and Hopwood, 1987). A fluorometric substrate (4-methyl-umbelliferyl-α-D-N-acetyglucosaminide-6-sulfate) is convenient and requires the subsequent action of α-N-acetylglucosaminidase to release fluorescence (He et al., 1993). This sulfatase can also be assayed using special substrates able to be separated from their products by affinity chromatography/electrospray ionization mass spectrometry (Gerber et al., 2001).

3.6.4.2 Mutations

Glucosamine-6-sulfatase has been purified to homogeneity from a number of human tissues (Freeman, Clements, and Hopwood, 1987); amino acid sequence used to isolate cDNA encoding this sulfatase predicts a 552 amino acid polypeptide with 13 potential N-glycosylation sites, with 10 likely to be occupied (Robertson et al., 1992). The gene is located on 12q14 (Robertson et al., 1988) and a number of mutations have been described (Hopwood and Morris, 1990; Beesley et al., 2003; Mok, Cao, and Hegele, 2003).

3.7 *Exo*-Hydrolase Activity Towards Uronic Acid Residues in HS

These enzymes are also required for the lysosomal degradation of the uronic acid species present in dermatan sulfate. Therefore, when deficient, both HS and dermatan sulfate accumulate in lysosomes and are found in excess in urine. Specifically, iduronate-2-sulfatase, α-L-iduronidase, and β-D-glucuronidase are required to act on the nonreducing end α-linked iduronate-2-sulfate and β-D-glucuronide residues present on HS and dermatan sulfate. Deficiencies in these enzymes that lead to storage and a clinical phenotype are known as MPS II (Hunter syndrome), MPS I (Hurler–Scheie syndrome), and MPS VII (Sly syndrome), respectively.

4 ANIMAL MODELS

A number of animal models for Sanfilippo syndrome have been characterized. These models have been used to investigate the pathophysiology and natural history of Sanfilippo syndrome. It is important to note that these models have also been extremely valuable to investigate/evaluate the effectiveness of various therapies for the treatment/management of Sanfilippo patients.

4.1 MPS III Type A

A mouse model has been described where a naturally occurring mutation (D31N) causes the loss of an essential aspartic acid residue needed for divalent metal ion binding at the active site of sulfamidase; this leads to a model with approximately 3% of normal sulfamidase activity, storage of HS in the brain, and Sanfilippo-like symptoms, for example, aggression and loss of learned behavior (Bhaumik et al., 1999; Gliddon and Hopwood, 2004; Hemsley and Hopwood, 2005; Gliddon et al., 2004). These mice have storage vacuoles in neurons containing membranous and floccular materials (Bhaumik et al., 1999; Bhattacharyya et al., 2001). Sanfilippo syndrome has been identified in wire-haired dachshunds and New Zealand Huntaways (Jolly et al., 2000). The mutation in the Huntaway was shown to lead to early termination of the sulfamidase protein and the development of progressive ataxia, hypermetria, and loss of learned behavior (Jolly et al., 2000; Yogalingam et al., 2002). The mutation in the dachshunds causes a deletion of a threonine residue and possibly a less severe phenotype than seen in the Huntaway, with progressive neurologic disease without apparent somatic involvement (Aronovich et al., 2000).

4.2 MPS III Type B

A mouse model has been produced by targeted disruption of exon 6 in the NAGLU gene (Li et al., 1999). These mice were healthy and fertile while young, survived for 12 months, stored HS in liver and kidney, and had elevated GM2 and GM3 gangliosides in brain. Large pleiomorphic inclusions were seen in some neurons and pericytes in the brain. At four to five months of age abnormal behavior in an open field test was observed in affected mice. MPS IIIB has been described in a Schipperke dog that results from a mutation reducing NAGLU activity to between 4 and 9% of normal, causing excess HS in the urine, and cerebellar disease including dysmetria, hind-limb ataxia, and a wide-based stance with truncal swaying (Ellinwood et al., 2003). A 2 bp deletion in the avian NAGLU gene causes elongation past the original stop codon to produce NAGLU with extra 37 amino acids together with 387 altered amino acids, and MPS IIIB in an emu (Aronovich et al., 2000).

4.3 MPS III Type D

A goat model was identified at post-mortem, which had similar clinical and histological properties to humans with Sanfilippo syndrome (Thompson et al., 1992). The mutation was shown to be R102X, which caused progressive dementia, storage of HS, and secondary storage of lipids leading to a severe clinical outcome (Cavanagh et al., 1995).

5 MECHANISMS OF NEUROPATHOGENESIS IN SANFILIPPO SYNDROME

Lysosomal storage vacuoles first appear in the CNS of MPS IIIA during the second trimester of fetal development (Ceuterick et al., 1980), with a progressive increase in vacuolization with gestational age. Although studies of the nature and rate of development of neuropathology in Sanfilippo individuals provide valuable insights into the consequence of HS accumulation at individual time points, a large amount of critical information is missing. Tamagawa et al. (1985) reported increased expression of a glial fibrillary acidic protein in the brain of MPS IIIB patients. This is where systematic studies of the brain in animal models become extremely valuable.

Much research is focused on studies of the nature of pathological cascades that are triggered in response to the primary storage of HS, in the case of Sanfilippo syndrome. Once HS starts to build up within the lysosome and reaches a predetermined level of burden other components begin to accumulate.

The similarity between the spectrum of pathology (for example, inflammatory response, intracellular inclusions, and cell loss) seen in Sanfilippo syndrome, other LSD involving the brain and neurodegenerative conditions such as multiple sclerosis, and Huntington's and Parkinson's diseases, has suggested the presence of cascade events that occur within the CNS following the initiation of neuropathology by an initial insult. In the case of Sanfilippo syndrome, this is reduced activity of lysosomal enzymes resulting in the storage of the primary substrate, HS.

Similar to other LSD, intracellular inclusions and axonal spheroids have been shown to occur in MPS IIIA mouse brain. In the MPS IIIA mouse it has been reported that, as HS storage increases, accumulation of GM2/GM3 gangliosides, cholesterol, and autofluorescent material is found to be present (McGlynn, Dobrenis, and Walkley, 2004). Their report indicates that, although both gangliosides accumulate in the same neuron, they were consistently located in separate populations of cytoplasmic vesicles. It is important to note that GM3 ganglioside only partially colocalized with primarily stored HS. Furthermore, the stored cholesterol likewise only partially colocalized with the GM2/GM3 gangliosides. These findings raise questions about the mechanisms that lead to the poor colocalization of primary storage of HS and secondary storage with ganglioside/cholesterol membrane rafts. These lipid rafts are believed to be critical in signal transduction events in neurons and therefore the nature of the storage relationships between these storage compounds is probably critical to the mechanism to explain neuronal dysfunction in Sanfilippo syndrome.

6 THERAPIES

Therapy for neurodegenerative LSD requires the introduction of active enzyme into the CNS such that the introduced enzyme can act faster than the progression of the disease. Bone marrow transplantation (allogenetic hematopoietic stem-cell transplantation) has been used to benefit LSD patients. Donor stem cells repopulate various tissues and deliver enzyme to correct storage in host cells. Bone marrow transplantation has been tried as a therapy in a few symptomatic Sanfilippo syndrome patients. As these patients had neurological symptoms at the time of the bone marrow transplantation interpreting the effect of the therapy was difficult (Hoogerbrugger et al., 1995). It was suggested that bone marrow transplantation should be considered in Sanfilippo patients if they have no evidence of neurological disease. Sivakumur and Wraith (1999) reported results of bone marrow transplantation in an asymptomatic MPS IIIA infant at ten months of age, and compared his progress to his older but untreated, affected sibling. Despite successful biochemical correction his neurological deterioration was not prevented over eight years and was no different to that of his untreated sibling. Like the untreated sibling the treated patient lost development quotients at the same rate over this time period and became immobile with swallowing dysfunction (Sivakumur and Wraith, 1999).

Bone marrow has traditionally been sourced to provide donor stem cells for transplantation. However, recruitment of unrelated adult donors to find a matched donor can take too long for rapidly progressing disorders. Banked umbilical cord blood is readily available from unrelated donors and has successfully been transplanted in a number of LSD patients (Kurtzberg et al., 1996; Staba et al., 2004; Escolar et al., 2005). One hypothesis posits that transplantation of umbilical cord blood from unrelated donors before the development of symptoms would favorably alter the natural history of the disease in affected newborns. Detection of affected newborns would result from prenatal diagnosis in the presence of a family history of a particular disorder, or alternately, from the introduction of newborn screening programs.

A study of umbilical cord blood transplantation from unrelated donors has been carried out in Krabbe disease: it compared the effect of transplantation in presymptomatic Krabbe newborns with affected infants who had been transplanted after symptoms had developed. The results were associated with improved neurological outcome and survival in the transplanted presymptomatic neonates (Escolar et al., 2005). These results encourage the hypothesis that Sanfilippo patients may also respond if transplanted in the first month of birth, particularly if relatively large numbers of stem cells are able to enter the brain and provide sufficient enzyme to correct or prevent storage in the host CNS cells before irreversible pathology occurs.

A number of therapies have been trialed in animal models of Sanfilippo syndrome. These include enzyme replacement and gene replacement therapies. Gliddon and Hopwood (2004) reported the impact of intravenously introducing recombinant sulfamidase into the CNS of newborn MPS IIIA mice when mannose-6-phosphate receptors are present in the blood–brain barrier (Urayana et al., 2004) so that enzyme is readily transported from circulation into the brain. Here, enzyme was able to delay the onset of pathological, biochemical, and behavioral changes, indicating that sulfamidase given early is able to improve behavioral function.

Another therapeutic approach is the reduction of storage burden by reducing the synthesis of the substrate for the defective enzyme. Substrate reduction therapy has been proposed for Sanfilippo syndrome through the application of HS synthesis inhibitors (Roberts et al., 2004). The advantages of this therapy are that the inhibitors may be taken orally, and may pass through the blood–brain barrier. One disadvantage is that in the absence of residual enzyme activity this will be of limited benefit. However, it may be an advantage in combination with other therapies. Furthermore, there needs to be a balance between substrate inhibition and ensuring that sufficient HS proteoglycan is present to satisfy functional needs.

Chemical chaperone therapy should also be considered for the treatment of Sanfilippo syndrome. Here, small molecules may act by stabilizing the mutant enzyme. For example, high concentrations of infused galactose have been reported to lead to clinical improvement in a Fabry patient, from increased stability of the mutant α-galactosidase involved in this patient's phenotype. It may be possible that the glucosamine therapy being trialed in Sanfilippo patients is affecting outcome via this mechanism.

Gene therapy is under intense study for Sanfilippo syndrome. Recent reports of the injection of adeno-associated virus vectors (serotype 2 or 5) encoding NAGLU into a single site in the putamen of 6-week old MPS IIIB mice showed that NAGLU activity was maintained over the entire brain at 38 weeks of age. Improvement was only partial in some mice (Cressant et al., 2004). Similarly, DiNatale et al. (2005) have injected a lentiviral-NAGLU vector intravenously into 8- to 10-week-old MPS IIIB mice. Here, NAGLU activity was able to maintain reduced levels of GAGs in target organs (liver, lung, and spleen).

7 EARLY DETECTION

Observations have led to the conclusion that therapies for LSD with CNS pathology are more likely to benefit patients if commenced presymptomatically (e.g., Escolar et al., 2005). This conclusion has supported the need to establish newborn screening for these disorders. Technology for LSD screening in the newborn population has been under considerable research for the past few years. Lysosomal enzyme activities and protein are measurable in rehydrated dried blood spots. Fluorometric, radiometric, immunochemical (protein profiling), and electrospray ionization tandem mass spectrometry (ESI-MS/MS) assays have been developed. The latter two offer the capability of assaying the products of several enzymes simultaneously (multiplexing). Both protein profiling and ESI-MS/MS assays that directly measure both the amount of lysosomal protein and reaction velocities in rehydrated dried blood spots have been reported and are readily adaptable to the process of newborn screening (Umapathysivam, Hopwood, and Meikle, 2001; Li et al., 2004; Wang et al., 2005; Meikle et al., 2005).

REFERENCES

Aronovich EL, Carmichael KP, Morizono H, Koutlas IG, Deanching M, Hoganson G, Fischer A, Whitley CB (2000) Canine heparan sulfate sulfamidase and the molecular pathology underlying Sanfilippo syndrome type A in Dachshunds. Genomics 68: 80-4.

Ausseil J, Loredo-Osti JC, Verner A, Darmond-Zwaig C, Maire I, Poorthuis B, van Diggelen OP, Hudson TJ, Fujiwara TM, Morgan K, Pshezhetsky AV (2004) Localisation of a gene for mucopolysaccharidosis IIIC to the pericentromeric region of chromosome 8. J Med Genet. 41: 941-5.

Barone R, Fiumara A, Villani GR, Di Natale P, Pavone L (2001) Extraneurologic symptoms as presenting signs of Sanfilippo disease. Pediatr Neurol. 2001 Sep;25(3): 254-7.

Beesley CE, Burke D, Jackson M, Vellodi A, Winchester BG, Young EP (2003) Sanfilippo syndrome type D: identification of the first mutation in the N-acetylglucosamine-6-sulphatase gene J Med Genet. 40: 192-4.

Bhattacharyya R, Gliddon B, Beccari T, Hopwood JJ, Stanley P (2001) A novel missense mutation in lysosomal sulfamidase is the basis of MPS III A in a spontaneous mouse mutant. Glycobiology 11: 99-103.

Bhaumik M, Muller VJ, Rozaklis T, Johnson L, Dobrenis K, Bhattacharyya R, Wurzelmann S, Finamore P, Hopwood JJ, Walkley SU, Stanley P (1999) A mouse model for mucopolysaccharidosis type III A (Sanfilippo syndrome). Glycobiology 9: 1389-96.

Bunge S, Ince H, Steglich C, Kleijer WJ, Beck M, Zaremba J, van Diggelen OP, Weber B, Hopwood JJ, Gal A (1997) Identification of 16 sulfamidase gene mutations including the common R74C in patients with mucopolysaccharidosis type IIIA (Sanfilippo A). Hum Mutat. 10: 479-85.

Byers S, Rozaklis T, Brumfield LK, Ranieri E, Hopwood JJ (1998) Glycosaminoglycan accumulation and excretion in the mucopolysaccharidoses: characterization and basis of a diagnostic test for MPS.Mol Genet Metab 65: 282-290.

Cavanagh KT, Leipprandt JR, Jones MZ, Friderici K (1995) Molecular defect of caprine N-acetylglucosamine-6-sulphatase deficiency. A single base substitution creates a stop codon in the 5'-region of the coding sequence. J Inherit Metab Dis. 18: 96.

Ceuterick C, Martin JJ, Libert J, Farriaux JP (1980) Sanfilippo A disease in the fetus--comparison with pre- and postnatal cases. Neuropadiatrie 11: 176-85.

Cressant A, Desmaris N, Verot L, Brejot T, Froissart R, Vanier MT, Maire I, Heard JM (2004) Improved behavior and neuropathology in the mouse model of Sanfilippo type IIIB disease after adeno-associated virus-mediated gene transfer in the striatum. J Neurosci. 24: 10229-39.

Dasgupta F, Masada RI (2002) Synthesis of 7-O-(2-deoxy-2-sulfamido-alpha-D-glucopyranosyl)-4-methylcoumarin sodium salt: a fluorogenic substrate for sulfamidase. Carbohydr Res. 337: 1055-8.

Date Y, Ohi T, Shioya K, Sukegawa K, Matsukura S (1998) Clinical and neuroradiological evaluation of the long-term surviving siblings of Sanfilippo syndrome A type Brain Nerve 50: 165-9.

Dhallan R, Au WC, Mattagajasingh S, Emche S, Bayliss P, Damewood M, Cronin M, Chou V, Mohr M (2004) Methods to increase the percentage of free fetal DNA recovered from the maternal circulation JAMA. 291:1114-9.

Di Natale P, Balzano N, Esposito S, Villani GR (1998) Identification of molecular defects in Italian Sanfilippo A patients including 13 novel mutations. Hum Mutat 11: 313-20.

Di Natale P, Di Domenico C, Gargiulo N, Castaldo S, Gonzalez Y, Reyero E, Mithbaokar P, De Felice M, Follenzi A, Naldini L, Villani GR (2005) Treatment of the mouse model of mucopolysaccharidosis type IIIB with lentiviral-NAGLU vector. Biochem J. 388: 639-46.

Ellinwood NM, Wang P, Skeen T, Sharp NJ, Cesta M, Decker S, Edwards NJ, Bublot I, Thompson JN, Bush W, Hardam E, Haskins ME, Giger U (2003) A model of mucopolysaccharidosis IIIB (Sanfilippo syndrome type IIIB): N-acetyl-alpha-D-glucosaminidase deficiency in Schipperke dogs. J Inherit Metab Dis 26: 489-504.

Escolar ML, Poe MD, Provenzale JM, Richards KC, Allison J, Wood S, Wenger DA, Pietryga D, Wall D, Champagne M, Morse R, Krivit W, Kurtzberg J (2005) Transplantation of umbilical-cord blood in babies with infantile Krabbe's disease. N England J Med 352: 2069-81.

Freeman C, Hopwood JJ (1986) Human liver sulphamate sulphohydrolase. Determinations of native protein and subunit Mr values and influence of substrate agylcone structure on catalytic properties. Biochem J. 234(1): 83-92.

Freeman C, Hopwood JJ (1987) Human liver N-acetylglucosamine-6-sulphate sulphatase. Catalytic properties. Biochem J. 246: 355-65.

Freeman C, Clements PR, Hopwood JJ (1987) Human liver N-acetylglucosamine-6-sulphate sulphatase. Purification and characterization. Biochem J. 246: 347-54.

Freeman C, Hopwood JJ (1991) Glucuronate-2-sulphatase activity in cultured human skin fibroblast homogenates. Biochem J. 279: 399-405.

Fuller M, Meikle PJ, Hopwood JJ (2004a) Glycosaminoglycan degradation fragments in mucopolysaccharidosis I. Glycobiology 14: 443-50.

Fuller M, Rozaklis T, Ramsay SL, Hopwood JJ, Meikle PJ (2004b) Disease-specific markers for the mucopolysaccharidoses. Pediatr Res. 56: 733-8.

Fuller M, Chau A, Nowak RC, Hopwood JJ, Meikle PJ (2006) A defect in exodegradative pathways provides insight into endodegradation of heparan and dermatan sulfates. Glycobiology 16:318-325.

Gabrielli O, Coppa GV, Bruni S, Villani GR, Pontarelli G, Di Natale P (2005) An adult Sanfilippo type A patient with homozygous mutation R206P in the sulfamidase gene. Am J Med Genet A. 133(1): 85-9.

Gerber SA, Scott CR, Turecek F, Gelb MH (2001) Direct profiling of multiple enzyme activities in human cell lysates by affinity chromatography/electrospray ionization mass spectrometry: application to clinical enzymology. Anal Chem. 73: 1651-7.

Gerber SA, Turecek F, Gelb MH (2001) Bioconjug Chem. 12: 603-15.

Gliddon BL, Hopwood JJ (2004) Enzyme-replacement therapy from birth delays the development of behavior and learning problems in mucopolysaccharidosis type IIIA mice. Pediatr Res. 56: 65-72.

Gliddon BL, Yogalingam G, Hopwood JJ (2004) Purification and characterization of recombinant murine sulfamidase. Mol Genet Metab. 83: 239-45.

He W, Voznyi YaV, Boer AM, Kleijer WJ, van Diggelen OP (1993) A fluorimetric enzyme assay for the diagnosis of Sanfilippo disease type D (MPS IIID). J Inherit Metab Dis. 16: 935-41.

He W, Voznyi YaV, Huijmans JG, Geilen GC, Karpova EA, Dudukina TV, Zaremba J, Van Diggelen OP, Kleijer WJ (1994) Prenatal diagnosis of Sanfilippo disease type C using a simple fluorometric enzyme assay. Prenat Diagn. 14: 17-22.

Hemsley KM, Hopwood JJ (2005) Development of motor deficits in a murine model of mucopolysaccharidosis type IIIA (MPS-IIIA). Behav Brain Res. 158: 191-9.

Hoogerbrugge PM, Brouwer OF, Bordigoni P, Ringden O, Kapaun P, Ortega JJ, O'Meara A, Cornu G, Souillet G, Frappaz D, et al (1995) Allogeneic bone marrow transplantation for lysosomal storage diseases. The European Group for Bone Marrow Transplantation. Lancet 345: 1398-402.

Hopwood JJ (2005) Prenatal diagnosis of Sanfilippo syndrome. Prenat Diagn. 25:148-50.

Hopwood JJ, Elliott (1981) The diagnosis of the Sanfilippo C syndrome, using monosaccharide and oligosaccharide substrates to assay acetyl-CoA: 2-amino-2-deoxy-alpha-glucoside N-acetyltransferase activity Clin Chim Acta. 112: 67-75.

Hopwood JJ, Elliott H (1982a) Detection of the Sanfilippo type B syndrome using radiolabelled oligosaccharides as substrates for the estimation of alpha-N-acetylglucosaminidase. Clin Chim Acta. 120: 77-86.

Hopwood JJ, Elliott H (1982b) Diagnosis of Sanfilippo type A syndrome by estimation of sulfamidase activity using a radiolabelled tetrasaccharide substrate. Clin Chim Acta. 123: 241-50.

Hopwood JJ, Harrison JR (1982) High-resolution electrophoresis of urinary glycosaminoglycans: an improved screening test for the mucopolysaccharidoses. Anal Biochem. 119: 120-7.

Hopwood JJ, Morris CP (1990) The mucopolysaccharidoses. Diagnosis, molecular genetics and treatment. Mol Biol Med. 7: 381-404.

Hopwood JJ, Ballibio A (2001) In: *The Metabolic and Molecular Bases of Inherited Disease.* (Scriver CR, Beaudet AL, Sly WS and Valle D, eds.), 8th edition. McGraw-Hill, New York, pp. 3725-32.

Jolly RD, Allan FJ, Collett M, Rozaklis T, Muller V, Hopwood JJ (2000) Mucopolysaccharidosis IIIA (Sanfilippo syndrome) in a New Zealand Huntaway dog with ataxia. N Z Vet J. 48: 144-8.

Kaplan D (1969) Classification of the mucopolysaccharidoses based on the pattern of mucopolysacchariduria.Amer. J Med 47: 721.

Karageorgos LE, Guo XH, Blanch L, Weber B, Anson DS, Scott HS, Hopwood JJ (1996) Structure and sequence of the human sulphamidase gene. DNA Res. 3: 269-71.

Karpova EA, Voznyi YaV, Keulemans JL, Hoogeveen AT, Winchester B, Tsvetkova IV, van Diggelen OP (1996) A fluorimetric enzyme assay for the diagnosis of Sanfilippo disease type A (MPS IIIA). J Inherit Metab Dis.19(3): 278-85.

King B, Savas P, Fuller M, Hopwood J, Hemsley K (2006) Validation of a heparan sulfate-derived disaccharide as a marker of accumulation in murine mucopolysaccharidosis type IIIA. Mol Genet Metab 87: 107-112.

Kresse H, Wiesman U, Cantz M, Hall CW, Neufeld EF (1971) Biochemical heterogeneity of the Sanfilippo syndrome: preliminary characterization of two deficient factors. Biochem Biophys Res Commun 42: 892-8.

Kresse H and Neufeld EF (1972) J Biol Chem 247: 2164-70.

Kurtzberg J, Laughlin M, Graham ML, Smith C, Olson JF, Halperin EC, Ciocci G, Carrier C, Stevens CE, Rubinstein P (1996) Placental blood as a source of hematopoietic stem cells for transplantation into unrelated recipients. N England J Med 335: 157-66.

Li HH, Yu WH, Rozengurt N, Zhao HZ, Lyons KM, Anagnostaras S, Fanselow MS, Suzuki K, Vanier MT, Neufeld EF (1999) Mouse model of Sanfilippo syndrome type B produced by targeted disruption of the gene encoding alpha-N-acetylglucosaminidase. Proc Natl Acad Sci U S A. 96: 14505-10.

Li Y, Scott CR, Chamoles NA, Ghavami A, Pinto BM, Turecek F, Gelb MH (2004) Direct multiplex assay of lysosomal enzymes in dried blood spots for newborn screening. Clin Chem. 50: 1785-96.

Lindor NM, Hoffman A, O'Brien JF, Hanson NP, Thompson JN (1994) Sanfilippo syndrome type A in two adult sibs. Am J Med Genet 53: 241-4.

Mason KE, Meikle PJ, Hopwood JJ, Fuller M (2006) Characterization of sulfated oligosaccharides in mucopolysaccharidosis type IIIA by electrospray ionization mass spectrometry. Anal Chem; 78:4534-42.

Marsh J, Fensom, AH (1985) 4-Methylumbelliferyl alpha-N-acetylglucosaminidase activity for diagnosis of Sanfilippo B disease. Clin. Genet. 27: 258-62.

McGlynn R, Dobrenis K, Walkley SU (2004) Differential subcellular localization of cholesterol, gangliosides, and glycosaminoglycans in murine models of mucopolysaccharide storage disorders. J Comp Neurol. 480: 415-26.

McKusick VA (1966) Heritable disorders of connective tissue, Ed 3, p 325, CV Mosby, St Louis

Meikle PJ, Whittle AM, Hopwood JJ (1995) Human acetyl-coenzyme A:alpha-glucosaminide N-acetyltransferase. Kinetic characterization and mechanistic interpretation. Biochem J. 308: 327-33.

Miyazaki T, Musuda N, Waragai M, Motoyoshi Y, Kurokawa K, Yuasa T (2002) An adult Japanese Sanfilippo A patient with novel compound heterozygous S347F and D444G mutations in the sulphamidase gene J Neurol NNeurosurg Psych 73:777-8.

Meikle PJ, Hopwood JJ, Clague AE, Carey WF (1999) Prevalence of lysosomal storage disorders. JAMA 281: 249-54.

Meikle PJ, Dean CJ, Grasby D, Bockmann MR, Whittle AM, Lang DL, Fuller M, Brooks DA, Hopwood JJ (2005) J Inherit Met Dis 28:14.

Mok A, Cao H, Hegele RA (2003) Genomic basis of mucopolysaccharidosis type IIID (MIM 252940) revealed by sequencing of GNS encoding N-acetylglucosamine-6-sulfatase. Genomics 81: 1-5.

Montfort M, Vilageliu L, Garcia-Giralt N, Guidi S, Coll MJ, Chabas A, Grinberg D (1998) Mutation 1091delC is highly prevalent in Spanish Sanfilippo syndrome type A patients. Hum Mutat.12: 274-9.

Perkins KJ, Byers S, Yogalingam G, Weber B, Hopwood JJ (1999) Expression and characterization of wild type and mutant recombinant human sulfamidase. Implications for Sanfilippo (Mucopolysaccharidosis IIIA) syndrome. J Biol Chem. 274: 37193-9.

Perkins KJ, Muller V, Weber B, Hopwood JJ (2001) Prediction of Sanfilippo phenotype severity from immunoquantification of heparan-N-sulfamidase in cultured fibroblasts from mucopolysaccharidosis type IIIA patients.Mol Genet Metab. 73: 306-12.

Poorthuis BJ, Wevers RA, Kleijer WJ, Groener JE, de Jong JG, van Weely S, Niezen-Koning KE, van Diggelen OP (1999) The frequency of lysosomal storage diseases in The Netherlands. Hum Genet. 105(1-2): 151-6.

Roberts ALK, Thomas BJ, Wilkinson AS, Byers S (2004) 8th International Symposium of Mucopolysaccharide and Related Desease, June 10 – 13 abstract V30, Mainz, Germany.

Robertson DA, Freeman C, Nelson PV, Morris CP, Hopwood JJ (1988) Human glucosamine-6-sulfatase cDNA reveals homology with steroid sulfatase Biochem Biophys Res Commun. 157: 218-24.

Robertson DA, Freeman C, Morris CP, Hopwood JJ (1992) A cDNA clone for human glucosamine-6-sulphatase reveals differences between arylsulphatases and non-arylsulphatases. Biochem J. 288: 539-44.

Scott HS, Blanch L, Guo XH, Freeman C, Orsborn A, Baker E, Sutherland GR, Morris CP, Hopwood JJ (1995) Cloning of the sulphamidase gene and identification of mutations in Sanfilippo A syndrome.Nat Genet. 11: 465-7.

Sivakumur P, Wraith JE (1999) Bone marrow transplantation in mucopolysaccharidosis type IIIA: a comparison of an early treated patient with his untreated sibling. J Inherit Metab Dis. 22: 849-50.

Staba SL, Escolar ML, Poe M, Kim Y, Martin PL, Szabolcs P, Allison-Thacker J, Wood S, Wenger DA, Rubinstein P, Hopwood JJ, Krivit W, Kurtzberg J (2004) Cord-blood transplants from unrelated donors in patients with Hurler's syndrome N England J Med 350: 1960-9.

Tamagawa K, Morimatsu Y, Fujisawa K, Hara A, Taketomi T (1985) Neuropathological study and chemico-pathological correlation in sibling cases of Sanfilippo syndrome type B. Brain Dev. 7: 599-609.

Thompson JN, Jones MZ, Dawson G, Huffman PS (1992) N-acetylglucosamine 6-sulphatase deficiency in a Nubian goat: a model of Sanfilippo syndrome type D (mucopolysaccharidosis IIID). J Inherit Metab Dis. 15: 760-8.

Thornhill AR, Snow K (2002) Molecular diagnostics in preimplantation genetic diagnosis. J Mol Diagn. 4: 11-29.

Tsuji D, Kuroki A, Ishibashi Y, Itakura T, Kuwahara J, Yamanaka S, Itoh K (2005) Specific induction of macrophage inflammatory protein 1-alpha in glial cells of Sandhoff disease model mice associated with accumulation of N-acetylhexosaminyl glycoconjugates. J Neurochem 92: 1497-507.

Umapathysivam K, Hopwood JJ, Meikle PJ (2001) Determination of acid alpha-glucosidase activity in blood spots as a diagnostic test for Pompe disease. Clin Chem. 47: 1378-83.

Urayana A, Grubb JH, Sly WS, Banks WA (2004) Proc Natl Acad Sci U S A. 101:12658-63.

Van Hove JL, Wevers RA, Van Cleemput J, Moerman P, Sciot R, Matthijs G, Schollen E, de Jong JG, Carey WF, Muller V, Nicholls C, Perkins K, Hopwood JJ (2003) Late-Onset visceral presentation with cardiomyopathy and without neurological symptoms of adult Sanfilippo A syndrome. Am J Med Genet A118: 382-7.

Wang D, Eadala B, Sadilek M, Chamoles NA, Turecek F, Scott CR, Gelb MH (2005) Tandem mass spectrometric analysis of dried blood spots for screening of mucopolysaccharidosis I in newborns. Clin Chem. 51: 898-900.

Weber B, Blanch L, Clements PR, Scott HS, Hopwood JJ (1996) Cloning and expression of the gene involved in Sanfilippo B syndrome (mucopolysaccharidosis III B). Hum Mol Genet. 5: 771-7.

Weber B, van de Kamp JJ, Kleijer WJ, Guo XH, Blanch L, van Diggelen OP, Wevers R, Poorthuis BJ, Hopwood JJ (1998) Identification of a common mutation (R245H) in Sanfilippo A patients from The Netherlands. J Inherit Metab Dis. 21: 416-22.

Wisniewski K, Rudelli R, Laure-Kamionowska M, Sklower S, Houck GE Jr, Kieras F, Ramos P, Wisniewski HM, Braak H (1985) Sanfilippo disease, type A with some features of ceroid lipofuscinosis. Neuropediatrics 16: 98-105.

Yogalingam G, Hopwood JJ (2001) Molecular genetics of mucopolysaccharidosis type IIIA and IIIB: Diagnostic, clinical, and biological implications. Hum Mutat. 18: 264-81.

Yogalingam G, Pollard T, Gliddon B, Jolly RD, Hopwood JJ (2002) Identification of a mutation causing mucopolysaccharidosis type IIIA in New Zealand Huntaway dogs. Genomics. 79: 150-3.

Zhao HG, Li HH, Bach G, Schmidtchen A, Neufeld EF (1996) The molecular basis of Sanfilippo syndrome type B. Proc Natl Acad Sci U S A. 93: 6101-5.

MUCOPOLYSACCHARIDOSIS IV (Morquio Syndrome; MPS IV)

Shunji Tomatsu,[1*] Adriana M. Montaño[1], Tatsuo Nishioka,[1] Tadao Orii,[2]

1 BACKGROUND

Morquio syndrome (mucopolysaccharidosis IV: MPS IV) is an autosomal recessive disease classified in the group of mucopolysaccharide storage diseases. Two forms are recognized, type A and type B. MPS IVA is characterized by the absence of the enzyme N-acetyl-galactosamine 6-sulfate sulfatase (GALNS). MPS IVB results from deficiency of the enzyme β-galactosidase. Both types excrete keratan sulfate (KS) in urine. In 1929, Morquio, a pediatrician in Uruguay, described cases of Morquio syndrome. In 1952, Brante isolated the stored mucopolysaccharides in Morquio patients. In 1976, the enzyme deficiency in MPS IVA (GALNS deficiency) was identified (Singh et al. 1976). Shortly thereafter, the enzyme deficiency in MPS IVB was described (β-galactosidase deficiency). Historically, MPS IVA was considered to have more severe manifestations than MPS IVB. However, with the ability to differentiate between types A and B by enzyme analysis, variability in clinical expression in both groups becomes apparent.

MPS IV causes skeletal dysplasia through excessive storage of KS. The unique clinical manifestations of this disorder are attributable to the restricted tissue distribution of KS (corneas and cartilage), in contrast to the much wider distribution of dermatan sulfate (DS) and heparan sulfate (HS). The ELISA-sandwich assay by a monoclonal antibody specific to KS has been developed. It has been found that blood and urine KS level varies with age and clinical severity, indicating that the new assay for KS may be suitable for early diagnosis and longitudinal assessment of disease severity in MPS IV (Tomatsu et al., 2004a).

The broad range of clinical phenotypes seen in MPS IVA is presumed to be the result of many different GALNS mutations. The isolation and characterization of the full-length cDNA encoding the human GALNS protein and genomic sequence (Tomatsu et al., 1991; Nakashima et al., 1994) have facilitated investigations into the molecular genetic heterogeneity in MPS IVA. GALNS is a member of the sulfatase gene family of which 13 sulfatase genes in humans have been cloned. All sulfatase gene products are closely related, showing 20–35% similarity at the amino acid level. C79 residue of human GALNS is conserved among all sulfatase proteins from many species. Characterization of sulfatase proteins by structural analysis and homology comparisons suggested C79 as an active site residue (Bond et al., 1997; Sukegawa et al., 2000). The posttranslational

[1]Department of Pediatrics, Cardinal Glennon Children's Hospital, Saint Louis University, USA; [2]Department of Pediatrics, Gifu University School of Medicine, Japan;
*Address for correspondence:
S. Tomatsu:
Department of Pediatrics, Saint Louis University Pediatric Research Institute
3662 Park Ave. St. Louis, MO 63110-2586
Phone: 314-577-5623, ext. 6213; Fax: 314-577-5398
E-mail: tomatsus@slu.edu

modification of the highly conserved cysteine residue is required for generating catalytically active sulfatases (Schmidt et al., 1995).

Recently, three different MPS IVA mice models have been established (Tomatsu et al., 2003b; unpublished data). Clinical manifestations in the affected mice are milder than those in human patients.

Only palliative measures are currently available for treatment of patients with MPS IV. Potential strategies, currently at different levels of development, include enzyme replacement, gene therapy, and allogenic bone marrow transplantation in which engrafted cells provide the normal enzyme. Those treatments are expected to enable dramatic improvements in visceral organs but little or no improvements in bone because the enzymes are not delivered to the bone effectively. In this context, the challenge still is to maximize the clinical efficacy to the bones, especially to systematic bone disease, MPS IV.

In this chapter, we review the clinical manifestations, incidence, diagnosis, and treatments for MPS IV, mainly focusing on IVA.

2 CLINICAL ASPECTS

2.1 Medical History

Patients with Morquio syndrome appear healthy at birth. Morquio-specific radiographic changes occurring before phenotypic changes are obvious, have been observed (Figure 1). The initial symptoms in most MPS IVA patients have been noticed by the age of three years. The MPS IVA patient is usually evaluated during the second or third year of life for unusual skeletal features. These include short trunk dwarfism, pectus carinatum, kyphosis, gibbus, scoliosis, genu valgus, flaring of the lower ribs, hypermobile joints, and abnormal gait with a tendency to fall (Figure 2).

Patients with MPS IVA can usually be clinically distinguished from patients with other MPSs because they preserve intelligence and have additional skeletal manifestations derived from a unique spondyloepiphyseal dysplasia and ligamentous laxity. These skeletal manifestations include odontoid hypoplasia, a striking short trunk dwarfism with a short neck, coxa valga, and genu valgus. Odontoid hypoplasia is the most critical skeletal feature to recognize in any patient with MPS IV. This in combination with ligamentous laxity and extradural mucopolysaccharide deposition, results in atlantoaxial subluxation, with consequential quadriparesis or even death.

Other potential complications include cervical myelopathy, pulmonary compromise, valvular (aortic and/or mitral regurgitation) and coronary heart disease, hearing loss, hepatomegaly, fine corneal clouding, coarse facial features (milder than in either MPS I or MPS II), and widely spaced teeth with abnormally thin enamel and frequent caries formation. A history of exercise intolerance in patients with MPS IV often predicts the presence of occult cervical myelopathy, which can also cause bowel and bladder dysfunction and compression of the spinal cord leading to weakness or paralysis. Patients with a severe form, primarily related to cervical instability and pulmonary compromise, often do not survive beyond the second or third decade of life. Paralysis from the myelopathy, restrictive chest wall movement, and valvular heart disease all contribute to their shortened life span. Patients with mild manifestations of MPS IVA have been reported to survive into the seventh decade of life. Length of survival may improve with the improved comprehensive care available to these patients today. Further natural history of MPS IVA is available through www.morquio.com.

MPS IVB was initially considered to be the attenuated form of Morquio syndrome because the progression of skeletal dysplasias and stunting of growth were less pronounced than in MPS IVA. However, subsequent reports have described patients with MPS IVB whose clinical disease was as severe as that seen in the severe form of MPS IV A. In addition, patients with an attenuated form of MPS IVA have been reported, having almost normal stature, mild skeletal abnormalities with dysplastic hips, corneal clouding, and absent keratosulfaturia (Fujimoto and Horwitz, 1983). Thus, the severe and attenuated forms of MPS IV are not caused by unique enzyme deficiencies, but each enzyme deficiency has a wide spectrum of clinical manifestations.

2.2 Pathogenesis

KS is predominantly found in cartilage and cornea, the major organs affected in Morquio syndrome. The pathogenesis of the bone dysplasia is largely unknown but it is believed that the accumulation of KS could be toxic to osteoblasts (Fang-Kircher et al, 1997). The specific mechanism(s) by which excess storage of KS results in the skeletal dysplasia unique to MPS IV remains unknown. The biology of KS is under investigation. Numerous KS-containing proteins have been identified, and the elucidation of their functional roles will provide a better understanding of the pathophysiology of MPS IV. HS and DS have a more generalized tissue distribution. Their normal metabolism in patients with Morquio syndrome spares these patients from mental retardation and disease manifestations observed in other types of MPS.

3 BIOCHEMICAL AND MOLECULAR DIAGNOSIS

3.1 GAG Assay

Urine spot tests are readily available to screen for mucopolysaccharides (GAGs). These tests are associated with false-negative results in MPS IV patients because mildly affected patients do not always excrete KS fragments largely; therefore, testing more than one urine sample is recommended. More accurate semiquantification of urinary GAG can be obtained by spectrophotometric assays with dimethylmethylene blue although some overlaps with normal values (10–20%) are observed (Whitley et al., 2002; Tomatsu et al., 2004a).

3.2 KS Assay

KS is the only glycosaminoglycan that does not contain uronic acid. Instead, galactose residues, mostly sulfated, alternate with sulfated N-acetylglucosamine residues. Inability to degrade KS gives rise to MPS IV. As with the other GAGs, KS is degraded sequentially from the nonreducing end by the action of glycosidases and sulfatases, as shown in Figure 3.

The ELISA for KS in blood and urine distinguished MPS IVA patients from the controls with the exception of a few cases, showing that blood KS level varied with age and clinical severity and that blood KS levels in both MPS IVA and controls peaked between five and ten years of age and declined with age (Figure 4). Urine KS level also varied with age and clinical severity and the severe MPS IVA phenotype was associated with greater urine KS excretion than the attenuated one (Figure 5; Tomatsu et al., 2004a).

Patients with high levels of KS in the blood and urine likely represent severe cases of MPS IVA; in these patients, cartilage is overloaded with undegraded KS. Between ages five and ten, the mean blood KS concentration in MPS IVA patients was the highest. Using the ELISA method, the incidence of nonkeratosulfaturic MPS IVA would probably be lower than reported previously. These findings indicate that the assay for blood or urine KS may be suitable for early diagnosis and longitudinal assessment of disease severity in MPS IVA

3.3 Enzyme Assay

In view of the specificity and reliability of enzyme assay, it is generally more efficient to omit extensive analysis of urinary GAGs. Definitive diagnosis of the MPS is established by enzyme assays in leukocytes or fibroblasts. The fluorogenic (4-methylumbelliferyl) substrates relevant to MPS IVA and B are now available, reproducible, and reliable (Lund-Hansen, Hoyer, and Andersen, 1984; van Diggelen et al., 1990) although there is no definite correlation between the residual activity and clinical severity. For prenatal diagnosis, the enzyme activity can be measured in amniocytes or chorionic villi. Determination of the carrier state by enzyme analysis is not always possible because the range of enzyme activity in noncarrier and carrier individuals overlaps. Detection of mutations in the GALNS and β-galactosidase genes can facilitate carrier testing if the family desires.

The sulfatase that cleaves the sulfate from galactose residues is deficient in patients with MPS IVA. The enzyme and the deficiency were discovered by use of oligosaccharide substrates that had been prepared from chondroitin 6-sulfate (C6S) and therefore contained sulfated N-acetylgalactosamine; for that reason, the enzyme was originally named N-acetylgalactosamine 6-sulfate sulfatase. Subsequently, it was shown that the enzyme was specific for the 6-sulfated galactose residues of KS and the 6-sulfated N-acetylgalactosamine residues of C6S. The degradation of the latter polymer is also impaired in MPS IVA. The enzyme has been purified to homogeneity (Masue et al., 1991); its cDNA encodes 522 amino acids, including a 26 amino acid signal sequence, and bears marked similarity to other sulfatases (Tomatsu et al., 1991).

The galactose residues of keratan sulfate are removed by β-galactosidase. The sequence of its cDNA predicts a protein of 677 amino acids, including a 23 amino acid signal peptide (Oshima et al., 1988). Newly made β-galactosidase undergoes the processing characteristic of lysosomal enzymes, but in addition undergoes aggregation to a multimer of ca. ~600 kDa. The normal enzyme hydrolyzes terminal β-linked galactose residues found in GM1 ganglioside, glycoproteins, and oligosaccharides, as well as in KS. Total absence of β-galactosidase activity results in GM1 gangliosidosis, whereas MPS IVB results from mutations that selectively impair catalytic activity toward KS.

3.4 Clinical Tests

A full skeletal survey should be obtained in a patient thought to have MPS IV. The following radiographic studies are recommended: standing anteroposterior (AP) and lateral views of flexion and extension radiographs of the cervical spine, the odontoid process, the chest, entire spine, pelvis view with visualization of the femoral heads articulating with the acetabulum, the lower extremities including the entire femur, articulation with tibia, and ankles, AP views of both hands, forearms, elbows in extension, humerus, and shoulder.

MRI of the brain stem and cervical spine should be performed to evaluate odontoid hypoplasia and cord compression. An ophthalmology examination with slit lamp should be performed at the time of initial evaluation to look for corneal clouding.

4 GENETICS

4.1 Incidence

MPS IV is a rare disorder, and epidemiological data are scarce. The earlier British Columbia survey in 1971 reported the incidence of MPS IV to be 1 per 300,000 total live births and later in 1990 increased to 1 per 200,000 live births. It was reported to be 1 per 75,000 live births in Northern Ireland, based on ascertained cases over 30 years. In Japan, the incidence was 1 per 100,000 live births. Discrepancy of incidence in different ethnic populations has been derived partly from the common founder effect. Majority of MPS IV patients are MPS IVA (over 95%) although incidence of MPS IVB is unknown.

4.2 Genes and Mutations

The GALNS gene is located on chromosome arm 16q24.3 and encodes a 522–amino acid protein that is stabilized in a complex with two other lysosomal enzymes (β-galactosidase and α-neuraminidase) and the protective protein cathepsin A. The assembly of these four components is necessary for correct posttranslation processing and stability of the component enzymes and for the efficient catabolism of KS. Investigations of the allelic heterogeneity in MPS IVA were facilitated by the isolation and characterization of the full-length cDNA encoding the human GALNS protein and the genomic GALNS gene (Tomatsu et al., 1991; Nakashima et al., 1994).

To date, about 130 different mutations and seven polymorphisms causing an amino acid change have been identified. This heterogeneity in GALNS gene mutations accounts for an extensive clinical variability within MPS IVA (Bunge et al., 1997; Fukuda et al., 1992; Hori et al., 1995; Kato et al., 1997; Ogawa et al., 1995; Tomatsu et al., 1995ab, 1997; Yamada et al., 1998; Montaño et al., 2003; Terzioglu et al., 2002). Genotype/ phenotype correlation exists for some of these mutations.

The GALNS structural modeling was designed based on homology modeling of related sulfatases. The tertiary structural model of the human GALNS protein indicates that the severe mutations are associated with the peptide(s) located at the core of the structure leading to destruction of the hydrophobic domain, modification of the packing, or modification of the active site. In contrast, mild mutations are mostly associated with peptide abnormalities located on the surface of the GALNS native protein (Sukegawa et al., 2000). Several common mutations have been found in Irish (pI113F, pT312S), Italian (pM1V, pW10X), Japanese (pN204K, double gene deletion), and panethnic (pR386C). A large proportion (over 70%) of known lesions derive from missense mutations.

The MPS IVB is caused by a deficiency of β-galactosidase that affects the degradation of KS without affecting the degradation of GM1 ganglioside in neuronal cells, as the latter would cause GM1 gangliosidosis. The nature of the mutations that permit such selectivity is an important and as yet unanswered question. The missense mutation Trp273Leu, which appears to be the cause of Morquio syndrome in patients from two unrelated families, was identified (Oshima et al., 1991).

5 MANAGEMENT

5.1 General

Typical skeletal anomalies of Morquio syndrome include platyspondyly, odontoid hypoplasia, kyphosis, hyperlordosis, scoliosis, ovoid deformities of the vertebrae, genu valgum, ulnar deviation of the wrist, valgus deformity of the elbow, inclinations of distal ends of radium and ulna toward each other, deformities of metacarpals and short phalanges, epiphyseal deformities of the tubular bones, widened metaphyses, and osteoporosis (Figure 1). Because most MPS IV patients are not candidates for specific therapies, management consists of supportive care and treatment of complications, even as specific therapies are being developed (see below). The progressive nature of bone involvement in MPS IV dictates the need for continual evaluation of their clinical status. Systematic evaluation of hearing, vision, and joint function coupled with treatment of specific problems can lead to improved quality of life by minimizing the handicapping effects of diffuse systemic disease.

5.2 Vision

Corneal clouding is common in MPS IV, and can lead to significant visual disability. Corneal transplantation has been performed, but the long-term outcome is not always successful.

5.3 Hearing

Deafness, usually of combined conductive and neurosensory origin, occurs commonly in MPS IV. The deafness has been attributed to three causes: frequent middle ear infections, deformity of the ossicles, and probable abnormalities of the inner ear. Auditory brainstem response is abnormal in a nonspecific way, probably reflecting a mixture of middle-ear, cochlear, eighth-nerve, and lower-brain-stem anomalies. Ventilating tubes can minimize the long-term sequelae of the frequent episodes of acute otitis media and chronic middle ear effusions. The majority of MPS IV patients have significant hearing loss and would benefit from hearing aids.

5.4 Hyperlaxity of Joints

Morquio syndrome is unique in having ligamentous laxity unlike other MPS diseases. The origin of abnormal joint function still remains unknown, presumably derived from a combination of metaphyseal deformities, hypoplasia of the bones, and degradation of connective tissues near the joint secondary to GAG accumulation. Range-of-motion exercises, swimming, and computer typing appear to offer some benefits in preserving joint function fine motor skills and should be started early in the clinical course. The indication of physical therapy and its benefits in MPS IV should be further studied.

5.5 Obstructive Airway Disease

Obstructive airway disease is probably common in the severely involved MPS IV patients, but sleep studies to determine the extent of airway compromise are not routinely performed. A narrowed trachea, thickened vocal cords, redundant tissue in the upper airway, and an enlarged tongue all contribute to airway obstruction (Dullenkopf et al.,

2002). Intermittent obstruction in severely involved MPS IV patients is common and may lead to sleep apnea. Obstructive apnea has been reported in MPS IV patients who had loud snoring and daytime sleepiness. Tracheostomy can produce a dramatic symptomatic improvement in patients with obstructive apnea, but experience is limited. Obstructive sleep apnea was successfully treated with high-pressure nasal continuous positive airway pressure and supplemental oxygen. Tonsillectomy and adenoidectomy are frequently performed to correct eustachian tube dysfunction and decrease airway obstruction.

5.6 Anesthesia

Patients with MPS IV present major anesthesia risks. In addition to chest wall deformities, MPS IV patients have hypoplasia of the dens (odontoid process) as a consistent finding, putting them at considerable risk of anterior dislocation of the vertebral axis with resultant spinal cord compression. This can occur during head positioning for tracheal intubation. Thus, head extension should be avoided by having an assistant hold the head in the neutral position during laryngoscopy. However, manual in-line stabilization of the head and neck can make direct laryngoscopic intubation difficult. Induction of anesthesia can be difficult because of inability to maintain an adequate airway. Visualization can be limited during intubation. A new device to facilitate this task is the angulated video-intubation laryngoscope (AVIL). The AVIL may become a helpful device to aid endotracheal intubation in patients when cervical spine immobilization impairs direct laryngoscopy (Dullenkopf et al., 2002). Recovery from anesthesia may be slow, and postoperative airway obstruction is common. Death has been reported as a result of anesthesia complication.

5.7 Cardiovascular Disease

Clinical evidence of heart disease occurs in adult MPS IV patients. Valvular disease, myocardial thickening, systemic and pulmonary hypertension, and narrowing of the coronary arteries with ischemia contribute to congestive heart failure. Aortic valvular disease is more likely to occur in MPS IV. Cardiac evaluation at regular intervals with echocardiography is useful in the treatment of patients through serial monitoring of ventricular function and size. Bacterial endocarditis prophylaxis should be advised for the MPS IV patient with cardiac abnormalities.

5.8 Dental Issues

Attention to daily oral hygiene and professional dental cleaning and evaluation every 6–12 months is necessary to minimize the effects of thin dental enamel.

5.9 Diet

No specific dietary restrictions exist for patients with MPS IV. On a practical level, these patients should avoid excess body weight to minimize pulmonary compromise caused by the skeletal deformities.

5.10 Activity

MPS IV patients can participate in activities as tolerated with a few important restrictions. Contact sports could damage the cervical spine and should be avoided. Repetitive motions at work or with sports could strain abnormal joints and should also be avoided. Swimming and computer activity are recommended.

6 MORTALITY/MORBIDITY

Mortality and morbidity rates are primarily related to the atlantoaxial instability and subsequent cervical myelopathy. A minor fall or extension of the neck can result in cord transection and subsequent quadriparesis or death. The cervical myelopathy can cause bowel and bladder dysfunction and apnea. Obstructive sleep apnea can cause prolonged periods of hypoxia, pulmonary hypertension, and even death. Airway obstruction also occurs secondary to thickening of tissue in the upper airway from mucopolysaccharide deposition. Patients with MPS IV have a predisposition to pulmonary infection because of progressive truncal deformity and immobility, namely, restricted lung capacity. Early-onset coronary heart disease and valve thickening (aortic and mitral) with resultant cardiac dysfunction are described, and endocarditis prophylaxis is recommended. Corneal clouding can cause visual disturbance and photophobia. Enamel abnormalities predispose them to dental caries.

7 ANIMAL MODELS

Recently, the MPS IVA knockout mouse has been established (Tomatsu et al., 2003a). Homozygous *Galns*$^{-/-}$ mice have no detectable GALNS enzyme activity and show increased urinary glycosaminoglycan (GAGs) levels. These mice accumulate GAGs in multiple tissues including liver, kidney, spleen, heart, brain, and bone marrow. At 2 months old, lysosomal storage is present primarily within reticuloendothelial cells such as Kupffer cells and cells of the sinusoidal lining of the spleen. In addition, by 12 months old, vacuolar change is observed in the visceral epithelial cells of glomeruli, cells at the base of heart valves, and the chondrocytes and osteoblasts in bone but it is not present in parenchymal cells such as hepatocytes and renal tubular epithelial cells. In the brain, hippocampal and neocortical neurons and meningeal cells had lysosomal storage. KS and C6S were more abundant in the cytoplasm of corneal epithelial cells of *Galns*$^{-/-}$ mice compared with wild-type mice by immunohistochemistry. Radiographs revealed no change in the skeletal bones of mice up to 12 months old. Thus, targeted disruption of the murine *Galns* gene has produced a murine model that shows storage of GAGs in visceral organs and the bone. Overall, clinical manifestations in the affected mice are milder than those in human patients.

The complete absence of GALNS in mutant mice enables us to study pharmacokinetics and tissue targeting of recombinant GALNS designed for enzyme replacement. However, cellular and humoral immune responses to the injected enzymes have been recognized as major impediments to evaluating experimental ERT on those knockout murine models. To solve this immunologic problem and to study the long-term effectiveness and side effects of therapies without immune responses, MPS IVA mice tolerant to human GALNS have been created as MPS VII-tolerant mice (Tomatsu et al., 2003b; Tomatsu et al., unpublished). The C79S mutation at the human *GALNS* cDNA and the corresponding C76S mutation at the exon 2 of the mouse *GALNS* gene by in vitro mutagenesis were introduced into the targeting vector.

Availability of these MPS IVA murine models is of particular value for testing new therapies, such as bone marrow transplantation, enzyme replacement, and gene transfer.

8 DEVELOPMENT OF THERAPIES

Only palliative measures are currently available for treatment of patients with MPS IV. Potential strategies for treatment of patients with the other MPSs, currently at different levels of development, include enzyme replacement, gene therapy, and allogenic bone marrow transplantation in which engrafted cells provide the normal enzyme. No medications are available to prevent, treat, or cure Morquio syndrome, therefore, supportive measures are used to treat the manifestations of this disorder. These include nonsteroidal anti-inflammatory drugs (NSAIDs) for joint pain, antibiotics for pulmonary infections, and oxygen for pulmonary compromise or obstructive sleep apnea. Medications for supportive care, such as NSAIDs for joint pain, antibiotics for pulmonary infections, and oxygen for pulmonary compromise and obstructive sleep apnea, can be used to treat the manifestations of this disorder.

8.1 Surgical Treatment

Admission for surgical interventions may be required. These procedures include femoral osteotomies, corrective knee surgery for severe genu valgus deformity, and cervical spinal fusion. Because all patients with MPS IV have odontoid hypoplasia that can lead to atlantoaxial subluxation, many physicians recommend cervical spine fusion for all of their affected patients. Some physicians recommend fusion of C1 and C2, whereas others recommend fusion of C1 and C2 and occipital fusion. Patients undergoing cervical fusion wear a halo brace for an extended period after the surgery. Ideally, the surgery should be performed before signs and symptoms of cervical myelopathy occur. However, because of the high risk of this procedure, some elect not to have this surgery.

Other potential operations for patients with Morquio syndrome include femoral osteotomies and corrective knee surgery for genu valgus deformity. Total joint replacement of hips and/or knees may be necessary. The early use of a back brace may delay or prevent surgical intervention for scoliosis.

A preoperative evaluation should be pursued, especially if evidence of cardiac dysfunction, obstructive apnea, or pulmonary insufficiency is present.

8.2 Therapies by Exogenously Supplied Enzymes

The glycosylation sites of lysosomal enzymes contain mannose 6-phosphate residues (M6P), which target the protein into the lysosome. The M6P moiety binds to a specific M6P receptor in the Golgi and is thus directed to prelysosomal compartments. Enzymes that escape this routing system are secreted by the cell via the constitutive secretory pathway and are often recaptured by cell surface M6P receptors that return it to the lysosome by the endocytic pathway (Kornfeld and Mellman, 1989). Thereby, the MPS and other lysosomal storage diseases (LSDs) have been considered potentially amenable to therapy by exogenously supplied enzymes. The modest success in altering the course of some MPS by bone marrow transplantation (BMT) suggests that enzyme produced by cells from the donor bone marrow can reduce the storage of GAG.

8.3 Bone Marrow Transplantation (BMT)

The clinical outcome of BMT in MPS patients has varied considerably. Factors that affect the outcome of BMT include the type of the MPS disorder, the type of donor, the type of preparative regimen, the degree of clinical involvement, and the age at the time of transplantation. Several studies on patients with MPS IVA by allogeneic BMT have been done. Previous studies showed that allogeneic BMT had a limited benefit in the treatment of Morquio's disease. Most MPS IVA patients had serious skeletal deformities and bad muscular habits before the transplant, and replaced enzyme could not correct such deformities or corneal clouding, even if normal range of peripheral blood GALNS activity was achieved.

However, the quality of life in some young MPS IVA patients is markedly better compared to older patients (personal communication with Dr. Yabe, Tokai University). Therefore, earlier transplant may be able to control the deterioration of skeletal deformity. The allogeneic BMT in other lysosomal storage diseases abated the visceral features such as hepatosplenomegaly, cardiac hypertrophy, and upper airway obstruction. However, regarding the skeletal symptoms, none of the patients showed improvement although most of the recipients achieved stabilization. Further study and long-term observation are necessary to assess the efficacy of allogeneic BMT in the treatment of Morquio's disease. The ideal age at which BMT should be performed is not known for MPS IVA, but the best outcome would be anticipated if the BMT were performed prior to significant skeletal dysplasia.

BMT is a major procedure that carries a high risk of mortality and morbidity from graft-versus-host disease and other complications. Its degree of efficacy has not yet been analyzed. The three-year survival rate in MPS patients after BMT is over 90%

8.4 Gene Therapy

The high risk of allogeneic BMT provides an impetus for the development of somatic gene therapy, using retroviral vectors to introduce cDNA encoding the missing lysosomal enzyme into the patient's fibroblasts. Only one report for MPS IVA has been described by using a recombinant retroviral vector containing a full-length human wild-type GALNS cDNA (Toietta et al., 2001). GALNS activity in MPS IVA transduced fibroblasts was severalfold higher than normal values. To measure the variability of GALNS expression among different transduced cells, lymphoblastoid B cells, human keratinocytes, murine myoblasts, and rabbit synoviocytes were transduced as well. In all cases, an increase of GALNS activity after transduction was confirmed. By the experiment of MPS IVA cells cocultivated with enzyme-deficient transduced cells, the enzyme uptake was mannose-6-phosphate dependent. Furthermore, retrovirus transduction into MPS IVA fibroblasts led to correction of the metabolic defect. These results provide the evidence that GALNS may be delivered either locally or systematically by various cells in an ex vivo gene therapy of MPS IVA. Further experiments using MPS IVA mice models are needed.

8.5 Enzyme Replacement Therapy

ERTs are in clinical trials for several lysosomal storage diseases including MPS I (Kakkis et al., 2001), MPS II, and MPS VI. Those ERTs revealed dramatic improvements in visceral organs but little or no improvements in bone and brain because the enzymes are not delivered to these tissues effectively. To improve quality of life, the enzyme must be maximally delivered to the bones to clear the storage materials.. Efficiency of GALNS delivery to the bone was compared among native (simple recombinant), sulfatase modifier factor (SUMF1) coexpressed (fully active sulfatase enzyme; Cosma et al., 2003; Dierks et al., 2003), and N-terminal bone targeting (NBT: six glutamines were attached to N-terminus of mature protein) enzymes. Pharmacokinetics and tissue distribution of purified enzymes and pathological improvements were assessed by using MPS IVA knockout mice (Tomatsu et al., 2004b, unpublished). All three enzymes cleared storage materials in visceral organs. SUMF 1 coexpressed GALNS provided 3–4 times more specific activity than other two enzymes. Native and SUMF-1 coexpressed GALNS had the same biphasic pattern of enzyme clearance in blood and tissue distribution, whereas NBT GALNS had ten times more prolonged clearance in blood in a monophasic manner. NBT GALNS changed tissue distribution of enzyme with a long retention of enzyme activity in tissues including the bone. These results suggest that a newly designed bone-targeting system is a potential treatment for MPS IVA.

REFERENCES

Bond, C.S., Clements, P.R., Ashby, S.J., Collyer, C.A., Harrop, S.J., Hopwood, J.J., and Guss, J.M., 1997, Structure of a human lysosomal sulfatase, *Structure* **5**:277–289.

Brante, G., 1952, Gargoylism, a mucopolysaccharidosis, *Scand. J. Clin. Lab. Invest.* **4**:43.

Bunge, S., Kleijer, W.J., Tylki-Szymanska, A., Steglich, C., Beck, M., Tomatsu, S., Fukuda, S., Poorthuis, B.J., Czartoryska, B., Orii, T., and Gal, A., 1997, Identification of 31 novel mutations in the N-acetylgalactosamine-6-sulfatase gene reveals excessive allelic heterogeneity among patients with Morquio A syndrome, *Hum. Mutat.* **10**:223–232.

Cosma, M.P., Pepe, S., Annunziata, I., Newbold, R.F., Grompe, M., Parenti, G., and Ballabio, A., 2003, The multiple sulfatase deficiency gene encodes an essential and limiting factor for the activity of sulfatases, *Cell.* **113**:445–456.

Dierks, T., Schmidt, B., Borissenko, L.V., Peng, J., Preusser, A., Mariappan, M., and von Figura, K., 2003, Multiple sulfatase deficiency is caused by mutations in the gene encoding the human C (alpha)-formylglycine generating enzyme, *Cell.* **113**:435–444.

Dullenkopf, A., Holzmann, D., Feurer, R., Gerber, A., and Weiss, M., 2002, Tracheal intubation in children with Morquio syndrome using the angulated video-intubation laryngoscope, *Can J Anaesth.* **49**:198–202.

Fang-Kircher, S.G., Herkner, K., Windhager, R., and Lubec, G., 1997, The effects of acid glycosaminoglycans of neonatal calvarian cultures- A role of keratan sulfate in Morquio syndrome?, *Life Sci.* **61**:771–775.

Fujimoto, A., and Horwitz, A.L., 1983, Biochemical defect of non-keratan-sulfate-excreting Morquio syndrome, *Am. J. Med. Genet.* **15**:265–273.

Fukuda, S., Tomatsu, S., Masue, M., Sukegawa, K., Iwata, H., Ogawa, T., Nakashima, Y., Hori, T., Yamagishi, A., Hanyu, Y., Morooka, K., Kiman, T., Hashimoto, T., and Orii, T., 1992, Mucopolysaccharidosis type IVA: N-Acetylgalactosamine-6-sulfate sulfatase exonic point mutations in classical Morquio and mild cases, *J. Clin. Invest.* **90**:1049–1053.

Hori, T., Tomatsu, S., Nakashima, Y., Uchiyama, A., Fukuda, S., Sukegawa, K., Shimozawa, N., Suzuki, Y., Kondo, N., and Horiuchi, T., Ogura, S., and Orii, T., 1995, Mucopoly-accharidosis type IVA: Common double deletion at the N-acetylgalactosamine-6-sulfate sulfatase gene, *Genomics.* **26**:535–542.

Kakkis, E.D., Muenzer, J., Tiller, G.E., Waber, L., Belmont, J., Passage, M., Izykowski, B., Phillips, J., Doroshow, R., Walot, I., Hoft, R., and Neufeld, E.F., 2001, Enzyme-replacement therapy in mucopolysaccharidosis I, *N Engl J Med.* **344**:182–188.

Kato, Z., Fukuda, S., Tomatsu, S., Vega, H., Yasunaga, T., Yamagishi, A., Yamada, N., Valencia, A., Barrera, L.A., Sukegawa, K., Orii, T., and Kondo, N., 1997, A novel common missense mutation G301C in the N-acetylgalactosamine-6-sulfate sulfatase gene in mucopolysaccharidosis IVA, *Hum. Genet.* **101**:97–101.

Kornfeld, S., and Mellman, I., 1989, The biogenesis of lysosomes, *Annu Rev Cell Biol.* **5**:483–525.

Lund-Hansen, T., Hoyer, P.E., Andersen, H., 1984, A quantitative cytochemical assay of beta-galactosidase in single cultured human skin fibroblasts, *Histochemistry* **81**:321–330.

Masue, M., Sukegawa, K., Orii, T. and Hashimoto, T., 1991, N-acetylgalactosamine-6-sulfate sulfatase in human placenta: purification and characteristics, *J Biochem (Tokyo).* **110**:965–970.

Montaño, A.M., Kaitila, I., Sukegawa, K., Tomatsu, S., Kato, Z., Nakamura, H., Fukuda, S., Orii, T., and Kondo, N., 2003, Mucopolysaccharidosis IVA: characterization of a common mutation found in Finnish patients with attenuated phenotype, *Hum. Genet.* **113**:162–169.

Morquio, L., 1929, Sur une forme de dystrophie osseuse familiale, *Bull. Soc. Pediat. Paris* **27**:145.

Nakashima, Y., Tomatsu, S., Hori, T., Fukuda, S., Sukegawa, K., Kondo, N., Suzuki, Y., Shimozawa, Y., and Orii, T., 1994, Mucopolysaccharidosis IVA: molecular cloning of the human N-acetylgalactosamine 6-sulfatase (GALNS) gene and analysis of the 5'-flanking region, *Genomics.* **20**:99–104.

Ogawa, T., Tomatsu, S., Fukuda, S., Yamagishi, A., Rezvi, G.M., Sukegawa, K., Kondo, N., Suzuki, Y., Shimozawa, N., and Orii, T., 1995, Mucopolysaccharidosis IVA: screening and identification of mutations of the N-acetyl galactosamine-6-sulfate sulfatase gene, *Hum. Mol. Genet.* **4**:341– 349.

Oshima, A., Tsuji, A., Nagao, Y., Sakuraba, H., Suzuki, Y., 1988, Cloning, sequencing, and expression of cDNA for human beta-galactosidase, *Biochem. Biophys. Res. Commun.* 1988, **157**:238–244.

Oshima, A., Yoshida, K., Shimmoto, M., Fukuhara, Y., Sakuraba, H., and Suzuki, Y., 1991, Human beta-galactosidase gene mutations in morquio B disease, *Am J Hum Genet.* **49**:1091–1093.

Schmidt, B., Selmer, T., Ingendoh, A., and von Figura, K., 1995, A novel amino acid modification in sulfatases that is defective in multiple sulfatase deficiency, *Cell* **82**:271–278.

Singh, J., DiFerrante, N.M., Niebes, P., and Travella, D., 1976, N-acetylgalactosamine-6-sulfate sulfatase in man: absence of the enzyme in Morquio disease, *J. Clin. Invest.* **57**:1036.

Sukegawa, K., Nakamura, H., Kato, Z., Tomatsu, S., Montaño, A.M., Fukao, T., Toietta, G., Tortora, P., Orii. T., and Kondo, N., 2000, Biochemical and structural analysis of missense mutations in N-acetylgalactosamine-6-sulfate sulfatase causing mucopolysaccharidosis IVA phenotypes, *Hum. Mol. Genet.* **22**:1283–1290.

Terzioglu, M., Tokatli, A., Coskun, T., and Emre, S., 2002, Molecular analysis of Turkish mucopolysaccharidosis IVA (Morquio A) patients: Identification of novel mutations in the N-acetylgalactosamine-6-sulfate sulfatase (GALNS) gene, *Hum. Mutat.* **20**:477– 478.

Toietta, G., Severini, G.M., Traversari, C., Tomatsu, S., Sukegawa, K., Fukuda, S., Kondo, N., Tortora, P., and Bordignon, C., 2001, Various cells retrovirally transduced with N-acetylgalactosoamine-6-sulfate sulfatase correct Morquio skin fibroblasts in vitro, *Hum Gene Ther.* **12**:2007–2016.

Tomatsu, S., Fukuda, S., Cooper, A., Wraith, J.E., Ferreira, P., Di Natale, P., Tortora, P., Fujimoto, A., Kato, Z., Yamada, N., Isogai, K., Yamagishi, A., Sukegawa, K., Suzuki, Y., Shimozawa, N., Kondo, N., Sly, W.S., and Orii, T., 1997, Fourteen novel mucopolysaccharidosis IVA producing mutations in GALNS gene, *Hum. Mutat.* **10**:368– 375.

Tomatsu, S., Fukuda, S., Cooper, A., Wraith, J.E., Rezvi, G.M., Yamagishi, A., Yamada, N., Kato, Z., Isogai, K., Sukegawa, K., Kondo, N., Suzuki, Y., Shimozawa, N., and Orii, T., 1995a, Mucopolysaccharidosis IVA: Identification of a common missense mutation I113F in the N-acetylgalactosamine-6-sulfate sulfatase gene, *Am. J. Hum. Genet.* **57**:556–563.

Tomatsu, S., Fukuda, S., Cooper, A., Wraith, J.E., Rezvi, G.M., Yamagishi, A., Yamada, N., Kato, Z., Isogai, K., and Sukegawa, K., Kondo, N., Suzuki, Y., Shimozawa, N., and Orii, T., 1995b, Mucopolysaccharidosis type IVA: Identification of six novel mutations among non-Japanese patients, *Hum. Mol. Genet.* **4**:741– 743.

Tomatsu, S., Fukuda, S., Masue, M., Sukegawa, K., Fukao, T., Yamagishi, A., Hori, T., Iwata, H., Ogawa, T., Nakashima, Y., Hanyu, Y., Hashimoto, T., Titani, K., Oyama, R., Suzuki, M., Yagi, K., Hayashi, Y., and Orii, T., 1991, Morquio disease: Isolation, characterization and expression of full-length cDNA for human N-acetylgalactos-amine-6-sulfate sulfatase, *Biochem. Biophys. Res. Commun.* **181**:677–683.

Tomatsu, S., Gutierrez, M.A., Nishioka, T., Pena, O., 2004b, Enzyme replacement therapy for mucopolysaccharidosis IVA: Development of bone targeting system, *The American Society of Human Genetics, 54th Annual Meeting, Toronto, Ontario,Canada* October 26–30: p. 71.

Tomatsu, S., Okamura, K., Taketani, T., Orii, K.O., Nishioka, T., Gutierrez, M.A., Velez-Castrillon, S., Fachel, A.A., Grubb, J.H., Cooper, A., Thornley, M., Wraith, E., Barrera, L.A., Giugliani, R., Schwartz, I.V., Frenking, G.S., Beck, M., Kircher, S.G., Paschke, E., Yamaguchi, S., Ullrich, K., Isogai, K., Suzuki, Y., Orii, T., Kondo, N., Creer, M., and Noguchi, A., 2004a, Development and testing of new screening method for keratan sulfate in mucopolysaccharidosis IVA, *Pediatr Res.* **5**:592–597.

Tomatsu, S., Orii, K.O., Vogler, C., Grubb, J.H., Snella, E.M., Gutierrez, M., Dieter, T., Holden, C.C., Sukegawa, K., Orii, T., Kondo, N., and Sly, W.S., 2003a, Production of MPS VII mouse (Gus(tm(hE540A x mE536A)Sly)) doubly tolerant to human and mouse beta-glucuronidase, *Hum Mol Genet.* **12**:961–973.

Tomatsu, S., Orii, K.O., Vogler, C., Nakayama, J., Levy, B., Grubb, J.H., Gutierrez, M.A., Shim, S., Yamaguchi, S., Nishioka, T., Montaño, A.M., Noguchi, A., Orii, T., Kondo, N., and Sly, W.S., 2003b, Mouse model of N-acetylgalactosamine-6-sulfate sulfatase deficiency (Galns-/-) produced by targeted disruption of the gene defective in Morquio A disease, *Hum Mol Genet.* **12**:3349–3358.

van Diggelen, O.P., Zhao, H., Kleijer, W.J., Janse, H.C., Poorthuis, B.J., vanPelt, J., Kamerling, J.P., and Galjaard, H., 1990, A fluorimetric enzyme assay for the diagnosis of Morquio disease type A (MPS IV A), *Clin. Chim. Acta* **187**:131–139.

Whitley, C.B., Spielmann, R.C., Herro, G., and Teragawa, S.S., 2002, Urinary glycos-aminoglycan excretion quantified by an automated semimicro method in specimens conveniently transported from around the globe, *Mol Genet Metab.* **75**:56–64.

Yamada, N., Fukuda, S., Tomatsu, S., Muller, V., Hopwood, J.J., Nelson, J., Kato, Z., Yamagishi, A., Sukegawa, K., Kondo N., and Orii, T., 1998, Molecular hetero-geneity in mucopolysaccharidosis IVA in Australia and Northern Ireland: Nine novel mutations including T312S, a common allele that confers a mild phenotype, *Hum. Mutat.* **11**:202–208.

MUCOPOLYSACCHARIDOSIS TYPE VI (MPS VI, Maroteaux–Lamy Syndrome)

J. E. Wraith

1 INTRODUCTION

Mucopolysaccharidosis type VI (MPS VI), or Maroteaux–Lamy syndrome (OMIM 253200) is a rare, autosomal recessive disorder of glycosaminoglycan (GAG) storage resulting from a deficiency of the lysosomal enzyme N-acetylgalactosamine-4-sulfatase (also known as arylsulfatase B, ASB, E.C.3.1.6.1). Affected individuals are unable to catabolise the GAG dermatan sulfate (DS, Figure 1) and this results in the intracellular accumulation of partially degraded GAG in the lysosomes of a wide variety of tissues. This accumulation, by an unknown method, causes a chronic progressive disorder involving multiple organs that can lead to death in early adult life.

The disorder was first described by Maroteaux and colleagues in a 13-year-old child (Maroteaux et al., 1963) as a Hurler-type syndrome with normal intelligence and the excretion of dermatan sulfate (which they called chondroitin sulfate B) alone. Since this initial description many further cases have been described and a very wide clinical phenotype has been described ranging from a severe disorder presenting in the first year of life with marked skeletal and multisystem disease to a disorder presenting in adulthood with very mild symptoms.

2 INCIDENCE AND INHERITANCE

As with many other rare genetic disorders the true incidence of MPS VI is unknown. Estimates have ranged from 1:320,000 in Western Australia (Nelson et al., 2003) to 1: 1,300,000 in British Columbia (Lowry et al., 1990). It is likely that the true prevalence will only be established when advances in therapy result in the need for newborn screening programmes for MPS VI and other MPS disorders.

The gene encoding for N-acetylgalactosamine-4-sulfatase (ARSB, http://uwcmml1s.uwcm.ac.uk/uwcm/mg/search/119008.html) comprises eight exons and is localized to chromosome 5q13-q14 (Modaressi et al., 1993; Litjens et al., 1989). Over 40 mutations have been identified, the majority of which are missense/nonsense nucleotide substitutions. The vast majority of MPS VI mutant alleles are either unique to a patient or are present in only a small number of patients. So far, no common mutations have been described and genotype/phenotype correlation in MPS VI is not currently possible.

J.E. Wraith MB ChB FRCPCH, Consultant Paediatrician, Royal Manchester Children's Hospital, Manchester M27 4HA, United Kingdom.

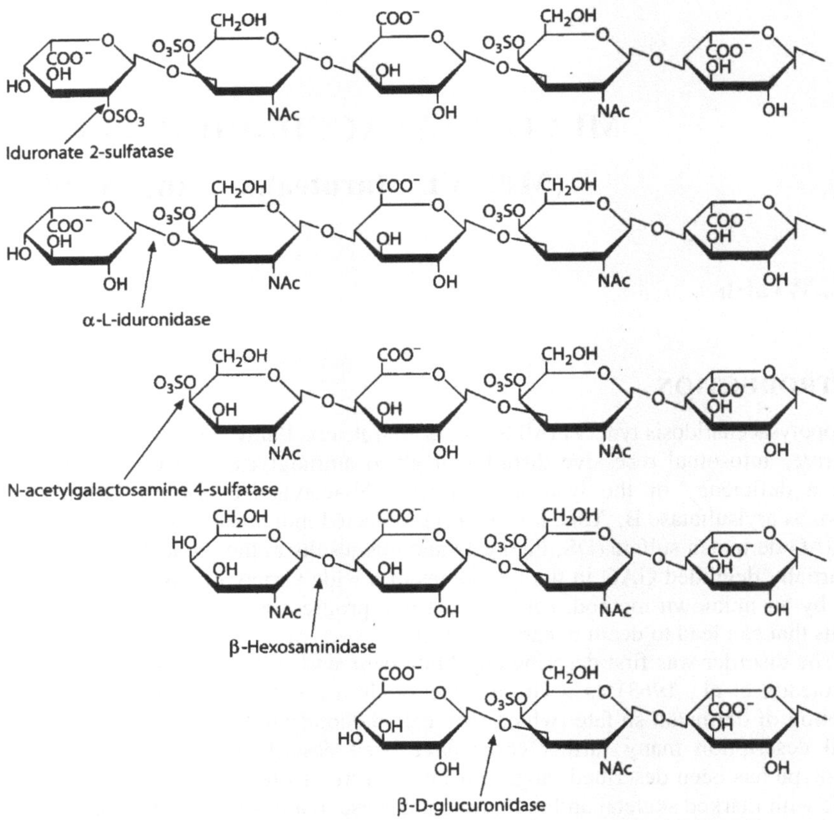

Figure 1. The catabolic pathway of dermatan sulphate.

3 ANIMAL MODELS

There are both feline (Jezyk et al., 1977) and murine (rat; Yoshida et al., 1993) models of MPS VI. The cat model has been extremely well studied and appears to be an excellent model of the human disease. Early trials of enzyme replacement therapy (ERT) in the cat model (Crawley et al., 1996) demonstrated widespread tissue uptake of recombinant N-acetylgalactosamine-4-sulfatase (rh4S) and subsequent work by the same group confirmed a dose-responsive effect of rh4S (Crawley et al., 1997). These studies (and others) were one of the driving forces behind the development of enzyme replacement therapy for humans affected by MPS VI by demonstrating proof of concept in this approach.

Both the cat (Gasper et al., 1984; Norrdin et al., 1995) and the rat (Simonaro et al., 1997) have also been used to study the effects of bone marrow transplantation (BMT) in MPS VI especially the effects on the skeleton which in the human remains resistant to correction following BMT.

4 CLINICAL PRESENTATION IN HUMANS

Irrespective of the age at which signs and symptoms of MPS VI become apparent the disease is present from conception and after birth is characterized by relentless progression.

Patients at the severe end of the clinical spectrum are usually diagnosed in the first 12–18 months of life because of the severe skeletal dysplasia which is a feature of this form of MPS VI. Older children or adults with more attenuated forms of the disease may present with joint stiffness, a cardiac murmur, or mild corneal clouding. Many medical specialties ultimately become involved in the multidisciplinary management of these patients and the effects of the disease on each of the major body systems is briefly presented.

4.1 Cognitive Development

Intellectual impairment is not a direct feature of MPS VI although severe physical limitations and complications such as blindness and deafness can impair an individual patient's ability to learn and develop.

4.2 Facial Dysmorphism

In patients at the severe end of the clinical spectrum facial features resemble those of Hurler disease with coarsening of the facial features secondary to a combination of storage in the soft tissues of the orofacial region and underlying facial bone dysostosis. Thickening of the alae nasi, lips, ear lobules, and tongue becomes progressively more evident and enlargement of the calvarium results in macrocephaly. By the age of 18 months the clinical diagnosis of a mucopolysaccharide disease is usually obvious. In patients with more attenuated forms of the disease the facial abnormalities may be more subtle and between these two extremes there is considerable variation. Figure 2 illustrates the range of facial abnormality that can be seen in this disorder.

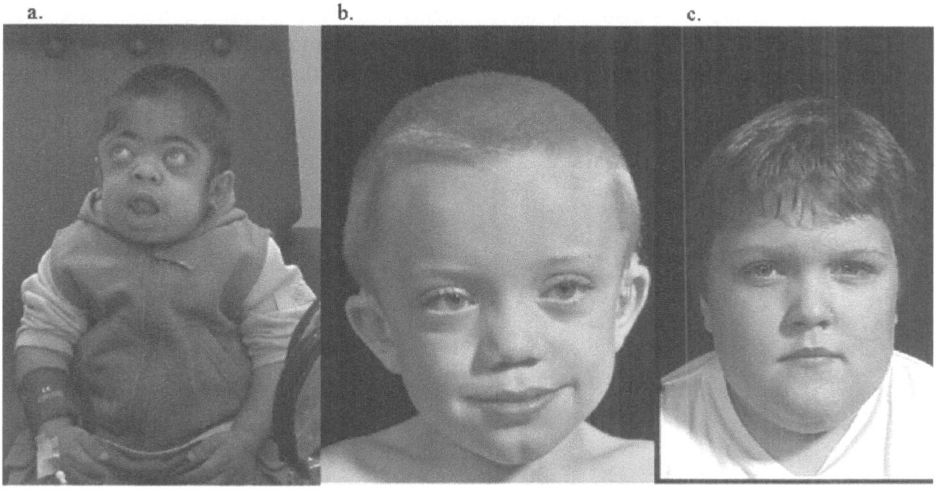

Figure 2. The facial appearance of a severe (2a), intermediate (2b) and a mildly affected patient (2c).

4.3 Ophthalmology

A wide range of ocular pathology has been described in MPS VI (Kenyon et al., 1972). This includes corneal opacity secondary to an alteration in corneal stromal matrix organization (Alroy, Haskins, and Birk, 1999) and both acute and chronic glaucoma (Cantor, Disseler, and Wilson, 1989). Corneal clouding can be treated by corneal transplantation, but in some patients reaccumulation of GAG can occur (Schwartz, Werblin, and Green, 1985). It is important to remember that in some patients the visual impairment in this disease is secondary to optic atrophy and raised intracranial pressure secondary to communicating hydrocephalus. A significant number of severely affected patients become blind and this loss of vision can be very rapid.

4.4 Hearing

Significant hearing loss is common and usually results from a combination of both conductive and sensorineural hearing loss. Most patients will have required ventilation tubes for middle ear fluid from an early age and a significant number will require hearing aids to provide appropriate sound amplification. All patients require regular audiological assessment and should be under regular review by an ear, nose, and throat surgeon.

4.5 Dental Disease

Macroglosia and gingival hyperplasia make caring for the teeth difficult. Delayed tooth eruption (Smith et al., 1995) and multiple dentigerous cysts (Roberts et al., 1984) lead to a high incidence of dental complications including dental abscess. It is important that early basic dental care is established and that patients undergo regular dental review to minimize the risk of infective endocarditis.

4.6 Upper Airways Obstruction, Anaesthesia, and the Lower Airway and Lungs

As with other severe MPS disorders many patients with MPS VI present major anaesthetic risks that can prove fatal (Walker et al., 1994). It is recommended that patients undergo procedures that require general anaesthesia only at centres that have access to anaesthetists

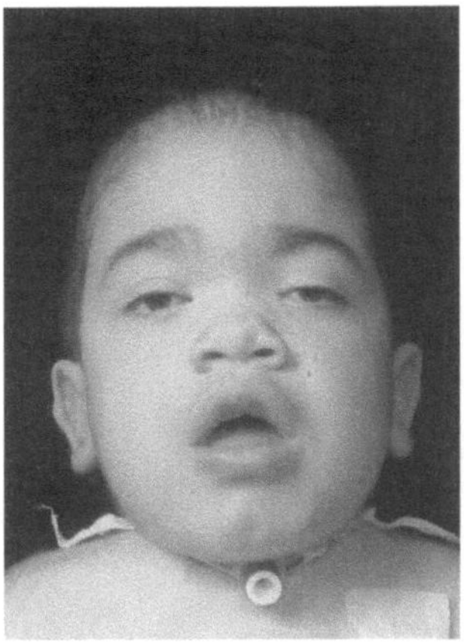

Figure 3. Severe upper airway obstruction treated by tracheostomy.

who are experienced in MPS disorders and their complications. These difficulties are a result of the severe upper airways obstruction that most severely affected MPS VI patients develop in the early years of life. Midface hypoplasia secondary to the skeletal dysplasia and gag deposition in the tongue, gums, pharynx, and upper respiratory tract greatly compromise airway patency. Tracheostomy for severe obstructive sleep apnoea and safe anaesthesia becomes necessary in some patients (Figure 3). In patients at the less severe end of the clinical spectrum noninvasive respiratory support and ventilation may become necessary.

As patients get older restrictive respiratory disease, secondary to the small thoracic cage, becomes more prominent and is a major cause of morbidity and mortality.

4.7 Cardiac Disease

Aortic and mitral valve thickening and dysfunction are the most common cardiac defects in MPS VI (Tan et al., 1992). Some severely affected patients present with severe cardiomyopathy indistinguishable from endocardial elastosis (Fong et al., 1987) and involvement of the conduction system has also been reported in some patients (Keller et al., 1987). Coronary artery disease has not been reported as a particular problem even in severely affected patients.

4.8 Abdomen

Enlargement of the liver is usual in older severely affected patients but splenomegaly is very variable. Most male patients develop inguinal hernias and umbilical hernias are common in both sexes. Most affected patients do not have the same tendency to diarrhea that occurs in MPS I, II, and III.

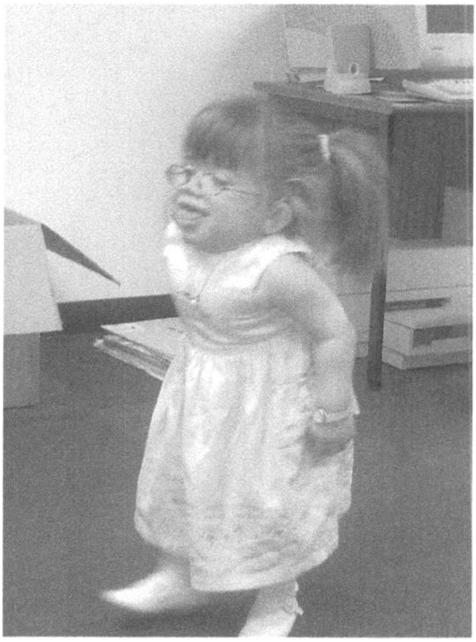

Figure 4. Severe dysostosis multiplex resulting in severe short stature.

4.9 Skeletal Dysplasia and Joint Disease

In severely affected patients there is florid dysostosis multiplex which results in extreme short stature (adult height 95–105 cm, Figure 4). Progressive contractures in the lower limb from soft tissue involvement further limit mobility and the development of severe claw-hand deformity and carpal tunnel syndrome (Musharbash 2002) affects dexterity often from a very young age.

Pathological studies of affected joints demonstrate premature chondrocyte death and increased levels of inflammatory cytokines (Simonaro, Haskins, and Schuchman, 2001).

4.10 Spinal Cord

Spinal cord compression due to a combination of dural hyperplasia and abnormalities at the craniocervical junction (Figure 5) can lead to cervical myelopathy (Young et al., 1980). Flexion and extension X-rays of the cervical spine are essential to exclude instability at the craniocervical junction. Compression is treated surgically by decompressive laminectomy accompanied by duraplasty in most affected patients. Fusion is also necessary in those patients with instability

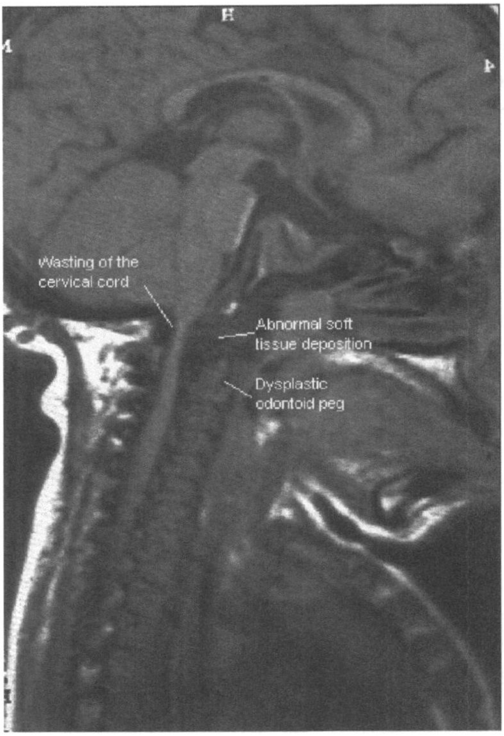

Figure 5. MRI of the cranio-cervical junction.

5 DIAGNOSIS

Diagnosis is suggested by the pattern of urinary GAG excreted in urine and confirmed by a diagnostic enzyme assay usually in leukocytes or cultured skin fibroblasts. Prenatal diagnosis can be performed by enzyme assay on chorion villus biopsy. Carrier testing is not routinely indicated and can only be done accurately by molecular analysis

6 DISEASE MANAGEMENT

Patients with this rare, progressive disorder are best managed in centres with experience in the multidisciplinary support that the patients require. Palliative therapies have been discussed in the clinical section above and in this section attempts at curative treatments are discussed.

BMT has been performed in a number of patients with MPS VI. Follow up of successfully treated patients has demonstrated an improvement in facial appearance, a stabilization of cardiac disease and an improvement in airways. Skeletal changes have persisted and even progressed in some patients although improvements in soft tissue have resulted in better mobility (Herskhovitz et al 1999). The use of bone marrow transplantation is limited by lack of suitable donors in a number of cases and the associated significant morbidity and mortality resulting from the procedure.

ERT has been developed for patients with MPS VI. A Phase I/II, randomized, two-dose, double-blind study has been performed in which patients were randomized to weekly infusions of either high (1.0 mg/kg) or low (0.2 mg/kg) doses of rhASB. Selected patients for the trial were severely affected at baseline and the study demonstrated an improvement in both clinical and biochemical parameters without significant side effects. In this study 6 patients were investigated and results after 48 weeks of treatment can be summarized as follows:

No drug-related serious adverse events, significant laboratory abnormalities, or allergic reactions were observed in the study. The high-dose group experienced a more rapid and larger relative reduction in urinary glycosaminoglycan that was sustained through week 48. Improvements in the 6-minute walk test were observed in all patients with dramatic gains in those walking <100 meters at baseline. Shoulder range of motion improved in all patients at week 48 and joint pain improved in patients with significant pain at baseline (Harmatz et al 2004).

A phase III study completed in June 2004 was designed to evaluate the safety and efficacy of RhASB, enrolled 39 patients, ranging in age from 5 to 29 years, at six sites located throughout the world. The study was double-blind and placebo-controlled. Approximately equal numbers of patients were randomized to receive either enzyme 1mg/kg intravenously or placebo solution weekly for 24 consecutive weeks. The results of the study can be summarized as follows:

(http://www.biomarinpharm.com/BM_ClinicalAndDevelopmentPrograms_Aryplase.html)

There was a statistically significant improvement in endurance ($p=0.025$) in enzyme treated patients as compared to placebo as measured by the distance walked in 12 minutes, the primary end point in the trial. The three minute stair climb demonstrated a positive trend ($p=0.053$) in favor of the enzyme treated group. And there was a statistically significant reduction in urine GAG ($p<0.001$) in enzyme treated patients compared to the placebo group.

In addition to efficacy the study confirmed that treatment with enzyme was safe and that the adverse events that occurred during the study were generally mild. The frequency of serious adverse events was more common in the placebo group and was disease related. As a result of these studies Naglazyme™ (galsulfase) is approved both in the United States (May 2005) and the European Union (January 2006) for the treatment of MPS VI.

7 CONCLUSIONS AND FUTURE THERAPY

MPS VI is a chronic progressive disorder that encompasses a wide spectrum of phenotypes. New developments in treatment (especially ERT) have improved the outlook for affected patients but considerable residual challenges will remain. Early diagnosis, potentially by newborn screening, will be necessary in order for the majority to benefit from these advances.

It is likely that in the future combination therapies (ERT and BMT) will be tried in this disease. In addition direct enzyme injection into affected joints may produce a better therapeutic response and this approach is currently under investigation in the MPS VI cat.

Gene therapy remains a distant goal in MPS VI although early work has confirmed transfection in skin fibroblasts (Fillat et al., 1996) and local delivery to the retina using adeno-associated virus (AAV)-mediated delivery has demonstrated local correction of the disease in affected cats (Ho et al., 2002).

REFERENCES

Alroy, J., Haskins, M., Birk, D.E. (1999), Altered corneal stromal matrix organization is associated with mucopolysaccharidosis I, III and VI, Exp Eye Res. 68: 523.

Cantor, B., Disseler, J.A., Wilson, F.M., 1989, Glaucoma in the Maroteaux– Lamy syndrome, Am J Ophthalmol. 108: 426.

Crawley, A.C., Brooks, D.A., Muller, V.J., Petersen, B.A., Isaac, E.L., Bielicki, J., King, B.M., Boulter, C.D., Moore, A.J., Fazzalari, N.L., Anson, D.S., Byers, S., Hopwood, J.J., 1996, Enzyme replacement therapy in a feline model of Maroteaux–Lamy syndrome, J Clin Invest. 97: 1864.

Crawley, A.C., Niedzielski, K.H., Isaac, E.L., Davey, R.C.A., Byers, S., Hopwood, J.J., 1997, Enzyme replacement therapy from birth in a feline model of mucopolysacc-haridosis type VI, J Clin Invest. 99: 651.

Fillat, C., Simonaro, C.M., Yeyati, P.L., Abkowitz, J.L., Haskins, M.E., Schuchman, E.H., 1996, Arylsulfatase B activities and glycosaminoglycan levels in retrovirally transduced mucopolysaccharidosis type VI cells. Prospects for gene therapy, J Clin Invest. 98: 497.

Fong, L.V., Menahem, S., Wraith, J.E., Chow, C.W., 1987, Endocardial fibroelastosis in mucopolysaccharidosis type VI, Clin Cardiol. 10: 362.

Gasper, P.W., Thrall, M.A., Wenger, D.A., Macy, D.W., Ham, L., Dornsife, R.E., McBiles, K., Quackenbush, S.L., Kesel, M.L., Gillette, E.L., Hoover, E., 1984, Correction of feline arylsulphatase B deficiency (mucopolysaccharidosis VI) by bone marrow transplantation, Nature. 312: 467.

Harmatz, P., Whitley, C.B., Waber, L., Pais, R., Steiner, R., Plecko, B., Kaplan, P., Simon, J., Butensky, E., Hopwood, J.J., 2004, Enzyme replacement therapy in mucopolysaccharidosis VI (Maroteaux–Lamy syndrome), J Pediatr. 144: 574.

Herskhovitz, E., Young, E., Rainer, J., Hall, C.M., Lidchi, V., Chong, K., Vellodi, A., (1999), Bone marrow transplantation for Maroteaux–Lamy syndrome (MPS VI): Long-term follow-up, J Inherit Metab Dis. 22: 50.

Ho, T.T., Maguire, A.M., Aguirre, G.D., Surace, E.M., Anand, V., Zeng, Y., Salvetti, A., Hopwood, J.J., Haskins, M.E., Bennett, J., 2002, Phenotypic rescue after adeno-associated virus-mediated delivery of 4-sulfatase to the retinal pigment epithelium of feline mucopolysaccharidosis VI, J Gene Med. 4: 613.

Jezyk, P.F., Haskins, M.E., Patterson, D.F., Mellman, W.J., Greenstein, M., 1977, Mucopolysaccharidosis in a cat with arylsulfatase B deficiency: A model of Maroteaux-Lamy syndrome, Science. 198: 834.

Keller, C., Briner, J., Schneider, J., Spycher, M., Rampini, S., Gitzelmann, R., 1987, Mucopolysaccharidosis 6-A (Maroteaux-Lamy disease): Comparison of clinical and pathologico-anatomic findings in a 27-year-old patient, Helv Paediatr Acta. 42: 317.

Kenyon, K.R., Topping, T.M., Green, W.R., Maumenee, A.E., 1972, Ocular pathology of the Maroteaux-Lamy syndrome (systemic mucopolysaccharidosis type VI). Histologic and ultrastructural report of two cases, Am J Ophthalmol. 73: 718.

Litjens, T., Baker, E.G., Beckmann, K.R., Morris, C.P., Hopwood, J.J., Callen, D.F., 1989, Chromosomal localization of ARSB, the gene for human N-acetylgalactos-amine-4-sulphatase, Hum Genet. 82: 67.

Lowry, R.B., Applegarth, D.A., Toone, J.R., MacDonald, E., Thunen, N.Y., 1990, An update on the frequency of mucopolysaccharide disorders in British Columbia, [Letter], Hum Genet. 85: 389.

Maroteaux, P., Leveque, B., Marie, J., Lamy, M., 1963, Une nouvelle dysostose avec elimination urinaire de chondroitine-sulfate B, Presse Med. 71: 1849.

Modaressi, S., Rupp, K., von Figura, K., Peters, C., 1993, Structure of the human arylsulfatase B gene, Biol Chem Hoppe Seyler, 374: 327.

Musharbash, A., 2002, Carpal tunnel syndrome in a 28-month-old child, Pediatr Neurosurg. 37: 32.

Nelson, J., Crowhurst, J., Carey, B., Greed, L., 2003, Incidence of the mucopolysaccharidoses in Western Australia, Am J Med Genet. 123A(3): 310.

Neufeld, E.F., and Muenzer, J., 2001, The Mucopolysaccharidoses, in: *The Metabolic & Molecular Bases Of Inherited Disease,* 8th edition, C.R. Scriver, A.L. Beaudet, W.S. Sly, D. Valle, Eds, McGraw-Hill, New York, pp. 3421–3452.

Norrdin, R.W., Simske, S.J., Gaarde, S., Schwardt, J.D., Thrall, M.A., 1995, Bone changes in mucopolysaccharidosis VI in cats and the effects of bone marrow trans-plantation: mechanical testing of long bones, Bone, 17: 485.

Roberts, M.W., Barton, N.W., Constantopoulos, G., Butler, D.P., Donahue, A.H., 1984, Occurrence of multiple dentigerous cysts in a patient with the Maroteaux-Lamy syndrome (mucopolysaccharidosis, type VI), Oral Surg Oral Med Oral Pathol. 58: 169.

Schwartz, M.F, Werblin, T.P., Green, W.R., 1985, Occurrence of mucopolysaccharide in corneal grafts in the Maroteaux-Lamy syndrome, Cornea. 4: 58.

Simonaro, C.M., Haskins, M.E., Kunieda, T., Evans, S.M., Visser, J.W., Schuchman, E.H., 1997, Bone marrow transplantation in newborn rats with mucopolysaccharidosis type VI: Biochemical, pathological, and clinical findings, Transplantation, 63: 1386.

Simonaro, C.M., Haskins, M.E., Schuchman, E.H., 2001, Articular chondrocytes from animals with a dermatan sulfate storage disease undergo a high rate of apoptosis and release nitric oxide and inflammatory cytokines: A possible mechanism underlying degenerative joint disease in the mucopolysaccharidoses, Lab Invest. 81: 1319.

Smith, K.S., Hallett, K.B., Hall, R.K., Wardrop, R.W., Firth, N., 1995, Mucopolysaccharidosis: MPS VI and associated delayed tooth eruption, Int J Oral Maxillofac Surg. 24: 176.

Tan, C.T., Schaff, H.V., Miller, F.A. Jr., Edwards, W.D., Karnes, P.S., 1992, Valvular heart disease in four patients with Maroteaux-Lamy syndrome, Circulation, 85: 188.

Walker, R.W., Darowski, M., Morris, P., Wraith, J.E., 1994, Anaesthesia and mucopolysaccharidoses. A review of airway problems in children, Anaesthesia. 49: 1078.

Yoshida, M., Ikadai, H., Maekawa, A., Takahashi, M., Nagase, S., 1993, Pathological characteristics of mucopolysaccharidosis VI in the rat, J Comp Pathol, 109: 141.

Young, R., Kleinman, G., Ojemann, R.G., Kolodny, E., Davis, K., Halperin, J., Zalneraitis, E., DeLong, G.R., 1980, Compressive myelopathy in Maroteaux-Lamy syndrome: Clinical and pathological findings, Ann Neurol. 8: 336.

MUCOPOLYSACCHARIDOSIS TYPE VII (SLY DISEASE): CLINICAL, GENETIC DIAGNOSIS AND THERAPIES

Denise J. Norato

Sly syndrome, or Mucopolysaccharidosis type VII (MPS VII), is a very rare autosomal-recessive lysosomal storage disorder (LSD) and about 30 cases have been described (www.orpha.net; Neufeld and Muenzer, 2001; Lee et al., 1985; Sewell et al., 1982; Nelson et al., 1982; Machin, 1989).

Sly syndrome is due to β-glucoronidase deficiency and shows a wide range of severity and system involvement heterogeneity similar to MPS I and II (Neufeld and Muenzer, 2001), with phenotypic extremes from the very severe fetal hydrops to the oligo-symptomatic variant. This chapter gives an overall view of the clinical aspects, pathophysiology, molecular genetics, and the recent advances towards an effective therapy.

1 HISTORY

Sly syndrome, (MPS VII) (OMIM+253220) was first reported by Sly and coworkers in 1973 (Sly et al., 1973). The patient presented hepatosplenomegaly, dysostosis multiplex, and granular inclusions in granulocytes compatible with a mucopolysacharidosis and showed deficiency of beta-glucuronidase (EC 3.2.1.31) in fibroblasts (Sly et al., 1973). Although almost the last MPS to be described, MPS VII was the first autosomal mucopolysaccharidosis for which chromosomal assignment was achieved (Knowles et al., 1977).

Beta-glucuronidase deficiency as a cause of hydrops fetalis was first reported in 1982 by Nelson and coworkers (1982) and in 1992, Kagie and collaborators demonstrated the deficiency in the amniotic fluid obtained at 25 weeks gestation of a suspected fetus (Kagie et al., 1992). The first prenatal diagnosis by enzymatic assay of chorionic villi was reported in 1998 by van Eyndhoven and coworkers. Birkenmeier's group described the first animal model in 1989, an autosomal recessive mutant mouse with beta-glucuronidase deficiency (Birkenmeier et al., 1989).

2 CLINICAL DIAGNOSIS

MPS VII, SLY syndrome, due to β-glucuronidase deficiency, has a wide range of severity and system involvement heterogeneity similar to MPS I and II (Neufeld and Muenzer,

Medical Sciences School, Life Sciences Center, Pontifical Catholic University, Campinas (PUC-CAMP).Av.John Boyd Dunlop, s/n, Campinas São Paulo Brasil
Mail Address: Rua dos Guaicurus 790, CEP- 13087-763; Phone/ Fax 55 19 32426206 Mobile 55 19 81760752.
E-mail: dnorato@uol.com.br

2001) and a severe neonatal form is seen in nonimmune hydrops fetalis cases (Kagie et al., 1992; van Dorpe et al., 1996; Stangenberg et al., 1992; van Eyndhoven et al., 1998).

Clinical symptoms are heterogeneous and are progressively limiting in nature, compromising several systems and organs and leading to early death in severe cases. (Hopwood and Morris, 1990; Neufeld and Muenzer, 2001).

The original patient with β–glucuronidase deficiency described by Sly had a clinical picture similar to MPS I patients with hepatosplenomegaly, umbilical hernia, and dysostosis with progressive kyphoscoliosis, odontoid hypoplasia, hip dysplasia, and moderate, nonprogressive mental retardation (Sly et al., 1973).

Milder forms of Sly syndrome are manifested in patients older than four years, but in the severe form, signs and symptoms are similar to the ones observed in patients with other severe forms of MPS, such as MPS I-H (Hurler syndrome) and may be already evident in the neonatal period. (Neufeld and Muenzer, 2001) Severe MPS VII may be present at birth as hydrops fetalis and patients with the severe neonate form may have neonatal jaundice, coarse facial features, and dysostosis multiplex (Lissens et al., 1991; Kagie et al., 1992; Stangenberg et al., 1992; Cheng et al., 2003).

On the other end of the severity spectrum, very mild cases of Sly syndrome have been reported. In these cases asymptomatic thoracic kyphosis and mild scoliosis were the main clinical features and no hepatosplenomegaly, corneal clouding, hernia, or short stature were present. In other cases severe skeletal dysplasia with no hepatosplenomegaly, hernia, corneal clouding, or neurologic abnormalities was observed. Patients with the less severe form retain trace residual amounts of enzyme that ameliorate the phenotype to varying degrees (Gitzelmann et al., 1978; de Kremer et al., 1992).

MPS VII has an autosomal recessive inheritance pattern and carriers are asymptomatic. Affected siblings in a family will have the same genotype, but can show phenotype variations in disease manifestations. Major clinical manifestations of MPS VII include hepatomegaly, skeletal abnormalities and joint stiffness, facial dysmorphism, communicating hydrocephalus, developmental delay, hearing loss, obstructive airway disease, and cardiac complications. Mental retardation is a common feature of Sly syndrome, but it usually is moderate and nonprogressive and begins in patients older than three years. Variable levels of corneal clouding are seen in MPS VII patients, and may develop at any time in patients older than one year, but usually begin in patients at approximately age eight years (Neufeld and Muenzer, 2001, Lee et al., 1985, Sewell et al., 1982). The details regarding these clinical manifestations of MPS VII are the same that have been reported for MPS I by several authors and extensively described by Neufeld and Muenzer (2001).

3 ENZYMOLOGY AND URINARY GLYCOSAMINOGLYCANS

The mucopolysaccharidosis (MPS) are characterized by glycosaminoglycan storage in several tissues (Leroy, 1979), due to the deficiency in one of 11 enzymes involved in the sequential degradation of glycosaminoglycans (GAGs), dermatan sulfate, heparan sulfate, and keratan sulfate that result from the breaking of proteoglycans of the fundamental extracelular matrix of connective tissue. As GAG degradation by hepatic endoglycosidases, sulfatases, and transferases is limited, the fragments that escape from the tissues are accumulated and excreted in urine, where they can be detected by qualitative and quantitative methods (McKusick and Neufeld, 1983; Hopwood and Morris, 1990).

The first of the lysosomal enzymes of glycosaminoglycan degradation to be studied at a molecular level was β̃ Glucuronidase (GUSB) and full-length cDNAs encoding proteins of 651 (human) and 648 (rodent) amino acids have been sequenced and expressed (Nishimura et al., 1986; Oshima et al., 1987; Powell et al., 1988).

The enzyme, a glycosidase, deficient in MPS VII, is responsible for one of the last steps for the degradation of most glycosaminoglycans removing the glucuronic acid residues present in dermatan sulfates, heparan sulfates and chondroitin sulfates (Stahl et al., 1971; Himeno et al., 1976; Brot et al., 1978). Although a lysosomal enzyme, it is also associated with the protein egasyn in microsomes in some tissues (Tomin et al., 1975; Stawser et al., 1979; Medda et al., 1989). The lysosomal enzyme transport recognition marker mannose 6-phosphate was first identified for β̃ glucuronidase (Kaplan et al., 1977).

Urine of MPS VII patients shows dermatan and heparan sulfates and also some traces of chondroitin sulfates. The action of lysosomal hyaluronidase and endohexosaminidase is thought to be responsible for the modest elevation of chondroitin 4- and 6-sulfate excretion seen in MPS VII (Lee et al., 1985).

4 MOLECULAR GENETICS

GUSB locus is 7q11.21-q11.22, by dosage analysis of chromosomal aberrations (Knowles et al., 1977; Ward et al., 1983; Frydman et al., 1986; Allanson et al., 1988; Fagan et al., 1989; Schwartz et al., 1991) and in situ hybridization (Speleman et al., 1996). This locus contains the GUSB gene which is 21 kb long and contains 12 exons ranging from 85 to 376 bp, and through alternate splicing two types of cDNAs arise; exon 6 corresponds to the 153-bp deletion in the shorter of the two types (Miller et al., 1990). Gene expression was first studied in human cDNA clones in *E. coli* (Guise et al., 1985) and studies of cDNA sequence for human placental beta-glucuronidase in transfected cultured cells demonstrated the existence of 2 populations of mRNA, only one of which specifies a catalytically active enzyme (Oshima et al., 1987).

More than 45 different mutations have been identified, approximately 90% of which were point mutations. The leu176-to-phe L176F mutation that was first identified in a Mennonite family (Wu et al., 1994) is associated with a mild phenotype, and accounts for approximately 20% of mutant alleles of the GUSB gene. Cells from L176F patients have less than 1% of normal GUSB activity, but expression of the L176F cDNA in COS cells produces nearly as much enzyme activity as the wildtype control cDNA (Tomatsu et al., 1991, 2002; Schwartz et al., 1991).

The first severely affected patient with MPS VII presented nonimmune hydrops fetalis and was a compound heterozygote for two GUSB mutations: a C-to-T transition at position 1061 of the cDNA in exon 6, giving rise to an ala354-to-val substitution (A354V); and a C-to-T transition at position 1831 in exon 12 producing an arg611-to-trp substitution (R611W). Cultured fibroblasts from this patient showed less than 1% of residual activity for the enzyme beta-glucuronidase. Transient expression in COS-7 cells demonstrated that both mutant enzymes were synthesized as normal-size precursors in normal quantities, but both exhibited accelerated turnover (Nelson et al., 1982; Wu and Sly, 1993).

In the severe form of MPS VII, in two patients, both compound heterozygotes, with a severe reduction of beta-glucuronidase activity, one had alleles P148S and Y495C and the other the allele W507X and a deletion at position 1642-1679 in exon 10: 1642del38nt, caused by a C-to-T transition which together with the previous guanine, created the GT of a new, premature 5-prime splice site (Yamada et al., 1995).

A compound heterozygote patient with mild MPS VII had a maternal allele for a 1363G-A transition in the GUSB gene, predicting a trp446-to-ter substitution in exon 8 that analysis of mRNA structure by RT/PCR and direct sequencing revealed the inclusion of a new exon derived from an antisense Alu-repeat in intron 8 and the skipping of exon 9 in a large proportion of the mRNA, and a paternal gene presenting a 2-bp deletion creating a strong 5-prime splice site, IVS8+0.6kb delTC. With a sensitive RT-PCR assay, they demonstrated that both the inclusion of the Alu-cassette and the skipping of exon 9 were minor events in control samples and that mRNA with both alterations was found only in the carrier of the intronic 2-bp deletion. The increased proportion of exon 9 skipping seemed to be related to the premature termination of translation (Vervoort et al., 1998a, b).

A 37-year-old patient who had a relatively mild MPS VII phenotype, described as the longest known survivor of MPS VII, (Pfeiffer et al., 1977; Storch et al., 2003) presented compound heterozygosity for two mutations in the GUSB gene: lys350 to asn (K350N) in exon 6 and arg577 to leu (R577L) in exon 11. The mild phenotype was attributed to the residual catalytic activity provided by the K350N mutant but expression of either the R577L or the R577L/K350N mutation resulted in rapid degradation of the enzyme in early biosynthetic compartments and a total loss of enzymatic activity (Storch et al., 2003).

A point mutation 2154A-G in the 3-prime noncoding region of the gene in the neighborhood of two consensus polyadenylation sites was found in five MPS VII patients and in normals and appears to represent a harmless polymorphism (Vervoort et al., 1997).

5 PATHOLOGY

Proteoglycans have a widespread distribution, therefore MPS disorders are associated with multiple system involvement. GAGs accumulate in the lysosomes of cells and multiple distended lysosomes can be seen in hepatocytes and Kupffer cells and in almost all tissues. The tissue accumulation of undegraded substrate affects cellular function and leads to the clinical manifestations of the disease (McKusick and Neufeld, 1983; Hopwood and Morris, 1990).

MPS VII shows the typical Alder–Reilly bodies in polymorphonuclear leukocytes that are seen in MPS II and VI. These granulations may be observed in peripheral blood smears, but are more likely to be found in bone marrow studies, and are a morphological aberration of all leukocytes in which there is a decreased degradation of mucopolysaccharides and the subsequent aggregation within the WBC to form deposits of mucopolysaccharides and glycogen, that stain varying shades of purple-red with toluidine blue O stain (Gitzelmann et al., 1978).

Pathological findings in autopsy samples from the first patient reported with mucopolysaccharidosis VII were dysostosis multiplex and extensive cardiovascular lesions including arterial stenosis, and marked fibrous thickening of the atrioventricular and aortic valves. Microscopic evidence of lysosomal storage was found in bone, cartilage, arteries, and cardiac valves, liver, spleen, lymph nodes, eyes, adrenal, pituitary, and the central nervous system. In the brain, storage was localized to specific regions, primarily intraneuronal, and appeared ultrastructurally as delicate whorled filamentous accumulations in lysosomes. Similar filamentous storage also occurred in medial cells of the aorta. Multiple postmortem tissues contained only trace amounts of beta-glucuronidase and elevated glycosaminoglycans, predominantly chondroitin 4- and 6-sulfate (Vogler et al., 1994).

In fetal hydrops MPS VII cases, microscopic study revealed prominently vacuolated Hofbauer cells in the placenta and foamy macrophages in liver, spleen, bone marrow, and other organs (Irani et al., 1983; van Dorpe et al., 1996; Wilson et al., 1982).

Mice with a recessively inherited deficiency of β-glucuronidase present mucopolysaccharidosis type VII (MPS VII) have a shortened lifespan, are dysmorphic, dwarfed, and have clinical evidence of behavioral and memory deficiencies. Widespread lysosomal distention with glycosaminoglycan accumulation affects most viscera and show progressive accumulation of lysosomal storage in neurons, glia, and mesenchymal tissue. The morphological character and the amount of lysosomal storage vary among neuronal groups. In the hippocampus, regional variation in the abundance of lysosomal storage in the MPS VII mice correlates with variation in the amount of beta-glucuronidase activity in normal mice (Levy et al., 1996).

Analysis of storage in very young MPS VII mice showed that pups have storage in fixed tissue macrophages of the liver and cartilage as soon as 12 days postcoitus (dpc), was present in central nervous system on the 15 dpc, and in osteoblast and primitive neocortical at 18 to 19 dpc. Lysosomal distention in leukocytes was observed 2 days after birth (Vogler et al., 2005).

Further studies of MPS VII brains. Pathologic findings of storage in other MPS show that besides glycosaminoglycans, several gangliosides (GM2, GM3, and GD3) accumulate in the brains of MPS patients (Constantopoulos et al., 1978a, b, 1980). The lipids are probably stored in zebra bodies, which are reminiscent of the lysosomal inclusions seen in the sphingolipidoses. The accumulation of gangliosides has been puzzling, because their catabolism does not require the enzymes of glycosaminoglycan degradation. However, the activity of several additional lysosomal enzymes is reduced in the MPS, probably as a result of inhibition by accumulated glycosaminoglycans (Kint et al., 1973; Baumkotter et al., 1983). The inhibition of ganglioside neuraminidase may be particularly relevant. It may be significant that lipid accumulation was found in brains of several patients with mental retardation (MPS I H, MPS II severe, and MPS III A and III B) but not in the brain of a patient with MPS I S, whose intelligence was normal (Constantopoulos et al., 1978a, b, 1980).

6 LABORATORY DIAGNOSIS, CARRIER DETECTION, AND PRENATAL DIAGNOSIS

Definitive diagnosis of MPS VII is established using an enzyme assay that measures enzyme activity in leukocytes, cultured skin fibroblasts, serum, or plasma or dried blood spots on filter paper. The amount of enzyme activity measured in vitro from cellular extracts does not always correlate with disease severity. Mutation analysis is usually reserved for carrier detection, however, certain genotypes can predict disease severity, and knowing a patient's genotype may help the patient's family and physician to make the most informed decision regarding treatment. MPS VII has an autosomal recessive inheritance pattern and carriers are asymptomatic. Enzyme assay and molecular analysis are possible for carrier detection and prenatal diagnosis (Hall et al., 1978; Neufeld and Muenzer, 2001; Chamoles et al., 2001).

Prenatal diagnosis by using chorionic villi samples, cultured CV cells, or amniocentesis for MPS VII is possible (Kagie et al.,1992; Diukman and Goldberg, 2003) and in the severe hydrops fetalis cases reports of very early prenatal diagnosis using chorionic villi samples have been published (van Eyndhoven et al., 1998).

Family screening is indicated for counseling, and early diagnosis of presymptomatic siblings should become mandatory when MPS VII can be treated. Symptomatic therapies, supportive care, and treatment of complications have been crucial for survival and a better quality of life, and can greatly improve the quality of life for patients and their families. While looking for the definitive treatment, it can't be forgotten that patients should have excellent health care and psychological support and that the multisystem involvement in MPS patients requires communication and collaboration among medical specialists.

7 RESEARCH ON THERAPIES AND ANIMAL MODELS

Therapy for MPS VII is not available, but promising studies are been done in animal models. Several mutant mice, cats, and dogs, animal models for Sly syndrome, have been studied and been subject to different therapies such as gene transfer, recombinant enzymes, bone marrow transplantation, and modified fibroblast implants.

7.1 Bone Marrow Transplantation

As no other therapy is available, BMT should be the treatment of choice for MPS VII patients. Although only one transplanted patient has been reported, a 12-year-old Japanese girl homozygous for the ala619-to-val mutation received an allogeneic bone marrow transplant that resulted in improved motor function and activities of daily living, decreased upper respiratory and ear infections, but no improvement in cognitive function (Yamada et al., 1998).

Research in animal models also suggests that BMT reverses the systhemic pathology but does not prevent or correct central nervous system manifestations.

In an autosomal recessive mutant mouse with a beta-glucuronidase gene complex site mutation on the distal end of chromosome 5 presenting beta-glucuronidase deficiency, syngeneic bone marrow transplantation increased the lifespan of affected animals to a value approaching that seen in normal mice and corrected widespread lysosomal storage (Birkenmeier et al., 1989, 1991).

Cardiac manifestations in β-glucuronidase (GUSB) null mice were prevented by bone marrow transplant in neonates without myeloablative pretreatment. β-hexosamindase increased in tissues and reduced storage in the stroma of heart valves, adventitial cells of the aortic root, perivascular and interstitial cells of the myocardium, and interstitial cells of the conduction tissue (Schuldt et al., 2004).

Behavioral deficits (grooming and spatial learning abilities) in GUSB mice were not restored to normal whereas similarly treated normal mice showed significant functional deterioration, indicating the detrimental consequence of bone marrow transplantation in the neonatal period (Bastedo et al., 1994).

7.2 Enzyme Replacement Therapy

Enzyme replacement therapy for human patients with MPS VII has not been yet tested but ERT in animal models showed good results for systemic involvement and there are indications that when started early the enzyme reaches the central nervous system.

Recombinant mouse beta-glucuronidase given to newborn MPS VII mice is rapidly cleared from the circulation and localized in many tissues and significantly reduces or prevents the accumulation of lysosomal storage during the first six weeks of life. Mice that received enzyme replacement at birth had less severe skeletal dysplasia and less lysosomal storage in neurons compared to those that started at six weeks of age. Animals that received enzyme at birth followed by syngeneic BMT at five weeks of age had reduced lysosomal distention in corneal fibroblasts, bone, and meninges. Mice from the three groups lived at least up to one year of age, had liver and spleen beta-glucuronidase activity, and decreased lysosomal storage (Sands et al., 1994, 1997).

MPS VII mice receiving weekly intravenous injections of recombinant beta-glucuronidase at birth showed a less severe phenotype and improvements in the histopathology of the brain and ear, in a spatially oriented learning test, and in hearing deficits. The central nervous system was virtually cleared of disease showing that neonatal treatment provided access to the central nervous system via an intravenous route (O'Connor et al., 1998).

A new tagged enzyme, a chimeric protein containing a portion of mature human IGF II fused to the C terminus of human beta-glucuronidase, was delivered effectively to storage sites in MPS VII mice and was effective in reversing the disease pathology in glomerular podocytes and osteoblasts, through mannose 6-phosphate independent delivery (LeBowitz et al., 2004).

7.3 Cell and Gene Therapy

Researchers in gene and cell therapies for lysosomal diseases have been using MPS VII animal models since the beginning of the 1990s, with incredible results in preventing and correcting disease manifestations, but no human subject has yet received gene transfer or stem cell implants as a result of the controversial results in other diseases concerning adverse effects (viral infections and cancer) as well as the ethical aspects of changing human genome or using stem cells.

Initially tested in human and canine fibroblasts deficient for beta-glucuronidase the retroviral vector-mediated beta-glucuronidase gene transfer is expressed and restores normal processing of specific glycosaminoglycans in the lysosomal compartment. (Wolfe et al., 1990) This retroviral vector-mediated transfer of the beta-glucuronidase gene when used in mutant stem cells in the MPS VII mouse model with a 1-bp deletion that creates a frameshift within exon 10 (Sands and Birkenmeier, 1993) resulted in long-term expression of low levels of beta-glucuronidase that partially corrected the disease by reducing lysosomal storage in liver and spleen (Wolfe et al., 1992).

Mucopolysaccharidosis type VII in the dog closely resembles the severe MPS VII clinical picture of human patients (Haskins et al., 1984, 1991). GUSB cDNA derived from MPS VII dogs show a single nucleotide substitution G-to-A base change at nucleotide position 559 that causes an arg-to-his substitution at amino acid position 166. A retroviral vector expressing the full-length canine beta-glucuronidase cDNA corrected the deficiency in MPS VII cells (Ray et al., 1998) and gene therapy prevented the clinical manifestations of MPS VII in dogs that were transduced with a GUSB-expressing retroviral vector as neonates (Ponder et al., 2002).

Experiments with the human GUSB gene have shown that it can be introduced in transgenic MPS VII mice and it expresses high levels of human beta-glucuronidase activity correcting the mouse disease (Kyle et al., 1990) and in another experiment, mutant mouse skin fibroblasts grown in primary culture had human beta-glucuronidase cDNA introduced with a retroviral vector and were implanted intraperitoneally (in collagen lattices) resulting in accumulated enzyme in tissues and disappearance of the lysosomal storage in the mice liver and spleen (Moullier et al., 1993).

An interesting approach was the injection of recombinant adenovirus carrying the human beta-glucuronidase cDNA coding region under the control of a nontissue-specific promoter intravitreally or subretinally into the eyes of mutant MPS VII mice that corrected the storage granules in the retinal pigment epithelium (Li and Davidson, 1995). MPS VII mice receiving intraocular AAV-mediated therapy for retinal disease showed nearly normal levels of beta-glucuronidase activity, preservation of cells in the outer nuclear layer of the retina, decreased lysosomal storage in retinal pigmented epithelial (RPE) cells in the eye, and significantly increased dark-adapted ERG amplitudes compared to untreated MPS VII at 16 weeks, an age at which untreated MPS VII mice have advanced histologic lesions and significantly reduced ERG amplitudes (Hennig et al., 2004).

Interspecies correction was also obtained between cats and rats. Cats with mucopolysaccharidosis VII present a clinical picture very similar to human MPS VII (Gitzelmann et al., 1994; Fyfe et al., 1999). Affected cats are homozygous and cats with half-normal beta-glucuronidase activity are heterozygous for the missense mutation, a G-to-A transition that predicted a glu351-to-lys substitution and destroyed a BssSI site. The beta-glucuronidase activity was restored by retroviral gene transfer of rat beta-glucuronidase cDNA (Fyfe et al., 1999).

Gene transfer mediated by a recombinant adeno-associated virus (rAAV) type 2 vector that carries the murine beta-glucuronidase cDNA under the transcriptional direction of the human elongation factor-1alpha promoter resulted in stable hepatic beta-glucuronidase expression for at least one year postinjection and widespread distribution of vector genomes and beta-glucuronidase within extrahepatic organs and is capable of reducing lysosomal storage within the liver, spleen, kidney, heart, lung, and brain and reversing the progression of storage in the adult MPS VII mouse (Sferra et al., 2004).

Neonatal intravenous injection of a gamma retroviral vector (RV) expressing canine GUSB resulted in transduction of hepatocytes, high levels of GUSB modified with mannose 6-phosphate in blood, and reduction in disease manifestations in the heart, bone, and eye. In the heart significant levels of GUSB were detected in myocardium and aorta and dogs that had mild or moderate mitral regurgitation at 4 to 5 months after birth showed improvement or normalization when evaluated at 9 to 11 and at 24 months, and the mitral valve thickening present early in some animals disappeared over time. (Sleeper) GUSB activity was also restored in osteocytes and bone-lining cells but not on chondrocytes and there was marked reduction in lysosomal storage in bone and at the bone:growth plate interface, resulting in improvements in facial morphology and long bone lengths, reducing erosions of the femoral head and increase of the internal area of the trachea (Mango et al., 2004). Furthermore, blood cells had and expressed substantial copies of RV DNA that may synergize with uptake of GUSB from blood to reduce storage in organs, suggesting that both secretion of enzyme into blood by hepatocytes, and expression in blood cells that migrate into organs, may contribute to correction of disease (Wang et al., 2006).

A challenge for new therapies in neurodegenerative disease is how to correct central nervous system pathology, and animal models of MPS VII are important tools in the developing of new strategies.

Newborn MPS VII mice received a single intravenous injection of recombinant adeno-associated virus encoding the human GUSB cDNA and therapeutic levels of expression were achieved by 1 week in several organs, including neurons, microglia, and meninges, and persisted for 16 weeks (Daly et al., 1999) and adult MPS VII mice that received recombinant feline immunodeficiency virus (FIV)-based vectors expressing beta-glucuronidase injected unilaterally into the striatum, showed bihemispheric correction of the characteristic cellular pathology and recovery of behavioral function (Brooks et al., 2002).

Treatment with a recombinant viral vector to correct the enzymatic defect quantitatively reversed the neurodegenerative lesions (ubiquitin and neurofilament inclusions, reactive astrogliosis) to normal levels in targeted regions vulnerable to neurodegeneration in brain of the mouse model of MPS VII, predominantly in the hippocampus and cerebral cortex (Heuer et al., 2002).

Affected MPS VII mice with established disease show impaired conditioned fear response and context discrimination that was reversed six weeks after gene transfer by recombinant AAV4 vectors encoding beta-glucuronidase that were injected unilaterally into the lateral ventricle that resulted in penetration of secreted enzyme in cerebral and cerebellar structures and brainstem (Liu et al., 2005).

A single adeno-associated virus serotype 1 vector injection into the ventricle of MPS type VII mouse at 15.5 days of gestation, resulted in widespread distribution and lifelong expression of the normal gene in the brain and spinal cord, and prevented the development of storage lesions throughout the central nervous system (CNS). A small amount of enzyme was also present in visceral tissues, consistent with transfer from cerebrospinal fluid to venous circulation (Karolewski et al., 2006).

Adult MPS VII mice, treated by injection of a herpes simplex virus type 1 (HSV-1) vector into a single site on each side of the brain, presented axonal transport of vector and GUSB. GUSB enzymatic activity levels were normal in several brain regions and storage lesions were corrected in a large volume of the brain (Berges et al., 2006).

The use of genetically modified cells or stem cells has brought new hope for degenerative central nervous diseases. In the MPS animal models the results have proved to be worth investing in developing treatment for humans. There was a marked improvement of cognitive function in MPS VII mice after brain transplantation of genetically modified bone marrow stromal (BMS) cells to lateral ventricle of newborn mucopolysaccharidosis VII (MPS VII), and lysosomal distention was not found in the treated animal brain. Transplanted cells were identified in olfactory bulb, striatum, and cerebral cortex after 2 weeks; the enzymatic activities increased and GAGs were reduced to near normal level after 4 weeks but this was temporary as activity declined after 8 weeks and the reduction of GAGs persisted only for 16 weeks (Sakurai et al., 2004).

A functional assessment with novel-object recognition tests confirmed improvements in behavioral patterns after neural stem cells, derived from embryonic mouse striata and expanded in vitro by neurosphere, transplantation into the lateral ventricles of newborn MPS VII mice. Donor cells migrated far beyond the site of injection within 24 h, GUSB activity was present in the brain at least for three weeks, and there was a decrease in lysosomal storage in hippocampus, cortex, and ependyma (Fukuhara et al., 2006).

New animal models useful for preclinical trials evaluating the effectiveness of enzyme and/or gene therapy have been produced as the new murine models for MPS VII produced by targeted mutagenesis: mice with no Gusb activity and a severe phenotype (glu536-to-ala E536A) and two mice models with low levels of residual enzyme activity and milder phenotypes (glu536-to-gln E536Q mutation, corresponding to active-site nucleophile replacements glu540 to ala (E540A) and glu540 to gln (E540Q) in the human GUSB gene, as well as leu175 to phe (L175F), corresponding to the most common human mutation, L176F (Tomatsu et al., 2002) and the transgenic mice expressing the human beta-glucuronidase cDNA with an amino acid substitution at the active site nucleophile (E540A) were bred onto the MPS VII (gus mps/mps) background that retained the clinical, morphologic, biochemical, and histopathological characteristics of the original MPS VII mouse and were tolerant to immune challenge with human beta-glucuronidase (Sly et al., 2001). New natural occurring animals have been identified, as the MPS VII mice due to spontaneous insertion of an intracisternal A particle element into intron 8 of the gus structural gene, which has less than 1% of normal beta-glucuronidase activity and a less severe disease are fertile and breed to produce litters, all of which are MPS VII pups (Vogler et al., 2001).

The research done up to date in MPS VII animal models will also help further research for treating patients with other types of MPS. Although large steps have been taken to achieve the goal of curing MPS VII, at the moment, only bone marrow and hematopoietic stem cells are available as specific therapy for MPS VII patients. There is an urgent necessity of a global effort in finding and registering these patients, gathering data on the natural history and finding financial support for the clinical tests and the commercial production of human recombinant beta-glucuronidase or even making gene or cell therapy protocols accessible for clinical use.

REFERENCES

Allanson JE, Gemmill RM, Hecht BK, Johnsen S, Wenger DA. Deletion mapping of the beta-glucuronidase gene. *Am. J. Med. Genet.* 29:517–522, 1988.

Bastedo L, Sands MS, Lambert DT, Pisa MA, Birkenmeier E, Chang PLJ. Behavioral consequences of bone marrow transplantation in the treatment of murine mucopolysaccharidosis type VII. *Clin Invest.* September; 94(3):1180–1186, 1994.

Baumkotter J, Cantz M. Decreased ganglioside neuraminidase activity in fibroblasts from mucopolysaccharidosis patients. *Biochim Biophys Acta* 761:163, 1983.

Berges BK, Yellayi S, Karolewski BA, Miselis RR, Wolfe JH, Fraser NW. Widespread correction of lysosomal storage in the mucopolysaccharidosis type VII mouse brain with a herpes simplex virus type 1 vector expressing beta-glucuronidase. *Mol Ther.* 13(5):859–869, 2006.

Birkenmeier EH, Barker JE, Vogler CA, Kyle JW, Sly WS, Gwynn B, Levy B, Pegors C. Increased life span and correction of metabolic defects in murine mucopolysaccharidosis type VII after syngeneic bone marrow transplantation. *Blood* 78: 3081–3092, 1991.

Birkenmeier EH, Davisson MT, Beamer WG, Ganschow RE, Vogler CA, Gwynn B, Lyford KA, Maltais LM, Wawrzyniak CJ. Murine mucopolysaccharidosis type VII: Characterization of a mouse with beta-glucuronidase deficiency. *J. Clin. Invest.* 83:1258–1266, 1989.

Brooks AI, Stein CS, Hughes SM, Heth J, McCray PM Jr, Sauter SL, Johnston JC, Cory-Slechta DA, Federoff HJ, Davidson BL. Functional correction of established central nervous system deficits in an animal model of lysosomal storage disease with feline immunodeficiency virus-based vectors. *Proc. Nat. Acad. Sci.* 99:6216–6221, 2002.

Brot FE, Bell CE, Sly WS. Purification and properties of b-glucuronidase from human placenta. *Biochemistry* 17:385, 1978.

Chamoles NA, Blanco M, Gaggioli D. Diagnosis of a-l-iduronidase deficiency in dried blood spots on filter paper: the possibility of newborn diagnosis. *Clin. Chem.* 47(4): 780–781, 2001.

Cheng Y, Verp MS, Knutel T, Hibbard JU. Mucopolysaccharidosis type VII as a cause of recurrent non-immune hydrops fetalis. *J Perinat Med.* 31(6):535–537, 2003.

Constantopoulos G, Dekaban AS. Neurochemistry of the mucopolysaccharidoses: Brain lipids and lysosomal enzymes in patients with four types of mucopolysaccharidosis and in normal controls. *J Neurochem.* 30:965, 1978a.

Constantopoulos G, Eiben RM, Schafer IA. Neurochemistry of the mucopolysaccharidoses: Brain glycosaminoglycans, lipids and lysosomal enzymes in mucopolysaccharidosis IIIB (a-N-acetylglucosaminidase deficiency). *J Neurochem.* 31:1215, 1978b.

Constantopoulos G, Iqbal K, Dekaban AS. Mucopolysaccharidosis types IH, IS, II and IIIA: Glycosaminoglycans and lipids of isolated brain cells and other fractions from autopsied tissues. *J Neurochem.* 34:1399, 1980.

Daly TM, Vogler C, Levy B, Haskins ME, Sands MS. Neonatal gene transfer leads to widespread correction of pathology in a murine model of lysosomal storage disease. *Proc. Nat. Acad. Sci.* 96:2296–2300, 1999.

de Kremer RD, Givogri I, Argarana CE, Hliba E, Conci R, Boldini CD, Capra AP. Mucopolysaccharidosis type VII (beta-glucuronidase deficiency): A chronic variant with an oligosymptomatic severe skeletal dysplasia. *Am J Med Genet.* 44:145, 1992.

Diukman R, Goldberg JD. Prenatal diagnosis of inherited metabolic diseases. *West J Med.* 159(3):374–381, 1993.

Frydman M, Steinberger J, Shabtai F, Steinherz R. Interstitial 7q deletion [46,XY,del(7)(pter-cen::q112-qter)] in a retarded quadriplegic boy with normal beta glucuronidase. *Am. J. Med. Genet.* 25:245–249, 1986.

Fukuhara Y, Li XK, Kitazawa Y, Inagaki M, Matsuoka K, Kosuga M, Kosaki R, Shimazaki T, Endo H, Umezawa A, Okano H, Takahashi T, Okuyama T. Histopathological and behavioral improvement of murine mucopolysaccharidosis type VII by intracerebral transplantation of neural stem cells. *Mol Ther.* 13(3):548–555, 2006.

Fyfe JC, Kurzhals RL, Lassaline ME, Henthorn PS, Alur PRK, Wang P, Wolfe JH, Giger U, Haskins ME, Patterson DF, Sun H, Jain S, Yuhki N. Molecular basis of feline beta-glucuronidase deficiency: An animal model of mucopolysaccharidosis VII. *Genomics* 58:121–128, 1999.

Gitzelmann R, Bosshard NU, Superti-Furga A, Spycher MA, Briner J, Wiesmann U, Lutz H, Litschi B. Feline mucopolysaccharidosis VII due to beta-glucuronidase deficiency. *Vet. Path.* 31:435–443, 1994.

Gitzelmann R, Wiesmann UN, Spycher MA, Herschkowitz N, Giedion A. Unusually mild course of beta-glucuronidase deficiency in two brothers (mucopolysaccharidosis VII). *Helv. Paediat. Acta* 33:413–428, 1978.

Guise KS, Korneluk RG, Waye J, Lamhonwah A-M, Quan F, Palmer R, Ganschow RE, Sly WS, Gravel RA. Isolation and expression in Escherichia coli of a cDNA clone encoding human beta-glucuronidase. *Gene* 34:105–110, 1985.

Haskins ME, Aguirre GD, Jezyk PF, Schuchman EH, Desnick RJ, Patterson DF. Mucopolysaccharidosis type VII (Sly syndrome): beta-glucuronidase-deficient mucopolysaccharidosis in the dog. *Am. J. Path.* 138:1553–1555, 1991.

Haskins ME, Desnick RJ, DiFerrante N, Jezyk PF, Patterson DF. Beta-glucuronidase deficiency in a dog: a model of human mucopolysaccharidosis VII. *Pediat. Res.* 18:980–984, 1984.

Hennig AK, Ogilvie JM, Ohlemiller KK, Timmers AM, Hauswirth WW, Sands MS.AAV-mediated intravitreal gene therapy reduces lysosomal storage in the retinal pigmented epithelium and improves retinal function in adult MPS VII mice. *Mol Ther.* 10(1):106–116, 2004.

Heuer GG, Passini MA, Jiang K, Parente MK, Lee VM-Y, Trojanowski JQ, Wolfe JH. Selective neurodegeneration in murine mucopolysaccharidosis VII is progressive and reversible. *Ann. Neurol.* 52:762–770, 2002.

Himeno M, Nishimura Y, Tsuji H, Kato K. Purification and characterization of microsomal and lysosomal b-glucuronidase from rat liver by use of immuno-affinity chromatography. *Eur J Biochem.* 70:349, 1976.

Hopwood JJ, Morris CP. The mucopolysaccharidoses: diagnosis, molecular genetics and treatment. *Mol Biol Med.* 7:381–404, 1990.

Irani D, Kim HS, El-Hibri H, Dutton RV, Beaudet A, Armstrong D. Post mortem observations on beta-glucuronidase deficiency presenting as hydrops fetalis. *Ann Neurol.* 14:486, 1983

Kagie MJ, Kleijer WJ, Huijmans JGM, Maaswinkel-Mooy P, Kanhai HHH. Beta-glucuronidase deficiency as a cause of fetal hydrops. *Am. J. Med. Genet.* 42: 693–695, 1992.

Kaplan A, Achord DT, Sly WS. Phosphohexosyl components of a lysosomal enzyme are recognized by pinocytosis receptors on human fibroblasts. *Proc Natl Acad Sci USA* 74:2026, 1977

Karolewski BA, Wolfe JH. Genetic correction of the fetal brain increases the lifespan of mice with the severe multisystemic disease mucopolysaccharidosis type VII. *Mol Ther.* 14(1):14–24, 2006

Kint JA, Dacremont G, Carton D, Orye E, Hooft C. Mucopolysaccharidosis: Secondarily induced abnormal distribution of lysosomal isoenzymes. *Science* 181:352, 1973.

Knowles BB, Solter D, Trinchieri G, Maloney KM, Ford SR, Aden DP. Complement-mediated antiserum cytotoxic reaction to human chromosome 7 coded antigen(s): Immunoselection of rearranged human chromosome 7 in human-mouse somatic cell hybrids. *J. Exp. Med.* 145:314–326, 1977.

Kyle JW, Birkenmeier EH, Gwynn B, Vogler C, Hoppe PC, Hoffmann JW, Sly WS. Correction of murine mucopolysaccharidosis VII by a human beta-glucuronidase transgene. *Proc. Nat. Acad. Sci.* 87:3914–3918, 1990.

LeBowitz JH, Grubb JH, Maga JA, Schmiel DH, Vogler C, Sly WS. Lycosylation-independent targeting enhances enzyme delivery to lysosomes and decreases storage in mucopolysaccharidosis type VII mice. *Proc. Nat. Acad. Sci.* 101:3083–3088, 2004.

Lee JES, Falk RE, Ng WG, Donnell GN. b-Glucuronidase deficiency: A heterogeneous mucopolysaccharidosis. *Am J Dis Child.* 139:57, 1985.

Leroy GL. Management of the mucopolysaccharidosis and allieddisorders. In: *The Management of Genetic Disorders*, New York, Alan R. Liss, pp.133–155, 1979.

Levy B, Galvin N, Vogler C, Birkenmeier EH, Sly WS. Neuropathology of murine mucopolysaccharidosis type VII.*Acta Neuropathol (Berl).* 92(6):562–568, 1996.

Li T, Davidson BL. Phenotype correction in retinal pigment epithelium in murine mucopolysaccharidosis VII by adenovirus-mediated gene transfer. *Proc. Nat. Acad. Sci.* 92:7700–7704, 1995.

Lissens W, Dedobbeleer G, Foulon W, De Catte L, Charels K, Goossens A, Liebaers I. Beta-glucuronidase deficiency as a cause of prenatally diagnosed non-immune hydrops fetalis. *Prenatal Diag.* 11:405–410, 1991.

Liu G, Martins I, Wemmie JA, Chiorini JA, Davidson BL. Functional correction of CNS phenotypes in a lysosomal storage disease model using adeno-associated virus type 4 vectors. *J Neurosci.* 12; 25(41):9321–9327, 2005.

Machin GA. Hydrops revisited: Literature review of 1414 cases published in the 1980s. *Am J Med Genet.* 34:366, 1989.

Mango RL, Xu L, Sands MS, Vogler C, Seiler G, Schwarz T, Haskins ME, Ponder KP. Neonatal retroviral vector-mediated hepatic gene therapy reduces bone, joint, and cartilage disease in mucopolysaccharidosis VII mice and dogs. *Mol Genet Metab.* 82(1):4–19, 2004.

McKusick VA, Neufeld EF. The mucopolysaccharide storage diseases, in Stanbury JB, Wyngaarden JB, Fredrickson DS, Goldstein JL, Brown MS (Eds.). *The Metabolic Basis of Inherited Disease*, 5th ed. New York, McGraw-Hill, p. 751, 1983.

Medda S, Chemelli RM, Martin JL, Pohl LR, Swank RT. Involvement of the carboxyl-terminal propeptide of b-glucuronidase in its compartmentalization within the endoplasmic reticulum as determined by a synthetic peptide approach. *J Biol Chem.* 264:15824, 1989.

Miller RD, Hoffmann JW, Powell PP, Kyle JW, Shipley JM, Bachinsky DR, Sly WS. Cloning and characterization of the human beta-glucuronidase gene. *Genomics* 7:280–283, 1990.

Moullier P, Bohl D, Heard J-M, Danos O. Correction of lysosomal storage in the liver and spleen of MPS VII mice by implantation of genetically modified skin fibroblasts. *Nature Genet.* 4:154–159, 1993.

Natowicz MR, Isman F, Prence EM, Cedrone P, Allen JJ. Rapid prenatal testing for human beta-glucuronidase deficiency (MPS VII).*Genet Test.* Fall; 7(3):241–243, 2003.

Nelson A, Peterson L, Frampton B, Sly WS. Mucopolysaccharidosis VII (b-glucuronidase deficiency) presenting as nonimmune hydrops fetalis. *J Pediatr.* 101:574, 1982.

Neufeld EF, Muenzer J. The mucopolysaccharidoses. In: *The Metabolic and Molecular Bases of Inherited Disease.* Scriver CR, Beaudet AL, Sly WS, Valle D, Childs B, Kinzler KW, and Vogelstein B (Eds.). 8th edition, Vol. III. New York, McGraw-Hill, Medical, p. 3421, 2001.

Nishimura Y, Rosenfeld MG, Kreibich G, Gubler U, Sabatini DD, Adesnik M, Andy R. Nucleotide sequence of rat preputial gland b-glucuronidase cDNA and in vitro insertion of its encoded polypeptide into microsomal membranes. *Proc Natl Acad Sci USA* 83:7292, 1986.

O'Connor LH, Erway LC, Vogler CA, Sly WS, Nicholes A, Grubb J, Holmberg SW, Levy B, Sands MS. Enzyme replacement therapy for murine mucopolysaccharidosis type VII leads to improvements in behavior and auditory function. *J. Clin. Invest.* 101:1394–1400, 1998.

Oshima A, Kyle JW, Miller RD, Hoffmann JW, Powell PP, Grubb JH, Sly WS, Tropak M, Guise KS, Gravel RA. Cloning, sequencing, and expression of cDNA for human beta-glucuronidase. *Proc. Nat. Acad. Sci.* 84:685–89, 1987.

Ponder KP, Melniczek JR, Xu L, Weil MA, O'Malley TM, O'Donnell PA, Knox VW, Aguirre GD, Mazrier H, Ellinwood NM, Sleeper M, Maguire AM, Volk SW, Mango RL, Zweigle J, Wolfe JH, Haskins ME. Therapeutic neonatal hepatic gene therapy in mucopolysaccharidosis VII dogs. *Proc. Nat. Acad. Sci.* 99:13102–13107, 2002.

Powell PP, Kyle JW, Miller RD, Pantano J, Grubb JH, Sly WS, Rat liver b-glucuronidase: cDNA cloning, sequence comparisons and expression of a chimeric protein in COS cells. *Biochem J.* 250:547, 1988.

Ray J, Bouvet A, DeSanto C, Fyfe JC, Xu D, Wolfe JH, Aguirre GD, Patterson DF, Haskins ME, Henthorn PS. Cloning of the canine beta-glucuronidase cDNA, mutation identification in canine MPS VII, and retroviral vector-mediated correction of MPS VII cells. *Genomics* 48:248–253, 1998.

Sakurai K, Iizuka S, Shen JS, Meng XL, Mori T, Umezawa A, Ohashi T, Eto Y. Brain transplantation of genetically modified bone marrow stromal cells corrects CNS pathology and cognitive function in MPS VII mice. *Gene Ther.* 11(19):1475–1481, 2004.

Sands MS, Birkenmeier EH. A single-base-pair deletion in the beta-glucuronidase gene accounts for the phenotype of murine mucopolysaccharidosis type VII. *Proc. Nat. Acad. Sci.* 90:6567–6571, 1993.

Sands MS, Vogler C, Kyle JW, Grubb JH, Levy B, Galvin N, Sly WS, Birkenmeier EH. Enzyme replacement therapy for murine mucopolysaccharidosis type VII. *J. Clin. Invest.* 93:2324–2331, 1994.

Sands MS, Vogler C, Torrey A, Levy B, Gwynn B, Grubb J, Sly WS, Birkenmeier EH. Murine mucopolysaccharidosis type VII: Long term therapeutic effects of enzyme replacement and enzyme replacement followed by bone marrow transplantation. *J. Clin. Invest.* 99:1596–1605, 1997.

Schwartz CE, Stanislovitis P, Phelan MC, Klinger K, Taylor HA, Stevenson RE. Deletion mapping of plasminogen activator inhibitor, type I (PLANH1) and beta-glucuronidase (GUSB) in 7q21-q22. *Cytogenet. Cell Genet.* 51:152–153, 1991.

Sewell AC, Gehler J, Mittermaier G, Meyer E. Mucopolysaccharidosis type VII (b-glucuronidase deficiency): A report of a new case and a survey of those in the literature. *Clin Genet.* 21:366, 1982.

Schuldt AJT, Hampton TJ, Chu V, Vogler CA, Galvin N, Lessard MD, Barker JE. Electrocardiographic and other cardiac anomalies in β-glucuronidase-null mice corrected by nonablative neonatal marrow transplantation *PNAS*, 101:603–608, 2004.

Sferra TJ, Backstrom K, Wang C, Rennard R, Miller M, Hu Y. Widespread correction of lysosomal storage following intrahepatic injection of a recombinant adeno-associated virus in the adult MPS VII mouse.Mol Ther. 10(3):478–491, 2004.

Sleeper MM, Fornasari B, Ellinwood NM, Weil MA, Melniczek J, O'Malley TM, Sammarco CD, Xu L, Ponder KP, Haskins ME. Gene therapy ameliorates cardiovascular disease in dogs with mucopolysaccharidosis VII.Circulation. 17; 110(7): 815–820, 2004.

Sly WS, Quinton BA, McAlister WH, Rimoin DL. b-Glucuronidase deficiency: Report of clinical, radiologic and biochemical features of a new mucopolysaccharidosis. *J Pediatr* 82:249, 1973.

Sly WS, Vogler C, Grubb JH, Zhou M, Jiang J, Zhou XY, Tomatsu S, Bi Y, Snella EM. Active site mutant transgene confers tolerance to human beta-glucuronidase without affecting the phenotype of MPS VII mice. *Proc. Nat. Acad. Sci.* 98:2205–2210, 2001.

Speleman F, Vervoort R, Van Roy N, Liebaers I, Sly WS, Lissens W. Localization by fluorescence in situ hybridization of the human functional beta-glucuronidase gene (GUSB) to 7q11.21-q11.22 and two pseudogenes to 5p13 and 5q13. *Cytogenet. Cell Genet.* 72:53–55, 1996.

Stahl PD, Touster O. b-Glucuronidase of rat liver lysosomes. J Biol Chem 246:5398, 1971.

Stangenberg M, Lingman G, Roberts G, Ozand P. Mucopolysaccharidosis VII as cause of fetal hydrops in early pregnancy. *Am. J. Med. Genet.* 44:142–144, 1992.

Stawser LD, Touster O. Demonstration of rat liver microsomal binding protein specific for b-glucuronidase. *J Biol Chem.* 254:3716, 1979.

Storch S, Wittenstein B, Islam R, Ullrich K, Sly WS, Braulke T. Mutational analysis in longest known survivor of mucopolysaccharidosis type VII. *Hum. Genet.* 112:190–194, 2003.

Tomatsu S, Fukuda S, Sukegawa K, Ikedo Y, Yamada S, Yamada Y, Sasaki T, Okamoto H, Kuwahara T, Yamaguchi S, Kiman T, Shintaku H, Isshiki G, Orii T. Mucopolysaccharidosis type VII: Characterization of mutations and molecular heterogeneity. *Am. J. Hum. Genet.* 48:89–96, 1991.

Tomatsu S, Orii KO, Vogler C, Grubb JH, Snella EM, Gutierrez MA, Dieter T, Sukegawa K, Orii T, Kondo N, Sly WS. Missense models [Gus(tm(E536A)Sly), Gus(tm(E536Q)Sly), and Gus(tm(L175F)Sly)] of murine mucopolysaccharidosis type VII produced by targeted mutagenesis. *Proc. Nat. Acad. Sci.* 99:14982–14987, 2002.

Tomin S, Paigen K. Egasyn, a protein complexed with microsomal b-glucuronidase. *J Biol Chem.* 250:1146, 1975.

van Dorpe J, Moerman P, Pecceu A, van den Steen P, Fryns JP. Non-immune hydrops fetalis caused by beta-glucuronidase deficiency (mucopolysaccharidosis VII): Study of a family with 3 affected siblings. *Genet. Counsel.* 7:105–112, 1996.

van Eyndhoven HWF, ter Brugge HG, van Essen AJ, Kleijer WJ. Beta-glucuronidase deficiency as cause of recurrent hydrops fetalis: The first early prenatal.diagnosis by chorionic villus sampling. *Prenatal Diag.* 18:959–962, 1998.

Vervoort R, Buist NRM, Kleijer WJ, Wevers R, Fryns J-P, Liebaers I, Lissens W. Molecular analysis of the beta-glucuronidase gene: Novel mutations in mucopolysaccharidosis type VII and heterogeneity of the polyadenylation region. *Hum. Genet.* 99:462–468, 1997.

Vervoort R, Gitzelmann R, Bosshard N, Maire I, Liebaers I, Lissens W. Low beta-glucuronidase enzyme activity and mutations in the human beta-glucuronidase gene in mild mucopolysaccharidosis type VII, pseudodeficiency and a heterozygote. *Hum. Genet.* 102:69–78, 1998a.

Vervoort R, Gitzelmann R, Lissens W, Liebaers I. A mutation (IVS8+0.6kbdelTC) creating a new donor splice site activates a cryptic exon in an Alu-element in intron 8 of the human beta-glucuronidase gene. *Hum. Genet.* 103:686–693, 1998b.

Vogler C, Levy B, Galvin N, Lessard M, Soper B, Barker J. Early onset of lysosomal storage disease in a murine model of mucopolysaccharidosis type VII: undegraded substrate accumulates in many tissues in the fetus and very young MPS VII mouse. *Pediatr Dev Pathol.* 8(4):453–462, 2005.

Vogler C, Levy B, Galvin N, Sands MS, Birkenmeier EH, Sly WS, Barker J. A novel model of murine mucopolysaccharidosis type VII due to an intracisternal A particle element transposition into the beta-glucuronidase gene: clinical and pathologic findings. *Pediat. Res.* 49:342–348, 2001.

Vogler C, Levy B, Kyle JW, Sly WS, Williamson J, Whyte MP. Mucopolysaccharidosis VII: Postmortem biochemical and pathological findings in a young adult with beta-glucuronidase deficiency *Mod Pathol.* 7(1):132–137, 1994.

Wang B, O'Malley TM, Xu L,Vite C, Wang P, O'Donnell PA, Ellinwood NM, Haskins ME, Ponder KP. Expression in blood cells may contribute to biochemical and pathological improvements after neonatal intravenous gene therapy for mucopolysaccharidosis VII in dogs. *Mol Genet Metab.* 87(1):8–21, 2006.

Ward JC, Sharpe CR, Luthardt FW, Martens PR, Palmer CG. Regional gene mapping of human beta-glucuronidase (GUSB) by dosage analysis: assignment to region 7q11.23-7q21. (Abstract) *Am. J. Hum. Genet.* 35:56A, 1983.

Wilson D, Melnik E, Sly W, Makesbery WR. Neonatal b-glucuronidase-deficiency mucopolysaccharidosis (MPS VII): Autopsy findings. *J Neuropathol Exp Neurol* 41:344, 1982.

Wolfe JH, Sands MS, Barker JE, Gwynn B, Rowe LB, Vogler CA, Birkenmeier EH. Reversal of pathology in murine mucopolysaccharidosis type VII by somatic cell gene transfer. *Nature* 360:749–753, 1992.

Wolfe JH, Schuchman EH, Stramm LE, Concaugh EA, Haskins ME, Aguirre GD, Patterson DF, Desnick RJ, Gilboa E. Restoration of normal lysosomal function in mucopolysaccharidosis type VII cells by retroviral vector-mediated gene transfer. *Proc. Nat. Acad. Sci.* 87:2877–2881, 1990.

Wu BM, Sly WS. Mutational studies in a patient with the hydrops fetalis form of mucopolysaccharidosis type VII. *Hum. Mutat.* 2:446–457, 1993.

Wu BM, Tomatsu S, Fukuda S, Sukegawa K, Orii T, Sly WS. Overexpression rescues the mutant phenotype of L176F mutation causing beta-glucuronidase deficiency mucopolysaccharidosis in two Mennonite siblings. *J. Biol. Chem.* 269:23681–23688, 1994.

Yamada S, Tomatsu S, Sly WS, Islam R, Wenger DA, Fukuda S, Sukegawa K, Orii T. Four novel mutations in mucopolysaccharidosis type VII including a unique base substitution in exon 10 of the beta-glucuronidase gene that creates a novel 5-prime-splice site. *Hum. Molec. Genet.* 4:651–655, 1995.

Yamada Y, Kato K, Sukegawa K, Tomatsu S, Fukuda S, Emura S, Kojima S, Matsuyama T, Sly WS, Kondo N, Orii T. Treatment of MPS VII (Sly disease) by allogeneic BMT in a female with homozygous A619V mutation. *Bone Marrow Transplant.* 21:629–634, 1998.

POMPE DISEASE-GLYCOGENOSIS TYPE II: ACID MALTASE DEFICIENCY

Arnold Reuser, Marian Kroos

1 INTRODUCTION

Pompe disease (OMIM #232300) is caused by acid α-glucosidase deficiency leading to lysosomal glycogen storage in numerous tissues but primarily affecting skeletal muscle function (Hirschhorn and Reuser 2001, Engel and Hirschhorn 2004). Several findings related to Pompe disease have contributed to the understanding of lysosomal storage disorders in general. It concerns: the detection of storage 'vacuoles' (Pompe 1932), the definition of a lysosomal enzyme deficiency, and the subsequent definition of a lysosomal storage disorder (Hers et al., 1963; Lejeune, 1963), the correlation between residual activity and clinical phenotype (Mehler and DiMauro, 1977; Reuser, 1978), the occurrence of posttranslational modification of lysosomal proteins and mannose 6-phosphorylation (Kaplan, 1977; Hasilik and Neufeld, 1980a,b), as well as the concept of enzyme replacement therapy (Baudhuin et al., 1964).

In 1932, the available histochemical staining procedure for glycogen allowed for the identification of the storage compound and revealed the difference between cytoplasmic glycogen accumulation, as in Von Gierke disease (Glycogenosis type I), and 'vacuolar' storage, as in Pompe disease (Pompe, 1932; Putschar, 1932). Pompe disease became a 'vacuolar' storage disorder long before the lysosomes were identified as intracellular organelle (De Duve, 1955). Pompe disease also attracted attention by its peculiarity of glycogen storage in virtually all tissue types whereas there were no abnormalities in the structure of the glycogen nor in any of the enzymes then known to be involved in glycogen synthesis and degradation. In 1954, the clinical condition earlier described by Drs. Pompe and Putschar, independently, was placed second (type II) on the list of known glycogenoses (Cori, 1954).

Nine years later, when the substrate specificities of acid α-glucosidase were already known, Hers and colleagues reported that patients with Pompe disease have acid α-glucosidase deficiency concurrent with lysosomal glycogen storage (Hers, 1963; Lejeune et al., 1963). This became a crucial finding as it set off a hunt for missing lysosomal enzyme activities in known vacuolar c.q. lysosomal pathologies with so far unknown cause. For several decades, identification of the proteins involved in the degradation of the lysosomal storage compounds remained the most successful approach to establish the primary cause of lysosomal storage disorders (Hers and Van Hoof, 1973).

Department of Clinical Genetics, Erasmus MC, Rotterdam, The Netherlands. E-mail: a.reuser@erasmusmc.nl

The simplicity of measuring acid α-glucosidase activity with both natural (glycogen or maltose) and artificial substrates (4-methylumbelliferyl-α-D-glucopyranoside) led soon to the notion that the clinical phenotype of Pompe disease is primarily determined by the level of residual acid α-glucosidase activity (Mehler and DiMauro, 1977; Reuser et al., 1978, 1995). In 1983, similar correlations were demonstrated to apply to several other lysosomal storage disorders (Conzelmann and Sandhoff, 1983).

Thanks to the availability of one of the very first substrate affinity purification matrices, antibodies could be raised against the purified enzyme and used to compare properties of the normal and the pathologic forms of acid α-glucosidase. This way it was shown that patients with classic-infantile Pompe disease mostly produce very little, catalytically inactive enzyme whereas low levels (7–28%) of otherwise normal α-glucosidase are frequently found in adult Pompe disease (Reuser et al., 1978). These early findings were corroborated by an elegant series of experiments by Hasilik and Neufeld who introduced pulse-chase immunoprecipitation procedures to visualize the various steps of lysosomal enzyme biosynthesis. Acid α-glucosidase was amongst the first few enzymes used to demonstrate the power of the technology that is to date indispensable for in-depth investigation of phenotype–genotype correlations (Hasilik and Neufeld, 1980a,b).

Pompe disease also played a pivotal role in the development of enzyme replacement therapy. In 1964, the first attempt was made employing an acid α-glucosidase extract from the fungus Aspergillus niger (Baudhuin et al., 1964). This attempt was not successful but followed by worldwide initiatives to apply enzyme replacement therapy in lysosomal storage disorders. These early efforts with enzyme preparations from various human and nonhuman sources – according to present day standards given in a relatively low dose and for a relatively short period – continued for about 15 years with little encouraging results. But now, another 20 years later, enzyme replacement therapy is back on stage and has proven its beneficial potential for several of the lysosomal storage diseases (Barton et al., 1990; Eng, 2001; Kakkis, 2001).

Ongoing clinical studies on the safety and efficacy of enzyme replacement therapy for Pompe disease are a challenging activity. Time was invested in producing recombinant human α-glucosidase in the milk of transgenic rabbits (Bijvoet et al., 1996, 1998, 1999). The method is feasible and valuable as proven by the six-year-long safe and promising clinical applications (Van den Hout et al., 2000, 2004; Winkel et al., 2004; Klinge et al., 2005). In addition, the clinical studies with the rabbit milk enzyme have led to improved understanding of the molecular requirements, the enzyme dose and the dosing regimen needed for successful outcome of enzyme replacement therapy in Pompe disease. The current aim is to obtain market approval for recombinant human α-glucosidase produced in genetically engineered CHO cells (Fuller et al., 1995; Van Hove et al., 1996; Amalfitano et al., 2001).

This chapter does not intend to supply an exhaustive review of all recent literature in the field of Pompe disease but addresses a number of selected items.

2 CLINICAL SPECTRUM

Pompe disease can manifest at any age leading to a truly continuous spectrum of disease phenotypes. Muscle weakness is the common and usually also the first clinical symptom. Affected newborns present as floppy babies, toddlers have developmental delays, children and teenagers can't keep up with their classmates while running or sporting, and adults experience difficulties with rising from a chair, climbing stairs, or lifting objects (Felice, 1995; Wokke, 1995; Hirschhorn and Reuser, 2001; Engel and Hirschhorn, 2004;

Kishnani and Howell, 2004; Van den Hout et al., 2003; Hagemans et al., 2005; Winkel, 2005). In a minority of cases, respiratory problems are the first sign of muscle weakness (Moufarrej and Bertorini, 1993; Margolis et al., 1994; Mellies et al., 2001, Hagemans et al., 2005). Once started, the symptoms gradually worsen whereby the rate of disease progression is higher in infants with early disease manifestations than in adults with late-onset of symptoms (Hagemans et al., 2005a,b). Cardiac manifestations are, next to skeletal muscle weakness, a salient feature of Pompe disease in severely affected newborns (the classic-infantile phenotype), but are rarely if at all encountered in teenagers and adults Infants with a 'nonclassic' or 'nontypical' infantile phenotype can have a borderline cardiomyopathy (Slonim et al., 2000).

Although the clinical signs are very much restricted to the heart and skeletal muscles, the storage of lysosomal glycogen is truly generalized. For instance, Schwann cells and motor neurons were found to accumulate glycogen, but patients with Pompe disease do not have neurological symptoms (Hirschhorn and Reuser, 2001; Winkel et al., 2003; Engel and Hirschhorn, 2004; Winkel, 2005). The theory that affected infants die before they develop such symptoms does not hold. Infants whose natural lifespan was extended by more than five years through the application of enzyme replacement therapy do not show signs of CNS involvement and the therapeutic enzyme is unlikely to cross the blood–brain barrier (Van den Hout et al., 2004). Among the problems possibly related to glycogen storage in smooth muscle cells are complaints about digestive and bladder dysfunctions, as well as the occurrence of aneurisms of the basal arteries (Hirschhorn and Reuser, 2001; Anneser et al., 2005; Winkel, 2005). The recent discovery of a hearing deficit in severely affected infants was a surprise finding. The symptom is probably due to cochlear dysfunction (Van den Hout et al., 2001, 2004; Kamphoven et al., 2004). Knockout mice with Pompe disease store glycogen in the inner and outer hair cells.

3 PATHOGENESIS

The pathogenic mechanism is largely unknown except that the process starts with the lysosomal accumulation of glycogen when the rate of glycogen import exceeds the rate of lysosomal glycogen degradation. Lysosomes are being filled with glycogen and the lysosomal compartment expands by the formation of more lysosomes. The size of the lysosomes also increases as small lysosomes fuse to form larger ones. The lysosomes that are normally located at the periphery of the skeletal muscle fibres are also seen between the contractile filaments, hamper contractility, and cause fibre splitting. Lakes of cyto-plasmic glycogen form in severely affected fibres. This is possibly due to mechanical forces shearing the limiting membrane of the rigid glycogen-loaded lysosomal structure. Alternatively, the autophagic capacity of the muscle cell can be compromised by the inaccessibility of the overloaded lysosomal system resulting in diminished turnover of cytoplasmic glycogen and excessive formation of autophagic vacuoles.

At the end, the architecture of the muscle fibre is totally disturbed so that normal metabolic and contractile functions are lost. Without the possibility of repair, the fibre is wasted and replaced by scar tissue. This is the picture that emerges when affected muscle sections are microscopically inspected, in particular when consecutive biopsies are taken or when the disease process is followed in aging knockout mice (Bijvoet, 1998; Raben et al., 1998; Bijvoet et al., 1999; Winkel et al., 2003; Engel and Hirschhorn, 2004; Lynch et al., 2005). The true pathogenic process is certainly more complicated but this very simple model forms a useful working hypothesis. Aspects of the model were addressed in two recent publications (Hesselink et al., 2002, 2003).

4 NOMENCLATURE

The continuity of the clinical spectrum complicates the subclassification of patients. The presently used nomenclature is far from uniform. The name Pompe disease was given to the condition described in 1932. In those days it was quite common to call a disease after the person presenting the first detailed description. Other lysosomal storage disorders obtained their name this way, as for instance Tay–Sachs disease, Anderson–Fabry disease, and Gaucher disease, and kept their name. The nomenclature for Pompe disease took a different turn. In 1954, the disease was listed as number two (Glycogenosis type II) among the several known glycogen storage diseases. A third name, 'Acid Maltase Deficiency' was introduced in 1963 when the acid α-glucosidase deficiency was discovered and maltose was used as one of the substrates to assay the enzymatic activity. The name shows up in 'Acid Maltase Deficiency Association', the largest Pompe patient organisation in the United States.

The title of this chapter was chosen so to emphasise that Pompe disease, Glycogenosis type II, and Acid Maltase Deficiency are one and the same disease entity that occurs in a variety of phenotypes constituting the clinical spectrum. It is irrational to apply the name Pompe disease uniquely to cases that mimic the originally described most severe, early infantile, form of Pompe disease, also termed classic (infantile-onset) Pompe disease. Adjectives given to milder phenotypes are: 'nonclassic' or 'nontypical' for infantile-onset disease with prolonged survival and modest cardiac hypertrophy, and 'childhood', 'juvenile', or 'adult', depending on the patient's age at start of disease manifestation or the patient's current age. The term 'muscular variant' applies to all forms of Pompe disease without cardiac manifestations and is nowadays rarely used (Zellweger et al., 1965).

The parallel use of 'infantile-onset' next to 'infantile' Pompe disease is really confusing. A patient with the infantile form of Pompe disease has a life expectancy of less than 1 year, whereas a 20-year-old patient may have had onset of symptoms in the first year of life and can therefore be called a patient with infantile-onset Pompe disease. The 'late-onset' phenotype is not well defined either. The term was originally used to refer to juvenile and adult forms of Pompe disease, but is occasionally also used to indicate that a patient did not present symptoms in the first year of life. A simple bipartite subdivision in either infantile-onset or late-onset Pompe disease masks rather than clarifies the patients' condition and prognosis (Kishnani et al., 2005).

It is most informative to classify patients according to their current age with annotation of the age at onset, such as, for instance, an infant with classic infantile-onset, disease; a child with either infantile- or childhood-onset disease; or an adult with onset of symptoms during early childhood or late adulthood. Precise information on the clinical course is necessary for a correct classification of the disease phenotype and essential for an educated prognosis. DNA, protein, and enzymatic analyses can be a tremendous help for classifying a patient accurately.

4.1 Salient Features of the Various Phenotypes

4.1.1 Classic-Infantile

Symptoms start at a median age of 1.6 months, and the median life expectancy is 6–8 months. Five to eight percent of 153 cases reported in the literature lived longer than one year, and only two patients longer than 1.5 years. The longest survival was 2.9 years (Slonim, 2000; Amalfitano, 2001; Van den Hout, 2003). A massive cardiac hypertrophy and a rapidly progressive loss of muscle function are the two key features that distinguish these patients from all other subtypes. Major developmental milestones are not reached. The patients have virtually no residual acid α-glucosidase activity and a set of two fully deleterious mutations in the acid α-glucosidase gene (GAA) (Kroos et al., 1995; Reuser et al., 1995; Hermans, 2004).

4.1.2 Nonclassic or Nontypical Infantile and Childhood

In contrast to classic-infantile Pompe disease, these patients have moderate cardiac involvement which can be transient. Muscle weakness dominates the clinical picture and pulmonary infections followed by respiratory insufficiency are the major life-threatening events. It concerns a heterogeneous group of patients who are usually diagnosed in the first year of life with either a rapid or more slowly deteriorating course (Slonim et al., 2000; Haley et al., 2003, 2004; Hagemans et al., 2005).

4.1.3 Juvenile and Adult

There are no criteria to delineate these subtypes separately other than by the age of the patient and the age at onset of symptoms. A teenager with a more than 10 years history of disease is at risk to develop a scoliosis in puberty and has obviously a more severe disease phenotype than a 60-year-old senior who experiences at that age difficulty climbing stairs. Yet, it is difficult to define clinical subclasses. When the group of patients with onset of symptoms after the first year of life is subdivided by age, in age categories from 1–6 y, 6–18 y, and over 18 y, there is no statistical difference between the groups in age of onset or specific clinical features (Winkel, 2005). All patients experience problems with walking when their disease progresses. Forty-four percent of the investigated patient population uses a wheelchair (Hagemans et al., 2005a,b). Respiratory problems are common and can lead to ventilator dependency (45% of patients uses respiratory support; Mellies et al., 2001; Iranzo, 2002; Hagemans et al., 2005). A recent study on the quality of life shows that these patients are markedly affected in the physical health domain but score only slightly lower than the general population in the mental health domain (Hagemans, 2004).

5 ENZYMATIC AND MOLECULAR DIAGNOSIS

Acid α-glucosidase (EC 3.2.1.3/20) belongs to the family 31hydrolases (Henrissat, 1991). The acid α-glucosidase gene (GAA) contains 19 coding exons in 20 kb of genomic DNA (Martiniuk, 1986, 1990; Hoefsloot, 1988, 1990). The locus is quite polymorphic with 36 nonpathogenic base changes identified so far. The number of pathogenic mutations has passed 100 and is still growing. A database with mutations is listed at www.pompecenter.nl. This Web site also supplies information about the occurrence of common mutations and the geographic distribution of mutations. There are only few truly common mutations; these are: IVS1(-13T>G; or c.-32-13T>G) frequently observed among adult Caucasian patients, Arg854X (c.2560C>T) among African Americans, and c.1935C>A in Chinese patients from Taiwan and the coastal areas of China (Martiniuk, 1991; Hermans, 1993; Huie, 1994; Shieh, 1994; Boerkoel, 1995; Lin and Shieh, 1996; Shieh and Lin, 1998; Becker et al., 1998). The latter mutation gives rise to the amino acid substitution Asp645Glu. Interestingly, there are several mutations known at this site.

Other common mutations are: delT525 (c.525del) and del exon 18 (c.2481+102_2646+31del), both not unique for, but frequent in, the Dutch patient population (Huie, 1994; Van der Kraan, 1994; Kroos, 1995; Ausems, 1999; Hirschhorn and Huie, 1999; Dagnino et al., 2000; Ausems et al., 2001). With present-day ease to sequence the entire GAA gene it rarely pays off to prescreen patients for the presence of common mutations. Instead, all exons are amplified and sequenced directly and if necessary also the mRNA in search for the pathogenic mutations. This approach has the additional advantage that it helps to define the GAA haplotypes, which in turn may be informative for deciphering the correlation between genotype and phenotype.

The genotype and phenotype in Pompe disease correlate quite well (Hirschhorn and Reuser et al., 2001; Raben et al., 2002). Any combination of two alleles carrying fully deleterious mutations is associated with the classic-infantile phenotype. And, IVS1(-13T>G) in combination with any other mutation predisposes for a nonclassic phenotype, usually juvenile or adult. The effect of each novel mutation needs to be assessed at the DNA, the protein, and the functional level, and the phenotype of the patient also needs to be taken into account in order to establish the correlation of genotype and phenotype (Hermans et al., 2004). In the lower range of residual enzyme activities it can be difficult to determine whether a given mutation is fully or almost fully deleterious (Castro-Gago et al., 1999; Hermans, 2004; Kroos, 2004). Secondary genetic, epigenetic, and environmental factors modulate the disease phenotype and complicate the overall good correlation of genotype and phenotype. Exceptional findings require thorough investigations before they are accepted as examples that contradict the rule (Beratis, 1978; Willemsen, 1993; Hermans, 1994; Kroos, 1997; Vorgerd, 1998).

The different levels of residual α-glucosidase activity in the different clinical subtypes of Pompe disease were first demonstrated in 1977–1978 and later confirmed by larger series of data (Mehler and DiMauro, 1977; Reuser et al., 1978, 1985, 1987, 1995). Occasionally, however, a total absence of acid α-glucosidase activity is reported in adults (Laforet et al., 2000). When such finding is contradicted by both the phenotype as well as the genotype of the patient there is reason to critically review the method of enzyme assay (Ausems et al., 2001). Published case reports show that the acid α-glucosidase assay in leukocytes is not a consistently reliable diagnostic procedure (Laforet et al., 2000; Ausems et al., 2001; Whitaker et al., 2004).

In contrast, the assay is error prone. It requires the use of antibodies to specifically extract lysosomal acid α-glucosidase or the use of glycogen as natural substrate, because neutral maltases interfere unpredictably when the artificial 4-methylumbelliferyl substrate is being used. Using glycogen as substrate one has to be aware that 1 in 16 Caucasians is a carrier of the GAA2 allele encoding an acid α-glucosidase isozyme with reduced activity for glycogen but not causing lysosomal glycogen storage (Swallow et al., 1975, 1989; Martiniuk, 1990). Another complicating factor in the leukocyte assay is the expression of glucoamylase (GANC) also releasing glucose from glycogen (Delque Bayer, 1989; Nichols, 1998, 2003). The GANC locus is polymorphic, and approximately 30% of the white population carries a 'null' allele (Martiniuk and Hirschhorn, 1981).

A muscle biopsy is not ideal either as a routine diagnostic tool. The procedure is invasive and the α-glucosidase activity in the muscle is relatively low. The procedure provides, however, the possibility to obtain a second, independent, diagnostic parameter via light and electron microscopy (Engel and Hirschhorn, 2004).

Fibroblasts grown out of a skin biopsy are the best material for (enzyme) diagnostic and investigational purposes. The acid α-glucosidase activity can be measured conveniently and sensitively at low pH using the 4-methylumbelliferyl substrate. However, one has to be aware that the α-glucosidase activity depends considerably on the cell density. The cells additionally can be used to investigate the biosynthesis of acid α-glucosidase variants. Furthermore, they can serve as DNA and mRNA sources and be stored in a cell bank for follow-up experiments (Hirschhorn and Reuser, 2001; Reuser et al., 1995).

6 PRENATAL DIAGNOSIS

Prenatal diagnosis is routinely performed by chorionic villi sampling around week 12 of gestation. The advantage of this method over amniocentesis is the earlier moment of intervention during pregnancy and the shorter time from biopsy to diagnostic report. An additional advantage is the higher activity of acid α-glucosidase in the chorionic villi sample compared with the cultured amniocytes making the enzymatic assay more sensitive. The risk of maternal contamination is a disadvantage of chorionic villi sampling. In practice, the risk is low when the samples are collected, selected, and prepared by experienced hands. In case of doubt, maternal contamination can be confirmed or excluded by DNA fingerprinting (Kleijer et al., 1995; Hirschhorn and Reuser, 2001).

The acid α-glucosidase activity in the chorionic villi is routinely measured with the artificial substrate 4-methylumbelliferyl-α-D-glucopyranoside at pH 4, but maltose can be used as well (Park et al., 1992; Kleijer et al., 1995). The assay is simple, reliable, and almost always conclusive for the diagnosis of classic infantile-onset Pompe disease. Complications arise when the index patient has residual activity or a parent has very low activity. In these two situations it is difficult to distinguish affected individuals from carriers and better to perform DNA analysis. The disease-causing mutations or the GAA haplotypes of the index patient and the parents need to be known in advance.

There are publications on the imaging of lysosomal glycogen storage in ultrathin sections of amniocytes and chorionic villi as a diagnostic tool, but the method is to our knowledge not applied as a routine procedure (Hug et al., 1984, 1991).

7 NEWBORN SCREENING

Several methods for newborn screening have recently been published. One method measures the activity of acid α-glucosidase in blot spots after immune-capture and is in essence suitable for the detection of Pompe disease as it excludes contaminating gluco-amylase and neutral maltase activities (Umapathysivam et al., 2000). In practise the method shows to be 100% sensitive and 100% specific, and to be better than a similar method based on the detection of the acid α-glucosidase protein (Umapathysivam et al., 2000). The latter is not surprising considering the different effects that GAA mutations can have on the biosynthesis of acid α-glucosidase. Deficiency of catalytic activity is not necessarily associated with deficiency of enzyme protein.

Two other methods developed for the assay of acid α-glucosidase activity in blood spots try to circumvent the contaminating glucoamylase and neutral maltase activities with inhibitors. One method measures both at acidic and neutral pH and employs maltose as acid α-glucosidase inhibitor (Chamoles et al., 2004). The diagnostic conclusions are reached by calculating the difference in activity between maltose inhibited and uninhibited samples while taking the neutral maltase activity measured at pH 6.5 as reference point. The method seems complicated at first sight and takes no account of the polymorphic variations in the neutral maltase activities, but appears to work reasonably well in practise. The last method applies two parallel incubations, one with and one without acarbose as inhibitor of the interfering maltases from the neutrophils. The novelty of this method is the application of a new artificial substrate for acid α glucosidase designed in such a way that five lysosomal enzyme activities can be measured simultaneously (Li et al., 2004). Each specific reaction product is detected by tandem mass spectroscopy, considered to be a versatile tool in newborn screening (Rinaldo et al., 2004). All three methods for newborn screening of Pompe disease will be validated on larger series of samples.

7.1 Acid a-Glucosidase

Acid α-glucosidase (EC.3.2.1.3/20) belongs to the family 31 hydrolases, consists of 952 amino acids, and has typical features of a lysosomal glycoprotein (Hoefsloot, 1988; Henrissat et al., 1991). All seven N-linked glycosylation sites contain carbohydrate side chains (Hermans et al., 1993). Starting from the pioneering work of Hasilik and Neufeld, insight was obtained in the posttranslational modifications of acid α-glucosidase, and transport defects due to pathogenic mutations were clarified (Hasilik and Neufeld, 1980a,b; Reuser 1985, 1987; Hermans et al., 2004; Moreland et al., 2004).

The acid α-glucosidase precursor (see Figure 1) has a relative molecular mass of 110 kD and an N-terminal signal peptide. Sequence comparison within the family 31 hydrolases predicts the existence of a P-type domain signature in the N-terminal propeptide of α-glucosidase with consensus pattern: 'RFDCAPDKAITQEQCEARGCCY' from amino acid position 89 to 110 (Prosite: PDOC00024; Hauser and Hoffmann, 1992; Hauser et al., 1993).

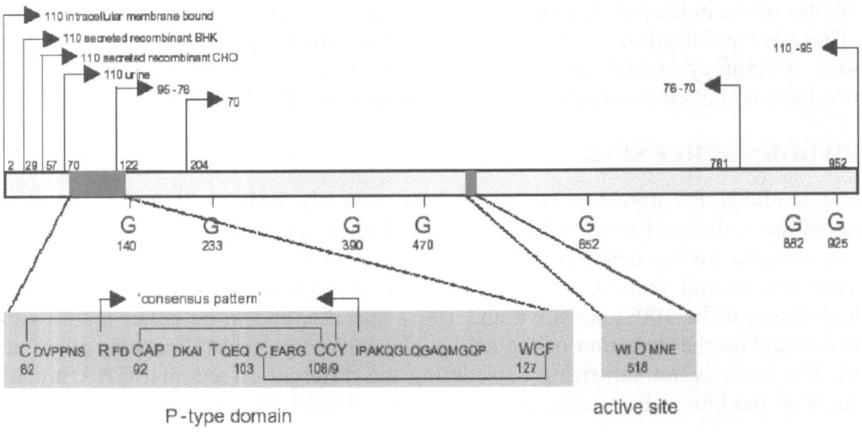

Figure 1. Schematic presentation of the posttranslational modification of acid α-glucosidase. The various sites of proteolytic cleavage are indicated by the numbered amino acid positions at the bottom of the arrows. The arrows mark the limits of the peptides that show by SDS-PAGE under reducing conditions. The figures at the arrow heads refer to the relative molecular mass of the peptides in kD units. The seven glycosylation sites are indicated by the letter G and the number of the asparagine residue involved. The amino acid composition of the P-type domain and the active site domain (D518) are shown in grey boxes.

The 110 kD precursor is cleaved at N- and C- terminal sites located around a 76–70 kD centre piece. This occurs stepwise in the trans-Golgi network, late endosomes and lysosomes (Hoefsloot et al., 1988; Wisselaar et al., 1993). When the biosynthetic process is visualized by pulse-chase labelling and SDS-PAGE under reducing conditions, a change of molecular mass is seen in time: the 110 kD precursor is converted to a 95 kD intermediate which is in turn converted to 76 and 70 kD mature catalytically active enzyme. The exact position of the cleavage sites was recently established, and it was demonstrated that all fragments remain associated with one another (Moreland et al., 2004). This explains how the frequent nonpathogenic Asp91Asn substitution located in the N-terminal propeptide can decrease the affinity of the mature lysosomal acid α-glucosidase for glycogen.

In fibroblasts, an estimated 95% of the 110 kD precursor pool is captured by the mannose 6-phosphate receptor and transported to the lysosomes and approximately 5% is secreted (Oude Elferink et al., 1985). A majority of missense mutations leads to reduced acid-α-glucosidase synthesis or premature degradation of the precursor protein, probably through misfolding and interception by the ER-Golgi complex quality control system. But, some missense mutations are compatible with low-level production of partially functional enzyme (Hermans et al., 2004). Deletions, insertions, and premature stop codons result often in unstable mRNA. Low-level synthesis of otherwise normal acid α-glucosidase is encountered in patients with splice site mutations; IVS1(-13T>G) is the best known example.

8 THERAPY

The options for Pompe disease are the same as for other lysosomal storage disorders: gene therapy to supply a healthy copy of the GAA gene, enzyme therapy to supplement the missing activity, and substrate deprivation to limit the entry of glycogen into the lysosomal system. The apparent absence of CNS involvement is an advantage. The third option is really intriguing as lysosomal glycogen degradation is not needed for glycogen turnover nor required to maintain the cellular energy balance. But, investigational steps in this direction have not yet been taken. The last two sections of this chapter summarize the recent literature on gene therapy and enzyme replacement therapy.

8.1 Gene Therapy

The key issue in somatic gene therapy is finding the right vector for delivery and long term expression of the transgene in the target tissues. Vector development and application are subject to rapidly changing insights and technical possibilities (Cheng and Smith, 2003). Seven years ago, gene therapy approaches were for the first time directed to Pompe disease (Zaretsky et al., 1997). Cultured myoblasts and fibroblasts from patients were transduced with an MLV vector harbouring human acid α-glucosidase cDNA, and this led to correction of the enzyme deficiency and clearance of the lysosomal glycogen. It was the first proof of principle, but the method seems far from optimal for in vivo correction of the muscle cells in Pompe disease as retroviruses transduce only dividing cells.

On the other hand, a recent experiment in MPS VII dogs demonstrated transduction of the liver by a retrovirus (RV) expressing canine ß-glucuronidase when administered intravenously in the neonatal period. The liver started to secrete ß-glucuronidase and supplied enzyme to distant organs such as heart. Mitral regurgitation improved or disappeared in 24 months time, and aortic dilation and aortic valve thickening were prevented (Sleeper et al., 2004). In the future, dividing myoblast progenitor cells can perhaps be transduced ex vivo and used for transplantation in Pompe disease (LaBarge and Blau, 2002). MLV-like vectors have proven their value mainly for treatment of blood cell related diseases such as X-linked SCID, by ex vivo transduction and reimplantation of bone marrow stem cells, but the application is not without risk (Kohn et al., 2003).

Following publications on gene therapy for Pompe disease described the in vitro and in vivo use of adenovirus (Ad), adeno-associated virus (AAV), and hybrid Ad-AAV vector systems. This started with the successful in vitro transduction of fibroblasts and myoblasts with Ad viral vectors (Nicolino et al., 1998; Pauly, 1998). Direct intramuscular and intracardiac injections of an Ad viral vector containing α-glucosidase cDNA in rat, and in acid α-glucosidase-deficient knockout mice and quail resulted in high-level expression and correction of the lysosomal glycogen storage close to the site of injection. But, the effect did not spread to adjacent muscle bundles (Pauly, 1998, Tsujino et al., 1998).

Systemic effects were obtained by intravenous administrations of GAA cDNA containing Ad viral vectors with a CMV promoter. A hundredfold normal acid α-glucosidase activities were achieved in the liver of treated Gaa -/- knockout mice and high plasma levels of the 110 kD precursor through hepatic secretion (Amalfitano et al., 1999; Ding et al., 2001; Pauly et al., 2001). This way, distant organs were cross-corrected, including skeletal muscle, heart, and diaphragm, the primary sites of functional impairment in Pompe disease. The lysosomal glycogen storage in these organs decreased. The evidence for cross-correction was based on the finding of mature 76 and 70 kD forms of α-glucosidase in the target organs, whereas copies of the viral genome were undetectable with sensitive PCR methods. Similar results were obtained in the quail model of Pompe disease (McVie-Wylie et al., 2003). But, the CMV promoter did not sustain expression in the liver for more than few weeks, and antibody formation against both α-glucosidase and Ad viral proteins occurred.

Modifications of vector dose and transgene promoter were introduced to counteract some of these effects. For instance, the CMV promoter was replaced by liver-specific promoters such as the modular liver-specific enhancer promoter (LSP) and the albumine promoter. The elongation factor 1-α enhancer/promoter was used too, but none of these modifications resulted in measurable levels of α-glucosidase in the plasma, and distant organs were not cross-corrected. As low-level expression of the GAA transgene driven by the LSP persisted in liver for six months it was reasoned that the lack of α-glucosidase in plasma was due to rising antibody titres (Ding et al., 2002).

The bottom line is that Ad viral vectors can be used for one-time liver transduction. Very high expression is obtained using the CMV promoter/enhancer and this results in hepatic secretion of α-glucosidase followed by cross-correction of distant organs. The heart profits more than the diaphragm and the skeletal muscle is difficult to cross-correct. However, the CMV promoter is rapidly down-regulated, and expression from liver-specific promoters is too low to achieve cross-correction. Ad viral vectors are prone to elicit an immune response that eventually leads to loss of therapeutic effect.

In the last ten years, broad attention was given to the application of adeno-associated (AAV) viral vectors. They proved powerful vehicles to functionally correct genetic defects of heart and skeletal muscle (Kessler et al., 1996; Herzog et al., 1999; Su et al., 2002), but also liver and other organs in animal models (Daly et al., 1999; Snyder et al., 1999; Jung et al., 2001). The first applications to Pompe disease appeared three years ago. They demonstrated efficient in vitro transduction of fibroblasts and myoblasts from patients resulting in normalization of the acid α-glucosidase activity (Fraites et al., 2002; Lin et al., 2002). Animal model experiments in which the vector was injected directly into the muscle came out equally positive. Four days after injection into the deep pectoral muscle of affected quail, the lysosomal glycolgen storage had disappeared and there were signs of muscle regeneration. But the effect did not spread to distant locations, T lymphocyte infiltrations were noted, and the local effect lasted for only 14 days (Lin et al., 2002). Notably, the CMV promoter was used.

Intramuscular and intracardiac administrations of AAV2- and AAV1-GAA cDNA vector constructs in α-glucosidase-deficient knockout mice resulted in near normal (AAV2) to eight times normal (AAV1) activities, depending on the particle dose, the site of injection, and the time of measurement. The effect was local but encouraging as the glycogen content of the muscle reversed to normal within two weeks time and the contractility of the soleus muscle was partially restored (Fraites et al., 2002). Persistent expression of α-glucosidase was observed in normal Balb/c mice for up to six months, despite the application of the CMV promoter.

Hybrid Ad-AAV vectors with improved packaging features and alternative promoters (CMV enhancer/chicken beta-actin promoter) were used in similar experimental settings. The effect of transduction varied by the vector dose. Intravenous administration of 2 × 1010 Ad-AAV particles via the retro-orbital sinus resulted after two weeks time in a slight increase of the α-glucosidase activity in the liver of Gaa -/- knockout mice but the effect subsided within six weeks. By contrast 1 × 1012 particles led to near normalization of the liver activity after six weeks time (Sun et al., 2003). To optimize the effect, the same high dose of AAV2 or AAV6 particles was infused via the portal vein in immuno-deficient double knockout Gaa -/- /SCID mice. This led to persistent expression of acid α-glucosidase in the liver at 10 times (AAV2) and two times (AAV6) the normal activity as measured after three months.

Moreover, there was at this time point also increased activity in spleen, heart, diaphragm, and gastrocnemius muscle. For heart it was demonstrated that the glycogen content diminished (Sun et al., 2003). The effects were ascribed to transduction of the liver followed by cross-correction of the distant organs in the immune-deficient knockout mice. The same Ad-AAV vector construct was also injected directly into the gastro-cnemius of three-day-old mice. Even after six months, the activity in the injected muscle was close to 50-fold normal, and the lysosomal glycogen storage was substantially less than in untreated animals. In this experiment, there also was increased activity in the hamstring, the diaphragm, and the heart as sign of a systemic effect (Sun et al., 2003). The procedure led to anti acid α-glucosidase and anti Ad antibody formation, but apparently not to such extent that it annihilated the therapeutic effect.

Further efforts to develop gene therapy for Pompe disease focused on optimizing the tissue targeting by cross-packaging of AAV2 vectors with non-AAV2 capsids. For instance, AAV2/7 was considered more suitable for transduction of muscle and AAV2/8 would be more suitable for liver (Gao et al., 2002; Sarkar et al., 2004). Aiming for transduction of the liver and cross-correction of heart and muscle, the AAV2/8 vector with the CMV enhancer and chicken beta-actin promoter was employed. This approach resulted in correction of the acid α-glucosidase-deficiency and reduction of the glycogen storage in the muscle of immune-deficient double knockout Gaa -/- /SCID mice (Sun et al., 2005). Six months after a single intravenous injection of virus particles, the mice showed improved muscle function on a rota rod. Indeed, the effect was brought about by liver transduction and not by muscle transduction. The α-glucosidase activity in the liver of male and female mice exceeded the normal activity by 25- and 10-fold, respectively. For heart, diaphragm, quadriceps, and gastrocnemius of male mice the respective values were 30%, 700%, 100%, and 100% of normal. In female mice there was a response in liver, heart, and diaphragm, but not in other muscle groups. The vector copy number in muscle and heart was in all animals very low.

The picture emerging from the total set of experiments reviewed above is that of efficient transduction of the liver and poor transduction of muscle and heart after systemic delivery. But, after transduction, the liver can in principle function as a continuous endogenous source of therapeutic enzyme and cross-correct distant organs. Raben et al. (2001) demonstrated in a different way that expression of an α-glucosidase transgene in the liver under a liver-specific, tetracycline-controlled, promoter works far better for systemic correction than expression of the transgene in the muscle under a muscle-specific promoter. After all, the liver is a gland and the muscle isn't.

Clinical applications of AAV-based therapies in humans are underway such as, for instance, two FDA-approved studies in which AAV-Factor IX is applied intramuscular and systemic for the treatment of hemophilia B (High, 2004).

A last publication about the use of Ad and AAV viral vectors to be mentioned in this section demonstrated selective transduction of the diaphragm when GAA cDNA containing AAV1 and AAV2 vectors were injected in utero in the fetal liver or in the intraperitoneal cavity (Rucker et al., 2004). The up to ten times normal acid α-glucosidase activity achieved this way in the diaphragm of one-month-old mice appeared sufficient to prevent lysosomal glycogen storage and loss of contractile function.

A totally different gene therapy approach aims to correct gene defects in vivo. Modified single-stranded oligonucleotides were employed to replace the DNA stretch containing the mutation. The method was shown to work in vitro by correcting the α-glucosidase deficiency in cultured fibroblasts of patients with Pompe disease to 0.5–4% of normal. The principle of in vivo use was demonstrated by the introduction of a Gaa mutation in the liver of mice (Lu et al., 2003).

There are several more hurdles to be taken before gene therapy for Pompe disease will find its way into the clinic, but there are options and new ones to come that may eventually lead to success.

8.2 Enzyme Replacement Therapy

In January 1999, recombinant human acid α-glucosidase was applied for the first time to treat Pompe disease. The real novelty of this first study on the safety and efficacy of the procedure was the production of recombinant human α-glucosidase in the milk of transgenic rabbits (Van den Hout et al., 2000). At that time other transgenic milk products were being tested in the clinic, but none had already obtained market approval. As a matter of fact, the situation is still the same. Decades of preclinical research preceded this trial. The work can now be summarized in a few lines. First it was demonstrated that mannose 6-phosphate containing α-glucosidase is taken up by cultured fibroblasts and muscle cells of patients with Pompe disease via the mannose 6-phosphate receptor and corrects the lysosomal glycogen storage in a matter of days (Reuser, 1984; Van der Ploeg, 1987, 1988a,b; Van den Hout et al., 2000). Uptake by cardiomyocytes was also demonstrated in a rat heart perfusion system and in heart and skeletal muscle of healthy mice after intravenous injection (Van der Ploeg et al., 1990, 1991). The mannose 6-phosphate content of α-glucosidase was shown to be important for uptake efficiency, but less so in vivo than in vitro (Van der Ploeg et al., 1991). A recent study describes that artificial conjugation of mannose 6-phosphate containing oligosaccharides to acid α-glucosidase improves the clearance of glycogen from tissues of treated *Gaa* $^{-/-}$ knockout mice (Zhu et al., 2004).

The feasibility of enzyme replacement therapy in Pompe disease was strengthened by the outcome of subsequent series of experiments wherein quail and acid α-glucosidase deficient knockout mice with Pompe disease were treated with recombinant human enzyme produced in CHO cells and milk of transgenic mice and rabbits (Fuller, 1995; Van Hove et al., 1996; Kikuchi et al., 1998; Bijvoet et al., 1998, 1999). Production in CHO cells was achieved by cloning the acid α-glucosidase cDNA either behind the human elongation factor 1α gene promoter or, alternatively, behind the CMV promoter in tandem with the dihydrofolate reductase gene to cause gene amplification. To produce α-glucosidase in the mammary gland of transgenic rabbits, the entire human GAA gene was cloned behind the bovine α-S_1 casein promoter to assure mammary-gland-specific, high-level expression of α-glucosidase in the secretory epithelium.

These preclinical studies showed that intravenously supplied enzyme reaches many tissues, except brain, and that liver and spleen are far more responsive than heart and skeletal muscles in terms of enzyme uptake and glycogen degradation. The minimal dose to obtain uptake in skeletal muscle was estimated to be in the range of 10–20 mg/kg body weight and the dose response was almost linear up to 100 mg/kg (Reuser, unpublished results). During the course of the clinical trials, these estimates appeared to be realistic.

Four patients were initially enrolled in a study testing enzyme replacement therapy with recombinant human α-glucosidase from rabbit milk. They were proven cases of classic-infantile Pompe disease by established clinical, enzymatic, and genetic criteria. The youngest two were 2.5 and 3 months old when their treatment started, and the oldest two were 7–8 months old. All four had cardiomegaly.

The youngest patient showed signs of cardiac failure, respiratory distress, and was fed by nasogastric tube from birth on. The second youngest did not manifest these signs but demonstrated axial hypotonia and head lag like the youngest. The two older infants were both in an end stage of disease at the time of inclusion. They were oxygen-dependent, could barely move their arms and legs, and were close to cardiorespiratory failure. One of them became respirator-dependent before the first enzyme infusion. The initial dose given was 15–20 mg/kg per week, but the dose was raised to 40 mg/kg per week after 12 weeks of treatment to optimize the effect. The weekly high doses were well tolerated. Antihistamine premedication was given for only a short period in the beginning, but later stopped. Occasionally occurring side effects were managed by temporarily lowering of the infusion rate.

The results obtained in the first 36 and 72 weeks of treatment were published (Van den Hout et al., 2000, 2001, 2004). After half a year of treatment, the muscle acid α-glucosidase activity of all infants had reached the normal range. After 72 weeks of treatment the activity of the youngest two patients was still high, but tended to drop again below the normal range in the older two patients. A decrease of the glycogen content was only measured at the 72 weeks time point in muscle of the second youngest patient who had the best condition at inclusion. In microscopic specimens of the muscle prepared in week 12, the staining intensity with PAS (glycogen) was less than in week 0, but the number of vacuoles had increased.

In week 24, there seemed to be some muscle fibre regeneration in three of the four infants but further improvement of the muscle architecture was only seen in the patient performing best at start (Winkel et al., 2003). He learned to walk, does not require a ventilator, attends school, and is at present seven years old. Despite his remarkable development he has residual disease in the form of a myopathic face, a somewhat unsturdy gait, and a hearing deficit.

The youngest patient responded very well in the beginning. She learned to sit unsupported and her respiratory function improved substantially till the age of two. From then on her condition slowly declined and she became ventilator-dependent after a bout of pneumonia. Her death came quite unexpectedly. She attracted a very high fever, went into coma, and died at the age of four years. The oldest two patients are still alive at the age of seven, but are fully ventilator-dependent and have lost almost all muscle function, but they are bright and do not show any sign of mental decline (Van den Hout et al., 2004). The latter is an important finding indicating that there is no immediate risk of developing brain disease when the life of patients with classic-infantile Pompe disease is prolonged by enzyme replacement therapy. The prolonged survival of all four patients in this study was considered the best proof of clinical efficacy and was largely attributed to the good response of cardiac muscle. A rapid decline of the left ventricular wall thickness was recorded after initiation of the therapy (Van den Hout et al., 2000).

Two more infants were included in a parallel study conducted in another centre. They too responded well to the treatment with recombinant human α-glucosidase from rabbit milk. The study report over the first 48 weeks of treatment describes an overall improvement of the left ventricular mass, the cardiac function, and the skeletal muscle morphology and function (Klinge et al., 2005).

Microscopic images of muscle biopsies taken during treatment gave the impression that severely affected muscle fibres cannot be rescued. Thus, the condition of the patients at start of treatment is largely decisive for the long-term clinical effects of treatment (Winkel et al., 2003). Similarly, it was demonstrated in a knockout mouse model of Pompe disease that the effect of enzyme replacement therapy diminishes when the disease duration increases. Furthermore, type I fibres appeared to respond better than type II fibres (Raben et al., 2005).

In the same year that the trials with recombinant human α-glucosidase from rabbit milk started, another three infants with Pompe disease were enrolled in a study investigating the safety and efficacy of recombinant human α-glucosidase from genetically engineered CHO cells (Amalfitano et al., 2001). Two of these patients were diagnosed prenatally and the third at two months of age. Two had severe cardiomegaly after birth and one a cardiomyopathy with cardiac dimensions within the upper limit of the normal range.

After start of treatment with 2 × 5 mg/kg per week, the left ventricular mass index (LVMI) of the two patients with cardiomegaly decreased substantially over a 57-week-long treatment period. The LVMI of the third patient remained within the normal range. Muscle biopsies were taken at start and after four months of treatment. There was a 2- to 3-fold increase of the α-glucosidase activity in the patients with cardiomegaly and an 18-fold increase in the patient without. Only the latter patient had a lowering of the glycogen content of the muscle. The AIMS scores of the two severely affected infants improved slightly within the first one or two months of treatment but then declined. The pulmonary function showed a similar trend, and full ventilator support was required after three months of treatment. The third patient had a slightly lower than normal AIMS score at inclusion, hit the normal range after the first few infusions, and kept following the normal curve till the last measuring point at 16 months. The pulmonary function of this third patient was normal at start of treatment and remained normal till the end of the study.

The different response of the patients was ascribed to the fact that the poor-responders had no endogenous α-glucosidase synthesis and therefore developed high antibody titres, whereas the good responder had immunologically detectable α-glucosidase and a low antibody titre (Amalfitano et al., 2001). Escalating doses of α-glucosidase, up to five times the intended dose, were administered for a long period to counteract the adverse effect of antibody formation. This dosing regimen caused nephrotic syndrome in the end. The problem gradually resolved after the dose was lowered again (Hunley et al., 2004). The best responding patient is still alive.

The potential problem of antibody formation led to the start of a new clinical trial with recombinant human α-glucosidase from CHO cells whereby the presence of immunologically detectable endogenous enzyme became one of the inclusion criteria. Dr M. Nicolino gave a preliminary account of the latter study at the June 2004 meeting of the MPS society in Mainz. Prolonged survival and reduction of the LVMI in this study were also clear signs of therapeutic efficacy, but there were good and poor responders in this study as well.

The patients who originally received recombinant human α-glucosidase from rabbit milk were transitioned to a new formula of recombinant enzyme from CHO cells. The limited data that could be collected indicate that the two preparations have a similar safety profile and a similar effective dose range (Winkel, personal communication). In current clinical studies with recombinant human α-glucosidase from CHO cells doses of 20–40 mg/kg per two weeks are applied.

Only three patients with juvenile-adult forms of Pompe disease have participated in an enzyme replacement therapy trial. By now, they have been treated for about six years. The results were described in a recent report covering the first three years (Winkel et al., 2004). The patients were at the time of inclusion 16, 32, and 11 years old, and used a wheelchair. One patient (16 y) was partially, and another (32 y) fully, ventilator-dependent. All three had a long history of slowly progressive muscle weakness. They were treated with α-glucosidase from rabbit milk in an initial dose of 10 mg/kg per week for the first 12 weeks. The dose was thereafter increased to 20 mg/kg per week. The least affected patient (11y) showed a dramatic gain of muscle strength and function as measured with a handheld dynamometer (over tenfold) and by GFM-score (from 57% to normal).

After 72 weeks of treatment the patient managed with difficulty to walk ten meters on tiptoe. The walking ability improved further after release of the Achillus tendon in week 75. After three years of treatment he could run and keep his balance while standing on one leg and kicking a ball. He had normal respiratory function at the start of treatment and a normal age-related increase of function during the six years treatment period.

The 16-year-old patient lost muscle strength during the first 53 weeks of treatment and a severe scoliosis developed causing lumbar pain and Babinsky reflexes. The ability to walk without assistance was lost in this period, but the GMFM score improved due to improved kneeling, crawling, and sitting. Between week 63 and 65, the scoliosis was largely corrected by surgical intervention, and the patient slowly gained muscle strength and function. After six years of treatment the patient can walk between parallel bars, uses a wheelchair for daily activities, and ventilator support during the night (Van der Ploeg, personal communication) the third patient was out of bed for only 2 h per day at the time of inclusion and had lost most muscle function. During treatment, the leg and neck muscles became a little bit stronger, and the PEDI scores for self-care items improved. The patient's vital capacity improved significantly. Now, after six years of treatment, the patient is no longer bedridden but participates in daily family life.

Although the results are fragmentary, they give a strong signal that not only infants but also patients with childhood, juvenile, or adult forms of Pompe disease can benefit from enzyme replacement therapy. The production of acid α-glucosidase in the milk of transgenic rabbits stopped, and all the patients initially participating in studies with this enzyme switched to the new version of recombinant human α-glucosidase from CHO cells. They all responded well to the switch and are now receiving doses of 20–40 mg/kg per week (infants) or the same dose per two weeks (adults). In December 2004, the new product was filed for approval by the European authority (EMEA).

The positive outcome of preclinical and clinical studies, the participation of industry, and the support of patient organizations give a positive outlook on the future treatment of Pompe disease. Sophisticated molecular tools and clinical parameters are at hand to diagnose and position the patients in the continuous clinical spectrum. Early diagnosis is essential for optimal guidance and for early intervention. Raising greater awareness is one way to reach this goal. Newborn screening has great potential. The present-day experience with enzyme replacement therapy is promising and the development of gene therapy is aggressively pursued.

ACKNOWLEDGEMENTS

We would like to acknowledge the pleasant and productive collaboration with our colleague Dr. Ans Van der Ploeg, paediatrician and principal investigator of clinical studies on the safety and efficacy of enzyme replacement therapy, and all who work with us in the PompeCenter at ErasmusMC, Rotterdam, The Netherlands. We appreciate the support of patients and patient organizations and our contacts with Genzyme Corp., Boston, MA and Naarden, The Netherlands.

REFERENCES

Amalfitano A, Bengur AR, Morse RP, Majure JM, Case LE, Veerling DL, Mackey J, Kishnani P, Smith W, McVie-Wylie A, Sullivan JA, Hoganson GE, Phillips JA, 3rd, Schaefer GB, Charrow J, Ware RE, Bossen EH and Chen YT (2001) Recombinant human acid alpha-glucosidase enzyme therapy for infantile glycogen storage disease type II: Results of a phase I/II clinical trial. Genet Med, **3,** 132–138.

Amalfitano A, McVie-Wylie AJ, Hu H, Dawson TL, Raben N, Plotz P and Chen YT (1999) Systemic correction of the muscle disorder glycogen storage disease type II after hepatic targeting of a modified adenovirus vector encoding human acid-alpha-glucosidase. Proceedings of the National Academy of Sciences of the United States of America, **96,** 8861–8866.

Anneser JM, Pongratz DE, Podskarbi T, Shin YS and Schoser BG (2005) Mutations in the acid alpha-glucosidase gene (M. Pompe) in a patient with an unusual phenotype. Neurology, **64,** 368–70.

Ausems MG, ten Berg K, Sandkuijl LA, Kroos MA, Bardoel AF, Roumelioti KN, Reuser AJ, Sinke R and Wijmenga C (2001) Dutch patients with glycogen storage disease type II show common ancestry for the 525delT and del exon 18 mutations. J Med Genet, **38,** 527–9.

Ausems MG, Verbiest J, Hermans MP, Kroos MA, Beemer FA, Wokke JH, Sandkuijl LA, Reuser AJ and van der Ploeg AT (1999) Frequency of glycogen storage disease type II in The Netherlands: Implications for diagnosis and genetic counselling. Euro J Hum Gen, **7,** 713–6.

Ausems MG, Wokke JH, Reuser AJ and van Diggelen OP (2001) Juvenile and adult-onset acid maltase deficiency in France: genotype-phenotype correlation. Neurology, **57,** 1938.

Barton NW, Furbish FS, Murray GJ, Garfield M and Brady RO (1990) Therapeutic response to intravenous infusions of glucocerebrosidase in a patient with Gaucher disease. Proc Natl Acad Sci U S A, **87,** 1913–6.

Baudhuin P, Hers HG and Loeb H (1964) An electron microscopic and biochemical study of type II glycogenosis. Lab Invest, **13,** 1139–1152.

Becker JA, Vlach J, Raben N, Nagaraju K, Adams EM, Hermans MM, Reuser AJ, Brooks SS, Tifft CJ, Hirschhorn R, Huie ML, Nicolino M and Plotz PH (1998) The African origin of the common mutation in African American patients with glycogen-storage disease type II [letter]. Am J Hum Genet, **62,** 991–4.

Beratis NG, LaBadie GU and Hirschhorn K (1978) Characterization of the molecular defect in infantile and adult acid alpha-glucosidase deficiency fibroblasts. Journal of Clinical Investigation, **62,** 1264–74.

Bijvoet AG, Kroos MA, Pieper FR, de Boer HA, Reuser AJ, van der Ploeg AT and Verbeet MP (1996) Expression of cDNA-encoded human acid alpha-glucosidase in milk of transgenic mice. Biochimica et Biophysica Acta, **1308,** 93–6.

Bijvoet AG, Kroos MA, Pieper FR, Van der Vliet M, De Boer HA, Van der Ploeg AT, Verbeet MP and Reuser AJ (1998) Recombinant human acid alpha-glucosidase: High level production in mouse milk, biochemical characteristics, correction of enzyme deficiency in GSDII KO mice. Human Molecular Genetics, **7,** 1815–24.

Bijvoet AG, van de Kamp EH, Kroos MA, Ding JH, Yang BZ, Visser P, Bakker CE, Verbeet MP, Oostra BA, Reuser AJ and van der Ploeg AT (1998) Generalized glycogen storage and cardiomegaly in a knockout mouse model of Pompe disease. Human Molecular Genetics, **7,** 53–62.

Bijvoet AG, Van Hirtum H, Kroos MA, Van de Kamp EH, Schoneveld O, Visser P, Brakenhoff JP, Weggeman M, van Corven EJ, Van der Ploeg AT and Reuser AJ (1999) Human acid alpha-glucosidase from rabbit milk has therapeutic effect in mice with glycogen storage disease type II. Human Molecular Genetics, **8,** 2145–53.

Bijvoet AG, Van Hirtum H, Vermey M, Van Leenen D, Van Der Ploeg AT, Mooi WJ and Reuser AJ (1999) Pathological features of glycogen storage disease type II highlighted in the knockout mouse model. J Pathol, **189,** 416–424.

Boerkoel CF, Exelbert R, Nicastri C, Nichols RC, Miller FW, Plotz PH and Raben N (1995) Leaky splicing mutation in the acid maltase gene is associated with delayed onset of glycogenosis type II. American Journal of Human Genetics, **56,** 887–97.

Castro-Gago M, Eiris-Punal J, Rodriguez-Nunez A, Pintos-Martinez E, Benlloch-Marin T and Barros-Angueira F (1999) Severe form of juvenile type II glycogenosis in a compound-heterozygous boy (Tyr-292--> Cys/Arg-854-->Stop) [Forma grave de glucogenosis tipo II juvenil en un nino heterocigoto compuesto (Tyr-292-->Cys/Arg-854-->Stop)]. Revista de Neurologia, **29,** 46–9.

Chamoles NA, Niizawa G, Blanco M, Gaggioli D and Casentini C (2004) Glycogen storage disease type II: Enzymatic screening in dried blood spots on filter paper. Clin Chim Acta, **347,** 97–102.

Cheng SH and Smith AE (2003) Gene therapy progress and prospects: Gene therapy of lysosomal storage disorders. Gene Ther, **10,** 1275-81.

Cori GT (1954) Glycogen structure and enzyme deficiencies in glycogen storage disease. Harvey Lectures, **8,** 145.

Conzelmann E and Sandhoff K (1983) Partial enzyme deficiencies: Residual activities and the development of neurological disorders. Dev Neurosci, **6,** 58–71.
Dagnino F, Stroppiano M, Regis S, Bonuccelli G and Filocamo M (2000) Evidence for a founder effect in Sicilian patients with glycogen storage disease type II. Hum Hered, **50,** 331–3.
Daly TM, Vogler C, Levy B, Haskins ME and Sands MS (1999) Neonatal gene transfer leads to widespread correction of pathology in a murine model of lysosomal storage disease. Proc Natl Acad Sci U S A, **96,** 2296–300.
De Duve C, Pressman BC, Gianetto R, Wattiaux R and Appelmans F (1955) Tissue fractionation studies. 6. Intracellular distribution patterns of enzymes in rat-liver tissue. Biochem J, **60,** 604–617.
Delque Bayer P, Vittori C, Sudaka P and Giudicelli J (1989) Purification and properties of neutral maltase from human granulocytes. Biochem J, **263,** 647–652.
Ding E, Hu H, Hodges BL, Migone F, Serra D, Xu F, Chen YT and Amalfitano A (2002) Efficacy of gene therapy for a prototypical lysosomal storage disease (GSD-II) is critically dependent on vector dose, transgene, promoter, and the tissues targeted for vector transduction. Mol Ther, **5,** 436–46.
Ding EY, Hodges BL, Hu H, McVie-Wylie AJ, Serra D, Migone FK, Pressley D, Chen YT and Amalfitano A (2001) Long-term efficacy after [E1-, polymerase-] adenovirus-mediated transfer of human acid-alpha-glucosidase gene into glycogen storage disease type II knockout mice. Hum Gene Ther, **12,** 955-65.
Eng CM, Guffon N, Wilcox WR, Germain DP, Lee P, Waldek S, Caplan L, Linthorst GE and Desnick RJ (2001) Safety and efficacy of recombinant human alpha-galactosidase A--replacement therapy in Fabry's disease. N Engl J Med, **345,** 9–16.
Engel AG and Hirschhorn R (Eds.) (2004) Acid maltase deficiency.
Felice KJ, Alessi AG and Grunnet ML (1995) Clinical variability in adult-onset acid maltase deficiency: report of affected sibs and review of the literature. Medicine, **74,** 131–5.
Fraites TJ, Jr., Schleissing MR, Shanely RA, Walter GA, Cloutier DA, Zolotukhin I, Pauly DF, Raben N, Plotz PH, Powers SK, Kessler PD and Byrne BJ (2002) Correction of the enzymatic and functional deficits in a model of Pompe disease using adeno-associated virus vectors. Mol Ther, **5,** 571–8.
Fuller M, Van der Ploeg A, Reuser AJ, Anson DS and Hopwood JJ (1995) Isolation and characterisation of a recombinant, precursor form of lysosomal acid alpha-glucosidase. Eur J Biochem, **234,** 903–9.
Gao GP, Alvira MR, Wang L, Calcedo R, Johnston J and Wilson JM (2002) Novel adeno-associated viruses from rhesus monkeys as vectors for human gene therapy. Proc Natl Acad Sci U S A, **99,** 11854–9.
Hagemans M (2005) Severity in Pompe's disease related to age and disease duration. Neurology, in press.
Hagemans ML, Janssens AC, Winkel LP, Sieradzan KA, Reuser AJ, Van Doorn PA and Van der Ploeg AT (2004) Late-onset Pompe disease primarily affects quality of life in physical health domains. Neurology, **63,** 1688–92.
Hagemans ML, Winkel LP, Van Doorn PA, Hop WJ, Loonen MC, Reuser AJ and Van der Ploeg AT (2005) Clinical manifestation and natural course of late-onset Pompe's disease in 54 Dutch patients. Brain.
Haley SM, Fragala MA and Skrinar AM (2003) Pompe disease and physical disability. Dev Med Child Neurol, **45,** 618–23.

Haley SM, Fragala-Pinkham MA, Ni PS, Skrinar AM and Kaye EM (2004) Pediatric physical functioning reference curves. Pediatr Neurol, **31,** 333–41.

Hasilik A and Neufeld EF (1980a) Biosynthesis of lysosomal enzymes in fibroblasts. Synthesis as precursors of higher molecular weight. J Biol Chem, **255,** 4937–4945.

Hasilik A and Neufeld EF (1980b) Biosynthesis of lysosomal enzymes in fibroblasts. Phosphorylation of mannose residues. J Biol Chem, **255,** 4946–50.

Hauser F and Hoffmann W (1992) P-domains as shuffled cysteine-rich modules in integumentary mucin C.1 (FIM-C.1) from Xenopus laevis. Polydispersity and genetic polymorphism. J Biol Chem, **267,** 24620–4.

Hauser F, Poulsom R, Chinery R, Rogers LA, Hanby AM, Wright NA and Hoffmann W (1993) hP1.B, a human P-domain peptide homologous with rat intestinal trefoil factor, is expressed also in the ulcer-associated cell lineage and the uterus. Proc Natl Acad Sci U S A, **90,** 6961–5.

Henrissat B (1991) A classification of glycosyl hydrolases based on amino acid sequence similarities. Biochem J, **280 (Pt 2),** 309–16.

Hermans MM, De Graaff E, Kroos MA, Mohkamsing S, Eussen BJ, Joosse M, Willemsen R, Kleijer WJ, Oostra BA and Reuser AJ (1994) The effect of a single base pair deletion (delta T525) and a C1634T missense mutation (pro545leu) on the expression of lysosomal alpha-glucosidase in patients with glycogen storage disease type II. Hum Molec Gen, **3,** 2213–8.

Hermans MM, de Graaff E, Kroos MA, Wisselaar HA, Willemsen R, Oostra BA and Reuser AJ (1993) The conservative substitution Asp-645-->Glu in lysosomal alpha-glucosidase affects transport and phosphorylation of the enzyme in an adult patient with glycogen-storage disease type II. Biochem J, **289,** 687–93.

Hermans MM, van Leenen D, Kroos MA, Beesley CE, Van Der Ploeg AT, Sakuraba H, Wevers R, Kleijer W, Michelakakis H, Kirk EP, Fletcher J, Bosshard N, Basel-Vanagaite L, Besley G and Reuser AJ (2004) Twenty-two novel mutations in the lysosomal alpha-glucosidase gene (GAA) underscore the genotype-phenotype correlation in glycogen storage disease type II. Hum Mutat, **23,** 47–56.

Hermans MMP, Wisselaar HA, Kroos MA, Oostra BA and Reuser AJJ (1993) Human lysosomal a-glucosidase: Functional characterization of the glycosylation sites. Biochem J, **289,** 681–686.

Hers HG (1963) alpha-Glucosidase deficiency in generalized glycogen storage disease (Pompe's disease). Biochem J, **86,** 11–16.

Hers HG and Van Hoof F (1973) Lysosomes and Storage Diseases. Academic Press, New York.

Herzog RW, Yang EY, Couto LB, Hagstrom JN, Elwell D, Fields PA, Burton M, Bellinger DA, Read MS, Brinkhous KM, Podsakoff GM, Nichols TC, Kurtzman GJ and High KA (1999) Long-term correction of canine hemophilia B by gene transfer of blood coagulation factor IX mediated by adeno-associated viral vector. Nat Med, **5,** 56–63.

Hesselink RP, Gorselink M, Schaart G, Wagenmakers AJ, Kamphoven J, Reuser AJ, Van Der Vusse GJ and Drost MR (2002) Impaired performance of skeletal muscle in alpha-glucosidase knockout mice. Muscle Nerve, **25,** 873–83.

Hesselink RP, Wagenmakers AJ, Drost MR and Van der Vusse GJ (2003) Lysosomal dysfunction in muscle with special reference to glycogen storage disease type II. Biochim Biophys Acta, **1637,** 164–70.

High KA (2004) Clinical gene transfer studies for hemophilia B. Semin Thromb Hemost, **30,** 257–67.

Hirschhorn R and Huie ML (1999) Frequency of mutations for glycogen storage disease type II in different populations: the delta525T and deltaexon 18 mutations are not generally "common" in white populations [letter; comment]. J Med Gen, **36,** 85–6.

Hirschhorn R and Reuser AJJ (2001) Glycogen storage disease type II (GSDII). In Scriver, C.R., Beaudet, A.L., Sly, W.S. and Valle, D. (Eds.), The Metabolic and Molecular Bases of Inherited Disease. 8 ed. McGraw-Hill, NY, pp. 3389–3420.

Hoefsloot LH, Hoogeveen-Westerveld M, Kroos MA, van Beeumen J, Reuser AJ and Oostra BA (1988) Primary structure and processing of lysosomal alpha-glucosidase; homology with the intestinal sucrase-isomaltase complex. EMBO J, **7,** 1697–704.

Hoefsloot LH, Hoogeveen-Westerveld M, Reuser AJJ and Oostra BA (1990) Characterization of the human lysosomal alpha-glucosidase gene. Biochem J, **272,** 493–497.

Hug G, Chuck G, Chen YT, Kay HH and Bossen EH (1991) Chorionic villus ultrastructure in type II glycogen storage disease (Pompe's disease) [letter]. New England Journal of Medicine, **324,** 342–3.

Hug G, Soukup S, Ryan M and Chuck G (1984) Rapid prenatal diagnosis of glycogen-storage disease type II by electron microscopy of uncultured amniotic-fluid cells. N Engl J Med, **310,** 1018–22.

Huie ML, Chen AS, Brooks SS, Grix A and Hirschhorn R (1994) A de novo 13 nt deletion, a newly identified C647W missense mutation and a deletion of exon 18 in infantile onset glycogen storage disease type II (GSDII). Hum Mol Gen, **3,** 1081–7.

Huie ML, Chen AS, Tsujino S, Shanske S, DiMauro S, Engel AG and Hirschhorn R (1994) Aberrant splicing in adult onset glycogen storage disease type II (GSDII): molecular identification of an IVS1 (-13T-->G) mutation in a majority of patients and a novel IVS10 (+1GT-->CT) mutation. Hum Mol Gens, **3,** 2231–6.

Hunley TE, Corzo D, Dudek M, Kishnani P, Amalfitano A, Chen YT, Richards SM, Phillips JA, 3rd, Fogo AB and Tiller GE (2004) Nephrotic syndrome complicating alpha-glucosidase replacement therapy for Pompe disease. Pediatrics, **114,** e532–5.

Iranzo A (2002) Article reviewed: Sleep-disordered breathing and respiratory failure in acid maltase deficiency. Sleep Med, **3,** 179–80.

Jung SC, Han IP, Limaye A, Xu R, Gelderman MP, Zerfas P, Tirumalai K, Murray GJ, During MJ, Brady RO and Qasba P (2001) Adeno-associated viral vector-mediated gene transfer results in long-term enzymatic and functional correction in multiple organs of Fabry mice. Proc Natl Acad Sci U S A, **98,** 2676–81.

Kakkis ED, Muenzer J, Tiller GE, Waber L, Belmont J, Passage M, Izykowski B, Phillips J, Doroshow R, Walot I, Hoft R and Neufeld EF (2001) Enzyme-replacement therapy in mucopolysaccharidosis I. N Engl J Med, **344,** 182–8.

Kamphoven JH, de Ruiter MM, Winkel LP, Van den Hout HM, Bijman J, De Zeeuw CI, Hoeve HL, Van Zanten BA, Van der Ploeg AT and Reuser AJ (2004) Hearing loss in infantile Pompe's disease and determination of underlying pathology in the knockout mouse. Neurobiol Dis, **16,** 14–20.

Kaplan A, Achord DT and Sly WS (1977) Phosphohexosyl components of a lysosomal enzyme are recognized by pinocytosis receptors on human fibroblasts. Proc Natl Acad Sci USA, **74,** 2026–2030.

Kessler PD, Podsakoff GM, Chen X, McQuiston SA, Colosi PC, Matelis LA, Kurtzman GJ and Byrne BJ (1996) Gene delivery to skeletal muscle results in sustained expression and systemic delivery of a therapeutic protein. Proc Natl Acad Sci U S A, **93,** 14082–7.

Kikuchi T, Yang HW, Pennybacker M, Ichihara N, Mizutani M, Van Hove JLK and Chen YT (1998) Clinical and metabolic correction of Pompe disease by enzyme therapy in acid maltase-deficient quail. J Clin Invest, **101,** 827–33.

Kishnani P, Byrne, B., Vanderploeg A., Mueller-Felber, W. (2005) The Pompe Registry: Centralized data collection to outline the natural course of Pompe disease. Gen Medicine.

Kishnani PS and Howell RR (2004) Pompe disease in infants and children. J Pediatr, **144,** S35–43.

Kleijer WJ, van der Kraan M, Kroos MA, Groener JE, van Diggelen OP, Reuser AJ and van der Ploeg AT (1995) Prenatal diagnosis of glycogen storage disease type II: Enzyme assay or mutation analysis? Pediatr Res, **38,** 103–6.

Klinge L, Straub V, Neudorf U, Schaper J, Bosbach T, Gorlinger K, Wallot M, Richards S and Voit T (2005) Safety and efficacy of recombinant acid alpha-glucosidase (rhGAA) in patients with classical infantile Pompe disease: Results of a phase II clinical trial. Neuromuscul Disord, **15,** 24–31.

Kohn DB, Sadelain M and Glorioso JC (2003) Occurrence of leukaemia following gene therapy of X-linked SCID. Nat Rev Cancer, **3,** 477–88.

Kroos MA, Kirschner J, Gellerich FN, Hermans MM, Van Der Ploeg AT, Reuser AJ and Korinthenberg R (2004) A case of childhood Pompe disease demonstrating phenotypic variability of p.Asp645Asn. Neuromuscul Disord, **14,** 371–4.

Kroos MA, Van der Kraan M, Van Diggelen OP, Kleijer WJ and Reuser AJ (1997) Two extremes of the clinical spectrum of glycogen storage disease type II in one family: A matter of genotype. Hum Mutat, **9,** 17–22.

Kroos MA, Van der Kraan M, Van Diggelen OP, Kleijer WJ, Reuser AJJ, Van den Boogaard MJ, Ausems MGEM, Ploos van Amstel HK, Poenaru L, Nicolino M and Wevers R (1995) Glycogen storage disease type II: Frequency of three common mutant alleles and their associated clinical phenotypes studied in 121 patients. J Med Genet, **32,** 836–7.

LaBarge MA and Blau HM (2002) Biological progression from adult bone marrow to mononucleate muscle stem cell to multinucleate muscle fiber in response to injury. Cell, **111,** 589–601.

Laforet P, Nicolino M, Eymard PB, Puech JP, Caillaud C, Poenaru L and Fardeau M (2000) Juvenile and adult-onset acid maltase deficiency in France: Genotype-phenotype correlation. Neurology, **55,** 1122–8.

Lejeune N, Thinès-Sempoux D and Hers HG (1963) Tissue fractionation studies: Intracellular distribution and properties of a-glucosidases in rat liver. Biochem J, **86,** 16–21.

Li Y, Scott CR, Chamoles NA, Ghavami A, Pinto BM, Turecek F and Gelb MH (2004) Direct multiplex assay of lysosomal enzymes in dried blood spots for newborn screening. Clin Chem.

Lin CY and Shieh JJ (1996) Molecular study on the infantile form of Pompe disease in Chinese in Taiwan. Chung-Hua Min Kuo Hsiao Erh Ko i Hsueh Hui Tsa Chih, **37,** 115–21.

Lin CY, Ho CH, Hsieh YH, and Kikuchi T (2002) Adeno-associated virus-mediated transfer of human acid maltase gene results in a transient reduction of glycogen accumulation in muscle of Japanese quail with acid maltase deficiency. Gene Ther, **9,** 554–63.

Lu IL, Lin CY, Lin SB, Chen ST, Yeh LY, Yang FY and Au LC (2003) Correction/mutation of acid alpha-D-glucosidase gene by modified single-stranded oligonucleotides: In vitro and in vivo studies. Gene Ther, **10,** 1910–6.

Lynch CM, Johnson J, Vaccaro C and Thurberg BL (2005) High-resolution light microscopy (HRLM) and digital analysis of Pompe disease pathology. J Histochem Cytochem, **53,** 63–73.

Margolis ML, Howlett P, Goldberg R, Eftychiadis A, and Levine S (1994) Obstructive sleep apnea syndrome in acid maltase deficiency. Chest, **105,** 947–9.

Martiniuk F and Hirschhorn R (1981) Characterization of neutral isozymes of human alpha-glucosidase: Differences in substrate specificity, molecular weight and electrophoretic mobility. Biochim Biophys Acta, **658,** 248–61.

Martiniuk F, Bodkin M, Tzall S and Hirschhorn R (1990) Identification of the base-pair substitution responsible for a human acid a-glucosidase allele with lower "affinity" for glycogen (GAA 2) and transient gene expression in deficient cells. Am J Hum Genet, **47,** 440–445.

Martiniuk F, Mehler M, Bodkin M, Tzall S, Hirschhorn K, Zhong N and Hirschhorn R (1991) Identification of a missense mutation in an adult-onset patient with glycogenosis type II expressing only one allele. DNA Cell Biol, **10,** 681–7.

Martiniuk F, Mehler M, Pellicer A, Tzall S, La Badie G, Hobart C, Ellenbogen A and Hirschhorn R (1986) Isolation of a cDNA for human acid alpha-glucosidase and detection of genetic heterogeneity for mRNA in three alpha-glucosidase-deficient patients. Proc Natl Acad Sci U S A, **83,** 9641–4.

Martiniuk F, Mehler M, Tzall S, Meredith G and Hirschhorn R (1990) Sequence of the cDNA and 5'-flanking region for human acid alpha-glucosidase, detection of an intron in the 5' untranslated leader sequence, definition of 18-bp polymorphisms, and differences with previous cDNA and amino acid sequences. Dna Cell Biol, **9,** 85–94.

McVie-Wylie AJ, Ding EY, Lawson T, Serra D, Migone FK, Pressley D, Mizutani M, Kikuchi T, Chen YT and Amalfitano A (2003) Multiple muscles in the AMD quail can be "cross-corrected" of pathologic glycogen accumulation after intravenous injection of an [E1-, polymerase-] adenovirus vector encoding human acid-alpha-glucosidase. J Gene Med, **5,** 399–406.

Mehler M and DiMauro S (1977) Residual acid maltase activity in late-onset acid maltase deficiency. Neurology, **27,** 178–84.

Mellies U, Ragette R, Schwake C, Baethmann M, Voit T and Teschler H (2001) Sleep-disordered breathing and respiratory failure in acid maltase deficiency. Neurology, **57,** 1290–5.

Moreland RJ, Jin X, Zhang XK, Decker RW, Albee KL, Lee KL, Cauthron RD, Brewer K, Edmunds T and Canfield WM (2004) Lysosomal acid a-glucosidase consists of four different peptides processed from a single-chain precursor. J Biol Chem.

Moufarrej NA and Bertorini TE (1993) Respiratory insufficiency in adult-type acid maltase deficiency. South Med J, **86,** 560–7.

Nichols BL, Avery S, Sen P, Swallow DM, Hahn D and Sterchi E (2003) The maltase-glucoamylase gene: Common ancestry to sucrase-isomaltase with complementary starch digestion activities. Proc Natl Acad Sci U S A, **100,** 1432–7.

Nichols BL, Eldering J, Avery S, Hahn D, Quaroni A and Sterchi E (1998) Human small intestinal maltase-glucoamylase cDNA cloning. Homology to sucrase-isomaltase. J Biol Chem, **273,** 3076–81.

Nicolino MP, Puech JP, Kremer EJ, Reuser AJ, Mbebi C, Verdiere-Sahuque M, Kahn A and Poenaru L (1998) Adenovirus-mediated transfer of the acid alpha-glucosidase gene into fibroblasts, myoblasts and myotubes from patients with glycogen storage disease type II leads to high level expression of enzyme and corrects glycogen accumulation. Hum Molec Gen, **7,** 1695–702.

Oude Elferink RPJ, Van Doorn-Van Wakeren J, Strijland A, Reuser AJJ and Tager JM (1985) Biosynthesis and intracellular transport of a-glucosidase and cathepsin D in normal and mutant human fibroblasts. Eur J Biochem, **153,** 55–63.

Park HK, Kay HH, McConkie-Rosell A, Lanman J and Chen YT (1992) Prenatal diagnosis of Pompe's disease (type II glycogenosis) in chorionic villus biopsy using maltose as a substrate. Prenatal Diagnosis, **12,** 169–73.

Pauly DF, Fraites TJ, Toma C, Bayes HS, Huie ML, Hirschhorn R, Plotz PH, Raben N, Kessler PD and Byrne BJ (2001) Intercellular transfer of the virally derived precursor form of acid alpha-glucosidase corrects the enzyme deficiency in inherited cardioskeletal myopathy Pompe disease. Hum Gene Ther, **12,** 527–38.

Pauly DF, Johns DC, Matelis LA, Lawrence JH, Byrne BJ and Kessler PD (1998) Complete correction of acid alpha-glucosidase deficiency in Pompe disease fibroblasts in vitro, and lysosomally targeted expression in neonatal rat cardiac and skeletal muscle. Gene Ther, **5,** 473–80.

Pompe JC (1932) Over idiopathische hypertrofie van het hart. Ned Tijdsch Geneesk, **76,** 304–311.

Putschar W (1932) Uber angeborne Glycogenspeicherkrankheit des herzens: Thesaurismosis glycogenica (v.Gierke). Beitr Pathol Anat, **90,** 222..

Raben N, Fukuda T, Gilbert AL, de Jong D, Thurberg BL, Mattaliano RJ, Meikle P, Hopwood JJ, Nagashima K, Nagaraju K and Plotz PH (2005) Replacing acid alpha-glucosidase in Pompe disease: Recombinant and transgenic enzymes are equipotent, but neither completely clears glycogen from type II muscle fibers. Mol Ther, **11,** 48–56.

Raben N, Lu N, Nagaraju K, Rivera Y, Lee A, Yan B, Byrne B, Meikle PJ, Umapathysivam K, Hopwood JJ and Plotz PH (2001) Conditional tissue-specific expression of the acid alpha-glucosidase (GAA) gene in the GAA knockout mice: Implications for therapy. Hum Mol Genet, **10,** 2039–2047.

Raben N, Nagaralu K, Lee E, Kessler P, Byrne B, Lee L, LaMarca M, King C, Ward J, Sauer B and Plotz P (1998) Targeted disruption of the acid a-glucosidase gene in mice causes an illness with critical features of both infantile and adult human glycogen storage disease type II. J Biol Chem, **273,** 19086–19092.

Raben N, Plotz P and Byrne BJ (2002) Acid alpha-glucosidase deficiency (glycogenosis type II, Pompe disease). Curr Mol Med, **2,** 145–66.

Reuser AJ, Kroos M, Oude Elferink RP and Tager JM (1985) Defects in synthesis, phosphorylation, and maturation of acid alpha-glucosidase in glycogenosis type II. J Biol Chem, **260,** 8336–41.

Reuser AJ, Kroos MA, Ponne NJ, Wolterman RA, Loonen MC, Busch HF, Visser WJ and Bolhuis PA (1984) Uptake and stability of human and bovine acid alpha-glucosidase in cultured fibroblasts and skeletal muscle cells from glycogenosis type II patients. Experimental Cell Research, **155,** 178–89.

Reuser AJ, Kroos M, Willemsen R, Swallow D, Tager JM and Galjaard H (1987) Clinical diversity in glycogenosis type II. Biosynthesis and in situ localization of acid alpha-glucosidase in mutant fibroblasts. J Clin Invest, **79,** 1689–99.

Reuser AJJ, Koster JF, Hoogeveen A and Galjaard H (1978) Biochemical, immunological, and cell genetic studies in glycogenosis type II. Am J Hum Genet, **30,** 132–143.

Reuser AJJ, Kroos MA, Hermans MMP, Bijvoet AGA, Verbeet MP, Van Diggelen OP, Kleijer WJ and Van der Ploeg AT (1995) Glycogenosis type II (acid maltase deficiency). Muscle Nerve, **3,** S61–S69.

Rinaldo P, Tortorelli S and Matern D (2004) Recent developments and new applications of tandem mass spectrometry in newborn screening. Curr Opin Pediatr, **16,** 427–33.

Rucker M, Fraites TJ, Jr., Porvasnik SL, Lewis MA, Zolotukhin I, Cloutier DA and Byrne BJ (2004) Rescue of enzyme deficiency in embryonic diaphragm in a mouse model of metabolic myopathy: Pompe disease. Development, **131,** 3007–19.

Sarkar R, Tetreault R, Gao G, Wang L, Bell P, Chandler R, Wilson JM and Kazazian HH, Jr. (2004) Total correction of hemophilia A mice with canine FVIII using an AAV 8 serotype. Blood, **103,** 1253–60.

Shieh JJ and Lin CY (1998) Frequent mutation in Chinese patients with infantile type of GSD II in Taiwan: Evidence for a founder effect. Hum Mutation, **11,** 306–12.

Shieh JJ, Wang LY and Lin CY (1994) Point mutation in Pompe disease in Chinese. J Inherited Metab Disease, **17,** 145–8.

Sleeper MM, Fornasari B, Ellinwood NM, Weil MA, Melniczek J, O'Malley TM, Sammarco CD, Xu L, Ponder KP and Haskins ME (2004) Gene therapy ameliorates cardiovascular disease in dogs with mucopolysaccharidosis VII. Circulation, **110,** 815–20.

Slonim AE, Bulone L, Ritz S, Goldberg T, Chen A and Martiniuk F (2000) Identification of two subtypes of infantile acid maltase deficiency. J Pediatr, **137,** 283–5.

Snyder RO, Miao C, Meuse L, Tubb J, Donahue BA, Lin HF, Stafford DW, Patel S, Thompson AR, Nichols T, Read MS, Bellinger DA, Brinkhous KM and Kay MA (1999) Correction of hemophilia B in canine and murine models using recombinant adeno-associated viral vectors. Nat Med, **5,** 64–70.

Su H, Arakawa-Hoyt J and Kan YW (2002) Adeno-associated viral vector-mediated hypoxia response element-regulated gene expression in mouse ischemic heart model. Proc Natl Acad Sci U S A, **99,** 9480–5.

Sun B, Chen YT, Bird A, Amalfitano A and Koeberl DD (2003) Long-term correction of glycogen storage disease type II with a hybrid Ad-AAV vector. Mol Ther, **7,** 193–201.

Sun B, Chen YT, Bird A, Xu F, Hou YX, Amalfitano A and Koeberl DD (2003) Packaging of an AAV vector encoding human acid alpha-glucosidase for gene therapy in glycogen storage disease type II with a modified hybrid adenovirus-AAV vector. Mol Ther, **7,** 467–77.

Sun B, Zhang H, Franco LM, Young SP, Schneider A, Bird A, Amalfitano A, Chen YT and Koeberl DD (2005) Efficacy of an adeno-associated virus 8-pseudotyped vector in glycogen storage disease type II. Mol Ther, **11,** 57–65.

Swallow DM, Corney G, Harris H and Hirschhorn R (1975) Acid a-glucosidase: a new polymorphism in man demonstrable by 'affinity' electrophoresis. Ann Hum Genet, **1975,** 391–406.

Swallow DM, Kroos M, Van der Ploeg AT, Griffiths B, Islam I, Marenah CB and Reuser AJ (1989) An investigation of the properties and possible clinical significance of the lysosomal a-glucosidase GAA*2 allele. Ann Hum Genet, **53,** 177–84.

Tsujino S, Kinoshita N, Tashiro T, Ikeda K, Ichihara N, Kikuchi H, Hagiwara Y, Mizutani M, Kikuchi T and Sakuragawa N (1998) Adenovirus-mediated transfer of human acid maltase gene reduces glycogen accumulation in skeletal muscle of Japanese quail with acid maltase deficiency. Hum Gene Ther, **9,** 1609–16

Umapathysivam K, Whittle AM, Ranieri E, Bindloss C, Ravenscroft EM, van Diggelen OP, Hopwood JJ and Meikle PJ (2000) Determination of acid alpha-glucosidase protein: Evaluation as a screening marker for Pompe disease and other lysosomal storage disorders. Clin Chem, **46,** 1318–1325.

Van den Hout H, Reuser AJ, Vulto AG, Loonen MC, Cromme-Dijkhuis A and Van der Ploeg AT (2000) Recombinant human alpha-glucosidase from rabbit milk in Pompe patients. Lancet, **356,** 397–8.

Van den Hout HM, Hop W, Van Diggelen OP, Smeitink JA, Smit GP, Poll-The BT, Bakker HD, Loonen MC, de Klerk JB, Reuser AJJ and Van der Ploeg AT (2003) The natural course of infantile Pompe's disease: 20 original cases compared with 133 cases from the literature. Pediatrics, **112,** 332–340.

Van den Hout JM, Kamphoven JH, Winkel LP, Arts WF, De Klerk JB, Loonen MC, Vulto AG, Cromme-Dijkhuis A, Weisglas-Kuperus N, Hop W, Van Hirtum H, Van Diggelen OP, Boer M, Kroos MA, Van Doorn PA, Van der Voort E, Sibbles B, Van Corven EJ, Brakenhoff JP, Van Hove J, Smeitink JA, de Jong G, Reuser AJ and Van der Ploeg AT (2004) Long-term intravenous treatment of Pompe disease with recombinant human alpha-glucosidase from milk. Pediatrics, **113,** e448–57.

Van den Hout JM, Reuser AJ, de Klerk JB, Arts WF, Smeitink JA and Van der Ploeg AT (2001) Enzyme therapy for Pompe disease with recombinant human alpha-glucosidase from rabbit milk. J Inherit Metab Dis, **24,** 266–74.

Van der Kraan M, Kroos MA, Joosse M, Bijvoet AG, Verbeet MP, Kleijer WJ and Reuser AJ (1994) Deletion of exon 18 is a frequent mutation in glycogen storage disease type II. Biochem Biophys Res Commun, **203,** 1535–41.

van der Ploeg AT, Kroos M, van Dongen JM, Visser WJ, Bolhuis PA, Loonen MC and Reuser AJ (1987) Breakdown of lysosomal glycogen in cultured fibroblasts from glycogenosis type II patients after uptake of acid alpha-glucosidase. J Neurol Sci, **79,** 327–36.

Van der Ploeg AT, Bolhuis PA, Wolterman RA, Visser JW, Loonen MC, Busch HF and Reuser AJ (1988) Prospect for enzyme therapy in glycogenosis II variants: A study on cultured muscle cells. J Neurol, **235,** 392–6.

Van der Ploeg AT, Kroos MA, Willemsen R, Brons NH and Reuser AJ (1991) Intravenous administration of phosphorylated acid alpha-glucosidase leads to uptake of enzyme in heart and skeletal muscle of mice. J Clin Invest, **87,** 513–8.

Van der Ploeg AT, Loonen MC, Bolhuis PA, Busch HM, Reuser AJ and Galjaard H (1988) Receptor-mediated uptake of acid alpha-glucosidase corrects lysosomal glycogen storage in cultured skeletal muscle. Pediatr Res, **24,** 90–4.

Van der Ploeg AT, Van der Kraaij AM, Willemsen R, Kroos MA, Loonen MC, Koster JF and Reuser AJ (1990) Rat heart perfusion as model system for enzyme replacement therapy in glycogenosis type II. Pediatr Res, **28,** 344–7.

Van Hove JL, Yang HW, Wu JY, Brady RO and Chen YT (1996) High-level production of recombinant human lysosomal acid alpha-glucosidase in Chinese hamster ovary cells which targets to heart muscle and corrects glycogen accumulation in fibroblasts from patients with Pompe disease. Proc Nat Acad Sci USA, **93,** 65–70.

Vorgerd M, Burwinkel B, Reichmann H, Malin JP and Kilimann MW (1998) Adult-onset glycogen storage disease type II: Phenotypic and allelic heterogeneity in German patients. Neurogenetics, **1,** 205–211.

Whitaker CH, Felice KJ and Natowicz M (2004) Biopsy-proven alpha-glucosidase deficiency with normal lymphocyte enzyme activity. Muscle Nerve, **29,** 440–2.

Willemsen R, van der Ploeg AT, Busch HF, Zondervan PE, Van Noorden CJ and Reuser AJ (1993) Synthesis and in situ localization of lysosomal alpha-glucosidase in muscle of an unusual variant of glycogen storage disease type II. Ultrastruct Pathol, **17,** 515–27.

Winkel LP (2005) The natural course of non classic Pompe disease; review of 225 published cases. in press.

Winkel LP, Kamphoven JH, Van Den Hout HJ, Severijnen LA, Van Doorn PA, Reuser AJ and Van Der Ploeg AT (2003) Morphological changes in muscle tissue of patients with infantile Pompe's disease receiving enzyme replacement therapy. Muscle Nerve, **27,** 743–51.

Winkel LP, Van den Hout JM, Kamphoven JH, Disseldorp JA, Remmerswaal M, Arts WF, Loonen MC, Vulto AG, Van Doorn PA, De Jong G, Hop W, Smit GP, Shapira SK, Boer MA, van Diggelen OP, Reuser AJ and Van der Ploeg AT (2004) Enzyme replacement therapy in late-onset Pompe's disease: A three-year follow-up. Ann Neurol, **55,** 495–502.

Wisselaar HA, Kroos MA, Hermans MM, van Beeumen J and Reuser AJ (1993) Structural and functional changes of lysosomal acid alpha-glucosidase during intra-cellular transport and maturation. J Biol Chem, **268,** 2223–31.

Wokke JH, Ausems MG, van den Boogaard MJ, Ippel EF, van Diggelene O, Kroos MA, Boer M, Jennekens FG, Reuser AJ and Ploos van Amstel HK (1995) Genotype-phenotype correlation in adult-onset acid maltase deficiency. Ann Neurol, **38,** 450–454.

Zaretsky JZ, Candotti F, Boerkoel C, Adams EM, Yewdell JW, Blaese RM and Plotz PH (1997) Retroviral transfer of acid alpha-glucosidase cDNA to enzyme-deficient myoblasts results in phenotypic spread of the genotypic correction by both secretion and fusion [published erratum appears in Hum Gene Ther 1998 Apr 10;9(6):930]. Hum Gene Ther, **8,** 1555–63.

Zellweger H, Illingworth Brown B, McCormick WF and Tu JB (1965) A mild from of muscular glycogenosis in two brothers with a-1,4-glucosidase deficiency. Ann Paediat, **205,** 412–437.

Zhu Y, Li X, Kyazike J, Zhou Q, Thurberg BL, Raben N, Mattaliano RJ and Cheng SH (2004) Conjugation of mannose 6-phosphate-containing oligosaccharides to acid alpha -glucosidase improves the clearance of glycogen in Pompe mice. J Biol Chem. 279: 50336–41.

LYSOSOMAL FREE SIALIC ACID STORAGE DISORDERS: SALLA DISEASE AND ISSD

Amanda Helip-Wooley, Robert Kleta, William A. Gahl

Salla disease and infantile free sialic acid storage disease (ISSD) are rare disorders of free sialic acid storage characterized by accumulation of the monosaccharide sialic acid (N-acetylneuraminic acid) in lysosomes. These disorders should be distinguished from sialuria in which free sialic acid accumulates in the cytoplasm. All three disorders present with increased excretion of free sialic acid in the urine. Only Salla disease and ISSD are discussed here in detail.

1 SIALIC ACID METABOLISM

Sialic acids are a family of negatively charged monosaccharides comprised of approximately 50 compounds derived from neuraminic acid. These compounds are important components of the complex carbohydrates found on many macromolecules. In normal tissues and fluids a small portion of sialic acid is found free, whereas the majority is bound to glycoconjugates. N-acetylneuraminic acid (Neu5Ac) is the predominant form in humans, and is referred to as sialic acid. One major biochemical difference between humans and the great apes is the lack of N-glycolylneuraminic acid (NeuGc) in humans, due to loss of the biosynthetic enzyme CMP-Neu5Ac hydroxylase in hominid evolution (Irie et al., 1998).

Sialic acid synthesis begins with cytoplasmic glucose, which undergoes several steps of processing to produce UDP-N-acetyl-D-glucosamine (UDP-GlcNAc), a key intermediate (Figure 1). The conversion of UDP-GlcNAc to N-acetylmannosamine-6-phosphate (ManNAc-6-P) comprises the rate-limiting steps in sialic acid biosynthesis, catalyzed by UDP-GlcNAc-2-epimerase/ManNAc kinase. The epimerase domain of this bifunctional enzyme is subject to feedback inhibition by cytidine monophosphate (CMP)-sialic acid (Kornfeld et al., 1964), and this feedback inhibition is defective in sialuria, resulting in massive overproduction of sialic acid (Weiss et al., 1989). In the course of glycoconjugate production, sialic acid becomes charged with CMP in the nucleus to form CMP-sialic acid, which is transported into the trans-Golgi for transfer to glycoconjugates by a sialyltransferase.

Degradation of sialylglycoconjugates occurs in lysosomes where terminal sialic acids are removed by acid sialidase, or neuraminidase. Deficiency of sialidase results in storage of bound sialic acid, in the form of sialylglycoconjugates (for review see Seyrantepe et al., 2003). Once cleaved, free sialic acid is transported out of the lysosome and into the

Section on Human Biochemical Genetics, Medical Genetics Branch, National Human Genome Research Institute, NIH, Bethesda MD, USA. Address correspondence to: William A. Gahl, M.D., Ph.D.Medical Genetics Branch, NHGRI, NIH. 10 Center Drive, MSC 1851, Building 10, Room 10C-103. Bethesda, Maryland 20892-1851. Telephone: 301-402-2739 FAX 301-402-2740. e-mail: bgahl@helix.nih.gov

cytosol where it can be degraded by sialate-pyruvate lyase or recycled (Schauer et al., 1999). Defects in the transport of free sialic acid out of the lysosome result in the allelic lysosomal free sialic acid storage disorders, Salla disease, and ISSD.

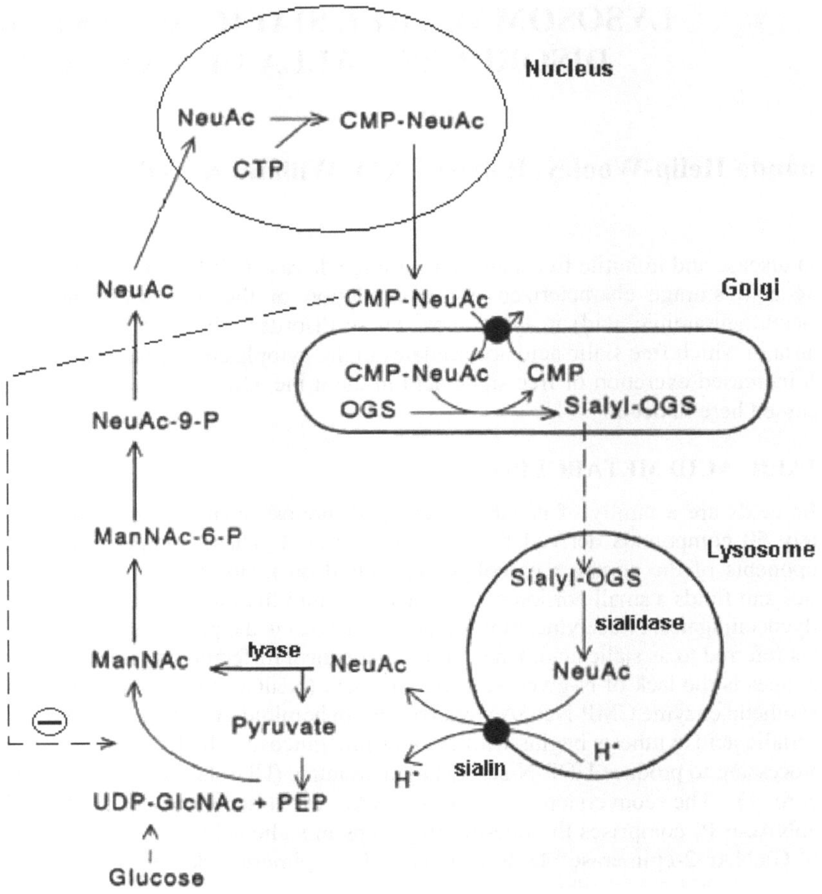

Figure 1. Pathway of sialic acid metabolism. UDP-GlcNAc = Uridine-diphosphate-N-acetylglucosamine; PEP = phosphoenolpyruvate; ManNAc = N-acetylmannosamine; CTP = cytidine-triphosphate; CMP = cytidine-monophosphate; OGS = oligosaccharide. (Adapted from Aula and Gahl, 2001.)

2 SALLA DISEASE AND ISSD

2.1 History

In 1979 four adult patients, three brothers and a female cousin, were described with "a new lysosomal storage disorder" (Aula et al., 1979). The name Salla disease was ascribed for the geographic region of northeastern Finland where the family resided. All four patients presented with severe mental retardation, coarse facial features, dysarthric speech, and ataxia. Large vacuoles in lymphocytes, skin biopsies, and cultured fibroblasts were suggestive of a storage disorder. However, the activities of several lysosomal hydrolases were normal, as were studies of urine amino acids, organic acids, oligosaccharides, and glycosaminoglycans. The urinary excretion of sialic acid was two to three times normal in three of the four patients. Nine additional Salla disease patients were subsequently described with increased sialic acid excretion in the urine (Renlund et al., 1979).

A disorder with similar ultrastructural and biochemical findings, but a much more severe clinical course, was described in 1982 (Tondeur et al., 1982). Affected patients presented in infancy with failure to thrive, hepatosplenomegaly, edema, coarse facies, psychomotor retardation, and death in infancy or early childhood. The disease was termed infantile free sialic acid storage disease (ISSD) and was later shown to result from the same biochemical defect as Salla disease (Tietze et al., 1989).

2.2 The Basic Biochemical Defect

Early attempts to understand the basic biochemical defect in Salla disease focused on the metabolic turnover of sialic acid. Several enzymes involved in sialic acid metabolism were studied, including sialate-pyruvate lyase, sialidase, CMP-N-acetylneuraminate N-acylneuraminohydrolase, and CTP:N-acylneuraminate cytidylyltransferase, but no abnormalities were found (Hancock et al., 1983; Renlund et al., 1983b). Salla disease patients' fibroblasts, incubated with radiolabeled sialic acid precursors [3H]-N-acetylmannosamine (ManNAc) or [3H]-glucosamine, yielded a marked accumulation of labeled free sialic acid, compared to normal cells (Hancock, Horwitz, and Dawson,, 1983; Thomas et al., 1983). In contrast, the distribution and amount of labeled sialic acid bound to glycoconjugates was normal. Affected patients' lysosomes, loaded with free sialic acid by incubation with [3H]-sialic acid-methylester or tritium-labeled glycoproteins, demonstrated a defect in the clearance of free sialic acid from lysosomes, compared to normal controls (Mancini, Verheijen, and Galjaard, 1986; Mendla et al., 1988; Renlund et al., 1986a). These results suggested a defect in the egress of free sialic acid from lysosomes, although initial velocity measurements were lacking.

Free sialic acid egress from normal and Salla disease fibroblast lysosomes was studied in greater detail by Renlund and co-workers (Renlund, Tietze, and Gahl, 1986b). Normal and Salla disease fibroblasts were loaded with equivalent concentrations of sialic acid by incubation with nonradioactive ManNAc, a sialic acid precursor, and the rate of sialic acid egress from a lysosome enriched granular fraction was measured. In normal cells, the initial velocity of free sialic acid egress increased linearly with sialic acid loading and temperature ($Q_{10} = 2.4$), whereas fibroblasts from five Salla disease patients exhibited negligible egress (Renlund, Tietze, and Gahl, 1986b; Tietze et al., 1989). Deficient egress of sialic acid from loaded lysosomes of five ISSD patient fibroblasts was later demonstrated, supporting the concept that Salla disease and ISSD result from the same basic defect, but with variable degrees of clinical severity (Tietze et al., 1989). At very high levels of sialic acid loading a minimal amount of egress was measured in Salla disease, suggesting some residual transport activity, but no sialic acid egress was measured in ISSD granular fractions, regardless of the level of loading (Renlund, Tietze, and Gahl, 1986b; Tietze et al., 1989).

The lysosomal sialic acid transporter was further characterized using resealed lysosomal membrane vesicles isolated from rat liver (Mancini et al., 1989). These studies demonstrated that a specific carrier with a K_m of 0.24 mM was responsible for sialic acid egress from lysosomes. Transport was dependent on a proton gradient across the membrane. Other monocarboxylic sugars, such as glucuronic acid, were ligands for this carrier, as demonstrated by their ability to both competitively inhibit and transstimulate labeled sialic acid uptake. The lysosomal sialic acid transporter was later purified to apparent homogeneity and its activity was reconstituted in phospholipid vesicles (Havelaar et al., 1998). This purified protein retained the same ligand specificity, affinity and proton gradient-dependent transport described earlier (Mancini et al., 1992; Mancini et al., 1989). In addition, L-lactate, but not mevalonate, was recognized by the sialic acid transporter, with a K_m of 0.4 mM (Havelaar et al., 1998). The properties of the lysosomal sialic acid transporter resemble those of plasma membrane monocarboxylate transporters.

3 CLINICAL FINDINGS

3.1 Salla Disease

The pregnancy and neonatal periods are unremarkable for the majority of Salla disease patients. Muscular hypotonia and ataxia are usually the first clinical signs and typically are evident between 6 and 12 months of age. Nystagmus is observed in about 25% of cases, and has been seen as early as three weeks of age (Varho et al., 2002). Motor development is always delayed, with an average age at walking of 4 years (Aula and Gahl, 2001). Approximately 33% of patients never learn to walk unsupported. As with most lysosomal storage disorders, developmental milestones are lost over time. This occurs more slowly in Salla disease than in ISSD. The mean age at death in Salla disease is 34.6 years ($N = 12$), and the oldest known patient is over 70 years old (Aula and Gahl, 2001).

Somatic findings are limited in classical Salla disease. Growth is usually normal in the first years of life but height eventually falls to one to two standard deviations below the mean for age (Varho et al., 2002). Head circumference is normal and there is no hepatosplenomegaly. Coarse facial features generally develop later in adulthood. Scoliosis is present in less than half of the patients and no skeletal dysplasia is observed. A thickened calvarium is present in many patients.

The cognitive ability of Salla disease patients is severely impaired. Adults with Salla disease have severe mental delay, with IQs of 20 to 40. The majority of patients (76%) are able to say single words, but few (27%) learn to speak in sentences (Varho et al., 2002). In general, speech deteriorates gradually, and older patients eventually become nonverbal. Dysarthria and dyspraxia further contribute to speech difficulties, but Salla disease patients have pleasant cheerful dispositions.

Neurologic features in childhood include primarily hypotonia and ataxia. In older patients athetosis and spasticity are more prominent. Epilepsy occurs in 32% of Salla disease patients (Varho et al., 2002). Brain MRI scans of Salla disease patients demonstrate defective myelination and an abnormally thin, hypoplastic corpus callosum (Aula and Gahl, 2001). Cortical and cerebellar atrophy also occur, especially in older or more severely affected patients. Abnormal nerve conduction velocities were observed in 10 of 21 Salla disease patients examined, indicating myelination defects in the peripheral nervous system (Varho et al., 2000). Magnetic resonance spectroscopy studies of Salla disease patients revealed a 34% increase in *N*-acetylaspartate signal and a 35% decrease in choline signal in the parietal white matter, compared with age-matched controls (Varho et al., 1999). The increased *N*-acetylaspartate signal may reflect increased sialic acid content of the cells, whereas the decreased choline signal may indicate hypomyelination. An increased *N*-acetylaspartate signal is commonly seen in Canavan's disesase.

3.2 Infantile Sialic Acid Storage Disease

The clinical course of ISSD is more severe than that of Salla disease and usually leads to death in early infancy (mean age at death, 13.1 months; Lemyre et al., 1999). In one study, approximately half of 27 ISSD cases presented in utero or in the newborn period with hydrops fetalis and/or ascites (Lemyre et al., 1999), findings also seen in MPS IVA, MPS VII, type 2 Gaucher disease, sialidosis, galactosialidosis, GMI gangliosidosis, - Niemann-Pick type C, Farber disease, and mucolipidosis II (Daneman, Stringer, and Reillly, 1983; Gillan et al., 1984; Stone and Sidransky, 1999).

All ISSD patients demonstrate failure to thrive, hypotonia, hepatosplenomegaly, growth retardation, and markedly delayed psychomotor development in early infancy (Aula and Gahl, 2001). Hypopigmentation of the skin and hair, mild skeletal dysplasias, and coarse facial features are also seen. Nephrosis has been reported in four cases of ISSD (Lemyre et al., 1999; Pueschel et al., 1988; Sperl et al., 1990). Electron microscopy of a renal biopsy from one patient revealed swollen epithelial, endothelial, and mesangial cells filled with membrane-bound, electron-lucent vacuoles (Lemyre et al., 1999). Cardiomegaly was reported in nine of twelve patients with ISSD and cardiac failure was documented in five of these nine cases.

Examination of brain pathology in ISSD demonstrates severe involvement of the central nervous system. Storage material accumulated in neurons and astrocytes, associated with severe demyelination, gliosis, and marked hypoplasia of the corpus callosum (Lemyre et al., 1999; Pueschel et al., 1988; Stevenson et al., 1983). Staining of neurons, endothelial cells, and Kupffer cells using the sialic acid-specific lectin wheat germ agglutinin, identified the storage material as sialic acid (Pueschel et al., 1988).

3.3 Intermediate Phenotypes

Several cases of sialic acid storage disease display an earlier age of onset and a more relentless course than that of classical Salla disease, yet are not as severe as ISSD (Aula et al., 2000; Kleta et al., 2003, 2004). These variants are called intermediate or severe Salla disease. Patients with the intermediate phenotype demonstrate early-onset (three to six months of age) hypotonia, spasticity, growth retardation, and developmental delay.

4 LABORATORY FINDINGS

Urinary excretion of free sialic acid is increased in both Salla disease and ISSD. In Salla disease, the levels of sialic acid excretion range from 5 to 20 times normal (280–2100 nmol/mg creatinine), whereas in ISSD urine sialic acid excretion is increased up to 200-fold (1071–14230 nmol/mg creatinine; Aula and Gahl, 2001). Free sialic acid excretion varies with age. Normal newborns may have appreciable quantities of sialic acid in their urine, so it is important that age-matched controls be used to establish reference ranges. Quantification of free sialic acid (in the form of N-acetylneuraminic acid) is most commonly performed using HPLC.

In addition to elevated sialic acid in urine, increased free sialic acid can be measured in cultured fibroblasts from Salla disease and ISSD patients. Free sialic acid in fibroblasts is approximately 10 to 30 times normal in Salla disease (4–46 nmol/mg protein) and 20 to two hundred times normal in ISSD (10–269 nmol/mg protein; Aula and Gahl, 2001). On subcellular fractionation, stored free sialic acid in these disorders cofractionates with lysosomal markers. Bound sialic acid, measured in either urine or cultured cells from patients, is within the normal range.

Electron microscopic studies of skin, kidney, and liver from sialic acid storage disease patients reveal numerous electron-lucent membrane-bound vacuoles in various cell types (Aula et al., 1979; Biancheri et al., 2002; Kleta et al., 2004; Lemyre et al., 1999; Pueschel et al., 1988; Tondeur et al., 1982). Vacuoles are also present in cultured fibroblasts from patients and can sometimes be observed in peripheral blood lymphocytes (Aula et al., 1979).

5 GENETICS

Salla disease and ISSD are allelic disorders inherited in an autosomal recessive manner. ISSD is panethnic. Salla disease is more common but is found almost exclusively in Finland, due to a founder mutation in the population. Linkage analysis using 27 Finnish families localized the Salla disease locus to an approximately 200 kb region of chromosome 6q (Haataja et al., 1994). This region was refined, using linkage disequilibrium data from a total of 50 Finnish families, to a critical region of 80 kb on chromosome 6q14-q15 (Schleutker et al., 1995).

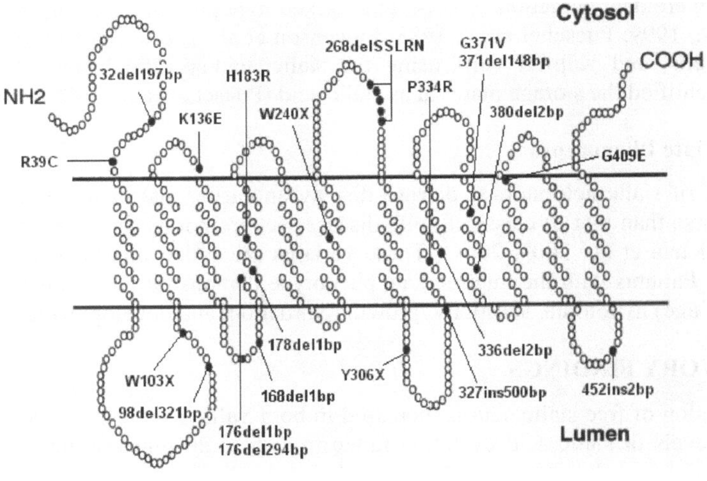

Figure 2. Predicted structure of sialin with the location of free sialic acid storage disease patients' mutations (described in Table 1) indicated. The starting points and size (in base pairs) of deletions and insertions are indicated; splice site mutations are not shown. (Adapted from Aula et al., 2000.)

The gene responsible for Salla disease and ISSD was identified in 1999 using positional cloning and homology with known transporter genes (Verheijen et al., 1999). EST clones found to hybridize to the critical region were used as probes to screen multiple cDNA libraries, and a 2.5-kb cDNA with an ORF of 1485 bp was identified. The gene, *SLC17A5*, is expressed ubiquitously in human tissues. The predicted protein product, designated sialin, consists of 495 amino acids and has homology to the anion/cation symporter (ACS) family of transporters. It is predicted to have 12 transmembrane (TM) domains and contains a characteristic motif in the fourth TM region that is conserved in all members of the ACS family (Figure 2).

Mutations in *SLC17A5* were found in both Salla disease and ISSD patients, confirming that the two disorders are allelic. A homozygous missense mutation (115C>T; R39C) in *SLC17A5* was found in all five of the Finnish Salla disease patients examined, accounting for the founder mutation in the Finnish population (Verheijen et al., 1999). This mutation, often referred to as $Salla_{FIN}$, changes a highly conserved arginine to a cysteine just prior to the first TM domain of sialin (Figure 2). Six different mutations in *SLC17A5* were identified in six ISSD patients, including two deletions leading to frameshifts (533delC and 1112-1259del 148 bp), two missense mutations (548A>G and 1001C>G), an insertion (978ins 500 bp), and a 15 bp in-frame deletion (802-816del 15 bp) (Table 1; Figure 2). The 15 bp deletion was found in an apparently homozygous state in three of the six ISSD patients. To date, more than 100 cases of Salla disease and more than 20 cases of ISSD have been reported. A summary of known mutations is presented in Table 1and their locations in the predicted protein are depicted in Figure 2.

Table 1. Mutations in the SLC17A5 gene

Nucleotide Change	Amino Acid/Protein Alteration	Reference
Salla Disease		
115C>T [1]	R39C; before TM domain 1	Verheijen et al., 1999
ISSD		
802-816del15bp	In-frame del [SSLRN] 268-272	Verheijen et al., 1999
533delC	Frameshift aa 178; PTC	Verheijen et al., 1999
1112-1259del148bp	Frameshift aa 371; PTC	Verheijen et al., 1999
548A>G	H183R; in TM domain 4	Verheijen et al., 1999
1001C>G	P334R; in TM domain 8	Verheijen et al., 1999
978-979ins500bp	Insertion after aa 327; PTC	Verheijen et al., 1999
1112G>T	G371V; in TM domain 9	Aula, et al., 2000
1355-1356insAA	Frameshift aa 452; read-through stop	Aula, et al., 2000
918T>G	Y306X; between TM domains 7 and 8	Aula, et al., 2000
526delG	Frameshift aa 176; PTC	Aula, et al., 2000

1138-1139delGT	Frameshift aa 380; PTC	Parazzini, et al., 2003
IVS9+1G>A	Splice site mutation; del exon 9; PTC	Kleta, et al., 2003
Intermediate		
1138-1139delGT [2]	Frameshift aa 380; PTC	Aula, et al., 2000
292-611del [2]	98del321bp; del exons 3 and 4	Aula, et al., 2000
526-819del [2]	176del294bp; del exons 4-6	Aula, et al., 2000
406A>G [2]	K136E; before TM domain 3	Aula, et al., 2000
1007-1008del [2]	Frameshift aa 336; PTC	Aula, et al., 2000
309G>A [2]	W103X; before TM domain 2	Aula, et al., 2000
719G>A [2]	W240X; in TM domain 6	Aula, et al., 2000
95-1G>C	Splice site mutation; del exon 2	Aula, et al., 2000
507delA [2]	Frameshift aa 168; PTC	Kleta, et al., 2003
802-816del15bp [2]	In-frame del [SSLRN] 268-272	Kleta, et al., 2003
291G>A	Splice site mutation; del exon 2	Kleta, et al., 2004
1226G>A	G409E; in TM domain 10	Kleta, et al., 2004

[1] $\text{Salla}_{\text{FIN}}$ mutation
[2] Compound heterozygous with 115C>T
PTC = premature termination codon

A study of 80 Finnish Salla disease patients revealed that 91% of affected individuals are homozygous for the R39C founder mutation, whereas the remainder are compound heterozygotes with the R39C mutation on one allele (Aula et al., 2000). These compound heterozygotes exhibit a more severe phenotype and are classified as having intermediate or severe Salla disease (see below). The carrier frequency of the R39C mutation across all regions of Finland is estimated at approximately 1:200. In the northeast region of Finland it approximates 1%.

In one report, sequence analysis of ten patients with ISSD revealed ten different mutations (Aula et al., 2000). A 15-bp in-frame deletion (del 802-816; del SSRLN) was found in five unrelated patients of Canadian, English, and French ancestry. The remaining mutations included insertions or deletions leading to frameshifts and premature termination codons or missense mutations leading to alterations of conserved amino acids (Table 1). Two case reports of patients with ISSD describe additional mutations including a 2-bp deletion (1138delGT) and a splice site mutation (IVS9+1G>A) (Kleta et al., 2003; Parazzini et al., 2003). No patients with ISSD have been reported with the R39C mutation.

A third phenotype of sialic acid storage disease includes patients presenting with a more severe form of Salla disease, that is, intermediate between ISSD and Salla disease. Patients with severe Salla disease, or intermediate type sialic acid storage disease, usually possess one mild mutation, such as R39C, and one severe mutation, as might be seen in ISSD (Table 1).

6 CELL BIOLOGY

The intracellular localization of wild-type sialin and sialin containing either the Salla$_{FIN}$ founder mutation (R39C) or the 15-bp deletion observed in several ISSD patients (del SSRLN) has been studied (Aula et al., 2002). The wild-type sialin targeted to lysosomes as expected, whereas an appreciable portion of the Salla$_{FIN}$ and the majority of the ISSD mutant proteins were found in the Golgi compartment. These findings account for the milder phenotype observed in Salla disease, because a portion of the mutant protein reaches lysosomes where it may express some residual activity. The severe ISSD phenotype reflects the observation that essentially none of the ISSD mutant protein is trafficked to lysosomes. Because the transporter never reaches its site of action, it cannot function, despite the small size of its deletion, that is, five amino acids.

Lysosomal membrane proteins such as cystinosin, the cystine carrier protein defective in cystinosis, and the LAMPs contain either a tyrosine-based sorting signal YXXϕ (where ϕ is a bulky hydrophobic amino acid) or a di-leucine motif in their C-terminus to direct them to lysosomes (Bonifacino and Dell'Angelica, 1999; Cherqui et al., 2001; Rapoport et al., 1998). Sialin does not appear to contain either of these traditional sorting signals in its C-terminal tail. Instead, a functional di-leucine motif was recently identified in the N-terminal portion of sialin (Morin, Sagne, and Gasnier, 2004; Wreden, Wlizla, and Reimer, 2005). The N-terminal DRTPLL motif (corresponding to amino acids 18–23) appeared necessary to target sialin to lysosomes; disruption of this motif by deletion or by mutation of the two leucine residues resulted in its mislocalization to the plasma membrane.

The mutant sialin expressed at the cell surface was utilized to quantitate sialin function by measuring cellular uptake rather than lysosomal efflux (Morin, Sagne, and Gasnier, 2004; Wreden, Wlizla, and Reimer, 2005). The transport activity of sialin in this system was dependent upon a proton gradient and had kinetics and specificities similar to those of sialic acid transport in native lysosomal membranes or in reconstituted purified protein systems. In addition, ISSD patient mutations (H183R, P334R, del SSLRN) introduced into the recombinant sialin abolished all transport activity, whereas the Salla$_{FIN}$ mutation (R39C) and an intermediate mutation (K136E) retained residual activity (Morin, Sagne, and Gasnier, 2004; Wreden, Wlizla, and Reimer, 2005). These assays demonstrate that the molecular defect in sialin directly correlates with the clinical phenotype.

It has been reported that sialin does not localize to lysosomes in primary cultured neurons as it does in nonneuronal cells (Aula et al., 2004). Instead, sialin was found in the plasma membrane and in a punctate pattern along neuronal processes. This raises the possibility that sialin may have additional or distinct functions in neurons, and is especially interesting in light of the severe neurological involvement in Salla disease and ISSD.

7 DIAGNOSIS AND TREATMENT

The diagnosis of Salla disease and ISSD is based on clinical presentation and the finding of increased free sialic acid in the urine. The presence of storage vacuoles in cultured cells or tissue biopsies supports the diagnosis, but these findings also occur in other lysosomal storage disorders. Identification of the disease-causing gene means that the diagnosis can be confirmed at the molecular level, particularly in Finnish Salla disease patients. Non-Finnish patients may also have the R39C sialin mutation, but the nonspecific nature of their clinical findings would make diagnosis difficult. It is likely that more cases of Salla disease exist outside of Finland than are currently reported.

The early presentation and severe course of ISSD make it more likely to be investigated as a possible lysosomal storage disorder. Appropriate diagnostic studies include EM of a skin biopsy, which reveals enlarged lysosomal vacuoles, generally described as electron-lucent and containing some granular material. The vacuoles are distinct from the membranous whorls and lamellar inclusions seen in the mucolipidoses and gangliosidoses, but their morphology cannot easily be distinguished from that present in several other lysosomal storage disorders. Vacuoles may or may not be present in lymphocytes of Salla disease and ISSD patients.

Free sialic acid in the urine is most often detected by thin-layer chromatography using resorcinol stain (Renlund et al., 1983a). Demonstration of a large resorcinol-positive spot that is appreciably greater than that of age-matched controls and runs with the same Rf as a free sialic acid standard is essentially diagnostic of a free sialic acid storage disorder. This finding does not, however, rule out cytoplasmic sialic acid accumulation, as seen in sialuria. Salla disease and ISSD are usually distinguished from sialuria based on clinical findings, but the presence of vacuoles in skin biopsies or in cultured fibroblasts can also be used to rule out sialuria. Urinary free sialic acid can be quantitated by HPLC but this is generally unnecessary for diagnostic purposes (Renlund et al., 1986a). There is some correlation between the severity of the clinical course and the amount of free sialic acid excreted in the urine, with ISSD patients excreting approximately ten times as much sialic acid as Salla disease patients and intermediate phenotypes falling somewhere in between (Aula and Gahl, 2001). A similar relationship is seen with the amount of free sialic acid stored in tissues.

Salla disease and ISSD have been diagnosed prenatally based on quantitation of free sialic acid in uncultured first trimester chorionic villus samples (CVS) or on cultured amniocytes (Clements, Taylor, and Hopwood, 1988; Lake, Young, and Nicolaides, 1989; Renlund and Aula, 1987; Salomaki et al., 2001; Vamos et al., 1986). The grossly elevated (up to 70-fold) levels of free sialic acid in ISSD amniocytes make prenatal diagnosis more straightforward than in Salla disease, in which the sialic acid levels of amniocytes may only be five times normal. For this reason, prenatal diagnosis on CVS is preferred in Salla disease. Free sialic acid measured in cell free amniotic fluid was in the normal range in one pregnancy affected with Salla disease and was only mildly elevated in ISSD, so this method should not be used for prenatal diagnosis (Renlund and Aula, 1987; Vamos et al., 1986). EM of CVS has demonstrated numerous vacuoles in ISSD but not in Salla disease (Lake, Young, and Nicolaides, 1989). In families where the mutation has been identified, molecular analysis of CVS DNA provides a simple means of diagnosis.

To date, no specific therapy is available to Salla disease or ISSD patients and there have been no reports of attempts at bone marrow transplantation. Further investigations into the function of the sialic acid transporter and the pathophysiology of the disorder may yield new therapeutic targets.

REFERENCES

Aula, N., Jalanko, A., Aula, P., and Peltonen, L., 2002, Unraveling the molecular pathogenesis of free sialic acid storage disorders: altered targeting of mutant sialin, *Mol Genet Metab* **77**: 99.

Aula, N., Kopra, O., Jalanko, A., and Peltonen, L., 2004, Sialin expression in the CNS implicates extralysosomal function in neurons, *Neurobiol Dis* **15**: 251.

Aula, N., Salomaki, P., Timonen, R., Verheijen, F., Mancini, G., Mansson, J. E., Aula, P., and Peltonen, L., 2000, The spectrum of SLC17A5-gene mutations resulting in free sialic acid-storage diseases indicates some genotype-phenotype correlation, *Am J Hum Genet* **67**: 832.

Aula, P., and Gahl, W., 2001, Disorders of Free Sialic Acid Storage. In The Metabolic and Molecular Bases of Inherited Disease, C. Scriver, A. Beaudet, W. Sly, and D. Valle, Eds. (New York, McGraw-Hill), pp. 5109–5120.

Aula, P., Autio, S., Raivio, K., Rapola, J., Thoden, C., Koskela, S., and Yamashina, I., 1979, "Salla Disease". A new lysosomal storage disorder, *Arch Neurol* **36**: 88.

Biancheri, R., Verbeek, E., Rossi, A., Gaggero, R., Roccatagliata, L., Gatti, R., van Diggelen, O., Verheijen, F. W., and Mancini, G. M., 2002, An Italian severe Salla disease variant associated with a SLC17A5 mutation earlier described in infantile sialic acid storage disease, *Clin Genet* **61**: 443.

Bonifacino, J., and Dell'Angelica, E., 1999, Molecular bases for the recognition of tyrosine-based sorting signals, *J Cell Biol* **145**: 923.

Cherqui, S., Kalatzis, V., Trugnan, G., and Antignac, C., 2001, The targeting of cystinosin to the lysosomal membrane requires a tyrosine-based signal and a novel sorting motif, *J Biol Chem* **276**: 13314.

Clements, P., Taylor, J., and Hopwood, J., 1988, Biochemical characterization of patients and prenatal diagnosis of sialic acid storage disease for three families., *J Inherit Metab Dis* **11**: 30.

Daneman, A., Stringer, D., and Reillly, B., 1983, Neonatal ascites due to lysosomal storage disease, *Radiology* **149**: 463.

Gillan, J. E., Lowden, J. A., Gaskin, K., and Cutz, E., 1984, Congenital ascites as a presenting sign of lysosomal storage disease, *J Pediatrics* **104**: 225.

Haataja, L., Schleutker, J., Laine, A. P., Renlund, M., Savontaus, M. L., Dib, C., Weissenbach, J., Peltonen, L., and Aula, P., 1994, The genetic locus for free sialic acid storage disease maps to the long arm of chromosome 6, *Am J Hum Genet* **54**: 1042.

Hancock, L., Horwitz, A., and Dawson, G., 1983, N-acetylneuraminic acid and sialoglycoconjugate metabolism in fibroblasts from a patient with generalized N-acetylneuraminic acid storage disease, *Biochem Biophys Acta* **760**: 42.

Havelaar, A. C., Mancini, G. M., Beerens, C. E., Souren, R. M., and Verheijen, F. W., 1998, Purification of the lysosomal sialic acid transporter. Functional characteristics of a monocarboxylate transporter, *J Biol Chem* **273**: 34568.

Irie, A., Koyama, S., Kozutsumi, Y., Kawasaki, T., and Suzuki, A., 1998, The molecular basis for the absence of N-glycolylneuraminic acid in humans, *J Biol Chem* **273**: 15866.

Kleta, R., Aughton, D. J., Rivkin, M. J., Huizing, M., Strovel, E., Anikster, Y., Orvisky, E., Natowicz, M., Krasnewich, D., and Gahl, W. A., 2003, Biochemical and molecular analyses of infantile free sialic acid storage disease in North American children, *Am J Med Genet* **120A**: 28.

Kleta, R., Morse, R., Orvisky, E., Krasnewich, D., Alroy, J., Ucci, A., Bernardini, I., Wenger, D., and Gahl, W., 2004, Clinical, biochemical, and molecular diagnosis of a free sialic acid storage disease patient of moderate severity, *Mol Genet Metab* **82**: 137.

Kornfeld, S., Kornfeld, R., Neufeld, E., and O'Brien, P., 1964, The feedback control of sugar nucleotide biosynthesis in liver, *Proc Natl Acad Sci USA* **52**: 371.

Lake, B., Young, E., and Nicolaides, K., 1989, Prenatal diagnosis of infantile sialic acid storage disease in a twin pregnancy, *J Inherit Metab Dis* **12**: 152.

Lemyre, E., Russo, P., Melancon, S. B., Gagne, R., Potier, M., and Lambert, M., 1999, Clinical spectrum of infantile free sialic acid storage disease, *Am J Med Genet* **82**: 385.

Mancini, G., Beerens, C., Galjaard, H., and Verheijen, F., 1992, Functional reconstitution of the lysosomal sialic acid carrier into proteoliposomes, *Proc Natl Acad Sci USA* **89**: 6609.

Mancini, G., de Jonge, H., Galjaard, H., and Verheijen, F., 1989, Characterization of a proton-driven carrier for sialic acid in the lysosomal membrane. Evidence for a group-specific transport system for acidic monosaccharides, *J Biol Chem* **264**: 15247.

Mancini, G., Verheijen, F., and Galjaard, H., 1986, Free N-acetylneuraminic acid storage disorders: evidence for defective NANA transport across the lysosomal membrane, *Hum Genet* **73**: 214.

Mendla, K., Baumkotter, J., Rosenau, C., Ulrich-Bott, B., and Cantz, M., 1988, Defective lysosomal release of glycoprotein-derived sialic acid in fibroblasts from patients with sialic acid storage disease, *Biochem J* **250**: 261.

Morin, P., Sagne, C., and Gasnier, B., 2004, Functional characterization of wild-type and mutant human sialin, *EMBO J* **23**: 4560.

Parazzini, C., Arena, S., Marchetti, L., Menni, F., Filocamo, M., Verheijen, F., Mancini, G., Triulzi, F., and Parini, R., 2003, Infantile sialic acid storage disease: Serial ultrasound and magnetic resonance imaging features, *AJNR Am J Neuroradiol* **24**: 398.

Pueschel, S., O'Shea, P., Alroy, J., Ambler, M., Dangond, F., Daniel, P., and Kolodny, E., 1988, Infantile sialic acid storage disease associated with renal disease, *Pediatr Neurol* **4**: 207.

Rapoport, I., Chen, Y., Cupers, P., Shoelson, S., and Kirchhausen, T., 1998, Dileucine-based sorting signals bind to the beta chain of AP-1 at a site distinct and regulated differently from the tyrosine-based motif-binding site., *EMBO J* **17**: 2148.

Renlund, M., and Aula, P., 1987, Prenatal detection of Salla disease based upon increased free sialic acid in amniocytes, *Am J Med Genet* **28**: 377.

Renlund, M., Aula, P., Raivio, K. O., Autio, S., Sainio, K., Rapola, J., and Koskela, S. L., 1983a, Salla disease: A new lysosomal storage disorder with disturbed sialic acid metabolism, *Neurology* **33**: 57.

Renlund, M., Chester, M. A., Lundblad, A., Aula, P., Raivio, K. O., Autio, S., and Koskela, S. L., 1979, Increased urinary excretion of free N-acetylneuraminic acid in thirteen patients with Salla disease, *Eur J Biochem* **101**: 245.

Renlund, M., Chester, M. A., Lundblad, A., Parkkinen, J., and Krusius, T., 1983b, Free N-acetylneuraminic acid in tissues in Salla disease and the enzymes involved in its metabolism, *Eur J Biochem* **130**: 39.

Renlund, M., Kovanen, P. T., Raivio, K. O., Aula, P., Gahmberg, C. G., and Ehnholm, C., 1986a, Studies on the defect underlying the lysosomal storage of sialic acid in Salla disease. Lysosomal accumulation of sialic acid formed from N-acetyl-mannosamine or derived from low density lipoprotein in cultured mutant fibroblasts, *J Clin Invest* **77**: 568.

Renlund, M., Tietze, F., and Gahl, W. A., 1986b, Defective sialic acid egress from isolated fibroblast lysosomes of patients with Salla disease, *Science* **232**: 759.

Salomaki, P., Aula, N., Juvonen, V., Renlund, M., and Aula, P., 2001, Prenatal detection of free sialic acid storage disease: Genetic and biochemical studies in nine families, *Prenat Diagn* **21**: 354.

Schauer, R., Sommer, U., Kruger, D., van Unen, H., and Traving, C., 1999, The terminal enzymes of sialic acid metabolism: Acylneuraminate-pyruvate-lyases, *Biosci Rep* **19**: 373.

Schleutker, J., Laine, A. P., Haataja, L., Renlund, M., Weissenbach, J., Aula, P., and Peltonen, L., 1995, Linkage disequilibrium utilized to establish a refined genetic position of the Salla disease locus on 6q14-q15, *Genomics* **27**: 286.

Seyrantepe, V., Poupetova, H., Froissart, R., Zabot, M., Maire, I., and Pshezhetsky, A., 2003, Molecular pathology of NEU1 gene in sialidosis, *Hum Mutat* **22**: 343.

Sperl, W., Gruber, W., Quatacker, J., Monnens, L., Thoenes, W., Fink, F. M., and Paschke, E., 1990, Nephrosis in two siblings with infantile sialic acid storage disease, *Eur J Pediatr* **149**: 477.

Stevenson, R., Lubinsky, M., Taylor, H., Wenger, D., Schroer, R., and Olmstead, P., 1983, Sialic acid storage disease with sialuria: Clinical and biochemical features in the severe infantile type, *Pediatrics* **72**: 441.

Stone, D. L., and Sidransky, E., 1999, Hydrops fetalis: Lysosomal storage disorders in extremis, *Adv Pediatr* **46**: 409.

Thomas, G., Scocca, J., Libert, J., Vamos, E., Miller, C., and Reynolds, L., 1983, Alterations in cultured fibroblasts of sibs with an infantile form of free (unbound) sialic acid storage disorder, *Pediatr Res* **17**: 307.

Tietze, F., Seppala, R., Renlund, M., Hopwood, J. J., Harper, G. S., Thomas, G. H., and Gahl, W. A., 1989, Defective lysosomal egress of free sialic acid (N-acetylneuraminic acid) in fibroblasts of patients with infantile free sialic acid storage disease, *J Biol Chem* **264**: 15316.

Tondeur, M., Libert, J., Vamos, E., Van Hoof, F., Thomas, G. H., and Strecker, G., 1982, Infantile form of sialic acid storage disorder: Clinical, ultrastructural, and biochemical studies in two siblings, *Eur J Pediatr* **139**: 142.

Vamos, E., Libert, J., Elkhazen, N., Jauniaux, E., Hustin, J., Wilkin, P., Baumkotter, J., Mendla, K., Cantz, M., and Strecker, G., 1986, Prenatal diagnosis and confirmation of infantile sialic acid storage disease, *Prenat Diagn* **6**: 437.

Varho, T., Jaaskelainen, S., Tolonen, U., Sonninen, P., Vainionpaa, L., Aula, P., and Sillanpaa, M., 2000, Central and peripheral nervous system dysfunction in the clinical variation of Salla disease, *Neurology* **55**: 99.

Varho, T., Komu, M., Sonninen, P., Holopainen, I., Nyman, S., Manner, T., Sillanpaa, M., Aula, P., and Lundbom, N., 1999, A new metabolite contributing to N-acetyl signal in 1H MRS of the brain in Salla disease. *Neurology* **52**: 1668.

Varho, T. T., Alajoki, L. E., Posti, K. M., Korhonen, T. T., Renlund, M. G., Nyman, S. R., Sillanpaa, M. L., and Aula, P. P., 2002, Phenotypic spectrum of Salla disease, a free sialic acid storage disorder, *Pediatr Neurol* **26**: 267.

Verheijen, F. W., Verbeek, E., Aula, N., Beerens, C. E., Havelaar, A. C., Joosse, M., Peltonen, L., Aula, P., Galjaard, H., van der Spek, P. J., and Mancini, G. M., 1999, A new gene, encoding an anion transporter, is mutated in sialic acid storage diseases, *Nat Genet* **23**: 462.

Weiss, P., Tietze, F., Gahl, W., Seppala, R., and Ashwell, G., 1989, Identification of the metabolic defect in sialuria, *J Biol Chem* **264**: 17635.

Wreden, C. C., Wlizla, M., and Reimer, R. J., 2005, Varied mechanisms underlie the free sialic acid storage disorders, *J Biol Chem* **280**: 1408.

CYSTINOSIS

Robert Kleta, Amanda Helip-Wooley, William A. Gahl

1 INTRODUCTION

Most lysosomal storage disorders result from deficiencies of acid hydrolases that degrade lipids, carbohydrates, or proteins. As a consequence of these defects, macromolecules accumulate within lysosomes. This stands in stark contrast to lysosomal storage disorders resulting from deficiency of small molecule transporters, in which an amino acid, sugar, or vitamin fails to exit the lysosome. The resulting diseases, corresponding to storage of cystine, sialic acid, or cobalamin, are cystinosis, Salla disease, and cobalamin F disease, respectively. This chapter deals with the first and most common of these lysosomal membrane transport disorders.

Cystinosis, an autosomal recessive disease, occurs throughout the world with an incidence estimated at 1:100,000 to 1:200,000 live births (Gahl, Thoene, and Schneider, 2001, 2002). This disorder of lysosomal cystine transport must be distinguished from cystinuria, in which cystine fails to be transported across the plasma membrane of renal tubular cells, leading to nephrolithiasis. Although rare, cystinosis is the single most common identifiable cause of renal Fanconi syndrome in children. Early recognition and diligent treatment of cystinosis unquestionably confers substantial medical benefit to affected individuals.

2 CYSTINE METABOLISM

Cystine consists of two cysteine molecules linked by a disulfide bond. Within proteins, this linkage generates and maintains functional structure. Cysteine itself can be formed from serine and methionine, the only other significant sulfur-containing amino acid in a normal diet. Cysteine is almost always present in its reduced form, maintained by the high glutathione content of the cytoplasm of cells. Cystine, on the other hand, can accumulate in lysosomes because of their acidic, nonreducing milieu, in which hydrolases generate the disulfide via protein degradation. A specific transporter, cystinosin, subsequently allows the egress of lysosomal cystine into the cytoplasm where it is reduced to cysteine and either participates in metabolic pathways or is excreted by the kidneys.

3 HISTORY

In 1903 Abderhalden described a patient who died in early childhood of progressive anorexia with a unique crystal deposition in several organs (Figure 1). Two siblings in

Section on Human Biochemical Genetics, Medical Genetics Branch, National Human Genome Research Institute, NIH, Bethesda MD, USA. Address correspondence to: William A. Gahl, M.D., Ph.D.Medical Genetics Branch, NHGRI, NIH. 10 Center Drive, MSC 1851, Building 10, Room 10C-103. Bethesda, Maryland 20892-1851. Telephone: 301-402-2739 FAX 301-402-2740. e-mail: bgahl@helix.nih.gov

this family had previously died in infancy under similar circumstances (Abderhalden, 1903). Elaborate chemical analyses in similar cases proved that the crystals in these patients consisted of cystine (Lignac, 1924). Studies by De Toni, Debre, and Fanconi in the 1930s showed that glucosuria and proteinuria were associated with this disease, which also led to intractable rickets (De Toni, 1933; Debre et al., 1934; Fanconi, 1936). In the 1950s, Bickel pointed to a renal origin for the rickets, and distinguished cystinosis from cystinuria (Bickel, 1955).

The 1960s brought the ability to measure amino acids in small quantities and, conesquently the finding of elevated cystine in lysosomes of cystinosis patients (Schneider, Bradley, and Seegmiller, 1967; Schulman, Bradley, and Seegmiller, 1969; Patrick et al., 1968). In the 1970s, Thoene et al. discovered that certain aminothiols, like cysteamine, were able to lower the cystine content of cystinotic lysosomes (Thoene et al., 1976). Careful biochemical studies documented the inability of cystinotic lysosomes to export cystine (Gahl et al., 1982a,b, 1983; Jonas et al., 1982,). Subsequent clinical studies proved the usefulness of oral cysteamine therapy in maintaining renal function and improving growth in patients with cystinosis (Gahl et al., 1987; Markello, Bernardini, and Gahl, 1993). Cystinosis entered the era of molecular biology in the late 1990s, when the cystinosis gene, *CTNS*, was mapped, isolated, sequenced, and characterized for its disease-causing mutations (The Cystinosis Collaborative Research Group, 1995; Town et al., 1998; Kleta et al., 2002).

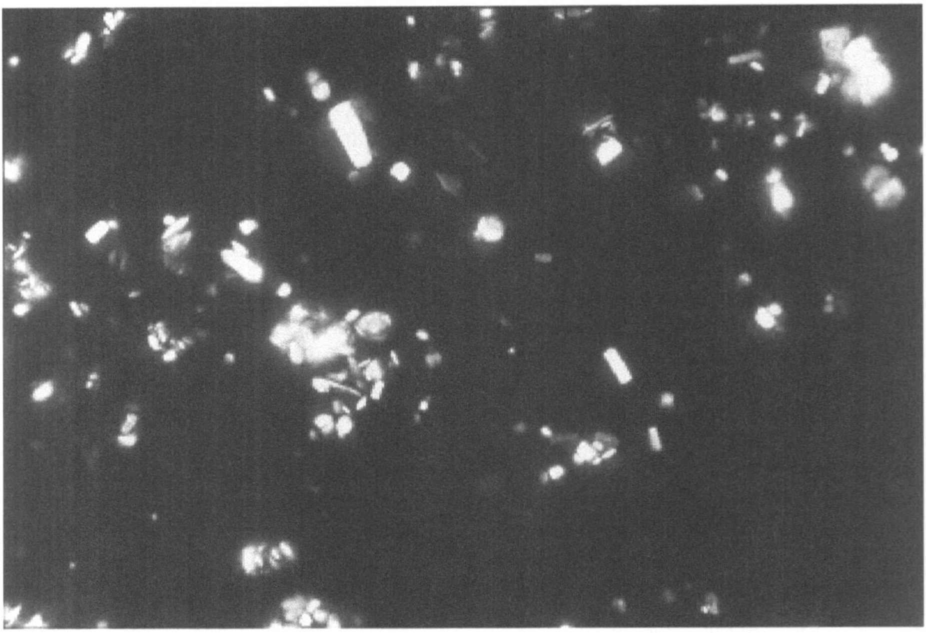

Figure 1. Cystine crystals. Light microscopy showing abundant cystine crystals in the spleen of a cystinosis patient under cross-polarizing light.

4 THE BASIC BIOCHEMICAL DEFECT

The basic biochemical defect in cystinosis is the inability of lysosomes to export cystine to the cytoplasm. Early attempts to understand the pathophysiology of cystinosis focused on the cytoplasmic metabolism of cystine, its possible enzymatic degradation, and its role in crystal formation and tissue destruction. Dietary measures to reduce the intake of sulfur-containing amino acids provided no clinical or biochemical improvement in affected patients. Studies involving reducing agents such as vitamin C also failed. In the 1960s, a decade after the lysosome was identified as a distinct subcellular compartment (De Duve et al., 1955), this vesicle was finally recognized as the locus for the pathological accumulation of cystine in cystinosis (Schneider, Bradley, and Seegmiller, 1967), explaining the failure of previous treatment strategies.

The elucidation of the lysosomal location of cystine storage in cystinosis also led to investigation of defective transport as the basic defect. In fact, cystinosis patients showed no transport of cystine out of cellular lysosomes, whereas cystine could be transported out of normal lysosomes at a measurable rate (Gahl et al., 1982a,b; Jonas et al., 1982). The cystine transporter exhibited ligand specificity, stereospecificity, and countertransport (Gahl et al., 1983), all characteristics of carrier-mediated transport. Heterozygotes showed half the maximal velocity of cystine transport (Gahl et al., 1984), reflecting a gene-dosage effect.

5 CLINICAL FINDINGS

Children affected with nephropathic (i.e., classical or infantile) cystinosis are generally born at term after uneventful pregnancies. They appear healthy without recognizable signs or symptoms. Cystinosis patients are often blonde, but this largely reflects the northern European background of the majority of patients due to a founder mutation in *CTNS*. Cystinosis does occur in African-American and other ethnic groups with black hair, so hair color or skin pigmentation should never exclude this diagnosis (Figure 2).

Beginning at approximately 6 months of age, cystinosis patients develop failure to thrive, polyuria, and polydipsia; by 12 to 18 months of age, they manifest pronounced growth retardation with rickets. Corneal cystine crystals can be appreciated after approximately 18 months of age, and this leads to photophobia later in childhood. Unrecognized and untreated cystinosis patients develop end-stage renal disease at approximately 10 years of age (Gretz et al., 1983; Gahl, Thoene, and Schneider, 2001, 2002).

Before cystinosis was recognized as a distinct entity, affected children would often die in infancy due to fluid and electrolyte losses. Later, but before kidney transplantation became available, these children would die of end-stage renal disease before puberty. With the advent of solid organ transplantation in the 1960s, children with cystinosis survived into adulthood but experienced other long-term effects of cystinosis (Theodoropoulus et al., 1993). These included severely short stature, hypothyroidism (Kimonis et al., 1995), muscle wasting (Gahl et al., 1988; Charnas et al., 1994), pulmonary dysfunction (Anikster et al., 2001), swallowing disorders (Sonies et al., 1990, 2004; Trauner et al., 2001), pancreatic insufficiency (Fivush et al., 1987; Fivush, Flick, and Gahl, 1988), liver involvement (O'Brien et al., 2006), neurological complications (Ehrich et al., 1979, Trauner et al., 1988; Fink et al., 1989, Dogulu et al., 2004), and testicular dysfunction (Chik et al., 1993). A variety of eye complications also increases with age in untreated

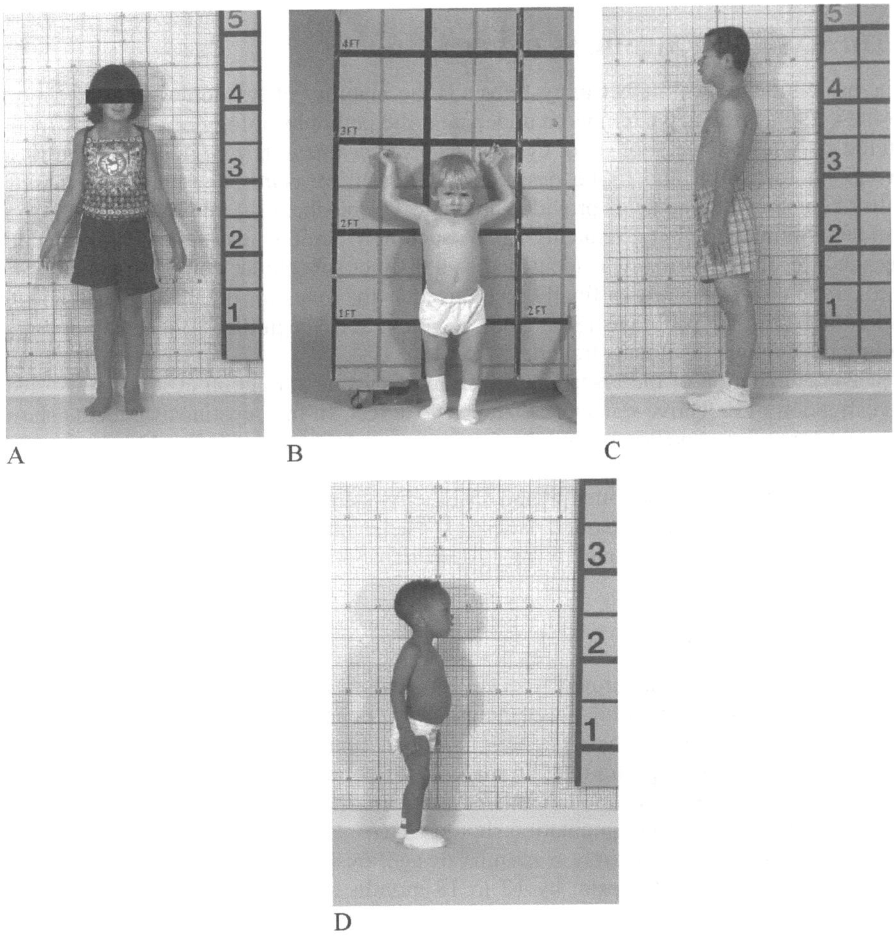

Figure 2. Nephropathic cystinosis patients of different ethnic backgrounds. (A) Eight-year-old Caucasian female treated diligently with cysteamine from age 2 months. She never received exogenous growth hormone. Note normal height and the dark complexion of her hair. (B) Three-year 7-month-old Caucasian male. Note short stature. (C) Twenty-year-old Mexican male. Note short stature and dark complexion of his hair. (D) Three-year 10-month-old African-American male. Note short stature.

cystinosis patients (Tsilou et al., 2002). Females are fertile and successful pregnancies have been reported (Reiss et al., 1988; Andrews et al., 1994). The occurrence of nearly all of the organ damage of cystinosis can be effectively delayed or prevented by early and rigorous treatment with cysteamine (see below; Gahl, Thoene, and Schneider, 2001, 2002; Kleta et al., 2004a b).

Although cystinosis patients display a spectrum of severities, two particular variants of the classical disease have been emphasized. Intermediate (i.e., late-onset or adolescent) cystinosis presents with kidney failure in early adulthood (Thoene et al., 1999). Ocular (i.e., benign, adult, or nonnephropathic) cystinosis has no renal involvement and presents with corneal and bone-marrow crystals only (Anikster et al., 2000).

6 LABORATORY FINDINGS

Cystinosis patients generally present in infancy with polyuria, polydipsia, and dehydration due to their renal Fanconi syndrome. Upon investigation, they exhibit a low urine specific gravity, hypokalemia, metabolic acidosis, glucosuria, phosphaturia, generalized aminoaciduria, and tubular proteinuria. Fractional excretions of phosphate and amino acids provide the most sensitive indicators of proximal tubular injury. The proteinuria of cystinosis can reach levels commonly seen in other kidney disorders, for example, nephrotic syndrome. Some patients display a picture resembling Bartter syndrome (juxtaglomerular hyperplasia, hyperaldosteronism, and hypokalemic alkalosis), except that children with Bartter syndrome show pathology from birth. Other children with cystinosis may be considered to have diabetes mellitus because of their glucosuria, but this diagnosis can be excluded by the absence of hyperglycemia. Timed urines are more accurate and more reliable than spot urines, and can provide the first documentation of a compromised glomerular filtration rate.

7 GENETICS

Cystinosis is inherited in an autosomal recessive manner. Parents are not affected and would only be recognized as carriers by molecular analysis, transport studies of lysosomal cystine egress, or measuring polymorphonuclear leukocyte cystine content. All these findings reflect defective cystinosin, the single lysosomal membrane protein that transports cystine out of lysosomes and into the cytoplasm.

The gene for cystinosin, *CTNS*, resides on chromosome 17p13 and consists of 12 exons, 10 of which code for an 1104-bp mRNA producing a 367 amino acid protein. The gene was isolated in 1998 (Town et al., 1998). To date, more than 80 missense, nonsense, insertion, deletion, splice site, and promoter mutations have been reported (Town et al., 1998; Shotelersuk et al., 1998; Thoene et al., 1999; Attard et al., 1999; McGowan-Jordan et al., 1999; Forestier et al., 1999; Anikster et al., 2000; Kleta et al., 2001; Phornphutkul et al., 2001; Kiehntopf et al., 2002; Kalatzis et al., 2002; Mason et al., 2003). Certain ethnic groups may show a predominance of specific mutations (Rupar et al., 2001). For example, a common 57-kb deletion is present in approximately 50% of all northern European patients' alleles. This deletion, which represents a founder mutation arising approximately 1500 years ago (Shotelersuk et al., 1998), also destroys an adjacent gene, *CARKL* (Touchman et al., 2000), whose function is unknown. Conventional PCR techniques are used to document mutations in *CTNS*, with the caveat that the presence or absence of the common 57-kb deletion should first be determined in any affected individual (Anikster et al., 1999; Forestier et al., 1999). Recently, a FISH technique was developed to detect the common 57-kb deletion and a less common 11.7-kb deletion (Bendavid et al., 2004).

All forms of cystinosis (nephropathic, intermediate, and ocular) are allelic and result from mutations in *CTNS*. There is some correlation between genotype and phenotype. For instance, ocular cystinosis patients carry a severe mutation on one allele and a mild mutation on the other. Their intracellular cystine content is slightly higher than that of heterozygotes but always well below nephropathic cystinosis levels (Anikster et al., 2000).

8 CELL BIOLOGY

Early studies in the 1980s predicted the properties and the function of the lysosomal cystine transporter as a proton coupled carrier (Oude Elferink et al., 1983). Modern molecular biology essentially confirmed earlier biochemical work by documenting the cystine-carrying capacity of the *CTNS* gene product, cystinosin. Specifically, investigators redirected the cystinosin protein from lysosomes to the plasma membrane by deleting its C-terminal sorting motif, GYDQL, thereby exposing the intralysosomal side of cystinosin to the extracellular medium. COS cells expressing cystinosin-ΔGYDQL selectively took up L-cystine from the extracellular medium at acidic pH. Disruption of the transmembrane pH gradient or incubation of the cells at neutral pH strongly inhibited the uptake. Cystinosin-ΔGYDQL was directly involved in the observed cystine transport, because this activity was highly reduced when the GYDQL motif was restored. It was concluded from these experiments that cystinosin represents a proton-driven transporter that is responsible for cystine export from lysosomes (Kalatzis et al., 2001).

Cystinosin has seven predicted transmembrane domains, is highly glycosylated at its N-terminus, and carries a lysosomal-targeting motif in its carboxy tail. Two lysosomal sorting motifs have been identified on cystinosin, including the well-described C-terminal targeting motif GYDQL and a novel YFPQA motif in the third cytoplasmic loop. This YFPQA motif was able to partially direct *CTNS* to lysosomes in the absence of the GYDQL signal. (Cherqui et al., 2001). Using fusion proteins, the effects of certain *CTNS* mutations on trafficking were demonstrated. Although the normal *CTNS*-GFP fusion protein colocalized almost exclusively with Lysotracker red (a lysosomal marker), a mutated GFP fusion product, lacking the GYDQL motif, was found in the plasma membrane and cytoplasm, as well as lysosomes (Helip-Wooley et al., 2002). Transport studies using mutated *CTNS* demonstrated the effect of particular mutations on the trafficking and transport of cystine. Most of the mutations did not alter the lysosomal localization of cystinosin, although three partially mislocalized the protein independently of its C-terminal sorting motif, thus confirming the presence of an additional sorting mechanism (Kalatzis et al., 2004).

Antibodies to cystinosin showed colocalization with other lysosomal markers. In immunohistochemical analyses, cystinosin localized to the tubular epithelia of normal human kidneys, with a pattern similar to that of LAMP 2. Cystinosin immunoreactivity was absent from kidneys of patients homozygous for the 57-kb *CTNS* deletion (Haq et al., 2002).

A *CTNS* knockout mouse, created using a promotor trap approach, surprisingly showed no kidney involvement (Cherqui et al., 2002). Even though the truncated cystinosin protein was mislocalized and nonfunctional, *CTNS* knockout mice showed no renal Fanconi syndrome or glomerular involvement. This discrepancy still requires explanation. The mice did show increased intracellular cystine content in all organs tested, and cysteamine was able to lower the lysosomal cystine content.

It is still not fully understood how an elevated intralysosomal cystine content leads to multiorgan failure in untreated cystinosis (Cetinkaya et al., 2002). A possible explanation for this pathophysiology could be the initiation of apoptosis through the elevated intralysosomal cystine content. Park et al. showed enhanced apoptosis associated with lysosomal cystine storage and speculated that this may lead to inappropriate cell death and decreased cell numbers in many tissues and hence contribute to the nephropathic cystinosis phenotype (Park, Helip-Wooley, and Thoene, 2002; Park and Thoene, 2005).

9 DIAGNOSIS

The diagnosis of cystinosis often relies heavily on the physician, who needs to recognize the signs and symptoms of defective renal tubular reabsorption that are present in affected individuals. This should prompt investigation of the cause of the renal Fanconi syndrome, which is cystinosis until proven otherwise. Even today, this fact is often overlooked, and occasionally the diagnosis of cystinosis is not made until end-stage renal disease has occurred or kidney transplantation is performed.

The specific diagnosis of cystinosis can be confirmed biochemically by determination of lysosomal cystine in polymorphonuclear neutrophils, a subset of leukocytes (Smolin, Clark, and Schneider, 1987). All cystinosis patients show an elevation of this value, with typical levels of 3–23 nmol half-cystine per mg protein (normal, <0.2). Several different techniques can be employed for this measurement; the cystine-binding protein assay is the most sensitive (Oshima et al., 1974). However, this test should be performed only in centers with expertise. From 16 months of age an experienced ophthalmologist can also make the diagnosis of cystinosis, by documenting typical corneal crystals (Gahl et al., 2000), although other diseases can mimic cystinosis in this respect (Kleta et al., 2004c). Prenatal diagnosis is available using amniocytes or chorionic villi, and can be performed in families where an index case is present (Smith et al., 1987). The use of any form of biopsy (kidney, cornea, bone-marrow) to make the diagnosis of cystinosis is obsolete. Newborn screening for cystinosis is not available.

10 TREATMENT

Treatment consists of symptomatic medications and a specific cystine-depleting compound. Symptomatic treatments include replacement therapy for the renal tubular Fanconi syndrome, such as free access to water and prescribed doses of sodium, potassium, bicarbonate, and phosphate (Gahl, 1986). When glomerular kidney failure occurs, vitamin D or erythropoietin and ultimately some form of kidney replacement therapy is required. Growth hormone has proven beneficial in cases of significant growth retardation prior to initiation of cystine-depleting therapy (Wuhl et al., 2001).

Specific therapy follows from the work of Thoene et al., who demonstrated in 1976 that certain aminothiols (e.g., cysteamine) lowered intralysosomal cystine content in cystinosis cells (Thoene et al., 1976). This major breakthrough provided the prospect for meaningful treatment of cystinosis. Cysteamine can freely enter the lysosome and react in a disulfide interchange reaction with cystine to generate free cysteine and the mixed disulfide cysteamine-cysteine. Both these compounds can leave the cystinotic lysosome by other transport mechanisms intact in cystinosis (Gahl et al., 1985; Pisoni, Thoene, and Christensen, 1985). Cysteamine essentially makes it possible for cystine to change its identity, bypassing the genetic defect of a nonfunctional or missing cystine transporter (Figure 3).

Based upon clinical studies (Gahl et al., 1992; Gahl, Thoene, and Schneider 2001, 2002), cysteamine was approved for use in pretransplant cystinosis patients in the United States in 1994 (Med. Lett. Drugs Ther., 1994; Schneider, 1995). Early and well-treated patients can grow at a normal rate without the need for growth hormone or thyroid hormone replacement, maintain their renal glomerular function (Kleta, Thoene, and Schneider, 2004a,b), and sometimes show an improvement of their renal Fanconi syndrome. The major drawbacks of this medication are gastrointestinal side effects, a foul taste and smell, and the need for a six-hour dosing regimen; other side effects are rare (Corden et al., 1981), especially with the use of incremental dosing. Proton pump inhibitors may be helpful for some of the gastrointestinal side effects (Wenner et al., 1997; Dohil et al., 2003, 2005). Nevertheless, treatment with cysteamine poses a challenge to the affected individual (Schneider, 2004).

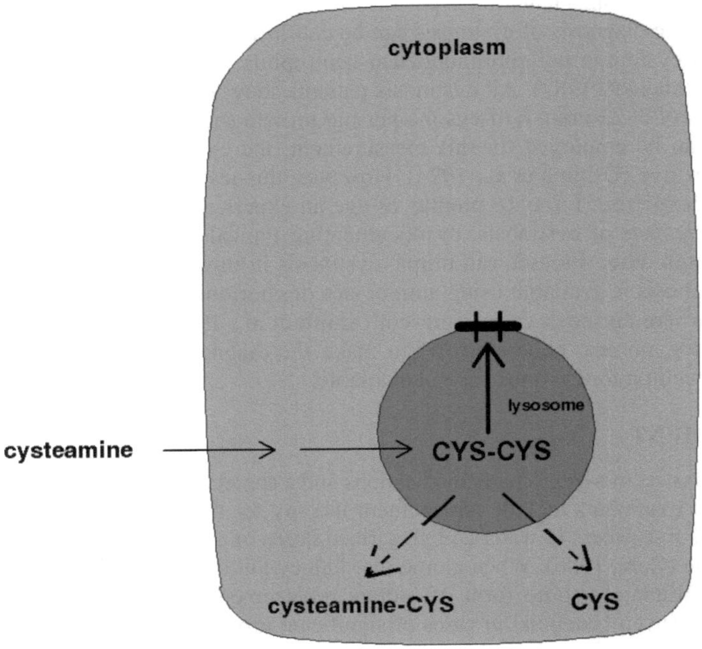

Figure 3. Mechanism by which cysteamine enables cystine to exit the lysosome in cystinosis. Cystine (CYS-CYS) cannot leave the lysosome because the cystine transporter, cystinosin, is defective. Cysteamine enters the cell and the lysosome, where it is trapped as a charged quaternary amine by the acidic milieu. An intralysosomal disulfide interchange reaction generates free cysteine (CYS) and the mixed disulfide cysteamine-cysteine (cysteamine-CYS), both of which can leave the lysosome by different transporters, which are intact in cystinosis. This lowers the intralysosomal cystine content, achieving the therapeutic goal for cystinosis.

Treatment with cysteamine in pregnancy has not been evaluated. Pregnant rats treated with cysteamine showed fetal pathology only in doses well above those recommended and used in humans (Assadi, Mullin, and Beckman, 1998; Beckman, Mullin, and Assadi, 1998, Assadi et al., 1999). Cysteamine's pharmacology has been well studied (Belldina et al., 2003; Tenneze et al., 1999; Kleta et al., 2004b) and its use has been demonstrated to be economically beneficial (Soohoo, Schneider, and Kaplan, 1997). Although systemic cysteamine treatment is unable to reach the cornea of the eye, corneal crystals can be dissolved with cysteamine eyedrops (Kaiser-Kupfer et al., 1986, 1987, 1990; Iwata et al., 1998; Gahl et al., 2000; Tsilou et al., 2003).

Two decades after the start of the first trials of cysteamine in cystinosis patients, it is becoming more and more apparent that higher doses of cysteamine lead to a better outcome (Markello, Bernardini, and Gahl, 1993; Van't Hoff et al., 1995, Kleta et al., 2004b, Kleta et al., 2005). Early initiation of therapy also predicts a better outcome (Da Silva et al., 1985; Gahl, 2003; Kleta et al., 2004a, Kleta et al., 2005).

11 CONCLUSION

Cystinosis has been associated with a number of "firsts" in biomedical research. It is the first lysosomal storage disorder to be discovered to result from a transport defect. This defect revealed, for the first time, the existence of any lysosomal transport system for small molecules. Cystinosis represents the first treatable lysosomal storage disease, and cysteamine eyedrops comprise the first topical therapy to dissolve any corneal crystals. Finally, cystinosis should be the first disorder than comes to mind when physicians diagnose renal tubular Fanconi syndrome in infants, because early diagnosis and treatment are absolutely critical for a good outcome.

REFERENCES

Abderhalden, E., 1903, Familiäre Cystindiathese, *Hoppe Seylers Zeitschr. f. physiol. Chemie.* **38**:557–561.

Andrews, P. A., Sacks, S. H., and Van't Hoff, W., 1994, Successful pregnancy in cystinosis, *JAMA* **272**:1327–1328.

Anikster, Y., Lacbawan, F., Brantly, M., Gochuico, B. L., Avila, N. A., Travis, W., and Gahl, W. A., 2001, Pulmonary dysfunction in adults with nephropathic cystinosis, *Chest.* **119**:394–401.

Anikster, Y., Lucero, C., Guo, J., Huizing, M., Shotelersuk, V., Bernardini, I., McDowell, G., Iwata, F., Kaiser-Kupfer, M. I., Jaffe, R., Thoene, J., Schneider, J. A., and Gahl, W. A., 2000, Ocular nonnephropathic cystinosis: Clinical, biochemical, and molecular correlations, *Pediatr. Res.* **47**:17–23.

Anikster, Y., Lucero, C., Touchman, J. W., Huizing, M., McDowell, G., Shotelersuk, V., Green, E. D., and Gahl, W. A., 1999, Identification and detection of the common 65–kb deletion breakpoint in the nephropathic cystinosis gene (CTNS), *Mol. Genet. Metab.* **66**:111–116.

Assadi, F. K., McCue, P., Jefferis, S., Shi, M., and Beckman, D. A., 1999, Effects of pre- and postnatal cysteamine exposure on renal function in the rat, *Pediatr. Nephrol.* **13**:812–815.

Assadi, F. K., Mullin, J. J., and Beckman, D. A., 1998, Evaluation of the reproductive and developmental safety of cysteamine in the rat: Effects on female reproduction and early embryonic development, *Teratology* **58**:88–95.

Attard, M., Jean, G., Forestier, L., Cherqui, S., van't Hoff, W., Broyer, M., Antignac, C., and Town, M., 1999, Severity of phenotype in cystinosis varies with mutations in the CTNS gene: predicted effect on the model of cystinosin, *Hum. Mol. Genet.* **8**:2507–2514.

Beckman, D. A., Mullin, J. J., and Assadi, F. K., 1998, Developmental toxicity of cysteamine in the rat: effects on embryo-fetal development, *Teratology* **58**:96–102.

Belldina, E. B., Huang, M. Y., Schneider, J. A., Brundage, R. C., and Tracy, T. S., 2003, Steady-state pharmacokinetics and pharmacodynamics of cysteamine bitartrate in paediatric nephropathic cystinosis patients, *Br. J. Clin. Pharmacol.* **56**:520–525.

Bendavid, C., Kleta, R., Long, R., Ouspenskaia, M., Muenke, M., Haddad, B. R., and Gahl, W. A., 2004, FISH diagnosis of the common 57-kb deletion in CTNS causing cystinosis, *Human Genet.* **115**:510–514.

Bickel, H., 1955, Die Entwicklung der biochemischen Läsion bei der Lignac-Fanconischen Krankheit, *Helv. Paed. Acta* **10**:259–268.

Cetinkaya, I., Schlatter, E., Hirsch, J. R., Herter, P., Harms, E., and Kleta, R., 2002, Inhibition of Na(+)-dependent transporters in cystine-loaded human renal cells: electrophysiological studies on the Fanconi syndrome of cystinosis, *J. Am. Soc. Nephrol.* **13**:2085–2093.

Charnas, L. R., Luciano, C. A., Dalakas, M.,Gilliatt, R. W., Bernardini, I., Ishak, K., Cwik, V. A., Fraker, D., Brushart, T. A., and Gahl, W. A., 1994, Distal vacuolar myopathy in nephropathic cystinosis, *Ann. Neurol.* **35**:181–188.

Cherqui, S., Kalatzis, V., Trugnan, G., and Antignac, C., 2001, The targeting of cystinosin to the lysosomal membrane requires a tyrosine-based signal and a novel sorting motif, *J. Biol. Chem.* **276**:13314–13321.

Cherqui, S., Sevin, C., Hamard, G., Kalatzis, V., Sich, M., Pequignot, M. O., Gogat, K., Abitbol, M., Broyer, M., Gubler, M. C., and Antignac, C., 2002, Intralysosomal cystine accumulation in mice lacking cystinosin, the protein defective in cystinosis, *Mol. Cell. Biol.* **22**:7622–7632.

Chik, C. L., Friedman, A., Merriam, G. R., and Gahl, W. A., 1993, Pituitary-testicular function in nephropathic cystinosis, *Ann. Intern. Med.* **119**:568–575.

Corden, B. J., Schulman, J. D., Schneider, J. A., and Thoene, J. G., 1981, Adverse reactions to oral cysteamine use in nephropathic cystinosis, *Dev. Pharmacol. Ther.* **3**:25–30.

Da Silva, V. A., Zurbrugg, R. P., Lavanchy, P., Blumberg, A., Suter, H., Wyss, S. R., Luthy, C. M., and Oetliker, O. H., 1985, Long-term treatment of infantile nephropathic cystinosis with cysteamine, *N. Engl. J. Med.* **313**:1460–1463.

Debre, R., Marie, J., Cleret, F., Messimy, R., 1934, Rachitisme tardif coexistant avec une nephrite chronique et une glycosurie. *Archives de Médecine des Enfants*, **37**:597–606.

De Duve, C., Pressman, B. C., Gianetto, R., Wattiaux, R., and Appelmans, F., 1955, Tissue fractionation studies. 6. Intracellular distribution patterns of enzymes in rat-liver tissue, *Biochem. J.* **60**:604–617.

De Toni, G., 1933, Remarks on the relations between renal rickets (renal dwarfism) and renal diabetes, *Acta Paediatrica*, **16**:479–484.

Dogulu, C. F., Tsilou, E., Rubin, B., Fitzgibbon, E. J., Kaiser-Kupfer, M. I., Rennert, O. M., and Gahl, W. A., 2004, Idiopathic intracranial hypertension in cystinosis, *J. Pediatr.* **145**:673–678.

Dohil, R., Fidler, M., Barshop, B., Newbury, R., Sellers, Z., Deutsch, R., and Schneider, J., 2005, Esomeprazole therapy for gastric acid hypersecretion in children with cystinosis, *Pediatr. Nephrol.* **20**:1786–1793.

Dohil, R., Newbury, R. O., Sellers, Z. M., Deutsch, R., and Schneider, J. A., 2003, The evaluation and treatment of gastrointestinal disease in children with cystinosis receiving cysteamine, *J. Pediatr.* **143**:224–230.

Ehrich, J. H., Stoeppler, L., Offner, G., and Brodehl, J., 1979, Evidence for cerebral involvement in nephropathic cystinosis, *Neuropadiatrie* **10**:128–137.

Fanconi, G., 1936, Der frühinfantile nephrotisch-glykosurische Zwergwuchs mit hypophosphatämischer Rachitis, *Jahrbuch für Kinderheilkunde*, **147**:299–338.

Fink, J. K., Brouwers, P., Barton, N., Malekzadeh, M. H., Sato, S., Hill, S., Cohen, W. E., Fivush, B., and Gahl, W. A., 1989, Neurologic complications in long-standing nephropathic cystinosis, *Arch. Neurol.* **46**:543–548.

Fivush, B., Flick, J. A., and Gahl, W. A., 1988, Pancreatic exocrine insufficiency in a patient with nephropathic cystinosis, *J. Pediatr.* **112**:49–51.

Fivush, B., Green, O. C., Porter, C. C., Balfe, J. W., O'Regan, S., and Gahl, W. A., 1987, Pancreatic endocrine insufficiency in posttransplant cystinosis, *Am. J. Dis. Child.* **141**:1087–1089.

Forestier, L., Jean, G., Attard, M., Cherqui, S., Lewis, C., van't Hoff, W., Broyer, M., Town, M., and Antignac, C., 1999, Molecular characterization of CTNS deletions in nephropathic cystinosis: Development of a PCR-based detection assay, *Am. J. Hum. Genet.* **65**:353–359.

Gahl, W. A., 1986, Cystinosis coming of age, *Adv. Pediatr.* **33**:95–126.

Gahl, W. A., 2003, Early oral cysteamine therapy for nephropathic cystinosis, *Eur. J. Pediatr.* **162**(Suppl. 1):S38–S41.

Gahl, W. A., Bashan, N., Tietze, F., and Schulman, J. D., 1984, Lysosomal cystine counter-transport in heterozygotes for cystinosis, *Am J. Hum. Genet.* **36**:277–282.

Gahl, W. A., Bashan, N., Tietze, F., Bernardini, I., and Schulman, J. D., 1982a, Lysosomal cystine transport is defective in cystinosis, *Science* **217**:1263–1265.

Gahl, W. A., Charnas, L., Markello, T. C., Bernardini, I., Ishak, K. G., and Dalakas, M. C., 1992, Parenchymal organ cystine depletion with long-term cysteamine therapy, *Biochem. Med. Metab. Biol.* **48**:275–285.

Gahl, W. A., Dalakas, M. C., Charnas, L., Chen, K. T., Pezeshkpour, G. H., Kuwabara, T., Davis, S. L., Chesney, R. W., Fink, J., and Hutchison, H. T., 1988, Myopathy and cystine storage in muscles in a patient with nephropathic cystinosis, *N. Engl. J. Med.* **319**:1461–1464.

Gahl, W. A., Kuehl, E. M., Iwata, F., Lindblad, A., and Kaiser-Kupfer, M. I., 2000, Corneal crystals in nephropathic cystinosis: Natural history and treatment with cysteamine eyedrops, *Mol. Genet. Metab.* **71**:100–120.

Gahl, W. A., Reed, G. F., Thoene, J. G., Schulman, J. D., Rizzo, W. B., Jonas, A. J., Denman, D. W., Schlesselman, J. J., Corden, B. J., and Schneider, J. A., 1987, Cysteamine therapy for children with nephropathic cystinosis, *N. Engl. J. Med.* **316**:971–977.

Gahl, W. A., Thoene, J., and Schneider, J. A., 2001, Cystinosis: A Disorder of Lysosomal Membrane Transport, in: *The Metabolic & Molecular Bases of Inherited Disease*, C. R. Scriver, A. L. Beaudet, D. Valle, W. S. Sly, Eds., McGraw-Hill, New York, pp. 5085–5108.

Gahl, W. A, Thoene, J. G., and Schneider, J. A., 2002, Cystinosis, *N. Engl. J. Med.* **347**:111–121.

Gahl, W. A., Tietze, F., Bashan, N., Bernardini, I., Raiford, D., and Schulman, J. D., 1983, Characteristics of cystine counter-transport in normal and cystinotic lysosome-rich leucocyte granular fractions, *Biochem. J.* **216**:393–400.

Gahl, W. A., Tietze, F., Bashan, N., Steinherz, R., and Schulman, J. D., 1982b, Defective cystine exodus from isolated lysosome-rich fractions of cystinotic leucocytes, *J. Biol. Chem.* **257**:9570–9575.

Gahl, W. A., Tietze, F., Butler, J. D., and Schulman, J. D., 1985, Cysteamine depletes cystinotic leucocyte granular fractions of cystine by the mechanism of disulphide interchange, *Biochem. J.* **228**:545–550.

Gretz, N., Manz, F., Augustin, R., Barrat, T. M., Bender-Gotze, C., Brandis, M., Bremer, H. J., Brodehl, J., Broyer, M., Bulla, M., Callis, L., Chantler, C., Diekmann, L., Dillon, M. J., Egli, F., Ehrich, J. H., Endres, W., Fanconi, A., Feldhoff, C., Geisert, J., Gekle, D., Gescholl-Bauer, B., Grote, K., Gruttner, R., Hagge, W., Haycock, C. B., Hennemann, H., Klare, B., Leupold, D., Lohr, H., Michalk, D., Oliveira, A., Ott, F., Pistor, K., Rau, J., Scharer, K., Schindera, F., Schmidt, H., Schulte-Wissermann, H., Verrier-Jones, K., Weber, H. P., Willenbockel, U., and Wolf, H., 1983, Survival time in cystinosis. A collaborative study, *Proc. Eur. Dial. Transplant. Assoc.* **19**:582–589.

Haq, M. R., Kalatzis, V., Gubler, M. C., Town, M. M., Antignac, C., Van't Hoff, W. G., and Woolf, A. S., 2002, Immunolocalization of cystinosin, the protein defective in cystinosis, *J. Am. Soc. Nephrol.* **13**:2046–2051.

Helip-Wooley, A., Park, M. A., Lemons, R. M., and Thoene, J. G., 2002, Expression of CTNS alleles: subcellular localization and aminoglycoside correction in vitro, *Mol. Genet. Metab.* **75**:128–133.

Iwata, F., Kuehl, E. M., Reed, G. F., McCain, L. M., Gahl, W. A., and Kaiser-Kupfer, M. I., 1998, A randomized clinical trial of topical cysteamine disulfide (cystamine) versus free thiol (cysteamine) in the treatment of corneal cystine crystals in cystinosis, *Mol. Genet. Metab.* **64**:237–242.

Jonas, A. J., Greene, A. A., Smith, M. L., and Schneider, J.A., 1982, Cystine accumulation and loss in normal, heterozygous, and cystinotic fibroblasts, *Proc. Natl. Acad. Sci. USA* **79**:4442–4445.

Kaiser-Kupfer, M. I., Caruso, R. C., Minkler, D. S., and Gahl, W. A., 1986, Long-term ocular manifestations in nephropathic cystinosis, *Arch. Ophthalmol.* **104**:706–711.

Kaiser-Kupfer, M. I., Fujikawa, L., Kuwabara, T., and Gahl, W. A., 1987, Removal of corneal crystals by topical cysteamine in nephropathic cystinosis, *N. Engl. J. Med.* **316**:775–779.

Kaiser-Kupfer, M. I., Gazzo, M. A., Datiles, M. B., Caruso, R. C., Kuehl, E. M., and Gahl, W. A., 1990, A randomized placebo-controlled trial of cysteamine eyedrops in nephropathic cystinosis, *Arch. Ophthalmol.* **108**:689–693.

Kalatzis, V., Cherqui, S., Antignac, C., and Gasnier, B., 2001, Cystinosin, the protein defective in cystinosis, is a H(+)-driven lysosomal cystine transporter, *EMBO J.* **20**:5940–5949.

Kalatzis, V., Cohen-Solal, L., Cordier, B., Frishberg, Y., Kemper, M., Nuutinen, E. M., Legrand, E., Cochat, P., and Antignac, C., 2002, Identification of 14 novel CTNS mutations and characterization of seven splice site mutations associated with cystinosis, *Hum Mutat.* **20**:439–446.

Kalatzis, V., Nevo, N., Cherqui, S., Gasnier, B., and Antignac, C., 2004, Molecular pathogenesis of cystinosis: effect of CTNS mutations on the transport activity and subcellular localization of cystinosin, *Hum. Mol. Genet.* **13**:1361–1371.

Kiehntopf, M., Schickel, J., Gonne, B., Koch, H. G., Superti-Furga, A., Steinmann, B., Deufel, T., and Harms, E., 2002, Analysis of the CTNS gene in patients of German and Swiss origin with nephropathic cystinosis, *Hum. Mutat.* **20**:237.

Kimonis, V. E., Troendle, J., Yang, M. L., Rose, S. R., Markello, T. C., and Gahl, W. A., 1995, Effects of early cysteamine therapy on thyroid function and growth in nephropathic cystinosis, *J. Clin. Endocrinol. Metab.* **80**:3257–3261.

Kleta, R., and Gahl, W. A., 2002, Cystinosis: antibodies and healthy bodies, *J. Am. Soc. Nephrol.* **13**:2189–2191.

Kleta, R., and Gahl, W. A., 2004b, Pharmacological treatment of nephropathic cystinosis with cysteamine, *Expert Opin. Pharmacother.* **5**:2255–2262.

Kleta, R., Anikster, Y., Lucero, C., Shotelersuk, V., Huizing, M., Bernardini, I., Park, M., Thoene, J., Schneider, J., and Gahl, W. A., 2001, CTNS mutations in African American patients with cystinosis, *Mol. Genet. Metab.* **74**:332–337.

Kleta, R., Bernardini, I., Ueda, M., Varade, W. S., Phornphutkul, C., Krasnewich, D., and Gahl, W. A., 2004a, Long-term follow-up of well-treated nephropathic cystinosis patients, *J. Pediatr.* **145**:555–560.

Kleta, R., Blair, S. C., Bernardini, I., Kaiser-Kupfer, M. I., and Gahl, W. A., 2004c, Keratopathy of multiple myeloma masquerading as corneal crystals of ocular cystinosis, *Mayo Clin Proc.* **79**:410–412.

Kleta, R., Kaskel, F., Dohil, R., Goodyer, P., Guay-Woodford, L. M., Harms, E., Ingelfinger, J. R., Koch, V. H., Langman, C. B., Leonard, M. B., Mannon, R. B., Sarwal, M., Schneider, J. A., Skovby, F., Sonies, B. C., Thoene, J. G., Trauner, D. A., and Gahl, W. A., 2005, First NIH/Office of Rare Diseases Conference on Cystinosis: past, present, and future, *Pediatr. Nephrol.* **20**:452–454.

Lignac, G. O. E., 1924, Über Störung des Cystinstoffwechsels bei Kindern, *Deut. Arch. Klin. Med.* **145**:139–150.

Markello, T. C., Bernardini, I. M., and Gahl, W. A., 1993, Improved renal function in children with cystinosis treated with cysteamine. *N. Engl. J. Med.* **328**:1157–1162.

Mason, S., Pepe, G., Dall'Amico, R., Tartaglia, S., Casciani, S., Greco, M., Bencivenga, P., Murer, L., Rizzoni, G., Tenconi, R., and Clementi, M., 2003, Mutational spectrum of the CTNS gene in Italy, *Eur. J. Hum. Genet.* **11**:503–508.

McGowan-Jordan, J., Stoddard, K., Podolsky, L., Orrbine, E., McLaine, P., Town, M., Goodyer, P., MacKenzie, A., and Heick, H., 1999, Molecular analysis of cystinosis: probable Irish origin of the most common French Canadian mutation, *Eur. J. Hum. Genet.* **7**:671–678.

Med. Lett. Drugs Ther., 1994, Cysteamine for cystinosis, **36**:118.

O'Brien, K., Hussain, N., Warady, B. A., Kleiner, D. E., Kleta, R., Bernardini, I., Heller, T., and Gahl, W. A., 2006, Nodular regenerative hyperplasia and severe portal hypertension in cystinosis, *Clin. Gastroenterol. Hepatol.* **4**:387–394.

Oshima, R. G., Willis, R. C., Furlong, C. E., and Schneider, J. A., 1974, Binding assays for amino acids. The utilization of a cystine binding protein from Escherichia coli for the determination of acid-soluble cystine in small physiological samples, *J. Biol. Chem.* **249**:6033–6039.

Oude Elferink, R. P., Harms, E., Strijland, A., and Tager, J. M., 1983, The intralysosomal pH in cultured human skin fibroblasts in relation to cystine accumulation in patients with cystinosis, *Biochem. Biophys. Res. Commun.* **116**:154–161.

Park, M., Helip-Wooley, A., and Thoene, J., 2002, Lysosomal cystine storage augments apoptosis in cultured human fibroblasts and renal tubular epithelial cells, *J Am. Soc. Nephrol.* **13**:2878–2887.

Park, M. A., and Thoene, J. G., 2005, Potential role of apoptosis in development of the cystinotic phenotype, *Pediatr. Nephrol.* **20**:441–446.

Patrick, A. D., and Lake, B. D., 1968, Cystinosis: Electron microscopic evidence of lysosomal storage of cystine in lymph node, *J. Clin. Pathol.* **21**:571–575.

Phornphutkul, C., Anikster, Y., Huizing, M., Braun, P., Brodie, C., Chou, J. Y., and Gahl, W .A., 2001, The promoter of a lysosomal membrane transporter gene, CTNS, binds

Sp-1, shares sequences with the promoter of an adjacent gene, CARKL, and causes cystinosis if mutated in a critical region, *Am. J. Hum. Genet.* **69**:712–721.

Pisoni, R. L., Thoene, J. G., and Christensen, H. N., 1985, Detection and characterization of carrier-mediated cationic amino acid transport in lysosomes of normal and cystinotic human fibroblasts, *J. Biol. Chem.* **260**:4791–4798.

Reiss, R. E., Kuwabara, T., Smith, M. L., and Gahl, W. A., 1988, Successful pregnancy despite placental cystine crystals in a woman with nephropathic cystinosis, *N. Engl. J. Med.* **319**:223–226.

Rupar, C. A., Matsell, D., Surry, S., and Siu, V., 2001, A G339R mutation in the CTNS gene is a common cause of nephropathic cystinosis in the south western Ontario Amish Mennonite population, *J. Med. Genet.* **38**:615–616.

Schneider, J. A., 1995, Approval of cysteamine for patients with cystinosis, *Pediatr. Nephrol.* **9**:254.

Schneider, J. A., 2004, Treatment of cystinosis: Simple in principle, difficult in practice, *J. Pediatr.* **145**:436–438.

Schneider, J. A., Bradley, K., and Seegmiller, J. E., 1967, Increased cystine in leukocytes from individuals homozygous and heterozygous for cystinosis, *Science* **157**:1321–1322.

Schulman, J. D., Bradley, K. H., and Seegmiller, J. E., 1969, Cystine: Compartmentalization within lysosomes in cystinotic leukocytes. *Science* **166**:1152–1154.

Shotelersuk, V., Larson, D., Anikster, Y., McDowell, G., Lemons, R., Bernardini, I., Guo, J., Thoene, J., and Gahl, W. A., 1998, CTNS mutations in an American-based population of cystinosis patients, *Am. J. Hum. Genet.* **63**:1352–1362.

Smith, M. L., Pellett, O. L., Cass, M. M., Kennaway, N. G., Buist, N. R., Buckmaster, J., Golbus, M., Spear, G. S., and Schneider, J. A., 1987, Prenatal diagnosis of cystinosis utilizing chorionic villus sampling, *Prenat. Diagn.* **7**:23–26.

Smolin, L. A., Clark, K. F., and Schneider, J. A., 1987, An improved method for heterozygote detection of cystinosis, using polymorphonuclear leukocytes, *Am. J. Hum. Genet.* **41**:266–275.

Sonies, B. C., Almajid, P., Kleta, R., Bernardini, I., and Gahl, W. A., 2005, Swallowing dysfunction in 101 patients with nephropathic cystinosis; Benefit of long-term cysteamine therapy, *Medicine* **84**:137–146.

Sonies, B. C., Ekman, E. F., Andersson, H. C., Adamson, M. D., Kaler, S. G., Markello, T. C., and Gahl, W. A., 1990, Swallowing dysfunction in nephropathic cystinosis, *N. Engl. J. Med.* **323**:565–570.

Soohoo, N., Schneider, J. A., and Kaplan, R. M., 1997, A cost-effectiveness analysis of the orphan drug cysteamine in the treatment of infantile cystinosis, *Med. Decis. Making* **17**:193–198.

Tenneze, L., Daurat, V., Tibi, A., Chaumet-Riffaud, P., and Funck-Brentano, C., 1999, A study of the relative bioavailability of cysteamine hydrochloride, cysteamine bitartrate and phosphocysteamine in healthy adult male volunteers, *Br. J. Clin. Pharmacol.* **47**:49–52.

The Cystinosis Collaborative Research Group, 1995, Linkage of the gene for cystinosis to markers on the short arm of chromosome 17, *Nat. Genet.* **10**:246–248.

Theodoropoulos, D. S., Krasnewich, D., Kaiser-Kupfer, M. I., and Gahl, W. A., 1993, Classical nephropathic cystinosis as an adult disease, *JAMA* **270**:2200–2204.

Thoene, J., Lemons, R., Anikster, Y., Mullet, J., Paelicke, K., Lucero, C., Gahl, W., Schneider, J., Shu, S. G., and Campbell, H. T., 1999, Mutations of CTNS causing intermediate cystinosis, *Mol. Genet. Metab.* **67**:283–293.

Thoene, J. G., Oshima, R. G., Crawhall, J. C., Olson, D. L., and Schneider, J. A., 1976, Cystinosis. Intracellular cystine depletion by aminothiols in vitro and in vivo, *J. Clin. Invest.* **58**:180–189.

Touchman, J. W., Anikster, Y., Dietrich, N. L., Maduro, V. V., McDowell, G., Shotelersuk, V., Bouffard, G. G., Beckstrom-Sternberg, S. M., Gahl, W. A., and Green, E. D., 2000, The genomic region encompassing the nephropathic cystinosis gene (CTNS): complete sequencing of a 200-kb segment and discovery of a novel gene within the common cystinosis-causing deletion, *Genome Res.* **10**:165–173.

Town, M., Jean, G., Cherqui, S., Attard, M., Forestier, L., Whitmore, S. A., Callen, D. F., Gribouval, O., Broyer, M., Bates, G. P., van't Hoff, W., and Antignac, C., 1998, A novel gene encoding an integral membrane protein is mutated in nephropathic cystinosis, *Nat. Genet.* **18**:319–324.

Trauner, D. A., Chase, C., Scheller, J., Katz, B., and Schneider, J. A., 1988, Neurologic and cognitive deficits in children with cystinosis, *J. Pediatr.* **112**:912–914.

Trauner, D. A., Fahmy, R. F., and Mishler, D. A., 2001, Oral motor dysfunction and feeding difficulties in nephropathic cystinosis, *Pediatr. Neurol.* **24**:365–368.

Tsilou, E. T., Rubin, B. I., Reed, G. F., Iwata, F., Gahl, W., and Kaiser-Kupfer, M. I., 2002, Age-related prevalence of anterior segment complications in patients with infantile nephropathic cystinosis, *Cornea* **21**:173–176.

Tsilou, E. T., Thompson, D., Lindblad, A. S., Reed, G. F., Rubin, B., Gahl, W., Thoene, J., Del Monte, M., Schneider, J. A., Granet, D. B., and Kaiser-Kupfer, M. I., 2003, A multicentre randomised double masked clinical trial of a new formulation of topical cysteamine for the treatment of corneal cystine crystals in cystinosis, *Br. J. Ophthalmol.* **87**:28–31.

Van't Hoff, W. G., and Gretz, N., 1995, The treatment of cystinosis with cysteamine and phosphocysteamine in the United Kingdom and Eire, *Pediatr. Nephrol.* **9**:685–689.

Wenner, W. J., and Murphy, J. L., 1997, The effects of cysteamine on the upper gastrointestinal tract of children with cystinosis, *Pediatr. Nephrol.* **11**:600–603.

Wuhl, E., Haffner, D., Offner, G., Broyer, M., van't Hoff, W., Mehls, O, and European Study Group on Growth Hormone Treatment in Children with Nephropathic Cystinosis, 2001, Long-term treatment with growth hormone in short children with nephropathic cystinosis, *J. Pediatr.* **138**:880–887.

I-CELL DISEASE

Doug Brooks,[1,2] Chris Turner,[1] Viv Muller,[1] John Hopwood,[1,2] Peter Meikle[1,2]

The lysosomal storage disorders (LSD), I-cell disease (mucolipidosis type II, ML II), and mucolipidosis type III (ML III) represent the severe and attenuated phenotypes resulting from the dysfunction of *N*-acetylglucosamine 1-phosphotransferase. Like other LSD, these disorders can present with a spectrum of clinical phenotypes. However, unlike other LSD these two disorders are genetically distinct, affecting different gene products (subunits) of the same *N*-acetylglucosamine 1-phosphotransferase enzyme. This chapter gives a brief overview of the disorder, how it has been diagnosed, together with the molecular cause and its impact on lysosomal biogenesis. Potential treatment strategies for patients with I-cell disease are also discussed.

1 HISTORY OF I-CELL DISEASE

Mucolipidoses type II and III are inherited disorders in which lysosomal storage of undegraded lipids and oligosaccharides can result in severe pathology and early death. In fact, the term mucolipidosis was derived from a clinical presentation consistent with both the mucopolysaccharidoses and the sphingolipidoses (Spranger and Wiedemann, 1970). Mucolipidosis II (ML II, classic I-cell disease) was first described in 1967 with patients having phase-dense inclusion bodies in affected cells (Leroy and DeMars, 1967). This visualisation of the intracellular pathology gave rise to the pseudonym I-cell (for inclusion-cell) disease. A second disorder was subsequently identified with similar pathology and inclusion bodies, but patients presented later and tended to have an attenuated clinical phenotype (mucolipidosis III, ML III or pseudo-Hurler polydystrophy; Maroteaux and Lamy, 1966; Taylor et al., 1973).

2 MUCOLIPIDOSIS CLINICAL PRESENTATION

The symptoms of classic I-cell disease (Kornfeld and Sly, 1999) are often apparent at birth, indicative of a very severe lysosomal storage disorder with early onset and rapidly progressive pathogenesis. I-cell disease symptoms include severe psychomotor retardation, developmental delay, failure to thrive, coarse facial features, flat nasal bridge, corneal clouding, hepatosplenomegaly, cardiac problems, short stature, kyphosis, dysostosis multiplex (with similar radiological changes to that observed in archetypical Hurler syndrome),

---

[1]*Lysosomal Diseases Research Unit, Department of Genetic Medicine, Children Youth and Women's Health Service, 72 King William Rd, North Adelaide, South Australia 5006, Australia,*
[2]*Department of Paediatrics, University of Adelaide, Adelaide, South Australia 5005, Australia.*
Address for correspondence: Assoc. Prof. Doug Brooks, Lysosomal Diseases Research Unit, Department of Genetic Medicine, Children Youth and Women's Health Service, 72 King William Rd, North Adelaide, South Australia 5006, Australia. Phone: (61-8) 8161-7341; Fax: (61-8) 8161-7100; E-mail: douglas.brooks@adelaide.edu.au)

limb bowing, short neck, restricted joint mobility, hip dislocation, clawed hands, carpal tunnel syndrome, hypotonia, ataxia, hernias, respiratory infections, otitis media, and gingival hyperplasia. Death is usually within the first decade of life, often by five years of age, usually as a result of cardiorespiratory complications.

ML III is a similar clinical condition but is caused by a different molecular lesion (see below). The onset of ML III is delayed compared to ML II and the clinical course is slower but still progressive, and there are phenotypic similarities to MPS I and MPS VI patients (Kornfeld and Sly, 1999). Clinical symptoms can include mental retardation, failure to thrive, short stature, kyphosis, hip pathology with waddling gait, dysostosis multiplex, delayed bone age, carpal tunnel syndrome, coarse facial features, flat nasal bridge, corneal clouding, cardiac disease, stiff joints, and hernias. ML III patients may survive for up to four or five decades.

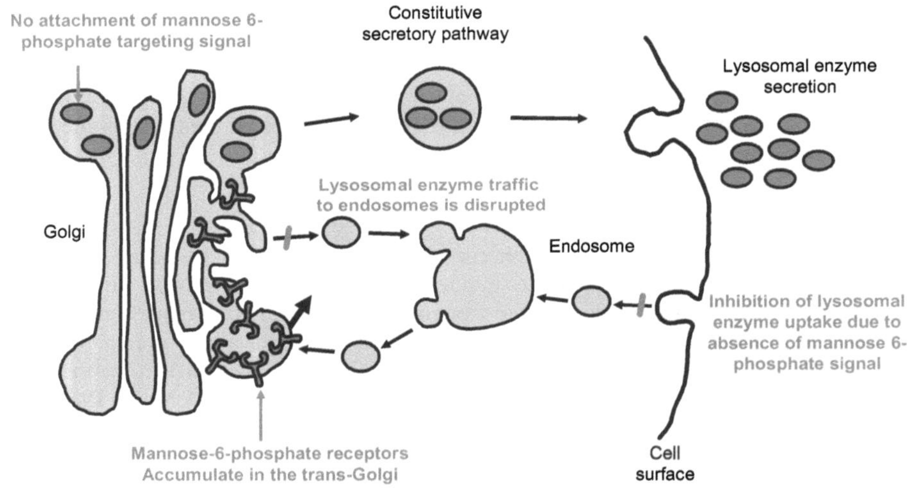

Figure 1. A schematic depicting the altered traffic of soluble lysosomal proteins in I-cell disease, which is caused by a failure to attach mannose 6-phosphate targeting signals to these proteins during Golgi processing. This inability to attach targeting signals to soluble lysosomal proteins is due to a deficiency of the enzyme *N*-acetylglucosamine 1-phosphotransferase.

3 CELL BIOLOGY AND MOLECULAR BASIS OF I-CELL DISEASE

I-cell disease and ML III are autosomal recessive inherited disorders involving a deficiency of *N*-acetylglucosamine 1-phosphotransferase activity. This enzyme is responsible for adding *N*-acetylglucosamine 1-phosphate to terminal mannose residues on newly synthesised lysosomal proteins, prior to the cleavage of the *N*-acetylglucosamine to leave the mannose 6-phosphate targeting residues. A deficiency of *N*-acetylglucosamine 1-phosphotransferase activity prevents lysosomal protein traffic and delivery to the endosomal-lysosomal compartment via the mannose 6-phosphate receptor (MPR) mediated trafficking pathway (Kornfeld and Sly, 1999). Instead, soluble lysosomal proteins are diverted to the default secretory pathway and exit the cell (Figure 1), and MPR are concentrated in the *trans*-Golgi (Figures 1 and 2; Brown and Farquhar, 1984). Consequently, the lysosome is unable to degrade various lipid and polysaccharide substrates, leading to the development of cellular pathology.

Fig. 2. ALTERED TRAFFIC IN I-CELL DISEASE

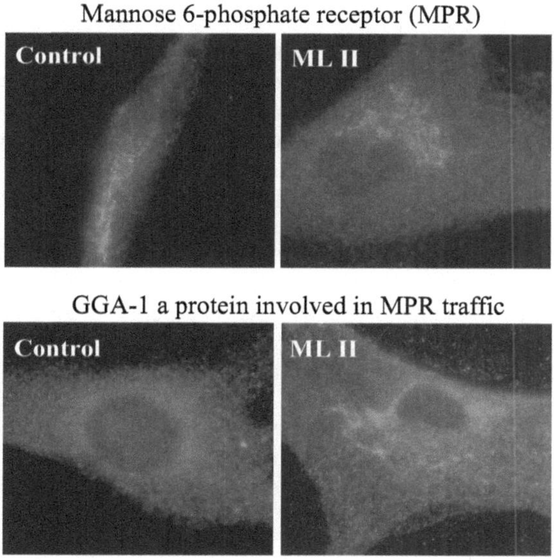

Figure 2. Fluorescence detection of mannose 6-phosphate receptors (top: MPR are receptors, for lysosomal protein ligands, which traffic between the *trans*-Golgi and endosomes) and GGA1 proteins (bottom: GGA proteins are part of the vesicular machinery that is involved in the traffic of MPR). The figure shows the altered intracellular distribution for these markers in I-cell compared to normal control human fibroblasts.

N-acetylglucosamine 1-phosphotransferase is a multisubunit enzyme with a $\alpha2\beta2\gamma2$ subunit structure, resulting from two gene products (4q21-q23 and 16p; http://www3.ncbi.nlm.nih.gov/entrez/query.fcgi?db=OMIM), the first encoding the α- and β-subunits and the other the γ-subunit (Little et al., 1986; Bao et al., 1996; Kornfeld et al., 1999; Raas-Rothschild et al., 2000, 2004). Molecular lesions in the γ-subunit appear to be the main reason for the onset of pathogenesis in ML III patients (Raas-Rothschild et al., 2000, 2004; Steet et al., 2005). These mutations are presumed to alter the structure of the multisubunit complex resulting in reduced but significant levels of *N*-acetylglucosamine 1-phosphotransferase activity. In contrast, I-cell patients with no residual *N*-acetylglucosamine 1-phosphotransferase activity develop an ML II clinical presentation. The absence of activity in ML II patients may be the result of mutations in the α- and β-subunits of *N*-acetylglucosamine 1-phosphotransferase, which contains the catalytic site. The mutation of different subunits of the *N*-acetylglucosamine 1-phosphotransferase protein may account for some of the genetic complementation groups observed in I-cell disease studies (Wright et al., 1979; Honey et al., 1982; Shows et al., 1982; Kornfeld and Sly, 1999; Sly, 2000).

ML II and ML III result in a generalised intracellular deficiency of lysosomal hydrolases which cause the widespread accumulation of oligosaccharides, mucopolysaccharides, and lipids leading to numerous electron-lucent inclusion bodies in affected cells. This deficiency is despite the normal synthesis of the complete range of lysosomal hydrolases, reflecting an inability to properly transport and localise these hydrolytic enzymes to the normal site of substrate degradation, the endosome–lysosome system (see Chapter 1, this book, and Figure 1).

Although *N*-acetylglucosamine 1-phosphotransferase is responsible for the attachment of the sugar structures required to generate the mannose 6-phosphate targeting signals, there are other theoretical ways to generate an I-cell phenotype. To expose the mannose 6-phophate targeting signal, an uncovering enzyme (*N*-acetylglucosamine 1-phosphodiester *α*-*N*-acetylglucosaminidase; Rohrer and Kornfeld, 2001) must operate in the *trans*-Golgi to remove the glucosamine residue revealing the mannose 6-phosphate structure. A defect in the activity of this enzyme could result in a defect similar to classic I-cell disease based on the inability of MPR to recognise the mannose 6-phosphate targeting signal. This could have a similar outcome to the combined mutation of MPR that gives rise to an I-celllike phenotype, as described below in an animal model. Other possible defects leading to an I-celllike phenotype could include mutations of the Golgi-localised, γ ear-containing ADP ribosylation factor-binding proteins (GGA; Boman et al., 2000; Dell'Angelica et al., 2000; Hirst et al., 2000) that are required in the transport of MPR. These could result in similar pathology due to the inability to transport lysosomal enzymes from the *trans*-Golgi to the endosome–lysosome system. To the authors' knowledge a clinical description of I-celllike disease has not been ascribed to a defect in the latter molecular machinery. It may be that there is a certain amount of redundancy in the machinery for this transport pathway.

4 DIAGNOSIS OF I-CELL DISEASE

The initial stage in the diagnosis of I-cell patients is clinical suspicion. The first laboratory testing is usually the assay of lysosomal enzyme activities in cultured cells (reduced levels of soluble hydrolases) and serum/plasma samples (increased levels of lysosomal enzyme activities) to indicate the presence of I-cell disease (e.g., Hwu et al., 1994). However, it should be noted that not all lysosomal enzymes are affected to the same degree and although the latter enzymology may be indicative of an I-cell phenotype, there are other theoretical ways of generating this pattern, as described above, leading to a potential misdiagnosis in terms of the molecular lesion. However, we have studied a number of I-cell patient samples ($n > 15$) but are yet to identify a molecular lesion in either of these genes (unpublished observations) and so the likelihood of this type of misdiagnosis is relatively low. The definitive diagnosis of I-cell disease involves the identification of a functional deficiency in *N*-acetylglucosamine 1-phosphotransferase enzyme activity (Varki, Reitman, and Kornfeld, 1981) and molecular genetic lesions associated with either one of the two subunits of *N*-acetylglucosamine 1-phosphotransferase. This is a difficult exercise, due mainly to the complexity and limited availability of *N*-acetylglucosamine 1-phosphotransferase enzymology.

Table 1. Analysis of lysosomal protein content in fibroblast and blood-spot samples from ML II and ML III patients

Protein	Skin Fibroblasts		Dried Blood Spots	
	Control ($n = 5$)	I-cell ($n = 8$)	Control ($n = 175$)	I-cell ($n = 10$)
Sulphamidase	27 ± 7	5 ± 3	9 ± 4	114 ± 36
Acid sphingomyelinase	27 ± 9	25 ± 8	17 ± 9	908 ± 151
α-Iduronidase	48 ± 30	8 ± 6	17 ± 7	107 ± 60
LAMP-1	105 ± 59	196 ± 70	124 ± 29	177 ± 60
α-Glucosidase	92 ± 11	80 ± 89	34 ± 13	125 ± 51
β-Glucosidase	70 ± 45	38 ± 19	10 ± 4	16 ± 6
Saposin C	21 ± 7	14 ± 6	23 ± 12	69 ± 29
α-Galactosidase	17 ± 8	4 ± 4	58 ± 24	55 ± 21
Arylsulphatase A	119 ± 87	38 ± 19	107 ± 35	973 ± 267
Iduronate-2-sulphatase	16 ± 8	4 ± 3	55 ± 16	756 ± 209
4-Sulphatase	11 ± 5	5 ± 3	6 ± 3	17 ± 13

All results in ng/mg of total protein for fibroblasts and ng/ml of blood for blood spots.

Table 2. Lysosomal enzymes and reference ranges used in the National Referral Laboratory

Protein	Skin Fibroblasts		Plasma	
	Control ($n = 5$)	I-cell ($n = 6$)	Control ($n = 100$)	I-cell ($n = 27$)
Arylsulphatase A	6–50	0.6–8.5	0.1–1.6	8–66
β-Glucosidase	2–30	0.4–1.7		
α-Galactosidase	2–60	0.1–3.1		
α-Fucosidase	0.8–5	0.01–0.8	2–25	7–450
β-Hexosaminidase A	30–700	6.4–19	0.5–3.1	7–14
β-Hexosaminidase B	6.3–70	1.5–3.3		
α–Mannosidase			0.5–3.0	15–106
β-Mannosidase	1–6	0.1–0.6		
Acid phosphatase	14–50	23–28	5–33	3–42
N-Acetylgalactos-Aminidase			0.1–0.6	3.1–12

All results for skin fibroblasts were in nmol/min/mg total protein and nmol/min/ml for plasma. Data from the National Referral Laboratory were kindly provided by Dr. Michael Fietz (Department of Genetic Medicine).

Although most lysosomal hydrolases are trafficked to the lysosome via the mannose 6-phosphate receptor system, not all are affected to the same degree in I-cell patients. For example, the level of selected lysosomal proteins in cultured skin fibroblasts from control and I-cell patients (Table 1) shows that of the ten proteins only sulphamidase and α-L-iduronidase are decreased below the control range in all patient cell lines, whereas acid sphingomyelinase and β-glucosidase are relatively unaffected. In the same table we can see that LAMP-1, a lysosomal membrane protein that does not use the mannose 6-phosphate targeting system, is elevated in cultured skin fibroblasts, reflecting the increased lysosomal load in these cells (Meikle et al., 1997; Sandoval et al., 1989).

In contrast to the decreased protein level in skin fibroblasts, increased amounts of proteins have been observed in plasma and whole blood from I-cell patients, resulting from secretion of mistargeted protein. In whole blood we observed elevations in sulphamidase and α-L-iduronidase corresponding to the decreased amounts observed in skin fibroblasts. Interestingly, we also observed elevations in arylsulphatase A, iduronate-2-sulphatase, and acid sphingomyelinase in whole blood; acid sphingomyelinase in particular was substantially elevated despite showing no corresponding decrease in cultured skin fibroblasts. β-Glucosidase was not elevated in blood-spots and not significantly reduced in skin fibroblasts, suggesting that this enzyme is retained in cells even in the absence of the mannose 6-phosphate moiety. Similarly, α-galactosidase and *N*-acetylgalactosamine-4-sulphatase were not elevated in blood spots from most I-cell patients. Consequently, laboratories must use care when interpreting the amount of lysosomal proteins in cultured fibroblasts, plasma, and blood spots. The panel of enzymes used in the National Referral Laboratory for the Diagnosis of Lysosomal, Peroxisomal and Related Genetic Diseases (Adelaide, Australia) is shown in Table 2 with reference ranges.

5. ANIMAL MODELS

An animal model of I-cell disease has been reported in domestic shorthair cats (Bosshard et al., 1996; Hubler et al., 1996; Mazrier et al., 2003). The clinical features of the affected cats were similar to those observed in human patients with I-cell disease, and included failure to thrive, behavioural dullness, facial dysmorphia, ataxia, dysostosis multiplex, and early death (ranging from several days to seven months, due mainly to respiratory or cardiac disease). Unlike human patients, affected cats had retinal degeneration leading to blindness by four months of age. The activities of several lysosomal enzymes were found to be elevated in serum and were correspondingly low in cultured fibroblast cells, indicative of the altered targeting of soluble lysosomal hydrolases in I-cell disease. The recessive mode of inheritance, pattern of lysosomal enzyme activities, morphological appearance, and radiology were all consistent with an I-cell phenotype. The deficiency of *N*-acetylglucosamine 1-phosphotransferase was confirmed in leukocytes and fibroblasts, which had the classical inclusion cell appearance (Bosshard et al., 1996), giving definitive confirmation to an I-cell defect.

Notably, a mouse model with a deficiency in MPR (MPR 300-, MPR 46-, and IGF-2-deficient) has been reported to have an I-cell disease phenotype (Dittmer et al., 1998). The mice displayed dwarfism, facial dysmorphia, waddling gait, dysostosis multiplex, elevated serum lysosomal enzyme activities, and histological signs of lysosomal storage. This was consistent with the ability to generate an I-celllike phenotype by several molecular mechanisms.

6 TREATMENT STRATEGIES

Hickman and Neufeld (1972) made the critical observation that I-cell fibroblasts were capable of internalising normal lysosomal enzymes, which contained the correct mannose 6-phosphorylation targeting signal. In contrast, lysosomal enzymes secreted by other I-cell fibroblasts were unable to be internalised by receptor-mediated uptake. Moreover, cell fusion was capable of correcting enzyme-deficient fibroblasts (D'Azzo et al., 1980). These observations were used as a basis to consider bone marrow transplantation as a therapeutic strategy for I-cell patients (Kurobane et al., 1986; Imaizumi et al., 1994; Grewal et al., 2003), which showed that some of the clinical symptoms, including cardiopulmonary complications, could be addressed by this strategy and that correction of the defect at the biochemical level was possible.

The use of the intravenous bisphosphonate, pamidronate, has also been evaluated in I-cell patients to address the severe bone pathology (Robinson et al., 2002). In a preliminary study, there was a "remarkable symptomatic improvement in bone pain" and increased mobility, despite incomplete suppression of the abnormal bone turnover. Pamidronate treatment may have a management role in I-cell patients, but further evaluation of this strategy is required.

REFERENCES

Bao, M., Booth, J.L., Elmendorf, B.J., and Canfield, W.M., 1996, Bovine UDP-*N*-acetylglucosamine:lysosomal-enzyme *N*-acetylglucosamine-1-phosphotransferase: I. Purification and subunit structure. *J. Biol. Chem.* **271**:31437.

Boman, A.L., Zhang, C., Zhu, X., and Kahn, R.A., 2000, A family of ADP-ribosylation factor effectors that can alter membrane transport through the trans-Golgi, *Mol. Biol. Cell.* **11**:1241.

Bosshard, N.U., Hubler, M., Arnold, S., Briner, J., Spycher, M.A., Sommerlade, H.J., von Figura, K., and Gitzelmann, R., 1996, Spontaneous mucolipidosis in a cat: An animal model of human I-cell disease, *Vet. Pathol.* **33**:1.

Brown, W.J., and Farquhar, M.G., 1984, Accumulation of coated vesicles bearing mannose 6-phosphate receptors for lysosomal enzymes in the Golgi region of I-cell fibroblasts, *Proc. Natl. Acad. Sci. U S A.* **81**:5135.

D'Azzo, A., Halley, D.J., Hoogeveen, A., and Galjaard, H., 1980, Correction of I-cell defect by hybridization with lysosomal enzyme deficient human fibroblasts, *Am. J. Hum. Genet.* **32**:519.

Dell'Angelica, E.C., Mullins, C., Caplans, S., and Bonifacino, J.S., 2000, Lysosome-related organelles, *FASEB J.* **14**:1265.

Dittmer, F., Hafner, A., Ulbrich, E.J., Moritz, J.D., Schmidt, P., Schmahl, W., Pohlmann, R., and von Figura, K., 1998, I-cell disease-like phenotype in mice deficient in mannose 6-phosphate receptors, *Transgenic. Res.* **7**:473.

Grewal, S., Shapiro, E., Braunlin, E., Charnas, L., Krivit, W., Orchard, P., and Peters, C., 2003, Continued neurocognitive development and prevention of cardiopulmonary complications after successful BMT for I-cell disease: A long-term follow-up report, *Bone Marrow Transplant.* **32**:957.

Hickman, S., and Neufeld, E.F., 1992, A hypothesis for I-cell disease: Defective hydrolases that do not enter lysosomes, *Biochem. Biophys. Res. Commun.* **49**:992.

Hirst, J., Lui, W.W., Bright, N.A., Totty, N., Seaman, M.N., and Robinson, M.S., 2000, A family of proteins with gamma-adaptin and VHS domains that facilitate trafficking between the trans-Golgi network and the vacuole/lysosome, *J. Cell Biol.* **149**:67.

Honey, N.K., Mueller, O.T., Little, L.E., Miller, A.L., and Shows, T.B., 1982, Mucolipidosis III is genetically heterogeneous, *Proc. Natl. Acad. Sci. U S A.* **79**:7420.

Hubler, M., Haskins, M.E., Arnold, S., Kaser-Hotz, B., Bosshard, N.U., Briner, J., Spycher, M.A., Gitzelmann, R., Sommerlade, H.J., and von Figura, K., 1996, Mucolipidosis type II in a domestic shorthair cat, *J. Small Anim. Pract.* **37**:435.

Hwu, W.L., Chuang, S.C., Wang, W.C., and Wang, T.R., 1994, Diagnosis of I-cell disease, *Zhonghua. Min. Guo. Xiao. Er. Ke. Yi. Xue. Hui. Za. Zhi.* **35**:508.

Imaizumi, M., Gushi, K., Kurobane, I., Inoue, S., Suzuki, J., Koizumi, Y., Suzuki, H., Sato, A., Gotoh, Y., Haginoya, K., et al., 1994, Long-term effects of bone marrow transplantation for inborn errors of metabolism: A study of four patients with lysosomal storage diseases, *Acta. Paediatr. Jpn.* **36**:30.

Kornfeld, S., and Sly, W.S., 1999, I-cell disease and pseudo-hurler polydystrophy: disorders of lysosomal phosphorylation and localisation, In Metabolic Basis of Inherited Diseases. Chapter 79 (McGraw-Hill, New York) pp. 2495–2508.

Kornfeld, R., Bao, M., Brewer, K., Noll, C., and Canfield, W., 1999, Molecular cloning and functional expression of two splice forms of human *N*-acetylglucosamine-1-phosphodiester alpha-*N*-acetylglucosaminidase, *J. Biol. Chem.* **274**:32778.

Kurobane, I., Inoue, S., Gotoh, Y., Kato, S., Tamura, M., Narisawa, K., and Tada, K., 1986, Biochemical improvement after treatment by bone marrow transplantation in I-cell disease, *Tohoku J. Exp. Med.* **150**:63.

Leroy, J.G., and DeMars, R.I., 1967, Mutant enzymatic and cytological phenotypes in cultured human fibroblasts, *Science.* **157**:804.

Little, L.E., Mueller, O.T., Honey, N.K., Shows, T.B., and Miller, A.L., 1986, Heterogeneity of *N*-acetyylglucosamine 1-phosphotransferase within mucolipidosis III, *J. Biol. Chem.* **261**:733.

Maroteaux, P., and Lamy, M., 1966, La pseudo-polydystrophie de Hurler, *Presse Med.* **74**:2889.

Mazrier, H., Van Hoeven, M., Wang, P., Knox, V.W., Aguirre, G.D., Holt, E., Wiemelt, S.P., Sleeper, M.M., Hubler, M., Haskins, M.E., and Giger, U., 2003, Inheritance, biochemical abnormalities, and clinical features of feline mucolipidosis II: The first animal model of human I-cell disease, *J. Hered.* **94**:363.

Meikle, P.J., Brooks, D.A., Ravenscroft, E.M., Yan, M., Williams, R.E., Jaunzems, A.E., Chataway, T.K., Karageorgos, L.E., Davey, R.C., Boulter, C.D., Carlsson, S.R., and Hopwood, JJ., 1997, Diagnosis of lysosomal storage disorders: Evaluation of lysosome-associated membrane protein LAMP-1 as a diagnostic marker, *Clin. Chem.* **43**:1325.

Raas-Rothschild, A., Bargal, R., Goldman, O., Ben-Asher, E., Groener, J.E., Toutain, A., Stemmer, E., Ben-Neriah, Z., Flusser, H., Beemer, F.A., Penttinen, M., Olender, T., Rein, A.J., Bach, G., and Zeigler, M., 2004, Genomic organisation of the UDP-*N*-acetylglucosamine-1-phosphotransferase gamma subunit (GNPTAG) and its mutations in mucolipidosis III, *J. Med. Genet.* **41**:e52.

Raas-Rothschild, A., Cormier-Daire, V., Bao, M., Genin, E., Salomon, R., Brewer, K., Zeigler, M., Mandel, H., Toth, S., Roe, B., Munnich, A., and Canfield, W.M., 2000, Molecular basis of variant pseudo-Hurler polydystrophy (mucolipidosis IIIC), *J. Clin. Invest.* **105**:673.

Robinson, C., Baker, N., Noble, J., King, A., David, G., Sillence, D., Hofman, P., and Cundy, T., 2002, The osteodystrophy of mucolipidosis type III and the effects of intravenous pamidronate treatment, *J. Inherit. Metab. Dis.* **25**:681.

Rohrer, J., and Kornfeld, R., 2001, Lysosomal hydrolase mannose 6-phosphate uncovering enzyme resides in the trans-Golgi network, *Mol. Biol. Cell.* **12**:1623.

Sandoval, I.V., Chen, J.W., Yuan, L., and August, J.T., 1989, Lysosomal integral membrane glycoproteins are expressed at high levels in the inclusion bodies of I-cell disease fibroblasts, *Arch. Biochem. Biophys.* **271**:157.

Shows, T.B., Mueller, O.T., Honey, N.K., Wright, C.E., and Miller, A.L., 1982, Genetic heterogeneity of I-cell disease is demonstrated by complementation of lysosomal enzyme processing mutants, *Am. J. Med. Genet.* **12**:343.

Sly, W.S., 2000, The missing link in lysosomal enzyme targeting, *J. Clin. Invest.* **105**:563.

Spranger, J.W., and Wiedemann, H.R., 1970, The genetic mucolipidoses: Diagnosis and differential diagnosis, *Humangenetik.* **9**:113.

Steet, R.A., Hullin, R., Kudo, M., Martinelli, M., Bosshard, N.U., Schaffner, T , Kornfeld, S., and Steinmann, B., 2005, A splicing mutation in the alpha/beta GlcNAc-1-phosphotransferase gene results in an adult onset form of mucolipidosis III associated with sensory neuropathy and cardiomyopathy, *Am. J. Med. Genet.* **132**:369.

Taylor, H.A., Thomas, G.H., Miller, C.S., Kelly, T.E., and Siggers, D., 1973, Mucolipidosis 3 (pseudo-Hurler polydystrophy): Cytological and ultrastructural observations of cultured fibroblast cells, *Clin. Genet.* **4**:388.

Varki, A.P., Reitman, M.L., and Kornfeld, S., 1981, Identification of a variant of mucolipidosis III (pseudo-Hurler polydystrophy): A catalytically active *N*-acetylglucosaminylphosphotransferase that fails to phosphorylate lysosomal enzymes, *Proc. Natl. Acad. Sci. U S A.* **78**:7773.

Wright, C.E., Miller, A.L., and Shows, T.B., 1979, Complementation analysis of the mucolipidoses demonstrates genetic heterogeneity, *Am. J. Hum. Genet.* **31**:66A.

Index

MIX
Papier aus verantwortungsvollen Quellen
Paper from responsible sources
FSC® C105338

If you have any concerns about our products,
you can contact us on
ProductSafety@springernature.com

In case Publisher is established outside the EU,
the EU authorized representative is:
Springer Nature Customer Service Center GmbH
Europaplatz 3, 69115 Heidelberg, Germany

Printed by Libri Plureos GmbH
in Hamburg, Germany